Lecture Notes in Computer Science   9644

Commenced Publication in 1973
Founding and Former Series Editors:
Gerhard Goos, Juris Hartmanis, and Jan van Leeuwen

Advanced Research in Computing and Software Science

Subline of Lecture Notes in Computer Science

More information about this series at http://www.springer.com/series/7407

Evangelos Kranakis · Gonzalo Navarro
Edgar Chávez (Eds.)

LATIN 2016:
Theoretical Informatics

12th Latin American Symposium
Ensenada, Mexico, April 11–15, 2016
Proceedings

Springer

Editors
Evangelos Kranakis
Carleton University
Ottawa, ON
Canada

Gonzalo Navarro
University Chile
Santiago
Chile

Edgar Chávez
Centro de Investigación Científica
    de Educación Superior de Ensenada
Ensenada
Mexico

ISSN 0302-9743   ISSN 1611-3349 (electronic)
Lecture Notes in Computer Science
ISBN 978-3-662-49528-5   ISBN 978-3-662-49529-2 (eBook)
DOI 10.1007/978-3-662-49529-2

Library of Congress Control Number: 2016932342

LNCS Sublibrary: SL1 – Theoretical Computer Science and General Issues

This Springer imprint is published by SpringerNature
The registered company is Springer-Verlag GmbH Berlin Heidelberg

Preface

This volume contains the papers presented at the 12th Latin American Theoretical Informatics Symposium (LATIN 2016) held during April 11–15, 2016, in Ensenada, Mexico. Previous editions of LATIN took place in Sao Paulo, Brazil (1992), Valparaiso, Chile (1995), Campinas, Brazil (1998), Punta del Este, Uruguay (2000), Cancun, Mexico (2002), Buenos Aires, Argentina (2004), Valdivia, Chile (2006), Buzios, Brazil (2008), Oaxaca, Mexico (2010), Arequipa, Peru (2012), and Montevideo, Uruguay (2014).

The conference received 131 submissions from around the world. Each submission was reviewed by at least three Program Committee members, and carefully evaluated on quality, originality, and relevance to the conference. Committee members wrote the reviews with the help of additional external referees. Based on an extensive electronic discussion, the committee selected 52 papers. In addition to the accepted contributions, the symposium featured distinguished lectures by Jin Akiyama (Tokyo University of Science), Allan Borodin (University of Toronto), José Correa (University of Chile), Alan Frieze (Carnegie Mellon University), and Héctor García-Molina (Stanford University).

The Imre Simon Test-of-Time Award started in 2012 and is given to the authors of the LATIN paper deemed to be most influential among all those published at least ten years prior to the current edition of the conference. Papers published in the LATIN proceedings up to and including 2006 were eligible for the 2016 award. This year the winner was Alistair Sinclair for his paper "Improved Bounds for Mixing Rates of Marked Chains and Multicommodity Flow," which appeared in LATIN 1992. This year the award was partially supported by Springer.

Many people helped to make LATIN 2016 possible. First, we would like to recognize the outstanding work of the members of the Program Committee. Their commitment contributed to a very detailed discussion on each of the submitted papers. The LATIN Steering Committee offered valuable advice and feedback; the conference benefitted immensely from their knowledge and experience.

The main organizer of the conference was the Centro de Investigación Científica y de Educación Superior de Ensenada (CICESE), located in northern Mexico. The conference was financially supported by CONACyT, CICESE, and the Mexican Mathematical Society. We are grateful for the facilities provided by EasyChair for paper evaluation and the preparation of the volume.

April 2016

Evangelos Kranakis
Gonzalo Navarro
Edgar Chávez

The Imre Simon Test-of-Time Award

For many fundamental sampling problems, the best and often the only known approach to solving them is to take a long enough random walk on a certain Markov chain and then return to the current state of the chain. Techniques to prove how long "long enough" is, i.e., the number of steps in the chain one needs to take in order to be sufficiently close to its stationary distribution, are crucial in obtaining estimates of running times of such sampling algorithms.

The mixing time of a Markov chain is quite tightly captured by the "spectral gap" of its underlying transition matrix. The spectral gap is closely related to a geometric parameter called "conductance," which is a measure of the edge-expansion of the Markov chain. Conductance also captures the mixing time up to square factors. Lower bounds on conductance, which give upper bounds on the mixing time, are typically obtained by a technique called "canonical paths" where the idea is to find a set of paths, one between every unequal source-destination pair, such that no edge is very heavily congested.

The method of canonical paths for bounding mixing time was introduced by Sinclair and Jerrum (1989), and then further developed by Diaconis and Stroock (1991). However, the canonical paths approach cannot always show rapid mixing of a rapidly mixing chain. In his LATIN 1992 paper, Sinclair establishes that this "drawback" disappears if one allows flow between a pair of states to be spread along multiple paths. Moreover, solutions to this multi-commodity flow problem are shown to capture the mixing rate closely. Thus, under fairly general conditions, we now know that a Markov chain is rapidly mixing if and only if it supports multicommodity flows of low cost.

In considering Sinclair's paper for the award, the selection committee was especially impressed by the elegance of the proposed technique, the quality of presentation, its general applicability, and its widespread recognition throughout the literature. This LATIN 1992 paper and its journal version (in the first volume of *Combinatorics, Probability and Computing*) has over 415 citations in Google Scholar. The areas that this paper has influenced include Markov chain Monte Carlo algorithms, random graphs, flows on graphs, approximation algorithms, statistical physics, and communication complexity, among others.

For all these reasons the committee selects "Improved Bounds for Mixing Rates of Markov Chains and Multicommodity Flow" by Alistair Sinclair (LATIN 1992, LNCS 583, 474–487) as the LATIN 2016 winner of the Imre Simon Test-of-Time Paper Award.

Michael Bender
Marcos Kiwi
Daniel Panario

Organization

Program Committee

Dimitris Achlioptas	UC Santa Cruz, USA
Amihood Amir	Bar-Ilan University, Israel and Johns Hopkins University, USA
Djamal Belazzougui	University of Helsinki, Finland
Michael Bender	Stony Brook University, USA
Edgar Chavez	CICESE, Mexico
Josep Diaz	UPC Barcelona, Spain
Martin Farach-Colton	Rutgers University, USA
Cristina Fernandes	University of São Paulo, Brazil
Esteban Feuerstein	University of Buenos Aires, Argentina
Fedor Fomin	University of Bergen, Norway
Leszek Gasieniec	University of Liverpool, UK
Joachim von zur Gathen	University of Bonn, Germany
Konstantinos Georgiou	Ryerson University, Canada
Roberto Grossi	University of Pisa, Italy
Giuseppe F. Italiano	University of Rome Tor Vergata, Italy
Christos Kaklamanis	University of Patras, Greece and CTI, The Netherlands
Marcos Kiwi	University of Chile, Chile
Evangelos Kranakis	Carleton University, Canada
Danny Krizanc	Wesleyan University, USA
Gregory Kucherov	CNRS/LIGM, France
Gad M. Landau	University of Haifa, Israel and NYU-Poly, USA
Lucia Moura	University of Ottawa, Canada
J. Munro	University of Waterloo, Canada
Lata Narayanan	Concordia University, Canada
Gonzalo Navarro	University of Chile, Chile
Yakov Nekrich	University of Waterloo, Canada
Jaroslav Opatrny	Concordia University, Canada
Daniel Panario	Carleton University, Canada
Pablo Pérez-Lantero	University of Valparaíso, Chile
Sergio Rajsbaum	National Autonomous University of Mexico, Mexico
Rajeev Raman	University of Leicester, UK
Ivan Rapaport	University of Chile, Chile
Jose Rolim	University of Geneva, Switzerland
Gelasio Salazar	Autonomous University of San Luis Potosi, Mexico
Nicola Santoro	Carleton University, Canada
Subhash Suri	UC Santa Barbara, USA

Dimitrios Thilikos   AlGCo project, CNRS, LIRMM, France and National
                            and Kapodistrian University of Athens, Greece
Jorge Urrutia   National Autonomous University of Mexico, Mexico
Peter Widmayer   ETH Zurich, Switzerland

Additional Reviewers

Alekseyev, Max
Alistarh, Dan
Alon, Noga
Alonso, Laurent
Alvarez, Carme
Ambainis, Andris
Amit, Mika
Aspnes, James
Bampas, Evangelos
Bampis, Evripidis
Bansal, Nikhil
Baste, Julien
Bodini, Oliver
Bohmova, Katerina
Bonomo, Flavia
Bravo, Mario
Brazdil, Tomas
Bringmann, Karl
Broutin, Nicolas
Bus, Norbert
Buss, Sam
Butman, Ayelet
Bärtschi, Andreas
Cao, Yixin
Carvajal, Rodolfo
Chan, Timothy M.
Chandran, L. Sunil
Chechik, Shiri
Cheng, Siu-Wing
Chitnis, Rajesh
Cicalese, Ferdinando
Conte, Alessio
Conway, Alexander
Cording, Patrick Hagge
Crochemore, Maxime
Cygan, Marek
Dabrowski, Konrad

Daigle, Alexandre
De Beaudrap, Jonathan
De Marco, Gianluca
de Pina, José Coelho
Diez Donoso, Yago
Dokka, Trivikram
Duch, Amalia
Durocher, Stephane
Dürr, Christoph
El-Zein, Hicham
Eppstein, David
Escoffier, Bruno
Feijao, Pedro
Fischer, Johannes
Fotakis, Dimitris
Freedman, Ofer
Gagie, Travis
Ganian, Robert
García-Colín, Natalia
Gawrychowski, Pawel
Geissmann, Barbara
Gekman, Efraim
Gelashvili, Rati
Giakkoupis, George
Giannopoulou, Archontia
Gonzalez-Aguilar, Hernan
Grabowski, Szymon
Graf, Daniel
Grant, Oliver
Grzesik, Andrzej
Hagerup, Torben
Hemaspaandra, Lane
Henning, Gabriela
Hernández-Vélez, César
Hwang, Hsien-Kuei
Jansen, Bart M.P.
Jeż, Artur

Kammer, Frank
Karakostas, George
Kempa, Dominik
Klein, Rolf
Koivisto, Mikko
Kolay, Sudeshna
Kolliopoulos, Stavros
Komusiewicz, Christian
Korman, Matias
Kostitsyna, Irina
Kowalik, Lukasz
Kuszner, Lukasz
Kärkkäinen, Juha
Laber, Eduardo
Lamprou, Ioannis
Lee, Orlando
Lewenstein, Noa
Lin, Min Chih
Liu, Chih-Hung
Löffler, Maarten
Maack, Marten
Madry, Aleksander
Mamageishvili, Akaki
Maneth, Sebastian
Maniatis, Spyridon
Marenco, Javier
Marino, Andrea
Martínez-Viademonte,
   Javier
Mastrolilli, Monaldo
Mayer, Tyler
Mayr, Richard
Mccauley, Samuel
Mcconnell, Ross
Mignot, Ludovic
Misra, Neeldhara
Mitsou, Valia

Mnich, Matthias
Moisset de Espanes, Pablo
Montanari, Sandro
Montealegre, Pedro
Moreno, Eduardo
Moura, Arnaldo
Moysoglou, Yannis
Mozes, Shay
Nebel, Markus
Nekrich, Yakov
Nicaud, Cyril
Nikoletseas, Sotiris
Nimbhorkar, Shriram
Nishimura, Naomi
Nisse, Nicolas
Ota, Takahiro
Panholzer, Alois
Panolan, Fahad
Papadopoulos, Charis
Pardini, Giovanni
Parotsidis, Nikos
Pedrosa, Lehilton L.C.
Peleg, David
Pelsmajer, Michael
Pietrzak, Krzysztof
Pilz, Alexander
Pizaña, Miguel
Ponty, Yann
Popa, Alexandru

Prencipe, Giuseppe
Pruhs, Kirk
Pröger, Tobias
Puleo, Gregory
Radoszewski, Jakub
Rampersad, Narad
Raymond, Jean-Florent
Rigo, Michel
Rojas, Javiel
Rozenberg, Liat
Rubinstein, Aviad
Sach, Benjamin
Salikhov, Kamil
Saptharishi, Ramprasad
Sau, Ignasi
Sauerwald, Thomas
Saurabh, Saket
Schabanel, Nicolas
Schmitz, Sylvain
Schouery, Rafael
Schutt, Andreas
Serna, Maria
Sitters, Rene
Soltys, Michael
Sorenson, Jonathan
Stojakovic, Milos
Strejilevich de Loma,
    Alejandro
Strømme, Torstein

Subramanya, Vijay
Suchan, Karol
Sulzbach, Henning
Suomela, Jukka
Svensson, Ola
Ta-Shma, Amnon
Talbot, Jean-Marc
Tani, Seiichiro
Thraves Caro, Christopher
Todinca, Ioan
Tschager, Thomas
Turowski, Krzysztof
Unger, Luise
Valicov, Petru
Versari, Luca
Verschae, José
Vialette, Stéphane
Viglietta, Giovanni
Wahlström, Magnus
Wakabayashi, Yoshiko
Weimann, Oren
Weinberg, S. Matthew
Xavier, Eduardo
Xiao, Mingyu
Yang, Siwei
Zabala, Paula
Zhang, Shaojie
Zito, Michele
Ziv-Ukelson, Michal

Abstracts

Reversible Figures and Solids

Jin Akiyama and Kiyoko Matsunaga

Tokyo University of Science
1-3 Kagurazaka, Shinjuku, Tokyo 162-8601, Japan
ja@jin-akiyama.com

An example of reversible (or hinge inside-out transformable) figures is Dudeney's Haberdasher's puzzle in which an equilateral triangle is dissected into four pieces, hinged like a chain, and then is transformed into a square by rotating the hinged pieces. Furthermore, the entire boundary of each figure goes into the inside of the other figure and becomes the dissection lines of the figure. Many intriguing results on reversibilities of figures have been found in the preceding research, but most of them are results on polygons. We generalize those results to general connected figures. It is shown that two nets obtained by cutting the surface of an arbitrary convex polyhedron along non-interesting dissection trees are reversible. Moreover, we generalize reversibility for 2D-figures to one for 3D-solids.

Definition (Reversible figures). A pair of hinged figures P and Q is said to be *reversible* (or *hinge inside-out transformable*) if P and Q satisfy the following conditions:

1. There exists a dissection of P into finite number of pieces $P_1, P_2, P_3, \ldots, P_n$. A set of dissection lines or curves forms a tree. Such a trees is called a *dissection tree*.
2. Pieces $P_1, P_2, P_3, \ldots, P_n$ can be joined by $n - 1$ hinges on the perimeter of P like a chain.
3. If one of the end-pieces of the chain is fixed and rotated, then the remaining pieces form Q when rotated clockwise and P when rotated counterclockwise.
4. The entire boundary of P goes into the inside of Q and the entire boundary of Q is composed of the edges of the dissection tree only.

Definition (trunk T, conjugate trunk T', (T, T')-chain). A trunk of P is a special kind of an inscribed region T of P. First, cut out an inscribed region T from P. For $i = 1, 2, \ldots, n$, let e_i be the perimeter part of T joining two vertices v_{i-1} and v_i of T, where $v_0 = v_n$. Denote by P_i the piece located outside of T that contains the perimeter part e_i. Some P_i may be empty (or just a part e_i). Then, hinge each pair P_i and P_{i+1} at their common vertex v_i for $(1 \leq i \leq n - 1)$; this gives us a chain of pieces P_i $(i = 1, 2, \ldots, n)$ of P. A chain and T are called a (T, T')-chain of P, a trunk of P, respectively, if an appropriate rotation of the chain forms T', which is one of the conjugate regions of T with all the pieces P_i packed inside T' without overlaps or gaps. T' is called a *conjugate trunk* of P.

Theorem A (Reversible Transformation between Figures). Let P be a figure with a trunk T and conjugate trunk T', and let Q have a trunk T' and conjugate trunk T. Then P is reversible to Q.

Theorem B (Reversible Transformation Between Nets of a Polyhedron). Let P be a polyhedron with n vertices $v_1, v_2, \ldots v_n$ and for $i = 1, 2$ let D_i be the dissection trees on the surface of P. Denote by N_i ($i = 1, 2$) the nets of P obtained by cutting P along D_i ($i = 1, 2$), respectively. If D_1 and D_2 don't intersect other than at the vertices of P, then a pair of nets N_1 and N_2 is reversible.

Theorem C. For any net N_1 of a polyhedron P with n vertices, there exist infinity many nets N_2 of P such that N_1 is reversible to N_2.

Theorem D. For any polyhedron P, there exist infinitely many pairs of non-self-overlapping nets of P that are reversible.

Theorem E (Reversible Transformation Between Nets of an Isotetrahedron). Let D_1 be an arbitrary dissection tree of an isotetrahedron T. Then there exists a dissection tree D_2 of T, which does not intersect D_1 other than vertices of T. A pair of nets N_i ($i = 1, 2$) obtained by cutting along D_1 is reversible, and each N_i tiles the plane.

Definition (Reversible solids). A pair of solids P, Q is said to be *hinge inside-out transformable* (or simply *reversible*) if P and Q satisfy these conditions:

(a) The solid P is dissected into several pieces by planes. Such a plane is called a *dissection* (or *cutting*) *plane*.
(b) The pieces are joined by piano hinges into a tree.
(c) If the pieces of P are reassembled inside out, you will get a solid Q.

We found a lot of reversible pairs of solids by using two different methods: the "chimera superimposition method" and "double-reversal-plates method".

Definition ((P, Q)-chimera superimposition). For a tessellative solid P, let $T(P)$ denote a tessellation by copies of P. A superimposition of $T(P)$ and $T(Q)$ is called a (P, Q)-chimera superimposition if $T(P)$ and $T(Q)$ satisfy these conditions:

1. Each copy of P in $T(P)$ is dissected into the same collection of pieces $P_1, P_2, \ldots, P_n$ by faces of copies of Q.
2. Each copy of Q in $T(Q)$ is dissected into the same collection of pieces $Q_1, Q_2, \ldots, Q_n$ by faces of copies of P.
3. P_i can be transferred to Q_i by rotations and translations for all $i = 1, 2, \ldots, n$ (by reordering $Q_1, Q_2, \ldots, Q_n$ appropriately).

Theorem 1. A (P, Q)-chimera superimposition of $T(P)$ and $T(Q)$ gives dissection planes such that P is reversible to Q.

Definition (Double-reversal-plates). A solid P is said a *double-reversal-plates solid* of T if P satisfies these conditions:

1. P contains an inscribed polyhedron T with n faces.
2. T is decomposed into n solids T_i ($i = 1, 2, \ldots, n$) each of which has one face f_i of P. If each T_i is glued on the face f_i of T, then the resultant solid is identical with P. Such an inscribed polyhedron T is called a *trunk* of P.

One example of double-reversal-plates solids is a rhombic dodecahedron. A rhombic dodecahedron P contains an inscribed cube T and the cube T can be decomposed into 6 congruent square pyramids T_i, each of which has one face f_i of T. A rhombic dodecahedron P can be constructed by putting a congruent square pyramid T on each face of a cube T.

Theorem 2. A pair of solids P and Q is reversible if both P and Q contain the identical trunk (inscribed polyhedron) T and are double-reversal-plates solids of T.

Theorem 3. A parallelohedron π is called *canonical* if it is axis-symmetric with respect to an orthogonal coordinate system, where the origin of the system is located at the center of π. Every canonical parallelohedron $S_i \in F_i$ ($i = 1, 2, ..., 5$) is reversible to the same canonical parallelohedron $S_i' \in F_i$. Moreover, for every canonical parallelohedron $S_{ij} \in F_i$ ($i = 1, 2, ..., 5$) there exists a canonical parallelohedron $S_{ji} \in F_j$ ($j = 1, 2, ..., 5$) such that S_{ij} is reversible to S_{ji}.

Simplicity Is in Vogue (again)

Allan Borodin

Abstract. Throughout history there has been an appreciation of the importance of simplicity in the arts and sciences. In the context of algorithm design, and in particular in apprxoximation algorithms and algorithmic game theory, the importance of simplicity is currently very much in vogue. I will present some examples of the current interest in the design of "simple algorithms". And what is a simple algorithm? Is it just "you'll know it when you see it", or can we benefit from some precise models in various contexts?

Subgame Perfect Equilibrium: Computation and Efficiency

José Correa

Department of Industrial Engineering, Universidad de Chile

The concept of Subgame Perfect Equilibrium (SPE) naturally arises in games which are played sequentially. In a simultaneous game the natural solution concept is that of a Nash equilibrium in which no players has an incentive to unilaterally deviate from her current strategy. However, if the game is played sequentially, i.e., there is a prescribed order in which the players make their moves, an SPE is a situation in which all players anticipate the full strategy of all other players contingent on the decisions of previous players. Although most research in algorithmic game theory has been devoted to understand properties of Nash equilibria including its computation and the so-called *price of anarchy* in recent years there has been an interest in understanding the computational properties of SPE and its corresponding efficiency measure, the *sequential price of anarchy*.

In this talk we will review some of these recent results putting particular emphasis on a very basic game, namely that of atomic selfish routing in a network [1–6]. In particular we will discuss some hardness results such as the PSPACE-completeness of computing an SPE and its NP-hardness even when the number of players fixed to two. We will also see that for interesting classes of games SPE avoid worst case Nash equilibria, resulting in substantial improvements for the price of anarchy. However, for the atomic network routing games with linear latencies, where the price of anarchy has long been known to be equal to 5/2, we prove that the sequential price of arachy is not bounded by any constant and can be as large as $\Omega(\sqrt{n})$, with n being the number of players.

References

1. Bhawalkar, K., Gairing, M., Roughgarden, T.: Weighted congestion games: the price of anarchy, universal worst-case examples, and tightness. ACM Trans. Econ. Comput. 2(4), 14 (2014)
2. Bilo, V., Flammini, M.,Monaco, G., Moscardelli, L.: Some anomalies of farsighted strategic behavior. In: WAOA 2012
3. Correa, J., de Keijzer, B., de Jong, J., Uetz, M.: The curse of sequentiality in routing games. In: WINE 2015

Partially supported by the Millennium Nucleus Information and Coordination in Networks ICM/FIC RC130003.

4. de Jong, J., Uetz, M.: The sequential price of anarchy for atomic congestion games. In: WINE 2014

5. Milchtaich, I.: Crowding games are sequentially solvable. Int. J. Game Theory **27**, 501–509 (1998)

6. Paes Leme, R., Syrgkanis, V., Tardos, É.: The curse of simultaneity. In: ITCS 2012

Buying Stuff Online

Alan Frieze and Wesley Pegden

Abstract. Suppose there is a collection x_1, x_2, ..., x_N of independent uniform [0, 1] random variables, and a hypergraph $\mathcal{F}$ of *target structures* on the vertex set $\{1, ..., N\}$. We would like to buy a target structure at small cost, but we do not know all the costs x_i ahead of time. Instead, we inspect the random variables x_i one at a time, and after each inspection, choose to either keep the vertex i at cost x_i, or reject vertex i forever.

In the present paper, we consider the case where $\{1, ..., N\}$ is the edge-set of some graph, and the target structures are the spanning trees of a graph; the spanning arborescences of a digraph; the Hamilton cycles of a graph; the prefect matchings of a graph; the paths between a fixed pair of vertices; or the cliques of some fixed size.

Data Crowdsourcing: Is It for Real?

Hector Garcia-Molina

Abstract. Crowdsourcing refers to performing a task using human workers that solve sub-problems that arise in the task. In this talk I will give an overview of crowdsourcing, focusing on how crowdsourcing can help traditional data processing and analysis tasks. I will also give a brief overview of some of the crowdsourcing research we have done at the Stanford University InfoLab.

Contents

Contents XXV

A Faster FPT Algorithm and a Smaller Kernel for BLOCK GRAPH VERTEX DELETION

Akanksha Agrawal[1(✉)], Sudeshna Kolay[2], Daniel Lokshtanov[1], and Saket Saurabh[1,2]

[1] University of Bergen, Bergen, Norway
{akanksha.agrawal,daniello}@uib.no
[2] Institute of Mathematical Sciences, Chennai, India
{skolay,saket}@imsc.res.in

Abstract. A graph G is called a *block graph* if every maximal 2-connected component of G is a clique. In this paper we study the BLOCK GRAPH VERTEX DELETION from the perspective of fixed parameter tractable (FPT) and kernelization algorithms. In particular, an input to BLOCK GRAPH VERTEX DELETION consists of a graph G and a positive integer k, and the objective to check whether there exists a subset $S \subseteq V(G)$ of size at most k such that the graph induced on $V(G) \setminus S$ is a block graph. In this paper we give an FPT algorithm with running time $4^k n^{\mathcal{O}(1)}$ and a polynomial kernel of size $\mathcal{O}(k^4)$ for BLOCK GRAPH VERTEX DELETION. The running time of our FPT algorithm improves over the previous best algorithm for the problem that runs in time $10^k n^{\mathcal{O}(1)}$, and the size of our kernel reduces over the previously known kernel of size $\mathcal{O}(k^6)$. Our results are based on a novel connection between BLOCK GRAPH VERTEX DELETION and the classical FEEDBACK VERTEX SET problem in graphs without induced C_4 and $K_4 - e$. To achieve our results we also obtain an algorithm for WEIGHTED FEEDBACK VERTEX SET running in time $3.618^k n^{\mathcal{O}(1)}$ and improving over the running time of previously known algorithm with running time $5^k n^{\mathcal{O}(1)}$.

1 Introduction

Deleting the minimum number of vertices from a graph such that the resulting graph belongs to a family $\mathcal{F}$ of graphs, is a measure on how close the graph is to the graphs in the family $\mathcal{F}$. In the problem of vertex deletion, we ask whether we can delete at most k vertices from the input graph G such that the resulting graph belongs to the family $\mathcal{F}$. Lewis and Yannakakis [12] showed that for any non-trivial and hereditary graph property Π on induced subgraphs, the vertex deletion problem is NP-complete. Thus these problems have been subjected to intensive study in algorithmic paradigms that are meant for coping with NP-completeness [7,8,13,14]. These paradigms among others include applying

The research leading to these results has received funding from the European Research Council under the European Union's Seventh Framework Programme (FP7/2007–2013) / ERC grant agreement no. 306992.

E. Kranakis et al. (Eds.): LATIN 2016, LNCS 9644, pp. 1–13, 2016.
DOI: 10.1007/978-3-662-49529-2_1

restrictions on inputs, approximation algorithms and parameterized complexity. The focus of this paper is to study one such problem from the viewpoint of parameterized algorithms.

Given a family $\mathcal{F}$, a typical parameterized vertex deletion problem gets as an input an undirected graph G and a positive integer k and the goal is to test whether there exists a vertex subset $S \subseteq V(G)$ of size at most k such that $G \setminus S \in \mathcal{F}$. In the parameterized complexity paradigm the main objective is to design an algorithm for the vertex deletion problem that runs in time $f(k) \cdot n^{\mathcal{O}(1)}$, where $n = |V(G)|$ and f is an arbitrary computable function depending only on the parameter k. Such an algorithm is called an FPT algorithm and such a running time is called FPT running time. We also design a polynomial time preprocessing algorithm that reduces the given instance to an equivalent one with size as small as possible. This is mathematically modelled by the notion of *kernelization*. A parameterized problem is said to admit a $h(k)$-*kernel* if there is a polynomial time algorithm (the degree of the polynomial is independent of k), called a *kernelization* algorithm, that reduces the input instance to an equivalent instance with size upper bounded by $h(k)$. In other words, let (x, k) be the input instance and (x', k') be the reduced instance. Then, $(x, k) \in \Pi$ if and only if $(x', k') \in \Pi$. Also, $|x'|, k' \leq h(k)$. If the function $h(k)$ is polynomial in k, then we say that the problem admits a polynomial kernel. For more background, the reader may refer to the following monograph [6].

A graph G is known as a *block graph* if every maximal 2-connected component in G is a clique. Equivalently, we can see a block graph as a graph obtained by replacing each edge in a forest by a clique. A *chordal graph* is a graph which has no induced cycles of length at least four. An equivalent characterisation of a block graph is a chordal graph with no induced $K_4 - e$ [2,9]. The class of block graphs is the intersection of the chordal and distance-hereditary graphs [9].

In this paper, we consider the problem which we call BLOCK GRAPH VERTEX DELETION (BGVD). Here, as an input we are given a graph G and an integer k, and the question is whether we can find a subset $S \subseteq V(G)$ of size at most k such that $G \setminus S$ is a block graph. The NP-hardness of the BGVD problem follows from [12].

BLOCK GRAPH VERTEX DELETION (BGVD)   **Parameter:** k
Input: An undirected graph $G = (V, E)$, and a positive integer k
Question: Is there a set $S \subseteq V$, of size at most k, such that $G \setminus S$ is a block graph?

Kim and Kwon [10] gave an FPT algorithm with running time $\mathcal{O}^\star(10^k)$ and a kernel of size $\mathcal{O}(k^6)$ for the BGVD problem. In this paper we improve both these results via a novel connection to FEEDBACK VERTEX SET problem.

Our Results and Methods. We start by giving the results we obtain in this article and then we explain how we obtain these results. Our three main results are:

Theorem 1. BGVD *has an* FPT *algorithm running in time* $\mathcal{O}^{\star}(4^k)$.

Theorem 2. BGVD *admits a factor four approximation algorithm.*

Theorem 3. BGVD *has a kernel of size* $\mathcal{O}(k^4)$.

Our two of the theorems improve both the results in [10]. That is, the running time of our FPT algorithm improves over the previous best algorithm for the problem that runs in time $10^k n^{\mathcal{O}(1)}$, and the size of our kernel reduces over the previously known kernel of size $\mathcal{O}(k^6)$.

Our results are based on a connection between the WEIGHTED-FVS and BGVD problems. In particular we show that if the given input graph does not have induced four cycles or diamonds ($K_4 - e$) then we can construct an auxiliary bipartite graph and solve WEIGHTED-FVS on it. This results in a faster FPT algorithm for BGVD. In the algorithm that we give for the BGVD problem, as a sub-routine we use the algorithm for the WEIGHTED-FVS problem. For obtaining a better polynomial kernel for BGVD, most of our Reduction Rules are same as those used in [10]. On the way to our result we also design a factor four approximation algorithm for BGVD.

Finally, we talk about WEIGHTED-FVS. For which, we also design a faster algorithm than known in the literature. The FEEDBACK VERTEX SET problem is one of the most well studied problems. Given an undirected graph $G = (V, E)$ and a positive integer k, the problem is to decide whether there is a set $S \subseteq V$ such that $G \setminus S$ is a forest. Thus, S is a vertex subset that intersects with every cycle of G. In the parameterized complexity setting, FEEDBACK VERTEX SET parameterized by k, has an FPT algorithm. The best known FPT algorithm runs in time $\mathcal{O}^{\star}(3.618^k)$ [4,11]. The problem also admits a kernel on $\mathcal{O}(k^2)$ vertices [15]. Another variant of FEEDBACK VERTEX SET that has been studied in parameterized complexity is WEIGHTED FEEDBACK VERTEX SET, where each vertex in the graph has some rational number as its weight.

WEIGHTED-FVS	**Parameter:** k
Input: An undirected graph $G = (V, E)$, a weight function $w : V \to \mathbb{Q}$, and a positive integer k	
Output: The minimum weighted set $S \subseteq V$ of size at most k, such that $G \setminus S$ is a forest.	

WEIGHTED-FVS is known to be in FPT with an algorithm of running time $5^k n^{\mathcal{O}(1)}$ [3]. We obtain a faster FPT algorithm for WEIGHTED-FVS. This algorithm uses, as a subroutine, the algorithm for solving WEIGHTED-MATROID PARITY [16]. In fact, this algorithm is very similar to the algorithm for FEEDBACK VERTEX SET given in [4,11]. Thus, our final new result is the following theorem.

Theorem 4 [⋆]. WEIGHTED-FVS *has an* FPT *algorithm running in time* $\mathcal{O}^{\star}(3.618^k)$.

Due to paucity of space, results stated without proof in the short version are marked with [⋆]. These proofs can be found in the full version of the paper.

2 Preliminaries

We start with some basic definitions and terminology from graph theory and algorithms. We also establish some of the notation that will be used in this paper.

We will use the $\mathcal{O}^\star$ notation to describe the running time of our algorithms. Given $f : \mathbb{N} \to \mathbb{N}$, we define $\mathcal{O}^\star(f(n))$ to be $\mathcal{O}(f(n) \cdot p(n))$, where $p(\cdot)$ is some polynomial function. That is, the $\mathcal{O}^\star$ notation suppresses polynomial factors in the running-time expression. We denote the set of rational numbers by $\mathbb{Q}$.

Graphs. A graph is denoted by $G = (V, E)$, where V and E are the vertex and edge sets, respectively. We also denote the vertex set and edge set of G by $V(G)$ and $E(G)$, respectively. All the graphs that we consider are finite graphs, possibly having loops and multi-edges. For any non-empty subset $W \subseteq V(G)$, the subgraph of G induced by W is denoted by $G[W]$; its vertex set is W and its edge set consists of all those edges of $E(G)$ with both endpoints in W. For $W \subseteq V(G)$, by $G \setminus W$ we denote the graph obtained by deleting the vertices in W and all edges which are incident to at least one vertex in W.

For a graph G, we denote the degree of vertex v in G by $d_G(v)$. A vertex $v \in V(G)$ is called as a *cut vertex* if the number of connected components in $G \setminus \{v\}$ is more than the number of connected components in G. For a vertex $v \in V(G)$, the neighborhood of v in G is the set $N_G(v) = \{u | (v, u) \in E(G)\}$. We drop the subscript G from $N_G(v)$, whenever the context is clear. Two vertices $u, v \in V(G)$ are called *true-twins* in G if $N(u) \setminus \{v\} = N(v) \setminus \{u\}$. For $A \subset V(G)$, an A-path in G is a path with at least one edge, whose end vertices are in A and all the internal vertices are from $V(G) \setminus A$.

A weighted undirected graph is a graph $G = (V, E)$, with a weight function $w : V(G) \to \mathbb{Q}$. For a subset $X \subseteq V(G)$, $w(X) = \sum_{v \in X} w(v)$.

A *feedback vertex set* is a subset $S \subseteq V(G)$ such that $G \setminus S$ is a forest. A minimum weight feedback vertex set of a weighted graph G is a subset $X \subseteq V(G)$, such that $G \setminus X$ is a forest and $w(X)$ is minimum among all possible weighted-fvs in G. In a graph with vertex weights, an FVS is called a weighted feedback vertex set (weighted-fvs). Similarly, for a given positive integer k, a minimum weighted-fvs of size k is a subset $X \subseteq V(G)$ such that $|X| \leq k$, $G \setminus X$ is a forest and $w(X)$ is minimum among all possible weighted-fvs in G that are of size at most k. Given a graph G and a vertex subset $S \subseteq V(G)$, we say that S is a *block vertex deletion set* if $G \setminus S$ is a block graph.

A maximal 2-connected subgraph of a graph G is called a *block*. By $K_4 - e$ we denote the graph obtained by removing an edge e from a complete graph on 4 vertices. For a graph G, let V_c denote the set of cut vertices of G, and $\mathcal{B}$ the set of its blocks. We then have a natural bipartite graph F on $V_c \cup \mathcal{B}$ formed by the edges (v, B) if and only if $c \in V(B)$. Note that for a block graph G, F is a forest [5]. The bipartite graph F is called as the block forest of G. We will arbitrarily root F at some vertex $B \in V(F)$.

A leaf block of a block graph G is a maximal 2-connected component with at most one cut vertex. For a maximal 2-connected component C in G a vertex $v \in V(C)$ is called as an *internal vertex* if v is not a cut vertex in G.

We refer the reader to [5] for details on standard graph theoretic notation and terminology we use in the paper.

3   FPT Algorithm for Block Graph Vertex Deletion

In this section, we present an FPT algorithm for the BGVD problem. First, we look at the special case, when the input graph does not have any small obstructions in the form of D_4's and C_4's. Here, $D_4 = K_4 - e$. We show that, in this case, BGVD reduces to WEIGHTED-FVS. Later, we solve the general problem, using the algorithm of the special case.

3.1   RESTRICTED BGVD

In this part, we solve the following special case of BGVD in FPT time.

RESTRICTED BGVD   **Parameter:** k
Input: A connected undirected graph G, which is $\{D_4, C_4\}$-free, and a positive integer k.
Question: Does there exist a set S such that $G \setminus S$ is a block graph?

Let G be the input graph. Let $\mathcal{C}$ be the set of maximal cliques in G. We start with the following simple observation about graphs without C_4 and D_4.

Lemma 1. *Let G be a graph that does not contain C_4 and D_4 as an induced subgraph then (a) any two maximal cliques intersect on at most one vertex and (b) the number of maximal cliques in G is at most n^2.*

Proof. Let C_1 and C_2 be two maximal cliques in $\mathcal{C}$. Since G is D_4-free, $V(C_1) \cap V(C_2)$ can have at most one vertex. Thus, each edge of G belongs to exactly one maximal clique. This gives a bound of n^2 on the number of maximal cliques. $\square$

We construct an auxiliary weighted bipartite graph $\hat{G}$ in the following way: $\hat{G}$ is a bipartite graph with vertex set bipartition $V(G) \cup V_c$, where V_c is the set where we add a vertex v_C corresponding to each $C \in \mathcal{C}$. Note that there is a bijective correspondence between the vertices of V_c and the maximal cliques in $\mathcal{C}$. A vertex v of a clique C is called *external* if it is part of at least two maximal cliques in $\mathcal{C}$. We add an edge between a vertex $v \in V(G)$ and a vertex $v_C \in V_c$ in $E(\hat{G})$ if and only if v is an external vertex of the clique $C \in \mathcal{C}$.

Lemma 2. *Let G be a graph without induced C_4 and D_4 and $S \subseteq V(G)$. Then S is block vertex deletion set of G if and only if $\hat{G} \setminus S$ is acyclic.*

Proof. First, let S be a block vertex deletion set solution for G. Suppose that $\hat{G} \setminus S$ has a cycle C. Notice that C cannot be a C_4, as this corresponds to two maximal cliques that share 2 vertices. Thus, C is an even cycle of length at least 6. Suppose C has length 6. This corresponds to maximal cliques C_1, C_2, C_3 such that $u = C_1 \cap C_2$, $v = C_2 \cap C_3$ and $w = C_1 \cap C_3$. Since C_1, C_2, C_3 are distinct maximal cliques, at least one of them must have a vertex other than u, v or w. Without loss of generality, let C_1 have a vertex $x \notin \{u, v, w\}$. Then, the set $\{x, u, v, w\}$ forms a D_4 in G. However, this is not possible, as G did not have a D_4 to start with. Hence, C must be an even cycle of length at least 8. However, this corresponds to a set of maximal cliques and external vertices, such that the external vertices form an induced cycle of length at least four. This contradicts that S was a block vertex deletion set for G. Thus, $\hat{G} \setminus S$ must be acyclic.

On the other hand, let $\hat{G} \setminus S$ be acyclic. Suppose $G \setminus S$ has an induced cycle C, of length at least four. As C is an induced cycle of length at least four, no two edges of C can belong to the same maximal clique. For an edge (u, v) of C, let $C_{(u,v)}$ be the maximal clique containing it. Also, let $c_{(u,v)}$ be the corresponding vertex in $\hat{G}$. We replace the edge (u, v) in C by two edges $(u, c_{(u,v)})$ and $(v, c_{(u,v)})$. In this way, We obtain a cycle C' of $\hat{G} \setminus S$, which is a contradiction. Thus, S must be a block vertex deletion set for G.   □

If the input graph G is without induced C_4 and D_4 then Lemma 2 tells us that to find block vertex deletion set of G of size at most k one can check whether there is a feedback vertex set of size at most k for $\hat{G}$ contained in $V(G)$. To enforce that we find feedback vertex set for $\hat{G}$ completely contained in $V(G)$ we solve an appropriate instance of WEIGHTED-FVS. In particular we give the weight function $w : V(\hat{G}) \to \mathbb{N}$ as follows. For $v \in V(G)$, $w(v) = 1$ and for $v_C \in V_C$, $w(v_C) = n^4$. Clearly, $V(G)$ is a feedback vertex set of $\hat{G}$ and thus the weight of a minimum sized feedback vertex set of $\hat{G}$ is at most n. This implies that running an algorithm for WEIGHTED-FVS on an instance $(\hat{G}, w, k)$ either returns a feedback vertex set contained inside $V(G)$ or returns that the given instance is a No instance.

Theorem 5. RESTRICTED BGVD *can be solved in* $\mathcal{O}^\star(3.618^k)$.

Proof. Given an instance (G, k) of Restricted BGVD. We apply the WEIGHTED-FVS on the instance $(\hat{G}, w, k)$, where $\hat{G}$ is obtained as described above. Let S be the weighted-fvs of size at most k in $\hat{G}$ returned by WEIGHTED-FVS (of course if there exists one). By the discussion above we know that if WEIGHTED-FVS does not return that the given instance is a No instance then $S \subseteq V(G)$. If it returns that the given instance is a No instance then we return the same. Else, assume that S is non-empty. Now we check whether $w(S)$ is at most k or not. Since every vertex in $V(G)$ has been assigned weight one we have that $w(S) = |S|$ and thus if $w(S) \leq k$ then we return S as block vertex deletion set of G. In the case when $w(S) > k$ we return that the given instance is a No instance for RESTRICTED BGVD. Correctness of these steps are guaranteed by Lemma 2. The running time of the algorithm is dominated by the running time of WEIGHTED-FVS and thus it is $\mathcal{O}^\star(3.618^k)$. This completes the proof.   □

3.2 Block Graph Vertex Deletion

We are now ready to describe an FPT algorithm for BGVD, and hence prove Theorem 1. We design the algorithm for the general case with the help of the algorithm for RESTRICTED BGVD.

Proof (of Theorem 1). Let O be a D_4 or C_4 present in the input graph G. For any potential solution S, at least one of the vertices of O must belong to S. Therefore, we branch on the choice of these vertices, and for every vertex $v \in O$, we recursively apply the algorithm to solve BGVD instance $(G \setminus \{v\}, k - 1)$. If one of these branches returns a solution X, then clearly $X \cup \{v\}$ is a block vertex deletion set of size at most k for G. Else, we return that the given instance is a No instance. On the other hand, if G is $\{D_4, C_4\}$-free, then we do not make any further recursive calls. Instead, we run the algorithm for RESTRICTED BGVD on G and return the output of the algorithm. Thus, the running time of this algorithm is upper bounded by $\mathcal{O}^*(4^k)$. □

4 An Approximation Algorithm for BGVD

In this section, we present a simple approximation algorithm $\mathcal{A}_1$ for BGVD. Given a graph G, we give a block vertex deletion set S of size at most $4 \cdot \mathsf{OPT}$, where OPT is the size of a minimum sized block vertex deletion set for G.

Proof (of Theorem 2). Let G be the given instance of BGVD and OPT be the size of a minimum sized block vertex deletion set for G and S_{OPT} be a minimum sized block vertex deletion set for G.

Let $\mathcal{S}$ be a maximal family of D_4 and C_4 such that any two members of $\mathcal{S}$ are pairwise disjoint. One can easily construct such a family $\mathcal{S}$ greedily in polynomial time. Let S_1 be the set of vertices contained in any obstruction in $\mathcal{S}$. That is, $S_1 = \bigcup_{O \in \mathcal{S}} O$. Since any block vertex deletion set must contain a vertex from each obstruction in $\mathcal{S}$ and any two members of $\mathcal{S}$ are pairwise disjoint, we have that $|S_{\mathsf{OPT}} \cap S_1| \geq |\mathcal{S}|$.

Let $G' = G \setminus S_1$. Observe that G' does not contain either D_4 or C_4 as an induced subgraph. Now we construct $\hat{G}'$, as described in Sect. 3.1. We apply the factor two approximation algorithm $\mathcal{A}$ given in [1] on the instance $(\hat{G}', w)$. This returns an fvs S_2 of $\hat{G}'$ such that $w(S_2)$ is at most twice the weight of a minimum weight feedback vertex set. By out construction $S_2 \subseteq V(G')$. Lemma 2 implies that S_2 is a factor two approximation for BGVD on G'. We return the set $S = S_1 \cup S_2$ as our solution. Since $S_{\mathsf{OPT}} \setminus S_1$ is also an optimum solution for G' we have that $|S_2| \leq 2|S_{\mathsf{OPT}} \setminus S_1|$.

It is evident that S is block vertex deletion set of G. To conclude the proof of the theorem we will show that $|S| \leq 4\mathsf{OPT}$. Towards this observe that

$$|S| = |S_1| + |S_2| \leq 4|\mathcal{S}| + 2|S_{\mathsf{OPT}} \setminus S_1|$$
$$\leq 4|S_{\mathsf{OPT}} \cap S_1| + 2|S_{\mathsf{OPT}} \setminus S_1|$$
$$\leq 4|S_{\mathsf{OPT}}| = 4\mathsf{OPT}.$$

This completes the proof. □

5   Improved Kernel for Block Graph Vertex Deletion

In this section, we give a kernel of $\mathcal{O}(k^4)$ vertices for BGVD. Let (G, k) be an instance of the BGVD problem. We start with some of the known reduction rules from [10].

Reduction Rule BGVD 1. *If G has a component H, where H is a block graph, then remove H from G.*

Reduction Rule BGVD 2. *If there is a vertex $v \in V(G)$, such that $G \setminus \{v\}$ has a component H, where $G[\{v\} \cup V(H)]$ is a connected block graph then, remove H from G.*

Reduction Rule BGVD 3. *Let $S \subseteq V(G)$, where each $u, v \in S$ are true-twins in G. If $|S| > k + 1$, then remove all the vertices from S except $k + 1$ vertices.*

Reduction Rule BGVD 4. *Let t_1, t_2, t_3, t_4 be an induced path in G. For $i \in \{1, 2, 3\}$, let $S_i \subseteq V(G) \setminus \{t_1, t_2, t_3, t_4\}$ be a clique in G such that the following holds.*

- *For $i \in \{1, 2, 3\}$, $v \in S_i$, $N_G(v) \setminus S_i = \{t_i, t_{i+1}\}$, and*
- *For $i \in \{2, 3\}$, $N_G(t_i) = \{t_{i-1}, t_{i+1}\} \cup S_{i-1} \cup S_i$.*

Remove S_2 from G and contract the edge (t_2, t_3).

Proposition 1 (Proposition 3.1 [10]). *Let G be a graph and k be a positive integer. For a vertex $v \in V(G)$, in $\mathcal{O}(kn^3)$ time, we can find one of the following.*

- *i. $k + 1$ pairwise vertex disjoint obstructions,*
- *ii. $k + 1$ obstructions whose pairwise intersection is exactly v,*
- *iii. $S'_v \subseteq V(G)$, such that $|S'_v| \leq 7k$ and $G \setminus S'_v$ has no block graph obstruction containing v.*

Reduction Rule BGVD 5. *Let $v \in V(G)$ and $G' = G \setminus \{v\}$. We remove the edges between $N_G(v)$ from G', i.e. $E(G') = E(G') \setminus \{(u, w) | u, w \in N_G(v)\}$. In G' if there are at least $2k + 1$ vertex-disjoint $N_G(v)$-paths in G' then we do one of the following.*

- *If G contains $k + 1$ vertex disjoint obstructions, then return that the graph is a no-instance.*
- *Otherwise, delete v from G and decrease k by 1.*

The Reduction rules BGVD 1 to BGVD 5 are safe and can be applied in polynomial time [10]. For sake of clarity we denote the reduced instance at each step by (G, k). We always apply the lowest numbered Reduction Rule, in the order that they have been stated, that is applicable at any point of time. For the rest of the discussion, we assume that Reduction rules BGVD 1 to BGVD 5 are not applicable.

For a vertex $v \in V(G)$, by Proposition 1, we may find $k + 1$ pairwise vertex-disjoint obstructions, and we can safely conclude that the graph is a No instance. Secondly, if we find $k + 1$ obstructions whose pairwise intersection is exactly v then the Reduction rule BGVD 5 will be applicable. Thus, we assume that for each vertex $v \in V(G)$, the third condition of Proposition 1 holds. In other words, we have a set S'_v of size at most $7k$, such that $G \setminus S'_v$ does not contain any obstruction passing through v. In fact, for each $v \in V(G)$, we can find a block vertex deletion set $S_v \subseteq V(G) \setminus \{v\}$ of bounded size.

Observation 1 [⋆]. *For every vertex $v \in V(G)$, we can find in $n^{\mathcal{O}(1)}$ time, a set $S_v \subseteq V(G) \setminus \{v\}$ such that $|S_v| \leq 11k$ and $G \setminus S_v$ is a block graph.*

For a vertex $v \in V(G)$, *component degree* of v is the number of connected components in $\mathcal{C}$, where $\mathcal{C}$ is the set of connected components in $G \setminus (S_v \cup \{v\})$ that have a vertex adjacent to v. We give a reduction rule that bounds the *component degree* of a vertex $v \in V(G)$, using *Expansion Lemma* [15].

A *q-star*, $q \geq 1$, is a graph with $q + 1$ vertices, one vertex of degree q and all other vertices of degree 1. Let $\mathcal{B}$ be a bipartite graph with the vertex bipartition as (X, Y). A set of edges $M \subseteq E(\mathcal{B})$ is called a *q-expansion* of X into Y if (i) every vertex of X is incident with exactly q edges of M and (ii) M saturates exactly $q|X|$ vertices in Y, i.e. edges in M are adjacent to exactly $q|X|$ vertices in Y.

Lemma 3 (Expansion Lemma). *Let q be a positive integer and $\mathcal{B}$ be a bipartite graph with vertex bipartition (X, Y) such that $|Y| \geq q|X|$ and there are no isolated vertices in Y. Then, there exist nonempty vertex sets $X' \subseteq X$ and $Y' \subseteq Y$ such that:*

1. *X' has a q-expansion into Y' and*
2. *no vertex in Y' has a neighbour outside X', i.e. $N(Y') \subseteq X'$.*

Furthermore, the sets X' and Y' can be found in polynomial time.

See [4] for the version of the Lemma 3 stated above. For a vertex $v \in V(G)$, let $\mathcal{C}_v$ be the set of connected components in $G \setminus (S_v \cup \{v\})$ that have a vertex adjacent to v. Consider a connected component $C \in \mathcal{C}_v$, such that no vertex $u \in V(C)$ is adjacent to any vertex in S_v. But then, $G \setminus \{v\}$ has a component which is a block graph (namely, the connected component C) therefore, Reduction rule BGVD 2 is applicable, a contradiction to the assumption that none of the previous Reduction rules are applicable. Therefore, for each $C \in \mathcal{C}$ there is a vertex $u \in V(C)$ and $s \in S_v$, such that $(u, s) \in E(G)$. Let $\mathcal{D}$ be a vertex set, with a vertex d corresponding to each component $D \in \mathcal{C}$. Consider the bipartite graph $\mathcal{B}_v$ with the vertex set bipartitioned as $(\mathcal{D}, S_v)$. There is an edge between $d \in \mathcal{D}$ and $s \in S_v$ if and only if the component D corresponding to which the vertex d was added to $\mathcal{D}$ has a vertex u_d such that $(u_d, s) \in E(G)$.

Reduction Rule BGVD 6. *For a vertex $v \in V(G)$ if $|\mathcal{C}_v| > 33k$, then we do the following.*

- *Let $\mathcal{D}' \subseteq \mathcal{D}$ and $S \subseteq S_v$ be the sets obtained after applying Lemma 3 with $q = 3$, $X = S_v$ and $Y = \mathcal{D}$;*
- *For each $d \in \mathcal{D}'$, let the component corresponding to d be $D \in \mathcal{C}_v$. Delete all the edges between (u, v), where $u \in V(D)$;*
- *For each $s \in S$, add two vertex disjoint paths between v and s.*

Safeness of the Reduction rule BGVD 6 follows from the safeness of Reduction rule 6 in [10].

5.1   Bounding the Number of Blocks in $G \setminus A$

Using the approximation algorithm for BGVD we compute an approximate solution A of size at most $4k$. Of course if $|A| > 4k$ then we can immediately return that G is a No instance. First, we bound the number of leaf blocks in $G \setminus A$, when none of the Reduction rules apply. Note that $G \setminus A$ is a block graph, since A is an approximate solution to BGVD. For $v \in A$, let S_v' be the set obtained from Proposition 1 and S_v be the set obtained from Observation 1. Let $\mathcal{C}_v$ be the set of connected components in $G \setminus (S_v \cup \{v\})$ which have a vertex adjacent to v. All the connected components in $G \setminus A$, which do not have a vertex that is adjacent to v, must be adjacent to some $v' \in A$. Otherwise, Reduction rule BGVD 1 will be applicable. Also, all the leaf blocks in $G \setminus A$ must have an internal vertex that is adjacent to some vertex in A, since the Reduction rules BGVD 1 and BGVD 2 are not applicable. The number of leaf blocks, in $G \setminus A$, whose set of internal vertices have a non-empty intersection with S_v', is at most $7k$. Therefore, it is enough to count, for each $v \in A$, the number of leaf blocks in $\mathcal{C}_v$. In the Observation 2, we give a bound on the number of leaf blocks in $G \setminus A$, not containing any vertex from S_v'.

Observation 2 [⋆]**.** *For $v \in A$, the number of leaf blocks in $G \setminus A$ not containing any vertex from S_v' is at most the number of leaf blocks in $G \setminus (S_v \cup \{v\})$.*

Therefore, for each $v \in A$ we count those leaf blocks in $\mathcal{C}_v$ which do not contain any vertex from S_v'.

Lemma 4. *Consider a vertex $v \in V(G)$ and its corresponding set S_v. Let $\mathcal{C}$ be the set of connected components in $G \setminus (S_v \cup \{v\})$. For each $C \in \mathcal{C}$, there is a block $\tilde{B}$ in C, such that $N_C(v) \subseteq V(\tilde{B})$.*

Proof. Let $\mathcal{C}$ be the set of connected components of $G \setminus (S_v \cup \{v\})$, $v \in V(G)$. By definition of S_v, for each $C \in \mathcal{C}$, $C \cup \{v\}$ is a block graph.

If for some $C \in \mathcal{C}$, $N_C(v) = \emptyset$, then the condition is trivially satisfied for that connected component C. Let $C \in \mathcal{C}$ be a connected component such that $N_C(v) \neq \emptyset$. Let t be a vertex in $N_B(v)$, where B is a block in C. Let B' be a block in C, where $B' \neq B$ and B' has a vertex $t' \in V(B') \setminus V(B)$ that is adjacent to v. Note that B, B' are in the same connected component C. Let P be the shortest path from t to t'.

We first argue for the case when $(t, t') \notin E(G)$. Therefore, the path P has at least 2 edges. We prove that we can find an obstruction, by induction on the length of the path (number of edges). If length of path P is 2, say $P = t, u, t'$. If $(u, v) \in E(G)$, then $\{t, t', u, v\}$ forms an induced D_4, otherwise they form an induced C_4, contradicting that $C \cup \{v\}$ is a block graph.

Let us assume that we can find an obstruction if the path length is l. We now prove it for paths of length $l + 1$. Let $P = t, x_1, x_2, \ldots, x_{l-1}, t'$ and y be the first vertex other than t in P such that $(y, v) \in E(G)$. If $y = t'$, then P along with v forms an induced cycle of length at least 5, contradicting that $C \cup \{v\}$ is a block graph. If $y = x_1$, then $\{t, x_1, x_2, v\}$ either forms a D_4, the case when $(x_2, v) \in E(G)$, or $\hat{P} = x_1, x_2, \ldots, t'$ is a path of shorter length with at least 2 edges and by induction hypothesis has an obstruction along with v. Otherwise, $P' = t, x_1, \ldots, y$ is a path of length less than l, with at least 2 edges, such that $(y, t) \in E(G)$. Therefore, by induction hypothesis there is an obstruction along with the vertex v, contradicting that $C \cup \{v\}$ is a block graph.

From the above arguments it follows that if v has a neighbour t in block B in C, then v cannot have a neighbour t' in block B', if the shortest path between t, t' has at least 2 edges.

If $(t, t') \in E(G)$, then t, t' are contained in some block $\hat{B}$. If v is adjacent to any other vertex u not in $V(\hat{B})$ then at most one of (t, u) or (t', u) can be an edge in G, since t, t' and u are in different blocks. If there is an edge, say (t, u), then t, t', u, v forms an induced D_4, contradicting that $C \cup \{v\}$ is a block graph. Otherwise, there is a path with at least two edges between u and t. Therefore, by the previous arguments we can find an obstruction along with the vertex v. Therefore, $N_C(v) \subseteq V(\hat{B})$ when $(t, t') \in E(G)$.

Hence, it follows that there is a block $\tilde{B}$ in C such that $N_C(v) \subseteq V(\tilde{B})$. $\square$

This leads us to the following Lemma.

Lemma 5 [⋆]. *For every $v \in A$, the number of leaf blocks in C_v is $\mathcal{O}(k)$.*

Observe that in $G \setminus A$, a vertex $v \in A$ can be adjacent to at most $\mathcal{O}(k)$ leaf blocks by Observation 2 and Lemma 5. Also, for a leaf block B in $G \setminus A$, there must be an internal vertex $b \in V(B)$, such that b is adjacent to some vertex in S_v, since the Reduction rule BGVD 2 is not applicable. Therefore, the number of leaf blocks in $G \setminus A$ is $\mathcal{O}(k^2)$.

Lemma 6 [⋆]. *The number of blocks B in $G \setminus A$ such that the vertex set of B intersects with the vertex set of at least three other block in $G \setminus A$ is $\mathcal{O}(k^2)$.*

Let $\mathcal{L}$ be the of leaf blocks in $G \setminus A$ and $\mathcal{T}$ be the set of blocks in $G \setminus A$ such that each block in $\mathcal{T}$ intersects with at least three other blocks in $G \setminus A$. By Lemmas 5 and 6, we have that $|\mathcal{L}| = \mathcal{O}(k^2)$ and $|\mathcal{T}| = \mathcal{O}(k^2)$.

Let B be a block in $G^* = G \setminus (S_v \cup \{v\})$ such that the vertex set of B has exactly two cut vertices, and intersects with exactly two blocks of G^*. Furthermore, the vertex set of B has an empty intersection with leaf blocks of G^* and those blocks in G^* which vertex set intersects with at least three other blocks of G^*. Also, B has a vertex that is neighbor to v. Such blocks are called *nice degree two blocks* of v. If a block satisfies the above conditions for some vertex $w \in A$, the block is called a *nice degree two block*. We denote the set of nice degree two blocks by $\mathcal{T}_1$.

Lemma 7 [⋆]. *Let $G^* = G \setminus (S_v \cup \{v\})$. Then G^* has at most $\mathcal{O}(k)$ nice degree two blocks of v.*

What remains is to bound the number of blocks which have exactly two cut vertices and are not nice degree two blocks.

Lemma 8 [⋆]. *The number of blocks in $G \setminus A$ with exactly two cut vertices is $\mathcal{O}(k^2)$.*

Now, we have a bound on the total number of blocks in $G \setminus A$.

Lemma 9. *Consider a graph G, a positive integer k and an approximate block vertex deletion set set A of size $\mathcal{O}(k)$. If none of the Reduction rules BGVD 1 to BGVD 6 is applicable then the number of blocks in $G \setminus A$ is bounded by $\mathcal{O}(k^2)$.*

Proof. Follows from Lemmas 5, 6 and 8.   □

5.2 Bounding the Number of Internal Vertices in a Maximal Clique of the Block Graph

We start by bounding the number of internal vertices in a maximal 2-connected component of $G \setminus A$. Consider a block B in $G \setminus A$. We partition the internal vertices $V_I(B)$ of block B into three sets $\mathcal{B}, \mathcal{R}$ and $\mathcal{I}$ depending on the neighborhood of A in block B. We also partition the vertices in A depending on the number of vertices they are adjacent to in B. In Lemma 10 we show that the number of internal vertices in a block B of $G \setminus A$ is upper bounded by $\mathcal{O}(k^2)$. We do so by partitioning the vertices into different sets and bounding each of these sets separately.

Lemma 10 [⋆]. *Let (G, k) be an instance to BGVD and let A be an approximate block vertex deletion set of G of size $\mathcal{O}(k)$. If none of the Reduction rules BGVD 1 to BGVD 6 is applicable then the number of internal vertices in a block B of $G \setminus A$ is bounded by $\mathcal{O}(k^2)$.*

We wrap up our arguments to show a $\mathcal{O}(k^4)$ sized vertex kernel for BGVD, and hence prove Theorem 3.

Proof (of Theorem 3). Let (G, k) be an instance to BGVD and let A be an approximate block vertex deletion set of G of size $\mathcal{O}(k)$. Also, assume that none of the Reduction rules BGVD 1 to BGVD 6 are applicable. By Theorem 9, the number of blocks in $G \setminus A$ is bounded by $\mathcal{O}(k^2)$. By Lemma 10 the number of internal vertices in a block of $G \setminus A$ is bounded by $\mathcal{O}(k^2)$. Also note that the number of cut-vertices in $G \setminus A$ is bounded by the number of blocks in $G \setminus A$, i.e. $\mathcal{O}(k^2)$. The number of vertices in $G \setminus A$ is sum of the internal vertices in $G \setminus A$ and the number of cut vertices in $G \setminus A$. Therefore, $|V(G)| = |V(G \setminus A)| + |A| = (\mathcal{O}(k^2) \cdot \mathcal{O}(k^2) + \mathcal{O}(k^2)) + \mathcal{O}(k) = \mathcal{O}(k^4)$.   □

References

1. Bafna, V., Berman, P., Fujito, T.: A 2-approximation algorithm for the undirected feedback vertex set problem. SIAM J. Discret. Math. **12**(3), 289–297 (1999)
2. Brandstädt, A., Le, V.B., Spinrad, J.P.: Graph Classes: A Survey. Society for Industrial and Applied Mathematics, Philadelphia (1999)
3. Chen, J., Fomin, F.V., Liu, Y., Lu, S., Villanger, Y.: Improved algorithms for feedback vertex set problems. J. Comput. Syst. Sci. **74**(7), 1188–1198 (2008)
4. Cygan, M., Fomin, F.V., Kowalik, L., Lokshtanov, D., Marx, D., Pilipczuk, M., Pilipczuk, M., Saurabh, S.: Parameterized Algorithms. Springer, Switzerland (2015)
5. Diestel, R.: Graph Theory. Graduate Texts in Mathematics, vol. 173, 4th edn. Springer, Heidelberg (2012)
6. Flum, J., Grohe, M.: Parameterized Complexity Theory. Texts in Theoretical Computer Science. An EATCS Series. Springer, New York (2006)
7. Fomin, F.V., Lokshtanov, D., Misra, N., Saurabh, S., F-deletion, P.: Approximation, kernelization and optimal FPT algorithms. In: FOCS (2012)
8. Fujito, T.: A unified approximation algorithm for node-deletion problems. Discrete Appl. Math. **86**, 213–231 (1998)
9. Howorka, E.: A characterization of ptolemaic graphs. J. Graph Theor. **5**(3), 323–331 (1981)
10. Kim, E.J., Kwon, O.: A polynomial kernel for block graph vertex deletion. CoRR, abs/1506.08477 (2015)
11. Kociumaka, T., Pilipczuk, M.: Faster deterministic feedback vertex set. Inf. Process. Lett. **114**(10), 556–560 (2014)
12. Lewis, J.M., Yannakakis, M.: The node-deletion problem for hereditary properties is NP-complete. J. Comput. Syst. Sci. **20**(2), 219–230 (1980)
13. Lund, C., Yannakakis, M.: On the hardness of approximating minimization problems. J. ACM **41**, 960–981 (1994)
14. Marx, D., O'Sullivan, B., Razgon, I.: Finding small separators in linear time via treewidth reduction. ACM Trans. Algorithms **9**(4), 30 (2013)
15. Thomassé, S.: A $4k^2$ kernel for feedback vertex set. ACM Trans. Algorithms **6**(2), 32: 1–32: 8 (2010)
16. Tong, P., Lawler, E.L., Vazirani, V.V.: Solving the weighted parity problem for gammoids by reduction to graphic matching. Technical report UCB/CSD-82-103, EECS Department, University of California, Berkeley, April 1982

A Middle Curve Based on Discrete Fréchet Distance

Hee-Kap Ahn[1], Helmut Alt[2], Maike Buchin[3], Eunjin Oh[1(✉)],
Ludmila Scharf[2], and Carola Wenk[4]

[1] Pohang University of Science and Technology, Pohang, Korea
{heekap,jin9082}@postech.ac.kr
[2] Free University of Berlin, Berlin, Germany
{alt,scharf}@mi.fu-berlin.de
[3] Ruhr University Bochum, Bochum, Germany
maike.buchin@rub.de
[4] Tulane University, New Orleans, USA
cwenk@tulane.edu

Abstract. Given a set of polygonal curves we seek to find a middle curve that represents the set of curves. We require that the middle curve consists of points of the input curves and that it minimizes the discrete Fréchet distance to the input curves. We present algorithms for three different variants of this problem: computing an ordered middle curve, computing an ordered and restricted middle curve, and computing an unordered middle curve.

1 Introduction

Sequential point data, such as time series and trajectories, are ever increasing due to technological advances, and the analysis of these data calls for efficient algorithms. An important analysis task is to find a "representative" or "middle" curve for a set of similar curves. For instance, this could be the route of a group of people or animals traveling together. Or it could be a representation of a handwritten letter for a class of similar handwritten letters. Such a middle curve provides a concise representation of the data, which is useful for data analysis and for reducing the size of the data, possibly by many magnitudes.

Since sampled locations are more reliable than positions interpolated in between those, we seek a middle curve consisting only of sampled point locations. The middle curve should then be as close as possible to the path of the individuals, hence we ask for it to minimize the discrete Fréchet distance to these. The Fréchet distance [1] and the discrete Fréchet distance [5] are well-known distance measures, which have been successfully used before in analyzing handwritten characters [7] and trajectories [2,9].

This work was partially supported by research grant AL 253/8-1 from Deutsche Forschungsgemeinschaft (German Science Association), and by the National Science Foundation under grant CCF-1301911. Work by Ahn and Oh was supported by the NRF grant 2011-0030044 (SRC-GAIA) funded by the government of Korea.

E. Kranakis et al. (Eds.): LATIN 2016, LNCS 9644, pp. 14–26, 2016.
DOI: 10.1007/978-3-662-49529-2_2

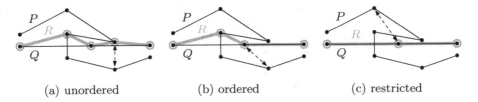

(a) unordered   (b) ordered   (c) restricted

Fig. 1. Illustration of the three different cases. The curve R is an optimal middle curve for each case. The two-way arrow which points to a point in $P \cup Q$ and a point in R indicates a mapping between two points realizing the discrete Fréchet distance.

We consider three variants of this problem, which we introduce now more formally for two curves. Given two point sequences P and Q, we wish to compute a *middle curve* R consisting of points from $P \cup Q$ that minimizes $\max(d_F(R, P),$ $d_F(R, Q))$, where d_F denotes the discrete Fréchet distance. In the following we assume that each point in R uniquely corresponds to a point in P or Q (in particular, if P and Q share points). We say a middle curve R is *ordered*, if any two points of P occurring in R have the same order as in P, likewise with points from Q. And we call an ordered middle curve R *restricted*, if points on R are mapped to themselves in a matching realizing the discrete Fréchet distance. Recall that points from R originate from P or Q, hence this seems a natural restriction. Furthermore, we distinguish whether points may occur *multiple times* or not (but still respecting the order/restriction if applicable).

Figure 1 illustrates the three cases we consider: the ordered, restricted, and unordered cases. Note how adding the restrictions (from unordered to restricted) changes the middle curve and increases the distance to the input curves. Requiring to respect the order of the input curves seems very natural. However, as we will see, the unordered case allows for the most efficient algorithm. Matching a vertex to itself on the middle curve is also natural. Furthermore, we will see that the restricted case allows for a more efficient algorithm.

Related Work. Several papers [3,6,8] study the problem of finding a middle curve but without the restriction that the middle curve should consist of points of the input curves. Buchin et al. [3] and Kreveld et al. [8] both require the middle curve to use parts of edges of the input. Buchin et al. aim to always "stay in the middle" in the sense of a median and give an $O(k^2 n^2)$-time algorithm, where k is the number of given curves and n is the number of vertices in each curve. Kreveld et al. aim to be as close as possible to all trajectories at any time and allow small jumps between different trajectories and give an $O(k^3 n)$-time algorithm. Note that neither of these approaches makes use of the Fréchet distance or its variants. Using neither input vertices nor input edges, Har-Peled and Raichel [6] show that a curve minimizing the Fréchet distance to k input curves can be computed in $O(n^k)$ time in the k-dimensional free space using the radius of the smallest enclosing disk as "distance".

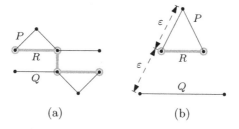

Fig. 2. (a) The middle curve may need to consist of vertices from both curves. (b) The 2-approximation is tight.

2-Approximation. A simple observation is that any of the input curves is a 2-approximation for minimizing the distance, which follows by triangle inequality. The 2-approximation is tight, as the example in Fig. 2 shows. We observe, however, that for an optimal middle curve we may need to choose a subset of vertices from both curves.

Our Results. We present algorithms for three variants of this problem for $k \geq 2$ curves of size at most n each:

1. **Ordered case:** An $O(n^{2k})$-time algorithm for computing an optimal ordered middle curve.
2. **Restricted case:** An $O(n^k \log^k n)$-time algorithm for computing an optimal restricted middle curve.
3. **Unordered case:** An $O(n^k \log n)$-time algorithm for computing an optimal unordered middle curve.

In the following sections, we present the algorithms for these cases. The algorithms for the restricted and the unordered cases allow points to appear multiple times. In the ordered case, we give algorithms for both.

Note that all algorithms run in time exponential in k, the number of trajectories. Hence these are practical only for small k. Other algorithms that compute variants of the Fréchet distance for k curves such as [4] and [6] also take time exponential in k due to the use of a k-dimensional free space diagram. Hence we do not expect faster algorithms for finding a middle curve based on the (discrete) Fréchet distance.

2 Algorithm for the Ordered Case

Here we present a dynamic programming algorithm for computing an ordered middle curve R. We first consider the case of two input curves $P = (p_1, \ldots, p_n)$ and $Q = (q_1, \ldots, q_m)$, and we do not allow multiple occurrences of the same point on R. Later we show how to generalize the algorithm to multiple input curves and to allow multiple point occurrences. Let $P_i, 1 \leq i \leq n$ denote the prefix $(p_1, ..., p_i)$ of P, where P_0 is defined as the empty sequence.

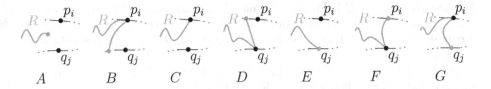

Fig. 3. Illustration of cases in the dynamic programming.

The dynamic programming algorithm operates with four-dimensional Boolean arrays of the form $X[i, j, k, l], 0 \le i, k \le n, 0 \le j, l \le m$, where $X[i, j, k, l]$ is **true** iff there exists an ordered sequence R from points in $P_i \cup Q_j$ with

$$\max(d_F(R, P_k), d_F(R, Q_l)) \le \varepsilon.$$

We say in this case that R *covers* P_k and Q_l (at distance ε). Clearly, the decision problem has a positive answer iff $X[n, m, n, m]$, or any $X[i, j, n, m]$, is true.

In order to determine the value of some $X[i, j, k, l]$ from entries of X with lower indices, we need more information, particularly, whether there is a covering sequence R in which the points p_i and q_j occur, and if they do, whether they occur in the interior or at the end of the sequence. To this end, we can represent the array X as the component-wise disjunction of seven Boolean arrays

$$X = A \vee B \vee C \vee D \vee E \vee F \vee G.$$

For each array defined below, a sequence R covering P_k and Q_l exists with the following properties, respectively:

$A[i, j, k, l]$: R contains neither p_i nor q_j.
$B[i, j, k, l]$: R contains p_i in its interior but does not contain q_j.
$C[i, j, k, l]$: R ends in p_i but does not contain q_j.
$D[i, j, k, l]$: R contains q_j in its interior but does not contain p_i.
$E[i, j, k, l]$: R ends in q_j but does not contain p_i.
$F[i, j, k, l]$: R contains q_j in its interior and ends in p_i.
$G[i, j, k, l]$: R contains p_i in its interior and ends in q_j.

Observe that R cannot contain both p_i and q_j in its interior. See Fig. 3 for an illustration of the seven different cases that can occur. Our dynamic programming algorithm is based on these recursive identities for $i, j, k, l \ge 0$:

$$
\begin{aligned}
A[0, 0, 0, 0] &= \text{true} \\
A[0, 0, k, l] &= \text{false for } k \ge 1 \text{ or } l \ge 1 \\
A[i, 0, k, l] &= X[i - 1, 0, k, l] \\
A[0, j, k, l] &= X[0, j - 1, k, l] \\
A[i, j, k, l] &= X[i - 1, j - 1, k, l] \\
B[i, 0, k, l] &= B[0, j, k, l] = \text{false} \\
B[i, j, k, l] &= G[i, j - 1, k, l] \vee B[i, j - 1, k, l]
\end{aligned}
$$

The first equality for B holds since p_i must be at the end of R if no points from Q are available. In the second equality, $G[i, j-1, k, l]$ accounts for the case that R contains q_{j-1} (which then must be at the end), and $B[i, j-1, k, l]$ for the case that it does not.

In the following, let $cl(p, q)$ denote the truth value of $\|p - q\| \le \varepsilon$, for two points p and q. The following identities hold for C.

$$C[i, j, 0, l] = C[i, j, k, 0] = C[0, j, k, l] = \texttt{false}$$
$$C[i, j, k, l] = cl(p_i, p_k) \wedge cl(p_i, q_l) \wedge$$
$$(A[i, j, k-1, l-1] \vee A[i, j, k-1, l] \vee A[i, j, k, l-1] \vee$$
$$C[i, j, k-1, l-1] \vee C[i, j, k-1, l] \vee C[i, j, k, l-1])$$

The first two equalities hold because only an empty middle curve can cover 0 points. The equality for $C[i, j, k, l]$ models the two cases of whether the final point p_i in R covers p_k and q_l only, or whether it also covers additional points that occur earlier in the sequences P_k and Q_l. The entries of D and E can be determined analogously to the ones of B and C with the roles of p_i and q_j exchanged. The identities of F have similar explanations as the ones of C:

$$F[0, j, k, l] = F[i, 0, k, l] = F[i, j, 0, l] = F[i, j, k, 0] = \texttt{false}$$
$$F[i, j, k, l] = cl(p_i, p_k) \wedge cl(p_i, q_l) \wedge$$
$$(D[i, j, k-1, l-1] \vee D[i, j, k-1, l] \vee D[i, j, k, l-1] \vee$$
$$E[i, j, k-1, l-1] \vee E[i, j, k-1, l] \vee E[i, j, k, l-1] \vee$$
$$F[i, j, k-1, l-1] \vee F[i, j, k-1, l] \vee F[i, j, k, l-1])$$

The entries of G can be determined analogously to the ones of F with the roles of p_i and q_j exchanged.

The dynamic programming algorithm runs in time $O(n^2 m^2)$, which is the size of each of the seven arrays. Not only the existence of a covering sequence R, but R itself can be computed by setting a pointer for each array entry of the form $Y[i, j, k, l]$, which is set to $\texttt{true}$, to the 4-tuple(s) of indices at the right hand side of an equality that has made it true. Note that there can be an exponential number of valid middle curves.

For the optimization problem, we can adapt a dynamic programming to compute the minimal value such that a covering middle curve exists. For this, X takes the minimum value of A to G; initialization is to $0|\infty$ instead of $\texttt{true}|\texttt{false}$; $\vee$ becomes min, and $\wedge$ becomes max. In this way we can solve the optimization problem in the same time as the decision problem.

The decision and optimization algorithms can be generalized to k sequences $P^1, ..., P^k$. The running time in this case is $O(n_1^2...n_k^2)$ for constant k, but the number of arrays is $2^{k-1}k + 2^k - 1$. The dynamic programming algorithm can also be modified to allow multiple occurrences of points on R, which requires distinguishing slightly more cases than before. Note that the length of a middle curve is at most nk if points may not appear multiple times, and at most $2nk$ if they may appear multiple times. The latter bound follows from a longest monotone path in the array of size n^{2k}.

Theorem 1. *For two polygonal curves with m and n vertices, the optimization problem for the ordered case can be solved in $O(m^2 n^2)$ time. An optimal covering sequence can be computed in the same time. For $k \geq 2$ curves of size at most n each, the optimization can be solved in $O(n^{2k})$ time.*

3   Algorithm for the Restricted Case

Now we consider the case where the reparameterizations are restricted to map every vertex of R to itself in the input curve it originated from. This case allows for a more efficient dynamic programming algorithm.

For this, we define arrays similar to Sect. 2. Let $X[i,j], 0 \leq i \leq n, 0 \leq j \leq m$, be **true** iff there exists an ordered sequence R from points in $P_i \cup Q_j$ with

$$\max(d_F(R, P_i), d_F(R, Q_j)) \leq \varepsilon,$$

with the restriction that any vertex of R is mapped to itself in the input curve it originated from. We say in this case that R *restrictively covers* P_i and Q_j. Clearly, the decision problem has a positive answer iff $X[n,m]$ is **true**. Similar to Sect. 2 we can write X as a disjunction of three Boolean arrays

$$X = A' \vee C' \vee E'.$$

For each array defined below, a sequence R covering P_i and Q_j exists with the following properties, respectively:

$A'[i,j]$: R ends in neither p_i nor q_j (but may contain one of them in its interior).

$C'[i,j]$: R ends in p_i (and may or may not contain q_j in its interior).

$E'[i,j]$: R ends in q_j (and may or may not contain p_i in its interior).

In contrast to Sect. 2, we now only distinguish the cases by the last point of R. Hence, we only distinguish three cases. (In comparison to the ordered case, A' combines A, B, D, and C' combines C, F, and E' combines E, G).

We compute all $X[i,j]$ incrementally for increasing j and increasing i using dynamic programming. Consider p_i being matched to q_j. We use the *upper* wedge $U_P(i,j)$ to describe the resulting coverage of P and Q. Specifically, $U_P(i,j)$ denotes the set of index pairs (i',j') such that $\|p_{i''} - p_i\| \leq \varepsilon$ and $\|q_{j''} - p_i\| \leq \varepsilon$ for all $i \leq i'' \leq i'$ and $j \leq j'' \leq j'$. That is, $U_P(i,j)$ consists of the connected set of index pairs $(i',j') \geq (i,j)$ that are covered by p_i. The *lower* wedge $L_P(i,j)$ denotes the set of index pairs (i',j') such that $\|p_{i''} - p_i\| \leq \varepsilon$ and $\|q_{j''} - p_i\| \leq \varepsilon$ for all $i' \leq i'' \leq i$ and $j' \leq j'' \leq j$. Furthermore, we define the *extended lower wedge* $\hat{L}_P(i,j)$ which, in addition to all points in the lower wedge $L_P(i,j)$ also contains (i',j') immediately to the left or below, i.e., for which $(i' + 1, j')$, $(i', j' + 1)$, or $(i' + 1, j' + 1)$ is contained in $L_P(i,j)$. The wedges $U_Q[i,j]$, $L_Q[i,j]$, and $\hat{L}_Q[i,j]$ are defined analogously, consisting of point pairs $(p_{i'}, q_{j'})$ for which $p_{i'}$ and $q_{j'}$

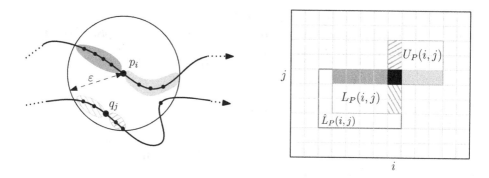

Fig. 4. Illustration of the wedges.

are both close to q_i. Figure 4 illustrates these wedges for a pair (i, j). Using this terminology we observe:

$$A'[i, j] \Leftrightarrow (\exists i' < i, j' \leq j : (C'[i', j'] \wedge (i, j) \in U_P(i', j')))$$
$$\vee (\exists i' \leq i, j' < j : (E'[i', j'] \wedge (i, j) \in U_Q(i', j')))$$
$$C'[i, j] \Leftrightarrow cl(p_i, q_j) \wedge (\exists i' \leq i, j' < j : (X[i', j'] \wedge (i', j') \in \hat{L}_P(i, j)))$$
$$E'[i, j] \Leftrightarrow cl(p_i, q_j) \wedge (\exists i' < i, j' \leq j : (X[i', j'] \wedge (i', j') \in \hat{L}_Q(i, j)))$$

During the dynamic programming, in order to efficiently compute the values $X[i, j] = A'[i, j] \vee C'[i, j] \vee E'[i, j]$ we maintain the upper envelope $\bar{X}$ of all true elements in X. More specifically, we define $\bar{X}[i] = \max\{j \mid X[i, j] = \texttt{true}\}$. Note that X as well as $\bar{X}$ change during the dynamic programming for increasing j and i.

We store $\bar{X}$ in an augmented balanced binary search tree sorted on i. Each leaf corresponds to an index i and stores $\bar{X}[i]$. Each internal node v represents the interval of indices stored in the leaves of the subtree rooted at v, and stores two key values $m[v]$ and $M[v]$. Here, $m[v]$ is the minimum of all $\bar{X}[i]$ over all leaves i in the subtree rooted at v, and $M[v]$ is the maximum.

We need the following two operations.

1. *Querying whether a rectangle intersects $\bar{X}$.* Given an extended lower wedge with bottom-left corner (i_B, j_B) and top-right corner (i_T, j_T), we need to check if there is an index pair (i, j) such that $j_B \leq \bar{X}[i]$ and $i_B \leq i \leq i_T$. This can be done as follows. Consider the search paths from the root to i_B and i_T. Let u_c be the lowest common ancestor of i_B and i_T. Whenever we descend into the right child at a node v on the path from u_c to the node i_T, we check the maximum key value of the left child v_L of v. The interval corresponding to v_L is fully contained in the interval $[i_B, i_T]$. Thus, if $M[v_L] \geq j_B$, the correct answer for the query is "yes". Otherwise, we do not need to consider the subtree rooted v_L further. Whenever we descend into the left child at a node on the path to i_B, we check the answer for the query analogously. Hence we can answer the query while we traverse the two paths, which takes logarithmic time.

2. *Updating* $\bar{X}$. We are given an upper wedge whose bottom-left corner is (i_B, j_B) and top-right corner is (i_T, j_T). We need to update $\bar{X}[i]$ to j_T for all $i_B \leq i \leq i_T$, if $\bar{X}[i] < j_T$.

We traverse the balanced binary search tree from the root as follows. Assume that we reach a node v. If $j_T \leq m[v]$ or the interval corresponding to v does not intersect $[i_B, i_T]$, then we do not need to update the values of the leaf nodes in the subtree rooted at v. Hence we do not traverse this subtree. If $m[v] < j_T$ and the interval corresponding to v intersects $[i_B, i_T]$, then we need to search further in the subtree rooted at v. So, we move to both children of v.

Finally we reach some leaf nodes, which will be updated. Then we go back to the root from those leaf nodes and update the key values for internal nodes lying on the paths. It is easy to see that the running time of the update is $O(c \log n)$, where c is the number of indices which are updated.

The algorithm consists of two parts: constructing all wedges and constructing the free space matrix X.

Constructing All Wedges. We construct the wedge $U_P(i, j)$ as follows: For fixed p_i, we first find the largest $k \geq i$ such that all $p_i, \ldots, p_k$ are in the disk of radius ε around p_i. Then we find the largest $l \geq j$ such that all $q_j, \ldots, q_l$ are in the disk of radius ε around p_i. This determines the upper right corner (k, l) of $U_P(i, j)$. Note that (k, l) is also the upper right corner for all $U_P(i, j')$ for $j \leq j' \leq l$. Hence, all wedges $U_P(i, j)$ can be computed in $O(m + n)$ time using two linear scans, one over P and one over Q. The wedges $U_Q(i, j)$, $L_P(i, j)$, $L_Q(i, j)$ are computed in a similar manner.

Constructing the Free Space Matrix X. First, initialize all $X[i, j]$ to `false`, except for $X[0, 0]$ which is set to `true`. Then compute $X[i, j]$ for $j = 1$ to m and for $i = 1$ to n. In each iteration, we process (p_i, q_j) only if they can be matched to each other, i.e., if $cl(p_i, q_j)$.

If $X[i, j]$ is `false`, i.e., we do not yet know of a middle curve covering P_i and Q_j, we first check whether adding p_i or q_j to a covering sequence extends the coverage to here. For this, we check if $\hat{L}_P(i, j)$ or $\hat{L}_Q(i, j)$ intersects $\bar{X}$. If $\hat{L}_P(i, j)$ intersects $\bar{X}$ then p_i can be added to a covering sequence, and we set $X[i, j]$ =`true`. Since in this case q_j can be added in addition, we set a flag in $X[i, j]$ to P, indicating that p_i has to be added first. Conversely, if $\hat{L}_Q(i, j)$ intersects $\bar{X}$, then q_j can be added to a covering sequence, and we do the same, setting a flag for q_j this time.

If $X[i, j]$ is `true`, then both p_i or q_j can be added to a covering sequence, hence we add the points covered by p_i or q_j, i.e., $U_P(i, j)$ and $U_Q(i, j)$, to X and $\bar{X}$. The wedge $U_P(i, j)$ is *added to X and $\bar{X}$* as follows: We update $\bar{X}$ with $U_P(i, j)$. During the update step we can identify all pairs $(i', j') \in U_P(i, j)$ with $\neg X[i', j']$; these are all (i', j') such that i' is a leaf in $\bar{X}$ that gets updated and

$\max(j_B, \bar{X}[i']) \leq j' \leq j_T$ where (i_B, j_B) is the lower left and (i_T, j_T) the upper right corner of $U_P(i,j)$. We set all $X[i',j'] =$ true and store a pointer from (i',j') to (i,j) that is labeled with P. Adding $U_Q(i,j)$ to X and $\bar{X}$ is done in a similar manner, but the pointers are labeled with Q. Note that the upper wedges are added to X in such a way that each $X[i,j]$ is touched only once, and at that time it is set to true.

The algorithm can now be summarized as follows.

Set $X[i,j] =$ false for all index pairs (i,j) except $X[0,0]$ which is set to true.
for $j = 1$ to m do
    for $i = 1$ to n do
        if $cl(p_i, q_j)$:
            if $\neg X[i,j]$: If $\hat{L}_P(i,j)$ or $\hat{L}_Q(i,j)$ intersects $\bar{X}$, set $X[i,j]$ to true and
               set the according flag in $X[i,j]$.
            if $X[i,j]$: Add $U_P(i,j)$ and $U_Q(i,j)$ to X and $\bar{X}$.

Analysis. For the correctness of the algorithm, observe that if $X[i,j]$ holds because of $A'[i,j]$, then it is marked when the last point of a covering is processed. If $X[i,j]$ holds by $C'[i,j]$ or $E'[i,j]$, then this is handled in the $\neg X[i,j] \wedge cl(p_i, q_j)$ case of the algorithm.

The running time for computing all wedges is $O((m+n)^2)$ since for each point $p_i \in P$ or $q_j \in Q$, we perform a constant number of linear scans. For the main part of the dynamic programming algorithm, when we consider an index pair (i,j), we may perform a query on $\bar{X}$ which takes $O(\log(mn))$ time, and we may add one or two upper wedges to X. The update operation that is part of adding a wedge takes $O(c \log n)$ time, where c is the number of indices that are updated. Note that $\bar{X}[i]$ is updated at most m times for each index i in total, and $X[i,j]$ is updated at most once for each index pair (i,j). Thus the running time for the decision algorithm is $O((m+n)^2 + mn\log(mn))$.

Lemma 1. *For two polygonal curves with m and n vertices, the decision problem for the restricted case can be solved in $O((m+n)^2 + mn\log(mn))$ time.*

Note that the algorithm allows multiple occurrences of vertices. However it restricts that if a vertex occurs multiple times, then all vertices of the other curve that occur in between are matched to that vertex in the discrete Fréchet matching. Figure 5 shows an example of this.

Optimization. The optimal distance will take one of the distances between pairs of points from $P \cup Q$, hence we first sort all distances in $O((m+n)^2 \log(mn))$ time and again search over them using the decision algorithm.

Lemma 2. *An optimal covering sequence for the restricted case can be computed in $O((m+n)^2 \log(mn) + mn\log^2(mn))$ time.*

Several Curves. For $k > 2$ curves the decision algorithm works the same with a $k-1$ dimensional range tree for $\bar{X}$ and runtime $O(n^k \log^{k-1} n)$. We again search over all distances between two points from any curves, so the optimal middle curve can be computed in $O(n^k \log^k n)$ time.

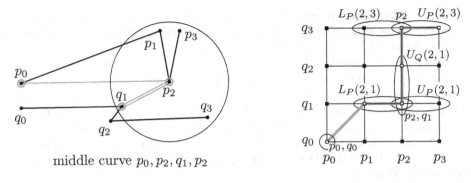

middle curve p_0, p_2, q_1, p_2

Fig. 5. An example of a middle curve R that uses a vertex (p_2) multiple times.

Output a Middle Curve. Using the pointers set by the algorithm, the algorithm can also output a middle curve. Note that a middle curve computed by the algorithm may have up to $2nk$ vertices. This follows from the algorithm because at each (i, j) at most two vertices (p_i and q_j) are added, and the length of a longest monotone path in the n^k grid is nk.

Theorem 2. *For two polygonal curves with m and n vertices, the decision problem for the restricted case can be solved in $O((m + n)^2 + mn \log(mn))$ time. An optimal covering sequence can be computed in $O((m + n)^2 \log(mn) + mn \log^2(mn))$ time. For $k \geq 2$ curves of size at most n each, the optimization can be solved in $O(n^k \log^k n)$ time.*

4   Algorithm for the Unordered Case

Let again $P = (p_1, \ldots, p_n)$ and $Q = (q_1, \ldots, q_m)$ be two input curves. To solve the decision problem for the unordered case, we modify the dynamic programming algorithm for computing the discrete Fréchet distance of two curves [5] as follows. We consider the $n \times m$ matrix X, which we call the *free space matrix*. Each entry $X[i, j]$ corresponds to the pair (p_i, q_j) of points. In contrast to the original algorithm, we color an entry $X[i, j]$ *free* if and only if there exists a point v from P or Q such that v has distance at most ε to both p_i and q_j. Then we search for a monotone path within the free entries in X.

4.1   Algorithm for the Decision Problem

We describe how to compute the labels more efficiently for the decision problem. Here, we use a circular sweep to determine for each point p_i all points q_j such that $X[i, j]$ is free, i.e., there is some point v of P or Q which has distance at most ε to both p_i and q_j. Let $U_{p_i}(\varepsilon)$ be the union of disks of radius ε centered at points in $P \cup Q$ and containing p_i. Then, for a point $q_j \in Q$ contained in $U_{p_i}(\varepsilon)$, $X[i, j]$ is free. To compute $X[i, j]$ for all $q_j \in Q$, we construct $U_{p_i}(\varepsilon)$ and perform a circular sweep around p_i for all points in Q. Once the circular arcs of the boundary $\partial U_{p_i}(\varepsilon)$ and all points $q_j \in Q$ are sorted along p_i in clockwise fashion, the circular sweep takes $O(m + n)$ time.

We design an algorithm that computes $U_{p_i}(\varepsilon)$ efficiently by constructing two data structures, called the *history* $\mathcal{H}_{p_i}$ and the *deletion list* $\mathcal{D}_{p_i}$. In the *pre-processing* phase, we gradually increase ε and construct the two data structures. When a fixed ε is given, we compute $U_{p_i}(\varepsilon)$ using the two data structures in the *construction* phase. This will allow us to solve the optimization problem efficiently in Sect. 4.2. The construction phase takes $O(m + n)$ time while the preprocessing phase takes $O(mn \log(mn))$ time. The space we use for the data structures is $O(mn)$.

In this extended abstract we give a sketch of the algorithm. The details will be presented in a full version of this paper.

The data structures for a point $p \in P$.

1. The history list $\mathcal{H}_{p_i} = \{x_1, \ldots, x_l\}$: This list represents the order of circular arcs of $\partial U_{p_i}(\varepsilon)$ for all $\varepsilon > 0$. For any three elements in $\mathcal{H}_{p_i}$, if all arcs corresponding to the elements appear on $\partial U_{p_i}(\varepsilon)$ for some $\varepsilon > 0$, then the order of them on $\partial U_{p_i}(\varepsilon)$ is the same as the one in $\mathcal{H}_{p_i}$.
2. The deletion list $\mathcal{D}_{p_i} = \{(\varepsilon_1, \varepsilon_1'), \ldots, (\varepsilon_t, \varepsilon_t')\}$: The k-th element of this list is assigned to the point in $P \cup Q$ that is the k-th closest to p_i. For any $\varepsilon > 0$, the disk of radius ε, centered at the k-th closest point, has at most two arcs appearing on $\partial U_{p_i}(\varepsilon)$. An arc of the disk disappears from $\partial U_{p_i}(\varepsilon)$ at $\varepsilon = \varepsilon_k$, and the other arc disappears from $\partial U_{p_i}(\varepsilon)$ at $\varepsilon = \varepsilon_k'$. Since $\mathcal{D}_{p_i}$ is an array of size $m + n$, we can access each element in $O(1)$ time.

Theorem 3. *For two polygonal curves with m and n vertices, the decision problem for the unordered case can be solved in $O(mn)$ time with $O(mn \log(mn))$ preprocessing time. A covering sequence can be computed in the same time.*

4.2 Algorithm for the Optimization Problem

We apply binary search on the set of distances between pairs of points from $P \cup Q$ involved in each step. Without loss of generality, assume that $n \leq m$. There are $O((m+n)^2)$ distances between pairs of points from $P \cup Q$, but we will show that we need only $O(mn)$ of them to compute the optimal distance ε^*.

1. Compute the set $\mathcal{D}$ of distances between pairs of points that are either both from P, or one from P and one from Q.
2. Sort the $O(mn)$ distances of $\mathcal{D}$ and apply binary search on the sorted list with the decision algorithm above. Let $[\varepsilon_1, \varepsilon_2]$ be the interval returned by the decision algorithm with $\varepsilon_1, \varepsilon_2 \in \mathcal{D}$. If $\varepsilon_1 \neq \varepsilon^*$ and $\varepsilon_2 \neq \varepsilon^*$, then ε^* is the distance of a pair of points in Q.
3. To find ε^*, for each point $p_i \in P$,
   (a) compute the set S_{p_i} of points in $P \cup Q$ that are at distance at most ε_2 from p_i, and construct the Voronoi diagram $\mathrm{VD}(S_{p_i})$.

(b) For each point q_j in $Q \setminus S_{p_i}$, locate the cell of $\text{VD}(S_{p_i})$ that contains q_j. If the site r associated with the cell is from Q and $\varepsilon_1 < \|q_j - r\| < \varepsilon_2$, then $\|q_j - r\|$ is a candidate for ε^*.

4. For a point pair (p_i, q_j), there exists at most one such point $r \in Q$, thus there are $O(mn)$ candidates in total, and we sort them and again apply binary search on the sorted list with the decision algorithm above.

Analysis. Let (p_i, q_j, r) be a tuple realizing ε^*, i.e., $\max(\|p_i - r\|, \|q_j - r\|) = \varepsilon^*$. Clearly, r is the point in $P \cup Q$ that minimizes $\max(\|p_i - r\|, \|q_j - r\|)$. If $r \in Q$, then r is the point in S_{p_i} that is closest to q_j. Thus, r is the point site associated with the Voronoi cell in $\text{VD}(S_{p_i})$ that contains q_j. This proves that ε^* is in the set of all candidates.

Let us analyze the running time of the optimization algorithm. The set $\mathcal{D}$ can be constructed in $O(mn)$ time. Sorting the distances in $\mathcal{D}$ takes $O(mn \log(mn))$ time. The binary search on the sorted list with the decision algorithm takes $O(mn \log(mn))$ time as the preprocessing phase is executed only once for each $p_i \in P$ and the history and deletion lists can be reused for different radii. In step 3, the Voronoi diagram $VD(S_{p_i})$ can be constructed in $O(m \log(mn))$ time for each $p_i \in P$, and the point location can be performed in the same time. Step 3(b) takes $O(m \log(mn))$ time for each $p_i \in P$.

Several Curves. The decision algorithm can be extended to k curves $P^1 = (p_1^1, \ldots, p_{n_1}^1), \ldots, P^k = (p_1^k, \ldots, p_{n_k}^k)$. If the outer loop iterates over all points $p_{i_1} \in P^1$ for $1 \leq i_1 \leq n_1$, then we determine which points $p_{i_2} \in P^2, \ldots, p_{i_k} \in P^k$ lie inside the disk of radius ε centered at p_{i_1}. For all tuples $(i_1, \ldots, i_k)$ the corresponding entries in the k-dimensional free space matrix are marked as free. The running time is $O(n_1 N \log N + M)$ where $N = \sum_{i=1}^k n_i$ and $M = \prod_{i=1}^k n_i$. If the history data structure has already been constructed, this running time can be reduced to $O(n_1 N + M)$ time. For a constant $k \geq 2$ curves of size at most n each, the running time becomes $O(n^k)$.

To compute an optimal middle curve, we sort all distances between point pairs from $P^1 \cup \ldots \cup P^k$ and search the optimal distance among them. Thus, we can compute an optimal covering sequence in $O(n^k \log n)$ time.

Theorem 4. *For two polygonal curves with m and n vertices, the optimization problem for the unordered case can be solved in $O(mn \log(mn))$ time. An optimal covering sequence can be computed in the same time. For $k \geq 2$ curves of size at most n each, the optimization can be solved in $O(n^k \log n)$ time.*

Acknowledgments. This work was initiated at the 17th Korean Workshop on Computational Geometry. We thank the organizers and all participants for the stimulating atmosphere. In particular we thank Fabian Stehn and Wolfgang Mulzer for discussing this paper.

References

1. Alt, H., Godau, M.: Computing the Fréchet distance between two polygonal curves. Int. J. Comput. Geom. Appl. **5**, 75–91 (1995)
2. Buchin, K., Buchin, M., Gudmundsson, J., Löffler, M., Luo, J.: Detecting commuting patterns by clustering subtrajectories. In: Hong, S.-H., Nagamochi, H., Fukunaga, T. (eds.) ISAAC 2008. LNCS, vol. 5369, pp. 644–655. Springer, Heidelberg (2008)
3. Buchin, K., Buchin, M., van Kreveld, M., Löffler, M., Silveira, R.I., Wenk, C., Wiratma, L.: Median trajectories. Algorithmica **66**(3), 595–614 (2013)
4. Dumitrescu, A., Rote, G.: On the Fréchet distance of a set of curves. In: Proceedings of the 16th Canadian Conference on Computational Geometry, CCCG 2004, Concordia University, Montréal, Québec, Canada, pp. 162–165, 9–11 August 2004
5. Eiter, T., Mannila, H.: Computing discrete Fréchet distance. Technical report, Technische Universität Wien (1994)
6. Har-Peled, S., Raichel, B.: The Fréchet distance revisited and extended. ACM Trans. Algorithms **10**(1), 3:1–3:22 (2014)
7. Sriraghavendra, E., Karthik, K., Bhattacharyya, C.: Fréchet distance based approach for searching online handwritten documents. In: Proceedings of the Ninth International Conference on Document Analysis and Recognition, ICDAR 2007, vol. 1, pp. 461–465. IEEE Computer Society (2007)
8. van Kreveld, M.J., Löffler, M., Staals, F.: Central trajectories. In: 31st European Workshop on Computational Geometry (EuroCG), Book of Abstracts, pp. 129–132 (2015)
9. Zhu, H., Luo, J., Yin, H., Zhou, X., Huang, J.Z., Zhan, F.B.: Mining trajectory corridors using Fréchet distance and meshing grids. In: Zaki, M.J., Yu, J.X., Ravindran, B., Pudi, V. (eds.) PAKDD 2010, Part I. LNCS, vol. 6118, pp. 228–237. Springer, Heidelberg (2010)

Comparison-Based FIFO Buffer Management in QoS Switches

Kamal Al-Bawani[1]([✉]), Matthias Englert[2], and Matthias Westermann[3]

[1] Department of Computer Science, RWTH Aachen University, Aachen, Germany
kbawani@cs.rwth-aachen.de
[2] DIMAP and Department of Computer Science, University of Warwick, Coventry, UK
englert@dcs.warwick.ac.uk
[3] Department of Computer Science, TU Dortmund, Dortmund, Germany
matthias.westermann@cs.tu-dortmund.de

Abstract. The following online problem arises in network devices, e.g., switches, with quality of service (QoS) guarantees. In each time step, an arbitrary number of packets arrive at a single FIFO buffer and only one packet can be transmitted. Packets may be kept in the buffer of limited size and, due to the FIFO property, the sequence of transmitted packets has to be a subsequence of the arriving packets. The differentiated service concept is implemented by attributing each packet with a non-negative value corresponding to its service level. A buffer management algorithm can reject arriving packets and preempt buffered packets. The goal is to maximize the total value of transmitted packets.

We study comparison-based buffer management algorithms, i.e., algorithms that make their decisions based solely on the relative order between packet values with no regard to the actual values. This kind of algorithms proves to be robust in the realm of QoS switches. Kesselman et al. [13] present a comparison-based algorithm that is 2-competitive. For a long time, it has been an open problem whether a comparison-based algorithm exists with a competitive ratio below 2. We present a lower bound of $1 + 1/\sqrt{2} \approx 1.707$ on the competitive ratio of any deterministic comparison-based algorithm and give an algorithm that matches this lower bound in the case of monotonic sequences, i.e., packets arrive in a non-decreasing order according to their values.

Keywords: Online algorithms · Competitive analysis · Network switches · Buffer management · Quality of service · Comparison-based

1 Introduction

We consider the following online problem which arises in network devices, e.g., switches, with quality of service (QoS) guarantees. In each time step, an arbitrary

---

The second and third author's work was supported by ERC Grant Agreement No. 307696.

E. Kranakis et al. (Eds.): LATIN 2016, LNCS 9644, pp. 27–40, 2016.
DOI: 10.1007/978-3-662-49529-2_3

number of packets arrive at a single buffer, i.e., a FIFO queue, of bounded capacity. Each packet has a non-negative value attributing its service level (also known as class of service (CoS)). Packets are stored in the buffer and only one packet can be transmitted in each time step. Due to the FIFO property, the sequence of transmitted packets has to be a subsequence of the arriving packets. A buffer management algorithm can reject arriving packets and preempt packets that were previously inserted into the buffer. The goal is to maximize the total value of transmitted packets.

In probabilistic analysis of network traffic, packet arrivals are often assumed to be Poisson processes. However, such processes are not considered to model network traffic accurately due to the fact that in reality packets have been observed to frequently arrive in bursts rather than in smooth Poisson-like flows (see, e.g., [15,17]). Therefore, we do not make any prior assumptions about the arrival behavior of packets, and instead resort to the framework of competitive analysis [16], which is the typical worst-case analysis used to assess the performance of online algorithms, i.e., algorithms whose input is revealed piece by piece over time, and the decision they make in each time step is irrevocable.

In competitive analysis, the benefit of an online algorithm is compared to the benefit of an optimal algorithm OPT which is assumed to know the entire input sequence in advance. An online algorithm ONL is called *c-competitive* if, for each input sequence σ, the benefit of OPT over σ is at most c times the benefit of ONL over σ. The value c is also called the *competitive ratio* of ONL.

Comparison-Based Buffer Management. In QoS networks, packet values are only an implementation of the concept of differentiated service. A packet value stands for the packet's service level, i.e., the priority with which this packet is transmitted, and does not have any intrinsic meaning in itself. However, just slight changes to the packet values, even though the relative order of their corresponding service levels is preserved, can result in substantial changes in the outcome of current buffer management algorithms. We aim to design new buffer management algorithms whose behavior is independent of how the service levels are implemented in practice. Therefore, we study *comparison-based* buffer management algorithms, i.e., algorithms that make their decisions based solely on the relative order between values with no regard to the actual values. Such algorithms are robust to order-preserving changes of packet values.

Kesselman et al. [13] present the following simple GREEDY algorithm: Accept any arriving packet as long as the queue is not full. If a packet arrives while the queue is full, drop the packet with the smallest value. Clearly, GREEDY is comparison-based, and Kesselman et al. show it is 2-competitive. Since the introduction of GREEDY, it has been an open problem to show whether a comparison-based algorithm exists with a competitive ratio below 2.

1.1   Related Work

In their seminal work, Mansour, Patt-Shamir and Lapid show that GREEDY is 4-competitive. Kesselman et al. [13] show that the exact competitive ratio of GREEDY is $2 - 1/B$, where B is the size of the buffer.

Azar and Richter [7] introduce the 0/1 principle for the analysis of comparison-based algorithms in a variety of buffering models. They show the following theorem.

Theorem 1 [7]. *Let* ALG *be a comparison-based switching algorithm (deterministic or randomized).* ALG *is c-competitive if and only if* ALG *is c-competitive for all input sequences of packets with values 0 and 1, under every possible way of breaking ties between equal values.*

For our model of a single FIFO queue, Andelman [4] employs the 0/1 principle to give a randomized comparison-based algorithm with a competitive ratio of 1.75. In fact, this is the only randomized algorithm known for this model.

In a related model with multiple FIFO queues, Azar and Richter [7] give a comparison-based deterministic algorithm with a competitive ratio of 3. In another related model, where the buffer is not FIFO and packet values are not known for the online algorithm, Azar et al. [6] use the 0/1 principle to show a randomized algorithm with a competitive ratio of 1.69. This algorithm is modified to a 1.55-competitive randomized algorithm, and a lower bound of 1.5 on the competitive ratio of any randomized algorithm is shown for that model [5].

Apart from comparison-based algorithms, the model of a single FIFO queue has been extensively studied. Kesselman, Mansour, and van Stee [12] give the state-of-the-art algorithm PG, and prove that PG is 1.983-competitive. Additionally, they give a lower bound of $(1 + \sqrt{5})/2 \approx 1.618$ on the competitive ratio of PG and a lower bound of 1.419 on the competitive ratio of any deterministic algorithm. Algorithm PG adopts the same preemption strategy of GREEDY and moreover, upon the arrival of a packet p, it proactively drops the first packet in the queue whose value is within a fraction of the value of p. This additional rule makes PG non-comparison-based. Bansal et al. [8] slightly modify PG and show that the modified algorithm is 1.75-competitive. Finally, Englert and Westermann [10] show that PG is in fact 1.732-competitive and give a lower bound of $1 + (1/\sqrt{2}) \approx 1.707$ on its competitive ratio.

In the case where packets take on only two values, 1 and $\alpha > 1$, Kesselman et al. [13] give a lower bound of 1.282 on the competitive ratio of any deterministic algorithm. Englert and Westermann [10] give an algorithm that matches this lower bound. In the non-preemptive model of this case, Andelman et al. [3] optimally provide a deterministic algorithm which matches a lower bound of $2 - 1/\alpha$ given by Aiello et al. [1] on the competitive ratio of any deterministic algorithm.

In the general-value case of the non-preemptive model, Andelman et al. [3] show a lower bound of $1 + \ln(\alpha)$ for any deterministic algorithm, where α is the ratio between the maximum and minimum packet values. This bound is achieved by a deterministic algorithm given by Andelman and Mansour [2].

The problem of online buffer management has also been studied under several other models. For example, the bounded delay model, where packets have deadlines besides their values [9,14]. A recent and comprehensive survey on this problem and most of its variants is given in [11].

1.2   Our Results

We present a lower bound of $1 + 1/\sqrt{2} \approx 1.707$ on the competitive ratio of any deterministic comparison-based algorithm. This lower bound is significantly larger than the lower bound of 1.419 known for general deterministic algorithms. We also give an algorithm, CPG, that matches our lower bound in the case of *monotonic sequences*, i.e., packets arrive in a non-decreasing order according to their values. Note that GREEDY remains 2-competitive in the case of monotonic sequences. For general sequences, we give a lower bound of 1.829 on the competitive ratio of CPG.

An intriguing question in this respect is whether a comparison-based algorithm with a competitive ratio close to $1 + 1/\sqrt{2} \approx 1.707$ could exist. If so, this would mean that we do not need to know the actual values of packets in order to compete with PG, the best non-comparison-based algorithm so far. If not, and in particular if 2 is the right lower bound for any comparison-based algorithm, the desired robustness of this kind of algorithms must come at a price, namely, a significantly degraded performance.

1.3   Model and Notations

We consider a single buffer that can store up to B packets. All packets are assumed to be of unit size, and each packet p is associated with a non-negative value, denoted by $v(p)$, that corresponds to its level of service. The buffer is implemented as a FIFO queue, i.e., packets are stored and sent in the order of their arrival.

Time is discretized into slots of unit length. An arbitrary number of packets arrive at fractional (non-integral) times, while at most one packet is sent from the queue, i.e., transmitted, at every integral time, i.e., at the end of each time slot. We denote the arrival time of a packet p by $\text{arr}(p)$. An arriving packet is either inserted into the queue or it is rejected, and an enqueued packet may be dropped from the queue before it is sent. The latter event is called preemption. Rejected and preempted packets are lost.

We denote the arrival of a new packet as an arrival event, and the sending of a packet as a send event. An input sequence σ consists of arrival and send events. The time that precedes the first arrival event of the sequence is denoted as time 0. We assume that the queue of any algorithm is empty at time 0.

The benefit that an algorithm ALG makes on an input sequence σ is denoted by $\text{ALG}(\sigma)$, and is defined as the total value of packets that ALG sends. We aim at maximizing this benefit. We denote by OPT an optimal (offline) algorithm that sends packets in FIFO order.

2 Lower Bound

The following theorem shows that no deterministic comparison-based algorithm can be better than $1 + 1/\sqrt{2} \approx 1.707$. Recall that the best lower bound for general deterministic algorithms is 1.419.

Theorem 2. *The competitive ratio of any deterministic comparison-based algorithm is at least $1 + 1/\sqrt{2} \approx 1.707$.*

Proof. Fix an online algorithm ONL. The adversary constructs a sequence of packets with non-decreasing values over a number of iterations. The 0/1 values corresponding to the packets' real values are revealed only when the sequence stops. In each iteration, the adversary generates a burst of B packets in one time slot followed by a number of individual packets, each in one time slot. We call a slot with B arrivals a *bursty* slot, and a slot with one arrival a *light* slot. A construction routine is repeated by the adversary until the desired lower bound is obtained. For $i \geq 0$, let f_i denote the i-th bursty slot, and let t_i denote the number of time slots that ONL takes to send and preempt all packets that it has in slot f_i.

As initialization, the adversary generates B packets in the first time slot. Thus, the first slot is f_0. After that, the adversary generates t_0 light slots, i.e., one packet arrives in each slot. Now, starting with $i = 0$, the adversary constructs the rest of the sequence by the following routine which is repeated until $t_i \geq B/\sqrt{2}$.

1. Generate the bursty slot f_{i+1}.
2. If $t_i \geq B/\sqrt{2}$, stop the sequence. At this point, all packets that arrive between f_0 and f_i (inclusive) are revealed as 0-packets and all packets after that are revealed as 1-packets, i.e., the 1-packets are those which arrive in the t_i light slots and in the bursty slot f_{i+1}. Clearly, the optimal algorithm, denoted as OPT, will send all the 1-packets while ONL will gain only the B 1-packets which it has in slot f_{i+1}. Notice that ONL sends only 0-packets in the t_i light slots. Hence, provided that $t_i \geq B/\sqrt{2}$,

$$\frac{\text{OPT}}{\text{ONL}} = \frac{t_i + B}{B}$$

$$\geq \frac{B/\sqrt{2} + B}{B} \ .$$

3. If $t_i < B/\sqrt{2}$, continue the sequence after f_{i+1} by generating t_{i+1} light slots.
   (a) If $t_{i+1} \leq t_i$, stop the sequence. At this point, all packets that arrive between f_0 and f_i (inclusive) are revealed as 0-packets and all packets after that are revealed as 1-packets, i.e., the 1-packets are those which arrive in the t_i and t_{i+1} light slots and in the bursty slot f_{i+1}. Clearly, OPT will send all the 1-packets while ONL will send only t_{i+1} packets of the B 1-packets which it has in slot f_{i+1} and also the t_{i+1} 1-packets which

it collects after f_{i+1}. Hence, provided that $B > \sqrt{2} \cdot t_i$ and $t_i \geq t_{i+1}$,

$$\frac{\text{OPT}}{\text{ONL}} = \frac{t_i + B + t_{i+1}}{t_{i+1} + t_{i+1}}$$

$$\geq \frac{t_i + \sqrt{2} \cdot t_i + t_{i+1}}{2 \cdot t_{i+1}}$$

$$\geq \frac{(1 + \sqrt{2})t_{i+1} + t_{i+1}}{2 \cdot t_{i+1}} .$$

(b) If $t_{i+1} > t_i$, set $i = i + 1$ and repeat the routine.

Obviously, the above routine terminates eventually, because a new iteration is invoked only when $t_{i+1} > t_i$, and thus the amount of t_i is strictly increased in each iteration. Therefore, there must exist i such that $t_i \geq B/\sqrt{2}$. □

3   Algorithm CPG

We present a comparison-based preemptive greedy (CPG) algorithm. This algorithm can be seen as the comparison-based version of the well-studied algorithm PG [8]. It follows a similar rule of preemption as PG, but without addressing the actual values of packets: Roughly speaking, once you have a set S of β packets in the queue with a packet r in front of them, such that r is less valuable than each packet in S, preempt r.

CPG is described more precisely in Algorithm 1. To avoid using the same set of packets to preempt many other packets, it associates with each arriving packet p a non-negative $credit$, denoted by $c(p)$. For a set S of packets, $c(S)$ will also denote the total credit of all packets in S. We now describe the above preemption rule in more details.

First, we present the notations of $preemptable$ packets and $preempting$ sets. Assume that a packet p arrives at time t. Let $Q(t)$ be the set of packets in CPG's queue immediately before t. For any packet $r \in Q(t)$, if there exists a set $S \subseteq (Q(t) \cup \{p\}) \setminus \{r\}$ such that (i) $p \in S$, (ii) $c(S) \geq \beta$, and (iii) for each packet $q \in (S)$, $\text{arr}(q) \geq \text{arr}(r)$ and $v(q) \geq v(r)$, then we say that r is $preemptable$ by p. Furthermore, we call S a $preempting$ set of r.

A packet r is preempted upon the arrival of another packet p if r is the first packet in the queue (in the FIFO order) such that r is preemptable by p and the value of r is less than the value of the packet that is behind r in the queue (if any). After a packet r is preempted, CPG invokes a subroutine CHARGE to deduct a total of β units from the credits of the preempting packets of r. This charging operation can be done arbitrarily, but subject to the non-negative constraint of credits, i.e., $c(p) \geq 0$, for any packet p. After that, the algorithm proceeds similarly to GREEDY: It inserts the arriving packet p into the queue if the queue is not full or p is more valuable than the packet with the least value in the queue. In the latter case, the packet with the least value is dropped. Otherwise, p is rejected. Finally, in send events, CPG simply sends the packet at the head of the queue.

Algorithm 1. CPG

---

arrival event. A packet p arrives at time t:

$c(p) \leftarrow 1$;

Let r be the first packet in the queue such that r is preemptable by p and the value of r is less than the value of the packet that is behind r (if any).

if r *exists* **then**
   | let S be a preempting set of r;
   | drop r;
   | CHARGE(S);

if *the queue is not full* **then**
   | insert p;
else
   | let q be the packet with the smallest value in the queue;
   | **if** $v(q) < v(p)$ **then**
   | | drop q and insert p;
   | **else**
   | | reject p;

Notice that CPG is a comparison-based algorithm. Hence, by Theorem 1, it is sufficient to show the competitive ratio of CPG for only $0/1$ sequences. We denote a packet of value 0 as 0-packet, and of value 1 as 1-packet.

Lost Packets. We distinguish between three types of packets lost by CPG:

1. Rejected packets: An arriving packet p is rejected if the queue is full and no packet in the queue is less valuable than p.
2. Evicted packets: An enqueued packet q is evicted by an arriving packet p if the queue is full and q is the least valuable among p and the packets in the queue.
3. Preempted packets: An enqueued packet r is preempted upon the arrival of another packet p if r is the first packet in the queue such that r is preemptable by p and the value of r is less than the value of the packet that is behind r (if any).

Notice that a 1-packet can only be evicted by a 1-packet. Also, if a 1-packet q is preempted, the preempting packets of q are all 1-packets.

3.1 Monotonic Sequences

In this section, we consider input sequences in which packets arrive with non-decreasing values, i.e., for any two packets p and q, $v(p) \leq v(q)$ if and only if p arrives before q. We observe that the 2-competitive greedy algorithm from [13] remains 2-competitive in this case.

Theorem 3. *Choosing* $\beta = \sqrt{2} + 1$, *the competitive ratio of* CPG *is at most* $1 + 1/\sqrt{2} \approx 1.707$.

For the rest of the analysis, we fix an event sequence σ of only 0- and 1-packets. Furthermore, let $Q(t)$ (resp. $Q^*(t)$) denote the set of 1-packets in the queue of CPG (resp. OPT) at time t.

Assumptions on the Optimal and the Online Algorithms. Notice that OPT, in contrast to CPG, can determine whether a packet of σ has value 0 or 1. Therefore, we can assume that OPT accepts arriving 1-packets as long as its queue is not full, and rejects all 0-packets. In send events, it sends 1-packets (in FIFO order) unless its queue is empty.

We further assume that no packets arrive after the first time in which the queue of CPG becomes empty. This assumption is also without loss of generality as we can partition σ into phases such that each phase satisfies this assumption and the queues of CPG and OPT are both empty at the start and the end of the phase. Then, it is sufficient to show the competitive ratio on any arbitrary phase. Consider for example the creation of the first phase. Let t be the first time in which the queue of CPG becomes empty. We postpone the packets arriving after t until OPT's queue is empty as well, say at time t', so that OPT and CPG are both empty at t'. This change can only increase the benefit of OPT. Clearly, t' defines the end of the first phase, and the next arrival event in σ defines the start of the second phase. The remaining of σ can be further partitioned in the same way.

Overflow Time Slot. We call a time slot in which CPG rejects or evicts 1-packets an overflow time slot. Assume for the moment that at least one overflow time slot occurs in σ. For the rest of the analysis, we will use f to denote the last overflow slot, and t_f to denote the time immediately before this slot ends. Obviously, rejection and eviction of 1-packets can happen only when the queue of CPG is full of 1-packets. Let t'_f be the point of time immediately before the first rejection or eviction in f takes place. Thus, the number of 1-packets in the queue at time t'_f is B. Thereafter, between t'_f and t_f, any 1-packet that is evicted or preempted is replaced by the 1-packet whose arrival invokes that eviction or preemption. Thus, the size of the queue does not change between t'_f and t_f, and hence the following observation.

Remark 1. $|Q(t_f)| = B$.

Furthermore, the following lemma shows that the B 1-packets in the queue at time t_f can be used to preempt at most one 1-packet in later arrival events.

Lemma 1. *Consider any arrival event e. Let t be the time immediately after e and let $D(t)$ denote the set of packets in the queue at time t except the head packet. Then, $c(D(t)) < \beta$.*

Proof. We show the lemma by contradiction. Let e be the first arrival event in σ, such that immediately after e, say at time t, $c(D(t)) \geq \beta$. Hence, immediately before e, say at time t', $\beta > c(D(t')) \geq \beta - 1$, since the total credit of the queue cannot increase by more than 1 in each arrival event.

Now, let p be the packet arriving in e and let q be the head packet at the arrival of p. Recall that σ is monotonic. Thus, the packets behind q in the queue and packet p are all at least as valuable as q. Hence, adding the credit of p to $c(D(t'))$, these packets would preempt q upon the arrival of p, and thus the total credit would decrease by 1. Therefore, $c(D(\cdot))$ does not change between t' and t which contradicts the definition of e. $\square$

Before we proceed, we introduce further notations. Let $\mathrm{ARR}(t, t')$ denote the set of 1-packets that arrive in σ between time t and t'. Furthermore, let $\mathrm{SENT}(t, t')$ and $\mathrm{LOST}(t, t')$ denote the set of 1-packets that CPG sends and loses, respectively, between time t and t'. Similarly, we define $\mathrm{SENT}^*(t, t')$ and $\mathrm{LOST}^*(t, t')$ for OPT.

Lemma 2. *It holds that*

$$|\mathrm{LOST}(0, t_f)| - |\mathrm{LOST}^*(0, t_f)| + |Q(t_f)| - |Q^*(t_f)| = |\mathrm{SENT}^*(0, t_f)| - |\mathrm{SENT}(0, t_f)| \ .$$

Proof. The lemma follows from this simple observation:

$$\begin{aligned}
|Q(t_f)| &+ |\mathrm{SENT}(0, t_f)| + |\mathrm{LOST}(0, t_f)| \\
&= |\mathrm{ARR}(0, t_f)| \\
&= |Q^*(t_f)| + |\mathrm{SENT}^*(0, t_f)| + |\mathrm{LOST}^*(0, t_f)| \ .
\end{aligned}$$

$\square$

The following lemma is crucial for the analysis of CPG. It essentially upper-bounds the number of 1-packets that CPG loses between the start of the sequence and the end of the overflow slot.

Lemma 3. $|\mathrm{LOST}(0, t_f)| - |\mathrm{LOST}^*(0, t_f)| + |Q(t_f)| - |Q^*(t_f)| \leq \frac{\beta}{\beta+1} B$.

Proof. First, we present further notations. If an algorithm ALG does not send anything in a sent event t, we say that ALG sends a $\emptyset$-packet in t. We call a send event in which OPT sends an x-packet and CPG sends a y-packet an x/y send event, where x and y take on values from $\{0, 1, \emptyset\}$. Furthermore, we denote by $\delta_{x/y}(t, t')$ the number of x/y send events that occur between time t and time t'.

Now, observe that

$$\begin{aligned}
|\mathrm{SENT}^*(0, t_f)| &= \delta_{1/0}(0, t_f) + \delta_{1/1}(0, t_f) + \delta_{1/\emptyset}(0, t_f) \ , \\
|\mathrm{SENT}(0, t_f)| &= \delta_{0/1}(0, t_f) + \delta_{1/1}(0, t_f) + \delta_{\emptyset/1}(0, t_f) \ .
\end{aligned}$$

Recall that OPT does not send 0-packets and that, by assumption, the queue of CPG does not get empty before t_f. Thus, $\delta_{0/1}(0, t_f) = \delta_{1/\emptyset}(0, t_f) = 0$, and therefore

$$|\mathrm{SENT}^*(0, t_f)| - |\mathrm{SENT}(0, t_f)| = \delta_{1/0}(0, t_f) - \delta_{\emptyset/1}(0, t_f) \leq \delta_{1/0}(0, t_f) \ .$$

Hence, given Lemma 2, it suffices to show that $\delta_{1/0}(0, t_f) \leq \lfloor \frac{\beta}{\beta+1} B \rfloor$.

Assume for the sake of contradiction that $\delta_{1/0}(0, t_f) > \lfloor \frac{\beta}{\beta+1} B \rfloor$. Let M_1 (resp. M_0) be the set of 1-packets (resp. 0-packets) that OPT (resp. CPG) sends in these 1/0 send events. Thus,

$$|M_1| = |M_0| \geq \lfloor \frac{\beta}{\beta+1} B \rfloor + 1 > \frac{\beta}{\beta+1} B \ . \tag{1}$$

Let p (resp. q) denote the first arriving packet in M_1 (resp. M_0). Furthermore, let r be the last arriving packet in M_0 and denote the time in which it is sent by t_r. Recall that σ is monotonic. Thus, all the 1-packets of M_1 arrive after r. Moreover, since CPG's buffer is FIFO, none of these 1-packets is sent before t_r. Also, since r, which is a 0-packet, is before them in the queue and is eventually sent, CPG does not either reject, evict or preempt any 1-packet from M_1 before t_r. Therefore, all the 1-packets of M_1 must be in the queue of CPG at time t_r.

Let's now look closely at the queue of CPG immediately after the arrival of p. Let that time be denoted as t_p. Since q is sent with p in the same 1/0 send event and since r is between q and p (by the above argument), q and r must be in the queue as well at time t_p. Moreover, since r is the last arriving 0-packet in M_0, the remaining 0-packets of M_0 must also be in the queue at t_p. Hence, the queue of CPG contains all the packets of M_0 along with p at time t_p.

Next, notice that all the 1-packets of M_1 are inserted in CPG's queue after r (which is a 0-packet) without preempting it. Since the credits of packets are used only in preemption, the credits of these 1-packets must be used to preempt other packets before r. Let R be the set of these preempted packets. Obviously,

$$|R| \geq \lfloor |M_1|/\beta \rfloor > |M_1|/\beta - 1 \ . \tag{2}$$

Since the packets of R cannot be preempted before the arrivals of the packets of M_1, all of them must be then before r in the queue at time t_p. Thus, the queue of CPG contains the packets of both M_0 and R along with p at time t_p. Clearly, $M_0 \cap R \cap \{p\} = \emptyset$. Hence, given Inequalities 1 and 2, the size of CPG's queue at t_p is at least

$$|M_0| + |R| + 1 > |M_0| + |M_1|/\beta = \frac{\beta+1}{\beta} M_0 > \frac{\beta+1}{\beta} \frac{\beta}{\beta+1} B = B \ ,$$

which is strictly larger than B, and hence a contradiction.   □

So far, our discussion has been focused on one half of the scene; namely, the one between the start of the sequence and the end of the last overflow slot. We shall now move our focus to the second half which extends from time t_f until the end of the sequence.

First, let t_0 be defined as follows: $t_0 = 0$ if no overflow slot occurs in σ, and $t_0 = t_f$ otherwise. Notice that in both cases, no 1-packet is rejected or evicted by CPG after t_0. Moreover, let T denote the first time by which the sequence stops and the queues of OPT and CPG are both empty. Thus, the benefits of OPT and CPG are given by $|\text{SENT}^*(0, T)|$ and $|\text{SENT}(0, T)|$, respectively.

The following lemma is the main ingredient of the proof of the competitive ratio.

Lemma 4. $|\text{SENT}(0,T)| \geq (\beta - 1)\left(|\text{LOST}(0,T)| - |\text{LOST}^*(0,T)|\right)$.

Proof. Obviously, we can write $|\text{SENT}(0,T)|$ as follows:

$$|\text{SENT}(0,T)| = |\text{SENT}(0,t_0)| + |Q(t_0)| + |\text{ARR}(t_0,T)| - |\text{LOST}(t_0,T)|$$
$$\geq |Q(t_0)| + |\text{ARR}(t_0,T)| - |\text{LOST}(t_0,T)| .$$

Due to the fact that no 1-packet is rejected or evicted by CPG after t_0, all packets in $\text{LOST}(t_0,T)$ are lost by preemption. We further notice that all these packets are preempted using packets that arrive after t_0. This is trivial in case $t_0 = 0$, and follows from Lemma 1 in case $t_0 = t_f$. (In fact, in the latter case, at most one packet of $\text{LOST}(t_0,T)$ can be preempted using the credits of packets that are in the queue at time t_f, but this anomaly can be covered by introducing an additive constant in the competitive ratio of CPG.) Since preempting a packet requires a credit of β, preempting the packets of $\text{LOST}(t_0,T)$ implies the arrival of at least new $\beta |\text{LOST}(t_0,T)|$ 1-packets that are inserted into the queue after t_0. Thus, $|\text{ARR}(t_0,T)| \geq \beta |\text{LOST}(t_0,T)|$, and hence we can rewrite $|\text{SENT}(0,T)|$ in the following way:

$$|\text{SENT}(0,T)| \geq |Q(t_0)| + \beta |\text{LOST}(t_0,T)| - |\text{LOST}(t_0,T)|$$
$$= |Q(t_0)| + (\beta - 1)|\text{LOST}(t_0,T)|$$
$$\geq |Q(t_0)| + (\beta - 1)\left(|\text{LOST}(t_0,T)| - |\text{LOST}^*(t_0,T)|\right) .$$

Now, if $t_0 = 0$, then $|Q(t_0)| = 0$ and thus the lemma follows immediately. If $t_0 = t_f$, we continue as follows:

$$|\text{SENT}(0,T)| \geq B + (\beta - 1)\Big(|\text{LOST}(t_f,T)| - |\text{LOST}^*(t_f,T)| - |Q(t_f)| + |Q^*(t_f)|\Big)$$
$$\geq \frac{\beta+1}{\beta}\Big(|\text{LOST}(0,t_f)| - |\text{LOST}^*(0,t_f)| + |Q(t_f)| - |Q^*(t_f)|\Big)$$
$$+ (\beta - 1)\Big(|\text{LOST}(t_f,T)| - |\text{LOST}^*(t_f,T)| - |Q(t_f)| + |Q^*(t_f)|\Big)$$
$$= (\beta - 1)\Big(|\text{LOST}(0,T)| - |\text{LOST}^*(0,T)|\Big) ,$$

where the first inequality follows from Remark 1, the second inequality from Lemma 3, and the equality from the fact that $\beta - 1 = (\beta+1)/\beta$, for $\beta = \sqrt{2}+1$. $\qquad\square$

Now, we use Lemma 4 to show that $|\text{SENT}^*(0,T)| \leq \frac{\beta}{\beta-1}|\text{SENT}(0,T)|$, which obviously completes the proof of Theorem 3:

$$|\text{SENT}^*(0,T)| = |\text{ARR}(0,T)| - |\text{LOST}^*(0,T)|$$
$$= |\text{SENT}(0,T)| + |\text{LOST}(0,T)| - |\text{LOST}^*(0,T)|$$
$$\leq |\text{SENT}(0,T)| + \frac{1}{\beta-1}|\text{SENT}(0,T)|$$
$$= \frac{\beta}{\beta-1}|\text{SENT}(0,T)| .$$

3.2   General Sequences

Theorem 3 shows that CPG is an optimal comparison-based algorithm in the case of monotonic sequences. In this section, we investigate how this algorithm performs on general sequences.

We notice that Lemma 1 does not necessarily hold for general sequences. Therefore, after an overflow of 1-packets takes place, the total credit of the 1-packets in the online buffer can significantly exceed β and thus some of these packets may be used in a subsequent time steps to preempt other packets from the same group, i.e., the group of the B 1-packets from the overflow slot. Consequently, the lower bound of B on the number of CPG's sent 1-packets may no longer hold in the general case, resulting in a competitive ratio worse than 1.707. Such a bad scenario for CPG is illustrated in the proof of the following theorem and leads to a lower bound of 1.829 on its competitive ratio.

Theorem 4. *For any value of β, CPG cannot be better than 1.829-competitive.*

Proof. The adversary generates one of the following two sequences based on the value of β:

Case 1. $\beta \leq 2.206$: In the first time slot, B 1-packets are generated in an increasing order (with respect to their original values). After that, no more packets arrive. Clearly, OPT sends all the B packets, while in CPG, every β packets preempt a packet from the front. Thus, CPG preempts B/β in total. Hence, its competitive ratio is given by

$$\frac{\text{OPT}}{\text{CPG}} = \frac{B}{B - B/\beta} = \frac{\beta}{\beta - 1} \geq 1.829.$$

Case 2. $\beta > 2.206$: In the first time slot, $(B-1)$ 0-packets are generated followed by a single 1-packet. Then, over the next $\beta B/(\beta + 1) - 1$ time slots, a single 1-packet is generated in each slot. Let M_1 denote the set of those 1-packets that arrive in the first $\beta B/(\beta+1)$ time slots. After that, in slot number $\beta B/(\beta+1)+1$, B 1-packets arrive at once. Let M_2 denote the set of these packets. Finally, in the next $B/(\beta(\beta + 1))$ time slots, a single 1-packet arrives in each slot. Let M_3 denote the set of these packets. After that, no more packets arrive.

Clearly, OPT sends all the 1-packets in the sequence. To minimize the number of 1-packets sent by CPG, the adversary can choose the original values of the 1-packets in the following malicious way. First, the values of packets in M_2 are all strictly less than the smallest value in M_1. Let M_2' denote the set of the first $B/(\beta + 1)$ packets in M_2. The packets of M_2' are ordered as follows. For each group of β packets, starting from the earliest, the first packet is strictly smaller than the $\beta - 1$ packets behind it, and all the β packets of this group are strictly smaller than all packets before them in M_2'. For example, for $\beta = 3$, theses groups may look like $|50, 51, 51|40, 41, 41|30, 31, 31| \cdots$. For the rest of M_2, i.e., the set $M_2 \setminus M_2'$, packets are given values that are strictly less than the smallest value in M_2'. Finally, the packets in M_3 are all assigned a value that is equal to the greatest value in M_2'.

Obviously, CPG accepts all the $\beta B/(\beta + 1)$ packets of M_1 and uses them to preempt $B/(\beta + 1)$ 0-packets. Meanwhile, the rest of the B 0-packets are sent in the first $\beta B/(\beta + 1)$ time slots. Thus, the packets of M_1 will be all in the queue of CPG when the packets of M_2 arrive. Clearly, this leads to an overflow of 1-packets and only the packets of M_2' can be accepted in this time slot. These packets are inserted with full credits into the queue, and thus when each packet from M_3 arrives, it groups with $\beta - 1$ packets from M_2' to preempt the first packet in one β-group of M_2', according to the above description of M_2'. Therefore, CPG sends a total of B 1-packets only, and hence its competitive ratio is given by

$$\frac{\text{OPT}}{\text{CPG}} = \frac{|M_1| + |M_2| + |M_3|}{B}$$

$$= \frac{\beta}{\beta + 1} + 1 + \frac{1}{\beta(\beta + 1)}$$

$$= \frac{\beta(\beta + 1) + \beta^2 + 1}{\beta(\beta + 1)} \geq 1.829 \ .$$

$\square$

4   Conclusions

Our main result is a lower bound of $1 + 1/\sqrt{2} \approx 1.707$ on the competitive ratio of any deterministic comparison-based algorithm, and an algorithm, CPG, that matches this lower bound in the case of monotonic sequences. For general sequences, CPG is shown to be no better than 1.829-competitive. However, for general sequences, the intriguing question of whether there exists a deterministic comparison-based online algorithm with a competitive ratio below 2 remains open.

References

1. Aiello, W., Mansour, Y., Rajagopolan, S., Rosén, A.: Competitive queue policies for differentiated services. J. Algorithms **55**(2), 113–141 (2005)
2. Andelman, N., Mansour, Y.: Competitive management of non-preemptive queues with multiple values. In: Fich, F.E. (ed.) DISC 2003. LNCS, vol. 2848, pp. 166–180. Springer, Heidelberg (2003)
3. Andelma, N., Mansour, Y., Zhu, A.: Competitive queueing policies for QoS switches. In: Proceedings of the 14th Annual ACM-SIAM Symposium on Discrete Algorithms (SODA), pp. 761–770 (2003)
4. Andelman, N.: Randomized queue management for DiffServ. In: Proceedings of the 17th ACM Symposium on Parallelism in Algorithms and Architectures (SPAA), pp. 1–10 (2005)
5. Azar, Y., Cohen, I.R.: Serving in the dark should be done non-uniformly. In: Halldórsson, M.M., Iwama, K., Kobayashi, N., Speckmann, B. (eds.) ICALP 2015. LNCS, vol. 9134, pp. 91–102. Springer, Heidelberg (2015)
6. Azar, Y., Cohen, IR., Gamzu, I.: The loss of serving in the dark. In: Proceedings of the 45th ACM Symposium on Theory of Computing (STOC), pp. 951–960 (2013)

7. Azar, Y., Richter, Y.: The zero-one principle for switching networks. In: Proceedings of the 36th ACM Symposium on Theory of Computing (STOC), pp. 64–71 (2004)
8. Bansal, N., Fleischer, L.K., Kimbrel, T., Mahdian, M., Schieber, B., Sviridenko, M.I.: Further improvements in competitive guarantees for QoS buffering. In: Díaz, J., Karhumäki, J., Lepistö, A., Sannella, D. (eds.) ICALP 2004. LNCS, vol. 3142, pp. 196–207. Springer, Heidelberg (2004)
9. Englert, M., Westermann, M.: Considering suppressed packets improves buffer management in QoS switches. In: Proceedings of the 18th Annual ACM-SIAM Symposium on Discrete Algorithms (SODA), pp. 209–218 (2007)
10. Englert, M., Westermann, M.: Lower and upper bounds on FIFO buffer management in QoS switches. Algorithmica 53(4), 523–548 (2009)
11. Goldwasser, M.H.: A survey of buffer management policies for packet switches. SIGACT News 41, 100–128 (2010)
12. Kesselman, A., Mansour, Y., Van Stee, R.: Improved competitive guarantees for QoS buffering. Algorithmica 43(1–2), 97–111 (2005)
13. Kesselman, A., Lotker, Z., Mansour, Y., Patt-Shamir, B., Schieber, B., Sviridenko, M.: Buffer overflow management in QoS switches. SIAM J. Comput. 33(3), 563–583 (2004)
14. Li, F., Sethuraman, J., Stein, C.: Better online buffer management. In: Proceedings of the 18th Annual ACM-SIAM Symposium on Discrete Algorithms (SODA), pp. 199–208 (2007)
15. Paxson, V., Floyd, S.: Wide-area traffic: the failure of Poisson modeling. IEEE/ACM Trans. Networking 3(3), 226–244 (1995)
16. Sleator, D., Tarjan, R.: Amortized efficiency of list update and paging rules. Commun. ACM 28(2), 202–208 (1985)
17. Veres, A., Boda, M.: The chaotic nature of TCP congestion control. In: Proceedings of IEEE INFOCOM, pp. 1715–1723 (2000)

Scheduling on Power-Heterogeneous Processors

Susanne Albers[1], Evripidis Bampis[2], Dimitrios Letsios[3],
Giorgio Lucarelli[4(✉)], and Richard Stotz[1]

[1] Fakultät für Informatik, Technische Universität München, Munich, Germany
{albers,stotz}@in.tum.de
[2] Sorbonne Universités, UPMC Univ. Paris 06, UMR 7606, LIP6, Paris, France
Evripidis.Bampis@lip6.fr
[3] Univ. Nice Sophia Antipolis, CNRS, I3S, UMR 7271,
06900 Sophia Antipolis, France
dletsios@unice.fr
[4] Université Grenoble-Alpes, INP, UMR 5217, LIG,Saint-Martin-d'Hères, France
giorgio.lucarelli@inria.fr

Abstract. We consider the problem of scheduling a set of jobs, each
one specified by its release date, its deadline and its processing volume,
on a set of heterogeneous speed-scalable processors, where the energy-
consumption rate is processor-dependent. Our objective is to minimize
the total energy consumption when both the preemption and the migration of jobs are allowed. We propose a new algorithm based on a compact
linear programming formulation. Our method approaches the value of the
optimal solution within any desired accuracy for a large set of continuous power functions. Furthermore, we develop a faster combinatorial
algorithm based on flows for standard power functions and jobs whose
density is lower bounded by a small constant. Finally, we extend and
analyze the AVerage Rate (AVR) online algorithm in the heterogeneous
setting.

1 Introduction

Nowadays energy consumption of computing devices is an important issue in
both industry and academia. One of the main technological alternatives in order
to take into account the energy consumed in modern computer systems is based
on the use of speed-scalable processors where the speed of a processor may be
dynamically changed over time. When a processor runs at speed s, then the
rate with which the energy is consumed (i.e., the power) is $f(s)$ with f a non-
decreasing function of the speed. According to the well-known cube-root rule

S. Albers—Work supported by the German Research Foundation, projects Al 464/
7-1 and Al 464/9-1.
E. Bampis—Research partially supported by projet GDR-RO AGaPe of CNRS.
D. Letsios—Research partially supported by ANR project Stint and ANR program
"Investments for the Future".
G. Lucarelli—Research supported by projet ANR Moebus.

E. Kranakis et al. (Eds.): LATIN 2016, LNCS 9644, pp. 41–54, 2016.
DOI: 10.1007/978-3-662-49529-2_4

for CMOS devices, the speed of a device is proportional to the cube-root of the power and hence $f(s) = s^3$. However, the standard model that is usually studied in the literature considers that the power is $f(s) = s^\alpha$ with $\alpha > 1$ a constant. Other works consider that the power is an arbitrary convex function [4,6].

The algorithmic study of this area started with the seminal paper of Yao et al. [16], where a set of jobs, each one specified by its work, its release date and its deadline, has to be scheduled preemptively on a single processor so that the energy consumption is minimized. In [16], an optimal polynomial-time algorithm has been proposed for this problem, while Li et al. [15] proposed an optimal algorithm with lower running time. The *homogeneous multiprocessor* setting in which the *preemption* and the *migration* of the jobs are allowed has been also studied. Chen et al. [9] proposed a greedy algorithm if all jobs have common release dates and deadlines. Bingham and Greenstreet [8] presented a polynomial-time algorithm for the more general problem with arbitrary release dates and deadlines. Their algorithm is based on solving a series of linear programs. Since the complexity of this algorithm can be high for practical applications, Albers et al. [1] and Angel et al. [2], independently, have been interested in the design of a combinatorial algorithm. Both works are based on the computation of several maximum flows in appropriate networks. Albers et al. [1] have also considered the online version of the multiprocessor problem and they studied two well-known algorithms, namely the *Optimal Available (OA)* and the *Average Rate (AVR)*, which have been proposed by Yao et al. in [16] for the single-processor setting. Specifically, they proved that OA is α^α-competitive and that AVR is $(2^{\alpha-1}\alpha^\alpha + 1)$-competitive. Note that, for the single-processor case, the competitive ratio of OA cannot be better than α^α [7], while the lower bound for AVR is $2^{\alpha-1}\alpha^\alpha$ [5].

In this paper, we consider the problem of scheduling a set of jobs on a set of *power-heterogeneous* processors when the preemption and the migration of the jobs are allowed. In our setting, each processor P_p is characterized by each own power function. This means that if a processor P_p runs at speed s, then its power is given by a non-decreasing function $f_p(s)$. The motivation to study power-aware scheduling problems is based on the need for more efficient computing. Indeed, parallel heterogeneous systems with multiple cores running at lower frequencies offer better performances than a single core. However, in order to exploit the opportunities offered by the heterogeneous systems, it is essential to focus on the design of new efficient power-aware algorithms taking into account the heterogeneity of these architectures. In this direction, some recent papers [3,13,14] have studied the impact of the introduction of the heterogeneity on the difficulty of some power-aware scheduling problems. Especially in [13], Gupta et al. show that well-known priority scheduling algorithms that are energy-efficient for homogeneous systems become energy inefficient for heterogeneous systems.

For the case where job migrations are allowed and the heterogeneous power functions are convex, an algorithm has been proposed in [3] that returns a solution within an additive factor of ϵ far from the optimal and runs in time polynomial to the size of the instance and to $1/\epsilon$. This result generalizes the results of $[1, 2, 4, 8]$ from the homogeneous setting to the heterogeneous one. However, the algorithm proposed in [3] is based on solving a configuration linear program using the Ellipsoid method. Given that this method may not be very efficient in practice, we focus on other approaches. We first propose a polynomial-time algorithm based on a compact linear programming formulation which solves the problem within any desired accuracy. Our algorithm does not need the use of the Ellipsoid method like in [3] and it applies for more general than convex power functions; it is valid for a large family of continuous non-decreasing power functions.

The above result leaves open a natural question: *is it possible to generalize the flow-based approach used in* [1, 2] *for the homogeneous multiprocessor problem to the power-heterogeneous case?* This question is interesting even for standard power functions of the form $f_p(s) = s^{\alpha_p}$. This last case is the goal of the second part of our paper. However, when power-heterogeneous processors are considered some structural properties of the optimal schedules of the homogeneous case are no more valid. For instance, in the heterogeneous setting, in any optimal schedule, the speed of a job is not necessarily unique, but it may change when parts of the same job are executed on different processors. A second difficulty comes from the fact that, while in the homogeneous case the processor on which a job is executed at a given time has no influence on the energy consumption, this is a crucial decision when scheduling on heterogeneous multiprocessors. Here, we overcome these subtle difficulties and we propose a max-flow based algorithm which is rather more complicated than its homogeneous counterpart (for example, the network formulation is more enhanced). In particular, we show that it produces a solution arbitrarily close to the optimal for jobs whose density is lower bounded by a small constant; this constant depends on the exponents of the power functions. The above assumption ensures that no job is processed with a speed less than one by any processor and allows us to solve the problem by performing maximum flow computations in a principled way. Note that this assumption is reasonable in practice because the speed of a processor is multiple CPU cycles per second.

The third part of our paper is devoted to the analysis of the well known online algorithm AVR. Our analysis simplifies the analysis in [1] for the homogeneous case and allows us to extend it in the power-heterogeneous setting. Specifically, we prove that Heterogeneous-AVR is $((1 + \epsilon)(\rho + 1))$-competitive algorithm for arbitrary power functions, where ρ is the worst competitive ratio of the single-processor AVR algorithm among all processors. This turns to be $((1 + \epsilon)(\alpha^\alpha 2^{\alpha-1} + 1))$-competitive algorithm for standard power functions of the form $f_p(s) = s^{\alpha_p}$, where α is the maximum power exponent among all processors.

In the following section we formally define our problem and we give the notation that we use. In Sect. 3, we present our LP-based algorithm, while in Sect. 4 we describe a flow-based combinatorial algorithm. Finally, the Heterogeneous-AVR and its analysis are given in Sect. 5. The missing proofs are given in the full version of this paper.

2 Problem Definition and Notations

An instance of the heterogeneous speed-scaling problem consists of a set of n jobs $\mathcal{J} = \{J_1, J_2, \ldots, J_n\}$ which have to be executed by a set of m parallel speed-scalable power-heterogeneous processors $\mathcal{P} = \{P_1, P_2, \ldots, P_m\}$. Each job J_j is specified by an amount of work w_j, a release time r_j and a deadline d_j. We say that J_j is *alive* during an interval of time I, if $I \subseteq [r_j, d_j)$. Moreover, we define the *density* of a job J_j as $\delta_j = \frac{w_j}{d_j - r_j}$.

The speed of a processor can be varied over time and it corresponds to the amount of work that the processor executes per unit of time. Furthermore, the power of processor P_p (i.e. its instantaneous energy consumption) is assumed to be a function $f_p(s)$ of its speed. We consider two classes of functions:

1. *Arbitrary Power Functions*: The function $f_p(s)$ of each processor P_p is an arbitrary and continuous function of s. However, we require an oracle for computing $f_p(s)$ in polynomial time, for any value of s.
2. *Standard Power Functions*: Each processor P_p satisfies the power function $f_p(s) = s^{\alpha_p}$, where $\alpha_p > 1$ is a small constant. This is the usual assumption in the speed-scaling literature. Note that, we denote by α the maximum power exponent, i.e., $\alpha = \max_{p \in \mathcal{P}} \{\alpha_p\}$.

During an interval of time I, the energy consumption of P_p is $\int_I f_p(s_{p,t}) dt$, where $s_{p,t}$ is the speed of P_p at $t \in I$. The objective is to find a minimum energy schedule such that every job J_j is executed during the interval $[r_j, d_j)$. Preemptions and migrations of jobs are allowed, which means that a job may be executed, suspended and resumed later from the point of suspension on the same or on a different processor. However, we do not allow parallel execution of a job, i.e., each job may be executed by at most one processor at each time.

We define the important times $t_1 < t_2 < \ldots < t_\ell < t_{\ell+1}$ which correspond to all the different possible release dates and deadlines of jobs, sorted in increasing order. Moreover, let $I_i = [t_i, t_{i+1})$, for $i = 1, 2, \ldots, \ell$ and $\mathcal{I}$ be the set of all I_i's. We denote by n_i the number of jobs which are alive during I_i. Then, for each interval $I_i \in \mathcal{I}$, we define $m_i = \min\{m, n_i\}$. Furthermore, we denote by $\mathcal{J}(t)$ and $\mathcal{J}(I)$ the set of the alive jobs at time t and during the interval I, respectively. At a given time t, we say that processor P_p is *occupied* if it executes some job, or we say that it is *idle*, otherwise. For a given schedule $\mathcal{S}$, we define by $E(\mathcal{S})$ the total energy consumption of $\mathcal{S}$. Finally, we denote by $\mathcal{S}^*$ an optimal schedule and by $OPT = E(\mathcal{S}^*)$ its energy consumption.

3 LP-Based Algorithm for Generalized Power Functions

In this section we present a linear program (LP) for the heterogeneous speed-scaling problem for a wide family of continuous power functions. Our formulation is more compact than the configuration LP proposed in [3] which contains an exponential number of variables and requires the use of the Ellipsoid method. Moreover, the formulation in [3] is polynomially solvable only for convex functions.

In order to define our LP, we discretize the possible speed values. Let s_{LB} and s_{UB} be a lower and an upper bound, respectively, on the speed of any processor in an optimal schedule. For example, we could choose $s_{LB} = w_{min}/[m\sum_i |I_i|]$ and $s_{UB} = \sum_j w_j d_j/|I_{min}|$. Given a constant $\epsilon > 0$, we geometrically discretize the interval $[s_{LB}, s_{UB}]$ and we define the set of discrete speeds $D = \{s_{LB}, s_{LB}(1+\epsilon), s_{LB}(1+\epsilon)^2, \ldots, s_{LB}(1+\epsilon)^k\}$, where $k = \min\{i : s_{LB}(1+\epsilon)^i \geq s_{UB}\}$. The set D contains $O(\frac{1}{\epsilon}\log(\frac{s_{UB}}{s_{LB}}))$ different speeds.

We consider a wide class of continuous power functions satisfying the following invariant: for any speed value $s \in [s_{LB}, s_{UB}]$ and small constant $\epsilon > 0$, there exists a sufficiently small constant $\epsilon' > 0$ such that $f((1+\epsilon)s) \leq (1+\epsilon')f(s)$. Note that ϵ' is a characteristic of the function f. For example, in the case of standard power functions, we have that $f((1+\epsilon)s) \leq (1+\epsilon')f(s)$ with $\epsilon' = (1+\epsilon)^\alpha - 1$. In the reminder of this section, we consider this kind of functions.

Lemma 1. *There exists a $(1+\epsilon')$-approximate schedule such that, at each time, the speed of every processor belongs to the discrete set D, where $|D| = O(\frac{1}{\epsilon}\log(\frac{s_{UB}}{s_{LB}}))$.*

The feasibility of our LP formulation is based on the following lemma.

Lemma 2. *Consider a schedule S and let $t_{i,j,p,s}$ be the total amount of time that job J_j is processed during the interval I_i by the processor P_p with speed s. Then, S is feasible if and only if all the following hold.*

- $\sum_{i,p,s} t_{i,j,p,s} \cdot s \geq w_j$, *for each job J_j,*
- $\sum_{p,s} t_{i,j,p,s} \leq |I_i|$, *for each interval I_i and job J_j, and*
- $\sum_{j,s} t_{i,j,p,s} \leq |I_i|$, *for each interval I_i and processor P_p.*

Let $E_{p,s} = f_p(s)$ be the power consumption of processor P_p if it runs with speed s. We introduce a variable $x_{i,j,p,s}$ which corresponds to the total amount of time that the job J_j is processed during the interval I_i by the processor P_p with speed s. Then, we obtain the following LP:

$$\min \sum_{i,j,p,s} x_{i,j,p,s} \cdot E_{p,s}$$

$$\sum_{i,p,s} x_{i,j,p,s} \cdot s \geq w_j \forall j$$

$$\sum_{p,s} x_{i,j,s,p} \leq |I_i| \forall i,j$$

$$\sum_{j,s} x_{i,j,p,s} \leq |I_i| \forall i,p$$

$$x_{i,j,p,s} \geq 0 \forall i,j,p,s$$

Given a solution of the above LP, we obtain an operation of job J_j on processor P_p with processing time $\sum_s x_{i,j,p,s}$ during each interval I_i. So, for each I_i, we obtain an instance of the preemptive open shop problem, which can be solved in polynomial time with the algorithm of Gonzalez and Sahni [11]. This observation implies an algorithm for our problem, and the following theorem holds.

Theorem 1. *There is an algorithm which produces an $(1 + \epsilon')$-approximate schedule in $O(poly(n, m, \frac{1}{\epsilon}, \log \frac{s_{UB}}{s_{LB}}))$ time.*

4   Flow-Based Algorithm for Standard Power Functions

In this section, we first characterize the structure of an optimal solution for the heterogeneous speed-scaling problem with power functions of the form $f_p(s) = s^{\alpha_p}$ and jobs whose density is lower bounded by a small constant, which is defined below. Then, we derive a combinatorial algorithm based on flow computations.

4.1   Structure of an Optimal Schedule

We elaborate on the structure of a specific optimal schedule and we derive a set of properties and lemmas which are always satisfied by this optimal schedule. Since we allow preemptions and migrations of jobs, more than one processors may execute part of one job J_j. Due to convexity of the power functions, in any minimum energy schedule, the part of job J_j assigned to processor P_p is executed (preemptively) with constant speed $s_{j,p}$. Of course, a job may be executed with different speeds by different processors. However, the following lemma shows that these speeds are related through the derivatives of the power functions.

Lemma 3. *For each job $J_j \in \mathcal{J}$ which is partially executed by the processors P_p and P_q with speeds $s_{j,p}$ and $s_{j,q}$, respectively, it holds that $f'_p(s_{j,p}) = f'_q(s_{j,q})$.*

The above lemma describes the relation of the speeds of a job on different processors. Based on this, we define the *hypopower* of a job $J_j \in \mathcal{J}$ as $Q_j = f'_p(s_{j,p})$, for every $P_p \in \mathcal{P}$. The following property is a corollary of Lemma 3.

Property 1. Each job $J_j \in \mathcal{J}$ is executed with constant hypopower Q_j.

The following property implies that the jobs which are executed at each time are the ones with the greatest hypopowers.

Property 2. For each pair of jobs $J_j, J_k \in \mathcal{J}$ and time $t \in [r_j, d_j) \cap [r_k, d_k)$ such that J_j is executed at t and J_k is not executed at t, it holds that $Q_j \geq Q_k$.

In the following lemma we set the minimum job density such that all speeds in the optimal schedule are at least one.

Lemma 4. *Assume that $\delta_j \geq \max_{p,q}\{(\frac{\alpha_p}{\alpha_q})^{1/(\alpha_q-1)}\}$ for every $J_j \in \mathcal{J}$. For every pair of job $J_j \in \mathcal{J}$ and processor $P_p \in \mathcal{P}$, it holds that $s_{j,p} \geq 1$.*

By using Lemma 4, we can define an order $P_1, P_2, \ldots, P_m$ of the processors such that for any value of speed $s \geq 1$, we have that $f_1(s) \leq f_2(s) \leq \ldots \leq f_m(s)$. Observe that this order is obtained by sorting the processors in non-decreasing order of their power exponent, i.e., $\alpha_1 \leq \alpha_2 \leq \ldots \leq \alpha_m$. Furthermore, it is not hard to verify that, for every speed s of a job in the optimal schedule, it also holds that $f'_1(s) \leq f'_2(s) \leq \ldots \leq f'_m(s)$. Based on the above, we say that $P_p \in \mathcal{P}$ is *cheaper* than $P_q \in \mathcal{P}$ if $p < q$; similarly, P_q is *more expensive* than P_p. The following lemma implies that cheap processors run, in general, with greater speeds than expensive processors in the optimal schedule.

Lemma 5. *For an interval I and any pair of jobs $J_j, J_k \in \mathcal{J}$ executed by the processors $P_p, P_q \in \mathcal{P}$ during whole I, respectively, if $p < q$ then $s_{j,p} \geq s_{k,q}$.*

The next property implies that cheap processors execute, in general, jobs with greater hypopowers compared to expensive processors.

Property 3. For an interval I and any pair of jobs $J_j, J_k \in \mathcal{J}$ executed by the processors $P_p, P_q \in \mathcal{P}$ during whole I, respectively, if $p < q$ then $Q_j \geq Q_k$.

The next property specifies the set of occupied processors at each time; these are the m_i cheapest ones. The remaining processors are idle. This means that, in the optimal schedule, the total processing time of all jobs is equal to $\sum_i \{m_i \cdot |I_i|\}$, i.e., the maximum possible that any feasible schedule may have.

Property 4. During an interval $I_i \in \mathcal{I}$, the processors in $\{P_1, P_2, \ldots, P_{m_i}\}$ are occupied, while the processors in $\{P_{m_i+1}, P_{m_i+2}, \ldots, P_m\}$ are idle.

The following corollary, which follows directly from Properties (1)-(4) implies that if we know the hypopowers of the jobs in the optimal schedule, then we know the speed of each processor at each time.

Corollary 1. *Consider an interval $I_i \in \mathcal{I}$ and let J_{j_k} be the alive job during I_i with the k-th greatest hypopower, breaking ties arbitrarily. Then, at each time $t \in I_i$, processors $P_1, P_2, \ldots, P_{m_i}$ run with hypopowers $Q_{j_1} \geq Q_{j_2} \geq \cdots \geq Q_{j_{m_i}}$, respectively. Moreover, processors $P_{m_i+1}, P_{m_i+2}, \ldots, P_m$ are idle.*

Theorem 2. *Properties (1)-(4) are necessary and sufficient for optimality.*

4.2   Presentation and Analysis of the Algorithm

Given the optimal structure presented in the previous section, we are now ready to describe a polynomial-time algorithm which is based on maximum flow computations. Initially, we present the high-level idea of the algorithm and, then, we describe its main components, in more detail, together with its analysis.

High-Level Idea. Initially, we define a slightly more general problem which is the one that the algorithm actually solves. An instance of this problem is specified by a triple $< \mathcal{J}, \mathcal{P}, \mathcal{I} >$. Specifically, there is a set $\mathcal{J}$ of n jobs which have to be executed by a set $\mathcal{P}$ of m parallel processors during a set $\mathcal{I}$ of disjoint time intervals. During each interval $I_i \in \mathcal{I}$ there is a subset $\mathcal{J}(I_i) \subseteq \mathcal{J}$ of alive jobs and a subset $\mathcal{P}(I_i) \subseteq \mathcal{P}$ of available processors. Every job $J_j \in \mathcal{J}(I_i)$ (and processor $P_p \in \mathcal{P}(I_i)$) is alive (resp. available) during the whole I_i. We denote by $n_i = |\mathcal{J}(I_i)|$ (and $a_i = |\mathcal{P}(I_i)|$) the number of alive jobs (resp. available processors) during I_i. Our original problem is a special case of the above; we observe that J_j is alive during every interval $I_i \in [r_j, d_j)$ and all the m processors are available in each interval. Moreover, the optimal structure of the previous section is extended in a straightforward way to this more general problem.

Let $\mathcal{S}^*$ be an optimal schedule with the structure presented in the previous section and consider an interval $I_i \in \mathcal{I}$. By Property 4, the $m_i = \min\{a_i, n_i\}$ cheapest processors are used during the entire I_i while the remaining ones are always idle during I_i. So, the property specifies the exact amount of time, say t_p, that each processor $P_p \in \mathcal{P}$ is used in $\mathcal{S}^*$ as well as the corresponding intervals. A similar argument with the one for proving Property 1 implies that the most energy-efficient, though not necessarily feasible, way to schedule the jobs is by executing them with the same hypopower Q such that

$$\sum_{p=1}^{m} t_p \left(\frac{Q}{\alpha_p} \right)^{\frac{1}{\alpha_p - 1}} = \sum_{J_j \in \mathcal{J}} w_j \tag{1}$$

In what follows, we assume that we can compute a solution to the above equation with arbitrary precision (we explain later how to treat errors occurred due to limited precision). If there is a feasible schedule in which all jobs are executed with equal hypopower Q, then this schedule is optimal and we are done. Note that, as we explain in the next subsection, this feasibility problem can be answered with a maximum flow computation. If such a feasible schedule does not exist, then $\mathcal{J}$ can be partitioned into two disjoint and non-empty subsets $\mathcal{J}_{\geq Q}$ and $\mathcal{J}_{<Q}$ containing the jobs executed with hypopower at least Q and smaller than Q, respectively, in $\mathcal{S}^*$. In each interval $I_i \in \mathcal{I}$, Properties 2 and 3 specify the subsets of available processors $\mathcal{P}_{\geq Q}(I_i), \mathcal{P}_{<Q}(I_i) \subseteq \mathcal{P}(I_i)$ dedicated to the execution of $\mathcal{J}_{\geq Q}$ and $\mathcal{J}_{<Q}$, respectively, which are disjoint. Specifically, let $\mathcal{J}_{\geq Q}(I_i)$ and $\mathcal{J}_{<Q}(I_i)$ be the subsets of jobs of $\mathcal{J}_{\geq Q}$ and $\mathcal{J}_{<Q}$, respectively, which are alive during I_i. The jobs in $\mathcal{J}_{\geq Q}$ occupy the cheapest $\min\{a_i, |\mathcal{J}_{\geq Q}(I_i)|\}$ processors during I_i while the jobs in $\mathcal{J}_{<Q}$ use the remaining occupied processors. Then, the problem $< \mathcal{J}, \mathcal{P}, \mathcal{I} >$ can be decomposed into the two independent subproblems $< \mathcal{J}_{\geq Q}, \mathcal{P}_{\geq Q}, \mathcal{I} >$ and $< \mathcal{J}_{<Q}, \mathcal{P}_{<Q}, \mathcal{I} >$. Therefore, $< \mathcal{J}, \mathcal{P}, \mathcal{I} >$ can be decomposed recursively as it is described in Algorithm 1.

Algorithm 1. $\text{OPT}(\mathcal{J}, \mathcal{P}, \mathcal{I})$

---

1 Compute the most energy-efficient hypopower Q for executing $(\mathcal{J}, \mathcal{P}, \mathcal{I})$;
2 $(\mathcal{J}_{\geq Q}, \mathcal{P}_{\geq Q}, \mathcal{I}), (\mathcal{J}_{<Q}, \mathcal{P}_{<Q}, \mathcal{I}) \leftarrow \text{BISEPARATION}(\mathcal{J}, \mathcal{P}, \mathcal{I}, Q)$;
3 **if** $\mathcal{J} = \mathcal{J}_{\geq Q}$ **then**
4 $\quad$ | $\quad$ **return** $\text{CONSTANTHYPOPOWERSCHEDULE}(\mathcal{J}, \mathcal{P}, \mathcal{I}, Q)$;
5 **else**
6 $\quad$ | $\quad$ $\mathcal{S}_{\geq Q} \leftarrow \text{OPT}(\mathcal{J}_{\geq Q}, \mathcal{P}_{\geq Q}, \mathcal{I})$;
7 $\quad$ | $\quad$ $\mathcal{S}_{<Q} \leftarrow \text{OPT}(\mathcal{J}_{<Q}, \mathcal{P}_{<Q}, \mathcal{I})$;
8 $\quad$ | $\quad$ **return** $\mathcal{S}_{\geq Q} \cup \mathcal{S}_{<Q}$;

In order to complete the presentation of our algorithm, it remains to describe a way of answering the *feasibility* of the problem $< \mathcal{J}, \mathcal{P}, \mathcal{I} >$ in which all jobs are executed with constant hypopower Q (which has been computed according to Eq. 1) and, in the case of infeasibility, the *biseparation* procedure.

Feasibility. Consider an interval $I_i \in \mathcal{I}$ and a processor $P_p \in \mathcal{P}(I_i)$. Recall that, if processor P_p runs with hypopower Q during I_i, then its speed is $s_{i,p} = (\frac{Q}{\alpha_p})^{1/(\alpha_p-1)}$. For simplicity, in what follows, we slightly abuse our notation: let $s_{i,p}$ be the speed of the p-th cheapest (and fastest) available processor during I_i, and $\mathcal{P}(I_i)$ be the set of the m_i cheapest available processors during I_i.

The feasibility of $< \mathcal{J}, \mathcal{P}, \mathcal{I} >$ w.r.t. the hypopower Q is based on a maximum flow computation in an appropriate network $\mathcal{N}(\mathcal{J}, \mathcal{P}, \mathcal{I}, Q)$ which is defined as follows (see Fig. 1). There is a source node u_0, a node u_j for each job $J_j \in \mathcal{J}$, a node $v_{i,p}$ for each pair of interval $I_i \in \mathcal{I}$ and processor $P_p \in \mathcal{P}(I_i)$, a node v_i for each interval $I_i \in \mathcal{I}$ and a destination node v_0. Moreover, the network contains the arc (u_0, u_j) with capacity w_j for each job $J_j \in \mathcal{J}$, the arc $(u_j, v_{i,p})$ with capacity $(s_{i,p} - s_{i,p+1})|I_i|$ for each interval $I_i \in \mathcal{I}$, job $J_j \in \mathcal{J}(I_i)$ and processor $P_p \in \mathcal{P}(I_i)$, the arc $(v_{i,p}, v_i)$ with capacity $p(s_{i,p} - s_{i,p+1})|I_i|$ for each interval $I_i \in \mathcal{I}$ and processor $P_p \in \mathcal{P}(I_i)$ as well as the arc (v_i, v_0) with infinite capacity for each $I_i \in \mathcal{I}$. By convention, let $s_{i,m_i+1} = 0$. This formulation was introduced by Federgruen and Groenevelt [10] and the following theorem is a corollary of [10].

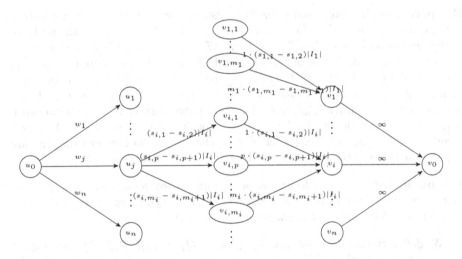

Fig. 1. The flow network $\mathcal{N}(\mathcal{J}, \mathcal{P}, \mathcal{I}, Q)$

Theorem 3. *There exists a feasible schedule of $< \mathcal{J}, \mathcal{P}, \mathcal{I} >$ with constant hypopower Q iff there exists a feasible flow in $\mathcal{N}(\mathcal{J}, \mathcal{P}, \mathcal{I}, Q)$ of value $\sum_{J_j \in \mathcal{J}} w_j$.*

Theorem 3 implies that any feasible schedule for $< \mathcal{J}, \mathcal{P}, \mathcal{I}, Q >$ can be transformed to a feasible flow of value $\sum_{J_j \in \mathcal{J}} w_j$ in the network $\mathcal{N}(\mathcal{J}, \mathcal{P}, \mathcal{I}, Q)$ and vice versa. In particular, consider a feasible schedule, an interval $I_i \in \mathcal{I}$ and assume that $w_{i,j,p} = t \cdot s_{i,p}$ units of work of job $J_j \in \mathcal{J}(I_i)$ are processed by the p-th fastest processor in $\mathcal{P}(I_i)$. Then, it holds that

$$w_{i,j,p} = t \cdot s_{i,p} = t \cdot (s_{i,p} - s_{i,p+1}) + t \cdot (s_{i,p+1} - s_{i,p+2}) + \ldots + t \cdot (s_{i,m_i} - s_{m_i+1})$$

Algorithm 2. BISEPARATION($\mathcal{J}, \mathcal{P}, \mathcal{I}, Q$)

1: $\mathcal{J}_{\geq Q} = \mathcal{J}$ and $\mathcal{J}_{<Q} = \emptyset$;
2: Find a maximum flow $\mathcal{F}$ in the network $\mathcal{N}(\mathcal{J}_{\geq Q}, \mathcal{P}, \mathcal{I}, Q)$;
3: **while** there is a path from v_0 to some job node u_j in $\widetilde{\mathcal{R}}_{\mathcal{F}}(\mathcal{J}_{\geq Q}, \mathcal{P}, \mathcal{I}, Q)$ **do**
4:    $\mathcal{J}_{\geq Q} = \mathcal{J}_{\geq Q} \setminus \{J_j\}$ and $\mathcal{J}_{<Q} = \mathcal{J}_{<Q} \cup \{J_j\}$;
5:    Remove the flow passing through u_j from $\mathcal{F}$ and u_j from $\mathcal{N}(\mathcal{J}_{\geq Q}, \mathcal{P}, \mathcal{I}, Q)$;
6: Compute $\mathcal{P}_{\geq Q}$ and $\mathcal{P}_{<Q}$ based on Corollary 1;
7: **return** $(\mathcal{J}_{\geq Q}, \mathcal{P}_{\geq Q}, \mathcal{I}), (\mathcal{J}_{<Q}, \mathcal{P}_{<Q}, \mathcal{I})$;

This observation shows the way for obtaining a feasible flow of value $\sum_{J_j \in \mathcal{J}} w_j$. Conversely, consider a feasible flow, an interval $I_i \in \mathcal{I}$, a job $J_j \in \mathcal{J}(I_i)$ and assume that $w_{i,j}$ units of flow cross the network induced by the nodes u_j, v_i and $v_{i,p}$ for each $p = 1, 2, \ldots, m_i$. By applying the algorithm of Gonzalez and Sahni [12] for scheduling a set of jobs (where job J_j has work $w_{i,j}$) with common release dates and deadlines on related machines, we obtain a feasible schedule.

Biseparation. If there is not a feasible schedule for $< \mathcal{J}, \mathcal{P}, \mathcal{I} >$ with constant hypopower Q computed by Eq. (1), we next show how to decompose the problem in the two subproblems $(\mathcal{J}_{<Q}, \mathcal{P}_{<Q}, \mathcal{I})$ and $(\mathcal{J}_{\geq Q}, \mathcal{P}_{\geq Q}, \mathcal{I})$. Initially, we introduce some notation. Consider an optimal schedule $\mathcal{S}^*$ with the structure presented in Sect. 4.1. We refer to every job $J_j \in \mathcal{J}_{\geq Q}$, i.e., which executed with hypopower at least Q, as *critical*. By Corollary 1, during interval $I_i \in \mathcal{I}$, the critical jobs occupy the $c_i = \min\{m_i, |\mathcal{J}_{\geq Q}(I_i)|\}$ fastest processors in $\mathcal{S}^*$. In the network $\mathcal{N}(\mathcal{J}, \mathcal{P}, \mathcal{I}, Q)$, we denote by $U(x, y)$ the capacity of the arc (x, y). Moreover, given a feasible (u_0, v_0)-flow $\mathcal{F}$, let $\mathcal{F}(x, y)$ the amount of flow crossing the arc (x, y). Our biseparation algorithm is based on the following lemma.

Lemma 6. *Let $\mathcal{J}' \subseteq \mathcal{J}_{<Q}$ be any subset of non-critical jobs. A job $J_j \in \mathcal{J} \setminus \mathcal{J}'$ is critical if and only if, in the network $\mathcal{N}(\mathcal{J} \setminus \mathcal{J}', \mathcal{P}, \mathcal{I}, Q)$, there exists a minimum (u_0, v_0)-cut which does not contain the arc (u_0, u_j).*

We define the *residual network* $\mathcal{R}_{\mathcal{F}}(\mathcal{J}, \mathcal{P}, \mathcal{I}, Q)$ of $\mathcal{N}(\mathcal{J}, \mathcal{P}, \mathcal{I}, Q)$ with respect to $\mathcal{F}$ as the network which contains the same nodes with $\mathcal{N}(\mathcal{J}, \mathcal{P}, \mathcal{I}, Q)$, the arc (x, y) with capacity $U(x, y) - \mathcal{F}(x, y)$, if (x, y) is not saturated by $\mathcal{F}$ in $\mathcal{N}(\mathcal{J}, \mathcal{P}, \mathcal{I}, Q)$ and the arc (y, x) with capacity $\mathcal{F}(x, y)$, if there is a positive amount of flow $\mathcal{F}(x, y) > 0$ crossing the arc (x, y) by $\mathcal{F}$ in $\mathcal{N}(\mathcal{J}, \mathcal{P}, \mathcal{I}, Q)$. Then, we define the *inverse residual network* $\widetilde{\mathcal{R}}_{\mathcal{F}}(\mathcal{J}, \mathcal{P}, \mathcal{I}, Q)$ which is the same as $\mathcal{R}_{\mathcal{F}}(\mathcal{J}, \mathcal{P}, \mathcal{I}, Q)$ except that all arcs are reversed. Algorithm 2 formally describes the biseparation procedure.

Lemma 7. *Algorithm 2 correctly identifies $\mathcal{J}_{<Q}$ and $\mathcal{J}_{\geq Q}$.*

Correctness and Running Time. The correctness of the algorithm follows from the fact that it produces a schedule satisfying Properties (1)-(4). Assume that by solving Eq. (1), we get a solution $Q+\epsilon$ instead of Q, where $\epsilon > 0$ is a small constant. If all jobs are executed with hypopower Q in the optimal schedule, then the algorithm will construct a feasible schedule in which all jobs are executed with hypopower $Q + \epsilon$. On the other hand, if there does not exists a feasible schedule of all jobs w.r.t. Q, then the algorithm will perform a biseparation w.r.t. $Q+\epsilon$. Even though this biseparation is performed w.r.t. $Q+\epsilon$, it is correct in the sense that a job is characterized as critical if and only if it is executed with hypopower at least $Q + \epsilon$ in the optimal schedule. Therefore, the algorithm will produce a $(1 + \epsilon')$-approximate schedule in which, at each time, the hypopower of a processor is at most an additive factor of ϵ more than its hypopower in the optimal schedule, where $\epsilon' > 0$ is a small constant.

Concerning its running time, it makes $O(n)$ recursive calls because there are $O(n)$ distinct values of hypopower in the optimal schedule. In every such call, it computes a hypopower value by solving Eq. (1) in $O(f(n, \frac{1}{\epsilon}))$ time, where ϵ is the desired accuracy, it computes a maximum flow in a graph with $O(nm)$ vertices in $O(g(nm))$ time and it performs $O(n)$ Breadth-First Searches in a graph with $O(n^2m)$ arcs in $O(n^3m)$.

Theorem 4. *Algorithm 1 produces a $(1+\epsilon')$-approximate schedule with running time $O(nf(n, \frac{1}{\epsilon}) + ng(nm) + n^4m)$.*

5 Online Scheduling with Heterogeneous AVR

For the single-processor case, Yao et al. [16] proposed the *AVerage Rate algorithm* (or simply AVR) and they showed that it is $\alpha^\alpha \cdot 2^{\alpha-1}$-competitive for standard power functions of the form $f(s) = s^\alpha$. AVR sets the processor's speed at each time t equal to the total density of the alive jobs at t, i.e., $\sum_{J_j \in \mathcal{J}(t)} \delta_j$. Then, it schedules the jobs according to the Earliest Deadline First (EDF) policy.

In order to generalize AVR to the multiprocessor setting, we consider a variation of the single-processor AVR algorithm which assigns exactly the same speed to the processor at each time t but it follows a different job selection policy. Without loss of generality, we assume that all release dates and deadlines are integers. Assume also that $r_{\min} = \min\{r_j : J_j \in \mathcal{J}\} = 0$ and let $d_{\max} = \max\{d_j : J_j \in \mathcal{J}\} = T$ be the maximum deadline among the released jobs. We can partition the time horizon into unit-size intervals of the form $I_t = [t, t+1), 0 \le t < T$. In particular, for each job $J_j \in \mathcal{J}(I_t)$, the algorithm assigns $\delta_j = \frac{w_j}{d_j - r_j}$ work of J_j to the interval I_t, and then it produces an arbitrary schedule of the total work assigned to I_t using constant speed $\sum_{J_j \in \mathcal{J}(I_t)} \delta_j$ during the whole I_t. The above variation achieves the same competitive ratio as the original AVR algorithm proposed in [16], since they both follow the same speed assignment rule and hence they have the same energy consumption.

We now turn our attention to the case of multiple heterogeneous processors. Based on the previous variation, we say that a schedule $\mathcal{S}$ is an *AVR-schedule* if for every job $J_j \in \mathcal{J}$ and interval $I_t \subseteq [r_j, d_j)$ the total amount of work of J_j executed during I_t on all processors in $\mathcal{S}$ is exactly equal to δ_j. The following lemma provides a lower bound on the optimal offline solution.

Lemma 8. *There exists a feasible AVR-schedule $\mathcal{S}_{AVR}$ for the heterogeneous speed-scaling problem with arbitrary power functions such that $E(\mathcal{S}_{AVR}) \leq (\max_{P_p \in \mathcal{P}} \{\rho_p\} + 1) OPT$, where ρ_p is the competitive ratio of the single-processor AVR algorithm when it is applied to the processor P_p with power function $f_p(s)$.*

Proof. Let $\mathcal{S}^*$ be an optimal offline schedule. We denote by $\mathcal{S}_p^*$ the part of $\mathcal{S}^*$ which corresponds to the processor P_p. In other words, $\mathcal{S}^*$ is the concatenation of $\mathcal{S}_p^*$'s. Let $w_{j,p}^*$ and $s_{j,p}^*$ be the amount of work and the corresponding speed of job J_j on processor P_p, respectively, in $\mathcal{S}^*$. For each $P_p \in \mathcal{P}$, we modify $\mathcal{S}_p^*$ to $\mathcal{S}_p$ by applying the variation of the single-processor AVR algorithm; the work executed for each $J_j \in \mathcal{J}$ in $\mathcal{S}_p$ is equal to $w_{j,p}^*$. Let $\mathcal{S}$ be the resulting schedule, i.e., the concatenation of $\mathcal{S}_p$'s. Moreover, let $w_{j,p,t}$ and $s_{j,p,t}$ be the amount of work and the corresponding speed of job J_j on processor P_p during I_t, respectively, in $\mathcal{S}$. Finally, we modify $\mathcal{S}$ by setting the speed of the piece of J_j executed by P_p during I_t equal to $\max\{s_{j,p}^*, s_{j,p,t}\}$ and we denote the obtained schedule by $\mathcal{S}_{AVR}$.

The total amount of work executed for J_j during I_t in $\mathcal{S}_{AVR}$ is equal to

$$\sum_{P_p \in \mathcal{P}} w_{j,p,t} = \sum_{P_p \in \mathcal{P}} \frac{w_{j,p}^*}{d_j - r_j} = \frac{w_j}{d_j - r_j} = \delta_j$$

Thus, $\mathcal{S}_{AVR}$ is an AVR-schedule.

The total processing time of all the pieces of J_j during I_t in $\mathcal{S}_{AVR}$ is equal to

$$\sum_{P_p \in \mathcal{P}} \frac{\frac{w_{j,p}^*}{d_j - r_j}}{\max\{s_{j,p}^*, s_{j,p,t}\}} \leq \frac{1}{d_j - r_j} \sum_{P_p \in \mathcal{P}} \frac{w_{j,p}^*}{s_{j,p}^*} \leq 1 = |I_t|$$

where the last inequality follows because $\mathcal{S}^*$ is feasible. By Lemma 2, we conclude that $\mathcal{S}_{AVR}$ can be constructed to be feasible.

In $\mathcal{S}_{AVR}$, the speed of the piece of J_j executed by P_p during I_t is equal either to the speed that the piece has in $\mathcal{S}^*$ or to the speed that it has in $\mathcal{S}$. Therefore, the energy consumption of $\mathcal{S}_{AVR}$ is

$$E(\mathcal{S}_{AVR}) = \sum_{P_p \in \mathcal{P}} \sum_{J_j \in \mathcal{J}} \sum_{t=0}^{T-1} \int_{I_t} \max\{f_p(s_{j,p}^*), f_p(s_{j,p,t})\}$$

$$\leq \sum_{P_p \in \mathcal{P}} \sum_{J_j \in \mathcal{J}} \sum_{t=0}^{T-1} \int_{I_t} f_p(s_{j,p}^*) + \sum_{P_p \in \mathcal{P}} \sum_{J_j \in \mathcal{J}} \sum_{t=0}^{T-1} \int_{I_t} f_p(s_{j,p,t})$$

$$= E(\mathcal{S}^*) + \sum_{P_p \in \mathcal{P}} E(\mathcal{S}_p)$$

For each $P_p \in \mathcal{P}$, let $\tilde{\mathcal{S}}_p$ be an optimal offline schedule for P_p in which for each job $J_j \in \mathcal{J}$ an amount of work $w_{j,p}^*$ is executed. Therefore, given that the single-processor AVR algorithm is ρ_p-competitive when it is applied to the processor P_p with power function $f_p(s)$, we have that

$$E(\mathcal{S}_{AVR}) \leq E(\mathcal{S}^*) + \sum_{P_p \in \mathcal{P}} \rho_p E(\tilde{\mathcal{S}}_p) \leq E(\mathcal{S}^*) + \max_{P_p \in \mathcal{P}}\{\rho_p\} \sum_{P_p \in \mathcal{P}} E(\tilde{\mathcal{S}}_p)$$

$$\leq E(\mathcal{S}^*) + \max_{P_p \in \mathcal{P}}\{\rho_p\} \sum_{P_p \in \mathcal{P}} E(\mathcal{S}_p^*) = E(\mathcal{S}^*) + \max_{P_p \in \mathcal{P}}\{\rho_p\} E(\mathcal{S}^*)$$

$$= (\max_{P_p \in \mathcal{P}}\{\rho_p\} + 1) E(\mathcal{S}^*)$$

where the last inequality follows by the optimality of $\tilde{\mathcal{S}}_p$ and the fact that the amount of work of each job $J_j \in \mathcal{J}$ is the same on both $\tilde{\mathcal{S}}_p$ and $\mathcal{S}_p^*$. ☐

We are now ready to describe our algorithm. The high level idea is that we create a $(1 + \epsilon)$-approximate AVR-schedule, by using the algorithm proposed in Sect. 3. More specifically, given the assignment of work into intervals implied by the definition of the AVR-schedules, for each interval $I_t = [t, t + 1)$ we create an offline $(1 + \epsilon)$-approximate schedule for this subinstance of the heterogeneous speed-scaling problem. We call this algorithm *Heterogeneous-AVR* (or simply H-AVR). Note that, if the time $t + 1$ does not correspond to a release date or a deadline then the schedules for the intervals I_t and I_{t+1} are the same, and hence we have to compute it only once. The following theorem follows.

Theorem 5. *H-AVR is a $((1+\epsilon)(\max_{P_p \in \mathcal{P}}\{\rho_p\}+1))$-competitive algorithm for the heterogeneous speed-scaling problem, where ρ_p is the competitive ratio of the single-processor AVR algorithm when it is applied to the processor P_p with power function $f_p(s)$.*

For the case of standard power functions of the form $f(s) = s^\alpha$, the single-processor AVR algorithm is $\alpha^\alpha 2^{\alpha-1}$-competitive [16]. Therefore, the following corollary holds.

Corollary 2. *H-AVR is a $((1 + \epsilon)(\alpha^\alpha 2^{\alpha-1} + 1))$-competitive algorithm for the heterogeneous speed-scaling problem for standard power functions of the form $f_p(s) = s^{\alpha_p}$, where $\alpha = \max_{P_p \in \mathcal{P}}\{\alpha_p\}$.*

References

1. Albers, S., Antoniadis, A., Greiner, G.: On multi-processor speed scaling with migration. J. Comput. Syst. Sci. **81**(7), 1194–1209 (2015)
2. Angel, E., Bampis, E., Kacem, F., Letsios, D.: Speed scaling on parallel processors with migration. In: Kaklamanis, C., Papatheodorou, T., Spirakis, P.G. (eds.) Euro-Par 2012. LNCS, vol. 7484, pp. 128–140. Springer, Heidelberg (2012)
3. Bampis, E., Kononov, A.V., Letsios, D., Lucarelli, G., Sviridenko, M.: Energy efficient scheduling and routing via randomized rounding. In: FSTTCS, vol. 24 of LIPIcs, pp. 449–460. Schloss Dagstuhl - Leibniz-Zentrum fuer Informatik (2013)
4. Bampis, E., Letsios, D., Lucarelli, G.: Green scheduling, flows and matchings. Theor. Comput. Sci. **579**, 126–136 (2015)

5. Bansal, N., Bunde, D.P., Chan, H.-L., Pruhs, K.: Average rate speed scaling. Algorithmica **60**(4), 877–889 (2011)
6. Bansal, N., Chan, H.-L., Pruhs, K.: Speed scaling with an arbitrary power function. ACM Trans. Algorithms **9**(2), 18 (2013)
7. Bansal, N., Kimbrel, T., Pruhs, K.: Speed scaling to manage energy and temperature. J. ACM **54**(1) (2007)
8. Bingham, B.D., Greenstreet, M.R.: Energy optimal scheduling on multiprocessors with migration. In: ISPA, pp. 153–161 (2008)
9. Chen, J.-J., Hsu, H.-R., Chuang, K.-H., Yang, C.-L., Pang, A.-C., Kuo, T.-W.: Multiprocessor energy-efficient scheduling with task migration considerations. In: ECRTS, pp. 101–108. IEEE Computer Society (2004)
10. Federgruen, A., Groenevelt, H.: Preemptive scheduling of uniform machines by ordinary network flow techniques. Manage. Sci. **32**(3), 341–349 (1986)   ·
11. Gonzalez, T., Sahni, S.: Open shop scheduling to minimize finish time. J. ACM **23**(4), 665–679 (1976)
12. Gonzalez, T., Sahni, S.: Preemptive scheduling of uniform processor systems. J. ACM **25**, 92–101 (1978)
13. Gupta, A., Im, S., Krishnaswamy, R., Moseley, B., Pruhs, K.: Scheduling heterogeneous processors isn't as easy as you think. In: SODA, pp. 1242–1253 (2012)
14. Gupta, A., Krishnaswamy, R., Pruhs, K.: Scalably scheduling power-heterogeneous processors. In: Abramsky, S., Gavoille, C., Kirchner, C., Meyer auf der Heide, F., Spirakis, P.G. (eds.) ICALP 2010. LNCS, vol. 6198, pp. 312–323. Springer, Heidelberg (2010)
15. Li, M., Yao, A.C., Yao, F.F.: Discrete and continuous min-energy schedules for variable voltage processors. PNAS **103**(11), 3983–3987 (2006)
16. Yao, F., Demers, A., Shenker, S.: A scheduling model for reduced CPU energy. In: FOCS, pp. 374–382 (1995)

Period Recovery over the Hamming and Edit Distances

Amihood Amir[1,2], Mika Amit[3]([✉]), Gad M. Landau[3,4], and Dina Sokol[5]

[1] Department of Mathematics and Computer Science,
Bar-Ilan University, Ramat Gan, Israel
amir@cs.biu.ac.il
[2] College of Computing, Georgia Tech, Atlanta, GA, USA
[3] Department of Computer Science, University of Haifa, Mount Carmel, Haifa, Israel
mika.amit2@gmail.com,landau@cs.haifa.ac.il
[4] Department of Computer Science and Engineering,
NYU Polytechnic School of Engineering, New York University,
Brooklyn, NY, USA
[5] Department of Computer and Information Science,
Brooklyn College of the City University of New York, Brooklyn, NY, USA
sokol@sci.brooklyn.cuny.edu

Abstract. A string S of length n has period P of length p if $S[i] = S[i+p]$ for all $1 \leq i \leq n-p$ and $n \geq 2p$. The shortest such substring, P, is called *the period* of S, and the string S is called *periodic* in P. In this paper we investigate the period recovery problem. Given a string S of length n, find the primitive period(s) P such that the distance between S and the string that is periodic in P is below a threshold τ. We consider the period recovery problem over both the Hamming distance and the edit distance. For the Hamming distance case, we present an $O(n \log n)$ time algorithm, where τ is given as $\frac{n}{(2+\epsilon)p}$, for $0 < \epsilon < 1$. For the edit distance case, $\tau = \frac{n}{(4+\epsilon)p}$, and we provide an $O(n^{4/3})$ time algorithm.

Keywords: Period recovery · Approximate periodicity · Hamming distance · Edit distance

A. Amir—Partially supported by the Israel Science Foundation grant 571/14, and grant No. 2014028 from the United States-Israel Binational Science Foundation (BSF).

M. Amit—Partially supported by the Israel Science Foundation grant 571/14, grant No. 2014028 from the United States-Israel Binational Science Foundation (BSF) and DFG.

G. M. Landau—Partially supported by the Israel Science Foundation grant 571/14, grant No. 2014028 from the United States-Israel Binational Science Foundation (BSF) and DFG.

D. Sokol—Partially supported by the United States-Israel Binational Science Foundation (BSF) grant No. 2014028.

© Springer-Verlag Berlin Heidelberg 2016
E. Kranakis et al. (Eds.): LATIN 2016, LNCS 9644, pp. 55–67, 2016.
DOI: 10.1007/978-3-662-49529-2_5

1 Introduction

The prevalence and importance of cyclic phenomena in nature is apparent in diverse areas, including astronomy, geology, earth science, oceanography, meteorology, biological systems, the genome, economics, and more. Assume that an instrument is taking measurements at fixed intervals. When the stream of measurements is analyzed, the question of whether the measurements represent a cycle is raised. The "cleanest" version of this question is whether the string of measurements is *periodic*.

Periodicity is one of the most important properties of a string and plays a key role in data analysis. As such, it has been extensively studied over the years [20], and linear time algorithms for exploring the periodic nature of a string were presented (e.g. [7,13,16]). However, realistic data may contain errors. Such errors may be caused by the process of gathering the data which might be prone to transient errors. Moreover, errors can also be an inherent part of the data because the periodic nature of the data represented by the string may be inexact. Thus, it is necessary to cope with periods that have errors.

In this paper, we present algorithms for the *period recovery problem*, defined in [1]. Informally, the problem is to recover a set of periods that are likely to be the underlying period of the corrupted periodic string, S. Given a sequence S, assume that S was originally a periodic string, which had been corrupted. Our task is to discover the original uncorrupted string, or more specifically, the exact period of the uncorrupted periodic string. Of course, too many errors can completely change the data, making it impossible to identify the original data and reconstruct the original cycle. However, it is intuitive that few errors should still preserve the periodic nature of the original string.

A related problem on which much work has been accomplished is finding all *runs* in a given string. Runs are substrings that contain two or more consecutive copies of a pattern. It has been shown in [14] that Fibonacci words contain only a linear number of runs. In [16] a conjecture that the maximal number of runs in a string of size n is at most n was given (see also [8], Chap. 8); this conjecture was proven recently in [3]. The problem of finding all approximate runs in a string was widely researched and many different measurements have been used in order to find such runs (see [2,17,18,23]).

Formally, a string S is *periodic* if it can be written as P^r, where P is some prefix of S, and $r \geq 2$ is a *real* number. A string P is *primitive* if it cannot be written as U^k for any *integer* $k \geq 2$, where U is some prefix of P. For example, the string *abcabca* is both periodic and primitive, while the string *abcabc* is periodic and *non*-primitive. Although a periodic string S may have many periods, when we refer to "the period" of S we always mean the shortest possible period, i.e. the period P of S such that P is primitive. Throughout the paper, when it is obvious from the context, we use lowercase p to represent the length of a string P.

If S can be written as P^2, it is called a *square*. In this case, the substring P is called the *root* of the square, and if P is primitive, then S is called a *primitively rooted* square. For example, the string $(ab)^4$ is a square with the non-primitive root *abab*, while *abab* is a primitively rooted square. P^3 is a *cube*, i.e. three consecutive copies of P, and in general P^k, i.e. k consecutive copies of P, is counted as $k - 1$ overlapping squares. The number of non-overlapping squares in a string P^k is equal to $\lfloor \frac{k}{2} \rfloor$. P^∞ denotes the periodic string in P that begins with P and extends infinitely to the right.

A Lyndon word is a string that is lexicographically smaller than all of its proper suffixes (see [21]). Any periodic string, $S = P^k$, for $k \geq 2$ contains a substring L, of length p, that is a Lyndon word. We call this substring the *L-root* of the periodic string.

In this paper we solve the following two problems, originally defined in [1]. Let ϵ be some real constant such that $0 < \epsilon < 1$.

Period Recovery over the Hamming Distance. Given a string S of length n defined over unbounded alphabet Σ, find all (primitive) periods P, such that the Hamming distance between a prefix of the string P^∞ and S is less than or equal to $\frac{n}{(2+\epsilon)p}$ (where p is the length of the period).

Period Recovery over the Edit Distance. Given a string S of length n defined over unbounded alphabet Σ, find each (primitive) period P such that S matches a prefix of P^∞ with at most $\frac{n}{(4+\epsilon)p}$ insertions, deletions, and mismatches (where p is the length of the period).

Remark. The parameter for the Hamming distance case, $\frac{n}{(2+\epsilon)p}$, was chosen to ensure that at least half of the occurrences of P in the input string are exact. This enables handling one candidate (or, in special cases, two candidates) per period length p. For the edit distance, [1] proved that when the number of allowed errors is bounded by $\frac{n}{(4+\epsilon)p}$, the number of answers is bounded by $O(\log n)$.

Two strings X, Y, are said to be *conjugate* if there exist strings U, V such that $X = UV$ and $Y = VU$, i.e. Y is a cyclic shift of X. Conjugacy defines an equivalence relation, and a primitive string of length p has exactly p distinct conjugates [20]. For the edit distance problem, often several conjugates of the same period P will satisfy the given constraints (due to the ability to delete leading characters). An algorithm can report all conjugates of P, but it is preferable to report only the *best* conjugate of P, i.e. the one that has the minimal distance to S. Henceforth, we assume that we are interested only in the best conjugate for each P in the solution set[1].

In this paper we use the same threshold as in [1] and greatly improve the time to find the candidates. We prove the following two theorems:

Theorem 1. *Given a string S of length n, the period recovery over the Hamming distance problem can be solved in $O(n \log n)$ time.*

Theorem 2. *Given a string S of length n, the period recovery over the edit distance problem can be solved in $O(n^{4/3})$ time.*

---

[1] In previous work, the lemmas state "up to cyclic rotations" which means that one conjugate is counted/reported for each set of cyclic permutations of a given period P. Here we clarify this language by always finding the single best conjugate.

We start in Sect. 2 with presenting a simple algorithm for period recovery over the Hamming distance, and prove Theorem 1. Our algorithm improves on [1] by a logarithmic factor. Then, in Sect. 3, we present an algorithm for period recovery over the edit distance, and prove Theorem 2. In this case the improvement is more extensive as the edit distance algorithm of [1] has $O(n^3)$ time complexity[2].

2   Period Recovery over the Hamming Distance

In the period recovery problem over the Hamming distance, the input is a string S of length n, and the output is all primitive periods P of length p, such that the Hamming distance between S and the string $P^{\frac{n}{p}}$ is less than or equal to $\frac{n}{(2+\epsilon)p}$, for some $0 < \epsilon < 1$.

We start with the following observations that reduce the number of candidate substrings for being an approximate period of the string S.

Observation 1. *For each solution P there are at least $\frac{(1+\epsilon)n}{(2+\epsilon)p} > \frac{n}{2p}$ positions i in S such that $i \in \{0, p, 2p, \ldots, \lfloor \frac{n}{p} \rfloor p\}$ and $S[i \ldots i + p - 1] = P$.*

The above observation follows immediately from the fact that there are at most $\frac{n}{(2+\epsilon)p}$ mismatches between S and $P^{\frac{n}{p}}$. In addition, since there are at least $\frac{n}{2p}$ exact copies of period P in S, at most one substring P can fulfill this requirement per each period length p. This leads to the following observation.

Observation 2. *For each period length p, $1 \le p \le \frac{n}{2}$ and $n = p^k$, if k is an integer then there can be at most 1 candidate of length p. In the case where k is a rational number, there can be at most 2 candidates of length p.*

Note that a substring, P, that has at least $\frac{n}{2p}$ exact copies in S is only a *candidate* substring for being an approximate period of S. The algorithm still needs to verify whether the Hamming distance between S and $P^{\frac{n}{p}}$ is not greater than $\frac{n}{(2+\epsilon)p}$.

The main idea of the algorithm comes from the above observations:

Algorithm Outline

Input: String S of length n.
Output: A set of primitive substrings that are approximate periods of S over the Hamming Distance.
for each period p from 1 to $\frac{n}{2}$ do

1. Let k be an integer such that $k = \lfloor \frac{n}{p} \rfloor$. Count the number of distinct substrings of length p that start at positions $0, p, 2p, \ldots, kp$ and take the one that occurs at least $\frac{k}{2}$ times. If no such substring occurs, there is no approximate period of length p in S.

---

[2] The paper actually states $O(n^3 \log n)$ time complexity. However, more recent work [4] for construction of a minimal augmented suffix tree can be used, reducing the time complexity of [1] to $O(n^3)$.

2. Let P be the majority substring of length p found in the previous step. Compute the Hamming distance between $P^{\frac{n}{p}}$ and S. If the distance is smaller than $\frac{n}{(2+\epsilon)p}$, then P is a candidate for being an approximate period of S.
3. Check whether the candidate P is a primitive substring. Report P if it is primitive.

Remark. Observe that in the special case where k is not an integer, there can be a situation where two substrings P have exactly $\lfloor \frac{k}{2} \rfloor$ exact occurrences in the input string. For example, for $S = abcabdab$, both abc and abd are candidates for being an approximate period of S. In these cases, both substrings will be verified by the algorithm. Additionally, if k is an integer and there exist two substrings occurring exactly $\lfloor \frac{k}{2} \rfloor$ times, none of the substrings will be verified by the algorithm since both have more than $\frac{n}{(2+\epsilon)p}$ mismatches with S.

2.1   Step 1: Finding a Candidate Substring of Length p

In this procedure we use the KMR naming technique of [15] in order to count the number of different substrings in specified positions. This technique renames all substrings of lengths $2^0, 2^1, 2^2, \ldots, 2^{\log n}$ in S, and for each position i in S computes a vector, $Names_i$, of size $\log n$. The entry $Names_i[j]$ contains a name of the substring of length 2^j that starts at position i. This naming procedure is run once for the entire algorithm: first, the characters of S are sorted lexicographic, and each $S[i]$ is given a "name" according to its rank in the sorted order. Then, for every position $i \in S$ and every power of 2, $j \in 2^k, k \in \{2, 4, \ldots, \lfloor \log n \rfloor\}$ a "name" is computed according to the lexicographical order the two names, $Name_i[j-1]$ and $Name_{i+j/2}[j-1]$, which have already been computed.

The product of the naming algorithm is the vectors $Names_i$, which are used in order to find the substring of length p that occurs the majority of times.

A substring P of length p that occurs more than $\frac{n}{2p}$ times at positions $0, p, 2p, \ldots, \lfloor \frac{n}{p} \rfloor p$ of S is denoted as a *candidate* substring of S. In order to find a candidate substring P for a specific period length p, an auxiliary list of size $\frac{n}{p}$ is computed in order to keep the names of the substrings of length p starting at positions $i \in \{0, p, 2p, \ldots, \lfloor \frac{n}{p} \rfloor p\}$ in S. Let P be the substring $S[i \ldots i + p - 1]$. The name of P is found in constant time using the $Names_i$ vector as follows. If p is of size 2^k then the name of P is equal to $Names_i[k]$. Otherwise, let k be the maximal integer such that $2^k < p$. P is split into two overlapping substrings such that P_1 is a prefix of P, P_2 is a suffix of P, and $|P_1| = |P_2| = 2^k$. The name of P is the concatenation of $Names_i[k]$ and $Names_{i+p-2^k}[k]$.

After the list of substring names is established, the names are sorted using radix sort. Then, for each name the number of its occurrences is summed, and the name of the substring that occurs more than half of the times corresponds to the winner substring. Denote the winner substring as P.

2.2   Step 2: Compute the Hamming Distance

The Hamming distance between $P^{\frac{n}{p}}$ and S must be smaller than $\frac{n}{2p}$ for P to be a valid candidate. We compute the Hamming distance using the technique of

"Kangaroo Jumps" [12] (i.e., using suffix tree and LCA algorithm for a constant time "jump" over equal substrings): we count the number of mismatches between all substrings of length p starting at positions $0, p, 2p, \ldots$ and P. If this number is smaller than $\frac{n}{2p}$, then P is a candidate substring for being an approximate period of S. A candidate substring P is represented as a pair (i, p), where i is a position in the string where P occurs and p is the length of the substring, such that $P = S[i \ldots i + p - 1]$.

2.3   Step 3: Primitivity Check

As in exact periodicity, we only consider a string P to be the approximate period of S if P is primitive. In order to decide whether a candidate substring is primitive, we use the algorithm presented in Crochemore et al. [9]. There, a string S of length n is preprocessed, such that given two indices, $0 \leq i \leq j \leq n - 1$, the algorithm returns $TRUE$ if the substring $S[i \ldots j]$ is primitive or $FALSE$, otherwise. This query is done in $O(\log n)$ time, and if P is indeed a primitive string, P is reported as an approximate period of S.

2.4   Time Complexity

The naming technique of [15] is done once in $O(n \log n)$ time. The preprocessing of S in order to support $O(\log n)$-time queries for substring primitivity is done once in $O(n \log^\gamma n)$ time, for an arbitrary positive real $0 < \gamma < 1$ (see [9])[3].
   In Step 1, for each period length p, the procedure of finding a candidate substring of length p is done in $O(\frac{n}{p})$ time: first, for each one of the positions $i \in \{0, p, \ldots, \lfloor \frac{n}{p} \rfloor p\}$, the name of the substring $S[i \ldots i + p - 1]$ is found and inserted to a list in constant time. Then, finding the majority substring is done by first sorting the list and then performing one pass over the sorted list, both done in time linear to the list size, $\lfloor \frac{n}{p} \rfloor$. In Step 2, we compute the total number of mismatches between P and the substrings $S[i \ldots i+p-1]$, where $i \in \{0, p, \ldots, \lfloor \frac{n}{p} \rfloor p\}$. Note that this procedure is run at most $\frac{3}{2} \cdot \frac{n}{p} = O(\frac{n}{p})$ times, since having more than $\frac{n}{2p}$ mismatches means that P is not a candidate substring. This gives a total of $\Sigma_{p=1}^{\frac{n}{2}} \frac{n}{p} = n\Sigma_{p=1}^{\frac{n}{2}} \frac{1}{p} = O(n \log n)$ time for finding all candidate substrings in S. Finally, in Step 3, primitivity checking costs $O(\log n)$ for each of the $n/2$ candidates, for a total of $O(n \log n)$. Thus, the total time complexity of the algorithm is bounded by $O(n \log n)$.

3   Period Recovery over the Edit Distance

In the period recovery problem over the edit distance, the input is a string S of length n, and the output is all primitive periods P of length p, such that the edit distance between S and any prefix of P^∞ is at most $\frac{n}{(4+\epsilon)p}$, for $0 < \epsilon < 1$.

---

[3] More precisely, this time complexity can be further improved to linear time preprocessing and $O(\log \log n)$ time query, by replacing, in Crochemore et al. [9], the 2D range minimum query algorithm of Chazelle [6] with the algorithm of Chan [5].

In [1] it was proven that there are at most $O(\log n)$ *solutions* for P. However, they computed the edit distance for $O(n \log n)$ candidates. In this paper we show how it is possible to narrow down the set of candidates for P *a priori*, so that verification is only necessary for $O(\log n)$ candidates (see Subsect. 3.1).

Our algorithm has three steps: as a first step, it finds the $O(\log n)$ candidate substrings in S (see Subsect. 3.3). In step 2, the algorithm verifies short candidates having $p < n^{1/3}$ (see Subsect. 3.4), and finally, in the third step, it verifies the long candidates with period length $p \geq n^{1/3}$ (see Subsect. 3.5).

For each candidate substring P, we compute the edit distance between *any* substring of P^∞ and S, and by this we find the best conjugate of P that has the minimum number of errors with S.

Remark. if the divisions do not yield whole numbers, we simply take the floor of the fractions without affecting the resulting complexities.

3.1 Reducing the Number of Candidates

In this subsection we prove that the initial number of possible candidates for approximate periods of S can be reduced to $O(\log n)$. In Corollary 1 it becomes apparent that for each possible period length p, $1 \leq p \leq \frac{n}{2}$, there can be at most 1 candidate substring for being an approximate period of S. We then proceed to prove that if U and V ($u > v$) are both candidates for being approximate periods of S, then the length of V must be a fraction of the length of U (Lemma 3). From both these facts we conclude that the total number of initial candidates cannot exceed $O(\log n)$.

We start with a definition of a zone. A *zone*, $z = (i, j, p)$, is a substring $S[i \ldots j]$ in which $S[x] = S[x + p]$ for every position $i \leq x < j - p$. If the size of the zone is greater than $2p$, the zone is actually a *run*, and it can be written as $z = U^k U'$ for some $k \geq 2$ an integer.

For a string S, if S contains x errors with a prefix of a string U^∞ then the string S can be partitioned into $x + 1$ zones with respect to U (see Example 1). It is easy to see that a zone z, $z = (i, j, p)$, in a string S contains exactly $\lfloor (j - i + 1)/2p \rfloor$ non-overlapping squares of the string $S[i \ldots i + p - 1]$. For simplicity of presentation, we further use the abbreviation $NOS(U)$ to refer to non-overlapping squares of some conjugate of the substring U.

Example 1. Let $S = ac\ ab\ ac\ ac\ ac\ ac\ a\ ac\ ac\ ac\ ac\ x\ ac\ ac$, and let $U = ac$ be a candidate substring for being an approximate period of S. The number of allowed errors is $n/(4 + \epsilon)u = 26/(4 + \epsilon)2 \leq 3$ for all $0 < \epsilon < 1$, and the errors are in the following positions in S: 4 (mismatch), between 13 and 14 (insertion), and 22 (deletion).

S has 4 zones with respect to U: $z_1 = S[1 \ldots 3]$, $z_2 = S[5 \ldots 13]$, $z_3 = S[14 \ldots 21]$ and $z_4 = S[23 \ldots 26]$. The first zone, z_1, does not contain a repetition of U or its conjugates, whereas the rest of the zones do: z_2 contains 2 $NOS(U)$ (of either ac or ca), z_3 contains 2 $NOS(U)$ (of the substring U) and z_4 contains 1 $NOS(U)$ (of the substring U).

In the following Lemma 1 we prove that a valid candidate for being an approximate period of S, U, must have at least $\frac{n}{4u}$ non-overlapping squares of conjugates of U in S.

In the proof of Lemma 1 we count the maximum number of characters in S that are not contained in $NOS(U)$ in S. Denote these characters as *"bad"* *characters*. In Example 1, there are 6 "bad" characters in S with respect to U: $S[1 \ldots 4]$, $S[21]$, and either $S[5]$ or $S[13]$.

Lemma 1. *Given a string S of length n, and a string U of length u, if S has at most $\frac{n}{(4+\epsilon)u}$ edit errors with a substring of U^∞, then S contains at least $\frac{n}{4u}$ non-overlapping squares, each having a conjugate of U as its root.*

Proof. Let S' be a prefix of U^∞ such that the alignment between S' and S contains at most $\frac{n}{(4+\epsilon)u}$ edit errors. The substrings between the alignment error positions are zones in S with respect to U. Therefore, there are at most $\frac{n}{(4+\epsilon)u}+1$ zones in S with respect to U.

Each zone contains at most $2u - 1$ characters that do not participate in $NOS(U)$. In addition, each error position can imply a character that does not participate in a $NOS(U)$.

Thus, the total number of "bad" characters in S with respect to U is at most $N_{bad} = (\frac{n}{(4+\epsilon)u} + 1) \cdot (2u - 1) + \frac{n}{(4+\epsilon)u} = \frac{2n}{(4+\epsilon)} + 2u - 1$.

The number of characters in S that do participate in a $NOS(U)$, is therefore at least: $N_{good} = n - N_{bad} = n - (\frac{2n}{(4+\epsilon)} + 2u - 1) = \frac{n(2+\epsilon)}{(4+\epsilon)} - 2u + 1$.

Dividing N_{good} by $2u$ gives a bound on the minimum number of non-overlapping squares of conjugates of U in S.

Consider the case where $\frac{N_{good}}{2u}$ does not yield a whole number. This means that there exists at least one zone, $z_i = (i, j, u)$, such that z_i contains $2ku + g$ "good" characters and $2u - 1$ "bad" characters, for $k \geq 0$ and $0 < g < 2u$ (i.e., the size of z_i is equal to $(2k + 2)u + g - 1$). The number of $NOS(U)$ in z_i is equal to $(2k + 2)/2 = k + 1$ (see Fig. 1).

Therefore, when counting the number of $NOS(U)$ in S, we take the *ceiling* of the value $\frac{N_{good}}{2u}$ as follows.

$$\#NOS(U) = \lceil \frac{N_{good}}{2u} \rceil = \lceil (\frac{n(2+\epsilon)}{(4+\epsilon)} - 2u + 1)/2u \rceil > \frac{n}{4u} - 1$$

for all $0 < \epsilon < 1$.

Since $\#NOS(U)$ is greater than $\frac{n}{4u} - 1$, S contains at least $\frac{n}{4u}$ non-overlapping squares of conjugates of U in S. □

Example 2. Let $S = (abcde\ abcdX)^{1000}(abcde\ abcde)^{1000}abcde$ and let $U = abcde$. In this example, $n = 20005$ and for $\epsilon = 0.01$, we get that the number of allowed errors is 1000, and the number of zones is 1001.

According to the equation above we get that the number of good characters in S is **at least** $20005 - (1001 \cdot 9 + 1000) = 9996$. This means that the number of $NOS(abcde) = \lceil \frac{N_{good}}{2u} \rceil = \lceil \frac{9996}{20} \rceil = 500 = \lfloor \frac{n}{4u} \rfloor = \lfloor \frac{20005}{40} \rfloor$. In this example, the actual number of "good" characters in S with respect to U is exactly 10000.

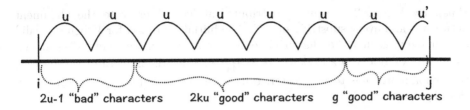

Fig. 1. A zone $z_i = (i, j, u)$ that contains $2ku+g$ "good" characters and was computed as having $2u - 1$ "bad" characters. The size of z_i is equal to $(2k + 2)u + g - 1$, which means that there are exactly $k + 1$ non-overlapping squares of conjugates of U in z_i.

Corollary 1. *For each period length p there can be at most 1 candidate substring for being an approximate period of S.*

Proof. Assume by contradiction that there are two candidate substrings, U and V, both of size p, such that V is not a conjugate of U. Each candidate has at least $\frac{n}{4p}$ non-overlapping squares of its conjugates in S. Each $NOS(U)$ contributes at least 2 errors to the alignment of S with V^∞. This is due to the fact that V is not a conjugate of U, hence a substring of S that equals U has to have at least 1 error with any substring of V^∞ of size p. This yields at least $\frac{n}{2p} > \frac{n}{(4+\epsilon)p}$ errors for each U and V, which means that both substrings are not valid candidates. $\square$

We now proceed with showing that this initial number of n candidates can be further reduced to $O(\log n)$.

Assume that a substring U contains $\frac{n}{4u}$ non-overlapping squares of its conjugates in S. Note that this property is not sufficient to render U a valid solution, but it can be used as a simple filter for initial candidates. We want to show that any shorter candidate substring V that may be a possible solution, must be significantly shorter than U. The main idea is to examine only the substrings of S that contain exact squares of conjugates of U and "count" the number of errors that must occur between the exact squares of the conjugates of U and any substring of V^∞.

Lemma 2 (Common Factor Lemma [10]). *For any primitive strings U and V, such that U is not a conjugate of V, a prefix of U^∞ and a prefix of V^∞ cannot have a common factor of length greater than or equal to $u + v - gcd(u, v)$ where gcd stands for the greatest common divisor.*

The common factor lemma implies that every square of a conjugate of U contributes at least one error to the alignment between S and every substring of V^∞ for every other possible candidate V, having $v \leq u$.

Lemma 3. *Let S be a string of length n, and $0 < \epsilon < 1$ be a constant. Let U be a primitive substring of S such that there are at least $\frac{n}{4u}$ exact non-overlapping squares of conjugates of U in S. A primitive substring V of S, such that $v \leq u$, can be a member of the solution set to the Period Recovery Problem, only if the following relation is true: $u \geq (1 + \frac{\epsilon}{4})v$.*

Proof. By Lemma 2, a square contributes at least 1 error to the alignment between S and every substring of V^∞. The initial condition for V being a valid answer requires that the alignment for V contains no more than $\frac{n}{(4+\epsilon)v}$ errors. Therefore, we bound u as follows: $\frac{n}{4u} \leq \frac{n}{(4+\epsilon)v}$. Since the numerator is the same, the denominator has to be greater, thus we have: $4u \geq (4+\epsilon)v$ dividing both sides by 4 yields: $u \geq (1 + \frac{\epsilon}{4})v$.   □

The above Lemma leads to the desired bound on the maximum number of possible candidates that needs to be considered in a string S of length n:

Lemma 4. *There are at most* $\log_{1+\frac{\epsilon}{4}} n = O(\log n)$ *possible candidates for being an approximate period over the edit distance in a string S of size n.*

Proof. The longest candidate U has length at most $\frac{n}{2}$. By Lemma 3, the next longest candidate must have length less than or equal to u divided by $1 + \frac{\epsilon}{4}$. Thus, the number of candidates equals the number of times $n/2$ can be divided by $1 + \frac{\epsilon}{4}$, which equals $\log_{1+\frac{\epsilon}{4}} n = O(\log n)$.   □

3.2   Algorithm Outline

Input: String S of length n.
Output: A set of at most $O(\log n)$ patterns that are approximate periods of S over the edit distance.

1. *Initialize candidates list:*
   (a) Find all runs in the string S. For each run, choose its L-root as a representative.
   (b) Sort the runs by their period size, and by their representative.
   (c) For each run r, $r = (i, j, p)$, compute the number of its non-overlapping squares by taking $\lfloor (j - i + 1)/2p \rfloor$.
   (d) For each representative, sum its non-overlapping conjugate squares (over all the runs it represents), and discard all representatives that have less than $\frac{n}{4u}$ non-overlapping squares.
   (e) For each representative U, starting from the longest candidate, discard all candidates V such that $u \leq (1 + \frac{\epsilon}{4})v$.
2. *Verify all $O(\log n)$ candidate strings:*
   (a) If the length of candidate U is $u < n^{1/3}$, use the verification procedure for short candidates.
   (b) Else (i.e., $u \geq n^{1/3}$), use the verification procedure for long candidates.

In the first part of the algorithm the initial list of candidates is computed. This part is described in Subsect. 3.3. At the end of this step, the list of candidates contains only substrings U such that U contains at least $n/4u$ non-overlapping squares over all the conjugates of U. In addition, for each two candidates in the list, U and V ($u > v$), we have that $u > (1 + \frac{\epsilon}{4})v$.

In the second part of the algorithm, an iteration over the list of candidates is done. For each candidate U, a verification procedure is called according to the size of the candidate: for candidates with $u < n^{1/3}$ the procedure for verifying short candidates is called (see Subsect. 3.4), and for candidates with $u \geq n^{1/3}$ the procedure for verifying long candidates is called (see Subsect. 3.5).

In the following subsections, the steps are further explained.

3.3 Finding the Candidate Substrings

As a preprocessing step, the suffix tree of S is built with a preparation for constant time LCA queries between suffixes of S. These data structures are built in linear time.

Step 1a of the algorithm uses the linear time algorithm of Bannai et al. [3] for finding all runs in a string S. For each run found by the algorithm, the algorithm marks a representative for the run, which is an occurrence of the L-root of the run. Step 1b is done in linear time using radix sort, since there are at most $n - 1$ runs in a string of length n. In addition, the comparison between two representatives is done in constant time using LCA queries.

In steps 1c and 1d of the algorithm, for each L-root, we compute the number of non-overlapping squares of all of its conjugates in S, and remove the ones that do not have sufficient number of non-overlapping squares. Since the number of L-roots is at most $n - 1$, the steps are done in linear time.

In the final step 1e, following Lemma 3, starting with the longest candidate substring U, all candidate substrings V such that $u \leq (1 + \frac{\epsilon}{4})v$ are discarded, since they cannot be a valid solution.

3.4 Verification of Short Candidates

In order to verify a given candidate P, it is necessary to perform an edit distance alignment between the input string S, and the periodic string P^∞. The goal is to report whether there is a match with fewer than $\frac{n}{(4+\epsilon)p}$ errors, and if there is, to report the best conjugate of P for which the match exists. We formally state the problem as follows. Given a string S of length n, and a string P of length p, such that $p < n^{1/3}$, find the best scoring edit distance alignment of the entire string S with any substring of P^∞. Note that if there is more than one alignment with the minimum distance, any alignment would be sufficient.

Wraparound dynamic programming (WDP) [11, 22] can be used directly to solve this problem. The idea of WDP is to align the string S with the period P, and to include an additional possibility in the dynamic programming calculation of the first value of each row. Each row is computed twice. The first time, the last value on the previous row is considered in the calculation of the first value of a row, and the second time, the last value in its own row is considered as well. In order to compute the edit distance between any conjugate of P and S, we set the first row of the table to zeros in step 2 of the algorithm.

Algorithm Short Candidate Verification

1. Construct an edit distance matrix with S on the left, and P on top.

2. Initialize the first row to all zero's.
3. Calculate the entire matrix using wraparound dynamic programming.
4. The lowest value in the last row is the minimum edit distance.

Time Complexity: WDP runs in $O(np)$ time. Since $p < n^{1/3}$, this results in $O(n^{4/3})$ time per short candidate.

3.5   Verification of Long Candidates

For a candidate P with $p \geq n^{1/3}$, we use the algorithm of Landau-Vishkin (LV) [19] to calculate the optimal edit distance alignment between S and substrings of P^∞. Given a fixed number of allowed errors, k, the LV algorithm calculates $O(k)$ elements on $2k$ diagonals using LCA queries on the suffix tree. We modify this algorithm, since our goal is to report the best conjugate of P that matches. We calculate $O(k)$ elements on $p + 2k$ diagonals, initializing the first $p - 1$ diagonals with a zero. This allows the algorithm to choose the best starting position from the first $p - 1$ positions, which in effect chooses the best conjugate of P.

Time Complexity: in our case, the $O(k^2)$ time algorithm of LV becomes $O(k(p + k)) = O(kp + k^2)$ since we calculate $p + 2k$ diagonals. The number of allowed errors for our algorithm is $k = n/(4 + \epsilon)p = O(n/p)$. Replacing this value for k we get $O(n + n^2/p^2)$. Recall that for long candidates, $p \geq n^{1/3}$, thus $p^2 \geq n^{2/3}$. This yields $O(n^2/n^{2/3}) = O(n^{4/3})$ time per long candidate.

3.6   Time Complexity

The initialization of the candidate list takes $O(n)$ time, as analyzed in Subsect. 3.3. In the verification part, $O(\log n)$ candidates are verified. Assume that the candidate sizes are $p_1 > p_2 > p_3 > \ldots > p_{\log n}$. Following Lemma 3, for every $1 \leq i \leq \log n$, $p_i - p_{i+1} > \frac{p_i}{1 + \frac{\epsilon}{4}}$.

For the short candidates (where $p_i < n^{1/3}$) the verification takes $O(np_i)$-time (see Subsect. 3.4). Therefore, for all short candidates the time complexity is expressed as the sum of the algebraic series: $n\Sigma_{i=0}^{\log n} \frac{n^{1/3}}{(1 + \frac{\epsilon}{4})^i} = O(n^{4/3})$.

For long candidates (where $p_i \geq n^{1/3}$), the verification takes $O(\frac{n^2}{p_i^2})$-time (see Subsect. 3.5). The time complexity for all long candidates is equal to: $\frac{n^2}{n^{2/3}}\Sigma_{i=0}^{\log n} \frac{1}{(1 + \frac{\epsilon}{4})^i} = O(n^{4/3})$. Thus, overall, the time complexity of the entire algorithm is bounded by $O(n^{4/3})$.

References

1. Amir, A., Eisenberg, E., Levy, A., Porat, E., Shapira, N.: Cycle detection and correction. ACM Trans. Algorithms **9**(1), 13:1–13:20 (2012)
2. Amit, M., Crochemore, M., Landau, G.M.: Locating all maximal approximate runs in a string. In: Fischer, J., Sanders, P. (eds.) CPM 2013. LNCS, vol. 7922, pp. 13–27. Springer, Heidelberg (2013)

3. Bannai, H.., Inenaga, T.I.S., Nakashima, Y., Takeda, M., Tsuruta, K.: The "runs" theorem. CoRR, abs/1406.0263v4 (2014)

4. Brodal, G.S., Lyngsø, R.B., Östlin, A., Pedersen, C.N.S.: Solving the string statistics problem in time $\mathcal{O}(n \log n)$. In: Widmayer, P., Triguero, F., Morales, R., Hennessy, M., Eidenbenz, S., Conejo, R. (eds.) ICALP 2002. LNCS, vol. 2380, pp. 728–739. Springer, Heidelberg (2002)

5. Chan, T.M.: Persistent predecessor search and orthogonal point location on the word ram. ACM Trans. Algorithms (TALG) **9**(3), 22 (2013)

6. Chazelle, B.: A functional approach to data structures and its use in multidimensional searching. SIAM J. Comput. **17**(3), 427–462 (1988)

7. Crochemore, M.: An optimal algorithm for computing the repetitions in a word. Inf. Process. Lett. **12**(5), 244–250 (1981)

8. Crochemore, M., Hancart, C., Lecroq, T.: Algorithms on Strings, 392 p. Cambridge University Press, Cambridge (2007)

9. Crochemore, M., Iliopoulos, C., Kubica, M., Radoszewski, J., Rytter, W., Waleń, T.: Extracting powers and periods in a string from its runs structure. In: Chavez, E., Lonardi, S. (eds.) SPIRE 2010. LNCS, vol. 6393, pp. 258–269. Springer, Heidelberg (2010)

10. Fine, N.J., Wilf, H.S.: Uniqueness theorems for periodic functions. Proc. Am. Math. Soc. **16**, 109–114 (1965)

11. Fischetti, V.A., Landau, G.M., Sellers, P.H., Schmidt, J.P.: Identifying periodic occurences of a template with applications to protein structure. Inf. Process. Lett. **45**(1), 11–18 (1993)

12. Galil, Z., Giancarlo, R.: Improved string matching with k mismatches. SIGACT News **17**(4), 52–54 (1986)

13. Gusfield, D., Stoye, J.: Linear time algorithms for finding and representing all the tandem repeats in a string. J. Comput. Syst. Sci. **69**(4), 525–546 (2004)

14. Iliopoulos, C.S., Moore, D., Smyth, W.F.: A characterization of the squares in a Fibonacci string. Theor. Comput. Sci. **172**(1–2), 281–291 (1997)

15. Karp, R.M., Miller, R.E., Rosenberg, A.L.: Rapid identification of repeated patterns in strings, trees, and arrays. In: STOC: ACM Symposium on Theory of Computing (STOC) (1972)

16. Kolpakov, R.M., Kucherov, G.: Finding maximal repetitions in a word in linear time. In: Proceedings of Symposium on Foundations of Computer Science (FOCS), pp. 596–604 (1999)

17. Kolpakov, R.M., Kucherov, G.: Finding approximate repetitions under Hamming distance. Theor. Comput. Sci **1**(303), 135–156 (2003)

18. Landau, G.M., Schmidt, J.P., Sokol, D.: An algorithm for approximate tandem repeats. J. Comput. Biol. **8**(1), 1–18 (2001)

19. Landau, G.M., Vishkin, U.: Fast parallel and serial approximate string matching. J. Algorithms **10**(2), 157–169 (1989)

20. Lothaire, M.: Applied Combinatorics on Words (Encyclopedia of Mathematics and its Applications). Cambridge University Press, New York (2005)

21. Lyndon, R.C.: On Burnside's problem. Trans. Am. Math. Soc. **77**(2), 202–215 (1954)

22. Myers, E.W., Miller, W.: Approximate matching of regular expressions. Bull. Math. Biol. **51**(1), 5–37 (1989)

23. Sim, J.S., Iliopoulos, C.S., Park, K., Smyth, W.F.: Approximate periods of strings. In: Crochemore, M., Paterson, M. (eds.) CPM 1999. LNCS, vol. 1645, pp. 123–133. Springer, Heidelberg (1999)

Chasing Convex Bodies and Functions

Antonios Antoniadis[1], Neal Barcelo[2], Michael Nugent[2], Kirk Pruhs[2],
Kevin Schewior[3]([✉]), and Michele Scquizzato[4]

[1] Max-Planck-Institut Für Informatik, Saarbrücken, Germany
`aantonia@mpi-inf.mpg.de`
[2] Department of Computer Science, University of Pittsburgh, Pittsburgh, USA
`{NCB30,mpn1,krp2}@pitt.edu`
[3] Technische Universität Berlin, Institut Für Mathematik, Berlin, Germany
`schewior@math.tu-berlin.de`
[4] Department of Computer Science, University of Houston, Houston, USA
`michele@cs.uh.edu`

Abstract. We consider three related online problems: Online Convex
Optimization, Convex Body Chasing, and Lazy Convex Body Chasing.
In Online Convex Optimization the input is an online sequence of convex
functions over some Euclidean space. In response to a function, the online
algorithm can move to any destination point in the Euclidean space. The
cost is the total distance moved plus the sum of the function costs at
the destination points. Lazy Convex Body Chasing is a special case of
Online Convex Optimization where the function is zero in some convex
region, and grows linearly with the distance from this region. And Convex Body Chasing is a special case of Lazy Convex Body Chasing where
the destination point has to be in the convex region. We show that these
problems are equivalent in the sense that if any of these problems have
an $O(1)$-competitive algorithm then all of the problems have an $O(1)$-
competitive algorithm. By leveraging these results we then obtain the
first $O(1)$-competitive algorithm for Online Convex Optimization in two
dimensions, and give the first $O(1)$-competitive algorithm for chasing linear subspaces. We also give a simple algorithm and $O(1)$-competitiveness
analysis for chasing lines.

1 Introduction

We consider the following three related online problems, all set in a d-dimensional
Euclidean space S, with some distance function ρ.

Convex Body Chasing: The input consists of an online sequence $\mathcal{F}_1, \mathcal{F}_2, \ldots, \mathcal{F}_n$
of convex bodies in S. In response to the convex body $\mathcal{F}_i$, the online algorithm
has to move to any destination/point $p_i \in \mathcal{F}_i$. The cost of such a feasible solution

K. Pruhs—Supported, in part, by NSF grants CCF-1115575, CNS-1253218, CCF-
1421508, and an IBM Faculty Award.

K. Schewior—Supported by the DFG within the research training group 'Methods
for Discrete Structures' (GRK 1408).

© Springer-Verlag Berlin Heidelberg 2016
E. Kranakis et al. (Eds.): LATIN 2016, LNCS 9644, pp. 68–81, 2016.
DOI: 10.1007/978-3-662-49529-2_6

is the total distance traveled by the online algorithm, namely $\sum_{i=1}^{n} \rho(p_{i-1}, p_i)$. The objective is to minimize the cost. If the convex bodies are restricted to be of a particular type $\mathcal{T}$, then we refer to the problem as $\mathcal{T}$ Chasing. So for example, Line Chasing means that the convex bodies are restricted to being lines.

Lazy Convex Body Chasing: The input consists of an online sequence of lazy convex bodies $(\mathcal{F}_1, \epsilon_1), (\mathcal{F}_2, \epsilon_2), \ldots, (\mathcal{F}_n, \epsilon_n)$, where each $\mathcal{F}_i$ is a convex body in S, and each slope ϵ_i is a nonnegative real number. In response to the pair $(\mathcal{F}_i, \epsilon_i)$, the online algorithm can move to any destination/point in the metric space S. The cost of such a feasible solution is

$$\sum_{i=1}^{n} \left(\rho(p_{i-1}, p_i) + \epsilon_i \rho(p_i, \mathcal{F}_i) \right),$$

where $\rho(p_i, \mathcal{F}_i)$ is the minimal distance of a point in $\mathcal{F}_i$ to p_i. So the online algorithm need not move inside each convex body, but if it is outside the convex body, in addition to paying the distance traveled, the online algorithm pays an additional cost that is linear in the distance to the convex body. The objective is to minimize the cost. Again if the convex bodies are restricted to be of a particular type $\mathcal{T}$, then we refer to the problem as Lazy $\mathcal{T}$ Chasing.

Online Convex Optimization: The input is an online sequence $F_1, F_2, \ldots, F_n$ of convex functions from S to $\mathbb{R}^+$. In response to the function F_i, the online algorithm can move to any destination/point in the metric space S. The cost of such a feasible solution is

$$\sum_{i=1}^{n} \left(\rho(p_{i-1}, p_i) + F_i(p_i) \right).$$

So the algorithm pays the distance traveled plus the value of the convex functions at the destinations points. The objective is to minimize the cost.

It is easy to see that a c-competitive algorithm for Online Convex Optimization implies a c-competitive algorithm for Lazy Convex Body Chasing, and a c-competitive algorithm for Lazy Convex Body Chasing implies a c-competitive algorithm for Convex Body Chasing. To see this, note that Convex Body Chasing is a special case of Lazy Convex Body Chasing where each ϵ_i is infinite (or, more formally, so large that any competitive algorithm would essentially have to move inside each convex body). Similarly, Lazy Convex Body Chasing is a special case of Online Convex Optimization in which the convex functions are zero on some convex set, and that grow linearly as one moves away from this convex set.

1.1   The History

Our initial interest in Online Convex Optimization arose from applications involving right-sizing data centers [1,14–18,21]. In these applications, there is a collection of d centrally-managed data centers, where each data center consists of a homogeneous collection of servers/processors which may be powered down. We represent the state of the data centers by a point in a d-dimensional space where coordinate i represents how many servers are currently powered-on in data center i (the assumption is that there are enough servers in each data center so that one may reasonably treat the number of servers as a real number instead of an integer). In response to a change in load, the number of servers powered-on in various data centers can be changed. Under the standard assumption that there is some fixed cost for powering a server on, or powering the server off, the Manhattan-distance between states represents the costs for powering on/off servers. The function costs represent the cost for operating the data-centers with the specified number of servers in each data center. The standard models of operating costs, such as those based on either queuing theoretic costs, and those based on energy costs for speed-scalable processors, are convex functions of the state.

Online Convex Optimization for $d = 1$: Essentially all the results in the literature for Online Convex Optimization are restricted to the case that the dimension is $d = 1$. [16] observed that the offline problem can be modeled as a convex program, which is solvable in polynomial time, and that if the line/states are discretized, then the offline problem can be solved by a straight-forward dynamic program. [16] also gave a 3-competitive deterministic algorithm that solves a (progressively larger) convex program at each time. [1] shows that there is an algorithm with sublinear regret, but that $O(1)$-competitiveness and sublinear regret cannot be simultaneously achieved. [1] gave a randomized online algorithm, RBG, and a 2-competitiveness analysis, but there is a bug in the analysis [22]. A revised 2-competitiveness analysis can be found in [2]. Independently, [4] gave a randomized algorithm and showed that it is 2-competitive. [4] also observed that any randomized algorithm can be derandomized, without any loss in the competitive ratio. [4] also gave a simple 3-competitive memoryless algorithm, and showed that this is optimally competitive for memoryless algorithms.

Convex Body Chasing: Convex Body Chasing and Lazy Convex Body Chasing were introduced in [10]. [10] assumed the standard Euclidean distance function, and observed that the optimal competitive ratio is $\Omega(\sqrt{d})$, where d is the dimension of the space. [10] gave a somewhat complicated algorithm and $O(1)$-competitiveness analysis for chasing lines in two dimensions, and observe that any $O(1)$-competitive line chasing algorithm for two dimensions can be extended to an $O(1)$-competitive line chasing algorithm for an arbitrary number of dimensions. [10] gave an even more complicated algorithm and $O(1)$-competitiveness analysis for chasing arbitrary convex bodies in two dimensions. [10] also observed that plane chasing in three dimensions is equivalent to lazy line chasing in two dimensions in the sense that one of these problems has an $O(1)$-competitive algorithm if and only if the other one does. [20] showed in a complicated analysis that the work function algorithm is $O(1)$-competitive for chasing lines and line segments in any dimension. [11] showed that the greedy algorithm is $O(1)$-competitive if $d = 2$ and the convex bodies are regular polygons with a constant number of sides.

Classic Online Problems: Online Convex Optimization is also related to several classic online optimization problems. It is a special case of the *metrical task system* problem in which the metric space is restricted to be a d-dimensional Euclidean space and the costs are restricted to be convex functions on that space. The optimal deterministic competitive ratio for a general metrical task system is $2n-1$, where n is the number of points in the metric [7], and the optimal randomized competitive ratio is $\Omega(\log n/\log\log n)$ [5,6] and $O(\log^2 n\log\log n)$ [9]. Online Convex Optimization is related to the *allocation problem* defined in [3], which arises when developing a randomized algorithm for the classic k-server problem using tree embeddings of the underlying metric space [3,8]. In fact, the algorithm RBG in [1] is derived from a similar algorithm in [8] for this allocation problem. The classic *ski rental* problem, where randomized algorithms are allowed, is a special case of Online Convex Optimization. The optimal competitive ratio for randomized algorithms for the ski rental problem is $e/(e-1)$ [12]. [4] showed that the optimal competitive ratio for Online Convex Optimization for $d = 1$ is strictly greater than the one for online ski rental. The k-server and CNN problems [13] can be viewed as chasing nonconvex sets.

1.2 Our Results

In Sect. 2 we show that all three of the problems that we consider are equivalent in the sense that if one of the problems has an $O(1)$-competitive algorithm, then they all have $O(1)$-competitive algorithms. More specifically, we show that if there is an $O(1)$-competitive algorithm for Lazy Convex Body Chasing in d dimensions then there is an $O(1)$-competitive algorithm for Online Convex Optimization in d dimensions. The crux of this reduction is to show that any convex function can be approximated to within a constant factor by a finite collection of lazy convex bodies. We then show that if there is an $O(1)$-competitive algorithm for Convex Body Chasing in d dimensions, then there is an $O(1)$-competitive algorithm for Lazy Convex Body Chasing (objects of the same type) in d dimensions. Intuitively in this reduction, each lazy convex body $(\mathcal{F}_i, \epsilon_i)$ is fed to the Convex Body Chasing algorithm with probability ϵ_i. As in [4], this algorithm can be derandomized by deterministically moving to the expected location of the randomized algorithm. The equivalence of these problems follows by combining these reductions with the obvious reductions in the other direction. Combining these reductions with the results in [10], most notably the $O(1)$-competitive algorithm for chasing halfspaces in two dimensions, we obtain an $O(1)$-competitive algorithm for Online Convex Optimization in two dimensions.

In Sect. 3 we give an online algorithm for Convex Body Chasing when the convex bodies are subspaces, in any dimension, and an $O(1)$-competitiveness analysis. In this context, subspace means a linear subspace closed under vector addition and scalar multiplication; So a point, a line, a plane, etc. The two main components of the algorithm are (1) A reduction from hyperplane chasing in d dimensions to lazy hyperplane chasing in $d-1$ dimensions, and (2) our reduction from Lazy Convex Body Chasing to Convex Body Chasing. The first reduction is the natural generalization of the continuous reduction from plane chasing in three dimesions to lazy line chasing in two dimensions given in [10]. Combining these two components gives an $O(1)$-approximation reduction from subspace chasing in d dimensions to subspace chasing in $d-1$ dimensions. One then obtains a $2^{O(d)}$-competitive algorithm by repeated applications of these reductions, and the use of any of the $O(1)$-competitive algorithms for Online Convex Optimization in one dimension. Within the context of right-sizing data centers, it is reasonable to assume that the number of data centers is a smallish constant, and thus this algorithm would be $O(1)$-competitive under this assumption.

In Sect. 4 we give an online algorithm for chasing lines and line segments in any dimension, and show that it is $O(1)$-competitive. The underlying insight of our online algorithm is the same as in [10], to be greedy with occasional adjustments toward the area where the adversary might have cheaply handled recent requests. However, our algorithm is cleaner/simpler than the algorithm in [10]. In particular our algorithm is essentially memoryless as the movement is based solely on the last two lines, instead of an unbounded number of lines as in [10]. Our analysis is based on a simple potential function: the distance between the location for the online algorithm and for the adversary, and is arguably cleaner than the analysis in [10], and is certainly cleaner than the analysis of the work function algorithm in [20].

While our results are not that technically deep, they do provide a much clearer picture of the algorithmic relationship of the various online problems in this area. Our results also suggest that the "right" problem to attack in this area is finding (if it exists) an $O(1)$-competitive algorithm for half-space chasing, as this is the simplest problem that would give an $O(1)$-competitive algorithm for all of these problems.

For concreteness we will assume ρ is the standard Euclidean distance function. Although as our focus is on $O(1)$-approximation, without being too concerned about the exact constant, our results will also hold for the Manhattan distance, and other standard normed distances.

2   Reductions

In this section we show in Lemma 1 that Lazy Convex Body Chasing is reducible to Convex Body Chasing, in Lemma 2 that Online Convex Optimization is reducible to Lazy Convex Body Chasing, and in Corollary 1 that these reductions give an $O(1)$-competitive algorithm for Online Convex Optimization in two dimensions.

Lemma 1. *If there is an $O(1)$-competitive algorithm A_C for Convex Body Chasing in d dimensions, then there is an $O(1)$-competitive algorithm A_L for Lazy Convex Body Chasing in d dimensions. The same result holds if the convex bodies in both problems are restricted to be of a particular type.*

Proof. We build a randomized algorithm A_L from A_C and then explain how to derandomize it. We first modify the input instance by replacing each lazy convex body $(\mathcal{F}_i, \epsilon_i)$ whose slope ϵ_i is greater than 1 by $\lceil \epsilon_i \rceil$ lazy convex bodies, each having $\mathcal{F}_i$ as the convex body. The first $\lfloor \epsilon_i \rfloor$ of these lazy convex bodies will have slope 1, and the potentially remaining convex body will have slope $\epsilon_i - \lfloor \epsilon_i \rfloor$. This modification does not affect the optimal cost, and will not decrease the online cost. From now on, we will assume that any input instance for Lazy Convex Body Chasing is of this modified form. It is easy to see how one can go back from a solution to the modified input to one to the original input without increasing the cost, since our algorithm will never "move away" from the line that just arrived.

Algorithm A_L: Upon the arrival of a new lazy convex body $(\mathcal{F}_i, \epsilon_i)$, the algorithm with (independent) probability $1 - \epsilon_i$ does not move, and with probability ϵ_i passes $\mathcal{F}_i$ to A_C and moves to the location to which A_C moves.

Notice that in the modified input instance every slope is a real number in $[0, 1]$, and thus probabilities ϵ_i and $1 - \epsilon_i$ are all well defined.

Analysis: Consider a particular input instance I_L of Lazy Convex Body Chasing, as defined before. Let I_C denote the random variable representing the sequence of convex bodies passed to A_C. Let OPT_L be the optimal solution for Lazy Convex Body Chasing on I_L. Let OPT_C be a random variable equal to the optimal solution for Convex Body Chasing on I_C. Let OPT_T be a random variable equal to the optimal solution for the Lazy Convex Body Chasing instance I_T derived from I_C by replacing each $\mathcal{F}_i \in I_C$ by the lazy convex body $(\mathcal{F}_i, 1)$. We will use absolute value signs to denote the cost of a solution.

The $O(1)$-competitiveness of A_L then follows from the following sequence of inequalities:

$$\mathbb{E}[|A_L(I_L)|] = \sum_i \mathbb{P}[\mathcal{F}_i \in I_C] \, \mathbb{E}[\text{Cost of } A_C \text{ on } \mathcal{F}_i \mid \mathcal{F}_i \in I_C]$$

$$+ \sum_i \mathbb{P}[\mathcal{F}_i \notin I_C] \, \mathbb{E}[\text{Cost of } A_L \text{ on } \mathcal{F}_i \mid \mathcal{F}_i \notin I_C] \qquad (1)$$

$$= \sum_i \epsilon_i \, \mathbb{E}[\text{Cost of } A_C \text{ on } \mathcal{F}_i \mid \mathcal{F}_i \in I_C]$$

$$+ \sum_i (1 - \epsilon_i) \, \mathbb{E}[\epsilon_i \, \rho(p_{i-1}, \mathcal{F}_i)] \qquad (2)$$

$$\leq \sum_i \epsilon_i \, \mathbb{E}[\text{Cost of } A_C \text{ on } \mathcal{F}_i \mid \mathcal{F}_i \in I_C]$$

$$+ \sum_i \epsilon_i \, \mathbb{E}[\rho(p_{i-1}, \mathcal{F}_i)] \qquad (3)$$

$$\leq 2 \sum_i \epsilon_i \, \mathbb{E}[\text{Cost of } A_C \text{ on } \mathcal{F}_i \mid \mathcal{F}_i \in I_C] \qquad (4)$$

$$= 2 \sum_i \mathbb{P}[\mathcal{F}_i \in I_C] \, \mathbb{E}[\text{Cost of } A_C \text{ on } \mathcal{F}_i \mid \mathcal{F}_i \in I_C] \qquad (5)$$

$$= 2 \, \mathbb{E}[|A_C(I_C)|] \qquad (6)$$

$$= O(\mathbb{E}[|\text{OPT}_C|]) \qquad (7)$$

$$= O(\mathbb{E}[|\text{OPT}_T|]) \qquad (8)$$

$$= O(|\text{OPT}_L|). \qquad (9)$$

Equality (1) follows from the definitions of expectation and conditional expectation, and linearity of expectation. Notice that all expectations involving $\mathcal{F}_i$ only depend upon the history up until $\mathcal{F}_i$ arrives. Equality (2) follows from the fact that $\mathcal{F}_i$ is added to I_C with probability ϵ_i, and if $\mathcal{F}_i$ is not added then A_L pays ϵ_i times the distance to $\mathcal{F}_i$. Inequality (3) follows from the linearity of expectation and since $1 - \epsilon_i \leq 1$. Inequality (4) holds since A_C has to move to each $\mathcal{F}_i \in I_C$ and thus in expectation has to pay at least $\mathbb{E}[\rho(p_{i-1}, \mathcal{F}_i)]$ (note that only by independence of the coin flips the expected position of A_C in case $\mathcal{F}_i \in I_C$ is identical to the expected position of A_L). Equality (5) holds since $\mathcal{F}_i \in I_C$ with probability ϵ_i. Equality (6) follows by linearity of expectation and the definition of conditional expectation. Inequality (7) follows by the assumption that A_C is $O(1)$-competitive. To prove Inequality (8) it is sufficient to construct a solution $\mathcal{S}$ for each possible instantiation of I_C that is at most a constant times more expensive than OPT_T. In response to a convex body $\mathcal{F}_i \in I_C$, $\mathcal{S}$ moves to the same destination point p_i as OPT_T, then moves to the closest point on $\mathcal{F}_i$, and then back to p_i. Thus the movement cost for $\mathcal{S}$ is at most the movement cost for OPT_T plus twice the function costs for OPT_T. To prove Inequality (9) it is sufficient to construct an algorithm B to solve Lazy Convex Body Chasing on I_T with expected cost $O(|\mathrm{OPT}_L|)$. For each convex body in $\mathcal{F}_i \in I_C$, Algorithm B first moves to the destination point p_i that OPT_L moves to after $\mathcal{F}_i$. Call this a basic move. Then algorithm B moves to the closest point in $\mathcal{F}_i$, and then back to p_i. Call this a detour move. Then by the triangle inequality the expected total cost of the basic moves for algorithm B is at most the movement cost of OPT_L. The probability that algorithm B incurs a detour cost for convex body $\mathcal{F}_i$ is ϵ_i, and when it incurs a detour cost, this detour cost is $2/\epsilon_i$ times the function cost incurred by OPT_L. Thus the expected cost for algorithm B on I_T is at most $3|\mathrm{OPT}_L|$.

Derandomization: As in [4], we can derandomize A_L to get a deterministic algorithm A_D with the same competitive ratio as A_L. A_D always resides in the expected position of A_L. More specifically, let x_i be a random variable denoting the position of A_L directly after the arrival of $\mathcal{F}_i$. Then A_D sets its position to $\mu_i := \mathbb{E}[x_i]$.

Then, we have that for each step i, A_L's expected cost is $\mathbb{E}[\rho(x_{i-1}, x_i)] + \epsilon_i \mathbb{E}[\rho(x_i, \mathcal{F}_i)]$. On the other hand, A_D's cost is $\rho(\mathbb{E}[x_i], \mathbb{E}[x_{i-1}]) + \epsilon_i \rho(\mathbb{E}[x_i], \mathcal{F}_i)$. By a generalization of Jensen's inequality (see for example Proposition B.1, page 343 in the book by Marshall and Olkin [19]), and by the convexity of our distance function ρ (the distance function is a norm, and therefore convexity follows by triangle inequality and absolute homogeneity), we have, for each i,

$$\mathbb{E}[\rho(x_i, x_{i-1})] \geq \rho(\mathbb{E}[x_i], \mathbb{E}[x_{i-1}])$$

and

$$\mathbb{E}[\rho(x_i, \mathcal{F}_i)] \geq \rho(\mathbb{E}[x_i], \mathcal{F}_i).$$

Summing over all i completes the analysis. □

Lemma 2. *If there is an $O(1)$-competitive algorithm A_L for Lazy Convex Body Chasing in d dimensions, then there is an $O(1)$-competitive algorithm A_O for Online Convex Optimization in d dimensions.*

Proof. Consider an arbitrary instance I_O of the convex optimization problem. We can without loss of generality ignore the prefix of the sequence of functions that can be handled with zero cost. So let $L > 0$ be the optimal cost for chasing function F_1. The algorithm will use L as a lower bound for the optimal cost.

For each function F_i that it sees, the algorithm A_O feeds the algorithm A_L a finite collection C_i of lazy convex bodies, and then moves to the final destination point that A_L moved to. Let I_L be the resulting instance of Lazy Convex Body Chasing. To define C_i assume without loss of generality that the minimum of F_i occurs at the origin. We can also assume without loss of generality that the minimum of F_i is zero.

Define $F_i'(x)$ to be the partial derivative of F_i at the point $x \in S = \mathbb{R}^d$ in the direction away from the origin. Now let C_j be the curve in $\mathbb{R}^{d+1}$ corresponding to the points $(x, F_i'(x))$ where $F_i'(x) = 2^j$ for integer $j \in (-\infty, +\infty)$. (Or more technically where the $F_i'(x)$ transitions from being less than 2^j to more than 2^j.) Let D_j be the projection of C_j onto S. Note that D_j is convex.

Let u be the minimum integer such that the location of A_O just before F_i is inside of D_u. Let ℓ be the maximum integer such that:

- the diameter of D_ℓ is less than $L/8^i$,
- the maximum value of $F_i(x)$ for an $x \in D_\ell$ is less than $L/8^i$, and
- $\ell < u - 10$.

Then C_i consists of the lazy convex bodies $(D_j, 2^j)$ for $j \in [\ell, u]$.

Let $G_j(x)$ be the function that is zero within D_j and grows linearly at a rate 2^j as one moves away from D_j. Now what we want to prove is that $\sum_j G_j(x) = \Theta(F_i(x))$ for all x outside of $D_{\ell+2}$. To do this consider moving toward x from the origin. Consider the region between D_j and D_{j+1} for $j \geq \ell+2$, and a point y in this region that lies on the line segment between the origin and x. Then we know that $2^j \leq F_i'(y) \leq 2^{j+1}$. Thus as we are moving toward x, the rate of increase of $\sum_j G_j(x)$ is within a constant factor of the rate of increase of $F_i(x)$, and thus $\sum_j G_j(x) = \Theta(F_i(x))$.

Let OPT_L be the optimal solution for the Lazy Line Chasing instance I_L and Let OPT_O be the optimal solution for the Online Convex Optimization instance I_O. Now the claim follows via the following inequalities:

$$|A_O(I_O)| = O(|A_L(I_L)|) \tag{10}$$
$$= O(|\text{OPT}_L|) \tag{11}$$
$$= O(|\text{OPT}_O|). \tag{12}$$

Inequality (10) holds since the movement cost for A_L and A_O are identical, and the function costs are within a constant of each other by the observation that $\sum_j G_j(x) = \Theta(F_i(x))$. Inequality (11) holds by the assumption that A_L is $O(1)$-competitive. Inequality (12) holds by the observation that $\sum_j G_j(x) = \Theta(F_i(x))$

and the observation that the maximum savings that OPT_L can obtain from being inside of each D_ℓ is at most $L/2$.   □

Corollary 1. *There is an $O(1)$-competitive algorithm for Online Convex Optimization in two dimensions.*

Proof. This follows from the reduction from Online Convex Optimization to Lazy Convex Body Chasing in Lemma 2, the reduction from Lazy Convex Body Chasing to Convex Body Chasing in Lemma 1, and the $O(1)$-competitive algorithm for Convex Body Chasing in two dimensions in [10].   □

3   Subspace Chasing

In this section we describe a $2^{O(d)}$-competitive algorithm for chasing subspaces in any dimension d. As noticed in [10], it suffices to give such an algorithm for chasing $(d-1)$-dimensional subspaces (hyperplanes). Essentially this is because every $f \leq d-1$-dimensional subspace is the intersection of $d-f$ hyperplanes, and by repeating these $d-f$ hyperplanes many times, any competitive algorithm can be forced arbitrarily close to their intersection.

Algorithm for Chasing Hyperplanes: The two main components of our algorithm for chasing hyperplanes in d dimensions are:

- A reduction from hyperplane chasing in d dimensions to lazy hyperplane chasing in $d-1$ dimensions. This is a discretized version of the continuous reduction given in [10] for $d = 3$. In this section we give an overview of the reduction, and the analysis, but defer the formal proof to the full version of the paper.
- Our reduction from Lazy Convex Body Chasing to Convex Body Chasing in the previous section.

Combining these two components gives a reduction from subspace chasing in d dimensions to subspace chasing in $d-1$ dimensions. One then obtains a $2^{O(d)}$-competitive algorithm by repeated applications of these reductions, and the use of any of the $O(1)$-competitive algorithms for lazy point chasing (or online convex optimization) when $d = 1$ [2,4,16].

Description of Reduction from Hyperplane Chasing in Dimension d to Lazy Hyperplane Chasing in Dimension $d-1$: Let A_{LHC} be the algorithm for lazy hyperplane chasing in dimension $d-1$. We maintain a bijective mapping from the $\mathbb{R}^{d-1}$ space S that the algorithm A_{LHC} moves in to the last hyperplane in $\mathbb{R}^d$. Initially, this mapping is an arbitrary one that maps the origin of $\mathbb{R}^{d-1}$ to the origin of $\mathbb{R}^d$. Call this hyperplane $\mathcal{F}_0$.

Each time a new hyperplane $\mathcal{F}_i$ in $\mathbb{R}^d$ arrives, the algorithm moves in the following way:

- If $\mathcal{F}_i$ is parallel to $\mathcal{F}_{i-1}$, then the algorithm moves to the nearest position in $\mathcal{F}_i$.

– If $\mathcal{F}_i$ is not parallel to $\mathcal{F}_{i-1}$, the two hyperplanes intersect in a $(d-2)$-dimensional subspace $\mathcal{I}_i$. Let $\alpha_i \leq \pi/2$ radians be the angle between hyperplanes $\mathcal{F}_{i-1}$ and $\mathcal{F}_i$. The algorithm then calls A_{LHC} with $(\mathcal{I}_i, \alpha_i)$. Let p_i be the point within $\mathcal{F}_{i-1}$ that A_{LHC} moves to. The bijection between $\mathcal{F}_i$ and S is obtained from the bijection between $\mathcal{F}_{i-1}$ and S by rotating α_i radians around $\mathcal{I}_i$. The algorithm then moves to the location of p_i in $\mathcal{F}_i$.

Analysis Overview: Consecutive parallel hyperplanes is the easy case. In this case, one can assume, at a loss of a factor of $\sqrt{2}$ in competitive ratio, that the optimal solution moves to the closest point on the new parallel hyperplane. Thus any competitive ratio c that one can prove under the assumption that consecutive hyperplanes are not parallel will hold in general as long as $c \geq \sqrt{2}$.

When there are no two consecutive non-parallel hyperplanes, we show that the cost for the reduction algorithm and the cost for A_{LHC} are within a constant of each other, and similarly the optimal cost for hyperplane chasing in d dimensions and the optimal cost for lazy hyperplane chasing are within a constant of each other. Intuitively this is because the additional movement costs incurred in (non-lazy) hyperplane chasing can be related to the angle between the last two hyperplanes, and thus to the distance cost (for not being on the hyperplane) that has to be paid in lazy hyperplane chasing. From this we can conclude that:

Theorem 1. *There is a $2^{O(d)}$-competitive algorithm for subspace chasing in d dimensions.*

We note that our algorithm can be implemented to run in time polynomial in d and n.

4   Line Chasing

We give an online algorithm for chasing lines and line segments in any dimension, and show that it is $O(1)$-competitive. Let $\mathcal{E}_i$ be the unique line that is an extension of the line segment $\mathcal{F}_i$.

Algorithm Description: Let $\mathcal{P}_i$ be a plane containing the line $\mathcal{E}_i$ that is parallel to the line $\mathcal{E}_{i-1}$ (note this is uniquely defined when $\mathcal{E}_i$ and $\mathcal{E}_{i-1}$ are not parallel). The algorithm first moves to the closest point $h_i \in \mathcal{P}_i$. The algorithm then moves to the closest point g_i in $\mathcal{E}_i$.

– If $\mathcal{E}_{i-1}$ and $\mathcal{E}_i$ are parallel, then the algorithm stays at g_i.
– If $\mathcal{E}_{i-1}$ and $\mathcal{E}_i$ are not parallel, let m_i be the intersection of $\mathcal{E}_i$ and the projection of $\mathcal{E}_{i-1}$ onto $\mathcal{P}_i$. Let $\beta \in (0,1)$ be some constant that we shall set later. The algorithm makes an adjustment by moving toward m_i along $\mathcal{E}_i$ until it has traveled a distance of $\beta \cdot \rho(h_i, g_i)$, or until it reaches m_i.
– Finally, the algorithm moves to the closest point p_i in $\mathcal{F}_i$.

Theorem 2. *This algorithm is $O(1)$-competitive for Line Chasing in any dimension.*

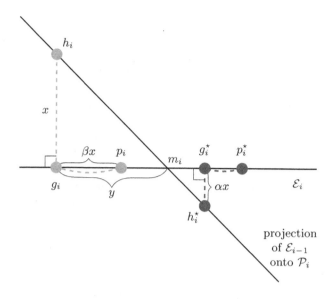

Fig. 1. An overview of the used points and distances in $\mathcal{P}_i$.

Proof. Initially assume that all the line segments are lines. Let $h_i^\star$ be the projection of the adversary's position, just before $\mathcal{F}_i$ arrives, onto $\mathcal{P}_i$. We will assume that the adversary first moves to $h_i^\star$, and then to the nearest point $g_i^\star$ on $\mathcal{F}_i$, and then to some arbitrary final point $p_i^\star$ on $\mathcal{F}_i$. This assumption increases the adversary's movement cost by at most a factor of $\sqrt{3}$ (a similar observation is made in [10]). We will charge the algorithm's cost for moving to $\mathcal{P}_i$ to the adversary's cost of moving to $\mathcal{P}_i$. Thus by losing at most a factor of $\sqrt{3}$ in the competitive ratio, we can assume that the adversary and the algorithm are at positions $h_i^\star$ and h_i right before $\mathcal{F}_i$ arrives.

Let ALG_i and OPT_i denote the movement cost of the algorithm and the adversary, respectively, in response to $\mathcal{F}_i$. Let the potential function Φ_i be $\delta \cdot \rho(p_i, p_i^\star)$ for some to be determined constant $\delta > 1$. To show that the algorithm is c-competitive it will be sufficient to show that, for each i,

$$\mathrm{ALG}_i + \Phi_i - \Phi_{i-1} \le c \cdot \mathrm{OPT}_i. \tag{13}$$

We will only initially consider the adversary's movement cost until it reaches $g_i^\star$. Equation (13) will continue to hold for any adversary's movement after $g_i^\star$ if

$$c \ge \delta. \tag{14}$$

We consider three cases. The notation that we will use is illustrated in Fig. 1.

In the first case assume $\rho(h_i^\star, g_i^\star) \ge \gamma x$, where $x = \rho(h_i, g_i)$, and $\gamma > 0$ is a constant that we define later. Intuitively, this is the easiest case as the adversary's cost will pay for both the algorithm's movement cost and the increase in the potential due to this movement. For Eq. (13) to hold, it is sufficient that

$$(1 + \beta)x + \delta((1 + \beta)x + \gamma x) \le c\gamma x,$$

or equivalently

$$c \geq \frac{(1 + \beta) + \delta((1 + \beta) + \gamma)}{\gamma}. \tag{15}$$

In the remaining two cases assume that $\rho(h_i^\star, g_i^\star) = \alpha x \leq \gamma x$, and let $y = \rho(m_i, g_i)$.

In the second case assume that $x \leq y$. Intuitively in this case the decrease in the potential due to the algorithm's adjustment on $\mathcal{F}_i$ decreases the potential enough to pay for the algorithm's movement costs. Since $\beta < 1$ and $x \leq y$, the algorithm will not have to stop at m_i when moving a distance of βx on $\mathcal{F}_i$ toward m_i. If

$$\gamma \leq 1 - \beta, \tag{16}$$

the algorithm will also not cross $g_i^\star$ while adjusting on $\mathcal{F}_i$, as this would contradict the assumption that $\rho(h_i^\star, g_i^\star) \leq \gamma x$. Equation (13) will be hardest to satisfy when $\Phi_i - \Phi_{i-1}$ is maximal. This will occur when $g_i^\star$ is maximally far from g_i, which in turn occurs when the points $g_i^\star$, m_i, and g_i lie on $\mathcal{F}_i$ in that order. In that case, Eq. (13) evaluates to

$$(1 + \beta)x + \delta((1 + \alpha)y - \beta x - (1 + \alpha)\sqrt{x^2 + y^2}) \leq c\alpha x.$$

Setting $L = \frac{1 + \beta - \delta\beta - c\alpha}{\delta(1 + \alpha)}$, this is equivalent to

$$Lx + y \leq \sqrt{x^2 + y^2}. \tag{17}$$

When $x \leq y$, Eq. (17) holds if $L \leq 0$. This in turn holds when

$$\delta \geq \frac{1 + \beta}{\beta}. \tag{18}$$

Finally we consider the third case that $x \geq y$. Intuitively in this case the decrease in the potential due to the algorithm's movement toward $\mathcal{F}_i$ decreases the potential enough to pay for the algorithms movement costs. First consider the algorithm's move from p_{i-1} to g_i. Again, the value of $\Phi_i - \Phi_{i-1}$ is maximized when $g_i^\star$ is on the other side of m_i as is g_i. In that case, the value of $\Phi_i - \Phi_{i-1}$ due to this move is at most

$$\delta((1 + \alpha)y - (1 + \alpha)\sqrt{x^2 + y^2}).$$

Note that the maximum possible increase in the potential due to the algorithm moving from g_i to p_i is $\delta\beta x$. Thus for Eq. (13) to hold, it is sufficient that

$$(1 + \beta)x + \delta\beta x + \delta((1 + \alpha)y - (1 + \alpha)\sqrt{x^2 + y^2}) \leq c\alpha x.$$

Setting $L = \frac{1 + \beta + \delta\beta - c\alpha}{\delta(1 + \alpha)}$, this is equivalent to

$$Lx + y \leq \sqrt{x^2 + y^2}. \tag{19}$$

When $x \geq y$, Eq. (19) holds if $L \leq \sqrt{2} - 1$. This in turn holds when

$$\delta \geq \frac{1 + \beta}{\sqrt{2} - 1 - \beta}. \tag{20}$$

We now need to find a feasible setting of β, γ, δ, and compute the resulting competitive ratio. Setting $\beta = \frac{\sqrt{2}-1}{2}$ and $\gamma = \frac{3-\sqrt{2}}{2}$, and $\delta = \frac{\sqrt{2}+1}{\sqrt{2}-1}$ one can see that Eqs. (16), (18), and (20) hold. The minimum c satisfying (15) is $c \approx 16.22$ and also satisfies (14). Then given that we overestimate the adversary's cost by a most a factor of $\sqrt{3}$, this gives us a competitive ratio for lines of approximately 28.1, approximately the same competitive ratio as obtained in [10].

If $\mathcal{F}_i$ is a line segment, then we need to account for the additional movement along $\mathcal{E}_i$ to reach $\mathcal{F}_i$. However, as we set $\delta > 1$, the decrease in the potential can pay for this movement cost.   □

Acknowledgement. We thank Nikhil Bansal, Anupam Gupta, Cliff Stein, Ravishankar Krishnaswamy, and Adam Wierman for helpful discussions. We also thank an anonymous reviewer for pointing out an important subtlety in one of our proofs.

References

1. Andrew, L.L.H., Barman, S., Ligett, K., Lin, M., Meyerson, A., Roytman, A., Wierman, A.: A tale of two metrics: simultaneous bounds on competitiveness and regret. In: Conference on Learning Theory, pp. 741–763 (2013)
2. Andrew, L.L.H., Barman, S., Ligett, K., Lin, M., Meyerson, A., Roytman, A., Wierman, A.: A tale of two metrics: simultaneous bounds on competitiveness and regret. CoRR, abs/1508.03769 (2015)
3. Bansal, N., Buchbinder, N., Naor, J.: Towards the randomized k-server conjecture: a primal-dual approach. In: ACM-SIAM Symposium on Discrete Algorithms, pp. 40–55 (2010)
4. Bansal, N., Gupta, A., Krishnaswamy, R., Pruhs, K., Schewior, K., Stein, C.: A 2-competitive algorithm for online convex optimization with switching costs. In: Workshop on Approximation Algorithms for Combinatorial Optimization Problems, pp. 96–109 (2015)
5. Bartal, Y., Bollobás, B., Mendel, M.: Ramsey-type theorems for metric spaces with applications to online problems. J. Comput. Syst. Sci. **72**(5), 890–921 (2006)
6. Bartal, Y., Linial, N., Mendel, M., Naor, A.: On metric Ramsey-type phenomena. In: ACM Symposium on Theory of Computing, pp. 463–472 (2003)
7. Borodin, A., Linial, N., Saks, M.E.: An optimal on-line algorithm for metrical task system. J. ACM **39**(4), 745–763 (1992)
8. Coté, A., Meyerson, A., Poplawski, L.: Randomized k-server on hierarchical binary trees. In: ACM Symposium on Theory of Computing, pp. 227–234 (2008)
9. Fiat, A., Mendel, M.: Better algorithms for unfair metrical task systems and applications. SIAM J. Comput. **32**(6), 1403–1422 (2003)
10. Friedman, J., Linial, N.: On convex body chasing. Discrete Comput. Geom. **9**, 293–321 (1993)
11. Fujiwara, H., Iwama, K., Yonezawa, K.: Online chasing problems for regular polygons. Inf. Process. Lett. **108**(3), 155–159 (2008)

12. Karlin, A.R., Manasse, M.S., McGeoch, L.A., Owicki, S.S.: Competitive randomized algorithms for nonuniform problems. Algorithmica **11**(6), 542–571 (1994)
13. Koutsoupias, E., Taylor, D.S.: The CNN problem and other k-server variants. Theor. Comput. Sci. **324**(2–3), 347–359 (2004)
14. Lin, M., Liu, Z., Wierman, A., Andrew, L.L.H.: Online algorithms for geographical load balancing. In: International Green Computing Conference, pp. 1–10 (2012)
15. Lin, M., Wierman, A., Andrew, L.L.H., Thereska, E.: Online dynamic capacity provisioning in data centers. In: Allerton Conference on Communication, Control, and Computing, pp. 1159–1163 (2011)
16. Lin, M., Wierman, A., Andrew, L.L.H., Thereska, E.: Dynamic right-sizing for power-proportional data centers. IEEE/ACM Trans. Netw. **21**(5), 1378–1391 (2013)
17. Lin, M., Wierman, A., Roytman, A., Meyerson, A., Andrew, L.L.H.: Online optimization with switching cost. SIGMETRICS Perform. Eval. Rev. **40**(3), 98–100 (2012)
18. Liu, Z., Lin, M., Wierman, A., Low, S.H., Andrew, L.L.H.: Greening geographical load balancing. In: ACM SIGMETRICS International Conference on Measurement and Modeling of Computer Systems, pp. 233–244 (2011)
19. Marshall, A.W., Olkin, I.: Inequalities: Theory of Majorization and Its Application. Academic Press, Cambridge (1979)
20. Sitters, R.: The generalized work function algorithm is competitive for the generalized 2-server problem. SIAM J. Comput. **43**(1), 96–125 (2014)
21. Wang, K., Lin, M., Ciucu, F., Wierman, A., Lin, C.: Characterizing the impact of the workload on the value of dynamic resizing in data centers. In: IEEE INFOCOM, pp. 515–519 (2013)
22. Wierman, A.: Personal Communication (2015)

Parameterized Lower Bounds and Dichotomy Results for the NP-completeness of H-free Edge Modification Problems

N.R. Aravind[1], R.B. Sandeep[1](✉), and Naveen Sivadasan[2]

[1] Department of Computer Science and Engineering,
Indian Institute of Technology Hyderabad, Hyderabad, India
{aravind,cs12p0001}@iith.ac.in
[2] TCS Innovation Labs, Hyderabad, India
naveen@atc.tcs.com

Abstract. For a graph H, the H-FREE EDGE DELETION problem asks whether there exist at most k edges whose deletion from the input graph G results in a graph without any induced copy of H. H-FREE EDGE COMPLETION and H-FREE EDGE EDITING are defined similarly where only completion (addition) of edges are allowed in the former and both completion and deletion are allowed in the latter. We completely settle the classical complexities of these problems by proving that H-FREE EDGE DELETION is NP-complete if and only if H is a graph with at least two edges, H-FREE EDGE COMPLETION is NP-complete if and only if H is a graph with at least two non-edges and H-FREE EDGE EDITING is NP-complete if and only if H is a graph with at least three vertices. Our result on H-FREE EDGE EDITING resolves a conjecture by Alon and Stav (2009). Additionally, we prove that, these NP-complete problems cannot be solved in parameterized subexponential time, i.e., in time $2^{o(k)} \cdot |G|^{O(1)}$, unless Exponential Time Hypothesis fails. Furthermore, we obtain implications on the incompressibility of these problems.

1 Introduction

Edge modification problems are to test whether modifying at most k edges makes the input graph satisfy certain properties. The three major edge modification problems are edge deletion, edge completion and edge editing problems. In edge deletion problems we are allowed to delete at most k edges from the input graph. Similarly, in completion problems, it is allowed to complete (add) at most k edges and in editing problems at most k editing (deletion or completion) are allowed. Edge modification problems come under the broader category of graph modification problems which have found applications in DNA physical mapping [11], numerical algebra [14], circuit design [9] and machine learning [3].

The focus of this paper is on H-free edge modification problems, in which we are allowed to modify at most k edges to make the input graph devoid of any

R.B. Sandeep—supported by TCS Research Scholarship.

© Springer-Verlag Berlin Heidelberg 2016
E. Kranakis et al. (Eds.): LATIN 2016, LNCS 9644, pp. 82–95, 2016.
DOI: 10.1007/978-3-662-49529-2_7

induced copy of H, where H is any fixed graph. Though these problems have been studied for four decades, a complete dichotomy result on the classical complexities of these problems are not yet found. We settle this by proving that H-FREE EDGE DELETION is NP-complete if and only if H is a graph with at least two edges, H-FREE EDGE COMPLETION is NP-complete if and only if H is a graph with at least two non-edges and H-FREE EDGE EDITING is NP-complete if and only if H is a graph with at least three vertices. Our result on H-FREE EDGE EDITING settles a conjecture by Alon and Stav [1]. Further, we obtain the parameterized lower bounds for these NP-complete problems. We obtain that these NP-complete problems cannot be solved in parameterized subexponential time (i.e., in time $2^{o(k)} \cdot |G|^{O(1)}$), unless Exponential Time Hypothesis (ETH) fails. Cai proved that these problems are in FPT and gave a branch and bound algorithm to solve these problems in time $|V(H)|^{O(k)} \cdot |G|^{O(1)}$. In this sense, our lower bounds are tight. Furthermore, we obtain implications on the incompressibility (non-existence of polynomial kernels) of these problems.

We build on our recent paper [2], in which we proved that H-FREE EDGE DELETION is NP-complete if H at least two edges and has a component with maximum number of vertices which is a tree or a regular graph. We also proved that these problems cannot be solved in parameterized subexponential time, unless ETH fails.

Related Work: In 1981, Yannakakis proved that H-FREE EDGE DELETION is NP-complete if H is a cycle [16]. Later in 1988, El-Mallah and Colbourn proved that the problem is NP-complete if H is a path of at least two edges [9]. Addressing the fixed parameter tractability of a generalized version of these problems, Cai proved that [4] H-FREE EDGE DELETION, COMPLETION and EDITING are fixed parameter tractable, i.e., they can be solved in time $f(k) \cdot |G|^{O(1)}$, for some function f. Polynomial kernelizability of these problems have been studied widely. Given an instance (G, k) of the problem the objective is to obtain in polynomial time an equivalent instance of size polynomial in k. Kratsch and Wahlström gave the first result on the incompressibility of H-free edge modification problems. They proved that [13] for a certain graph H on seven vertices, H-FREE EDGE DELETION and H-FREE EDGE EDITING do not admit polynomial kernels, unless NP $\subseteq$ coNP/poly. They use polynomial parameter transformation from an NP-complete problem and hence their results imply the NP-completeness of these problems. Later, Cai and Cai proved that H-FREE EDGE EDITING, DELETION and COMPLETION do not admit polynomial kernels if H is a path or a cycle with at least four edges, unless NP $\subseteq$ coNP/poly [5]. Further, they proved that H-FREE EDGE EDITING and DELETION are incompressible if H is 3-connected but not complete, and H-FREE EDGE COMPLETION is incompressible if H is 3-connected and has at least two non-edges, unless NP $\subseteq$ coNP/poly [5]. Under the same assumption, it is proved that H-FREE EDGE DELETION and H-FREE EDGE COMPLETION are incompressible if H is a tree on at least 7 vertices, which is not a star graph and H-FREE EDGE DELETION is incompressible if H is the star graph $K_{1,s}$, where $s \geq 10$ [6]. They also use polynomial parameter transformations and hence these problems are NP-complete.

Outline of the Paper: Sect. 2 gives the notations and terminology used in the paper. It also introduces a construction which is a modified version of the main construction used in [2]. Section 3 settles the case of H-FREE EDGE EDITING. Section 4 obtains results for H-FREE EDGE DELETION and COMPLETION. In the concluding section, we discuss the implications of our results on the incompressibility of H-free edge modification problems.

2   Preliminaries and Basic Tools

Graphs: For a graph G, $V(G)$ denotes the vertex set and $E(G)$ denotes the edge set. We denote the symmetric difference operator by $\triangle$, i.e., for two sets F and F', $F \triangle F' = (F \setminus F') \cup (F' \setminus F)$. For a graph G and a set $F \subseteq [V(G)]^2$, $G \triangle F$ denotes the graph $(V(G), E(G) \triangle F)$. A component of a graph is largest if it has maximum number of vertices. By $|G|$ we denote $|V(G)| + |E(G)|$. The disjoint union of two graphs G and G' is denoted by $G \cup G'$ and the disjoint union of t copies of G is denoted by tG. A simple path on t vertices is denoted by P_t. The graph t-diamond is $K_2 + tK_1$, the join of K_2 and tK_1. Hence, 2-diamond is the diamond graph. The minimum degree of a graph G is denoted by $\delta(G)$ and the maximum degree is denoted by $\Delta(G)$. Degree of a vertex v in a graph G is denoted by $\deg_G(v)$. We remove the subscript when there is no ambiguity. We denote the complement of a graph G by $\overline{G}$. For a graph H and a vertex set $V' \subseteq V(H)$, $H[V']$ is the graph induced by V' in H. A null graph is a graph without any edge.

For integers ℓ and h such that $h > \ell$, (ℓ, h)-degree graph is a graph in which every vertex has degree either ℓ or h. The set of vertices with degree ℓ is denoted by V_ℓ and the set of vertices with degree h is denoted by V_h. An (ℓ, h)-degree graph is called *sparse* if V_l induces a graph with at most one edge and V_h induces a graph with at most one edge.

The context determines whether H-FREE EDGE DELETION (COMPLETION/ EDITING) denotes the classical problem or the parameterized problem. For the parameterized problems, we use k (the size of the solution being sought) as the parameter. In this paper, edge modification implies either deletion, completion or editing.

Technique for Proving Parameterized Lower Bounds: Exponential Time Hypothesis (ETH) is a complexity theoretic assumption that 3-SAT cannot be solved in time $2^{o(n)}$, where n is the number of variables in the 3-SAT instance. A linear parameterized reduction is a polynomial time reduction from a parameterized problem A to a parameterized problem B such that for every instance (G, k) of A, the reduction gives an instance (G', k') such that $k' = O(k)$. The following result helps us to obtain parameterized lower bound under ETH.

Proposition 2.1. [7] *If there is a linear parameterized reduction from a parameterized problem A to a parameterized problem B and if A does not admit a parameterized subexponential time algorithm, then B does not admit a parameterized subexponential time algorithm.*

Two parameterized problems A and B are linear parameter equivalent if there is a linear parameterized reduction from A to B and there is a linear parameterized reduction from B to A. We refer the book [7] for various aspects of parameterized algorithms and complexity. The following are some folklore observations.

Proposition 2.2. H-FREE EDGE DELETION and $\overline{H}$-FREE EDGE COMPLETION are linear parameter equivalent. Similarly, H-FREE EDGE EDITING and $\overline{H}$-FREE EDGE EDITING are linear parameter equivalent.

Proposition 2.3. (i) H-FREE EDGE DELETION is NP-complete if and only if $\overline{H}$-FREE EDGE COMPLETION is NP-complete. Furthermore, H-FREE EDGE DELETION cannot be solved in parameterized subexponential time if and only if $\overline{H}$-FREE EDGE COMPLETION cannot be solved in parameterized subexponential time.

(ii) H-FREE EDGE EDITING is NP-complete if and only if $\overline{H}$-FREE EDGE EDITING is NP-complete. Furthermore, H-FREE EDGE EDITING cannot be solved in parameterized subexponential time if and only if $\overline{H}$-FREE EDGE EDITING cannot be solved in parameterized subexponential time.

Proposition 2.4. (i) H-FREE EDGE DELETION is polynomial time solvable if H is a graph with at most one edge.

(ii) H-FREE EDGE COMPLETION is polynomial time solvable if H is a graph with at most one non-edge.

(iii) H-FREE EDGE EDITING is polynomial time solvable if H is a graph with at most two vertices.

In this paper, we prove that these are the only polynomial time solvable H-free edge modification problems. For any fixed graph H, the H-free edge modification problems trivially belong to NP. Hence, we may state that these problems are NP-complete by proving their NP-hardness.

2.1   Basic Tools

The following construction is a slightly modified version of the main construction used in [2]. The modification is done to make it work for reductions of COMPLETION and EDITING problems. The input of the construction is a tuple (G', k, H, V'), where G' and H are graphs, k is a positive integer and $V' \subseteq V(H)$. In the old construction (Construction 1 in [2]), for every copy C of $H[V']$ in G', we introduced $k + 1$ copies of H such that the intersection of every pair of them is C. In the modified construction given below, we do the same for every copy C of $H[V']$ on a complete graph on $V(G')$.

Construction 1 Let (G', k, H, V') be an input to the construction, where G' and H are graphs, k is a positive integer and V' is a subset of vertices of H. Label the vertices of H such that every vertex gets a unique label. Let the labelling be ℓ_H. Consider a complete graph K' on $V(G')$. For every subgraph (not necessarily induced) C with a vertex set $V(C)$ and an edge set $E(C)$ in K' such that C is isomorphic to $H[V']$, do the following:

– *Give a labelling ℓ_C for the vertices in C such that there is an isomorphism f between C and $H[V']$ which maps every vertex v in C to a vertex v' in $H[V']$ such that $\ell_C(v) = \ell_H(v')$, i.e., $f(v) = v'$ if and only if $\ell_C(v) = \ell_H(v')$.*
– *Introduce $k+1$ sets of vertices $V_1, V_2, \ldots, V_{k+1}$, each of size $|V(H) \setminus V'|$.*
– *For each set V_i, introduce an edge set E_i of size $|E(H) \setminus E(H[V'])|$ among $V_i \cup V(C)$ such that there is an isomorphism h between H and $(V(C) \cup V_i, E(C) \cup E_i)$ which preserves f, i.e., for every vertex $v \in V(C)$, $h(v) = f(v)$.*

This completes the construction. Let the constructed graph be G.

We remark that the complete graph K' on $V(G')$ is not part of the constructed graph. The complete graph is only used to find where we need to introduce new vertices and edges. An example of the construction is shown in Fig. 1. We use the terminology used in [2]. We repeat it here for convenience. Let C be a copy of $H[V']$ in K'. Then, C is called a *base*. Let $\{V_i\}$ be the $k+1$ sets of vertices introduced in the construction for the base C. Then, each V_i is called a *branch* of C and the vertices in V_i are called the *branch vertices* of C. If V_j is a branch of C, then the vertex set of C is denoted by B_j. The vertex set of G' in G is denoted by $V_{G'}$. The copy of H formed by V_j, E_j and C is denoted by H_j. Since H is a fixed graph and k can safely be assumed to be at most $|V(G')| \cdot (|V(G')| - 1)/2$, the construction runs in polynomial time. The following two Lemmas are the generalized version of Lemma 2.3 and 3.5 of [2].

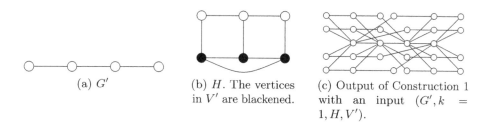

(a) G'

(b) H. The vertices in V' are blackened.

(c) Output of Construction 1 with an input $(G', k = 1, H, V')$.

Fig. 1. An example of Construction 1

Lemma 2.5. *Let G be obtained by Construction 1 on the input (G', k, H, V'), where G' and H are graphs, k is a positive integer and $V' \subseteq V(H)$. Then, if (G, k) is a yes-instance of H-FREE EDGE EDITING (DELETION/ COMPLETION), then (G', k) is a yes-instance of H'-FREE EDGE EDITING (DELETION/COMPLETION), where H' is $H[V']$.*

Proof. Let F be a solution of size at most k of (G, k). For a contradiction, assume that $G' \triangle F$ has an induced H' with a vertex set U. Hence there is a base C in G' isomorphic to H' with the vertex set $V(C) = U$. Since there are $k+1$ copies of H in G, where each pair of copies of H has the intersection C, and $|F| \leq k$, operating with F cannot kill all the copies of H associated with C. Therefore, since U induces an H' in $G' \triangle F$, there exists a branch V_i of C such that $U \cup V_i$ induces H in $G \triangle F$, which is a contradiction. □

Lemma 2.6. *Let H be any graph and d be any integer. Let V' be the set of vertices in H with degree more than d. Let H' be $H[V']$. Then, there is a linear parameterized reduction from H'-FREE EDGE EDITING (DELETION/COMPLETION) to H-FREE EDGE EDITING (DELETION/COMPLETION).*

Proof. Let (G', k) be an instance of H'-FREE EDGE EDITING (DELETION/ COMPLETION). Apply Construction 1 on (G', k, H, V') to obtain G. We claim that (G', k) is a yes-instance of H'-FREE EDGE EDITING (DELETION/COMPLETION) if and only if (G, k) is a yes-instance of H-FREE EDGE EDITING (DELETION/COMPLETION).

Let F' be a solution of size at most k of (G', k). For a contradiction, assume that $G \triangle F'$ has an induced H with a vertex set U. Since a branch vertex has degree at most d, every vertex in U with degree more than d in $(G \triangle F')[U]$ must be from $V_{G'}$. Hence there is an induced H' in $G' \triangle F'$, which is a contradiction. Lemma 2.5 proves the converse. □

3 H-FREE EDGE EDITING

In this section, we prove that H-FREE EDGE EDITING is NP-complete if and only if H is a graph with at least three vertices. We also prove that these problems cannot be solved in parameterized subexponential time unless ETH fails. We use the following known results.

Proposition 3.1. *The following problems are NP-complete. Furthermore, they cannot be solved in time $2^{o(k)} \cdot |G|^{O(1)}$, unless ETH fails.*

(i) P_3-FREE EDGE EDITING [12].
(ii) P_4-FREE EDGE EDITING [8].
(iii) C_ℓ-FREE EDGE EDITING, *for any fixed $l \geq 3$ [Follows from the proof for the corresponding DELETION problems in [16]].*
(iv) $2K_2$-FREE EDGE EDITING *[(3.1) and Proposition 2.3(ii)].*
(v) DIAMOND-FREE EDGE EDITING *[Follows from the proof for the corresponding DELETION problem in [15]].*

In our previous work [2], we proved that R-FREE EDGE DELETION is NP-complete if R is a regular graph with at least two edges. We also proved that these NP-complete problems cannot be solved in parameterized subexponential time, unless ETH fails. We observe that the results for R-FREE EDGE DELETION follow for R-FREE EDGE EDITING as well. The proofs are very similar except that we use Construction 1 instead of its ancestor in [2] and we reduce from EDITING problems instead of DELETION problems. We can use P_3-FREE EDGE EDITING, C_ℓ-FREE EDGE EDITING and $2K_2$-FREE EDGE EDITING as the base cases instead of their DELETION counterparts. We skip the proof as it will be a repetition of that in [2].

Lemma 3.2. *Let R be a regular graph with at least two edges. Then R-FREE EDGE EDITING is NP-complete. Furthermore, the problem cannot be solved in time $2^{o(k)} \cdot |G|^{O(1)}$, unless ETH fails.*

Now, we strengthen the above lemma by proving the same results for all regular graphs with at least three vertices.

Lemma 3.3. *Let R be a regular graph with at least three vertices. Then R-FREE EDGE EDITING is* NP-complete. *Furthermore, the problem cannot be solved in time $2^{o(k)} \cdot |G|^{O(1)}$, unless ETH fails.*

Proof. If R has at least two edges then the statements follows from Lemma 3.2. Assume that R has at most one edge and at least three vertices. It is straightforward to see that R must be the null graph. Then the complement of R is a complete graph with at least two edges. Now, the statements follows from Proposition 2.3(ii) and Lemma 3.2. □

Having these results in hand, we use Lemma 2.6 to prove the dichotomy result and the parameterized lower bound of H-FREE EDGE EDITING. Given a graph H with at least three vertices, we introduce a method Editing-Churn(H) to obtain a graph H' such that there is a linear parameterized reduction from H'-FREE EDGE EDITING to H-FREE EDGE EDITING and H' is a graph with at least three vertices and is a regular graph or a P_3 or a P_4 or a diamond.

Editing-Churn(H)

H is a graph with at least three vertices.

Step 1: If H is a regular graph, a P_3, a P_4 or a diamond, then return H.

Step 2: If H is a graph in which the number of vertices with degree more than $\delta(H)$ is at most two, then let $H = \overline{H}$ and goto Step 1.

Step 3: Delete all vertices with degree $\delta(H)$ in H and go to Step 1.

Observation 3.4 *Let H be a graph with at least three vertices. Then Editing-Churn(H) returns a graph H' which has at least three vertices and is a regular graph or a P_3 or a P_4 or a diamond. Furthermore, there is a linear parameterized reduction from H'-FREE EDGE EDITING to H-FREE EDGE EDITING.*

Proof. At any stage of the method, we make sure that the graph has at least three vertices. Let H' be an intermediate graph obtained in the method such that it is neither a regular graph nor a P_3 nor a P_4 nor a diamond. If Step 2 is applicable to both H' and $\overline{H'}$, then H hat at most four vertices. Hence H has either three or four vertices. It is straight-forward to verify that a graph (with three or four vertices) or its complement, satisfying the condition in Step 2, is either a regular graph or a P_3 or a P_4 or a diamond, which is a contradiction. The linear parameterized reduction from H'-FREE EDGE EDITING to H-FREE EDGE EDITING follows from Proposition 2.3(ii) and Lemma 2.6. □

Theorem 3.5. *H-FREE EDGE EDITING is* NP-complete *if and only if H is a graph with at least three vertices. Furthermore, these* NP-complete *problems cannot be solved in time $2^{o(k)} \cdot |G|^{O(1)}$, unless ETH fails.*

Proof. If H is a graph with at most two vertices, the statements follows from Proposition 2.4(iii). Let H be a graph with at least three vertices. Let H' be the graph returned by Editing-Churn(H). By Observation 3.4, H' is either a regular graph or a P_3 or a P_4 or a diamond and there is a linear parameterized reduction from H'-FREE EDGE EDITING to H-FREE EDGE EDITING. Now, the statements follows from the lower bound results for these graphs (Proposition 3.1(i), (ii), (v) and Lemma 3.3). □

4 H-FREE EDGE DELETION

In this section, we prove that H-FREE EDGE DELETION is NP-complete if and only if H is a graph with at least two edges. We also prove that these NP-complete problems cannot be solved in parameterized subexponential time, unless ETH fails. Then, from Proposition 2.3(i), we obtain a dichotomy result for H-FREE EDGE COMPLETION. We apply a technique similar to that we applied for EDITING in the last section.

Proposition 4.1. *The following problems are* NP-complete. *Furthermore, they cannot be solved in time* $2^{o(k)} \cdot |G|^{O(1)}$, *unless ETH fails.*

(i) P_3-FREE EDGE DELETION [12].
(ii) DIAMOND-FREE EDGE DELETION [10, 15].
(iii) H-FREE EDGE DELETION, if H is a graph with at least two edges and has a largest component which is a regular graph or a tree [2].

The following Lemma is a consequence of Lemma 2.6 and Proposition 2.3(i).

Lemma 4.2. *Let H be any graph. Then the following hold true:*

(i) Let H' be the subgraph of H obtained by removing all vertices with degree $\delta(H)$. Then there is a linear parameterized reduction from H'-FREE EDGE DELETION to H-FREE EDGE DELETION.

(ii) Let H' be the subgraph of H obtained by removing all vertices with degree $\Delta(H)$. Then there is a linear parameterized reduction from H'-FREE EDGE DELETION to H-FREE EDGE DELETION.

Proof. The first part directly follows from Lemma 2.6 by setting $d = \delta(H)$. To prove the second part, consider the problem $\overline{H}$-FREE EDGE COMPLETION. Let H'' be the graph obtained by removing all vertices with degree $\delta(\overline{H})$ from $\overline{H}$. Now, by Lemma 2.6, there is a linear parameterized reduction from H''-FREE EDGE COMPLETION to $\overline{H}$-FREE EDGE COMPLETION. We observe that H'' is $\overline{H'}$. Hence, by Proposition 2.3(i), there is a linear parameterized reduction from H'-FREE EDGE DELETION to H-FREE EDGE DELETION. □

Given a graph H, we keep on deleting either the minimum degree vertices or the maximum degree vertices by making sure that the resultant graph has at least two edges. We do this process until we obtain a graph in which vertices with degree more than $\delta(H)$ induce a graph with at most one edge and vertices with degree less than $\Delta(H)$ induce a graph with at most one edge. We call this method Deletion-Churn.

Deletion-Churn(H)

H is a graph with at least two edges.

1. If H is a graph in which the vertices with degree more than $\delta(H)$ induce a subgraph with at most one edge and the vertices with degree less than $\Delta(H)$ induce a subgraph with at most one edge, then return H.
2. If H is a graph in which the vertices with degree more than $\delta(H)$ induce a subgraph with at least two edges, then delete all vertices with degree $\delta(H)$ from H and goto Step 1.
3. If H is a graph in which the vertices with degree less than $\Delta(H)$ induce a subgraph with at least two edges, then delete all vertices with degree $\Delta(H)$ from H. Goto Step 1.

Observation 4.3 *Let H be a graph with at least two edges. If the vertices with degree more than $\delta(H)$ induce a graph with at most one edge and the vertices with degree less than $\Delta(H)$ induce a graph with at most one edge, then H is either a regular graph or a forest or a sparse (ℓ, h)-degree graph.*

Proof. Assume that H is not a regular graph. Since H has at least two edges and it satisfies the premises, $\delta(H) \geq 1$. If $\delta(H) = 1$, the premises imply that H is a forest. Assume that $\delta(H) \geq 2$. Then we prove that H is a sparse (ℓ, h)-degree graph. For a contradiction, assume that there exists a vertex $v \in V(H)$ such that $\delta(H) < deg(v) < \Delta(H)$. The premises imply that v has degree at most two, which is a contradiction. □

Lemma 4.4. *Let H be a graph with at least two edges. Then Deletion-Churn(H) returns a graph H' such that:*

(i) There is a linear parameterized reduction from H'-FREE EDGE DELETION to H-FREE EDGE DELETION.

(ii) H' has at least two edges and is either a regular graph or a forest or a sparse (ℓ, h)-degree graph.

Proof. In every step, we make sure that there are at least two edges in the resultant graph. Now, the first part follows from Lemma 4.2 and the second part follows from Observation 4.3. □

If the output of Deletion-Churn(H), H' is a regular graph or a forest, we obtain from Proposition 4.1(iii) that H-FREE EDGE DELETION is NP-complete and cannot be solved in parameterized subexponential time, unless ETH fails. Therefore, the only graphs to be handled now are the sparse (ℓ, h)-degree graphs with at least two edges. We do that in the next two subsections.

4.1 t-DIAMOND-FREE EDGE DELETION

We recall that t-diamond is the graph $K_2 + tK_1$ and that 2-diamond is the diamond graph (see Fig. 2). Clearly, t-diamond is a sparse (ℓ, h)-degree graph. In this subsection, we prove that t-DIAMOND-FREE EDGE DELETION is NP-complete. Further, we prove that the problem cannot be solved in parameterized subexponential time, unless ETH fails. We use an inductive proof where the base case is DIAMOND-FREE EDGE DELETION. For the proof, we introduce a simple construction, which is given below.

Fig. 2. A 2-diamond is isomorphic to a diamond graph.

Construction 2. *Let (G', k) be an input to the construction. For every edge $\{u, v\}$ in G', introduce a clique $C_{\{u,v\}}$ of $k + 1$ vertices such that every vertex in $C_{\{u,v\}}$ is adjacent to both u and v. This completes the construction. Let G be the resultant graph.*

Due to space constraints, the proof of the following lemma is moved to an extended version of this paper.

Lemma 4.5. *For any $t \geq 2$, t-DIAMOND-FREE EDGE DELETION is* NP-complete. *Furthermore, the problem cannot be solved in time $2^{o(k)} \cdot |G|^{O(1)}$, unless ETH fails.*

4.2   Handling Sparse (ℓ, h)-degree Graphs

We recall that for $h > \ell$, every vertex of a sparse (ℓ, h)-degree graph H is either of degree ℓ or of degree h and that V_ℓ induces a graph with at most one edge and V_h induces a graph with at most one edge. We have already handled t-diamond graphs. We handle the rest of the sparse (ℓ, h)-degree graphs in this subsection. Let H be any sparse (ℓ, h)-graph. There are four cases to be handled:

Case 1: V_h is an independent set; V_ℓ is an independent set
Case 2: V_h induces a graph with one edge; V_ℓ is an independent set
Case 3: V_h is an independent set; V_ℓ induces a graph with one edge
Case 4: V_h induces a graph with one edge; V_ℓ induces a graph with one edge

Observation 4.6. *Let H be a sparse (ℓ, h)-graph with at least two edges. Then the following hold true:*

(i) If $\ell = 1$, then H is a forest.
(ii) If $\ell \geq 2$, then $|V_\ell| \geq 2$ and the equality holds only when H is a diamond.

Proof. To prove the first part, we observe that $H \setminus V_\ell$ has at most one edge. To prove the second part, we observe that if $|V_\ell| \leq 2$ and if H is not a diamond, then $h \leq \ell$, which is a contradiction.   □

Since the case of forest is already handled in Proposition 4.1(iii), we can safely assume that $\ell \geq 2$ and hence $h \geq 3$. We start with handling Case 1. We use a slightly modified version of Construction 1. We recall that, in Construction 1, with an input (G', k, H, V'), For every copy C of $H[V']$ in K' (a complete graph on $V(G')$), we introduced $k + 1$ branches such that each branch along with C form a copy of H. In the modified construction, in addition to this, we make every pair of vertices from different branches mutually adjacent.

Construction 3. *Let (G', k, H, V') be an input to the construction, where G' and H are graphs, k is a positive integer and V' is a subset of vertices of H. Apply Construction 1 on (G', k, H, V') to obtain G''. For every pair of vertices $\{v_i, v_j\}$ such that $v_i \in V_i$ and $v_j \in V_j$, where $i \neq j$, make v_i and v_j adjacent. This completes the construction. Let the constructed graph be G.*

Now, we have a lemma similar to Lemma 2.5. We skip the proof as it is quite similar to that of Lemma 2.5.

Lemma 4.7. *Let G be obtained by Construction 3 on the input (G', k, H, V'), where G' and H are graphs, k is a positive integer and $V' \subseteq V(H)$. Then, if (G, k) is a yes-instance of H-FREE EDGE DELETION, then (G', k) is a yes-instance of H'-FREE EDGE DELETION, where H' is $H[V']$.*

The following lemma is proved by a reduction from P_3-FREE EDGE DELETION using Construction 3. Due to space constraints, the proof is moved to an extended version of this paper.

Lemma 4.8. *Let H be a sparse (ℓ, h)-graph, where $h > \ell \geq 2$ such that both V_ℓ and V_h are independent sets. Then H-FREE EDGE DELETION is NP-complete. Furthermore, the problem cannot be solved in time $2^{o(k)} \cdot |G|^{O(k)}$, unless ETH fails.*

Now we handle the cases in which V_ℓ induces a graph with one edge.

Lemma 4.9. *Let H be a sparse (ℓ, h)-graph with at least two edges such that V_l induces a graph with one edge. Let v_{ℓ_1} and v_{ℓ_2} be the two adjacent vertices in V_ℓ. Let H' be the graph induced by $V(H) \setminus \{v_{\ell_1}, v_{\ell_2}\}$. Then, there is a linear parameterized reduction from H'-FREE EDGE DELETION to H-FREE EDGE DELETION.*

Proof. Let (G', k) be an instance of H'-FREE EDGE DELETION. Apply Construction 1 on (G', k, H, V'), where V' is $V(H) \setminus \{v_{\ell_1}, v_{\ell_2}\}$. Let G be the graph obtained from the construction. We claim that (G', k) is a yes-instance of H'-FREE EDGE DELETION if and only if (G, k) is a yes-instance of H-FREE EDGE DELETION.

Let (G', k) be a yes-instance of H'-FREE EDGE DELETION and let F' be a solution of size at most k of (G', k). For a contradiction, assume that $G - F'$ has an induced H with a vertex set U. It is straight-forward to verify that If a branch vertex $v_1 \in V_1$ is in U, then its neighbor in the same branch $u_1 \in V_1$ must be in U and both acts as v_{ℓ_1} and v_{ℓ_2} in the H induced by U in $G - F'$. Hence $' - F'$ has an induced H', which is a contradiction. Lemma 2.5 proves the converse. □

Observation 4.10. *Let H be a sparse (ℓ, h)-graph with at least two edges where $h > \ell \geq 2$ such that V_l induces a graph with one edge. Let v_{ℓ_1} and v_{ℓ_2} be the two adjacent vertices in V_ℓ. Let H' be the graph induced by $V(H) \setminus \{v_{\ell_1}, v_{\ell_2}\}$. Then H' has at least two edges.*

Proof. By Observation 4.6(ii) since H is not a diamond, $|V_\ell| \geq 3$. This implies that $V \setminus \{v_{\ell_1}, v_{\ell_2}\}$ is nonempty. Now the observation follows from the fact that $\ell \geq 2$. □

Now we handle Case 2, i.e., V_h induces a graph with one edge and V_ℓ is an independent set.

Lemma 4.11. *Let H be a sparse (ℓ, h) graph where $h > \ell \geq 2$, V_h induces a graph with one edge and V_ℓ is an independent set. Let H be not a t-diamond. Let v_{h_1} and v_{h_2} be the two adjacent vertices in $H[V_h]$. Let V' be $V_\ell \cup \{v_{h_1}, v_{h_2}\}$. Let H' be $H[V']$. Then, there is a linear parameterized reduction from H'-FREE EDGE DELETION to H-FREE EDGE DELETION.*

Proof. For convenience, we give a reduction from $\overline{H'}$-FREE EDGE COMPLETION to $\overline{H}$-FREE EDGE COMPLETION. Then the statements follow from Proposition 2.3(i).

Let (G', k) be an instance of $\overline{H'}$-FREE EDGE COMPLETION. Apply Construction 1 on (G', k, H, V'), where V' is $V_\ell \cup \{v_{h_1}, v_{h_2}\}$. Let G be the graph obtained from the construction. We claim that (G', k) is a yes-instance of $\overline{H'}$-FREE EDGE COMPLETION if and only if (G, k) is a yes-instance of $\overline{H}$-FREE EDGE COMPLETION.

Let (G', k) be a yes-instance of $\overline{H'}$-FREE EDGE COMPLETION and let F' be a solution of size at most k of (G', k). For a contradiction, assume that $G + F'$ has an induced H with a vertex set U. It is straight-forward to verify that If a branch vertex $v_1 \in V_1$ is in U, then all its neighbors in the same branch are in U and V_1 acts as $V_h \setminus \{v_{h_1}, v_{h_2}\}$ of H in $\overline{H}$ induced by U in $G + F'$. Hence $G' + F'$ has an induced $\overline{H'}$, which is a contradiction. Lemma 2.5 proves the converse. □

Observation 4.12. *Let H be a sparse (ℓ, h) graph where $h > \ell \geq 2$, V_h induces a graph with one edge and V_ℓ is an independent set. Let H be not a t-diamond, for $t \geq 2$. Let v_{h_1} and v_{h_2} be the two adjacent vertices in $H[V_h]$. Let V' be $V_\ell \cup \{v_{h_1}, v_{h_2}\}$. Let H' be $H[V']$. Then H' has at least two edges and $|V(H')| < |V(H)|$.*

Proof. Follows from the facts that $h \geq 3$ and H is not a t-diamond. □

Lemma 4.13. *Let H be a sparse (ℓ, h)-degree graph with at least two edges. Then H-*FREE EDGE DELETION *is* NP-complete. *Furthermore, the problem cannot be solved in time $2^{o(k)} \cdot |G|^{O(1)}$, unless ETH fails.*

Proof. If V_ℓ induces a graph with an edge, then we apply the technique used in Lemma 4.9 and obtain a graph H' with at least two edges. Similarly, if H is not a t-diamond and V_h induces a graph with an edge, then we apply the technique used in Lemma 4.11 to obtain a graph H' with at least two edges. If the obtained graph H' is not a sparse (ℓ, h)-degree graph, then we apply Deletion-Churn(H') to obtain H''. We repeat this process until no more repetition is possible. Then, it is straight-forward to verify that we obtain a graph which is either a t-diamond, or a graph handled in Lemma 4.8 or a regular graph or a forest with at least two edges. □

4.3   Dichotomy Results

We are ready to state the dichotomy results and the parameterized lower bounds for H-FREE EDGE DELETION and H-FREE EDGE COMPLETION.

Theorem 4.14. *H-*FREE EDGE DELETION *is* NP-complete *if and only if H is a graph with at least two edges. Furthermore, the problem cannot be solved in time $2^{o(k)} \cdot |G|^{O(k)}$. H-*FREE EDGE COMPLETION *is* NP-complete *if and only if H is a graph with at least two non-edges. Furthermore, the problem cannot be solved in time $2^{o(k)} \cdot |G|^{O(k)}$.*

Proof. Consider H-FREE EDGE DELETION. The statements follow from Proposition 2.4(i), Lemma 4.4, Proposition 4.1(iii) and Lemma 4.13. Now the results for H-FREE EDGE COMPLETION follows from Proposition 2.3(i). □

5   Concluding Remarks

Our results have implications on the incompressibility of H-free edge modification problems. Polynomial parameter transformation (PPT) is a technique to prove the incompressibility of problems. It is a polynomial time reduction from a parameterized problem to another where the parameter blow-up is polynomial. To prove the incompressibility of a problem it is enough to to give a PPT from a problem which is already known to be incompressible, under some complexity theoretic assumption. All our reductions are PPTs. The following lemma is a direct consequence of Lemma 2.6.

Lemma 5.1. *Let H be a graph and d be any integer. Let H' be obtained from H by deleting vertices with degree d or less. Then, if H'-*FREE EDGE EDITING *(*DELETION/COMPLETION*) is incompressible, then H-*FREE EDGE EDITING *(*DELETION/COMPLETION*) is incompressible.*

We give a simple example to show an implication of this lemma. Consider an n-sunlet graph which is a graph in which a vertex with degree one is attached to each vertex of a cycle of n vertices. From the incompressibility of C_n-FREE EDGE EDITING, DELETION and COMPLETION, for any $n \geq 4$, it follows that n-SUNLET-FREE EDGE EDITING, DELETION and COMPLETION are incompressible for any $n \geq 4$. We believe that our result is a step towards a dichotomy result on the incompressibility of H-free edge modification problems. Another direction is to get a dichotomy result on the complexities of $\mathcal{H}$-free edge modification problems where $\mathcal{H}$ is a finite set of graphs.

References

1. Alon, N., Stav, U.: Hardness of edge-modification problems. Theor. Comput. Sci. **410**(47–49), 4920–4927 (2009)
2. Aravind, N.R., Sandeep, R.B., Sivadasan, N.: Parameterized lower bound and NP-completeness of some H-free edge deletion problems. In: Lu, Z., et al. (eds.) COCOA 2015. LNCS, vol. 9486, pp. 424–438. Springer, Heidelberg (2015). doi:10. 1007/978-3-319-26626-8_31
3. Bansal, N., Blum, A., Chawla, S.: Correlation clustering. Mach. Learn. **56**(1–3), 89–113 (2004)
4. Cai, L.: Fixed-parameter tractability of graph modification problems for hereditary properties. Inf. Process. Lett. **58**(4), 171–176 (1996)
5. Cai, L., Cai, Y.: Incompressibility of H-free edge modification problems. Algorithmica **71**(3), 731–757 (2015)
6. Yufei, C.: Polynomial kernelisation of H-free edge modification problems. M.phil thesis, Department of Computer Science and Engineering, The Chinese University of Hong Kong, Hong Kong SAR, China (2012)
7. Cygan, M., Fomin, F.V., Kowalik, L., Lokshtanov, D., Marx, D., Pilipczuk, M., Saurabh, S.: Parameterized Algorithms. Springer, Heidelberg (2015)
8. Drange, P.G.: Parameterized Graph Modification Algorithms. PhD dissertation, University of Bergen (2015)
9. El-Mallah, E.S., Colbourn, C.J.: The complexity of some edge deletion problems. IEEE Trans. Circuits Syst. **35**(3), 354–362 (1988)
10. Fellows, M.R., Guo, J., Komusiewicz, C., Niedermeier, R., Uhlmann, J.: Graph-based data clustering with overlaps. Discrete Optim. **8**(1), 2–17 (2011)
11. Paul, W., Goldberg, M.C., Kaplan, H., Shamir, R.: Four strikes against physical mapping of DNA. J. Comput. Biol. **2**(1), 139–152 (1995)
12. Komusiewicz, C., Uhlmann, J.: Cluster editing with locally bounded modifications. Discrete Appl. Math. **160**(15), 2259–2270 (2012)
13. Kratsch, S., Wahlström, M.: Two edge modification problems without polynomial kernels. Discrete Optim. **10**(3), 193–199 (2013)
14. Rose, D.J.: A Graph-theoretic Study of the Numerical Solution of Sparse Positive Definite Systems of Linear Equations. Academic Press, Cambridge (1972). pp. 183–217
15. Sandeep, R.B., Sivadasan, N.: Parameterized lower bound and improved Kernel for diamond-free edge deletion. In: Husfeldt, T., Kanj, I. (eds) IPEC Leibniz International Proceedings in Informatics (LIPIcs), vol. 43, pp. 365–376. Schloss Dagstuhl-Leibniz-Zentrum fuer Informatik (2015)
16. Yannakakis, M.: Edge-deletion problems. SIAM J. Comput. **10**(2), 297–309 (1981)

Parameterized Complexity of RED BLUE SET COVER for Lines

Pradeesha Ashok[1][(✉)], Sudeshna Kolay[1], and Saket Saurabh[1,2]

[1] Institute of Mathematical Sciences, Chennai, India
{pradeesha,skolay,saket}@imsc.res.in
[2] University of Bergen, Bergen, Norway
saket.saurabh@uib.no

Abstract. We investigate the parameterized complexity of GENERAL-IZED RED BLUE SET COVER (GEN-RBSC), a generalization of the classic SET COVER problem and the more recently studied RED BLUE SET COVER problem. Given a universe U containing b blue elements and r red elements, positive integers k_ℓ and k_r, and a family $\mathcal{F}$ of ℓ sets over U, the GEN-RBSC problem is to decide whether there is a subfamily $\mathcal{F}' \subseteq \mathcal{F}$ of size at most k_ℓ that covers all blue elements, but at most k_r of the red elements. This generalizes SET COVER and thus in full generality it is intractable in the parameterized setting, when parameterized by $k_\ell + k_r$. In this paper, we study GEN-RBSC-LINES, where the elements are points in the plane and sets are defined by lines. We study this problem for the parameters k_ℓ, k_r, and $k_\ell + k_r$. For all these cases, we either prove that the problem is W-hard or show that the problem is fixed parameter tractable (FPT). Finally, for the parameter $k_\ell + k_r$, for which GEN-RBSC-LINES admits FPT algorithms, we show that the problem does not have a polynomial kernel unless co-NP $\subseteq$ NP/poly. Further, we show that the FPT algorithm does not generalize to higher dimensions.

1 Introduction

The input to a covering problem consists of a universe U of size n, a family $\mathcal{F}$ of m subsets of U and a positive integer k, and the objective is to check whether there exists a subfamily $\mathcal{F}' \subseteq \mathcal{F}$ of size at most k satisfying some desired properties. A set S is said to *cover* a point $p \in U$ if $p \in S$. If $\mathcal{F}'$ is required to contain all the elements of U, then it corresponds to the classical SET COVER problem. The SET COVER problem is part of Karp's 21 NP-complete problems [10]. This, together with its numerous variants, is one of the most well-studied problems in the area of algorithms and complexity. It is one of the central problems in all the paradigms that have been established to cope with NP-hardness, including approximation algorithms, randomized algorithms and parameterized complexity.

S. Saurabh—The research leading to these results has received funding from the European Research Council under the European Union's Seventh Framework Programme (FP7/2007-2013) / ERC grant agreement no. 306992.

© Springer-Verlag Berlin Heidelberg 2016
E. Kranakis et al. (Eds.): LATIN 2016, LNCS 9644, pp. 96–109, 2016.
DOI: 10.1007/978-3-662-49529-2_8

Problems Studied, Context and Framework. The goal of this paper is to study a generalization of a variant of SET COVER namely the RED BLUE SET COVER problem.

RED BLUE SET COVER (RBSC)
Input: A universe $U = (R, B)$ where R is a set of r red elements and B is a set of b blue elements, a family $\mathcal{F}$ of ℓ subsets of U, and a positive integer k_r.
Question: Is there a subfamily $\mathcal{F}' \subseteq \mathcal{F}$ that covers all blue elements but at most k_r red elements?

RED BLUE SET COVER was introduced in 2000 by Carr et al. [2]. This problem is closely related to several combinatorial optimization problems such as the GROUP STEINER, MINIMUM LABEL PATH, MINIMUM MONOTONE SATISFYING ASSIGNMENT and SYMMETRIC LABEL COVER problems. This has also found applications in areas like fraud/anomaly detection, information retrieval and the classification problem. RED BLUE SET COVER is NP-complete, following from an easy reduction from SET COVER itself.

In this paper, we study the parameterized complexity, under various parameters, of a common generalization of both SET COVER and RED BLUE SET COVER, in a geometric setting.

GENERALIZED RED BLUE SET COVER (GEN-RBSC)
Input: A universe $U = (R, B)$ where R is a set of r red elements and B is a set of b blue elements, a family $\mathcal{F}$ of ℓ subsets of U, and positive integers k_ℓ, k_r.
Question: Is there a subfamily $\mathcal{F}' \subseteq \mathcal{F}$ of size at most k_ℓ that covers all blue elements but at most k_r red elements?

It is easy to see that when $k_\ell = |\mathcal{F}|$ then it is a RBSC instance, while it is a SET COVER instance when $k_\ell = k, R = \emptyset, k_r = 0$. Next we take a short detour and give a few essential definitions regarding parameterized complexity.

Parameterized Complexity. The goal of parameterized complexity is to find ways of solving NP-hard problems more efficiently than brute force: here the aim is to restrict the combinatorial explosion to a parameter that is hopefully much smaller than the input size. Formally, a *parameterization* of a problem is assigning a positive integer parameter k to each input instance and we say that a parameterized problem is *fixed-parameter tractable* (FPT) if there is an algorithm that solves the problem in time $f(k) \cdot |I|^{\mathcal{O}(1)}$, where $|I|$ is the size of the input and f is an arbitrary computable function depending only on the parameter k. If the problem had a set Γ of positive integers as parameters, then the problem is called FPT if there is an algorithm solving the problem in $f(\Gamma) \cdot |I|^{\mathcal{O}(1)}$, where $|I|$ is the size of the input and f is an arbitrary computable function depending only on the parameters in Γ. Equivalently, the problem can be considered to be parameterized by $k = \sum_{q \in \Gamma} q$. Such an algorithm is called an FPT algorithm and such a running time is called FPT running time. There is also an accompanying theory of parameterized intractability using which one can identify parameterized problems that are unlikely to admit FPT algorithms. These are essentially proved by showing that the problem is W-hard. A parameterized problem is said to admit a $h(k)$-*kernel* if there is a polynomial time algorithm (the degree of the polynomial is independent of k), called a *kernelization* algorithm, that reduces the input instance to an instance with size upper bounded by $h(k)$, while preserving the answer. If the function $h(k)$ is polynomial in k, then we say that the problem admits a polynomial kernel. While positive kernelization results have appeared regularly over the last two decades, the first results establishing infeasibility of polynomial kernels for specific problems have appeared only recently. In particular, Bodlaender et al. [1], and Fortnow and Santhanam [9] have developed a framework for showing that a problem does not admit a polynomial kernel unless co-NP $\subseteq$ NP/poly, which is deemed unlikely. For more background, the reader is referred to the following monograph [8].

In the parameterized setting, SET COVER, parameterized by k, is W[2]-hard [7] and it is not expected to have an FPT algorithm. The NP-hardness reduction from SET COVER to RED BLUE SET COVER implies that RED BLUE SET COVER is W[2]-hard parameterized by the size k_ℓ of a solution subfamily. However, the hardness result was not the end of the story for the SET COVER problem in parameterized complexity. In literature, various special cases of SET COVER have been studied. A few examples are instances with sets of bounded size, sets with bounded intersection [13,18], and instances where the bipartite incidence graph corresponding to the set family has bounded treewidth or excludes some graph H as a minor. Apart from these results, there has also been extended study on different parameterizations of SET COVER. A special case of SET COVER which is central to the topic of this paper, is the one where the *universe consists of a point set and sets in the family are defined by intersection of point set with some geometric object*. In the simplest geometric variant of SET COVER, called POINT LINE COVER, the elements of U are points in $\mathbb{R}^2$ and each set contains a maximal number of collinear points. This version of the problem was motivated by the problem of covering a rectilinear polygon with holes using rectangles [14] which in turn has applications in printing integrated circuits and image compression [4]. POINT LINE COVER is in FPT and in fact has a polynomial kernel [13]. Moreover, the size of these kernels has been proved to be tight, under standard assumptions, in [11]. Similarly, we can take our universe to be defined by n points in $\mathbb{R}^d$, for a fixed d, and each set to be defined by a maximal set of points that lie on the same hyperplane. A hyperplane in $\mathbb{R}^d$ is the affine hull of a set of d affinely independent points. SET COVER with hyperplanes is also known to be FPT with a polynomial kernel [13].

In this paper, we concentrate on the GENERALIZED RED BLUE SET COVER WITH LINES problem, parameterized by the size of the solution.

GENERALIZED RED BLUE SET COVER WITH LINES (GEN-RBSC-LINES)
Input: A universe $U = (R, B)$ where R is a set of r red points and B is a set of b blue points, in $\mathbb{R}^2$, a family $\mathcal{F}$ of ℓ subsets of U such that each set contains a maximal set of collinear points of U, and positive integers k_ℓ, k_r.
Question: Is there a subfamily $\mathcal{F}' \subseteq \mathcal{F}$ of size at most k_ℓ that covers all blue points but at most k_r red points?

We also study a generalization of this problem in higher dimensions. In this case each set is a maximal set of points that lie on a hyperplane of $\mathbb{R}^d$. A hyperplane in $\mathbb{R}^d$ is the affine hull of a set of $d+1$ affinely independent points [13].

GENERALIZED RED BLUE SET COVER WITH HYPERPLANES
Input: A universe $U = (R, B)$ where R is a set of r red points and B is a set of b blue points, in $\mathbb{R}^d$, a family $\mathcal{F}$ of ℓ subsets of U such that each set is a maximal set of points that lie on a hyperplane in $\mathbb{R}^d$., and positive integers k_ℓ, k_r.
Question: Is there a subfamily $\mathcal{F}' \subseteq \mathcal{F}$ of size at most k_ℓ that covers all blue points but at most k_r red points?

We finish this section with some related results. As mentioned earlier, the RED BLUE SET COVER problem in classical complexity is NP-complete. Interestingly, if the incidence matrix, built over the sets and elements, has the consecutive ones property then the problem is in P [5]. The problem has been studied in approximation algorithms as well [2,17]. Specially, the geometric variant, where every set is the space bounded by a unit square, has a polynomial time approximation scheme (PTAS) [3].

Our Contributions

1. We show that GEN-RBSC-LINES parameterized by k_r is para-NP-complete. This also shows that RBSC is para-NP-complete under standard parameterization.
2. We show that GEN-RBSC-LINES parameterized by k_ℓ is W[1]-hard.
3. We give an FPT algorithm for GEN-RBSC-LINES parameterized by $k_r + k_\ell$ that runs in $\mathcal{O}(k_\ell^{\mathcal{O}(k_\ell)} . k_r^{\mathcal{O}(k_r)})$ time. We further show that this problem does not admit a polynomial kernel unless co-NP $\subseteq$ NP/poly.
4. Finally, we show that GEN-RBSC for hyperplanes in $\mathbb{R}^d$, $d > 2$, parameterized by $k_\ell + k_r$ is W[1]-hard.

2   Preliminaries

In this paper an undirected graph is denoted by a tuple $G = (V, E)$, where V denotes the set of vertices and E the set of edges. For a set $S \subseteq V$, the *subgraph of G induced by S*, denoted by $G[S]$, is defined as the subgraph of G with vertex set S and edge set $\{(u, v) \in E : u, v \in S\}$. The subgraph obtained after deleting S is denoted as $G \setminus S$. All vertices adjacent to a vertex v are called neighbors of v and the set of all such vertices is called the neighborhood of v. Similarly, a non-adjacent vertex of v is called a non-neighbor and the set of all non-neighbors of v is called the non-neighborhood of v. The neighborhood of v is denoted by $N(v)$. A vertex in a connected graph is called a cut vertex if its deletion results in the graph becoming disconnected.

Recall that showing a problem W[1] or W[2] hard implies that the problem is unlikely to be FPT. One can show that a problem is W[1]-hard (W[2]-hard) by presenting a parameterized reduction from a known W[1]-hard problem (W[2]-hard) such as CLIQUE (SET COVER) to it. The most important property of a parameterized reduction is that it corresponds to an FPT algorithm that bounds the parameter value of the constructed instance by a function of the parameter of the source instance. A parameterized problem is said to be in the class para-NP if it has a nondeterministic algorithm with FPT running time. To show that a problem is para-NP-hard we need to show that the problem is NP-hard for some constant value of the parameter. For an example 3-COLORING is para-NP-hard parameterized by the number of colors. See [8] for more details.

Lower Bounds in Kernelization. In the recent years, several techniques have been developed to show that certain parameterized problems belonging to the FPT class cannot have any polynomial sized kernel unless some classical complexity assumptions are violated. One such technique that is widely used is the polynomial parameter transformation technique.

Definition 1. *Let Π, Γ be two parameterized problems. A polynomial time algorithm $\mathcal{A}$ is called a polynomial parameter transformation (or ppt) from Π to Γ if, given an instance (x, k) of Π, $\mathcal{A}$ outputs in polynomial time an instance (x', k') of Γ such that $(x, k) \in \Pi$ if and only if $(x', k') \in \Gamma$ and $k' \leq p(k)$ for a polynomial p.*

We use the following theorem together with ppt reductions to rule out polynomial kernels.

Proposition 1. *Let Π, Γ be two parameterized problems such that Π is NP-hard and $\Gamma \in$ NP. Assume that there exists a polynomial parameter transformation from Π to Γ. Then, if Π does not admit a polynomial kernel neither does Γ.*

For further details on lower bound techniques in kernelization refer to [1,9].

Generalized Red Blue Set Cover. A set S in a GENERALIZED RED BLUE SET COVER instance $(U, \mathcal{F})$ is said to *cover* a point $p \in U$ if $p \in S$. A *solution family* for the instance is a family of sets of size at most k_ℓ that covers all the blue points and at most k_r red points. In case of RED BLUE SET COVER, the solution family is simply a family of sets that covers all the blue points but at most k_r red points. Such a family will also be referred to as a *valid family*. A *minimal family of sets* is a family of sets such that every set contains a unique blue point. In other words, deleting any set from the family implies that a strictly smaller set of blue points is covered by the remaining sets. It is safe to assume that $r \geq k_r$, and $\ell \geq k_\ell$. Since it is enough to find a minimal solution family $\mathcal{F}'$, we can also assume that $b \geq k_\ell$. The sets of GENERALIZED RED BLUE SET COVER WITH LINES are also called *lines* in this paper. We now mention a key observation about lines that is crucial in many arguments in this paper.

Observation 1. *Given a set of points S, let $\mathcal{F}$ be the set of lines such that each line contains at least 2 points from S. Then $|\mathcal{F}| \leq \binom{|S|}{2}$.*

Definition 2. *An intersection graph $G_{\mathcal{F}} = (V, E)$ for an instance $(U, \mathcal{F})$ of* GENERALIZED RED BLUE SET COVER *is a graph with vertices corresponding to the sets in $\mathcal{F}$. We give an edge between two vertices if the corresponding sets have non-empty intersection.*

The following proposition is a collection of results on the SET COVER problem, that will be repeatedly used in the paper. The results are from [6,7]

Proposition 2. *The* SET COVER *problem is:*

  i. W[2] *hard when parameterized by the solution family size k.*
 ii. FPT *when parameterized by the universe size n, but does not admit polynomial kernels unless* co-NP $\subseteq$ NP/*poly.*
iii. FPT *when parameterized by the number of sets m in the instance, but does not admit polynomial kernels unless* co-NP $\subseteq$ NP/*poly.*

3   Hardness When Parameterized by k_r and by k_ℓ

In this section we show that GEN-RBSC-LINES parameterized by k_r and GEN-RBSC-LINES parameterized by k_ℓ are hard.

Theorem 1. [⋆][1] GEN-RBSC-LINES *is para-NP-complete parameterized by k_r.*

Theorem 1 follows from a reduction from POINT LINE COVER problem, which is NP-hard [12].

We now show that GEN-RBSC-LINES parameterized by k_ℓ is hard. We give a reduction to this problem from the MULTICOLORED CLIQUE problem, which is known to be W[1] hard even on regular graphs [16].

MULTICOLORED CLIQUE   **Parameter:** k
Input: A graph $G = (V, E)$ where $V = V_1 \uplus V_2 \uplus \ldots \uplus V_k$ with V_i being an independent set for all $1 \leq i \leq k$, and an integer k.
Question: Is there a multi-colored clique $C \subseteq G$ of size k such that $\forall 1 \leq i \leq k, C \cap V_i \neq \emptyset$.

Theorem 2. GEN-RBSC-LINES *parameterized by k_ℓ is* W[1]-*hard.*

Proof. We give a reduction from MULTICOLORED CLIQUE on regular graphs. Let $(G = (V, E), k)$ be an instance of MULTICOLORED CLIQUE, where G is a d-regular graph, $|V| = v$, $|E| = e$. We construct an instance of GEN-RBSC-LINES $(R \cup B, \mathcal{F})$, as follows. Let $V = V_1 \uplus V_2 \uplus \ldots \uplus V_k$.

1. For each vertex class $V_i, 1 \leq i \leq k$, add two blue points b_i at $(0, i)$ and b'_i at $(i, 0)$.

---

[1] All results marked with a ⋆ have their full proofs given in the full version.

2. For each vertex $v \in V_i$, we add a line l_v^1, which we call a *near-horizontal line*, such that all the near-horizontal lines corresponding to vertices in V_i intersect at b_i. Also, the lines are drawn such that for any two vertices $u \in V_i$ and $v \in V_j$, with $i \neq j$, the lines l_u^1 and l_v^1 do not intersect at a point with x-coordinate from the closed interval $[0, k]$.

3. Similarly, for each vertex $v \in V_i$, we add a line l_v^2, which we call a *near-vertical line*, such that all the near-vertical lines corresponding to vertices in V_i intersect at b_i'. Also, the lines are drawn such that for any two vertices $u \in V_i$ and $v \in V_j$, with $i \neq j$, the lines l_u^2 and l_v^2 do not intersect at a point with y-coordinate from the closed interval $[0, k]$. However, a near-horizontal line and a near-vertical line will intersect at a point with both x and y-coordinate from the closed interval $[0, k]$.

4. For each edge $e = (u, v) \in E$, add two red points, r_{uv} at the intersection of lines l_u^1 and l_v^2, and r_{vu} at the intersection of lines l_v^1 and l_u^2.

5. For each vertex $v \in V$, add a red point at the intersection of the lines l_v^1 and l_v^2.

Thus we have an instance $(R \cup B, \mathcal{F})$ of GEN-RBSC-LINES with $2v$ lines, $2k$ blue points and $2e + v$ red points. The construction ensures that no 3 lines in $\mathcal{F}$ intersect at a red point.

Claim. $G = (V, E)$ has a multi-colored clique of size k if and only if $(R \cup B, \mathcal{F})$ has a solution family of $2k$ lines, covering the $2k$ blue points and at most $2(d+1)k - k^2$ red points.

Proof. Assume there exists a multi-colored clique C of size k in G. Select the $2k$ lines corresponding to the vertices in the clique. That is, select the subset of lines $\mathcal{F}' = \{l_u^j \mid 1 \leq j \leq 2, u \in C\}$ in the GEN-RBSC-LINES instance. Since the clique is multi-colored, these lines cover all the blue points. Each line (near-horizontal or near-vertical) has exactly $d + 1$ red points. Thus, the number of red points covered by $\mathcal{F}'$ is at most $(d + 1)2k$. However, each red point corresponding to vertices in C and the two red points corresponding to each edge in C are counted twice. Thus, the number of red points covered by $\mathcal{F}'$ is at most $(d+1)2k - k - 2\binom{k}{2} = 2(d+1)k - k^2$. This completes the proof in the forward direction.

Now, assume there is a minimal solution family of size at most $2k$, containing at most $2(d + 1)k - k^2$ red points. As no two blue points are on the same line and there are $2k$ blue points, there exists a unique line covering each blue point. Let $\mathcal{L}^1$ and $\mathcal{L}^2$ represent the sets of near-horizontal and near-vertical lines respectively in the solution family. Observe that $\mathcal{L}^1$ covers $\{b_1, \ldots, b_k\}$ and $\mathcal{L}^2$ covers $\{b_1', \ldots, b_k'\}$. Let $C = \{v_1, \ldots, v_k\}$ be the set of vertices in G corresponding to the lines in $\mathcal{L}^1$. We claim that C forms a multicolored k-clique in G. Since b_i can only be covered by lines corresponding to the vertices in V_i and $\mathcal{L}^1$ covers $\{b_1, \ldots, b_k\}$ we have that $C \cap V_i \neq \emptyset$. It remains to show that for every pair of vertices in C there exists an edge between them in G. Let v_i denote the vertex in $C \cap V_i$.

Consider all the lines in $\mathcal{L}^1$. Each of these lines are near-horizontal and have exactly $d+1$ red points. Furthermore, no two of them intersect at a red point. Since the total number of red points covered by $\mathcal{L}^1 \cup \mathcal{L}^2$ is at most $2(d+1)k - k^2$, we have that the k lines in $\mathcal{L}^2$ can only cover at most $k(d+1) - k^2$ red points that are not covered by the lines in $\mathcal{L}^1$. That is, the k lines in $\mathcal{L}^2$ contribute at most $k(d+1) - k^2$ *new red points* to the solution. Thus, the number of red points that are covered by both $\mathcal{L}^1$ and $\mathcal{L}^2$ is k^2. Therefore, any two lines l_1 and l_2 such that $l_1 \in \mathcal{L}^1$ and $l_2 \in \mathcal{L}^2$ must intersect at a red point. This implies that either l_1 and l_2 correspond to the same vertex in V or there exists an edge between the vertices corresponding to them. Let $C' = \{w_1, \ldots, w_k\}$ be the set of vertices in G corresponding to the lines in $\mathcal{L}^2$. Since b'_i can only be covered by lines corresponding to the vertices in V_i and $\mathcal{L}^2$ covers $\{b'_1, \ldots, b'_k\}$ we have that $C' \cap V_i \neq \emptyset$. Let w_i denote the vertex in V_i such that $l^2_{w_i} \in \mathcal{L}^2$ covers b'_i. We know that $l^1_{v_i}$ and $l^2_{w_i}$ must intersect on a red point. However, by construction no two distinct vertices v_i and w_i belonging to the same vertex class V_i intersect at red point. Thus $v_i = w_i$. This means $C = C'$. This, together with the fact that two lines l_1 and l_2 such that $l_1 \in \mathcal{L}^1$ and $l_2 \in \mathcal{L}^2$ (now lines corresponding to C) must intersect at a red point, implies that C is a multicolored k-clique in G. $\square$

Since $k_\ell = 2k$, we have that GEN-RBSC-LINES is W[1]-hard parameterized by k_ℓ. This concludes the proof. $\square$

4   FPT Algorithm When Paremeterized by $k_\ell + k_r$

In this section, we describe an FPT algorithm for GEN-RBSC-LINES parameterized by $k_\ell + k_r$, which is our main technical/algorithmic contribution.

We start with a few preprocessing rules, after which we obtain an equivalent instance for the problem.

Reduction Rule 1: If there is a set $S \in \mathcal{F}$ with only red points then delete S from $\mathcal{F}$.

Reduction Rule 2: If there is a set $S \in \mathcal{F}$ with more than k_r red points in it then delete S from $\mathcal{F}$.

Reduction Rule 3: If there is a set $S \in \mathcal{F}$ with at least $k_\ell + 1$ blue points then reduce the budget of k_ℓ by 1 and the budget of k_r by $|R \cap S|$. The new instance is $(U \setminus S, \widetilde{\mathcal{F}})$, where $\widetilde{\mathcal{F}} = \{F \setminus S \mid F \in \mathcal{F}$ and $F \neq S\}$.

The last Rule is similar to a Reduction Rule used in [13], for the POINT LINE COVER problem. The following simple observation can be made after exhaustive application of Reduction Rule 3.

Observation 2. *There can be at most $b \leq k_\ell^2$ blue points in a YES instance. If there are more than k_ℓ^2 blue points remaining to be covered then we correctly say NO.*

At first, we consider a simpler case where any line in the input instance contains exactly 1 blue point. Here, no two blue points can be covered by the same line and therefore, any solution family contains at least b lines. Thus, $b \leq k_\ell$

or else, it is a NO instance. Also, a minimal solution family contains at most $b \leq k_\ell$ lines. Hence, a minimal solution family contains exactly b lines.

Definition 3. *Given a universe U of points and a family $\mathcal{F}$ of subsets of U, an intersection graph $G_\mathcal{F} = (V, E)$ is a graph with vertices corresponding to the sets in $\mathcal{F}$ and an edge between two vertices implies that the corresponding sets have non-empty intersection.*

Let $G_{\mathcal{F}'}$ be the intersection graph that corresponds to a minimal solution family $\mathcal{F}'$.

Definition 4. *Given an instance $(R, B, \mathcal{F})$ of GEN-RBSC-LINES we call a tuple $\left(b, p, s, P, \{I'_1, \ldots, I'_s\}, (k^1_r, k^2_r, \ldots, k^s_r)\right)$ good if the following hold.*
(a) Integers $p \leq k_r$ and $s \leq b \leq k_\ell$; Here b is the number of blue vertices in the instance.
(b) $P = P_1 \cup \cdots \cup P_s$ is an s-partition of B;
(c) For each $1 \leq i \leq s$, I'_i is an ordering for the blue points in part P_i;
(d) Integers $k^i_r, 1 \leq i \leq s$, are such that $\Sigma_{1 \leq i \leq s} k^i_r = p$.

Definition 5. *We say that the minimal solution family $\mathcal{F}'$ conforms with a good tuple $\left(b, p, s, P, \{I'_1, \ldots, I'_s\}, (k^1_r, k^2_r, \ldots, k^s_r)\right)$ if the following properties hold:*

1. *The components $C_1, \ldots, C_s$ of $G_{\mathcal{F}'}$ give the partition $P = P_1, \ldots, P_s$ on the blue points.*
2. *For each component C_i, $1 \leq i \leq s$, let $t_i = |P_i|$. Let $I'_i = b^i_1, \ldots, b^i_{t_i}$ be an ordering of blue points in P_i. Furthermore assume that $L^i_j \in \mathcal{F}'$ covers the blue point b^i_j. I'_i has the property that for all $j \leq t_i$, $G_{\mathcal{F}'}[\{L^i_1, \ldots, L^i_j\}]$ is connected. In other words for all $j \leq t_i$, L^i_j intersects with at least one of the lines from the set $\{L^i_1, \ldots, L^i_{j-1}\}$. Notice that, by minimality of $\mathcal{F}'$, the point of intersection for such a pair of lines is a red point.*
3. *$\mathcal{F}'$ covers $p \leq k_r$ red points.*
4. *In each component C_i, k^i_r is the number of red points covered by the lines in that component. It follows that $\Sigma_{1 \leq i \leq s} k^i_r = p$. In other words, the integers k^i_r form a combination of p.*

Basically, a good tuple provides a numerical representation of connected components of $G_{\mathcal{F}'}$.

Lemma 1. *Let $(U, \mathcal{F})$ be an input to GEN-RBSC-LINES parameterized by $k_\ell + k_r$, such that every line contains exactly 1 blue point. If there exists a solution subfamily $\mathcal{F}'$ then there is a conforming good tuple.*

Proof. Let $\mathcal{F}'$ be a minimal solution family of size $b \leq k_\ell$ that covers $p \leq k_r$ red points. Let $G_{\mathcal{F}'}$ have s components viz. $C_1, C_2, \cdots, C_s$, where $s \leq k_\ell$. For each $i \leq s$, let $\mathcal{F}_{C_i}$ denote the set of lines corresponding to the vertices of C_i. $P_i = B \cap \mathcal{F}_{C_i}$, $t_i = |P_i|$ and $k^i_r = |R \cap \mathcal{F}_{C_i}|$. In this special case and by minimality of $\mathcal{F}'$, $|\mathcal{F}_{C_i}| = t_i$. As C_i is connected, there is a sequence $\{L^i_1, L^i_2, \ldots L^i_{t_i}\}$ for the lines in $\mathcal{F}_{C_i}$ such that for all $j \leq t_i$ we have that $G_{\mathcal{F}'}[\{L_1^i, \ldots, L_j^i\}]$ is connected. This means that, for all $j \leq t_i$ L_j^i intersects with at least one of the lines from the set $\{L_1^i, \ldots, L_{j-1}^i\}$. By minimality of $\mathcal{F}'$, the point of intersection for such a pair of lines is a red point. For all $j \leq t_i$, let L_j^i cover the blue point b_j^i. Let $I_i' = b_1^i, b_2^i, \ldots, b_{t_i}^i$. The tuple $\left(b, p, s, P = P_1 \cup P_2 \ldots \cup P_s, \{I_1', \ldots, I_s'\}, (k_r^1, k_r^2, \ldots, k_r^s)\right)$ is a good tuple and it also conforms with $\mathcal{F}'$. This completes the proof.

The idea of the algorithm is to generate all good tuples and then check whether there is a solution subfamily $\mathcal{F}'$ that conforms to it.

Lemma 2. *For a good tuple* $(b, p, s, P, \{I_1', \ldots, I_s'\}, (k_r^1, k_r^2, \ldots, k_r^s))$, *we can verify in* $\mathcal{O}(b\ell p^b)$ *time whether there is a minimal solution family* $\mathcal{F}'$ *that conforms with this tuple.*

Proof. The algorithm essentially builds a search tree for each partition $P_i, 1 \leq i \leq s$. For each part P_i, we define a set of points R_i' which is initially an empty set.

For each $1 \leq i \leq s$, let $t_i = |P_i|$ and let $I_i' = b_1^i, \ldots, b_{t_i}^i$ be the ordering of blue points in P_i. Our objective is to check whether there is a subfamily $\mathcal{F}_i' \subseteq \mathcal{F}$ such that it covers $b_1^i, \ldots, b_{t_i}^i$, and at most k_r^i red points.

Initially, $\mathcal{F}_i' = \emptyset$, $R_i' = \emptyset$. At any point of the recursive algorithm, we represent the problem to be solved by the following tuple: $(\mathcal{F}_i', R_i', (b_j^i, \ldots, b_{t_i}^i), k_r^i - |R_i'|)$. Here, $\mathcal{F}_i'$ covers $b_1^i, \ldots, b_{j-1}^i$, and at most k_r^i red points of R_i'. In the next step we try to extend $\mathcal{F}_i'$ in such a way that it also covers b_j^i, but still covers at most k_r^i red points. Thus we follow the ordering given by I_i' to build $\mathcal{F}_i'$.

We start the process by guessing the line in $\mathcal{F}$ that covers b_1^i, say L_1^i. That is, for every $L \in \mathcal{F}$ such that b_1^i is contained in L we recursively check whether there is a solution to the tuple $(\mathcal{F}_i' := \mathcal{F}_i' \cup \{L\}, R_i' := R_i' \cup (R \cap L), (b_2^i, \ldots, b_{t_i}^i), k_r^i := k_r^i - |R_i'|)$. If any tuple returns YES then we return that there is a subset $\mathcal{F}_i' \subseteq \mathcal{F}$ which covers $b_1^i, \ldots, b_{t_i}^i$, and at most k_r^i red points.

Similarly, at an intermediate stage of the algorithm, let the tuple we have be $(\mathcal{F}_i', R_i', (b_j^i, \ldots, b_{t_i}^i), k_r^i)$. Let $\mathcal{L}$ be the set of lines that contain b_j^i and a red point from R_i'. Clearly, $|\mathcal{L}| \leq |R_i'| \leq k_r^i$. For every line $L \in \mathcal{L}$, we recursively check whether there is a solution to the tuple $(\mathcal{F}_i' := \mathcal{F}_i' \cup \{L\}, R_i' := R_i' \cup (R \cap L), (b_{j+1}^i, \ldots, b_{t_i}^i), k_r^i := k_r^i - |R_i'|)$. If any tuple returns YES then we return that there is a subset $\mathcal{F}_i' \subseteq \mathcal{F}$ which covers $b_1^i, \ldots, b_{t_i}^i$, and at most k_r^i red points.

Let $\mu = t_i$. At each stage μ drops by one and, except for the first step, the algorithm recursively solves at most k_r^i subproblems. This implies that the algorithm takes at most $\mathcal{O}(|\mathcal{F}|k_r^{t_i}) = \mathcal{O}(\ell k_r^{t_i})$ time.

Notice that the lines in the input instance are partitioned according to the blue points contained in it. Hence, the search corresponding to each part P_i is independent of those in other parts. In effect, we are searching for the components for $G_{\mathcal{F}'}$ in the input instance, in parallel. If for each P_i we are successful in finding a minimal set of lines covering exactly the blue points of P_i while covering at most k_r^i red points, we conclude that a solution family $\mathcal{F}'$ that conforms to the given tuple exists and hence the input instance is a YES instance.

The time taken for the described procedure in each part is at most $\mathcal{O}(\ell k_r^{t_i})$. Hence, the total time taken to check if there is a conforming minimal solution family $\mathcal{F}'$ is at most $\mathcal{O}(\ell \cdot \sum_{i=1}^{s} k_r^{t_i}) = \mathcal{O}(s\ell p^b) = \mathcal{O}(b\ell p^b)$. This concludes the proof. $\qquad\square$

We are ready to describe our FPT algorithm for this special case.

Lemma 3. *Let* $(U, \mathcal{F}, k_\ell, k_r)$ *be an input to* GEN-RBSC-LINES *such that every line contains exactly 1 blue point. Then we can check whether there is a solution subfamily $\mathcal{F}'$ to this instance in time* $k_\ell^{\mathcal{O}(k_\ell)} \cdot k_r^{\mathcal{O}(k_r)} \cdot (|U| + |\mathcal{F}|)^{\mathcal{O}(1)}$ *time.*

Proof. Lemma 1 implies that all we need to do is enumerate all possible good tuples $(b, p, s, P, \{I'_1, \dots, I'_s\}, (k_r^1, k_r^2, \dots, k_r^s))$, and for each tuple, check whether there is a conforming minimal solution family. We first give an upper bound on the number of tuples and how to enumerate them. There are k_ℓ choices for s and k_r choices for p. There can be at most b^{k_ℓ} choices for P which can be enumerated in $\mathcal{O}(b^{k_\ell} \cdot k_\ell)$ time. Recall that, for each $1 \le i \le s$, I'_i represents an ordering for blue points in P_i. If $|P_i| = t_i$, then the number of distinct orderings is upper bounded by $\prod_{i=1}^{s} t_i! \le \prod_{i=1}^{s} t_i^{t_i} \le \prod_{i=1}^{s} b^{t_i} = b^b$. Such orderings can be enumerated in $\mathcal{O}(b^b)$ time. For a fixed $p \le k_r, s \le k_\ell$, there are at most $\binom{p+s-1}{s-1}$ solutions for $k_r^1 + k_r^2 + \dots + k_r^s = p$ and this set of solutions can be enumerated in $\mathcal{O}(\binom{p+s-1}{s-1} \cdot ps)$ time. Notice that if $p \ge s$ then the time required for enumeration is $\mathcal{O}((2p)^p \cdot ps)$. Otherwise, the required time is $\mathcal{O}((2s)^s \cdot ps)$. As $p \le k_r$ and $s \le k_\ell$, the time required to enumerate the set of solutions is $\mathcal{O}(k_\ell^{\mathcal{O}(k_\ell)} k_r^{\mathcal{O}(k_r)} \cdot k_\ell k_r)$. Thus we can generate the set of tuples in time $k_\ell^{\mathcal{O}(k_\ell)} \cdot k_r^{\mathcal{O}(k_r)}$. Using Lemma 2, for each tuple, we can check whether there is a conforming solution family or not in $\mathcal{O}(k_r^{k_\ell} \cdot k_\ell \ell)$ time. If there is no tuple with a conforming solution family, we know that the input instance is a NO instance. The total time for this algorithm is $k_\ell^{\mathcal{O}(k_\ell)} k_r^{\mathcal{O}(k_r)} k_\ell^{\mathcal{O}(k_\ell)} \cdot (|U| + |\mathcal{F}|)^{\mathcal{O}(1)}$. Again, if $k_r \le k_l$ then $k_r^{\mathcal{O}(k_\ell)} = k_\ell^{\mathcal{O}(k_\ell)}$. Otherwise, $k_r^{\mathcal{O}(k_\ell)} = k_r^{\mathcal{O}(k_r)}$. Either way, it is always true that $k_r^{\mathcal{O}(k_\ell)} = k_\ell^{\mathcal{O}(k_\ell)} k_r^{\mathcal{O}(k_r)}$. Thus, we can simply state the running time to be $k_\ell^{\mathcal{O}(k_\ell)} \cdot k_r^{\mathcal{O}(k_r)} \cdot (|U| + |\mathcal{F}|)^{\mathcal{O}(1)}$. $\qquad\square$

We return to the general problem of GEN-RBSC-LINES parameterized by $k_\ell + k_r$. Instances in this problem may have lines containing 2 or more blue points. We use the results and observations described above to arrive at an FPT algorithm for GEN-RBSC-LINES parameterized by $k_\ell + k_r$.

Theorem 3. GEN-RBSC-LINES *parameterized by $k_\ell + k_r$ is FPT, with an algorithm that runs in* $k_\ell^{\mathcal{O}(k_\ell)} \cdot k_r^{\mathcal{O}(k_r)} \cdot (|U| + |\mathcal{F}|)^{\mathcal{O}(1)}$ *time.*

Proof. Given an input $(U, \mathcal{F}, k_\ell, k_r)$ for GEN-RBSC-LINES parameterized by $k_\ell + k_r$, we do some preprocessing to make the instance simpler. We exhaustively apply Reduction Rules 1, 2 and 3. After this, by Observation 2, the reduced equivalent instance has at most k_ℓ^2 blue points if it is a YES instance.

A minimal solution family can be broken down into two parts: the set of lines containing at least 2 blue points, and the remaining set of lines which contain exactly 1 blue point. Let us call these sets $\mathcal{F}_2$ and $\mathcal{F}_1$ respectively. We start with the following observation.

Observation 3. *Let $\mathcal{F}'' \subseteq \mathcal{F}$ be the set of lines that contain at least 2 blue points. There are at most $\binom{k_\ell^4}{k_\ell}$ ways in which a solution family can intersect with $\mathcal{F}''$.*

From Observation 3, there are $k_\ell^{4k_\ell}$ choices for the set of lines in $\mathcal{F}_2$. We branch on all these choices of $\mathcal{F}_2$. On each branch, we reduce the budget of k_ℓ by the number of lines in $\mathcal{F}_2$ and the budget of k_r by $|R \cap \mathcal{F}_2|$. Also, we make some modifications on the input instance: we delete all other lines containing at least 2 blue points from the input instance. We delete all points of U covered by $\mathcal{F}_2$ and all lines passing through blue points covered by $\mathcal{F}_2$. Our modified input instance in this branch now satisfies the assumption of Lemma 3 and we can find out in $k_\ell^{O(k_\ell)} k_r^{O(k_r)} \cdot (|U| + |\mathcal{F}|)^{O(1)}$ time whether there is a minimal solution family $\mathcal{F}_1$ for this reduced instance. If there is, then $\mathcal{F}_2 \cup \mathcal{F}_1$ is a minimal solution for our original input instance and we correctly say YES. Thus the total running time of this algorithm is $k_\ell^{O(k_\ell)} \cdot k_r^{O(k_r)} \cdot (|U| + |\mathcal{F}|)^{O(1)}$.

It may be noted here that for a special case where we can use any line in the plane as part of the solution, the second part of the algorithm becomes considerably simpler. Here for each blue point b, we can use an arbitrary line containing only b and no red point.   □

4.1   Kernelization for GEN-RBSC-LINES Parameterized by $k_\ell + k_r$

We give a polynomial parameter transformation from SET COVER parameterized by universe size n to GEN-RBSC-LINES parameterized by $k_\ell + k_r$.

Theorem 4. GEN-RBSC-LINES *parameterized by $k_\ell + k_r$ does not allow a polynomial kernel unless co-NP $\subseteq$ NP/poly.*

Proof. Let $(U, \mathcal{S})$ be a given instance of SET COVER. Let $|U| = n, |\mathcal{S}| = m$. We construct an instance $(R \cup B, \mathcal{F})$ of GEN-RBSC-LINES as follows. We assign a blue point $b_u \in B$ for each element $u \in U$ and a red point $r_S \in R$ for each set $S \in \mathcal{S}$. The red and blue points are placed such that no three points are collinear. We add a line between b_u and r_S if $u \in S$ in the SET COVER instance. Thus the new instance that we have constructed has $b = n$, $r = m$ and $\ell = \sum_{S \in \mathcal{S}} |S|$. We set $k_r = k$ and $k_\ell = n$.

Claim.[⋆] All the elements in $(U, \mathcal{S})$ can be covered by k sets if and only if there exist n lines in $(R \cup B, \mathcal{F})$ that contain all blue points but only k red points.

If $k > n$, then the SET COVER instance is a trivial YES instance. Hence, we can always assume that $k \leq n$. This completes the proof.   □

5   Hardness in Higher Dimensions When Parameterized by $k_\ell + k_r$

While we obtain an FPT algorithm for GEN-RBSC-LINES we also show that on generalizing the sets from lines to hyperplanes in higher Euclidean spaces, the problem is W[1]-hard under $k_\ell + k_r$.

Theorem 5. GEN-RBSC *for hyperplanes in* $\mathbb{R}^d$, $d > 2$, *parameterized by* $k_\ell + k_r$ *is* W[1] *-hard.*

Proof. The proof of hardness follows from a reduction from k-CLIQUE problem. The proof follows a framework given in [15].

Let $(G(V, E), k)$ be an instance of k-CLIQUE problem. Our construction consists of a $k \times k$ matrix of gadgets G_{ij}, $1 \leq i, j, \leq k$. Consecutive gadgets in a row are connected by horizontal connectors and consecutive gadgets in a column are connected by vertical connectors. Let us denote the horizontal connector connecting the gadgets G_{ij} and G_{ih} as $H_{i(jh)}$ and the vertical connector connecting the gadgets G_{ij} and G_{hj} as $V_{(ih)j}$, $1 \leq i, j, h \leq k$.

Gadgets: The gadget G_{ij} contains a blue point b_{ij} and a set R_{ij} of $d - 2$ red points. In addition there are n^2 sets $R'_{ij}(a, b), 1 \leq a, b \leq n$, each having two red points each.

Connectors: The horizontal connector $H_{i(jh)}$ has a blue point $b_{i(jh)}$ and a set $R_{i(jh)}$ of $d - 2$ red points. Similarly, the vertical connector $V_{(ih)j}$ a blue point $b_{(ih)j}$ and a set $R_{(ih)j}$ of $d - 2$ red points.

The points are arranged in general position i.e., no set of $d + 2$ points lie on the same d-dimensional hyperplane. In other words, any set of $d + 1$ points define a distinct hyperplane.

Hyperplanes: Assume $1 \leq i, j, h \leq k$ and $1 \leq a, b, c \leq n$. Let $P_{ij}(a, b)$ be the hyperplane defined by the $d + 1$ points of $b_{ij} \cup R_{ij} \cup R'_{ij}(a, b)$. Let $P^h_{i(jh)}(a, b, c)$ be the hyperplane defined by $d + 1$ points of $b_{i(jh)} \cup R_{i(jh)} \cup r_1 \cup r_2$ where $r_1 \in R'_{ij}(a, b)$ and $r_2 \in R'_{ih}(a, c)$. Let $P^v_{(ij)h}(a, b, c)$ be the hyperplane defined by $d + 1$ points of $b_{(ij)h} \cup R_{(ij)h} \cup r_1 \cup r_2$ where $r_1 \in R'_{ih}(a, c)$ and $r_2 \in R'_{jh}(b, c)$. For each edge $(a, b) \in E(G)$, we add $k(k-1)$ hyperplanes of the type $P_{ij}(a, b)$, $i \neq j$. Further, for all $1 \leq a \leq n$, we add k hyperplanes of the type $P_{ii}(a, a)$, $1 \leq i \leq k$. The hyperplane $P^h_{i(jh)}(a, b, c)$ containing the blue point $b_{i(jh)}$ in a horizontal connector is added to the construction if $P_{ij}(a, b)$ and $P_{ih}(a, c)$ are present in the construction. Similarly, the hyperplane $P^v_{(ij)h}(a, b, c)$ containing the blue point $b_{(ij)h}$ in a vertical connector is added to the construction if $P_{ih}(a, c)$ and $P_{jh}(b, c)$ are present in the construction.

Thus our construction has $k^2 + 2k(k - 1)$ blue points, $(k^2 + 2k(k - 1))(d - 2) + 2n^2k^2$ red points and $O((m^2k^2)$ hyperplanes.

Claim.[⋆] G has a k-clique if and only if all the blue points in the constructed instance can be covered by $k^2 + 2k(k - 1)$ hyperplanes covering at most $k^2d + 2k(k - 1)(d - 2)$ red points.   □

References

1. Bodlaender, H.L., Downey, R.G., Fellows, M.R., Hermelin, D.: On problems without polynomial kernels. J. Comput. Syst. Sci. **75**(8), 423–434 (2009)
2. Carr, R.D., Doddi, S., Konjevod, G., Marathe, M.V.: On the red-blue set cover problem. In: SODA, vol. 9, pp. 345–353 (2000)
3. Chan, T.M., Hu, N.: Geometric red-blue set cover for unit squares and related problems. In: CCCG (2013)
4. Ying Cheng, S., Iyengar, S., Kashyap, R.L.: A new method of image compression using irreducible covers of maximal rectangles. IEEE Trans. Software Eng. **14**(5), 651–658 (1988)
5. Dom, M., Guo, J., Niedermeier, R., Wernicke, S.: Red-blue covering problems and the consecutive ones property. J. Discrete Algorithms **6**(3), 393–407 (2008)
6. Dom, M., Lokshtanov, D., Saurabh, S.: Kernelization lower bounds through colors, ids. ACM Trans. Algorithms **11**(2), 1–20 (2014)
7. Downey, R.G., Fellows, M.R.: Parameterized Complexity, p. 530. Springer-Verlag, Heidelberg (1999)
8. Flum, J., Grohe, M.: Parameterized Complexity Theory. Texts in Theoretical Computer Science. An EATCS Series. Springer-Verlag, Heidelberg (2006)
9. Fortnow, L., Santhanam, R.: Infeasibility of instance compression and succinct pcps for NP. J. Comput. Syst. Sci. **77**(1), 91–106 (2011)
10. Karp, R.M.: Reducibility among combinatorial problems. In: Proceedings of a Symposium on the Complexity of Computer Computations, pp. 85–103 (1972)
11. Kratsch, S., Philip, G., Ray, S.: Point line cover: The easy kernel is essentially tight. In: SODA, pp. 1596–1606 (2014)
12. Kumar, V.S.A., Arya, S., Ramesh, H.: Hardness of set cover with intersection 1. In: Welzl, E., Montanari, U., Rolim, J.D.P. (eds.) ICALP 2000. LNCS, vol. 1853, pp. 624–635. Springer, Heidelberg (2000)
13. Langerman, S., Morin, P.: Covering things with things. Discrete Comput. Geom. **33**(4), 717–729 (2005)
14. Levcopoulos, C.: Improved bounds for covering general polygons with rectangles. In: Nori, K.V. (ed.) FSTTCS 1987. LNCS, vol. 287, pp. 95–102. Springer, Heidelberg (1987)
15. Marx, D.: Parameterized complexity of independence and domination on geometric graphs. In: Bodlaender, H.L., Langston, M.A. (eds.) IWPEC 2006. LNCS, vol. 4169, pp. 154–165. Springer, Heidelberg (2006)
16. Mathieson, L., Szeider, S.: The parameterized complexity of regular subgraph problems and generalizations. CATS **77**, 79–86 (2008)
17. Peleg, D.: Approximation algorithms for the label-cover$_{max}$ and red-blue set cover problems. J. Discrete Algorithms **5**(1), 55–64 (2007)
18. Raman, V., Saurabh, S.: Short cycles make W-hard problems hard: FPT algorithms for W-hard problems in graphs with no short cycles. Algorithmica **52**(2), 203–225 (2008)

Tight Bounds for Beacon-Based Coverage in Simple Rectilinear Polygons

Sang Won Bae[1], Chan-Su Shin[2], and Antoine Vigneron[3(✉)]

[1] Department of Computer Science, Kyonggi University, Suwon, Korea
swbae@kgu.ac.kr
[2] Division of Computer and Electronic Systems Engineering,
Hankuk University of Foreign Studies, Yongin, Korea
cssin@hufs.ac.kr
[3] Visual Computing Center, King Abdullah University of Science
and Technology (KAUST), Thuwal 23955-6900, Saudi Arabia
antoine.vigneron@kaust.edu.sa

Abstract. We establish tight bounds for beacon-based coverage problems. In particular, we show that $\lfloor \frac{n}{6} \rfloor$ beacons are always sufficient and sometimes necessary to cover a simple rectilinear polygon P with n vertices. When P is monotone and rectilinear, we prove that this bound becomes $\lfloor \frac{n+4}{8} \rfloor$. We also present an optimal linear-time algorithm for computing the beacon kernel of P.

1 Introduction

A *beacon* is a facility or a device that attracts objects within a given domain. We assume that objects in the domain, such as mobile agents or robots, know the exact location or the direction towards an activated beacon in the domain, even if it is not directly visible. More precisely, given a polygonal domain P, a beacon is placed at a fixed point in P. When a beacon $b \in P$ is activated, an object $p \in P$ moves along the ray starting at p and towards the beacon b until it either hits the boundary ∂P of P, or it reaches b (See Fig. 1a). If p hits an edge e of P, then it continues to move along e in the direction such that the Euclidean distance to b decreases. When p reaches an endpoint of e, it may move along the ray from the current position of p towards b, if possible, until it again hits the boundary ∂P of P. So, p is pulled by b in a greedy way, so that the Euclidean distance to b is monotonically decreasing, as an iron particle is pulled by a magnet. There are two possible outcomes: Either p finally reaches b, or it stops at a local minimum, called a *dead point*, where there is no direction in

Work by S.W.Bae was supported by Basic Science Research Program through the National Research Foundation of Korea (NRF) funded by the Ministry of Science, ICT & Future Planning (2013R1A1A1A05006927) and by the Ministry of Education (2015R1D1A1A01057220). Work by C.-S. Shin was supported by Research Grant of Hankuk University of Foreign Studies. Work by A. Vigneron was supported by KAUST base funding.

© Springer-Verlag Berlin Heidelberg 2016
E. Kranakis et al. (Eds.): LATIN 2016, LNCS 9644, pp. 110–122, 2016.
DOI: 10.1007/978-3-662-49529-2_9

which, locally, the distance to b strictly decreases. In the former case, p is said to be *attracted* by the beacon b.

This model of beacon attraction was recently suggested by Biro [1–3], and extends the classical notion of visibility. Biro et al. [1–3] introduced the *coverage* problem under this model: We want to place beacons in P so that any point $p \in P$ is attracted by at least one of the beacons. In this case, we say that the set of beacons *covers* or *guards* P.

In this paper, we are interested in combinatorial bounds on the number of beacons required for covering a simple rectilinear polygon P, called the *domain*. Our bounds are variations on visibility-based guarding results, such as the well-known *art gallery theorem* [4] and its relatives [5–7,9,10]. For the art gallery problem, it is known that $\lfloor \frac{n}{3} \rfloor$ point guards are sufficient, and sometimes necessary, to guard a simple polygon P with n vertices [4]. If P is rectilinear, then $\lfloor \frac{n}{4} \rfloor$ are necessary and sufficient [5,7,11]. Other related results are mentioned in the book [11] by O'Rourke or the surveys by Shermer [12] and Urrutia [14].

Biro et al. [2] initiated research on combinatorial bounds for beacon-based coverage, and for the corresponding routing problem. (In this routing problem, one wants to place beacons such that any object can be moved to any point of the domain by activating a sequence of beacons one at a time.) They gave several nontrivial bounds for different types of domains such as rectilinear or non rectilinear polygons, with or without holes. When the domain P is a simple rectilinear polygon with n vertices, they showed that $\lfloor \frac{n}{4} \rfloor$ beacons are sufficient to cover any rectilinear polygon with n vertices, while $\lfloor \frac{n+4}{8} \rfloor$ beacons are necessary to cover the polygon in Fig. 1, and conjectured that $\lfloor \frac{n+4}{8} \rfloor$ would be the tight bound. They also proved that $\lfloor \frac{n}{2} \rfloor - 1$ beacons are always sufficient for routing, and some domains, such as the domain depicted in Fig. 1a, require $\lfloor \frac{n}{4} \rfloor - 1$ beacons. Recently, Shermer closed the gap by showing that $\lceil \frac{n-4}{3} \rceil$ beacons are sufficient, and sometimes necessary, for beacon-based routing within a simple rectilinear polygon [13].

Our Results. We first present an optimal linear-time algorithm that computes the *beacon kernel* $\mathcal{K}(P)$ of a simple rectilinear polygon P (Sect. 3). The beacon kernel $\mathcal{K}(P)$ of P is defined to be the set of points $p \in P$ such that placing a single beacon at p is sufficient to completely cover P. Biro first presented an $O(n^2)$-time algorithm that computes the kernel $\mathcal{K}(P)$ of a simple polygon P in his thesis [1], and Kouhestani et al. [8] soon improved it to $O(n \log n)$ time with the observation that $\mathcal{K}(P)$ has a linear complexity. Our algorithm is based on a new, yet simple, characterization of the kernel $\mathcal{K}(P)$.

Our main result is presented in Sect. 4: We prove tight bounds on beacon-based coverage problems for simple rectilinear polygons. We first show how to construct a simple rectilinear polygon with n vertices that cannot be covered by less than $\lfloor \frac{n}{6} \rfloor$ beacons, and then we present a method for covering any simple rectilinear polygon with the same number of beacons. These results settle the open questions on the beacon-based coverage problem for simple rectilinear polygons

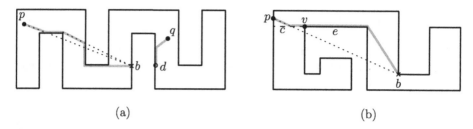

Fig. 1. (a) A lower bound construction P by Biro et al. [2]. A point $p \in P$ is attracted by a beacon b through the beacon attraction path depicted by the thick gray path, while $q \in P$ is not since it stops at the dead point d. (b) Another rectilinear polygon P. If one partitions P by the horizontal cut c (dashed segment) at v into two subpolygons P_c^+ and P_c^- and handle each separately, then it does not guarantee that P is guarded. In this case, $p \in P_c^+$ is attracted by b inside the subpolygon P_c^+ while it is not the case in the whole domain P.

posed by Biro et al. [2]. We also prove that $\lfloor \frac{n+4}{8} \rfloor$ beacons are always sufficient to cover a monotone rectilinear polygon, which matches the lower bound by Biro et al. [2], and proves their conjecture.

2   Preliminaries

A *simple rectilinear polygon* is a simple polygon whose edges are either horizontal or vertical. The internal angle at each vertex of a rectilinear polygon is always $90°$ or $270°$. We call a vertex with internal angle $90°$ a *convex vertex*, and a vertex with internal angle $270°$ is called a *reflex vertex*. For any simple rectilinear polygon P, we let $r = r(P)$ be the number of its reflex vertices. If P has n vertices in total, then $n = 2r + 4$, because the sum of the signed turning angles along ∂P is $360°$. An edge of P between two convex vertices is called a *convex edge*, and an edge between two reflex vertices is called a *reflex edge*. Each convex or reflex edge e shall be called *top*, *bottom*, *left* or *right* according to its orientation: If e is horizontal and the two adjacent edges of e are downwards from e, then e is a top convex or reflex edge. (The edge e in Fig. 1b is a top reflex edge.) For each edge e of P, we are often interested in the half-plane H_e whose boundary supports e and whose interior includes the interior of P locally at e. We shall call H_e the *half-plane supporting* e.

A rectilinear polygon P is called *x-monotone* (or *y-monotone*) if any vertical (resp., horizontal) line intersects P in at most one connected component. If P is both *x*-monotone and *y*-monotone, then P is said to be *xy-monotone*. It follows from this definition that:

Observation 1. *A rectilinear polygon P is x-monotone if and only if P has no vertical reflex edge. Hence, P is xy-monotone if and only if P has no reflex edge.*

Our approach to attain tight upper bounds relies on partitioning a given rectilinear polygon P into subpolygons by cuts. More precisely, a *cut* in P is a chord[1] of P that is horizontal or vertical. There is a unique cut at a point p on the boundary ∂P of P unless p is a vertex of P. If p is a reflex vertex, then there are two cuts at p, one of which is *horizontal* and the other is *vertical*, while there is no cut at p if p is a convex vertex. Any horizontal cut c in P partitions P into two subpolygons: one below c, denoted by P_c^-, and the other above c denoted by P_c^+. Analogously, for any vertical cut c, let P_c^- and P_c^+ denote the subpolygons to the left and to the right of c, respectively.

For a beacon b and a point $p \in P$, the *beacon attraction path* of p with respect to b, or simply the *b-attraction path* of p, is the piecewise linear path from p created by the attraction of b as described in Sect. 1 (See Fig. 1a). If the b-attraction path of p reaches b, then we say that p is *attracted* to b. As was done for the classical visibility notion [6,10], a natural approach would find a partition of P into smaller subpolygons of similar size, and handle them recursively. However, we must be careful when choosing a partition of P, because an attraction path within a subpolygon may not be an attraction path within P (See Fig. 1b). So P is not necessarily guarded by the union of the guarding sets of the subpolygons.

Thus, when applying a cut in P, we want to make sure that beacon attraction paths in a subpolygon Q of P do not *hit* the new edge of Q produced by c. To be more precise, we say that an edge e of P is *hit* by p with respect to b if the b-attraction path of p makes a bend along e.

Observation 2. *Let b be a beacon in P and $p \in P$ be any point such that p is attracted by b. If the b-attraction path of p hits an edge e of P, then $p \in H_e$ and $b \notin H_e$, where H_e denotes the half-plane supporting e. Therefore, no beacon attraction path hits a convex edge of P.*

Thus, if we choose a cut that becomes a convex edge on both sides, then we will be able to handle each subpolygon separately.

In this paper, we make the general position assumption that *no cut in P connects two reflex vertices*. This general position can be obtained by perturbing the reflex vertices of P locally, and such a perturbation does not harm the upper bounds on our problems in general. It will be discussed in the full version of this paper.

3 The Beacon Kernel

Before continuing to the beacon-based coverage problem, we consider simple rectilinear polygons that can be covered by a single beacon. This is related to the *beacon kernel* $\mathcal{K}(P)$ of a simple polygon P, defined to be the set of all points $p \in P$ such that a beacon placed at p attracts all points in P. Specifically, we give a characterization of rectilinear polygons P such that $\mathcal{K}(P) \neq \emptyset$. Our characterization is simple and constructive, resulting in a linear-time algorithm that computes the beacon kernel $\mathcal{K}(P)$ of any simple rectilinear polygon P.

---

[1] A *chord* c of a polygon is a line segment between two points on the boundary such that all points on c except the two endpoints lie in the interior of the polygon.

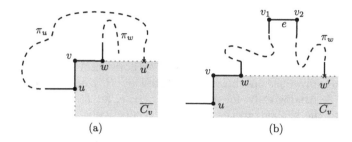

Fig. 2. Proof of Lemma 2.

Let R be the set of reflex vertices of P. Let $v \in R$ be any reflex vertex with two incident edges e_1 and e_2. For $i \in \{1, 2\}$, define N_i to be the closed half-plane whose boundary is the line orthogonal to e_i through v and whose interior excludes e_i. Let $C_v := N_1 \cup N_2$. Observe that C_v is a closed cone with apex v. Biro [1, Theorem 5.2.8] showed that the kernel $\mathcal{K}(P)$ of P is the set of points in P that lie in C_v for all reflex vertices $v \in R$:

Lemma 1 (Biro [1]). *For any simple polygon P with set R of reflex vertices, it holds that*

$$\mathcal{K}(P) = \left(\bigcap_{v \in R} C_v \right) \cap P = P \setminus \left(\bigcup_{v \in R} \overline{C_v} \right),$$

where $\overline{C_v} = \mathbb{R}^2 \setminus C_v$ denotes the complement of C_v.

Note that Lemma 1 holds for any simple polygon P. We now assume that P is a simple rectilinear polygon. Then, for any reflex vertex $r \in R$, the set C_v forms a closed cone with aperture angle $270°$ whose boundary consists of two rays following the two edges incident to v. Let $R_1 \subseteq R$ be the set of reflex vertices incident to at least one reflex edge, and let $R_2 := R \setminus R_1$. So a vertex in R_1 is adjacent to at least one reflex vertex that also belongs to R_1, and a vertex in R_2 is always adjacent to two convex vertices. We then observe the following.

Lemma 2. *For any simple rectilinear polygon P,*

$$\left(\bigcap_{v \in R_1} C_v \right) \cap P \subseteq \left(\bigcap_{v \in R_2} C_v \right) \cap P.$$

Proof. For a contradiction, suppose that there exists a point $p \in P$ that is included in $\bigcap_{v \in R_1} C_v$ but avoids $\bigcap_{v \in R_2} C_v$. Then, there must exist a reflex vertex $v \in R_2$ such that $p \notin C_v$, or equivalently, $p \in \overline{C_v}$. That is, $\bigcap_{v' \in R_1} C_{v'}$ and $\overline{C_v}$ have a nonempty intersection. Let u and w be the two vertices adjacent to v such that u, v, and w appear on ∂P in counterclockwise order. Note that both u and w are convex since $v \in R_2$.

Since $\overline{C_v} \cap P \neq \emptyset$ and $\overline{C_v}$ is an open set, the boundary ∂P of P crosses $\partial \overline{C_v}$ at some points other than the two edges uv and vw. Let w' be the first point in $\partial P \cap \partial \overline{C_v}$ that we encounter when traveling along ∂P counterclockwise, starting at w. Analogously, let u' be the first point in $\partial P \cap \partial \overline{C_v}$ that we encounter when traveling along ∂P clockwise, starting at u. Let $\pi_w \subset \partial P$ and $\pi_u \subset \partial P$ be the paths described above from w to w' and from u to u', respectively. As π_w and π_u are subpaths of ∂P, they do not intersect, and we have $w' \notin uv$ and $u' \notin vw$.

The boundary $\partial \overline{C_v}$ of $\overline{C_v}$ consists of two rays ρ_w and ρ_u, starting from v towards w and u, respectively. We claim that either w' lies on ρ_w or u' lies on ρ_u (See Fig. 2a). Indeed, suppose that $u' \notin \rho_u$. Then π_w should be contained in the region bounded by the simple closed curve $\pi_u \cup u'v \cup vu$, since π_w does not intersect $\overline{C_v} \cup uv$. This implies that w' must lie on $wu' \subset \rho_w$. Hence, our claim is true.

Without any loss of generality, we assume that $w' \in \rho_w$, the edge vw is horizontal, and the interior of P lies locally above vw, as shown in Fig. 2b. Then, the path π_w must contain at least one top reflex edge e lying above the line through w and w', since w and w' have the same y-coordinate and π_w avoids $\overline{C_v}$. Let v_1 and v_2 be the two endpoints of e, so $v_1, v_2 \in R_1$. Then $C_{v_1} \cap C_{v_2}$ is the half-plane H_e supporting e. Since e is a top reflex edge, $H_e \cap \overline{C_v} = \emptyset$. This is a contradiction to our assumption that $\bigcap_{v' \in R_1} C_{v'}$ intersects $\overline{C_v}$. $\qquad\square$

Let $\mathcal{R}(P)$ be the intersection of the half-planes H_e supporting e over all reflex edges e of P. We conclude the following.

Theorem 1. *Let P be a simple rectilinear polygon. A point $p \in P$ lies in its beacon kernel $\mathcal{K}(P)$ if and only if $p \in H_e$ for any reflex edge e of P. Therefore, it always holds that $\mathcal{K}(P) = \mathcal{R}(P) \cap P$, and the kernel $\mathcal{K}(P)$ can be computed in linear time.*

Proof. Recall that C_v for any $v \in R_1$ forms a cone with apex v and aperture angle $270°$. Since any $v \in R_1$ is adjacent to another reflex vertex $w \in R_1$ the intersection $C_v \cap C_w$ forms exactly the half-plane H_e supporting the reflex edge e with endpoints v and w. It implies that $\mathcal{R}(P) = \bigcap_{v \in R_1} C_v$. So by Lemma 2, we have

$$\mathcal{K}(P) = \bigcap_{v \in R} C_v \cap P = \bigcap_{v \in R_1} C_v \cap P = \mathcal{R}(P) \cap P.$$

The set $\mathcal{R}(P)$ is an intersection of axis-parallel halfplanes, so it is a (possibly unbounded) axis-parallel rectangle. In order to compute the kernel $\mathcal{K}(P)$, we identify the extreme reflex edge in each of the four directions to compute $\mathcal{R}(P)$, and then intersect it with P. This can be done in linear time. $\qquad\square$

4   Beacon-Based Coverage

In this section, we study the beacon-based coverage problem for rectilinear polygons. A set of beacons in P is said to *cover* or *guard* P if and only if every point $p \in P$ can be attracted by at least one of them.

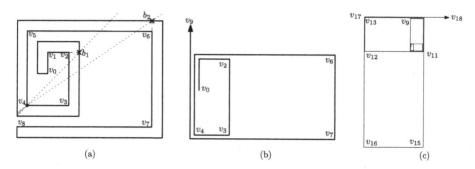

Fig. 3. Lower bound construction: (a) Placing two beacons b_1 and b_2 in P_7 near v_2 and v_6 is not enough to cover the shaded region near the reflex vertex v_4. (b)(c) The spine of our construction P_r for $r = 9$ and for $r = 18$.

Our main result is the following.

Theorem 2. *Let P be a simple rectilinear polygon P with $n \geqslant 6$ vertices and $r \geqslant 1$ reflex vertices. Then $\lfloor \frac{n}{6} \rfloor = \lceil \frac{r}{3} \rceil$ beacons are sufficient to guard P, and sometimes necessary. Moreover, all these beacons can be placed at reflex vertices of P.*

We now sketch the proof of Theorem 2. The lower bound construction is a rectangular spiral P_r consisting of a sequence of $r+1$ thin rectangles, as depicted in Fig. 3. The sequence of vertices $v_0 v_1 v_2 \ldots v_{r+1}$, where $v_1 \ldots v_r$ are the reflex vertices of P_r, forms a polyline called the *spine* of the spiral. The key idea is the following. Consider the case $r = 7$ (Fig. 3a). At first glance, it seems that the spiral can be covered by two beacons b_1 and b_2 placed near v_2 and v_6, respectively. However, at closer look, it appears that the small shaded triangular region on the bottom left corner is not covered. Hence, P_7 requires $3 = \lceil \frac{7}{3} \rceil$ beacons, as announced. More generally, we can prove that for a suitable choice of the edge lengths of the spine of P_r, an optimal coverings for P_r consists in placing a beacon at every third rectangle of P_r, which yields the bound $\lceil \frac{r}{3} \rceil$. The spine of P_r is depicted in Fig. 3b and c, where the aspect ratio of the rectangles is roughly $4 + r/2$.

The construction for the upper bound in Theorem 2 is more involved. We first prove that for any polygon with at most 3 reflex vertices, one beacon placed at a suitable reflex vertex is sufficient. For a larger number $r \geqslant 4$ of reflex vertices, we proceed by induction. So we partition P using a cut, and we handle each side recursively. As mentioned in Sect. 2, the difficulty is that in some cases, the union of the two guarding sets of the subpolygons does not cover P. So we first try to perform a *safe cut* c, that is, a cut c which is not incident to any reflex vertex, such that there is at least one reflex vertex on each side, and such that $\lceil r(P_c^-)/3 \rceil + \lceil r(P_c^+)/3 \rceil = \lceil r/3 \rceil$ (See Fig. 4a). If such a cut exists, then we can recurse on both side. By Observation 2, the union of the guarding sets of the two subpolygons guards P. Unfortunately, some polygons do not admit any safe cut.

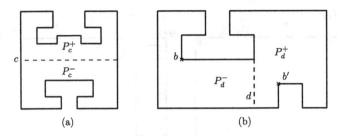

Fig. 4. Upper bound construction. (a) A safe cut c of a polygon with $r = 10$ reflex vertices. (b) This polygon admits no safe cut. We cut along d, and the polygon is guarded by any two beacons b and b' placed at reflex vertices of P_d^- and P_d^+, respectively.

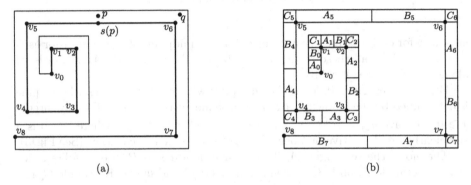

Fig. 5. (a) The spine (bold) of a spiral. The point $s(p)$ appears before $v_6 = s(q)$ along the spine, so $p \prec q$. (b) The partition of P_7 into rectangles A_i, B_i, C_i.

In this case, we show by a careful case analysis that we can always find a suitable cut (See the example in Fig. 4b). Due to space limitation, we omit the proof of this upper bound.

4.1 Proof of the Lower Bound

In this section, we prove the lower bound in Theorem 2. Our construction is a spiral-like rectilinear polygon P_r that cannot be guarded by less than $\lceil \frac{r}{3} \rceil$ beacons (See Fig. 3). More precisely, a rectilinear polygon is called a *spiral* if all its reflex vertices are consecutive along its boundary.

The *spine* of a spiral P with r reflex vertices is the portion of its boundary ∂P connecting $r + 2$ consecutive vertices $v_0, v_1, \ldots, v_{r+1}$ such that $v_1, \ldots, v_r$ are the reflex vertices of P (See Fig. 5a). Note that the two end vertices v_0 and v_{r+1} of the spine of a spiral are the only convex vertices that are adjacent to a reflex vertex. The spine can also be specified by the sequence of edge lengths $(a_0, \ldots, a_r)$ such that a_i is the length of the edge $v_i v_{i+1}$ for $i = 0, \ldots, r$.

We will use the following partition of a spiral P with r reflex vertices into $3r + 2$ rectangular subpolygons. It is obtained by applying the vertical and horizontal

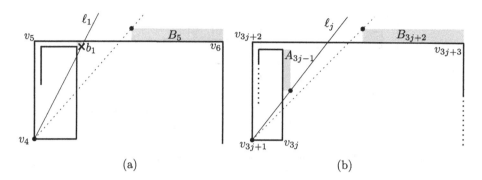

Fig. 6. Proof of Lemma 3.

cuts at v_i for each $i = 1, \ldots, r$ and the cut at the midpoint of edge $v_i v_{i+1}$ for each $i = 0, \ldots, r$. We call these rectangles $A_0, B_0, C_1, A_1, B_1, \ldots, C_r, A_r, B_r$, ordered along the spine (See Fig. 5b).

For any integer $r \geqslant 0$, let P_r be the spiral with r reflex vertices whose spine is determined by the following edge length sequence $(a_0, \ldots, a_r)$: For any integer $j \geqslant 0$, we have $a_{2j} = \rho^{2\lceil \frac{j}{3} \rceil} + j\epsilon$ and $a_{2j+1} = \rho^{1+2\lfloor \frac{j+1}{3} \rfloor} + j\epsilon$, where $\epsilon > 0$ is a sufficiently small positive number and $\rho > 2 + (r+2)\epsilon$ is a constant (See Fig. 3).

Therefore, the rectangles A_i, B_i, C_i corresponding to P_r are as follows, for any i. Rectangle A_i and B_i have side lengths $a_i/2$ and $w < \epsilon$. Rectangle C_i is a square with side length w.

We define an order $\prec$ among points in any spiral P as follows. Let p, q be two points in P. Let $s(p)$ and $s(q)$ denote the closest point to p and q along the spine, according to the geodesic distance within P (See Fig. 5a). Then we say that p precedes q, which we denote by $p \prec q$, if $s(p)$ precedes $s(q)$ along the spine, that is, $s(p)$ is on the portion of the spine between v_0 and $s(q)$.

Let $k = k(r)$ denote the smallest possible number of beacons that can guard P_r. We will say that a sequence of beacons $b_1, \ldots, b_k$ covering P_r is a *greedy placement* if $s(b_1) \prec \ldots \prec s(b_k)$, and the sequence $s(b_1), \ldots, s(b_k)$ is maximum in lexicographical order. So intuitively, we obtain the greedy placement by pushing the beacons as far as possible from the origin v_0 of the spiral. Clearly, b_1 must be placed in C_2. We then observe the following for $b_2, \ldots, b_{k-1}$.

Lemma 3. *For $2 \leqslant i \leqslant k - 1$, the i-th beacon b_i in a greedy placement for P_r is in A_{3i-1}.*

Proof. We prove the lemma by induction on i. We first verify the lemma for b_2. Without loss of generality, we assume that the edge $v_1 v_2$ is a top reflex edge. Let ℓ_1 be the line through v_4 and b_1. Observe that b_1 attracts all points in C_3, but not all of those in C_4. More precisely, b_1 attracts those in C_4 below ℓ_1 but misses those above ℓ_1. Hence, b_2 must be placed on ℓ_1 to cover the points in C_4 above ℓ_1. For our purpose, we compare the slopes of ℓ_1 and any line through v_4 and a point in B_5 (See Fig. 6a). Recall that $\rho > 2 + (r+2)\epsilon$ and $w < \epsilon$. The slope of ℓ_1 is at least

$$\frac{a_2}{a_3+w} > \frac{a_2}{a_3+\epsilon} = \frac{\rho^2+\epsilon}{\rho+2\epsilon} > \frac{\rho^2}{\rho+2\epsilon} > \frac{(2+2\epsilon)\rho}{\rho+2\epsilon} > 2.$$

On the other hand, the slope of any line through v_4 and a point in B_5 is at most

$$\frac{a_4+w}{a_5/2} < 2\frac{a_4+\epsilon}{a_5} = 2\frac{\rho^2+3\epsilon}{\rho^3+2\epsilon} < \frac{2\rho^2+6\epsilon}{\rho^3} < \frac{2+6\epsilon/\rho^2}{2+(r+2)\epsilon} < 1.$$

This implies that ℓ_1 cannot intersect B_5. Thus, if $b_2 \in B_5$, then b_2 fails to attract some points near v_4 and above ℓ_1, similarly as in Fig. 3a, so b_2 must lie in A_5.

For the inductive step, assume that $j \in \{2, \ldots, k-2\}$ and b_j lies in A_{3j-1}. Since b_j attracts all points in $C_{3j} \cup A_{3j} \cup B_{3j}$, we must have $r \geqslant 3j+1$, otherwise b_j would the last beacon in the greedy placement, contradicting our assumption that $j \leqslant k-2$. Then C_{3j+1} cannot be completely covered by b_j, so the next beacon b_{j+1} must cover C_{3j+1} partially. More precisely, b_{j+1} must lie on the line ℓ_j through v_{3j+1} and b_j (See Fig. 6b). Without loss of generality, we assume that the edge $v_{3j-1}v_{3j}$ is a right reflex edge. Also, note that $r \geqslant 3j+2$, since otherwise placing b_{j+1} completes the greedy placement and thus $k = j+1$, contradicting our assumption that $j \leqslant k-2$.

First assume that $3j-1$ is even, and hence $3j-1 = 2j'$ for some integer j'. It implies that $j' \equiv 1 \pmod 3$ and $r \geqslant 2j'+3$. Similarly to the above argument, the slope of ℓ_j is at least

$$\frac{a_{3j-1}/2}{a_{3j}+w} > \frac{a_{3j-1}/2}{a_{3j}+\epsilon} = \frac{1}{2}\frac{\rho^{2\lceil \frac{j'}{3}\rceil}+j'\epsilon}{\rho^{1+2\lfloor \frac{j'+1}{3}\rfloor}+(j'+1)\epsilon} = \frac{1}{2}\frac{\rho^{2\frac{j'+2}{3}}+j'\epsilon}{\rho^{1+2\frac{j'-1}{3}}+(j'+1)\epsilon}$$

$$> \frac{1}{2}\frac{\rho^{2\frac{j'+2}{3}}}{\rho^{1+2\frac{j'-1}{3}}+(j'+1)\epsilon} = \frac{\rho}{2}\frac{\rho^{\frac{2j'+1}{3}}}{\rho^{\frac{2j'+1}{3}}+(j'+1)\epsilon}$$

$$> \frac{\rho^{\frac{2j'+1}{3}}(1+r\epsilon/2+\epsilon)}{\rho^{\frac{2j'+1}{3}}+(j'+1)\epsilon} > \frac{\rho^{\frac{2j'+1}{3}}(1+r\epsilon/2+\epsilon)}{\rho^{\frac{2j'+1}{3}}(1+j'\epsilon+\epsilon)} > 1.$$

On the other hand, the slope of any line through v_{3j+1} and any point in B_{3j+2} is at most

$$\frac{a_{3j+1}+\epsilon}{a_{3j+2}/2} < 2\frac{\rho^{2\lceil\frac{j'+1}{3}\rceil}+(j'+2)\epsilon}{\rho^{1+2\lfloor\frac{j'+2}{3}\rfloor}+(j'+1)\epsilon} = 2\frac{\rho^{2\frac{j'+2}{3}}+(j'+2)\epsilon}{\rho^{1+2\frac{j'+2}{3}}+(j'+1)\epsilon}$$

$$< 2\frac{\rho^{2\frac{j'+2}{3}}+(j'+2)\epsilon}{\rho^{1+2\frac{j'+2}{3}}} = \frac{2+2(j'+2)\epsilon/\rho^{2\frac{j'+2}{3}}}{\rho} < \frac{2+(r+2)\epsilon}{\rho} < 1.$$

This implies that the next beacon b_{j+1} should also be placed in A_{3j+2}.

We now handle the remaining case, where $3j - 1 = 2j' + 1$ for some integer j'. It implies that $j' \equiv 2 \pmod 3$ and $r \geqslant 2j' + 4$. The slope of ℓ_j is at least

$$
\frac{a_{3j-1}/2}{a_{3j} + w} > \frac{a_{3j-1}/2}{a_{3j} + \epsilon} = \frac{1}{2} \frac{\rho^{1 + 2\lfloor \frac{j'+1}{3} \rfloor} + j'\epsilon}{\rho^{2\lceil \frac{j'+1}{3} \rceil} + (j'+2)\epsilon} = \frac{1}{2} \frac{\rho^{1 + 2\frac{j'+1}{3}} + j'\epsilon}{\rho^{2\frac{j'+1}{3}} + (j'+2)\epsilon}
$$

$$
> \frac{1}{2} \frac{\rho^{1 + 2\frac{j'+1}{3}}}{\rho^{2\frac{j'+1}{3}} + (j'+2)\epsilon} = \frac{\rho}{2} \frac{\rho^{2\frac{j'+1}{3}}}{\rho^{2\frac{j'+1}{3}} + (j'+2)\epsilon}
$$

$$
> \frac{\rho^{2\frac{j'+1}{3}}(1 + r\epsilon/2 + \epsilon)}{\rho^{2\frac{j'+1}{3}} + (j'+2)\epsilon} > \frac{\rho^{2\frac{j'+1}{3}}(1 + r\epsilon/2 + \epsilon)}{\rho^{2\frac{j'+1}{3}}(1 + j'\epsilon + 2\epsilon)} > 1.
$$

On the other hand, the slope of any line through v_{3j+1} and any point in B_{3j+2} is at most

$$
\frac{a_{3j+1} + \epsilon}{a_{3j+2}/2} < 2 \frac{\rho^{1 + 2\lfloor \frac{j'+2}{3} \rfloor} + (j'+2)\epsilon}{\rho^{2\lceil \frac{j'+2}{3} \rceil} + (j'+2)\epsilon} = 2 \frac{\rho^{1 + 2\frac{j'+1}{3}} + (j'+2)\epsilon}{\rho^{2 + 2\frac{j'+1}{3}} + (j'+2)\epsilon}
$$

$$
< 2 \frac{\rho^{1 + 2\frac{j'+1}{3}} + (j'+2)\epsilon}{\rho^{2 + 2\frac{j'+1}{3}}} = \frac{2 + 2(j'+2)\epsilon/\rho^{1 + 2\frac{j'+1}{3}}}{\rho} < \frac{2 + (r+2)\epsilon}{\rho} < 1
$$

Again, it implies that the next beacon b_{j+1} must be placed in A_{3j+2}.   □

It follows from Lemma 3 that the first $k - 1$ beacons in an optimal greedy placement cover the whole spiral until the block B_{3k-3}, and thus $r \geqslant 3k - 2$. In other words:

Lemma 4. *The spiral P_r defined above cannot be guarded by less than $\lceil \frac{r}{3} \rceil = \lfloor \frac{n}{6} \rfloor$ beacons, where n denotes the number of vertices of P_r.*

4.2 Monotone Polygons

Our last result is to show that in the worst case, monotone rectilinear polygons require fewer beacons than simple rectilinear polygons. It matches the lower bound by Biro [1].

Theorem 3. *For any rectilinear monotone polygon P with n vertices, r of which are reflex, $\lfloor \frac{n+4}{8} \rfloor = \lfloor \frac{r}{4} \rfloor + 1$ beacons are sufficient to guard P, and sometimes necessary.*

Proof. Without loss of generality, we assume that P is x-monotone. Thus, P has no vertical reflex edge by Observation 1. Our proof is by induction on the number r of reflex vertices. If P has at most one reflex edge e, then we observe that any point on e is contained in the kernel $\mathcal{K}(P)$ by Theorem 1. Thus, one beacon is sufficient to guard P.

Now, assume that P has at least two reflex edges. This implies that $r \geqslant 4$ since P is x-monotone. Let $v_1, v_2, \ldots, v_k$ be the right endpoints of the reflex edges sorted from left to right. Let e_1 and e_2 be the reflex edges that are incident to v_1 and v_2, respectively. Let c be the vertical cut at v_2. We partition P into P_c^+ and P_c^- by c. Then the left side subpolygon P_c^- has at most one reflex edge e_1, and thus can be guarded by a single beacon placed at any point on e_1. The right side subpolygon P_c^+ has $r(P_c^+) = r - 4$ reflex vertices. Thus, by induction, at most $\lfloor \frac{r-4}{4} \rfloor + 1$ beacons can guard P_c^+. The total number of beacons placed in P is at most

$$1 + \left\lfloor \frac{r-4}{4} \right\rfloor + 1 = \left\lfloor \frac{r}{4} \right\rfloor + 1,$$

as desired.

Finally, observe that cutting by c always makes a new convex edge in P_c^- and P_c^+ since there is no vertical reflex edge in P. This implies that separately guarding P_c^- and P_c^+ is sufficient to guard the whole P by Observation 2.   □

5   Concluding Remarks

In this paper, we gave tight bounds for beacon-based coverage in a simple rectilinear polygon, and in a monotone rectilinear polygon, which settles two open problems given by Biro et al. [2]. Furthermore, we presented an optimal linear time algorithm for computing the beacon-based kernel of a simple rectilinear polygon P. The problem of computing in subquadratic time the *inverse kernel* of P, which is defined as a set of points in P that are attracted to all the points in P, remains open.

Acknowledgments. We thank the anonymous referees for their helpful comments.

References

1. Biro, M.: Beacon-based routing and guarding. Dissertation, Stony Brook University (2013)
2. Biro, M., Gao, J., Iwerks, J., Kostitsyna, I., Mitchell, J.S.B.: Combinatorics of beacon-based routing and coverage. In: Proceedings of the 25th Canadian Conference on Computational Geometry (CCCG) (2013)
3. Biro, M., Iwerks, J., Kostitsyna, I., Mitchell, J.S.B.: Beacon-based algorithms for geometric routing. In: Dehne, F., Solis-Oba, R., Sack, J.-R. (eds.) WADS 2013. LNCS, vol. 8037, pp. 158–169. Springer, Heidelberg (2013)
4. Chavátal, V.: A combinatorial theorem in plane geometry. J. Combinat. Theory Series B **18**, 39–41 (1975)
5. Győri, E.: A short proof of the rectilinear art gallery theorem. SIAM J. Algebraic Discrete Methods **7**(3), 452–454 (1986)
6. Győri, E., Hoffmann, F., Kriegel, K., Shermer, T.: Generalized guarding, partitioning for rectilinear polygons. Comput. Geom. Theory Appl. **6**(1), 21–44 (1996)
7. Kahn, J., Klawe, M., Kleitman, D.: Traditional galleries require fewer watchmen. SIAM J. Algebraic Discrete Methods **4**(2), 194–206 (1983)

8. Kouhestani, B., Rappaport, D., Salmoaa, K.: Routing in a polygonal terrain with the shortest beacon watchtower. In: Proceedings of the 26th Canadian Conference on Computational Geometry (CCCG) (2014)

9. Michael, T.S., Pinciu, V.: Art gallery theorems for guarded guards. Comput. Geom. Theory Appl. **26**, 247–258 (2003)

10. O'Rourke, J.: An alternative proof of the rectilinear art gallery theorem. J. Geom. **21**, 118–130 (1983)

11. O'Rourke, J.: Art Gallery Theorems and Algorithms. International Series of Monographs on Computer Sciences. Oxford University Press, New York (1987)

12. Shermer, T.: Recent results in art galleries. IEEE Proc. **90**(9), 1384–1399 (1992)

13. Shermer, T.C.: A combinatorial bound for beacon-based routing in orthogonal polygons. In: Proceedings of CCCG 2015, pp. 213–219 (2015)

14. Urrutia, J.: Art gallery and illumination problems (chap. 22). In: Sack, J.-R., Urrutia, J. (eds.) Handbook of Computational Geometry, pp. 973–1027. North-Holland (2000)

On Mobile Agent Verifiable Problems

Evangelos Bampas$^{(\boxtimes)}$ and David Ilcinkas

CNRS and University of Bordeaux, LaBRI, UMR 5800, F-33400 Talence, France
{evangelos.bampas,david.ilcinkas}@labri.fr

Abstract. We consider decision problems that are solved in a distributed fashion by synchronous mobile agents operating in an unknown, anonymous network. Each agent has a unique identifier and an input string and they have to decide collectively a property which may involve their input strings, the graph on which they are operating, and their particular starting positions. Building on recent work by Fraigniaud and Pelc [LATIN 2012, LNCS 7256, pp. 362–374], we introduce several natural new computability classes allowing for a finer classification of problems below co-MAV or MAV, the latter being the class of problems that are verifiable when the agents are provided with an appropriate certificate. We provide inclusion and separation results among all these classes. We also determine their closure properties with respect to set-theoretic operations. Our main technical tool, which is of independent interest, is a new meta-protocol that enables the execution of a possibly infinite number of mobile agent protocols essentially in parallel, similarly to the well-known dovetailing technique from classical computability theory.

1 Introduction

1.1 Context and Motivation

The last few decades have seen a surge of research interest in the direction of studying computability- and complexity-theoretic aspects for various models of distributed computing. Significant examples of this trend include the investigation of unreliable failure detectors [5,6], as well as wait-free hierarchies [14]. A more recent line of work studies the impact of randomization and nondeterminism in what concerns the computational capabilities of the $\mathcal{LOCAL}$ model [9,12], as well as the impact of identifiers in the same model [10,11]. A different approach considers the characterization of problems that can be solved under various notions of termination detection or various types of knowledge about the network in message-passing systems [1–4,17]. Finally, a recent work focuses on the computational power of teams of mobile agents [13]. Our work lies in this latter direction.

This work was partially funded by the ANR projects DISPLEXITY (ANR-11-BS02-014) and MACARON (ANR-13-JS02-002). This study has been carried out in the frame of the "Investments for the future" Programme IdEx Bordeaux – CPU (ANR-10-IDEX-03-02).

© Springer-Verlag Berlin Heidelberg 2016
E. Kranakis et al. (Eds.): LATIN 2016, LNCS 9644, pp. 123–137, 2016.
DOI: 10.1007/978-3-662-49529-2_10

The mobile agent paradigm has been proposed since the 90's as a concept that facilitates several fundamental networking tasks including, among others, fault tolerance, network management, and data acquisition [15], and has been of significant interest to the distributed computing community (see, e.g., the recent surveys [7,16]). As such, it is highly pertinent to develop a computability theory for mobile agents, that classifies different problems according to their degree of (non-)computability, insofar as we are interested in really understanding the computational capabilities of groups of mobile agents.

In this paper, we consider a distributed system in which computation is performed by one or more deterministic mobile agents, operating in an unknown, anonymous network. Each agent has a unique identifier and is provided with an input string, and they have to collectively decide a property which may involve their input strings, the graph on which they are operating, and their particular starting positions. One may argue about the usefulness of developing a theory specifically for mobile agent decision problems. We believe that, apart from its inherent theoretical interest, such a study is bound to yield intermediate results, tools, intuitions, and techniques that will prove useful when one moves on to consider from a computability/complexity point of view other, perhaps more traditional, mobile agent problems, such as exploration, rendezvous, pattern formation, etc. One such tool is the protocol that we develop in this paper, which enables the interleaving of the executions of a possibly infinite number of mobile agent protocols.

1.2   Related Work

In [13], Fraigniaud and Pelc introduced two natural computability classes, MAD and MAV, as well as their counterparts co-MAD and co-MAV. The class MAD, for "Mobile Agent Decidable", is the class of all mobile agent decision problems which can be *decided*, i.e., for which there exists a mobile agent protocol such that all agents accept in a "yes" instance, while at least one agent rejects in a "no" instance. On the other hand, the class MAV, for "Mobile Agent Verifiable", is the class of all mobile agent decision problems which can be *verified*. More precisely, in a "yes" instance, there exists a certificate such that if each agent receives its dedicated piece of it, then all agents accept, whereas in a "no" instance, for every possible certificate, at least one agent rejects. Certificates are for example useful in applications in which repeated verifications of some property are required. Fraigniaud and Pelc proved in [13] that MAD is strictly included in MAV, and they exhibited a problem which is complete for MAV under an appropriate notion of oracle reduction.

In [8], Das et al. focus on the complexity of distributed verification, rather than on its computability. In fact, their model differs in several aspects. First of all, the networks in which the mobile agents operate are not anonymous, but each node has a unique identifier. This greatly facilitates symmetry breaking, a central issue in anonymous networks. On the other hand though, the memory of the mobile agents is limited. Indeed, in [8], the authors study the minimal amount of memory needed by the mobile agents to distributedly verify some

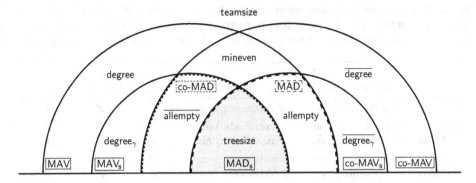

Fig. 1. Containments between classes below MAV and co-MAV with corresponding illustrative problems. Class and problem definitions are summarized in Tables 1 and 2, respectively.

classes of graph properties. Again, the studied properties are different from the ones studied here and in [13], since they do not depend on the mobile agents or their starting positions. However, they may depend on labels that nodes can possess in addition to their unique identifiers.

1.3 Our Contributions

We introduce several new mobile agent computability classes which play a key role in our endeavor for a finer classification of problems below MAV and co-MAV. The classes MAD$_s$ and MAV$_s$ are strict versions of MAD and MAV, respectively, in which unanimity is required in both "yes" and "no" instances. Furthermore, we consider the class co-MAV' (and its counterpart MAV') of mobile agent decision problems that admit a certificate for "no" instances, while retaining the system-wide acceptance mechanism of MAV.

We perform a thorough investigation of the relationships between the newly introduced and pre-existing classes. As a result, we obtain a complete Venn diagram (Fig. 1) which illustrates the tight interconnections between them. We take care to place natural decision problems (in the mobile agent context) in each of the considered classes. Among other results, we obtain a couple of fundamental, previously unknown, inclusions which concern pre-existing classes: MAD $\subseteq$ co-MAV and co-MAD $\subseteq$ MAV.

We complement our results with a complete study of the closure properties of these classes under the standard set-theoretic operations of union, intersection, and complement. The various class definitions together with the corresponding closure properties are summarized in Table 1.

The main technical tool that we develop and use in the paper is a new metaprotocol that enables the execution of a possibly infinite number of mobile agent protocols essentially in parallel. This can be seen as a mobile agent computing analogue of the well-known dovetailing technique from classical recursion theory.

Proofs are omitted due to lack of space.

Table 1. Overview of mobile agent decidability and verifiability classes and their closure properties. The notation yes (resp. no) means that all agents accept (resp. reject). Similarly, ȳes (resp. n̄o) means that at least one agent accepts (resp. rejects).

| | Definition | | Closure Properties | | |
	"yes" instances	"no" instances	Union	Intersec.	Compl.
MAD$_s$	(∀ certificate:) yes	(∀ certificate:) no	✓	✓	✓
MAD	(∀ certificate:) yes	(∀ certificate:) n̂o	✗	✓	✗
co-MAD	(∀ certificate:) ŷes	(∀ certificate:) no	✓	✗	✗
MAV$_s$	∃ certificate: yes	∀ certificate: no	✓	✓	✗
co-MAV$_s$	∀ certificate: yes	∃ certificate: no	✓	✓	✗
MAV	∃ certificate: yes	∀ certificate: n̂o	✗	✓	✗
co-MAV	∀ certificate: ŷes	∃ certificate: no	✓	✗	✗
MAV'	∃ certificate: ŷes	∀ certificate: no	✓	✓	✗
co-MAV'	∀ certificate: yes	∃ certificate: n̂o	✓	✓	✗

2 Preliminaries

The graphs in which the mobile agents operate are undirected, connected, and anonymous. The edges incident to each node v (ports) are assigned distinct local port numbers (also called labels) from $\{1, \ldots, d_v\}$, where d_v is the degree of node v. The port numbers assigned to the same edge at its two endpoints do not have to be in agreement.

We conventionally fix a binary alphabet $\Sigma = \{0, 1\}$. In view of the natural bijection between binary strings and $\mathbb{N}$ which maps a string to its rank in the quasi-lexicographic order of strings (shorter strings precede longer strings, the rank of the empty string ε being 0), we will occasionally treat strings and natural numbers interchangeably. If x and y are strings, then $\langle x, y \rangle$ stands for any standard encoding as a string of the pair of strings (x, y).

If $\boldsymbol{x}$ is a list, then $|\boldsymbol{x}|$ is the length of $\boldsymbol{x}$ and x_i is the i-th element of $\boldsymbol{x}$. If f is a function that can be applied to the elements of $\boldsymbol{x}$, then we will use the notation $f(\boldsymbol{x}) = (f(x_1), \ldots, f(x_{|\boldsymbol{x}|}))$. In the same spirit, if $\boldsymbol{x}$ and $\boldsymbol{y}$ are equal-length lists of strings, then $\langle \boldsymbol{x}, \boldsymbol{y} \rangle$ stands for the list $(\langle x_1, y_1 \rangle, \ldots, \langle x_{|\boldsymbol{x}|} y_{|\boldsymbol{y}|} \rangle)$.

We denote by Σ_1^0 the set of recursively enumerable (or Turing-acceptable) decision problems, $\Pi_1^0 = \text{co-}\Sigma_1^0$, and $\Delta_1^0 = \Sigma_1^0 \cap \Pi_1^0$. Δ_1^0 is exactly the set of Turing-decidable problems.

2.1 Mobile Agent Computations

A *mobile agent protocol* is modeled as a deterministic Turing machine. *Mobile agents* are modeled as instances of a mobile agent protocol (i.e., copies of the corresponding deterministic Turing machine) which move in an undirected, connected, anonymous graph with port labels. Each mobile agent is provided initially with two input strings: its ID, denoted by id, and its input, denoted by x.

By assumption, in any particular execution of the protocol, the ID of each agent is unique. The execution of a group of mobile agents on a graph G proceeds in synchronous steps. At the beginning of each step, each agent is provided with an additional input string, which contains the following information: (i) the degree of the current node u, (ii) the port label at u through which the agent arrived at u (or ε if the agent is in its first step or did not move in the previous step), and (iii) the configuration of all other agents which are currently on u. Then, each agent performs a local computation and eventually halts by accepting or rejecting, or it moves through one of the ports of u, or remains at the same node. We assume that all local computations take the same time and that edge traversals are instantaneous. Therefore, the execution is completely synchronous.

Let M be a mobile agent protocol, G be a graph, **id** be a list of distinct IDs, s be a list of nodes of G, and x be a list of strings such that $|\textbf{id}| = |s| = |x| = k > 0$. We denote by $M(\textbf{id}, G, s, x)$ the execution of k copies of M, the i-th copy starting on node s_i and receiving as inputs the ID id_i and the string x_i. The tuple $(\textbf{id}, G, s, x)$ is called the *implicit input*. Similarly, we denote by $M(\text{id}, x; \textbf{id}, G, s, x)$ the personal view of the execution of M on the implicit input, as experienced by the agent with ID id and input x. We distinguish between the *explicit* input (id, x), which is provided to the agent at the beginning of the execution, and the implicit input, which may or may not be discovered by the agent in the course of the execution.

Given an implicit input, we write $M(\text{id}, x; \textbf{id}, G, s, x) = \textsf{yes}$ (resp. no) if the agent with explicit input (id, x) accepts (resp. rejects) during $M(\textbf{id}, G, s, x)$. Furthermore, we write $M(\textbf{id}, G, s, x) \mapsto \textsf{yes}$ (resp. no), if $\forall i\ M(\text{id}_i, x_i; \textbf{id}, G, s, x) = \textsf{yes}$ (resp. no), and $M(\textbf{id}, G, s, x) \mapsto \widehat{\textsf{yes}}$ (resp. $\widehat{\textsf{no}}$), if all agents halt and for some $i\ M(\text{id}_i, x_i; \textbf{id}, G, s, x) = \textsf{yes}$ (resp. no).

2.2 Mobile Agent Decision Problems

Definition 1 [13]. *A mobile agent decision problem on anonymous graphs is a set Π of instances (G, s, x), where G is a graph, s is a non-empty list of nodes of G, and x is a list of strings with $|x| = |s|$, which satisfies the following closure property: For every G and for every automorphism α of G that preserves port numbers, $(G, s, x) \in \Pi$ if and only if $(G, \alpha(s), x) \in \Pi$.*[1]

We will refer to instances which belong to a problem Π as "yes" instances of Π. Instances that do not belong to Π will be called "no" instances of Π. The *complement* $\overline{\Pi}$ of a mobile agent decision problem Π is the problem $\overline{\Pi} = \{(G, s, x) : |s| = |x| \text{ and } (G, s, x) \notin \Pi\}$.[2] Some examples of decision problems are shown in Table 2.

---

[1] Note that this closure property is syntactically different from the one used in [13] due to notational differences, but the two are equivalent.

[2] It is easy to check that if Π is a decision problem, then $\overline{\Pi}$ also satisfies the closure property of Definition 1. Therefore, $\overline{\Pi}$ is also a decision problem.

Table 2. Definitions of some mobile agent decision problems that we use in the rest of the paper.

alone	$= \{(G, \boldsymbol{s}, \boldsymbol{x}) : |\boldsymbol{s}| = 1\}$
allempty	$= \{(G, \boldsymbol{s}, \boldsymbol{x}) : \forall i \; x_i = \varepsilon\}$
consensus	$= \{(G, \boldsymbol{s}, \boldsymbol{x}) : \forall i, j \; x_i = x_j\}$
degree	$= \{(G, \boldsymbol{s}, \boldsymbol{x}) : \forall i \; \exists v \; d_v = x_i\}$
degree$_\gamma$	$= \{(G, \boldsymbol{s}, \boldsymbol{x}) : G$ contains a node of degree $\gamma\}$ (for $\gamma \geq 1$)
mineven	$= \{(G, \boldsymbol{s}, \boldsymbol{x}) : \min_i x_i$ is even$\}$
path	$= \{(G, \boldsymbol{s}, \boldsymbol{x}) : G$ is a path$\}$
teamsize	$= \{(G, \boldsymbol{s}, \boldsymbol{x}) : \forall i \; x_i = |\boldsymbol{s}|\}$
treesize	$= \{(G, \boldsymbol{s}, \boldsymbol{x}) : \forall i \; G$ is a tree of size $x_i\}$

Definition 2 [13]. *A decision problem Π is mobile agent decidable if there exists a protocol M such that for all instances $(G, \boldsymbol{s}, \boldsymbol{x})$: if $(G, \boldsymbol{s}, \boldsymbol{x}) \in \Pi$ then $\forall \mathsf{id} \; M(\mathsf{id}, G, \boldsymbol{s}, \boldsymbol{x}) \mapsto$ yes, whereas if $(G, \boldsymbol{s}, \boldsymbol{x}) \notin \Pi$ then $\forall \mathsf{id} \; M(\mathsf{id}, G, \boldsymbol{s}, \boldsymbol{x}) \mapsto \widehat{\text{no}}$. The class of all decidable problems is denoted by* MAD.

Definition 3 [13]. *A decision problem Π is mobile agent verifiable if there exists a protocol M such that for all instances $(G, \boldsymbol{s}, \boldsymbol{x})$: If $(G, \boldsymbol{s}, \boldsymbol{x}) \in \Pi$ then $\exists \boldsymbol{y} \; \forall \mathsf{id} \; M(\mathsf{id}, G, \boldsymbol{s}, \langle \boldsymbol{x}, \boldsymbol{y}\rangle) \mapsto$ yes, whereas if $(G, \boldsymbol{s}, \boldsymbol{x}) \notin \Pi$ then $\forall \boldsymbol{y} \; \forall \mathsf{id} \; M(\mathsf{id}, G, \boldsymbol{s}, \langle \boldsymbol{x}, \boldsymbol{y}\rangle) \mapsto \widehat{\text{no}}$. The class of all verifiable problems is denoted by* MAV.

When there is no room for confusion, we will use the term *certificate* both for the string y provided to an agent and for the collection of certificates $\boldsymbol{y}$ provided to the group of agents. If we need to distinguish between the two, we will refer to $\boldsymbol{y}$ as a *certificate vector*. Finally, if X is a class of mobile agent decision problems, then co-$\mathsf{X} = \{\Pi : \overline{\Pi} \in \mathsf{X}\}$.

Remark 1. Note that in [13], only decidable (in the classical sense) mobile agent decision problems were taken into consideration. As a result, it was by definition the case that MAD and MAV were both subsets of Δ^0_1. For the purposes of this work, we do not impose this constraint.

3   Mobile Agent Decidability Classes

A problem Π is in co-MAD if and only if it can be decided by a mobile agent protocol in a sense which is dual to that of Definition 2: If the instance is in Π, then at least one agent must accept, whereas if the instance is not in Π, then all agents must reject. We will consider one more such variant in the form of the "strict" class MAD$_\mathsf{s}$. A problem belongs to this class if it can be solved in such a way that every agent always outputs the correct answer.

Definition 4. *A decision problem Π is in* MAD$_\mathsf{s}$ *if and only if there exists a protocol M such that for all instances $(G, \boldsymbol{s}, \boldsymbol{x})$: if $(G, \boldsymbol{s}, \boldsymbol{x}) \in \Pi$ then $\forall \mathsf{id} \; M(\mathsf{id}, G, \boldsymbol{s}, \boldsymbol{x}) \mapsto$ yes, whereas if $(G, \boldsymbol{s}, \boldsymbol{x}) \notin \Pi$ then $\forall \mathsf{id} \; M(\mathsf{id}, G, \boldsymbol{s}, \boldsymbol{x}) \mapsto$ no.*

By definition, MAD$_s$ is a subset of both MAD and co-MAD and it is easy to check that MAD$_s$ = co-MAD$_s$. Moreover, all of these classes are subsets of Δ_1^0, since a centralized algorithm, provided with an encoding of the graph and the starting positions, inputs, and IDs of the agents, can simulate the corresponding mobile agent protocol and decide appropriately. As mentioned in [13], path is an example of a mobile agent decision problem which is in $\Delta_1^0 \setminus$ MAD, since, intuitively, an agent cannot distinguish a long path from a cycle. In fact, this observation yields path $\in \Delta_1^0 \setminus ($MAD $\cup$ co-MAD$)$.

A nontrivial problem in MAD$_s$ is treesize. The problem was already shown to be in MAD in [13]. For the stronger property that treesize $\in$ MAD$_s$, we need a modification of the protocol given in [13].

Proposition 1. treesize $\in$ MAD$_s$.

We now show that MAD and co-MAD are strict supersets of MAD$_s$.

Proposition 2. allempty $\in$ MAD $\setminus$ MAD$_s$ *and* $\overline{\text{allempty}} \in$ co-MAD $\setminus$ MAD$_s$.

As we mentioned, MAD$_s$ is included in both MAD and co-MAD. In fact, MAD$_s$ = MAD $\cap$ co-MAD. We state this as a theorem without proof, since it can be obtained as a corollary of Theorems 2 and 3, which we will prove in Sect. 5.

Theorem 1. MAD$_s$ = MAD $\cap$ co-MAD.

By Theorem 1, if allempty was included in co-MAD, we would obtain allempty $\in$ MAD$_s$, which we know to be false. Thus, allempty $\notin$ co-MAD and we obtain a separation between MAD and co-MAD. Symmetrically, $\overline{\text{allempty}} \in$ co-MAD $\setminus$ MAD.

4   Interleaving Multiple Mobile Agent Protocols

It is important to have a tool that enables the execution of several mobile agent protocols on the same instance, and that also permits the mobile agents to make decisions based on the outcomes of these executions. For example, if one were to give a direct proof of Theorem 1 above, one would need a way for the agents to coordinate in order to execute both the MAD and the co-MAD protocol for a particular problem, and then, based on the outcome of these executions, to give a unanimous correct answer (in the spirit of MAD$_s$).

In classical computing, the well known *dovetailing* technique achieves this interleaving of different computations. Classical dovetailing proceeds in phases: in phase T, the first T steps of the first T programs are executed. At this point, an auxiliary function is executed, which decides, based on these executions, whether to accept, reject, or continue with the next phase. Correspondingly, the mobile agent meta-protocol which we propose in this section, proceeds in phases: in phase T, the agents execute the first T steps of the first T mobile agent protocols and then decide whether to accept, reject, or proceed to the next phase. In the mobile agent case, each agent decides independently by locally executing a function, which is given as a parameter to the meta-protocol. We call this function a *local decider*.

Still, it may happen that one or more agents halt as a result of executing the local decider, while others decide to continue. In such a case, the execution of the protocols in the next phase could be corrupted because the halted agents no longer follow the protocol. However, these halted agents can now be regarded as fixed tokens and the meta-protocol uses them in order to create a map of the graph. In fact, this is done in such a way as to ensure that all non-halted agents obtain not only the map of the graph but actually full knowledge of the implicit input. Based on this knowledge, each agent decides irrevocably whether to accept or reject by means of a second function which is given as a parameter to the meta-protocol, and which we call a *global decider*.

4.1 Ingredients of the Meta-Protocol

We propose a generic meta-protocol $\mathcal{P}_{\mathcal{N},f,g}$, which is parameterized by $\mathcal{N}, f, g$. The set $\mathcal{N}$ is a, possibly infinite, recursively enumerable set of mobile agent protocols. Let N_i, $i \geq 0$, denote the i-th protocol in such an enumeration. The functions f and g are computable functions which represent local computations with the following specifications:

Global decider: The function f maps pairs consisting of an explicit and an implicit input, i.e., tuples of the form $(\mathsf{id}, x; \mathbf{id}, G, \boldsymbol{s}, \boldsymbol{x})$, to the set $\{\mathbf{accept}, \mathbf{reject}\}$. In this case, we say that f is a *global decider*. When an agent executes f, it halts by accepting or rejecting according to the outcome of f.

Local decider: The function g takes as input an explicit input (id, x) and a list $(H_1, \ldots, H_\sigma)$ of arbitrary length σ, where each H_j is the history of the partial execution of $N_j(\mathsf{id}, x; \mathbf{id}, G, \boldsymbol{s}, \boldsymbol{x})$ for a certain number of steps and $(\mathbf{id}, G, \boldsymbol{s}, \boldsymbol{x})$ is an implicit input common for all histories $H_1, \ldots, H_\sigma$. The outcome of g is one of $\{\mathbf{accept}, \mathbf{reject}, \mathbf{continue}\}$. When an agent executes g, it halts in the corresponding state if the outcome is **accept** or **reject**, otherwise it continues without halting.

If for every implicit input $(\mathbf{id}, G, \boldsymbol{s}, \boldsymbol{x})$ and for every T_0, there exists a $T \geq T_0$ and some i such that the local computation $g(\mathsf{id}_i, x_i, H_1, \ldots, H_{\min(T,|\mathcal{N}|)})$ returns either **accept** or **reject**, where each H_j is an encoding of the execution of $N_j(\mathsf{id}_i, x_i; \mathbf{id}, G, \boldsymbol{s}, \boldsymbol{x})$ for T steps, then we say that g is a *local decider for $\mathcal{N}$*.

The meta-protocol uses the following procedures CREATE-MAP and RDV:

Procedure Create-Map(R): An agent executes this procedure only when it is on a node which contains at least one halted (or idle) agent. Starting from this node, and treating the halted agent as a fixed mark, it attempts to create a map of the graph assuming that the graph contains at most R nodes. More precisely, the agent first creates a map consisting in a single node corresponding to the marked node r, with d_r pending edges with port numbers from 1 to d_r. Then, while there remain some pending edges and there are at most R explored nodes, the agent explores some arbitrary pending edge as follows. The agent goes to the known extremity u of the pending edge by using the map and traverses it. It then determines whether its current position v corresponds to a node of its map, as follows: For every node w in its map, it computes a path in the map going from w to r and follows the corresponding sequence of port numbers in the unknown graph, starting from v. If it leads to the marked node, then $v = w$ and the agent updates its map by linking the pending edges of u and w with the appropriate port numbers. Otherwise, it retraces its steps to come back to v and tests a next node w. If all nodes turn out to be different from v, then the agent goes back to the marked node through u, and updates its map by adding a new node corresponding to v, linked to u, and with the appropriate number of new pending edges. At the end of the procedure, the agent either has a complete map of the graph, or knows that the graph has more than R nodes. This procedure takes at most $4R^4$ steps.

Procedure Rdv(R, id): This procedure guarantees that a group of k agents which (a) know the same upper bound R on the number of nodes in the graph, (b) have distinct id's $\{\text{id}_1, \ldots, \text{id}_k\}$, and (c) start executing RDV(R, id_i) at the same time from different nodes s_i, will all meet each other after finite time. Moreover, each agent knows when it has met all other agents executing RDV, even without initial knowledge of k.

The RDV procedure uses as a subroutine the following EXPLORE-BALL procedure: An agent executing EXPLORE-BALL(R) attempts to explore the ball of radius R around its starting node s_i, assuming an upper bound of R on the maximum degree of the graph. This is achieved by having the agent try every sequence of length R of port numbers from the set $\{1, \ldots, R\}$, retracing its steps backward after each sequence to return to s_i. If a particular sequence instructs the agent to follow a port number that does not exist at the current node (i.e., the port number is larger than the degree of the node), then the agent aborts that sequence and returns to s_i. Attempting all possible sequences takes at most $B(R) = 2R \cdot R^R$ steps. If an agent finishes earlier, it waits on s_i until $B(R)$ steps are completed. Therefore, a team of agents that start executing EXPLORE-BALL(R) at the same time from different nodes are synchronized and back at their starting positions after $B(R)$ steps.

Now, for each bit of id_i, the RDV procedure executes the following: If the bit is 0, the agent waits at s_i for $B(R)$ steps and then executes EXPLORE-BALL(R), whereas if the bit is 1, the agent first executes EXPLORE-BALL(R) and then waits on its starting position for $B(R)$ steps. After it exhausts the bits of id_i, the agent executes twice EXPLORE-BALL(R). This guarantees that, if the number of nodes is at most R, then after $2 \cdot (|\text{id}_i| + 1) \cdot B(R)$ steps, each agent i is located at s_i and has met all other agents executing RDV. Note that after every integer multiple of $B(R)$ steps, each agent is located at its initial node s_i.

4.2   Description of the Meta-Protocol

The meta-protocol $\mathcal{P}_{\mathcal{N}, f, g}$ works in phases, which correspond to increasing values of a presumed upper bound T on the number of nodes in the graph, the length

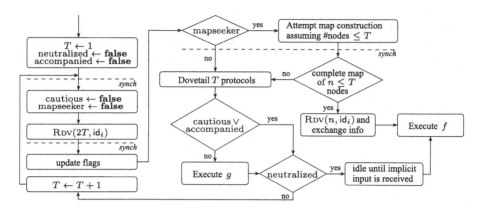

Fig. 2. High-level flowchart of the meta-protocol of Sect. 4.

of all agent identifiers, and the completion time of protocols $N_1, \ldots, N_T$. We will say that an agent is *idle* if it is waiting indefinitely on its starting node for some other agent to provide it with the knowledge of the full implicit input. We will say that an agent is *participating* if it is not halted and not idle. Note that an agent may halt only as a result of executing one of the decider functions f and g. In each phase T, the agents perform the following actions (see also Fig. 2):

Search for nearby starting positions and set flags. Each participating agent i first executes $\mathrm{RDV}(2T, \mathsf{id}_i)$ for at most $2(T+1)B(2T)$ steps. By design of RDV, this guarantees that agent i will explore its $2T$-neighborhood at least once and, in particular, if $T \geq |\mathsf{id}_i|$, then for each other participating agent, agent i will explore its $2T$-neighborhood at least once with that agent staying on its starting node. If, in the process, the agent meets any agent, then it sets its accompanied flag. It also sets its neutralized flag if the encountered agent is participating and it has a lexicographically larger ID. If the encountered agent is halted or idle, the agent sets its mapseeker flag. Finally, if the agent finds a node with degree larger than $2T$ or if the length of its ID is greater than T, it sets its cautious flag. All agents synchronize at this point.

Mapseeker agents attempt to create a map of the graph. Next, each agent i with the mapseeker flag set moves to a halted or idleagent which it has found previously, while executing RDV in the current phase. Then, it attempts to create a map of the graph by executing $\mathrm{CREATE\text{-}MAP}(T)$ and returns to s_i. Overall, this takes at most $4T^4 + 4T$ steps. Moreover, during the execution of $\mathrm{CREATE\text{-}MAP}$, mapseeker agents collect starting position and input information from allhalted and idle agents that they encounter. Meanwhile, non-mapseeker agents wait for $4T^4 + 4T$ steps. All agents synchronize at this point.

   So far, we have achieved that, if $T \geq n$, where n is the number of nodes in G, then either no agent is a mapseeker having the full map of G, or all participating agents have the mapseeker flag set and they have the full map of G (Lemma 1 below). If all mapseeker agents have the full map of G and $T \geq n$, then each such agent i executes $\text{RDV}(n, \text{id}_i)$, which guarantees that, finally, it is located at s_i and has met all other agents executing RDV. Therefore, after concluding the RDV procedure, each mapseeker executes f with full knowledge of the implicit input (Lemma 2).

Perform dovetailing. At this point, if no agent is a mapseeker having the full map of G, the agents execute each of the protocols $N_1, \ldots, N_{\min(T, |\mathcal{N}|)}$ for at most T steps, and then retrace backward to s_i (agents are synchronized after executing each protocol). If any of these protocols instructs an agent to halt, the agent instead waits until the T-step execution period has finished, and then returns to s_i. If the agent does not have the cautiousor accompanied flags set, it then executes $g(\text{id}, x, H_1, \ldots, H_{\min(T, |\mathcal{N}|)})$, where H_j is the history of the T-step execution of N_j with explicit input (id, x). Since this process takes at most $2T^2$ steps, all agents that do not halt as a result of executing g are synchronized at the end of the current phase. It is guaranteed that the histories fed to the local decider g correspond to correct executions of the corresponding protocols for implicit input (id, G, s, x), even though some of the agents may have halted or become idle in earlier phases (Lemma 3 and Corollary 1).

Neutralized agents become idle. Finally, at the end of the phase, neutralized agents start waiting for the implicit input (i.e., they become idle), and when they receive it (from some mapseeker agent), they execute the global decider f.

Lemma 1. *In each phase, either all or none of the participating agents (i.e., non-halted and non-idle) execute f.*

Lemma 2. *Any agent that executes f has full knowledge of the implicit input (id, G, s, x).*

Lemma 3. *If an agent i executes g during phase T, then no other agent's starting node is at distance $2T$ or less from s_i.*

By Lemma 3, we obtain the following corollary:

Corollary 1. *Any agent i that executes g has histories which correspond to the correct histories of $N_j(\text{id}_i, x_i; \text{id}, G, s, x)$ for T steps $(1 \le j \le \min(T, |\mathcal{N}|))$, even though some of the agents may have halted or become idle in earlier phases.*

In view of Corollary 1, we can show that all agents terminate and, in fact, they all terminate on their respective starting nodes.

Lemma 4. *Let f be a global decider and let g be a local decider for $\mathcal{N}$. Then, each agent halts under the execution $\mathcal{P}_{\mathcal{N}, f, g}(\text{id}, G, s, x)$ by executing either f or g. Moreover, each agent i halts on its starting node s_i.*

4.3   Application of the Meta-Protocol

To summarize, the meta-protocol is a generic tool that enables us to interleave the executions of a possibly infinite set of mobile agent protocols. Eventually, each agent accepts or rejects, based either on the histories of the executions of a number of these protocols (by means of the local decider), or on full knowledge of the implicit input (by means of the global decider).

We use the meta-protocol in order to place a particular problem in one of the mobile agent computability classes of Table 1. A common part of the proofs consists in defining the list of protocols $\mathcal{N}$ and suitable deciders f and g, and in showing that f and g indeed satisfy the global and local decider properties, respectively. This is followed by a part tailored to each particular result, where we use the properties of the meta-protocol (Lemmas 1–4 and Corollary 1) and the particular definitions of f and g, in order to show that agents that execute $\mathcal{P}_{\mathcal{N},f,g}$ always terminate in the desired state. The desired state is indicated by the class in which we wish to place the problem. For example, if we wish to show that a problem is in $\mathsf{MAD_s}$, we will have to show that all agents give the correct answer for all implicit inputs.

5   Mobile Agent Verifiability Classes

Definition 5. *A decision problem Π is in $\mathsf{MAV_s}$ if and only if there exists a protocol M such that for all instances (G, s, x): if $(G, s, x) \in \Pi$ then $\exists y \; \forall \mathsf{id} \; M(\mathsf{id}, G, s, \langle x, y \rangle) \mapsto$ yes, whereas if $(G, s, x) \notin \Pi$ then $\forall y \; \forall \mathsf{id} \; M(\mathsf{id}, G, s, \langle x, y \rangle) \mapsto$ no.*

By definition, $\mathsf{MAV_s} \subseteq \mathsf{MAV}$. Moreover, $\mathsf{MAV} \subseteq \Sigma_1^0$, since a centralized algorithm can simulate the MAV protocol for all possible certificate vectors (by classical dovetailing) and accept if it finds a certificate for which all agents accept. By taking complements, we obtain as well that $\mathsf{co\text{-}MAV_s} \subseteq \mathsf{co\text{-}MAV} \subseteq \Pi_1^0$.

There exist several nontrivial problems in $\mathsf{MAV_s}$ and $\mathsf{co\text{-}MAV_s}$ (Proposition 3). Furthermore, we can show that MAV is a strict superset of $\mathsf{MAV_s}$ and, as a corollary, $\mathsf{co\text{-}MAV}$ is a strict superset of $\mathsf{co\text{-}MAV_s}$ (Proposition 4).

Proposition 3. *For any fixed $\gamma \geq 1$, $\mathsf{degree}_\gamma \in \mathsf{MAV_s}$. Furthermore, $\mathsf{consensus} \in \mathsf{co\text{-}MAV_s}$ and $\mathsf{alone} \in \mathsf{co\text{-}MAV_s}$.*

Proposition 4. $\mathsf{degree} \in \mathsf{MAV} \setminus (\mathsf{MAV_s} \cup \mathsf{co\text{-}MAV})$.

Proposition 4 also separates MAV from $\mathsf{co\text{-}MAV}$. In order to separate Σ_1^0 from MAV and Π_1^0 from $\mathsf{co\text{-}MAV}$, we observe that the teamsize problem, which is clearly in $\Delta_1^0 = \Sigma_1^0 \cap \Pi_1^0$, is neither in MAV nor in $\mathsf{co\text{-}MAV}$.

Proposition 5. $\mathsf{teamsize} \in \Delta_1^0 \setminus (\mathsf{MAV} \cup \mathsf{co\text{-}MAV})$.

Decision problems with "no" certificates. In classical computability, the class $\Pi_1^0 = \mathsf{co\text{-}}\Sigma_1^0$ can be seen as the class of problems that admit a "no" certificate, i.e.: for "no" instances, there exists a certificate that leads to rejection, whereas for "yes" instances, no certificate can lead to rejection. In this respect, while MAV can certainly be considered as the mobile agent analogue of Σ_1^0, $\mathsf{co\text{-}MAV}$ is not quite the analogue of Π_1^0. Problems in $\mathsf{co\text{-}MAV}$ indeed admit a "no" certificate, but the acceptance mechanism is reversed: for "no" instances, there exists a certificate that leads all agents to reject. This motivates us to define and study co-MAV', the class of mobile agent problems that admit a "no" certificate while retaining the MAV acceptance mechanism, as well as its complement MAV'. We give the definition of MAV' below.

Definition 6. *A decision problem Π is in* MAV' *if and only if there exists a protocol M such that for all instances (G, s, x): if $(G, s, x) \in \Pi$ then $\exists y\ \forall \mathrm{id}\ M(\mathrm{id}, G, s, \langle x, y \rangle) \mapsto$ yes, whereas if $(G, s, x) \notin \Pi$ then $\forall y\ \forall \mathrm{id}\ M(\mathrm{id}, G, s, \langle x, y \rangle) \mapsto$ no.*

By definition, it holds that $\mathsf{MAV_s} \subseteq \mathsf{MAV'}$ and $\mathsf{co\text{-}MAV_s} \subseteq \mathsf{co\text{-}MAV'}$. To show $\mathsf{MAV'} = \mathsf{MAV_s}$ (and thus $\mathsf{co\text{-}MAV'} = \mathsf{co\text{-}MAV_s}$), we need to "boost" the MAV' protocol so that the agents answer unanimously even in "yes" instances. We achieve this by supplying an extra certificate, which is interpreted as the number of nodes of the graph. This enables the agents to meet and exchange information in "yes" instances, and therefore reach a unanimous decision. The meta-protocol from Sect. 4 essentially provides "for free" the necessary subroutines for meeting and exchanging information.

Theorem 2. $\mathsf{MAV'} = \mathsf{MAV_s}$ *and* $\mathsf{co\text{-}MAV'} = \mathsf{co\text{-}MAV_s}$.

In view of Theorem 2, it follows that $\mathsf{MAV_s} \subseteq \mathsf{MAV} \cap \mathsf{co\text{-}MAV}$ and $\mathsf{co\text{-}MAV_s} \subseteq \mathsf{MAV} \cap \mathsf{co\text{-}MAV}$. We separate $\mathsf{MAV} \cap \mathsf{co\text{-}MAV}$ from both of these classes with the problem mineven:

Proposition 6. mineven $\in (\mathsf{MAV} \cap \mathsf{co\text{-}MAV}) \setminus (\mathsf{MAV_s} \cup \mathsf{co\text{-}MAV_s})$.

Connections with the decidability classes. We explore the relationships among the decidability classes of Sect. 3 and the classes defined in this section. From the definitions we know that $\mathsf{MAD} \subseteq \mathsf{co\text{-}MAV'}$, therefore, by Theorem 2, $\mathsf{MAD} \subseteq \mathsf{co\text{-}MAV_s}$. Similarly, $\mathsf{co\text{-}MAD} \subseteq \mathsf{MAV_s}$. Therefore, since $\mathsf{MAD_s} \subseteq \mathsf{MAD} \cap \mathsf{co\text{-}MAD}$, we also have that $\mathsf{MAD_s} \subseteq \mathsf{MAV_s} \cap \mathsf{co\text{-}MAV_s}$.

We show in Theorem 3 that, in fact, $\mathsf{MAD_s} = \mathsf{MAV_s} \cap \mathsf{co\text{-}MAV_s}$. Furthermore, from the definitions and Theorem 2, we have $\mathsf{MAD} \subseteq \mathsf{MAV} \cap \mathsf{co\text{-}MAV_s}$ and $\mathsf{co\text{-}MAD} \subseteq \mathsf{MAV_s} \cap \mathsf{co\text{-}MAV}$. We show that these actually hold as equalities in Theorem 4 below. The proof of Theorem 3 (resp. Theorem 4) is based on trying all possible combinations of certificates for the $\mathsf{MAV_s}$ (resp. MAV) and $\mathsf{co\text{-}MAV_s}$ protocols. Here, we use the full power of the meta-protocol of Sect. 4 in order to interleave and synchronize this infinite number of executions.

Theorem 3. $\mathsf{MAD_s} = \mathsf{MAV_s} \cap \mathsf{co\text{-}MAV_s}$.

Theorem 4. $\mathsf{MAD} = \mathsf{MAV} \cap \mathsf{co\text{-}MAV_s}$ *and* $\mathsf{co\text{-}MAD} = \mathsf{MAV_s} \cap \mathsf{co\text{-}MAV}$.

Note that it was shown in [13] that, if we consider decision problems that are decidable or verifiable by a single agent (thus giving rise to the classes $\mathsf{MAD_1}$ and $\mathsf{MAV_1}$), then it holds that $\mathsf{MAD_1} = \mathsf{MAV_1} \cap \mathsf{co\text{-}MAV_1}$. Theorems 3 and 4 can be seen as generalizations of that result to multiagent classes.

Proposition 7. *For any fixed* $\gamma \geq 1$, degree$_\gamma \in$ MAV$_s$ \ co-MAD *and* $\overline{\text{degree}_\gamma} \in$ co-MAV$_s$ \ MAD.

In view of Theorem 4, Proposition 7 yields a separation between MAV$_s$ and co-MAV, as degree$_\gamma \in$ MAV$_s$ \ co-MAV, and a separation between co-MAV$_s$ and MAV, as $\overline{\text{degree}_\gamma} \in$ co-MAV$_s$ \ MAV.

By combining the results of this section with the results of Sect. 3, we obtain a picture of the relationships among the classes below MAV and co-MAV, as illustrated in Fig. 1.

References

1. Boldi, P., Vigna, S.: An effective characterization of computability in anonymous networks. In: Welch, J.L. (ed.) DISC 2001. LNCS, vol. 2180, pp. 33–47. Springer, Heidelberg (2001)
2. Boldi, P., Vigna, S.: Universal dynamic synchronous self-stabilization. Distrib. Comput. **15**(3), 137–153 (2002)
3. Chalopin, J., Godard, E., Métivier, Y.: Local terminations and distributed computability in anonymous networks. In: Taubenfeld, G. (ed.) DISC 2008. LNCS, vol. 5218, pp. 47–62. Springer, Heidelberg (2008)
4. Chalopin, J., Godard, E., Métivier, Y., Tel, G.: About the termination detection in the asynchronous message passing model. In: van Leeuwen, J., Italiano, G.F., van der Hoek, W., Meinel, C., Sack, H., Plášil, F. (eds.) SOFSEM 2007. LNCS, vol. 4362, pp. 200–211. Springer, Heidelberg (2007)
5. Chandra, T.D., Hadzilacos, V., Toueg, S.: The weakest failure detector for solving consensus. J. ACM **43**(4), 685–722 (1996)
6. Chandra, T.D., Toueg, S.: Unreliable failure detectors for reliable distributed systems. J. ACM **43**(2), 225–267 (1996)
7. Das, S.: Mobile agents in distributed computing: network exploration. Bull. Eur. Assoc. Theor. Comput. Sci. EATCS **109**, 54–69 (2013)
8. Das, S., Kutten, S., Lotker, Z.: Distributed verification using mobile agents. In: Frey, D., Raynal, M., Sarkar, S., Shyamasundar, R.K., Sinha, P. (eds.) ICDCN 2013. LNCS, vol. 7730, pp. 330–347. Springer, Heidelberg (2013)
9. Fraigniaud, P., Göös, M., Korman, A., Parter, M., Peleg, D.: Randomized distributed decision. Distrib. Comput. **27**(6), 419–434 (2014)
10. Fraigniaud, P., Göös, M., Korman, A., Suomela, J.: What can be decided locally without identifiers? In: PODC 2013, pp. 157–165. ACM (2013)
11. Fraigniaud, P., Halldórsson, M.M., Korman, A.: On the impact of identifiers on local decision. In: Baldoni, R., Flocchini, P., Binoy, R. (eds.) OPODIS 2012. LNCS, vol. 7702, pp. 224–238. Springer, Heidelberg (2012)
12. Fraigniaud, P., Korman, A., Peleg, D.: Towards a complexity theory for local distributed computing. J. ACM **60**(5), 35 (2013)
13. Fraigniaud, P., Pelc, A.: Decidability classes for mobile agents computing. In: Fernández-Baca, D. (ed.) LATIN 2012. LNCS, vol. 7256, pp. 362–374. Springer, Heidelberg (2012)
14. Herlihy, M.: Wait-free synchronization. ACM Trans. Program. Lang. Syst. **13**(1), 124–149 (1991)

15. Lange, D.B., Oshima, M.: Seven good reasons for mobile agents. Commun. ACM **42**(3), 88–89 (1999)
16. Markou, E.: Identifying hostile nodes in networks using mobile agents. Bull. Eur. Assoc. Theor. Comput. Sci. EATCS **108**, 93–129 (2012)
17. Yamashita, M., Kameda, T.: Computing functions on asynchronous anonymous networks. Math. Syst. Theory **29**(4), 331–356 (1996)

Computing Maximal Layers of Points in $E^{f(n)}$

Indranil Banerjee$^{(\boxtimes)}$ and Dana Richards

Department of Computer Science, George Mason University,
Fairfax, VA 22030, USA
{ibanerje,richards}@cs.gmu.edu

Abstract. In this paper we present a randomized algorithm for computing the collection of maximal layers for a point set in E^k ($k = f(n)$). The input to our algorithm is a point set $P = \{p_1, \ldots, p_n\}$ with $p_i \in E^k$. The proposed algorithm achieves a runtime of $O\left(kn^{2 - \frac{1}{\log k} + \log_k \left(1 + \frac{2}{k+1}\right)} \log n\right)$ when P is a random order and a runtime of $O(k^2 n^{3/2 + (\log_k (k-1))/2} \log n)$ for an arbitrary P. Both bounds hold in expectation. Additionally, the run time is bounded by $O(kn^2)$ in the worst case. This is the first non-trivial algorithm whose run-time remains polynomial whenever $f(n)$ is bounded by some polynomial in n while remaining sub-quadratic in n for constant k (in expectation). The algorithm is implemented using a new data-structure for storing and answering dominance queries over the set of incomparable points.

Keywords: Maximal layers · Random order · Complexity

1 Introduction

The problem of finding the maximal layers of a set $P = \{p_1, \ldots, p_n\}$ of n points[1] in $[0,1]^k$ (where $k = f(n)$) is analogous to the problem of finding the convex layers of P. Given P its first maximal layer is defined to be the set $\mathcal{M}_1$ of points $q \in P$ such that for any other $p \in P$, $p \not\succ q$. Here, $\succ$ is the usual domination order between two points. That is: $p \succ q$ if $p[j] \geq q[j]$ (where $p[j]$ is the jth coordinate of p) for all j. The first maximal set $\mathcal{M}_1$, which we simply refer to as the maximal set of P, has been well studied [4–6]. The lth maximal layer $\mathcal{M}_l$ is recursively defined as the first maximal layer of remainder of P upon removing from P all the elements of layers from 1 to $l-1$. Note that $\mathcal{M}_l$ could be empty. The maximal layers problem is to identify all the non-empty maximal layers of P and report them. We shall denote this problem as MAXLAYERS(P).

Related Work. We only have a tight bound for the runtime of MAXLAYERS(P) when $k \leq 3$, which is $\Theta(n \log n)$ [6,12]. However, we do not have any improved lower bound when $k > 3$. For fixed $k > 3$ best known upper-bound

---

[1] We have restricted the sampling set to $[0,1]^k$ in order to simplify our analysis. The results hold for any arbitrary compact subset of E^k.

© Springer-Verlag Berlin Heidelberg 2016
E. Kranakis et al. (Eds.): LATIN 2016, LNCS 9644, pp. 138–151, 2016.
DOI: 10.1007/978-3-662-49529-2_11

is $O(n(\log n)^{k-1})$ [3]. Interestingly, the upper bound to find only the first maximal set is $O(n(\log n)^{\max(1,k-2)})$. Both of these bounds hold in the worst-case. We see that, for fixed k, these algorithms can be regarded as almost optimal, as they only have a poly-logrithmic overhead over the theoretical lower bound. Conceptually, they implement multi-dimensional divide and conquer algorithms [7] on input P which introduces the poly-log factor in their runtimes. The point set P is partitioned into subsets based on ordering of points in some arbitrary dimension. Then the maximal sets are computed recursively and merged later.

Things get interesting if the number of dimensions is not bounded by a constant. When $k = \Omega(\log n)$, these poly-logarithmic upper-bounds above become quasi-polynomial (in n). However, there is a trivial algorithm (which compares each point against the other, and keeps track of the computed transitive relations) that requires in the worst case $O(kn^2)$ comparisons. When $k = n$ there is a deterministic algorithm that runs in $O(n^{(3+\omega)/2})$ to compute the first maximal layer [10], where $O(n^\omega)$ is the complexity of multiplying two $n \times n$ matrices. So we see that the algorithm runs in $\omega(n^2)$ time. In a recent paper [11], authors show that determining whether there exists a pair (u, v), where $u \in A$ and $v \in B$ (A and B are both sets of vector of size $O(n)$) such that $u \succ v$, can be done in sub-quadratic time provided $k = O(\log n)$.

Our Results. In this paper we propose a randomized algorithm for the MaxLayers(P) problem. When the point set P is also a random order the run-time of our algorithm is bounded by $O\left(kn^{2-\frac{1}{\log k}+\log_k\left(1+\frac{2}{k+1}\right)}\log n\right)$ in expectation. Otherwise, it is $O(k^2 n^{3/2+(\log_k (k-1))/2}\log n)$ also in expectation. Additionally, it takes $O(kn^2)$ time in the worst case. This is the first non-trivial algorithm for which the following two conditions holds simultaneously: (1) The worst case run time is polynomially bounded (in n) as long as k is bounded by some polynomial in n. (2) Whenever k is a constant the run time of the proposed algorithm is sub-quadratic in n (in expectation).

2   Preliminaries

We denote $P = \{p_1, \ldots, p_n\}$ as the input set of n points in E^k. The jth coordinate of a point p is denoted as $p[j]$. For any points $p, q \in P$, we define an ordering relation $\succ$, such that $p \succ q$ if $p[j] \; op \; q[j] \; \forall j \in [1 \ldots k]$. Where, op is a place holder for $\geq$ or $\leq$. Consequently, there are 2^k different ordering relations $\succ$ and for each such an ordering there is a unique set of maximal layers (of P). Without loss of generality, we assume that op is $\geq$ for all j, in this paper. Henceforth we will simply use $\geq$ in place of op. We will use the notation $S \succ p$, where S is a set of incomparable elements, to denote that $\exists q \in S$ such that $q \succ p$. If $S \succ p$ we say that S is "above" p. Furthermore, if $p \succeq q$ then either $p = q$ or $p \succ q$.

Clearly, $(P, \succ)$ defines a partial order. We shall simply use P to denote this poset when the context is clear. If $p \succ q$ then we say that p precedes (or dominates) q in the partial order and that they are comparable. We say that p and

q are incomparable (denoted by $p \parallel q$) if $p \not\succ q$ and $q \not\succ p$. If p and q belong to the same maximal layer then $p \parallel q$. Let the height h of P be defined as the number of non-empty maximal layers of P. We also define the width w of P as the size of the largest subset of P of mutually incomparable elements. Note, that the maximum size of any layer is $\leq w$.

Let $\mathcal{O} : E^k \times E^k \to \{0,1\}^k$, such that $\mathcal{O}(p,q)[j] = 1$ if $p[j] < q[j]$ and 0 otherwise. This definition, which might seem inverted, will make sense when we discuss it in the context of our data structures. We call $\mathcal{O}$ the orthant function as it computes the orthant with origin p in which q resides. Henceforth, the maximal layers will simply be referred to as layers. Let T be an ordering of the points in P such that for any two points $p, q \in P$, if $p \succ q$ then p precedes q in the ordering T. Then T is a *liner-extension* of P. Let $|S|$ denote the size of the set S. We are now ready to state the MaxLayers(P) problem formally:

Definition 1 (MaxLayers(P)). *Given a point set P along with an ordering relation defined above, label each point in P with rank of the maximal layer it belongs to.*

In our analysis we shall use the typical RAM model, where operation of the form $p[j] \overset{?}{\geq} q[j]$ takes constant time. Also we shall initially assume that the point set P is a random order. Then we will extend the result to arbitrary set of points. Below we define random orders formally according to its definition in [1].

Definition 2. *We pick a set of n points uniformly at random from $[0,1]^k$. Then the partial order generated by these points is a random order.*

This is equivalent to saying that $(P, \succ)$ is the intersection of k linear orders $T_1 \times \ldots \times T_k$ where the k-tuple $(T_1, \ldots, T_k)$ is chosen uniformly at random from $(n!)^k$ such tuples. Here, each T_j is a linear ordering (permutation) of $\{1, 2, \ldots, n\}$. Whenever we present our run-time results in terms of w or (and) h it is assumed that both are upper bounded by n, the number of points in P. To simplify our analysis we ignore the expected values of w and h, which could only have made our results stronger (for example, see [1,2]).

3   The Iterative Algorithm

We shall use MaxPartition(P) as the main procedure for solving an instance of MaxLayers(P). First we will describe a simpler algorithm and analyze it for a random order P. Then we extend it for an arbitrary set of points.

3.1   Data Structures

In this section we introduce the framework on which our algorithm is based. Let B be a self-balancing binary search tree (for example B could be realized as a red-black tree). Let $B(i)$ be the ith node in the in-order of B. Each node of B stores three pointers. One for each of its children (null in place of an empty child) and another pointer which points to an auxiliary data structure. If X is a node in B then left and right children of X are denoted as $l(X)$ and $r(X)$ respectively. We also denote by $L(X)$ the auxiliary data structure associate with X. When the context is clear, we shall simply use L in place of $L(X)$. We also let L be a placeholder for any data structure that can be used to store the set of points from a single layer of P. For example, L could be realized as a linked list. Additionally, L must support INSERT(L, p) and ABOVE(L, p). The ABOVE(L, p) operation takes a query point p, and answers the query $L \overset{?}{\succ} p$. The INSERT(L, p) operation inserts p into L, which assumes p is incomparable to the elements in L. So, we must ensure that L is the correct layer for p before calling INSERT(L, p).

We observe that the layers of P are themselves linearly ordered by their ranks from 1 to h. We can thus use B to store the layers in sorted order, where each node $B(i)$ would store the corresponding layer $\mathcal{M}_i$ (using $L(B(i))$). We endow B with INSERT(B, p) and SEARCH(B, p) (we do not need deletion) operations. The INSERT(B, p) procedure first calls the SEARCH(B, p) procedure to identify which node $B(i)$ of B the new point p should belong and then calls INSERT$(L(B(i)), p)$. If p does not belong to any layer currently in B then we create a new node in B. The SEARCH(B, p) procedure works as follows: we can think of B as a normal binary search tree, where the usual comparison operator $\geq$ has been replaced by the ABOVE(L, p) procedure. Furthermore, the procedure can only identify whether $L \succ p$ or $L \not\succ p$. This is exactly equivalent to the situation where we have replaced the comparison operator $\geq$ with $>$. So we must determine two successive nodes $B(i)$ and $B(i+1)$ such that $L(B(i)) \succ p$ and $L(B(i+1)) \not\succ p$. If such a pair of nodes does not exist then we return a null node.

3.2 MaxPartition(P)

We begin by first computing a linear extension T of P. We initialize B as an empty tree. We iteratively pick points from P in increasing order of their ranks in T and call INSERT(B, p), where p is the current point to be processed. INSERT(B, p) subsequently calls SEARCH(B, p). We have two possibilities:

CASE 1: SEARCH(B, p) returns a non-empty node $B(i)$. We then call INSERT$(L(B(i), p)$.

CASE 2: SEARCH(B, p) returns a null node. Then we create the node $B(m+1)$ in B, where m is the number of nodes currently in B. We first initialize $B(m+1)$ and then call INSERT$(L(B(m+1), p)$ on it. We note that, when we create a new node in B it must always be the right-most node of B. This follows from the order in which we process the points. Since p succeeds a processed point q in the linear extension T, hence $p \not\succ q$. Thus, if p does not belong to any of nodes currently in B then it must be the case that p is below all layers in B.

MAXPARTITION(P) terminates after all points have been processed. At termination $L(B(i))$ stores all of the points in $\mathcal{M}_i$ for $1 \leq i \leq h$. We make a couple of observations here. (1) When a point is inserted into a node $B(i)$ it will never be displaced from it by any point arriving after it. (2) Since, nodes are always added as the right-most node in B, for SEARCH(B,p) to be efficient, B must support self-rebalancing. If we assume that ABOVE(L,p) and INSERT(L,p) to work correctly, at once we see that SEARCH(B,p) and INSERT(B,p) are also correct. Hence, each point is correctly assigned to the layer it belongs to.

3.3 Runtime Analysis

Let ABOVE(L,p) take $t_a(|L|)$ time. As mentioned in Sect. 2, $|L| \le w$ for any layer in B. Hence, $t_a(w)$ is an upper bound on the runtime of ABOVE(L,p). Similarly, we bound the runtime of INSERT(L,p) with $t_i(w)$. Let p be the next point to be processed. At the time B will have at most h nodes. In order to process p the INSERT(B,p) will be invoked, which in turn calls the SEARCH(B,p) as discussed above. But the SEARCH(B,p) will employ a normal binary search on B with the exception that at each node of B it invokes the ABOVE(L,p) instead of doing a standard comparison. Since, B is self-balancing the height of B is bounded by $O(\log h)$. Hence, number of calls to ABOVE(L,p) is also bounded by $O(\log h)$, each of which takes $t_a(w)$ time. Also, for each point p, INSERT(L,p) is called only once. We also assume initializing a node in B takes constant time. So, processing of p takes $O(t_a(w)\log h + t_i(w))$ and this holds for any point.

Lemma 1. *We can compute a linear extension T of a random order P in $O(n\log n + kn)$ time in the worst case.*

Proof. We shall compute T as follows: Let $\mu(p) = \max_{1 \le j \le k} p[j]$. Then sorting the points in decreasing order of $\mu(p)$ will give us T. It is trivial to see that T is a linear extension of P when no two point in P share the same coordinate (see footnote 2). For a random order this is true almost surely. Furthermore this procedure takes $O(n\log n + kn)$ in the worst case. $\qquad\square$

The reason for computing T in this way will be clear when we get to the analysis of our algorithm. Later we shall see that the time bounds for ABOVE(L,p) and INSERT(L,p) will dominate the time it takes to compute T. So we shall ignore this term in our run-time analysis. The next theorem trivially follows from the discussion above.

Theorem 1. *The procedure MAXPARTITION(P) takes $O(n(t_a(w)\log h + t_i(w)))$ time and upon termination outputs a data structure consisting of the maximal layers of P in sorted order.*

4   Realization of L Using Half-Space Trees

In this section we introduce a new data structure for implementing L. We shall refer to it as Half-Space Tree (HST). The function $\mathcal{O}(p,q)$ computes which orthant q belongs to with respect to p as the origin. Clearly, there are 2^k such orthants, each having a unique label in $\{0,1\}^k$. Let $H_j(p)$ be a half space defined as: $H_j(p) = \{q \in [0,1]^k \mid \mathcal{O}(p,q) = \{0,1\}^{j-1}0\{0,1\}^{k-j}\}$ passing through origin

p whose normal is parallel to dimension j. Here, $\{0,1\}^{j-1}0\{0,1\}^{k-j}$ represents a 0–1 vector for which the jth component is 0. We shall use the notation $h_j(p)$ to denote the extremum orthant of $H_j(p)$ (w.r.t $\succ$), that is, $h_j(p) = 1^{j-1}01^{k-j}$. There are k such half spaces. An orthant whose label contains m 1's lies in the intersection of some $k - m$ such half spaces.

Lemma 2. *If $p, q \in P$ and $p \parallel q$ then $\mathcal{O}(p,q) \in \{0,1\}^k \setminus \{0^k, 1^k\}$. That is, q can only belong to orthants which lie in the intersection of at most $k - 1$ half spaces.*

Proof. Trivially follows from definitions. ☐

Corollary 1. *The above lemma holds if p and q belongs to the same layer. However, the converse of this statement is not true.*

4.1   Half-Space Tree

We define a k-dimensional HST recursively as follows:

Definition 3 (HST)

1. *A singleton node (root) storing a point p.*
2. *A root has a number of non-empty children nodes (up to k) each of which is a HST.*
3. *If node q is the jth child of node p then $h_j(p) \succeq \mathcal{O}(p,q)$.*

An HST stores points from a single layer. So Corollary 1 tells us that for any node p and a new point q at most $k - 1$ of the children nodes satisfy $h_j(p) \succeq \mathcal{O}(p,q)$. Hence, q can be inserted into any one out of these children nodes. Henceforth, we will also use w (the width of $(P, \succ)$) to bound the number of points currently stored inside L.

Above(L, p). Let us assume that L is realized by an HST. The Above(L,p) works as follows: First we compute $\mathcal{O}(r,p)$. Here, r is the root node. If $\mathcal{O}(r,p) = 0^k$ then we return $L \succ p$. Otherwise we call Above$(j(L),p)$ recursively on each non-empty child node j of root r, such that $h_j(r) \succeq \mathcal{O}(r,p)$. When all calls reach some leaf node, we stop and return $L \not\succ p$.

Proof of Correctness. CASE 1:$(L \succ p)$ Let q be some point in L such that $q \succ p$, prior to calling Above(L,p). Before reaching the node q, if we find some other node $q' \succ p$ then we are done. So we assume this is not the case. We claim that p will be compared with q. We show this as follows: Let the length of path from root r to q be $i + 1$. Let $u_0, \ldots, u_i$ be the sequence of nodes in this path (here $u_0 = r$ and $u_i = q$). Since, $q \succ p$, $\mathcal{O}(u_m, q) \succeq \mathcal{O}(u_m, p)$ for all $0 \leq m < i$. But, u_m is a predecessor node in the path from r to q, hence $h_{j_m}(u_m) \succeq \mathcal{O}(u_m, q)$ where u_{m+1} is the j_m^{th} child of u_m. Which implies $h_{j_m}(u_m) \succeq \mathcal{O}(u_m, p)$ (from transitivity of $\succeq$) for $0 \leq m < i$. Thus we will traverse this path at some point during our search.

CASE 2:$(L \not\succ p)$ Follows trivially from the description of Above(L,p). ☐

Insert(L, p). INSERT(L,p) is called with the assumption that $L \not\succ p$. If the root is empty then we make p as the root and stop. Otherwise, we pick one element uniformly at random from the set $S_r = \{j \in \{1,\ldots,k\} \mid h_j(r) \succeq \mathcal{O}(r,p)\}$ and recursively call INSERT($j(L),p$).

Proof of Correctness. It is easy to verify that the insert procedure maintains the properties of HST given in Definition 2. □

Although the insert procedure is itself quite simple, it is important that we understand the random choices it makes before moving further. These observations will be crucial to our analysis later. Let the current height of L be h_L. By L^* we denote the complete HST of height h_L, clearly L^* has k^{h_l} nodes. We color edges of L^* red if both of the nodes it is incident to are present in L, otherwise we color it blue. Unlike ABOVE, we can imagine that the INSERT procedure works with L^* instead of L. Upon reaching a node r in L^* the procedure samples uniformly at random from the set S_r as above. This set may contain edges of either color. If a blue edge have been sampled then we stop and insert p into the empty node incident to the blue edge in L. So we see that, despite not being in L, the nodes incident to blue edges effect the sampling probability equally.

4.2   Runtime Analysis

Here we compute $t_a(w)$ and $t_i(w)$ in expectation over the random order P and the internal randomness of the INSERT(L,p) procedure. From the discussion in Sect. 4.1 we clearly see that $t_i(w) = O(t_a(w))$. So it suffices to upper bound $t_a(w)$ in expectation. Furthermore, we only need to consider the case when ABOVE(L,p) returns $L \not\succ p$ as the other case would take fewer number of comparisons. Let this time be $u(w)$. We divide our derivations to compute $u(w)$ into two main steps:

i. Compute the expected number of nodes at depth d of L having a total of w nodes.
ii. Use that to put an upper bound on the number of nodes visited during a call to ABOVE(L,p) (when $L \not\succ p$).

We choose to process points according to T as detailed earlier. We denote this ordering by the ordered sequence $(p_1,\ldots,p_n)$.

Lemma 3. *For any two points p,q where p precedes q in T we have the probability that $p[j] > q[j]$ is $\eta_1(k) = 1 - \frac{1}{2}\frac{k-1}{k+1}$. Additionally, if p and q are incomparable then it is $\eta_2(k) = 1 - \frac{1}{k} - \frac{1}{2}\frac{k-2}{k+2}$.*

Proof. See appendix. □

Theorem 2. *After w insertions the expected number of nodes at depth d in L is given by:*

$$k^d \left(1 - \sum_{i=1}^{d} \frac{(1 - \frac{1}{k^i})^{w-1}}{\prod_{j=1, j\neq i}^{d} (1 - \frac{1}{k^{i-j}})} \right)$$

Proof. Let $X_{w,d}$ be the number of nodes at depth d of L after w insertions. Due to the second assertion of Lemma 3 we know that any new point to be inserted can belong to any of the k half-spaces with probability $\eta_2(k)$, which is constant over the half-spaces. The insert procedure selects one of these candidate half-spaces uniformly at random. Thus it follows from symmetry that a particular half-space will be chosen for insertion with probability $\frac{1}{k}$. If the subtree is non-empty then we do this recursively. We define an indicator random variable for the event that the tth insertion adds a node at depth d as $I_{t,d}$. Then,

$$X_{w,d} = \sum_{t=1}^{w} I_{t,d}$$

Taking expectation on both sides we get,

$$\mathbb{E}[X_{w,d}] = \sum_{t=1}^{w} \Pr[I_{t,d}]$$

Trivially, $\mathbb{E}[X_{w,0}] = 1$ for $t > 0$. When $d = 1$ and $t \geq 2$ then $\Pr[I_{t,1}] = 1 - \frac{X_{t-1,1}}{k}$. This is because there are $X_{t-1,1}$ nodes at depth 1 (nodes directly connected to the root) hence there are $k - X_{t-1,1}$ empty slots for the node to get inserted at depth 1, otherwise it will be recursively inserted to some deeper node. Hence we have,

$$\mathbb{E}[X_{w,1}] = \sum_{t=2}^{w} \left(1 - \frac{X_{t-1,1}}{k}\right)$$

For $d = 2$, we can similarly argue that the probability of insertion at depth 2 for some $t \geq 3$ is equal to probability of reaching a node at depth 1 times the probability of being inserted at depth 2. It is not difficult to see that this equals: $\left(\frac{X_{t-1,1}}{k}\right)\left(1 - \frac{X_{t-1,2}}{kX_{t-1,1}}\right)$. Hence,

$$\mathbb{E}[X_{w,2}] = \sum_{t=3}^{w} \left(\frac{X_{t-1,1}}{k}\right)\left(1 - \frac{X_{t-1,2}}{kX_{t-1,1}}\right)$$

Proceeding in this way we see that,

$$\mathbb{E}[X_{w,d}] = \sum_{t=1}^{w} \left(\frac{\mathbb{E}[X_{t-1,d-1}]}{k^{d-1}} - \frac{\mathbb{E}[X_{t-1,d}]}{k^d}\right)$$

Here we again take expectation on both sides and simplify the expression so that the sum starts from $t = 1$ since the terms $\mathbb{E}[X_{t,d}] = 0$ when $t \leq d$.

Let $a(w,d) = \mathbb{E}[X_{w,d}]$, we can then simplify the above equation to get the following recurrence,

$$a(w,d) = \frac{a(w-1,d-1)}{k^{d-1}} + \left(1 - \frac{1}{k^d}\right)a(w-1,d)$$

with $a(w,d) = 0$ for $w \le d$. The solution to this can be found by choosing an ordinary generating function $G_d(z)$ with parameter d, such that $G_d(z) = \sum_{t=0}^{\infty} a(t,d)z^t$. The solution [see appendix] completes the proof of the theorem. $\qquad\square$

Before moving on to the main theorem we need another lemma:

Lemma 4. *If $B = (b_0, b_1, \ldots, b_n)$ is a sequence such that $b_r \ge b_{r+1} \ge \ldots \ge b_n$, then the sum $S = \sum_{i=0}^{n} b_i m^i \le \sum_{i=0}^{r} b_i m^i + \frac{b_{r+1} m^{r+1}}{(1-m)}$ where $m < 1$.*

Proof. Follows from elementary algebra. Details are omitted. $\qquad\square$

Corollary 2. *If $m = 1 - \frac{1}{2}\frac{k-1}{k+1}$ and $k \ge 4$ then , $S \le \sum_{i=0}^{r} b_i m^i + \frac{7}{3}b_{r+1} m^r$.*

Theorem 3. *Expected number of nodes visited during an unsuccessful search $u(w)$ is bounded by $O\left(w^{1-\frac{1}{\log k}+\log_k\left(1+\frac{2}{k+1}\right)}\right)$.*

Proof. Before proving this we make the following observation. If for any $d = d_0$, the sequence $a(w,d)$ becomes decreasing, that is, $a(w,d_0) \ge a(w, d_0 - 1)$ and $a(w, d_0) > a(w, d_0 + 1)$, then afterwards it will stay decreasing. This is clear from the fact that $a(w,d)$ represents the expected number of nodes at depth d after w insertions. So the sequence $a(w,d)$ is unimodal since $a(w,0) \le a(w,1)$ trivially for $w \ge 2$. Let d_0 be the value that maximizes $a(w,d)$.

Let us compute the probability of visiting a node at depth d during a call to ABOVE when the query point is not below L. Let q be the current node being checked and p be the query point. According to Lemma 3 the probability $\Pr[p \in H_j(q)]$ is same for any j and is not dependent on the rank of q in T. Hence it is also not dependent on the depth of q in L. Furthermore, this probability is $\eta_1 = 1 - \frac{1}{2}\frac{k-1}{k+1}$, again from Lemma 3.

Thus the probability of visiting a node at depth d is the result of d independent moves each having probability η_1, hence it is η_1^d. Now we can find the expression for the expected number of nodes visited:

$$u(w) = \sum_{d=0}^{w-1} \eta_1^d a(w,d)$$

$$\le \sum_{d=0}^{d_0} \eta_1^d a(w,d) + \frac{7}{3}\eta_1^{d_0} a(w, d_0 + 1) \tag{1}$$

Here we use Theorem 2, Lemma 4 and its corollary and the fact that the sequence $a(w,d)$ is unimodal; to bound $u(w)$. Also note that $a(w,d) \le k^d$. Now we need to upper bound d_0. With some tedious algebra (details are omitted due to space constraints) we get, $d_0 \le \log_k w + 2$. Again, after some more algebra we finally get,

$$u(w) \le O\left(w^{1-\frac{1}{\log k}+\log_k\left(1+\frac{2}{k+1}\right)}\right) \tag{2}$$

This proves Theorem 3. $\qquad\square$

Corollary 3. *The algorithm runs in $O\left(kn^{2-\frac{1}{\log k}+\log_k\left(1+\frac{2}{k+1}\right)}\log n\right)$ time in expectation.*

Proof. From Theorem 1 and the first paragraph of Sect. 4.2 we see that the run-time of the algorithm is $O(knu(w)\log h)$. Since computing $\mathcal{O}(p,q)$ between pairs of vectors takes $O(k)$ time. Using the upper-bound of $u(w)$ and the fact that $w, h \leq n$ we get the runtime as claimed above. □

Corollary 4. *The worst case running time of the algorithm is bounded by $O(kn^2)$.*

Proof. The worst case occurs when each of the HSTs are just unbalanced chains. Hence, both ABOVE and INSERT takes linear time ($O(kn)$) per point. □

5 Extension to Arbitrary P

The previous algorithm would still be correct[2] if P is not a random order. However the expected runtime will no longer hold. In order to make our previous analysis work for any set of points we modify the way we store the layers. In this new setting the layers are still arranged using a self-balancing binary tree B, exactly as before. However, each layer is now maintained using a list of HSTs instead of just a single one. Let us call this data structure LIST-HST. We extend the ABOVE and INSERT procedure for HST as follows.

LIST-HST starts with an empty list. Attached to a LIST-HST is another list R in which newly arrived points are kept temporarily before they are ready to be inserted in the LIST-HST. Initially this list R is also empty. We take the maximum size of R as $\sqrt{w}$ ($\leq \sqrt{n}$). As long as R has less than $\sqrt{w}$ points we keep adding to it. Once R has been filled, we create an HST from the points in R and remove these points from R. This becomes the first HST in the list. We repeat these steps again when R is full. We describe the modified LIST-HST-ABOVE and LIST-HST-INSERT below.

List-HST-Insert(p). If R is not yet full then we just add p to R. Otherwise we create an HST from points in $P' = R \cup p$. We randomly permute elements in P' and pick the first element in this ordering as the root. We then build the HST iteratively by picking rest of the elements in this order. Next we prove a lemma similar to Lemma 3.

Lemma 5. *Assume that an HST is built by inserting points in a random order. Let X represent a point which is already inserted and Y a new point being compared to X. Then Y belongs to any of the children subtree of X with equal probability.*

---

[2] However, we need to modify the procedure for computing a linear extension since the assumption that no two point share a coordinate may longer hold. In this case we can simply take the sorted order of the points according to the sum of their coordinates.

Proof. Since the insertion order is random, X and Y are both random variables. We compute the probability $\Pr[X[i] > Y[i]]$ for some i. Now for some arbitrary pair of points $\{p, q\}$ the probability that p precedes q in the ordering is $1/2$. If $X, Y \in \{p, q\}$ then, $\Pr[X[i] > Y[i]|X, Y \in \{p, q\}] = (\Pr[X[i] > Y[i]|X = p, Y = q] + \Pr[X[i] > Y[i]|X = q, Y = p])/2$. $\Pr[p[i] > q[i]]$ is either 0 or 1 since the points p and q themselves are not random. Hence, $\Pr[X[i] > Y[i]|X, Y \in \{p, q\}] = 1/2$ for any pair $\{p, q\}$, which proves the claim above. $\qquad\square$

Using the above lemma and techniques used to prove Theorem 3 we can show that $d_0 \leq \log_k \sqrt{w} + 2$. Hence, building an HST takes $O(\sqrt{w} \log_k w)$ time in expectation and $O(w)$ in worst case. Since, there can only be $\sqrt{n}$ such steps where we build an HST and each takes $O(n)$ (since $w \leq n$) time hence the insert operation on LIST-HST adds $O(kn^{3/2})$ to the overall running time of our algorithm. This is insignificant compared to the total time.

List-HST-Above(p). For each HST L in the list we call ABOVE(L, p). If none of these calls find a point above p then we check the remaining points in R. However, since p is not random we cannot compute the probability η_1 as we did before. However, we can upper bound the fraction of subtrees that are visited from a node. We see that a point p can visit at most $k - 1$ subtrees of a node q otherwise we can conclude that $q \succ p$. The LIST-HST-INSERT procedure creates the jth subtree of q with equal probability for all j. Hence, during search the point p will visit a non-empty subtree of q is with probability $\leq (k - 1)/k$. This value can be substituted as an upper bound for η_1 in Eq. 1, which leads to $u(w) \leq O(k\sqrt{w}^{\log_k (k-1)})$. Since there are at most $\sqrt{w}$ HSTs in a layers, it takes $O(kw^{1/2+(\log_k (k-1))/2})$ time in expectation to search a list of HSTs. We ignore the time it takes to check the set R, which is $O(k\sqrt{w})$. Hence the total runtime is bounded by $O(k^2 n^{3/2+(\log_k (k-1))/2} \log n)$ in expectation.

5.1   A Summary of Results

We summarize the main results as follows:

i. *k is a constant.* From Corollary 3 we can easily verify that the algorithm has a runtime of $O(n^{2-\delta(k)})$ where $\delta(k) > 0$. This remains true even when P is not a random order.

ii. *k is some function of n.* We let $k = f(n)$. For any k the runtime of our algorithm is bounded by $O(kn^2)$ in the worst case. This bound does not hold for the divide-and-conquer algorithm in [3]. Also, the proposed algorithm never admits a quasi-polynomial runtime unlike any of the previously proposed non-trivial algorithms.

Concluding Remarks. In this paper we proposed a randomized algorithm for the MAXLAYERS(P) problem. Unlike previous authors we also consider the case when k is not a constant; this is often the case for many real-world data sets whose tuple dimensions are not insignificant with respect to its set size. In this setting we show that the expected runtime of our algorithm is $O\left(kn^{2-\frac{1}{\log k}+\log_k\left(1+\frac{2}{k+1}\right)}\log n\right)$ when P is a random order. For any arbitrary set of points in E^k it exhibits a runtime of $O(k^2 n^{3/2+(\log_k (k-1))/2}\log n)$ in expectation. It remains to be seen if there exists a deterministic algorithm that runs in $o(kn^2)$ for this problem. As a future work it would be interesting to know whether HST can be used for the unordered convex layers problem in higher dimensions. We know that unlike the maximal layers problem this problem is not decomposable [8]. So it would be interesting to know within our iterative framework whether we can extend HST to store the convex layers also.

Appendix

Proof of Lemma 3

Proof. Recall that T is a linear extension of P. Since p precedes q in T, $\mu(p) > \mu(q)$. Hence, $\exists\ j' \in \{1,\ldots,k\}$ such that $p[j'] > q[j']$. Let $j' = \mathrm{argmax}_{1\leq j\leq k} p[j]$. We compute the probability $\Pr[p[j] > q[j] \mid \mu(p) > \mu(q)]$ in two parts over the disjoint sets $\{j = j'\}$ and $\{j \neq j'\}$:

$$\Pr[p[j] > q[j] \mid \mu(p) > \mu(q)] = \Pr[p[j] > q[j] \mid j = j', \mu(p) > \mu(q)]\Pr[j = j']$$
$$+ \Pr[p[j] > q[j] \mid j \neq j', \mu(p) > \mu(q)]\Pr[j \neq j']$$
$$= 1\frac{1}{k} + \left(1 - \frac{1}{2}\frac{\mu(q)}{\mu(p)}\right)\left(1 - \frac{1}{k}\right) \qquad (3)$$

Since,

$$\Pr[p[j] > q[j] \mid j \neq j', \mu(p) > \mu(q)] = \frac{(\mu(p) - \mu(q))\mu(q) + \frac{\mu(q)^2}{2}}{\mu(p)\mu(q)} = 1 - \frac{1}{2}\frac{\mu(q)}{\mu(p)}$$

This follows from the fact that $p[j]$ and $q[j]$ are independent random variables uniformly distributed over $[0, \mu(p)]$ and $[0, \mu(q)]$ (given $\mu(p) > \mu(q)$) respectively. In the set $\{j = j'\}$ clearly $p[j] > q[j]$. However, in the set $\{j \neq j'\}$ the probability that $p[j] > q[j]$ is $\left(1 - \frac{1}{2}\frac{\mu(q)}{\mu(p)}\right)$. We note that $\mu(p), \mu(q)$ are themselves random variables. More importantly they are i.i.d random variables having the following distribution:

$$\Pr[\mu(p) < t] = t^k$$

on the interval $[0, 1]$. This follows from how points in P are constructed. We take the expectation of both side of Eq. 3 over the event space generated by $\mu(p), \mu(q)$ on the set $\{\mu(p) > \mu(q)\}$:

$$\Pr[p[j] > q[j] \mid \mu(p) > \mu(q)] = 1 - \frac{1}{2}\mathbb{E}\left[\frac{\mu(q)}{\mu(p)} \mid \mu(p) > \mu(q)\right]\left(1 - \frac{1}{k}\right)$$
$$= 1 - \frac{1}{2}\left(\frac{k}{k+1}\right)\left(1 - \frac{1}{k}\right) = 1 - \frac{1}{2}\left(\frac{k-1}{k+1}\right)$$

Now, let us compute, $\mathbb{E}\left[\frac{\mu(q)}{\mu(p)} \mid \mu(p) > \mu(q)\right]$. Recall that $\mu(p) = max_{1 \leq i \leq k} p[j]$. So the distribution function of $\mu(p)$ is,

$$F[\mu(p)] = \Pr[\mu(p) < t] = \Pr\left[\bigwedge_{1 \leq i \leq k} p[i] < t\right] = \prod_{1 \leq i \leq k} \Pr[p[i] < t] = t^k$$

Where the second equality comes from that fact that each component of p are independent and identically distributed on $[0, 1]$ with uniform probability. Hence,

$$\mathbb{E}\left[\frac{\mu(q)}{\mu(p)} \mid \mu(p) > \mu(q)\right] = \int_{\mu(p) > \mu(q)} \frac{\mu(q)}{\mu(p)} dF[\mu(p)] dF[\mu(q)]$$

$$= \frac{k^2}{\Pr[\mu(p) > \mu(q)]} \int_0^1 \int_0^{\mu(p)} \mu(p)^{k-2} \mu(q)^k d\mu(p) d\mu(p)$$

$$= \frac{k}{k+1}$$

A similar argument can be used to prove the second claim. $\qquad\square$

Solving $a(w, d)$

To simplify our calculations we modify the recurrence slightly: With $a(w, d) = k^d b(w, d)$, the recurrence equation becomes,

$$b(w, d) = \frac{1}{k^d} b(w-1, d-1) + (1 - \frac{1}{k^d}) b(w-1, d)$$

Let, $G_d(z) = \sum_{w=0}^{\infty} b(w, d) z^w$. We note that $b(w, d) = 0$ when $w \leq d$. Then we have,

$$G_d(z) = \frac{z}{k^d} G_{d-1}(z) + z(1 - \frac{1}{k^d}) G_d(z)$$

$$= \frac{z}{k^d(1 - (1 - \frac{1}{k^d})z)} G_{d-1}(z)$$

$$\cdots$$

$$= \frac{z^d}{\prod_{i=1}^{d} k^i(1 - (1 - \frac{1}{k^i})z)} G_0(z)$$

But, $G_0(z) = \sum_{w=0}^{\infty} b(w, 0) z^w = \sum_{w=1}^{\infty} z^w = \frac{z}{1-z}$ as $b(w, 0) = a(w, 0) = 1$ when $w \geq 1$. Hence,

$$b(w, d) = k^{-d(d+1)/2} [z^{w-d-1}] G_d(z)$$

$$= k^{-d(d+1)/2} [z^{w-d-1}] \frac{1}{(1-z) \prod_{i=1}^{d} (1 - (1 - k^{-i})z)} \tag{4}$$

where the notation $[z^i]p(z)$ means the coefficient of z^i in the polynomial $p(z)$ as usual. Using partial fractions: Let,

$$\frac{1}{(1-z)\prod_{i=1}^{d}\left(1-(1-k^{-1})z\right)} \equiv \frac{\beta_0}{1-z} + \sum_{i=1}^{d} \frac{\beta_i}{\left(1-(1-k^{-i})z\right)}$$

For which we get the following solution,

$$\beta_0 = k^{d(d+1)/2}$$

$$\beta_i = \frac{k^{d(d+1)/2}(1-k^{-i})^d}{\prod_{j\neq i, j\geq 1}^{d}\left(1-k^{j-i}\right)}$$

Substituting these in Eq. 4 above we get, $b(w,d) = 1 - \sum_{i=1}^{d} \frac{(1-k^{-i})^{w-1}}{\prod_{j=1, j\neq i}^{d}(1-k^{j-i})}$, which gives us the desired result for $a(w,d)$.

References

1. Winkler, P.: Random orders. Order **1–4**, 317–331 (1985)
2. Brightwell, G.: Random k-dimensional orders: width and number of linear extensions. Order **9**, 333–342 (1992)
3. Jensen, M.T.: Reducing the run-time complexity of multiobjective EAs: the NSGA-II and other algorithms. IEEE TEC **7–5**, 503–515 (2003)
4. Kung, H.T., Luccio, F., Preparata, F.P.: On finding the maxima of a set of vectors. J. ACM **22–4**, 469–476 (1975)
5. Bentley, J.L., Kung, H.T., Schkolnick, M., Thompson, C.D.: On the average number of maxima in a set of vectors and applications. J. ACM **25–4**, 536–543 (1978)
6. Yao, F.F.: On Finding the maximal elements in a set of plane vectors. Computer Science Department of Report, University of Illinois at Urbana-Champaign, Urbana, Illinois (1974)
7. Bentley, J.L.: Multidimensional divide-and-conquer. Commun. ACM **23–4**, 214–229 (1980)
8. Preparata, P., Shamos, M.: Computational Geometry: An Introduction. Texts and Monographs in Computer Science. Springer, Berlin (1985)
9. Overmars, M.H.: The Design of Dynamic Data Structures. LNCS, vol. 156. Springer, New York, Tokyo (1983)
10. Matoušek, J.: Computing dominance in E^n. Inf. Process. Lett. **38**, 277–278 (1991)
11. Impagliazzo, R., Lovett, S., Paturi, R., Schneider, S.: 0–1 Integer Linear Programming with a Linear Number of Constraints, eprint arXiv:1401.5512 (2014)
12. Nielsen, F.: Output-sensitive peeling of convex and maximal layers. Inf. Process. Lett. **59**, 255–259 (1996)

On the Total Number of Bends for Planar Octilinear Drawings

Michael A. Bekos$^{(\boxtimes)}$, Michael Kaufmann, and Robert Krug

Wilhelm-Schickard-Institut für Informatik, Universität Tübingen,
Tübingen, Germany
{bekos,mk,krug}@informatik.uni-tuebingen.de

Abstract. An *octilinear drawing* of a planar graph is one in which each edge is drawn as a sequence of horizontal, vertical and diagonal at 45° line-segments. For such drawings to be readable, special care is needed in order to keep the number of bends small. As the problem of finding planar octilinear drawings of minimum number of bends is NP-hard, in this paper we focus on upper and lower bounds. From a recent result of Keszegh et al. on the slope number of planar graphs, we can derive an upper bound of $4n - 10$ bends for 8-planar graphs with n vertices. We considerably improve this general bound and corresponding previous ones for triconnected 4-, 5- and 6-planar graphs. We also derive non-trivial lower bounds for these three classes of graphs by a technique inspired by the network flow formulation of Tamassia.

1 Motivation and Background

Octilinear drawings of graphs have a long history of research, which dates back to the early thirties of the last century, when an English technical draftsman, Henry Charles Beck (also known as Harry Beck), designed the first schematic map of London Underground. His map, the so-called Tube map, looked more like an electrical circuit diagram (consisting of horizontal, vertical and diagonal line segments) rather than a true map, as the underlying geographic accuracy was neglected. Laying out networks in such a way is called *octilinear graph drawing*. In particular, an octilinear drawing $\Gamma(G)$ of a graph $G = (V, E)$ is one in which each vertex occupies a point on an integer grid and each edge is drawn as a sequence of horizontal, vertical and diagonal at 45° line segments.

In planar octilinear graph drawing, an important goal is to keep the number of bends small, so that the produced drawings can be understood easily. However, the problem of determining whether a given embedded planar graph of maximum degree eight admits a bend-less planar octilinear drawing is NP-complete [16]. This motivated us to neglect optimality and study upper and lower bounds on the total number of bends of such drawings. Surprisingly enough, very few results were known, even if the octilinear model has been extensively studied in the areas of metro-map visualization and map schematization.

This work has been supported by DFG grant Ka812/17-1.

E. Kranakis et al. (Eds.): LATIN 2016, LNCS 9644, pp. 152–163, 2016.
DOI: 10.1007/978-3-662-49529-2_12

Table 1. A short summary of our results.

| Graph class | Lower bound | Ref. | Upper bounds | | | |
			Previous	Ref.	New	Ref.
3-con. 4-planar	$n/3 - 1$	Theorem 4	$2n$	[2]	$n + 5$	Theorem 1
3-con. 5-planar	$2n/3 - 2$	Theorem 4	$5n/2$	[2]	$2n - 2$	Theorem 2
3-con. 6-planar	$4n/3 - 6$	Theorem 4	$4n - 10$	[14]	$3n - 8$	Theorem 3

One can derive the first (non-trivial) upper bound on the required number of bends from a result on the planar slope number of graphs by Keszegh et al. [14], who proved that every k-planar graph (that is, planar of maximum degree k) has a planar drawing with at most $\lceil \frac{k}{2} \rceil$ different slopes in which each edge has two bends. For $3 \le k \le 8$, the drawings are octilinear, which yields an upper bound of $6n - 12$, where n is the number of vertices of the graph. This bound can be reduced to $4n - 10$ as follows. The edge that "enters" a vertex from its south port and the edge that "leaves" each vertex from its top port in the s-t ordering of the algorithm of Keszegh et al. can both be drawn with one bend each. This leads to a reduction by $2n - 2$ bends and the bound of $4n - 10$ follows.

On the other hand, it is known that every 3-planar graph with five or more vertices admits a planar octilinear drawing in which all edges are bend-less [8,13]. Recently, it was proved that 4- and 5-planar graphs admit planar octilinear drawings with at most one bend per edge [2], which implies that their total number of bends can be upper bounded by $2n$ and $5n/2$, respectively.

Octilinear drawings form a natural extension of the so-called *orthogonal drawings*, which allow for horizontal and vertical edge segments only. For such drawings, several bounds on the total number of bends are given, see e.g., [4,5,15,18]. It is also known that the bend minimization problem can be solved efficiently, when the input graph is embedded [17], while the corresponding minimizaton problem over all embeddings is NP-hard [11]. In [17], the author describes how one can extend his approach, so to compute a bend-optimal octilinear representation of any given embedded 8-planar graph. However, such a representation may not be realizable by a corresponding planar octilinear drawing [6].

The remainder of this paper is organized as follows. In Sect. 2, we recall basic definitions. In Sect. 3, we improve all aforementioned bounds for the classes of triconnected 4-, 5- and 6-planar graphs. In Sect. 4, we present lower bounds for these classes of planar graphs. We conclude in Sect. 5 with open problems. For a summary of our results refer to Table 1.

2 Preliminaries

Let $G = (V, E)$ be a triconnected planar graph and let $\Pi = (P_0, \dots, P_m)$ be a partition of V into paths, such that $P_0 = \{v_1, v_2\}$, $P_m = \{v_n\}$ and $v_2 \to v_1 \to v_n$ is a path on the outerface of G. For $k = 0, 1, \dots, m$, let G_k be the subgraph induced by $\cup_{i=0}^{k} P_i$. Partition Π is a *canonical order* [7,12] of G if the following hold: (i) G_k is biconnected, (ii) all neighbors of P_k in G_{k-1} are on the outer face of G_{k-1} and (iii) all vertices of P_k have at least one neighbor in P_j for some $j > k$. P_k is called *singleton* if $|P_k| = 1$ and *chain* otherwise.

To simplify the description of our algorithms, we direct and color the edges of G based on Π (similar to Schnyder colorings [9]) as follows. The first path P_0 of Π consists of exclusively one edge (that is, edge (v_1, v_2)), which we color blue and direct towards vertex v_1. For each path $P_k = \{v_i, \ldots, v_{i+j}\} \in \Pi$ later in the order, let v_ℓ and v_r be the leftmost and rightmost neighbors of P_k in G_{k-1}, respectively. In the case where P_k is a chain (that is, $j > 0$), we color edge (v_i, v_ℓ) and all edges of P_k blue and direct them towards v_ℓ. The edge (v_{i+j}, v_r) is colored green and is directed towards v_r (see Fig. 1a). In the case where P_k is a singleton (that is, $j = 0$), we color the edges (v_i, v_ℓ) and (v_i, v_r) blue and green, respectively and we direct them towards v_ℓ and v_r. We color the remaining edges of P_k that are incident to G_{k-1} (if any) red and we direct them towards v_i (see Fig. 1e). Given a vertex $v \in V$ of G, we denote by $\mathrm{indeg}_x(v)$ ($\mathrm{outdeg}_x(v)$, resp.) the in-degree (out-degree, resp.) of vertex v in color $x \in \{r, b, g\}$.

3   Upper Bounds

3.1   Triconnected 4-Planar Graphs

Let $G = (V, E)$ be a triconnected 4-planar graph. Before we proceed with the description of our approach, we need to define two useful notions. First, a *vertical cut* is a y-monotone continuous curve that crosses only horizontal segments and divides a drawing into a left and a right part; see e.g. [10]. Such a cut makes a drawing horizontally stretchable in the following sense: One can shift the right part of the drawing that is defined by the vertical cut further to the right while keeping the left part of the drawing in place and the result is a valid octilinear drawing. Similarly, a *horizontal cut* is defined. Since G has at most $2n$ edges, it follows that G has at most $n + 2$ faces. In order to construct a drawing $\Gamma(G)$ of G, which has roughly at most $n + 2$ bends, we also need to associate to each face of G a so-called *reference edge*. This is done as follows.

Let $\Pi = \{P_0, \ldots, P_m\}$ be a canonical order of G and assume that $\Gamma(G)$ is constructed incrementally by placing a new path of Π each time, so that the boundary of the drawing constructed so far is a x-monotone path. When placing a new path $P_k \in \Pi$, $k = 1, \ldots, m-1$, one or two bounded faces of G are formed (note that we treat the last partition P_m of Π separately). More precisely, if P_k is a chain or a singleton of degree 3 in G_k, then only one bounded face is formed. Otherwise (that is, P_k is a singleton of degree 4 in G_k), two new bounded faces are formed. In both cases, each newly-formed bounded face consists of at least two edges incident to vertices of P_k and at least one edge of G_{k-1}. In the former case, the reference edge of the newly-formed bounded face, say f, is defined as follows. If f contains at least one green edge that belongs to G_{k-1}, then the reference edge of f is the leftmost such edge (see Fig. 1a and c). Otherwise, the reference edge of f is the leftmost blue edge of f that belongs to G_{k-1} (see Fig. 1b and d). In the case where P_k is a singleton of degree 4 in G_k, the reference edge

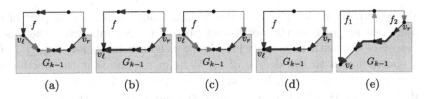

Fig. 1. Illustration of the reference edge (bold drawn) in the case of: (a-b) a chain, (c-d) a singleton of degree 2 in G_k and (e) a singleton of degree 3 in G_k.

of each of the newly formed faces is the edge of G_{k-1} that is incident to the endpoint of the red edge involved.

As already stated, we will construct $\Gamma(G)$ in an incremental manner by placing one partition of Π at a time. For the base, we momentarily neglect edge (v_1, v_2) of the first partition P_0 of Π and we start by placing the second one, say a chain $P_1 = \{v_3, \ldots, v_{|P_1|+2}\}$, on a horizontal line from left to right. Since by definition of Π, v_3 and $v_{|P_1|+2}$ are adjacent to the two vertices, v_1 and v_2, of the first partition P_0, we place v_1 to the left of v_3 and v_2 to the right of $v_{|P_1|+2}$. So, they form a single chain where all edges are drawn using horizontal line segments that are attached to the east and west port at their endpoints. The case where P_1 is a singleton is analogous. Assume now that we have already constructed a drawing for G_{k-1} which has the following invariant properties:

IP-1: The number of edges of G_{k-1} with a bend is at most equal to the number of reference edges in G_{k-1}.

IP-2: The north-west, north and north-east (south-west, south and south-east) ports of each vertex are occupied by incoming (outgoing) blue and green edges and by outgoing (incoming) red edges[1].

IP-3: If a horizontal port of a vertex is occupied, then it is occupied either by an edge with a bend (to support vertical cuts) or by an edge of a chain.

IP-4: A red edge is not on the outerface of G_{k-1}.

IP-5: A blue (green, resp.) edge of G_{k-1} is never incident to the north-west (north-east, resp.) port of a vertex of G_{k-1}.

IP-6: From each reference edge on the outerface of G_{k-1} one can devise a vertical cut through the drawing of G_{k-1}.

The base of our algorithm conforms with the aforementioned invariant properties. Next, we consider the three main cases.

C.1: $P_k = \{v_i\}$ *is a singleton of degree 2 in* G_k. Let v_ℓ and v_r be the leftmost and rightmost neighbors of v_i in G_{k-1}. We claim that the north-east port of v_ℓ and the north-west port of v_r cannot be simultaneously occupied. Assume to the contrary that the claim does not hold and denote by $v_\ell \rightsquigarrow v_r$ the path from v_ℓ to v_r at the outerface of G_{k-1} (neglecting the direction of

---

[1] Note, however, that not all of them can be simultaneously be occupied due to the degree restriction.

the edges). By IP-5, $v_\ell \rightsquigarrow v_r$ starts as blue from the north-east port of v_ℓ and ends as green at the north-west port of v_r. So, inbetween there is a vertex of the path $v_\ell \rightsquigarrow v_r$ which has a neighbor in P_j for some $j \geq k$; a contradiction to the degree of v_i. W.l.o.g. assume that the north-east port of v_ℓ is unoccupied. If (v_i, v_ℓ) is the reference edge of a face, then we draw (v_i, v_ℓ) as a horizontal-diagonal combination from the west port of v_i towards the north-east port of v_ℓ. Otherwise, (v_i, v_ℓ) is drawn bend-less from the south-west port of v_i towards the north-east port of v_ℓ. To draw the edge (v_i, v_r), again we distinguish two cases. If the north-west port at v_r is unoccupied, then (v_i, v_r) will use this port at v_r. Otherwise, (v_i, v_r) will use the north port at v_r. In addition, if (v_i, v_r) is the reference edge of a face, then (v_i, v_r) will use the east port at v_i. Otherwise, the south-east port at v_i. The port assignment described above conforms to IP-2–5. Clearly, IP-1 also holds. IP-6 holds because the newly introduced edges that are reference edges have a horizontal segment, which inductively implies that vertical cuts through them are possible.

C.2: $P_k = \{v_i\}$ *is a singleton of degree 3 in* G_k. This is the most involved case. However, due to space constraints we give the details in [3].

C.3: $P_k = \{v_i, \ldots v_{i+j}\}$; $j \geq 1$ *is a chain*. This case is similar to case C.1, as P_k has exactly two neighbors in G_{k-1}, which we denote by v_ℓ and v_r. The edges between $v_i, \ldots, v_{i+j}$ will be drawn as horizontal segments connecting the west and east ports of the respective vertices. Edges (v_i, v_ℓ) and (v_{i+j}, v_r) are drawn based on the rules of the corresponding case of a singleton with two neighbors in G_{k-1}. The port assignment still conforms to IP-2–IP-5. IP-1 and IP-6 hold, since all edges of the chain are horizontal.

Note that the coordinates of the newly introduced vertices are determined by the shape of the edges connecting them to G_{k-1}. If there is not enough space between v_ℓ and v_r to accommodate the new vertices, IP-6 allows us to stretch the drawing horizontally using the reference edge of the newly formed face.

If v_n is of degree 3, we cope with the last partition $P_m = \{v_n\}$ as being an ordinary singleton. Otherwise (i.e., v_n is of degree 4), we momentarily ignore (v_n, v_1) and proceed to draw the remaining edges incident to v_n. Edge (v_n, v_1) can be drawn afterwards using two bends. Since by construction v_1 and v_2 are horizontally aligned, we can draw the edge (v_1, v_2) with a single bend, emanating from the south-east port of v_1 towards the south-west port of v_2.

Theorem 1. *Let G be a triconnected 4-planar graph with n vertices. A planar octilinear drawing $\Gamma(G)$ with at most $n+5$ bends can be computed in $O(n)$ time.*

Proof. By IP-1, all bends of $\Gamma(G)$ are in correspondence with the reference edges of G, except for the bends of (v_1, v_2) and (v_n, v_1). Since the number of reference edges is at most $n + 2$ and the edges (v_1, v_2) and (v_n, v_1) require 3 additional bends, the total number of bends of $\Gamma(G)$ does not exceed $n + 5$. The linear running time follows by adopting the shifting method of Kant [13] to compute the actual coordinates of the vertices of G. □

3.2   Triconnected 5-Planar Graphs

Our algorithm for triconnected 5-planar graphs is an extension of an algorithm of Bekos et al. [2], which computes for a given triconnected 5-planar graph G on n vertices a planar octilinear drawing $\Gamma(G)$ of G with at most one bend per edge. Since G cannot have more than $5n/2$ edges, it follows that the total number of bends of $\Gamma(G)$ is at most $5n/2$. However, before we proceed with the description of our extension, we first provide some insights into this algorithm, which is based on a canonical order Π of G. Central are IP-2 and IP-4 of the previous section and the so-called *stretchability invariant*, according to which all edges on the outerface of the drawing constructed at some step of the canonical order have a horizontal segment and therefore one can devise corresponding vertical cuts to horizontally stretch the drawing. We claim that we can appropriately modify this algorithm, so that all red edges of Π are bend-less.

Since we seek to draw all red edges of Π bend-less, our modification is limited to singletons. So, let $P_k = \{v_i\}$ be a singleton of Π. The degree restriction implies that v_i has at most two incoming red edges (we also assume that P_k is not the last partition of Π, that is $k \neq m$). We first consider the case where v_i has exactly one incoming red edge, say $e = (v_j, v_i)$, with $j < i$. By construction, e must be attached to one of the northern ports of v_j (that is, north-west, north or north-east). On the other hand, e can be attached to any of the southern ports of v_i, as e is its only incoming red edge. This guarantees that e can be drawn bend-less. Due to space constraints, the more involved case, according to which v_i has exactly two incoming red edges, is discussed in [3].

Theorem 2. *Let G be a triconnected 5-planar graph with n vertices. A planar octilinear drawing $\Gamma(G)$ with at most $2n-2$ bends can be computed in $O(n)$ time.*

Proof. The only edges of $\Gamma(G)$ that have a bend are the blue and the green ones and possibly the third incoming red edge of vertex v_n of the last partition P_m of Π. Hence, $2n - 2$ at most. The running time remains linear since the shifting technique can still be applied. $\qquad\qquad\square$

3.3   Triconnected 6-Planar Graphs

We present an algorithm that based on a canonical order $\Pi = \{P_0, \ldots, P_m\}$ of a given triconnected 6-planar graph G results in a drawing $\Gamma(G)$ of G, in which each edge has at most two bends. So, in total $\Gamma(G)$ has $6n - 12$ bends. Then, we appropriately adjust $\Gamma(G)$ to reduce the total number of bends.

Algorithm 1 describes *rules* R1 - R6 to assign the edges to the ports of the corresponding vertices. It is not difficult to see that all port combinations implied by these rules can be realized with at most two bends, so that all edges have a horizontal segment (which makes the drawing horizontally stretchable): (i) a blue edge emanates from the west or south-west port of a vertex (by rule R4) and leads to one of the south-east, east, north-east, north or north-west ports of its other endvertex (by rule R1); see Fig. 2g and h, (ii) a green edge emanates

---

Algorithm 1. PortAssignment(v)

---

input : A vertex v of a triconnected 6-planar graph.
output: The port assignment of the edges around v.

R1: The **incoming blue edges** of v occupy consecutive ports in counterclockwise order around v starting from:

   a. the south-east port of v, if $\mathrm{indeg}_b(v) + \mathrm{outdeg}_r(v) = 5$; see Figure 2a.
   b. the east port of v, if $\mathrm{indeg}_b(v) + \mathrm{outdeg}_r(v) = 4$; see Figure 2b
   c. the east port of v, if $\mathrm{outdeg}_g(v)=0$ and (a),(b) do not hold; see Figure 2c
   d. the north-east port of v, otherwise; see Figure 2d

R2: The **outgoing red edge** occupies the counterclockwise next free port, if v has at least one incoming blue edge. Otherwise, the north-east port of v.

R3: The **incoming green edges** of v occupy consecutive ports in clockwise order around v starting from:

   a. the west port of v, if $\mathrm{indeg}_g(v)+\mathrm{outdeg}_r(v)+\mathrm{indeg}_b(v) \geq 4$; see Figure 2e
   b. the north-west port of v, otherwise; see Figure 2f

R4: The **outgoing blue edge** of v occupies the west port of v, if it is free; otherwise, the south-west port of v.

R5: The **outgoing green edge** of v occupies the east port of v, if it is free; otherwise, the south-east port of v.

R6: The **incoming red edges** of v occupy consecutively counterclockwisel the south-west, south and south-east ports of v starting from the first available.

---

from the east or south-east port of a vertex (by rule R5) and leads to one of the west, north-west, north or north-east ports of its other endvertex (by rule R3); see Fig. 2i and j, (iii) a red edge emanates from one of the north-west, north, north-east ports of a vertex (by rule R2) and leads to one of the south-west, south, south-east ports of its other endvertex (by rule R6); see Fig. 2k.

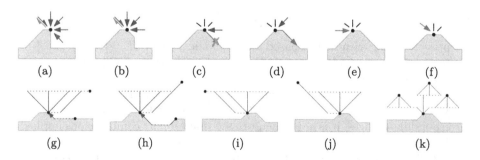

Fig. 2. (a)-(f) Illustration of the port assignment computed by Algorithm 1. (g)-(k) Different segment combinations with at most two bends (horizontal ones are drawn dotted)

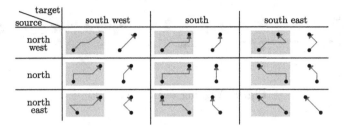

target source	south west	south	south east
north west			
north			
north east			

Fig. 3. Red edges can be redrawn with one bend (in boxes we show their initial shapes)

The shape of each edge is completely determined by the aforementioned rules. To compute the actual drawing $\Gamma(G)$ of G, we follow an incremental approach according to which one partition (that is, a singleton or a chain) of Π is placed at a time, similar to Kant's approach [12] and the 4- or 5-planar case. Each edge is drawn based on its shape, while the horizontal stretchability ensures that potential crossings can always be eliminated. Note additionally that we adopt the leftist canonical order [1].

We reduce the total number of bends in two steps. In the first step, we show that all red edges can be drawn with at most one bend each. Recall that a red edge emanates from one of the north-west, north, north-east ports of a vertex and leads to one of the south-west, south, south-east ports of its other-endvertex. So, in order to prove that all red edges can be drawn with at most one bend each, we have to consider in total nine cases, which are illustrated in Fig. 3. It is not difficult to see that in each of these cases, the red edge can be drawn with at most one bend. Note that the absence of horizontal segments at the red edges does not affect the stretchability of $\Gamma(G)$, since each face of $\Gamma(G)$ has at most two such edges (which both "point upward" at a common vertex). Since a red edge cannot be incident to the outerface of any intermediate drawing constructed during the incremental construction of $\Gamma(G)$, it follows that it is always possible to use only horizontal segments (of blue and green edges) to define vertical cuts, thus, avoiding all red edges.

The second step of our bend reduction is more involved. We seek to "save" two bends per vertex[2], which yields a reduction by roughly $2n$ bends in total. Consider an arbitrary vertex $u \in V$ of G. Our goal is to prove that there always exist two edges incident to u, which can be drawn with only one bend each. By rules R3 and R4, it follows that the west port of vertex u is always occupied, either by an incoming green edge (by rule R3) or by a blue outgoing edge (by rule R4; $u \neq v_1 \in P_0$). Analogously, the east port of vertex u is always occupied; either by a blue incoming edge (by rules R1 and R2) or by an outgoing green edge (by rule R5). Let $(u, v) \in E$ be the edge attached at the west port of u (symmetrically we cope with the edge at the east port of u). If edge (u, v) is attached to a non-horizontal port at v, then (u, v) is by construction drawn with one bend (regardless of its color; see Fig. 2g and i) and our claim follows.

---

[2] Except for vertex v_1 of the first partition P_0 of Π, which has no outgoing blue edge.

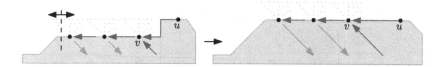

Fig. 4. Aligning vertices u and v.

It remains to consider the case where (u, v) is attached to a horizontal port at v. Assume first that (u, v) is blue (we discuss the case where (u, v) is green later). By Algorithm 1, it follows that (u, v) is either the first blue incoming edge attached at v (by rules R1b and R1c) or the second one (by rule R1a). We consider each of these cases separately. In rule R1c, observe that (u, v) is part of a chain (because $\text{outdeg}_g(u) = 0$). Hence, when placing this chain in the canonical order, we will place u directly to the right of v. This implies that (u, v) will be drawn as a horizontal line segment (that is, bend-less). Similarly, we cope with rule R1b, when additionally $\text{outdeg}_g(u) = 0$. So, there are still two cases to consider: rule R1a and rule R1b, when additionally $\text{outdeg}_g(u) = 1$; see the left part of Fig. 4. In both cases, the current degree of vertex u is 3 and vertex v (and other vertices that are potentially horizontally-aligned with v) must be shifted diagonally up, when u is placed based on the canonical order, such that (u, v) is drawn as a horizontal line segment (that is, bend-less; see the right part of Fig. 4). Note that when v is shifted up, vertex v and all vertices that are potentially horizontally-aligned with v are also of degree 3, since otherwise one of these vertices would not have a neighbor in some later partition of Π, which contradicts the definition of Π.

We complete our case analysis with the case where (u, v) is green. By rule R3a, it follows that (u, v) is the first green incoming edge attached at u. In addition, when (u, v) is placed based on the canonical order, there is no red outgoing edge attached at u. The leftist canonical order also ensures that there is no blue incoming edge at u drawn before (u, v). Hence, u is of degree two, when edge (u, v) is placed, which guarantees that v can be shifted up (potentially with other vertices that are horizontally-aligned with u), such that (u, v) is drawn as a horizontal line segment. We summarize our approach in the following theorem.

Theorem 3. *Let G be a triconnected 6-planar graph with n vertices. A planar octilinear drawing $\Gamma(G)$ with at most $3n-8$ bends can be computed in $O(n^2)$ time.*

Proof. Before the two bend-reduction steps, $\Gamma(G)$ contains at most $6n - 12$ bends. In the first reduction step, all red edges are drawn with one bend. Hence, $\Gamma(G)$ contains at most $5n - 9$ bends. In the second reduction step, we "save" two bends per vertex (except for $v_1 \in P_0$), which yields a reduction by $2n - 1$ bends. So, in total $\Gamma(G)$ contains at most $3n-8$ bends. On the negative side, we cannot keep the running time of our algorithm linear. The reason is the second reduction step, which yields changes in the y-coordinates of the vertices. In the worst case, however, quadratic time suffices. □

Note that there exist 6-planar graphs that do not admit planar octilinear drawings with at most one bend per edge [2]. Theorem 3 implies that on average one bend per edge suffices.

4 Lower Bounds

4.1 4-Planar Graphs

We start our study with the case of 4-planar graphs. Our main observation is that if a 3-cycle C_3 of a graph has at least two vertices, with at least one neighbor in the interior of C_3 each, then at least one edge of C_3 must contain a bend, since the sum of the interior angles at the corners of C_3 exceeds $180°$. In fact, elementary geometry implies that a k-cycle, say C_k with $k \geq 3$, whose vertices have $\sigma \geq 0$ neighbors in the interior of C_k requires (at least) $\max\{0, \lceil (\sigma - 3k + 8)/3 \rceil\}$ bends. Therefore, a bend is necessary. Now, refer to the 4-planar graph of Fig. 5a, which contains $n/3$ nested triangles, where n is the number of its vertices. It follows that this particular graph requires at least $n/3 - 1$ bends in total.

4.2 5- and 6-Planar Graphs

For these classes of graphs, we follow an approach inspired by Tamassia's min-cost flow formulation [17] for computing bend-minimum representations of embedded planar graphs of bounded degree. Since it is rather difficult to implement this algorithm in the case where the underlying drawing model is not the orthogonal model, we developed an ILP instead. Recall that a representation describes the "shape" of a drawing without specifying its exact geometry. This is enough to determine a lower bound on the number of bends, even if a bend-optimal octilinear representation may not be realizable by a corresponding (planar) octilinear drawing.

In our formulation, variable $\alpha(u, v) \cdot 45°$ corresponds to the angle formed at vertex u by edge (u, v) and its cyclic predecessor around vertex u. Hence, $1 \leq \alpha(u, v) \leq 8$. Since the sum of the angles around a vertex is $360°$, it follows that $\sum_{v \in N(u)} a(u, v) = 8$. Given an edge $e = (u, v)$, variables $\ell_{45}(u, v)$, $\ell_{90}(u, v)$ and $\ell_{135}(u, v)$ correspond to the number of left turns at $45°$, $90°$ and $135°$ when moving along (u, v) from vertex u towards vertex v. Similarly, variables $r_{45}(u, v)$, $r_{90}(u, v)$ and $r_{135}(u, v)$ are defined for right turns. All aforementioned variables are integer lower-bounded by zero. For a face f, we assume that its edges are directed according to the clockwise traversal of f. This implies that each (undirected) edge of the graph appears twice in our formulation. For reasons of symmetry, we require $\ell_{45}(u, v) = r_{45}(v, u)$, $\ell_{90}(u, v) = r_{90}(v, u)$ and $\ell_{135}(u, v) = r_{135}(v, u)$. Since the sum of the angles formed at the vertices and at the bends of a bounded face f equals to $180° \cdot (p(f) - 2)$, where $p(f)$ denotes the total number of such angles, it follows that $\sum_{(u,v) \in E(f)} \alpha(u, v) + (\ell_{45}(u, v) + \ell_{90}(u, v) + \ell_{135}(u, v)) - (r_{45}(u, v) + r_{90}(u, v) + r_{135}(u, v)) = 4a(f) - 8$, where $a(f)$ denotes the total number of vertex angles in f, and, $E(f)$ the directed arcs of f in its clockwise traversal.

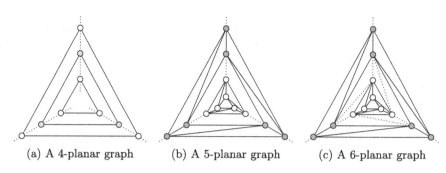

(a) A 4-planar graph   (b) A 5-planar graph   (c) A 6-planar graph

Fig. 5. Planar graphs of different degrees that require (a) $n/3 - 1$, (b) $2n/3 - 2$ and (c) $4n/3 - 6$ bends.

If f is unbounded, the respective sum is increased by 16. Of course, the objective is to minimize the total number of bends over all edges, or, equivalently $\min \sum_{(u,v) \in E} \ell_{45}(u, v) + \ell_{90}(u, v) + \ell_{135}(u, v) + r_{45}(u, v) + r_{90}(u, v) + r_{135}(u, v)$.

Now, consider the 5-planar graph of Fig. 5b and observe that each "layer" of this graph consist of six vertices that form an octahedron (solid-drawn), while octahedrons of consecutive layers are connected with three edges (dotted-drawn). Using our ILP formulation, we prove that each octahedron subgraph requires at least 4 bends, when drawn in the octilinear model (except for the innermost one for which we can guarantee only two bends). This implies that $2n/3 - 2$ bends are required in total to draw the graph of Fig. 5b. For the 6-planar case, we apply our ILP approach to a similar graph consisting of nested octahedrons that are connected by six edges each; see Fig. 5c. This leads to a better lower bound of $4n/3 - 6$ bends, as each octahedron except for the innermost one requires 8 bends. Summarizing we have the following theorem.

Theorem 4. *There exists a class $G_{n,k}$ of triconnected embedded k-planar graphs, with $4 \leq k \leq 6$, whose octilinear drawings require at least: (i) $n/3 - 1$ bends, if $k = 4$, (ii) $2n/3 - 2$ bends, if $k = 5$ and (iii) $4n/3 - 6$ bends, if $k = 6$.*

5   Conclusions

In this paper, we studied bounds on the total number of bends of octilinear drawings of triconnected planar graphs. We showed how one can adjust an algorithm of Keszegh et al. [14] to derive an upper bound of $4n - 10$ bends for general 8-planar graphs. Then, we improved this general bound and previously-known ones for the classes of triconnected 4-, 5- and 6-planar graphs. For these classes of graphs, we also presented corresponding lower bounds.

We mention two major open problems in this context. The first one is to extend our results to biconnected and simply connected graphs and to further tighten the bounds. Since our drawing algorithms might require superpolynomial area (cf. arguments from [2]), the second problem is to study trade-offs between the total number of bends and the required area.

References

1. Badent, M., Brandes, U., Cornelsen, S.: More canonical ordering. J. Graph Algorithms Appl. **15**(1), 97–126 (2011)
2. Bekos, M.A., Gronemann, M., Kaufmann, M., Krug, R.: Planar octilinear drawings with one bend per edge. J. Graph Algorithms Appl. **19**(2), 657–680 (2015)
3. Bekos, M.A., Kaufmann, M., Krug, R.: On the total number of bends for planar octilinear drawings. Arxiv report arxiv.org/abs/1512.04866 (2014)
4. Biedl, T.C.: New lower bounds for orthogonal graph drawings. In: Brandenburg, F.J. (ed.) GD 1995. LNCS, vol. 1027, pp. 28–39. Springer, Heidelberg (1996)
5. Biedl, T.C., Kant, G.: A better heuristic for orthogonal graph drawings. Comput. Geom. **9**(3), 159–180 (1998)
6. Bodlaender, H.L., Tel, G.: A note on rectilinearity and angular resolution. J. Graph Algorithms Appl. **8**(1), 89–94 (2004)
7. De Fraysseix, H., Pach, J., Pollack, R.: How to draw a planar graph on a grid. Combinatorica **10**(1), 41–51 (1990)
8. Di Giacomo, E., Liotta, G., Montecchiani, F.: The Planar Slope Number of Subcubic Graphs. In: Pardo, A., Viola, A. (eds.) LATIN 2014. LNCS, vol. 8392, pp. 132–143. Springer, Heidelberg (2014)
9. Felsner, S.: Schnyder woods or how to draw a planar graph? In: Geometric Graphs and Arrangements, pp. 17–42. Advanced Lectures in Mathematics, Vieweg/Teubner Verlag (2004)
10. Fößmeier, U., Heß, C., Kaufmann, M.: On improving orthogonal drawings: the 4M-algorithm. In: Whitesides, S.H. (ed.) GD 1998. LNCS, vol. 1547, pp. 125–137. Springer, Heidelberg (1999)
11. Garg, A., Tamassia, R.: On the computational complexity of upward and rectilinear planarity testing. SIAM J. Comput. **31**(2), 601–625 (2001)
12. Kant, G.: Drawing planar graphs using the lmc-ordering. In: FOCS, pp. 101–110. IEEE (1992)
13. Kant, G.: Hexagonal grid drawings. In: Mayr, E.W. (ed.) WG 1992. LNCS, vol. 657, pp. 263–276. Springer, Heidelberg (1993)
14. Keszegh, B., Pach, J., Pálvölgyi, D.: Drawing planar graphs of bounded degree with few slopes. SIAM J. Discrete Math. **27**(2), 1171–1183 (2013)
15. Liu, Y., Morgana, A., Simeone, B.: A linear algorithm for 2-bend embeddings of planar graphs in the two-dimensional grid. Discrete Appl. Math. **81**(1–3), 69–91 (1998)
16. Nöllenburg, M.: Automated drawings of metro maps. Technical Report 2005–25, Fakultät für Informatik, Universität Karlsruhe (2005)
17. Tamassia, R.: On embedding a graph in the grid with the minimum number of bends. SIAM J. Comput. **16**(3), 421–444 (1987)
18. Tamassia, R., Tollis, I.G., Vitter, J.S.: Lower bounds for planar orthogonal drawings of graphs. Inf. Process. Lett. **39**(1), 35–40 (1991)

Bidirectional Variable-Order de Bruijn Graphs

Djamal Belazzougui[1], Travis Gagie[2(✉)], Veli Mäkinen[2], Marco Previtali[3],
and Simon J. Puglisi[2]

[1] Center for Research on Technical and Scientific Information (CERIST),
Algiers, Algeria
[2] Department of Computer Science, Helsinki Institute for Information Technology,
University of Helsinki, Helsinki, Finland
travis.gagie@gmail.com
[3] Department of Computer Science, University of Milano-Bicocca, Milan, Italy

Abstract. Implementing de Bruijn graphs compactly is an important problem because of their role in genome assembly. There are currently two main approaches, one using Bloom filters and the other using a kind of Burrows-Wheeler Transform on the edge labels of the graph. The second representation is more elegant and can even handle many graph-orders at once, but it does not cleanly support traversing edges backwards or inserting new nodes or edges. In this paper we resolve the first of these issues and partially address the second.

1 Introduction

De Bruijn graphs are central to state-of-the-art methods for genome assembly [1,6,13,20,23], which is in turn fundamental to bioinformatics and many modern biological and medical projects [10,18,24,25]. The assembly process builds long contiguous DNA sequences, called contigs, from a set of much shorter DNA fragments, called reads. All de-Bruijn-graph-based assemblers follow the same general outline: extract the $(K + 1)$-mers from the reads, for some value K; construct the de Bruijn graph for the $(K + 1)$-mers; simplify the graph; and take contigs to be simple paths in the graph. The value of K can be, and is often required to be, specified by the user. Construction and navigation of the graph is a space and time bottleneck in practice and the main hurdle for assembling large genomes, so space-efficient representations of de Bruijn graphs have been the focus of intense research in recent years.

Simpson et al. [23] made one of the first attempts to reduce the space required by the graph via the use of a distributed hash table. For a set of reads from a human genome (HapMap: NA18507), this method requires 336 GB to store the graph. Conway and Bromage [9] reduced space requirements to 32 GB for the same data set by using a sparse bitvector (see [17]) to represent the edges

Supported by the Academy of Finland through grants 258308, 268324, and 284598, and Italian MIUR PRIN 2010–2011 grant Automi e Linguaggi Formali: Aspetti Matematici e Applicativi (code 2010LYA9RH). This work was done while the fourth author was visiting the University of Helsinki.

© Springer-Verlag Berlin Heidelberg 2016
E. Kranakis et al. (Eds.): LATIN 2016, LNCS 9644, pp. 164–178, 2016.
DOI: 10.1007/978-3-662-49529-2_13

(i.e. the $(K+1)$-mers). Efficient traversal of the graph was supported by rank and select operations over the bitvectors. Minia, by Chikhi and Rizk [8], uses a Bloom filter to store edges (with additional structures to avoid false positive edges that would affect the assembly). They traverse the graph by generating all possible outgoing edges[1] at each node and testing their membership in the Bloom filter. Using this approach, the graph was reduced to 5.7 GB on the same dataset. Salikov et al. [21] recently used cascading Bloom filters to further reduce the space requirements of Chikhi and Rizk's approach. Other authors have also made use of Bloom filters to implement de Bruijn graphs for metagenomic assembly [19] and pan-genomics [11].

Bowe, Onodera, Sadakane and Shibuya [4] developed a different succinct de Bruijn graph data structure based on the Burrows-Wheeler transform (BWT) [5]. We refer to this representation as BOSS, after the authors' initials. On the above mentioned data set the BOSS representation requires 2.5 GB. Boucher et al. [3] have since shown how to augment the BOSS representation to allow changing the order of the graph (denoted $k \leq K$) during traversal, while adding little space overhead to the original representation. Supporting variable-orders is useful when non-uniform sampling of the reads leads to some parts of the graph being sparser than others. Our results in this paper build on the BOSS and variable-order BOSS representations, and so we describe them briefly in Sect. 2 below. More recently, Chikhi et al. [7] implemented the de Bruijn graph using a combination of the BWT and minimizers. Their method uses 1.5 GB on the same NA18507 data.

The BOSS representation is asymmetric in the sense that, when visiting a node v, it takes much longer to follow the edge with a given label arriving at v than it does to follow the edge with that label leaving v. As we describe in Sect. 2.4, we can easily make the original BOSS representation bidirectional by storing a second BWT; to make the variable-order BOSS representation bidirectional in the same way, however, would require another BWT for each order. Our first contribution is a space-efficient bidirectional variable-order BOSS representation. Although we admit to being motivated partly by aesthetics — symmetry is beautiful — we also think that such a representation could be useful someday, either for bioinformatics or for some other application.

For example, given a long string containing relatively few distinct K-mers, we can build the BOSS representation of its de Bruijn graph and later, given a pattern of length at most K, we can quickly determine whether the string contains any occurrences of that pattern. Moreover, if we build a bidirectional variable-order BOSS representation, then we should be able to use it to list all the distinct substrings of length at most K that approximately match the pattern (see [2] for a discussion of how to use a bidirectional BWT for approximate pattern matching). We can even use the de Bruijn graph to compile a list of candidate approximate matches of length more than K, which contains all the approximate matches in the string and possibly some false positive matches

---

[1] Note that there are at most four possible outgoing edges, corresponding to each letter of the DNA alphabet {A, C, G, T}.

that are not contained in the string. Once we know all the distinct substrings approximately matching the pattern (and possibly some false positives), we can find all their occurrences in the string itself via exact pattern matching. At least in theory, therefore, we should be able to use a bidirectional variable-order BOSS representation to reduce approximate pattern matching to exact pattern matching. We are currently investigating this possibility.

The BOSS representation also lacks support for adding new nodes and edges, although this is possible with some representations based on Bloom filters. Our second contribution is to augment the BOSS representation such that we can both add and delete nodes and edges efficiently. It should be possible to adapt our approach to add support for deletions to implementation based on Bloom filters, but we leave this as future work. We make only the fixed-order BOSS representation dynamic in this paper; in the full version we will also show how to do the same for the variable-order representation. Our motivation for this contribution is mainly to provide a representation with complete functionality, without an immediate application. A dynamic bidirectional variable-order BOSS representation could be useful, however, for the approximate-matching problem discussed in the previous paragraph when the long string to be indexed is dynamic.

The rest of this paper is laid out as follows: in Sect. 2 we review the definition of de Bruijn graphs, the fixed- and variable-order BOSS representations and the bidirectional BWT; in Sect. 3 we describe our bidirectional variable-order BOSS representation; and in Sect. 4 we show how to make a fixed-order BOSS representation dynamic.

2   Preliminaries

2.1   De Bruijn Graphs

In bioinformatics, a de Bruijn graph of order K over some alphabet is a directed graph in which each node is labelled by a K-tuple from that alphabet, such that if there is an edge between nodes u and v then u's K-tuple can be changed into v's K-tuple by deleting the first character and appending a character. (Some authors assume that, if it is possible to change u's label into v's label this way, then the edge is always present.) As an aside, we note that such a graph is a subgraph (induced, under the assumption just mentioned) of the de Bruijn graph of order K as defined in combinatorics.

2.2   BOSS Representation

To construct the BOSS representation [4] of a Kth-order de Bruijn graph from a set of $(K+1)$-mers, we first add enough dummy $(K+1)$-mers starting with \$s so that if αa is in the set, then some $(K+1)$-mer ends with α (α a K-mer, a a symbol). We also add enough dummy $(K+1)$-mers ending with \$ that if $b\alpha$ is in the set, with α containing no \$ symbols, then some $(K+1)$-mer starts with α.

We then sort the set of $(K+1)$-mers into the right-to-left lexicographic order of their first K symbols (with ties broken by the last symbol) to obtain a matrix. If the ith through jth $(K+1)$-mers start with α, then we say node $[i,j]$ in the graph has label α, with $j-i+1$ outgoing edges labelled with the last symbols of the ith through jth $(K+1)$-mers. If there are n nodes in the graph, then there are at most σn rows in the matrix, i.e., $(K+1)$-mers. We store the last column W of the matrix, which functions as a BWT of the edge labels. Nodes correspond to the substrings of this BWT that contain the edge labels. We also store a bitvector L with 1 s marking the each character in W that labels the last outgoing edge of each node, and another bitvector marking how many incoming edges each node has. Bowe et al. described a number of queries for traversing the graph, all of which can be implemented in terms of the following three basic queries, with at most an $\mathcal{O}(\sigma)$-factor slowdown:

- forward(v, a) returns the node w reached from v by an edge labelled a, or NULL if there is no such node;
- backward(v) lists the nodes u with an edge from u to v;
- lastchar(v) returns the last character of v's label.

2.3 Variable-Order BOSS

In the BOSS representation, nodes correspond to the intervals of the edge-BWT. Boucher et al. [3] augment the BOSS representation of the original graph, by storing the length of the longest common suffix of each consecutive pair of nodes, to support the following three queries:

- shorter(v, k) returns the node whose label is the last k characters of v's label;
- longer(v, k) lists nodes whose labels have length $k \leq K$ and end with v's label;
- maxlen(v, a) returns some node in the original graph whose label ends with v's label, and that has an outgoing edge labelled a, or NULL if there is no such node.

Together, these operations allow the order (k) of the de Bruijn graph to be changed on the fly. The main addition to the BOSS representation is a wavelet tree over the array L^* storing the length of the longest suffix common to each row in the BOSS matrix and the preceding row.

2.4 Bidirectional BWT

Given a text string $T = t_1, t_2, \ldots, t_n$, where each $t_i \in \Sigma$, the *suffix array* of $T\$$ is an array $A[1..n+1]$ where $A[1] = n+1$ and $T[A[i]..n]\$$, for $i > 1$, is the lexicographically i-th smallest suffix of $T\$$, with \$ interpreted as a unique smallest character. A *lexicographic range of suffixes* is an interval in A. The *bidirectional Burrows-Wheeler transform* (bidirectional BWT) [14,22] allows us to move from a lexicographic range of text suffixes *prefixed by* a pattern string P to the range cP and to the range Pc, for any character $c \in \Sigma$, without needing an explicit suffix array to be available. In [2] an $O(n \log \sigma)$ bits data structure was developed to support these operations in constant time.

3 Bidirectional BOSS

BOSS allows us to move backward in the graph, accurately using rank and select on W and L (defined in Subsect. 2.2). This procedure has a major drawback, which is that is we cannot read which is the label of the edge we are traversing backward but can only read labels when we traverse edges forward.

A naïve yet inefficient solution would be to traverse k edges backward in order to retrieve the character in the first position of the current k-mer. Note that this way we *shift* the label of the current vertex until the first character becomes the label of the outgoing edge of the reached vertex. This procedure clearly requires $\mathcal{O}(k)$ for each incoming edge of the source node. Moreover, if the current node has j incoming edges we should perform $j \times k$ backward steps in order to find all the possible backward labels of those edges. Therefore, for an alphabet of size σ this procedure would require $\mathcal{O}(\sigma \times k)$.

For fixed-k BOSS we can avoid backtracking in the graph by storing the first character of each k-mer, although this approach is not viable in variable-order DBGs since it require storing the whole k-mers. Indeed, note that storing the first character of the k-mers allows us to gather the backward label of an edge in the graph of order k but we need to store the second character in order to gather the backward label of an edge in the graph of order $k - 1$, the third character for the graph of order $k - 2$, etc. We now introduce an elegant and efficient approach to move backward and forward in BOSS, namely bidirectional BOSS (biBOSS for short). This idea is loosely inspired by bidirectional BWT.

First, note that if we build the DBG of order k for a set of strings and their reverse, we obtain two isomorphic graphs; we refer to the former as DBG_k^f and to the latter as DBG_k^r. For each vertex v_i^f with label l_i in DBG_k^f there is a vertex v_i^r with label $\texttt{reverse}(l_i)$ in DBG_k^r and for each edge $e_h^f = (v_i^f, v_j^f)$ labeled with $\sigma^f = v_j^f[k]$ in DBG_k^f there is an edge $e_h^r = (v_j^r, v_i^r)$ labeled with $\sigma^r = v_i^r[k] = v_i^f[1]$ in DBG_k^r. Therefore, if we can maintain a link between the nodes and the edges in the two graphs we can easily retrieve the forward and backward labels simply by looking at e_h^f and e_h^r.

Moreover, note that outgoing edges from v_i^f in DBG_k^f are edges incoming to v_i^r in DBG_k^r and, conversely, edges outgoing from v_i^r are incoming to v_i^f. This remark clearly points out that we can simulate a backward step in DBG_k^f with a forward step in DBG_k^r without any need for further backtracking in neither DBG_k^f nor DBG_k^r.

A biBOSS for a set of strings S is therefore a data structure composed by two BOSSs. The first one, $BOSS^f$, is the BOSS data structure for S whereas the second one, $BOSS^r$, is the BOSS data structure for $S^r = \{\texttt{reverse}(s_i) \mid s_i \in S\}$. Each node v_i^f in DBG_k^f is defined in $BOSS^f$ as an interval $I_i^f = [i, j]$ over W, L, and L^* (W^f, L^f, and L^{*f} from now on). (Recall the definition of L^* from Subsect. 2.3.) The last vector is needed only when we want to augment BOSS to support variable order. Conversely, each node v_i^r in DBG_k^r is defined in $BOSS^r$ as an interval $I_i^r = [p, q]$ over W^r, L^r, and L^{*r}.

Therefore, in order to support forward and backward navigation of the DBG we propose maintaining a pair of intervals (I_i^f, I_i^r), one to describe the vertex v_i^f (I_i^f) and the other to describe the vertex v_i^r (I_i^r). From now on we will use v_i^f and v_i^r to define vertices and intervals interchangeably.

Boucher et al. [3] described three main functions for a variable order BOSS, namely shorter, longer, and maxlen in order to move upwards and downwards in the order of the DBGs and to move forward. Obviously we want to support the same set of functions so we will provide the bidirectional versions of the first two. We will not define maxlen for biBOSS since it is related to a specific direction of the graph. The examples in Figs. 1, 2 and 3 show a biBOSS graphical representation for a DBG of order 5. We will use this example to show the execution of the new procedures graphically.

Let v_i^f and v_i^r be two vertices of order k labeled by l_i and reverse(l_i) and let K be the maximum order of the variable order BOSS. We define bi-shorter as bi-shorter(v_i^f, v_i^r) = (v_j^f, v_j^r) such that the labels of v_j^f and v_j^r are respectively $l_i[2 : k]$ and reverse($l_i[2 : k]$) = reverse(l_i)[1 : k − 1]. This function returns the node in DBG_{k-1}^f which label is the last $k − 1$ character of l_i and its linked node in DBG_{k-1}^r. Note that we can swap the two vertices in order to move downward between the orders of DBG_*^r.

Computing v_j^f is straightforward since we only need to compute shorter ($v_i^f, k − 1$) as defined in Sect. 2.3. This procedure requires $\mathcal{O}(\log K)$ time [3].

Computing v_j^r is more challenging since we need to remove the last character of reverse(l_i). Nevertheless we can easily find a node in the graph with maximum order K that ends with reverse(l_i)[1 : k − 1] selecting any position in the interval v_i^r and applying backward as defined in Bowe et al. [4] and then moving downwards in the order of the graph. More formally we can say that $v_j^r = $ shorter(backward(maxlen($v_i^r, *$)), k − 1). This procedure requires $\mathcal{O}(\log K)$ time since we can compute maxlen and backward in constant time and shorter in $\mathcal{O}(\log K)$.

The example in Fig. 1 shows the computation of bi-shorter on the pair of linked vertices $v_i^f = AAGTA$ and $v_i^r = ATGAA$. The vertex $v_j^f = AGTA$ is computed by finding the L^{*f}-interval of order 4 that contains v_j^f. In order to compute v_j^r we first gather any node in the original graph contained in v_i^r (in this case the same node since v_i^r has maximum order 5) and, by applying backward to it, obtain a node in which the reversed label of v_j^f is a suffix ($CATGA$). Finally we move downwards in the order of $BOSS^r$ obtaining the vertex $v_j^r = ATGA$.

Let $\sigma_k \in \Sigma$ be a character in the alphabet; we define bi-longer as

$$\text{bi-longer}(v_i^f, v_i^r, \sigma_k) = (v_j^f, v_j^r)$$

such that the labels of v_j^f and v_j^r are respectively $\sigma_k \cdot l_i$ and reverse(l_i) $\cdot \sigma_k$, thus the two computed vertices have reversed labels. Note that bi-longer is slightly different than its corresponding function in variable order BOSS, namely longer. Indeed, applying the latter function (longer) to a vertex $v \in DBG_i^f$ in order to gather the nodes of order j, returns a list of vertices $V \in DBG_j^f$ such

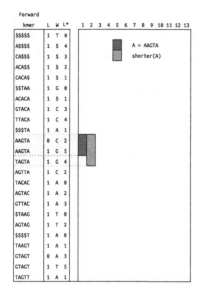 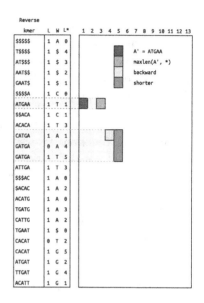

Fig. 1. A graphical example of `bi-shorter`.

that each vertex $v_h \in V$ has a label that ends with the label of v. `bi-longer`, instead, allow us to select the character we want to concatenate to the labels. Nevertheless, gathering all the vertices from `bi-longer` to simulate `longer` is straightforward.

Clearly, we cannot directly compute v_j^f using `longer`. Our goal is therefore to provide a method that allows us to select the correct vertex from the list produced by `longer`, i.e. the one labeled with $\sigma_k \cdot l_i$. First, note that if V is the list returned by `longer`($v_i^f, k+1$) then the labels of the vertices in V end with l_i and have length equal to $|l_i| + 1$. Moreover, the vertices in V are sorted by RLO (reverse lexicographic order) and for each vertex v_h in V the first character of its label is in the interval v_i^r. Indeed, since v_h is a backward extension of v_i^f its first character is a label of an edge outgoing from v_i^r. This remark hints that we can analyze v_i^r in order to correctly label the vertices in V. It is easy to prove that $|V|$ is equal to the cardinality of the set of the distinct characters in v_i^r and since the elements of V are sorted we can easily link each vertex with its first character and select the vertices that starts with σ_k (if it exists).

Computing v_j^r is straightforward, we need only to follow an edge labeled by σ_k (if it exists) and then find the vertex of order $k+1$. More formally, if t is the cardinality of the set $\{c \mid c \in v_i^r \cap c < \sigma_k\}_{\neq}$, v_j^f and v_j^r can be computed respectively as `longer`($v_i^f, k+1$)[$t+1$] and `shorter(forward(maxlen`(v_i^r, σ_k))$, k+1$).

This procedure requires $\mathcal{O}(\sigma \log \sigma + |V| \log K)$ time since computing t requires us to compute the rank values for each character in the alphabet at the beginning and at the end of the interval of v_i^r ($\mathcal{O}(\sigma \log \sigma)$), `longer` takes $\mathcal{O}(|V| \log K)$, `forward` takes constant time, `maxlen` takes $\mathcal{O}(\log \sigma)$, `shorter`

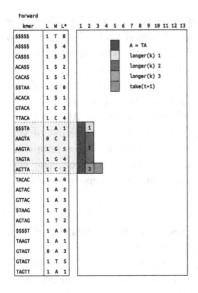 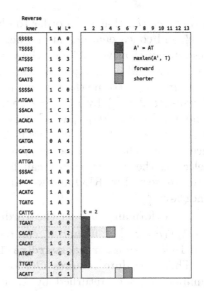

Fig. 2. A graphical example of `bi-longer`.

takes $\mathcal{O}(\log K)$, and selecting an element from a list takes constant time. When $\sigma = \mathcal{O}(1)$, therefore, `bi-longer` takes $\mathcal{O}(\log K)$ time.

The example in Fig. 2 shows the computation of `longer` on the pair of linked vertices $v_i^f = TA$ and $v_i^r = AT$ with $\sigma_k = T$. The interval v_i^f is first split into the 3 possible vertices of order 3 using L^{*f}. By analyzing v_i^r we find that the characters at the beginning of each 3-mer are respectively \$, G, and T. We therefore select the third interval since \$ $< G < T$. In order to compute v_j^r we first select an edge in v_i^r labeled with T (the one outgoing from $CACAT$) and follow it, obtaining a vertex in the graph of maximum order 5 that has the label of v_j^r as suffix ($ACATT$). As a last step we move upward in the order of the graph using the information stored in L^{*r} and obtaining the correct vertex in $BOSS^r$ (ATT, represented by the same interval).

Until now we have described how we can move between the different orders described by the variable order BOSS maintaining the linking between the nodes. We will now show how to move forward in either of the two graphs and maintain the link with the vertex in the other one with reversed label. Note that this actually means we move backward in one of the two graphs by selecting the correct graph in which to perform the forward step.

Let σ_k be a character in the alphabet; we define $\texttt{FwdBwd}(v_i^f, v_i^r, \sigma_k) = (v_j^f, v_j^r)$ such that the labels of v_j^f and v_j^r are respectively $l_i[2:k] \cdot \sigma_k$ and $\sigma_k \cdot \texttt{reverse}(l_i)[1:k-1] = \texttt{reverse}(l_i[2:k] \cdot \sigma_k)$.

Computing v_j^f is straightforward since we only need to compute `forward` (v_i^f, σ_k) as defined in Sect. 2.3. This step requires $\mathcal{O}(\log K)$.

Computing v_j^r is mostly a combination of the previous two functions. Indeed, if we consider the labels of v_i^r (l_i) and v_j^r (l_j) we can note that, for some $c_h \in \Sigma$,

$l_i \cdot c_k = c_h \cdot l_j$, that is, in order to obtain v_j^r we must remove the last character of v_i^r and concatenate c_k at the beginning of the obtained node. This two-step description clearly highlights the connections with `bi-shorter` (delete the last character) and `bi-longer` (concatenate a character at the beginning).

Our proposed method is as follows. First we compute the interval for `reverse`($l_i[2 : k]$) by applying `backward` to v_i^r; note that this step produces a node (v_h^r) in DBG_{k-1}^r which label is a suffix of the label of v_j^r. At this point it is easy to see that we can apply `longer` to v_h^r in order to find the list of vertices V that share the label of v_h^r as suffix; clearly v_j^r will be in V by definition. Selecting the correct vertex can be done similarly as in `bi-longer`, the vertices are sorted by RLO and we can access the different characters by analyzing `shorter`($v_i^f, k-1$).

More formally, if t is the cardinality of the set $|\{c \mid c \in$ `shorter`(v_i^f, $k-1$) $\cap c < \sigma_k\}_{\neq}|$, v_j^f and v_j^r can be computed respectively as `forward`(v_i^f, σ_k) and `longer`(`shorter`(`backward`(`maxlen`($v_i^r, *$)), $k-1$), $k+1$)[t].

This procedure requires $\mathcal{O}(\sigma \log \sigma + |V| \log K)$ time where $|V| \leq \sigma$ is the number of nodes returned by `longer`. Computing i requires us to compute `shorter` ($\mathcal{O}(\log K)$) and perform the same rank operation as for `bi-longer` ($\mathcal{O}(\sigma \log \sigma)$), computing v_j^f requires $\mathcal{O}(\log K)$, and computing v_j^r requires $\mathcal{O}(|V| \log K)$ since `maxlen` and `backward` can be computed in constant time, `shorter` requires $\mathcal{O}(\log K)$, and `longer` requires $\mathcal{O}(|V| \log K)$. When $\sigma = \mathcal{O}(1)$, therefore, $fwdbwd$ takes $\mathcal{O}(\log K)$ time.

The example in Fig. 3 shows the computation of `FwdBwd` on the pair of linked vertices $v_i^f = GTA$ and $v_i^r = ATG$ with $\sigma_k = G$. First we gather the index t, that we will use in the last step, by applying `shorter` to v_i^f (obtaining the vertex TA) and counting the number of distinct character smaller than σ_k. We then compute v_j^f by selecting an edge in v_i^f labeled with σ_k and following it, obtaining the vertex in the original graph labeled with $AGTAG$. We then apply `shorter` in order to gather the vertex v_j^f of order 3. In order to compute v_j^r we first select any edge in v_i^r and traverse it backward obtaining a vertex with suffix AT. We then compute v_j^r by applying `shorter` (order $k-1$) and `longer` (order k) and selecting the t-th interval computed. This example clearly shows why we cannot directly compute `shorter` of order k. When we select a random edge in v_i^r we cannot directly access the k-mers so we may concatenate a character $\sigma_h \neq \sigma_k$ at the beginning of the label of our vertex (in our example $\sigma_h = C$). We therefore need to get rid of this character by moving downward between the orders obtaining the node AT and then select the correct vertex by moving upward using the information of the edges outgoing from its reverse ($TA = $ `shorter`($v_i^f, k-1$)).

Theorem 1. *When $\sigma = \mathcal{O}(1)$, we can store a variable-order de Bruijn graph of maximum order K in $\mathcal{O}(n \log K)$ bits on top of the BOSS representation of the order-K graph, where n is the number of nodes in the order-K graph, such that incrementing or decrementing the order and forward or backward traversals take $\mathcal{O}(\log K)$ time.*

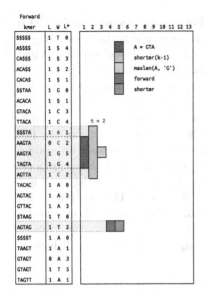
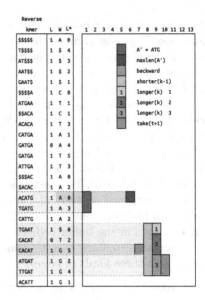

Fig. 3. A graphical example of FwdBwd.

4   Updating the Graph

Implementations based on Bloom filters can usually be made semi-dynamic (insertions only) without much difficulty. We can easily make our data structure dynamic by using dynamic sequences (see, e.g., [15]), but this would slow down our queries even on stable parts of the graph by an almost logarithmic factor.[2] In this section we show how we can achieve $O(\log^{2+\epsilon} m)$ time updates (deletes and insertions) while retaining the same $O(1)$ query time. Here m denotes the total number of edges in the graph. For simplicity and due to space constraints, here we describe only how to make a fixed-order BOSS representation dynamic; we will show how to make our variable-order representation dynamic in the full version of this paper.

The basic idea is to use a dynamic pointer-based structure to describe all new added nodes. We call this data structure a *buffer*. The buffer will also

---

[2] In the original (single-order) BOSS representation, one maintains a sequence of characters from alphabet 2σ and a bitvecor. The sequence is built by sorting the k-mers (nodes) in lexicographic order and for each k-mer storing the characters labelling its outgoing edges. The bitvector is used to mark the beginning of the labels of each k-mer in the sequence. We additionally use another bitvector to mark the beginning of each block of edges outgoing from a group of nodes whose labels have the same prefix of length $k - 1$. Queries, and insertions or deletions of nodes and edges can then easily be done by rank and select queries and insertions, deletions in the two bitvectors and the sequence (the reader can easily guess the details by looking at the original BOSS representation [4]). The total time for an update or query will be $O(\log m / \log \log m)$.

store the edges that connect new nodes. The buffer capacity will be limited to $O(m/\log^{1+\epsilon} n)$ edges. Since the description of an edge or a node takes $O(\log m)$ bits, the total space used by the data structure will be $O(m/\log^{\epsilon} m)$ bits. Whenever the buffer reaches its capacity we rebuild the whole succinct data structure in $O(m \log m/\log\log m)$ time and in succinct space. The amortized time will be $O(\log^{2+\epsilon} m)$ by edge or node. The process can be deamortized by standard techniques, e.g., keeping two versions of the static representation, one under construction and one on which queries are made. The construction of the first representation is advanced by $O(\log^{1+\epsilon} m)$ steps each time we have an update. We will store a dictionary for the new nodes stored in the buffer (henceforth the *new-node dictionary*). This dictionary will be based on Karp-Rabin hashes which are now computed on the full k-mers that label the nodes. This dictionary will work with high probability and will allow us at insertion time to check whether a node exists in the buffer or needs to be created.

It remains to show how to remove or add edges. For this, we will have to distinguish three kinds of edges:

1. an edge that connects an old node with another old node, both present in the static structure.
2. an edge that connects a new node with a new node, both present in the buffer.
3. an edge that connects a new node with an old node (one is present in the static structure and the other in the buffer).

We will use some additional information and data structures which will allow us to properly encode and handle insertions of the different kinds of edges.

In order to manage the first type of edge, we will use a hash-based dictionary which will store a set of pairs (key, value), where key is an identifier of the edge made of an integer describing one of the two ending nodes and the character labelling the edge, and value is satellite information that stores the identifier of the other ending node. Each edge will thus generate two entries in the dictionary, one for each of the two ending nodes. This dictionary will thus store information about old nodes, and the hash values used for building it are computed on the identifiers of the nodes (the field key). The field value will either store a pointer to the other node in the buffer, if the node is new, or an integer identifier if the node is old. We call this dictionary the *old-edge dictionary*.

In order to manage the second kind of edges, we will store another dictionary which will store a pair (key, value), where key is a pair of a pointer to a new node and a character that labels an edge outgoing from that node. The field value will an identifier of the other ending node, which will be either an integer if the node is old or a pointer in the pointer-based structure if the node is new. We will call this dictionary the *new-edge dictionary*.

Finally for each node in the graph, we will keep the locus of the longest common prefix of the $k-$mer that labels the node with the suffix tree of the static structure. This locus will be represented with a pair (STnodeID, depth), where STnodeID is the identifier of a node in the suffix tree of the static dBG representation and depth is the length of that longest common prefix. In order to be able to maintain the locus for all nodes, we will store the suffix tree topology of the static dBG representation as well as the support for the operation which provides the string depth of every node. While navigating the graph we will always keep the Karp-Rabin hash of the k-mer that labels the current node, even if the node exists in the static structure.

We are now ready to describe how updates and queries are supported. For queries, supposing we are at an old node and want to check for the existence of an outgoing edge labelled by character a. We first check whether that edge exists in static structure using the usual algorithm. If that is not the case, we check for the existence of the edge using the *old-edge dictionary*. If the edge exists in the dictionary, we will get the identifier of the other ending node of the edge. That edge might be in either the static structure or in the buffer. We then generate the Karp-Rabin hash of that node from the Karp-Rabin hash of the current node in constant time. We now describe the queries, whenever we are at a new node (a node in the buffer). In this case, we query the *old-edge dictionary*, using the pointer to the node in the buffer and the edge label. If the answer is positive, we get the other ending node and compute the Karp-Rabin hash of that node from the Karp-Rabin hash of current node in constant time.

In order to support insertions we will need to maintain the locus information. In particular for that, suppose we are at a node N_0 with locus ($\texttt{STnodeID}_0, \texttt{depth}_0$) and have an edge labelled with character a starting from that node and ending in a node N_1 (the edge N_1 might or might not exist yet). We can compute the locus of node N_1 denoted by ($\texttt{STnodeID}_1, \texttt{depth}_1$), by doing binary search on ancestors of $\texttt{STnodeID}$ in the suffix tree topology of the static dBG, and each time checking whether the node has a Weiner link labelled with a. The binary search will determine the deepest (closest) ancestor of $\texttt{STnodeID}_0$ with Weiner link labelled by a. Then $\texttt{depth}_1$ will be equal to the depth of that ancestor incremented by one and $\texttt{STnodeID}_1$ will be the target of the Weiner link that starts from that ancestor and is labelled by a. The total time needed to determine the locus will be $O(\log k \cdot t_{\texttt{ancestor}} + t_{\texttt{depth}})$, where $t_{\texttt{ancestor}}$ and $t_{\texttt{depth}}$ are respectively the times needed to support the $\texttt{ancestor}$ and $\texttt{depth}$ operations on the suffix tree topology of the static dBG. By setting $t_{\texttt{ancestor}}$ to $O(1)$ and $t_{\texttt{depth}}$ to $O(\log^{1+\epsilon} m)$ and by considering that $k \leq m$, we deduce that the total time for determining the locus is $O(\log^{1+\epsilon} m)$.

We now describe how updates are supported. We start with the description of insertions and, specifically, with the case of an insertion of an edge that starts from a node that exists in the static structure (evidently, we assume that the edge exists neither in the static structure nor in the buffer). Supposing we are at a node N_0 and want to determine the ending node N_1 of a new edge that connects N_0 to N_1 and is labelled by a. Since the k-mer that corresponds to that edge does not exist, we cannot directly use the navigation on the static structure to determine N_1 from N_0; instead we will use the locus computation as described above, starting from the suffix tree node of N_1 and then check whether the locus of N_1 has depth k. If that is the case, then the edge exists in the static structure and we insert it into the *old-edge dictionary* and we are done. Otherwise the node does not exist in the static structure and we check whether it exists in the buffer by querying the *new-node dictionary*. If the node is not found, we insert it into the *new-node dictionary*. We finally insert the edge N_0 to N_1 into the *old-edge dictionary*.

We now describe the case, where the edge starts at a node N_0 that exists in the buffer. This case is very similar to the previous one, the main difference being that now the locus of the node will have a depth $k' < k$. We can then compute the locus N_1 and as before, if the locus has depth k, then we know the node N_1 will be in the static structure. If it has depth less than k, then we query the node *new-edge dictionary*, and if it does not exist, we create the node N_1 and store it in the *new-node dictionary*. We finally insert the edge in the *old-edge dictionary* and we are done.

Another case would be the one in which we are at a node in the static structure and the node we want to go to does not exist in the static structure. We can check for its existence in the buffer by querying the *new-node dictionary*. This can be done since we can easily compute the Karp-Rabin hash of the new node from the Karp-Rabin hash of the current node (which, by assumption, we already have) and the edge label. If the node exists, then we just need to insert an element in the *old-edge dictionary* and an element in the *new-edge dictionary*.

Supporting deletions is straightforward. If the edge is new (it did not exist in the static structure), we can remove it by updating the three dictionaries. To remove edges that exist into old structure, we will need to use another dictionary which we call *removed-edge dictionary* and which will store the removed edges. Consequently queries will have to be modified so that an edge in the old structure is reported only if it is not found in the *removed-edge dictionary*.

Finally the reconstruction algorithm will use the dynamic algorithm of [16] which uses space $m \log \sigma + o(m \log \sigma)$ bits of space to construct the BWT in time $O(n \log m / \log \log m)$ and [12] to construct the suffix tree topology and the support for string depth operation in time $O(m \log^\epsilon m)$.

Theorem 2. *We can maintain a dynamic dBG representation in $O(m \log \sigma)$ bits of space with query time $O(1)$ and expected update time $O(\log^{2+\epsilon} m)$ (with no amortization) to insert or delete a node or edge, where m is the number of distinct k-mers. The data structure works with high probability (it has polynomially small probability of failure).*

We note that we can make the data structure error-free if we can afford to check the label of a node, by following $O(k)$ edges. We can thus remove the last sentence of the theorem if $k = O(\log^{2+\epsilon} n)$.

References

1. Bankevich, A., et al.: SPAdes: a new genome assembly algorithm and its applications to single-cell sequencing. J. Comput. Biol. **19**(5), 455–477 (2012)
2. Belazzougui, D., Cunial, F., Kärkkäinen, J., Mäkinen, V.: Versatile succinct representations of the bidirectional burrows-wheeler transform. In: Bodlaender, H.L., Italiano, G.F. (eds.) ESA 2013. LNCS, vol. 8125, pp. 133–144. Springer, Heidelberg (2013)

3. Boucher, C., Bowe, A., Gagie, T., Puglisi, S.J., Sadakane, K.: Variable-order de Bruijn graphs. In: Proceedings of the Data Compression Conference (DCC), pp. 383–392. IEEE (2015)
4. Bowe, A., Onodera, T., Sadakane, K., Shibuya, T.: Succinct de Bruijn graphs. In: Raphael, B., Tang, J. (eds.) WABI 2012. LNCS, vol. 7534, pp. 225–235. Springer, Heidelberg (2012)
5. Burrows, M., Wheeler, D.J.: A block sorting lossless data compression algorithm. Technical report 124, Digital Equipment Corporation (1994)
6. Butler, J., et al.: ALLPATHS: de novo assembly of whole-genome shotgun microreads. Genome Res. **18**(5), 810–820 (2008)
7. Chikhi, R., Limasset, A., Jackman, S., Simpson, J.T., Medvedev, P.: On the representation of de Bruijn graphs. In: Sharan, R. (ed.) RECOMB 2014. LNCS, vol. 8394, pp. 35–55. Springer, Heidelberg (2014)
8. Chikhi, R., Rizk, G.: Space-efficient and exact de Bruijn graph representation based on a Bloom filter. Algorithm. Mol. Biol. **8**(22) (2012)
9. Conway, T.C., Bromage, A.J.: Succinct data structures for assembling large genomes. Bioinformatics **27**(4), 479–486 (2011)
10. Haussler, D., et al.: Genome 10K: a proposal to obtain whole-genome sequence for 10,000 vertebrate species. J. Hered. **100**(6), 659–674 (2009)
11. Holley, G., Wittler, R., Stoye, J.: Bloom filter trie – a data structure for pangenome storage. In: Pop, M., Touzet, H. (eds.) WABI 2015. LNCS, vol. 9289, pp. 217–230. Springer, Heidelberg (2015)
12. Hon, W.-K., Sadakane, K.: Space-economical algorithms for finding maximal unique matches. In: Apostolico, A., Takeda, M. (eds.) CPM 2002. LNCS, vol. 2373, pp. 144–152. Springer, Heidelberg (2002)
13. Li, R., et al.: De novo assembly of human genomes with massively parallel short read sequencing. Genome Res. **20**(2), 265–272 (2010)
14. Li, R., Yu, C., Li, Y., Lam, T.-W., Yiu, S.-M., Kristiansen, K., Wang, J.: SOAP2. Bioinformatics **25**(15), 1966–1967 (2009)
15. Munro, J.I., Nekrich, Y.: Compressed data structures for dynamic sequences. In: Bansal, N., Finocchi, I. (eds.) ESA 2015. LNCS, vol. 9294, pp. 891–902. Springer, Heidelberg (2015)
16. Navarro, G., Nekrich, Y.: Optimal dynamic sequence representations. SIAM J. Comput. **43**(5), 1781–1806 (2014)
17. Okanohara, D., Sadakane, K.: Practical entropy-compressed rank/select dictionary. In: ALENEX, pp. 60–70 (2007)
18. Ossowski, S., et al.: Sequencing of natural strains of Arabidopsis thaliana with short reads. Genome Res. **18**(12), 2024–2033 (2008)
19. Pell, J., Hintze, A., Canino-Koning, R., Howe, A., Tiedje, J.M., Brown, C.T.: Scaling metagenome sequence assembly with probabilistic de Bruijn graphs. Proc. Nat. Acad. Sci. **109**(33), 13272–13277 (2012)
20. Peng, Y., Leung, H.C.M., Yiu, S.M., Chin, F.Y.L.: IDBA – a practical iterative de Bruijn graph de novo assembler. In: Berger, B. (ed.) RECOMB 2010. LNCS, vol. 6044, pp. 426–440. Springer, Heidelberg (2010)
21. Salikhov, K., Sacomoto, G., Kucherov, G.: Using cascading Bloom filters to improve the memory usage for de Bruijn graphs. Algorithms Mol. Biol. **9**(2) (2014)
22. Schnattinger, T., Ohlebusch, E., Gog, S.: Bidirectional search in a string with wavelet trees. In: Amir, A., Parida, L. (eds.) CPM 2010. LNCS, vol. 6129, pp. 40–50. Springer, Heidelberg (2010)

23. Simpson, J.T., et al.: ABySS: a parallel assembler for short read sequence data. Genome Res. **19**(6), 1117–1123 (2009)
24. The 1000 Genomes Project Consortium. An integrated map of genetic variation from 1,092 human genomes. Nature **491**(7422), 56–65 (2012)
25. Turnbaugh, P.J., et al.: The human microbiome project: exploring the microbial part of ourselves in a changing world. Nature **449**(7164), 804–810 (2007)

The Read/Write Protocol Complex
Is Collapsible

Fernando Benavides[1,2] and Sergio Rajsbaum[1(✉)]

[1] Instituto de Matemáticas, Universidad Nacional Autónoma de México,
Ciudad Universitaria, 04510 Mexico City, Mexico
rajsbaum@im.unam.mx
[2] Departamento de Matemáticas y Estadística, Universidad de Nariño,
San Juan de Pasto, Colombia
fandresbenavides@gmail.com

Abstract. The celebrated *asynchronous computability theorem* provides
a characterization of the class of decision tasks that can be solved in a
wait-free manner by asynchronous processes that communicate by writing and taking atomic snapshots of a shared memory. Several variations
of the model have been proposed (immediate snapshots and iterated
immediate snapshots), all equivalent for wait-free solution of decision
tasks, in spite of the fact that the protocol complexes that arise from the
different models are structurally distinct. The topological and combinatorial properties of these snapshot protocol complexes have been studied
in detail, providing explanations for why the asynchronous computability theorem holds in all the models.

In reality concurrent systems do not provide processes with snapshot
operations. Instead, snapshots are implemented (by a wait-free protocol)
using operations that write and read individual shared memory locations.
Thus, read/write protocols are also computationally equivalent to snapshot protocols. However, the structure of the read/write protocol complex has not been studied. In this paper we show that the read/write
iterated protocol complex is collapsible (and hence contractible). Furthermore, we show that a distributed protocol that wait-free implements
atomic snapshots in effect is performing the collapses.

1 Introduction

A *decision task* is the distributed equivalent of a function, where each process
knows only part of the input, and after communicating with the other processes,
each process computes part of the output. For example, in the *k-set agreement*
task processes have to agree on at most k of their input values; when $k = 1$ we
get the *consensus* task [8].

A central concern in distributed computability is studying which tasks are solvable in a distributed computing model, as determined by the type of communication mechanism available and the reliability of the processes. Early on it was shown

Full version in arXiv 1512.05427. Partially supported by UNAM-PAPIIT grant
IN107714.

© Springer-Verlag Berlin Heidelberg 2016
E. Kranakis et al. (Eds.): LATIN 2016, LNCS 9644, pp. 179–191, 2016.
DOI: 10.1007/978-3-662-49529-2_14

that consensus is not solvable even if only one process can fail by crashing, when asynchronous processes communicate by message passing [8] or even by writing and reading a shared memory [22]. A graph theoretic characterization of the tasks solvable in the presence of at most one process failure appeared soon after [3].

The *asynchronous computability theorem* [15] exposed that moving from tolerating one process failure, to any number of process failures, yields a characterization of the class of decision tasks that can be solved in a wait-free manner by asynchronous processes based on simplicial complexes, which are higher dimensional versions of graphs. In particular, n-set agreement is not wait-free solvable, even for $n + 1$ processes [4,15,24].

Computability theory through combinatorial topology has evolved to encompass arbitrary malicious failures, synchronous and partially synchronous processes, and various communication mechanisms [13]. Still, the original wait-free model of the asynchronous computability theorem, where crash-prone processes that communicate wait-free by writing and reading a shared memory is fundamental. For instance, the question of solvability in other models (e.g. f crash failures), can in many cases be reduced to the question of wait-free solvability [7,14].

More specifically, in the *AS model* of [13] each process can write its own location of the shared-memory, and it is able to read the whole shared memory in one atomic step, called a *snapshot*. The characterization is based on the *protocol complex,* which is a geometric representation of the various possible executions of a protocol. Simpler variations of this model have been considered. In the *immediate snapshot* (IS) version [2,4,24], processes can execute a combined write-snapshot operation. The *iterated immediate snapshot* (IIS) model [6] is even simpler to analyze, and can be extended (IRIS) to analyze partially synchronous models [23]. Processes communicate by accessing a sequence of shared arrays, through immediate snapshot operations, one such operation in each array. The success of the entire approach hinges on the fact that the topology of the protocol complex of a model determines critical information about the solvability of the task and, if solvable, about the complexity of solution [17].

All these snapshot models, AS, IS, IIS and IRIS can solve exactly the same set of tasks. However, the protocol complexes that arise from the different models are structurally distinct. The combinatorial topology properties of these complexes have been studied in detail, providing insights for why some tasks are solvable and others are not in a model.

Results. In reality concurrent systems do not provide processes with snapshot operations. Instead, snapshots are implemented (by a wait-free protocol) using operations that write and read individual shared memory locations [1]. Thus, read/write protocols are also computationally equivalent to snapshot protocols. However, the structure of the read/write protocol complex has not been studied. Our results are the following.

1. The one-round read/write protocol complex is collapsible to the IS protocol, i.e. to a chromatic subdivision of the input complex. The collapses can be performed simultaneously in entire orbits of the natural symmetric group

action. We use ideas from [21], together with distributed computing techniques of partial orders.

2. Furthermore, the distributed protocol that wait-free implements immediate snapshots of [5,9] in effect is performing the collapses.

3. Finally, also the multi-round iterated read/write complex is collapsible. We use ideas from [10], together with carrier maps e.g. [13].

All omitted proofs are in the full version in http://arxiv.org/abs/1512.05427

Related Work. The one-round immediate snapshot protocol complex is the simplest, with an elegant combinatorial representation; it is a chromatic subdivision of the input complex [13,19], and so is the (multi-round) IIS protocol [6]. The multi-round (single shared memory array) IS protocol complex is harder to analyze, combinatorially it can be shown to be a pseudomanifold [2]. IS and IIS protocols are homeomorphic to the input complex. An AS protocol complex is not generally homeomorphic to the underlying input complex, but it is homotopy equivalent to it [12]. The span of [15] provides an homotopy equivalence of the (multi-round) AS protocol complex to the input complex [12], clarifying the basis of the obstruction method [11] for detecting impossibility of solution of tasks.

Later on stronger results were proved, about the collapsibility of the protocol complex. The one-round IS protocol complex is collapsible [20] and homeomorphic to closed balls. The structure of the AS is more complicated, it was known to be contractible [12,13], and then shown to be collapsible (one-round) to the IS complex [21]. The IIS (multi-round) version was shown to be collapsible too [10].

There are several wait-free implementations of atomic snapshots starting with [1], but we are aware of only two algorithms that implement immediate snapshots; the original of [5], and its recursive version [9].

2   Preliminaries

2.1   Distributed Computing Model

The basic model we consider is the one-round *read/write* model (WR), e.g. [16]. It consists of $n+1$ processes denoted by the numbers $[n] = \{0, 1, \ldots, n\}$. A process is a deterministic (possibly infinite) state machine. Processes communicate through a shared memory array mem$[0 \ldots n]$ which consists of $n + 1$ single-writer/multi-reader atomic registers. Each process accesses the shared memory by invoking the atomic operations write(x) or read(j), $0 \leq j \leq n$. The write(x) operation is used by process i to write value x to register i, and process i can invoke read(j) to read register mem$[j]$, for any $0 \leq j \leq n$. Each process i has an input value, which may be its own id i. In its first operation, process i writes its input to mem$[i]$, then it reads each of the $n + 1$ registers, in an arbitrary order. Such a sequence of operations, consisting of a write followed by all the reads is abbreviated by WScan(x).

An *execution* consists of an interleaving of the operations of the processes, and we assume any interleaving of the operations is a possible execution. We may also consider an execution where only a subset of processes participate, consisting of an interleaving of the operations of those processes. These assumptions represent a wait-free model where any number of processes may fail by crashing.

In more detail, an execution is described as a set of atomic operations together with the irreflexive and transitive partial order given by: op precedes op' if op was completed before op'. If op does not precede op' and viceversa, the operations are called *concurrent*. The set of values read in an execution α by process i is called the *local view* of i which is denoted by $view(i, \alpha)$. It consists of pairs (j, v), indicating that the value v was read from the j-th register. The set of all local views in the execution α is called the *view* of α and it is denoted by $view(\alpha)$. Let $\mathcal{E}$ be a set of executions of the WR model. Consider the equivalence relation on $\mathcal{E}$ given by: $\alpha \sim \alpha'$ if $view(\alpha) = view(\alpha')$. Notice that for every execution α there exists an equivalent *sequential* execution α' with no concurrent operations. In other words, if op and op' are concurrent operations in α we can suppose that op was executed immediately before op' without modifying any views. Thus, we often consider only sequential executions α, consisting of a linear order on the set of all write and read operations.

Two other models can be derived from the WR model. In the *iterated* WR model, processes communicate through a sequence of arrays. They all go through the sequence of arrays mem_0, mem_1 ... in the same order, and in the r-th round, they access the r-th array, mem_r, exactly as in the one-round version of the WR model. Namely, process i executes one write to $mem_r[i]$ and then reads one by one all entries j, $mem_r[j]$, in arbitrary order. In the *non-iterated, multi-round* version of the WR model, there is only one array mem, but processes can execute several rounds of writing and then reading one by one the entries of the array. The *immediate snapshot* model (IS) [4,24], consists of a subset of executions of the WR one round model. Namely, all the executions where the operations can be organized in *concurrency classes*, each one consisting a set of writes by the set of processes participating in the concurrency class, followed by a read to all registers by each of these processes. See Sect. 3.1.

2.2   Algorithm IS

Consider the recursive algorithm IS of [9] for the iterated WR model, presented in Fig. 1. Processes go trough a series of disjoint shared memory arrays $mem_0, mem_1, \ldots, mem_n$. Each array mem_k is accessed by process i invoking WScan(i) in the recursive call IS($n + 1 - k$). Process i executes WScan(i) (line (1)), by performing first write(i), followed by read(j) for each $j \in [n]$, in an arbitrary order. The set of values read (each one with its location) is what the invocation of WScan(i) returns. In line (2) the process checks if $view$ contains $n + 1 - k$ id's, else IS($n - k$) is again invoked on the next shared memory in line (3). It is important to note that in each recursive call IS($n + 1 - k$) at least one process returns with $|view| = n + 1 - k$, given that $n + 1 - k$ processes invoked IS. For example, in the first recursive call IS($n + 1$) the last process to write reads $n + 1$ values and terminates the algorithm.

```
Algorithm IS(n + 1)
(1) view ← WScan(i)
(2) if |view| = n + 1 then return view
(3) else return IS(n).
```

Fig. 1. Code for process i

Every execution of the IS protocol can be represented by a finite sequence $\alpha = \alpha_0, \alpha_1, \ldots, \alpha_l$ with α_k an execution of the WR one round model where every process that takes a step in α_k invokes the recursive call with $IS(n + 1 - k)$. Since at least one process terminates the algorithm the length $l(\alpha) = l + 1$ is at most $n + 1$. The last returned local view in execution α for process i is denoted $view(i, \alpha)$, and the set of all local views is denoted by $view(\alpha)$.

Denote by $\mathcal{E}_l$ the set of views of all executions α with $l(\alpha) = l + 1$. Then $\mathcal{E}_n \subseteq \cdots \subseteq \mathcal{E}_0$. In particular, $\mathcal{E}_0$ corresponds to the views of executions of the one round WR of Sect. 2.1. Also, $\mathcal{E}_n$ corresponds to the views of the immediate snapshot model, see Theorem 1 of [9].

3   Definition and Properties of the Protocol Complex

Here we define the protocol complex of the write/read model and other models, which arise from the sets $\mathcal{E}_i$ mentioned in the previous section.

3.1   Additional Properties About Executions

Recall from Sect. 2.1 that an execution can be seen as a linear order on the set of write and read operations. For a subset $I \subseteq [n]$ let

$$\mathcal{O}_I = \{w_i, r_i(j) \ : \ i \in I, \ j \in [n]\}.$$

with $I = \mathcal{O}_i = \emptyset$. A *wr-execution on I* is a pair $\alpha = (\mathcal{O}_I, \to_\alpha)$ with $\to_\alpha$ a linear order on $\mathcal{O}_I$ such that $w_i \to_\alpha r_i(j)$ for all $j \in [n]$. The set I is called the Id set of α which is denoted by $\text{Id}(\alpha)$. Hence the view of i is $view(i, \alpha) = \{j \in I \ : \ w_j \to_\alpha r_i(j)\}$ and the view of α is $view(\alpha) = \{(i, view(i, \alpha)) \ : \ i \in I\}$. Note the chain $w_i \to_\alpha r_i(j_0) \to_\alpha \cdots \to_\alpha r_i(j_n)$ represents the invoking of WScan by the process i in the *wr*-execution α. Consider a *wr*-execution α and suppose that the order in which the process i reads the array $\text{mem}[0 \ldots n]$ is given by $r_i(j_0) \to_\alpha \cdots \to_\alpha r_i(j_n)$. If every write operation w_k satisfies $w_k \to_\alpha r_i(j_0)$ or $r_i(j_n) \to_\alpha w_k$ then $view(i, \alpha)$ corresponds to an atomic snapshot.

As a consequence, every execution in the snapshot model and immediate snapshot model corresponds to an execution in the write/read model. For instance in the *wr*-execution

$$\alpha : w_2 \to r_2(0) \to w_0 \to r_0(0) \to r_0(1) \to r_0(2) \to w_1 \to$$
$$\to r_1(0) \to r_2(1) \to r_1(1) \to r_2(2) \to r_1(2)$$

the $view(0, \alpha) = \{0, 2\}$ and $view(1, \alpha) = [2]$ are immediate snapshots, this means the processes 0 and 2 could have read the array instantaneously. In contrast, the $view(2, \alpha) = \{1, 2\}$ does not correspond to a snapshot. For the following consider the process j such that $w_i \rightarrow_\alpha w_j$ for all i.

Proposition 1. *Let α be a wr-execution on I. Then there exists $j \in I$ such that $view(j, \alpha) = I$.*

Let α be a *wr*-execution. For $0 \leq k \leq n$, define $\mathrm{Id}_k(\alpha) = \{j \in \mathrm{Id}(\alpha) : |view(j, \alpha)| = n + 1 - k\}$. An *IS-execution* is a finite sequence $\alpha = \alpha_0, \ldots, \alpha_l$ such that α_0 is a *wr*-execution on $[n]$, and α_{k+1} is a *wr*-execution on $\mathrm{Id}(\alpha_k) - \mathrm{Id}_k(\alpha_k)$. Given an *IS*-execution α, Proposition 1 implies $l(\alpha) \leq n + 1$. Moreover $\mathrm{Id}(\alpha_{k+1}) \subseteq \mathrm{Id}(\alpha_k)$ for all $0 \leq k \leq l - 1$. Hence $|\mathrm{Id}(\alpha_k)| \leq n + 1 - k$. Executions α, α' are *equivalent* if $view(\alpha) = view(\alpha')$, denoted $\alpha \sim \alpha'$.

Lemma 1. *Let α and α' be IS-executions with $l(\alpha) = l(\alpha)$. Given $0 \leq k \leq l$, (1) If $\alpha \sim \alpha'$ then $\mathrm{Id}(\alpha_k) = \mathrm{Id}(\alpha'_k)$. (2) If $\alpha_k \sim \alpha'_k$ then $\alpha \sim \alpha'$.*

According to the behavior of the protocol in Fig. 1, the local view of i is defined as $view(i, \alpha) = view(i, \alpha_k)$, if $i \in \mathrm{Id}(\alpha_k) - \mathrm{Id}(\alpha_{k+1})$ and $view(i, \alpha) = view(i, \alpha_l)$ for $k = l$. Hence the view of α is defined as $view(\alpha) = \{(i, view(i, \alpha)) : i \in [n]\}$.

Lemma 2. *Let $\alpha = \alpha_0, \ldots, \alpha_{l+1}$ be an IS-execution, $l(\alpha) = l + 2$. Then $view(\alpha) = view(\alpha')$ for some IS-execution α' such that $l(\alpha') = l + 1$.*

The *wr*-execution $\alpha' = \alpha_0, \ldots, \alpha_{l-1}, \alpha'_l$ of the lemma is obtained by, α'_l such that

$$view(i, \alpha'_l) = \begin{cases} view(i, \alpha_l), & \text{if } i \in \mathrm{Id}_l(\alpha_l) \\ view(i, \alpha_{l+1}), & \text{if } i \notin \mathrm{Id}_l(\alpha_l). \end{cases}$$

It follows $\mathcal{E}_l = \{view(\alpha) : \alpha = \alpha_0, \ldots, \alpha_l\}$. Thus, Lemma 2 implies $\mathcal{E}_{l+1} \subseteq \mathcal{E}_l$. For example consider the *IS*-execution $\alpha = \alpha_0, \alpha_1, \alpha_2$ where $\alpha_0 : w_0 \rightarrow r_0(0) \rightarrow r_0(1) \rightarrow r_0(2) \rightarrow w_1 \rightarrow r_1(0) \rightarrow r_1(1) \rightarrow r_1(2) \rightarrow w_2 \rightarrow r_2(0) \rightarrow r_2(1) \rightarrow r_2(2)$. $\alpha_1 : w_0 \rightarrow r_0(0) \rightarrow r_0(1) \rightarrow r_0(2) \rightarrow w_1 \rightarrow r_1(0) \rightarrow r_1(1) \rightarrow r_1(2)$. $\alpha_2 : w_0 \rightarrow r_0(0) \rightarrow r_0(1) \rightarrow r_0(2)$.
So $view(\alpha) = \{(0, \{0\}), (1, \{0, 1\}), (2, \{0, 1, 2\})\} \in \mathcal{E}_2 \subseteq \mathcal{E}_1 \subseteq \mathcal{E}_0$, Figs. 2 and 3.

3.2   Topological Definitions

The following are standard technical definitions, see [18, 21]. A (abstract) *simplicial complex* Δ on a finite set V is a collection of subsets of V such that for any $v \in V$, $\{v\} \in \Delta$, and if $\sigma \in \Delta$ and $\tau \subseteq \sigma$ then $\tau \in \Delta$. The elements of V are called *vertices* and the elements of Δ *simplices*. The *dimension* of a simplex σ is $\dim(\sigma) = |\sigma| - 1$. For instance the vertices are 0-simplices. For the purposes of this paper, we adopt the convention that the *void complex* $\Delta = \emptyset$ is a simplicial complex which is different from the *empty complex* $\Delta = \{\emptyset\}$. Given a positive integer n, Δ^n denotes the standard simplicial complex whose vertex set is $[n]$

and every subset of $[n]$ is a simplex. From now on we identify a complex Δ with its collection of subsets. For every simplex τ we denote by $I(\tau)$ the set of all simplices ρ, $\tau \subseteq \rho$. A simplex τ of Δ is called *free* if there exists a maximal simplex σ such that $\tau \subseteq \sigma$ and no other maximal simplex contains τ. The procedure of removing every simplex of $I(\tau)$ from Δ is called a *collapse*.

Let Δ and Γ be simplicial complexes, Δ *is collapsible to* Γ if there exists a sequence of collapses leading from Δ to Γ. The corresponding procedure is denoted by $\Delta \searrow \Gamma$. In particular, if the collapse is elementary with free simplex τ, it is denoted by $\Delta \searrow_\tau \Gamma$. If Γ is the void complex, Δ is *collapsible*. The next definition from [21] gives a procedure to collapse a simplicial complex, by collapsing simultaneously by entire orbits of the group action on the vertex set. Let Δ be a simplicial complex with a simplicial action of a finite group G. A simplex τ is called G-free if it is free and for all $g \in G$ such that $g(\tau) \neq \tau$, $I(\tau) \cap I(g(\tau)) = \emptyset$. If τ is G-free, the procedure of removing every simplex $\rho \in \bigcup_{g \in G} I(g(\tau))$ is called a G-collapse of Δ.

Since, if τ is G-free then $g(\tau)$ is free as well, the above definition guarantees that all collapses in the orbit of τ can be done in any order *i.e.* every G-collapse is a collapse. A simplicial complex Δ is G-*collapsible to* Γ if there exist a sequence of G-collapses leading from Δ to Γ, it is denoted by $\Delta \searrow_G \Gamma$. In similar way, if the G-collapse is elementary with G-free simplex τ, the notation $\Delta \searrow_{G(\tau)} \Gamma$ will be used. In the case Γ is the void complex, Δ is called G-collapsible. For instance consider a 2-simplex σ, τ a 1-face of σ and the action of $\mathbb{Z}_3$ over σ, then τ is free but not $\mathbb{Z}_3$-free.

3.3 Protocol Complex

Let n be a positive integer. The abstract simplicial complex $\mathsf{WR}_l(\Delta^n)$ with $0 \leq l \leq n$ consists of the set of vertices $V = \{(i, view_i) : i \in view_i \subseteq [n]\}$. A subset $\sigma \subseteq V$ forms a simplex if only if there exist an IS-execution α of length $l+1$ such that $\sigma \subseteq view(\alpha)$.

The complex $\mathsf{WR}_0(\Delta^n)$ is called the *protocol complex* of the write/read model and it will be denoted by $\mathsf{WR}(\Delta^n)$. Protocol complexes for the particular cases $n = 1$ and $n = 2$ are shown in Fig. 2. In [21] a combinatorial description of the protocol complex View^n associated to the snapshot model is given. There every simplex of View^n is represented as a $2 \times t$ matrix. Every simplex $\sigma \in \mathsf{WR}(\Delta^n)$ can be expressed as

$$\mathrm{W}(\sigma) = \begin{pmatrix} V_1 & \cdots & V_{t-1} & [n] \\ I_1 & \cdots & I_{t-1} & I_t \end{pmatrix}$$

where $I_i \cap I_j = \emptyset$ with $i \neq j$ and $I_i \subseteq V_j$ for all $i \leq j$. Moreover we can associate a matrix for every simplex in the complex $\mathsf{WR}_l(\Delta^n)$. This representation implies that $\chi(\Delta^n)$ and View^n are subcomplexes of $\mathsf{WR}(\Delta^n)$. Figure 3 shows the complex $\mathsf{WR}_l(\Delta^2)$. From now on we will write WR_l instead of $\mathsf{WR}_l(\Delta^n)$ unless we specify the standard complex. Lemma 2 implies that every maximal simplex of WR_{l+1} is a simplex of WR_l, which implies that WR_{l+1} is a subcomplex of WR_l. From now on σ will denote a simplex of WR_l. For $0 \leq k \leq l$ let $\sigma_k^< = \{(i, view_i) \in$

σ : $|view_i| < n + 1 - k\}$. In a similar way $\sigma_k^=$ and $\sigma_k = \sigma_k^< \cup \sigma_k^=$ are defined. Therefore, the set of processes in σ which participate in the $l + 1$ call recursive of Algorithm 1 is partitioned in those which read $n + 1 - l$ processes and those which read less than $n + 1 - l$ processes in the $l + 1$ layer shared memory. Let us define $I_\sigma^< = \bigcup\limits_{i \in Id(\sigma_l^<)} view_i$ and $I_\sigma = \bigcup\limits_{i \in Id(\sigma_l)} view_i$.

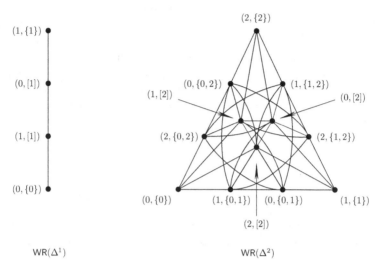

Fig. 2. Protocol complex for $n = 1$ and $n = 2$.

Theorem 1. $\sigma \in WR_{l+1}$ if only if $I_\sigma^< \cap Id(\sigma_l^=) = \emptyset$ and $|I_\sigma^<| < n + 1 - l$.

Proof. Suppose $I_\sigma^< \cap Id(\sigma_l^=) \neq \emptyset$ or $|I_\sigma^<| = n + 1 - l$ and there exists an *IS*-execution $\alpha = \alpha_0, \dots, \alpha_{l+1}$ such that $\sigma_l^< \subseteq view(\alpha_{l+1})$. Then there exist processes i and k such that $|view_i| < n + 1 - l$, $|view_k| = n + 1 - l$ and $k \in view_i$. This implies that k wrote in the $l + 1$ shared memory, a contradiction. For the other direction, since $I_\sigma^< \cap Id(\sigma_l^=) = \emptyset$ and $|I_\sigma^<| < n + 1 - l$ we can build an *IS*-execution $\alpha = \alpha_0, \dots, \alpha_{l+1}$ such that $\sigma \subseteq view(\alpha)$. $\square$

Notice that I_σ represents the set of processes which have been read in the $l + 1$ recursive call of the algorithm in Fig. 1.

Corollary 1. *If* $\sigma \notin WR_{l+1}$ *then*

1. $|I_\sigma| = n + 1 - l$.   2. $I_\sigma = I_\tau$ *for all* $\sigma \subseteq \tau$.

Let $inv(\sigma) = \{(k, I_\sigma) : k \in I_\sigma \backslash I_\sigma^<\}$ if $I_\sigma^< \neq I_\sigma$ else $inv(\sigma) = \{(k, I_\sigma) : k \in I_\sigma \backslash Id(\sigma_l^<)\}$. Notice that if $\sigma \notin WR_{l+1}$ then $inv(\sigma) \neq \emptyset$.
For the simplices $\sigma^- = \sigma - inv(\sigma)$ and $\sigma^+ = \sigma \cup inv(\sigma)$.

Proposition 2. *If $\sigma \notin \mathsf{WR}_{l+1}$ then*

1. $\sigma^+ = \sigma^- \cup inv(\sigma)$.
2. $\sigma^- \subseteq \sigma \subseteq \sigma^+$.
3. $\sigma^- \notin \mathsf{WR}_{l+1}$.
4. *If $\sigma^- \subseteq \tau \subseteq \sigma^+$ then $\sigma^- = \tau^-$ and $\sigma^+ = \tau^+$.*
5. $(\sigma^-)^- = \sigma^-$.

Consider $\mathrm{I}_-^+(\sigma) = \{\tau \in \mathsf{WR}_l \ : \ \sigma^- \subseteq \tau \subseteq \sigma^+\}$. Item (3) above implies that $\mathrm{I}_-^+(\sigma) \cap \mathsf{WR}_{l+1} = \emptyset$ if $\sigma \notin \mathsf{WR}_{l+1}$. Moreover from (4) it is obtained that $\mathrm{I}_-^+(\sigma) \cap \mathrm{I}_-^+(\tau) = \emptyset$ or $\mathrm{I}_-^+(\sigma) = \mathrm{I}_-^+(\tau)$.

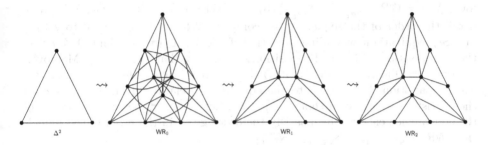

Fig. 3. Complexes WR_l.

4   Collapsibility

Let $S_{[n]}$ denote the permutation group of $[n]$. Notice that if the Id's of processes in a wr-execution on I are permuted according to $\pi \in S_{[n]}$ then we obtain a new linear order on $\pi(I)$. In other words if α is a wr-execution on I and $\pi \in S_{[n]}$ then $\alpha' = \pi(\alpha)$ is a wr-execution on $\pi(I)$. Moreover if $\sigma = view(\alpha)$ then $\pi(\sigma) = view(\pi(\alpha))$. This shows that there exists a natural group action on each simplicial complex WR_l.

Proposition 3. *Let $\sigma \in \mathsf{WR}_l$ be a simplex. Then*

1. $\pi(\sigma) \in \mathsf{WR}_l$.
2. $\pi(\sigma^-) = \pi(\sigma)^-$.

3. $\pi(\sigma^+) = \pi(\sigma)^+$.
4. $\pi(\mathrm{I}_-^+(\sigma)) = \mathrm{I}_-^+(\pi(\sigma))$.

For example in Fig. 2, $\sigma = \{(1, \{1,2\}), (2, \{0,2\}), (0, [2])\}$ and $\pi(0) = 1$, $\pi(1) = 2$ and $\pi(2) = 0$, then $\pi(\sigma) = \{(2, \{0,2\}), (0, \{0,1\}), (1, [2])\}$.

Theorem 2. *For every $0 \leq l \leq n+1$,*

1. WR_l *is collapsible to* WR_{l+1}.
2. WR_l *is $S_{[n]}$-collapsible to* WR_{l+1}.

Proof. Since $\sigma \in I_-^+(\sigma)$ for all simplices $\sigma \in \mathsf{WR}_l$, the intervals $I_-^+(\sigma)$ cover $L = \{\sigma : \sigma \in \mathsf{WR}_l, \sigma \notin \mathsf{WR}_{l+1}\}$. Also, Proposition 2 (4) implies that L can be decomposed as a disjoint union of intervals $I_-^+(\sigma_1), \ldots, I_-^+(\sigma_k)$ s.t. $\sigma_i = \sigma_i^+$ for all $1 \leq i \leq k$. Suppose $\dim(\sigma_{i+1})_l^< \leq \dim(\sigma_i)_l^<$ or if $\dim(\sigma_{i+1})_l^< = \dim(\sigma_i)_l^<$ then $\dim(\sigma_{i+1}) \leq \dim(\sigma_i)$. We will prove by induction on i, $1 \leq i \leq k$, that $\mathsf{WR}_l^i \searrow_{\sigma_i^-} \mathsf{WR}_l^{i+1}$ where $\mathsf{WR}_l^1 = \mathsf{WR}_l$ and $\mathsf{WR}_l^{k+1} = \mathsf{WR}_{l+1}$. If there exists a maximal simplex $\sigma \in \mathsf{WR}_l^i$ such that $\sigma_i \subseteq \sigma$ then $\sigma = \sigma_j$ for some $i \leq j \leq k$. Hence $(\sigma_i)_l^< \subseteq (\sigma_j)_l^<$ and therefore $\sigma_i = \sigma_j$. Now suppose there exists a maximal simplex $\sigma_j \in \mathsf{WR}_l^i$ with $i \leq j \leq k$ such that $\sigma_i^- \subseteq \sigma_j$. This implies that $(\sigma_i)_l^< = (\sigma_j)_l^<$ and $inv(\sigma_i) = inv(\sigma_j)$. Thus $\sigma_i = \sigma_i^- \cup inv(\sigma_i) \subseteq \sigma_j \cup inv(\sigma_j) = \sigma_j$ and therefore σ_i^- is free in WR_l^i. Therefore, $\mathsf{WR}_l = \mathsf{WR}_l^1 \searrow_{\sigma_1^-} \cdots \searrow_{\sigma_k^-} \mathsf{WR}_l^{k+1} = \mathsf{WR}_{l+1}$. Now if we specify in more detail the order of the sequence, the complex WR_l can be collapsed to WR_{l+1} in a $S_{[n]}$-equivariant way. First note that if $\pi(\sigma_i) \in I_-^+(\sigma_j)$ for some $1 \leq j \leq k$, then Proposition 3 (3) and Proposition 2 (4) imply that $\pi(\sigma_i) = \sigma_j$. Moreover, $\dim(\sigma_i)_l^< = \dim \pi(\sigma_i)_l^<$ and $\dim(\sigma_i) = \dim \pi(\sigma_i)$. Hence the set $\{\sigma_1, \ldots, \sigma_k\}$ can be partitioned according to the equivalence relation given by: $\sigma_i \sim \sigma_j$ if there exists $\pi \in S_{[n]}$ such that $\pi(\sigma_i) = \sigma_j$. Let $\tau_1, \ldots, \tau_p$ be representatives of the equivalence classes which satisfy the order given in the proof of the item 1, then $\mathsf{WR}_l \searrow_{S_{[n]}(\tau_1^-)} \cdots \searrow_{S_{[n]}(\tau_p^-)} \mathsf{WR}_{l+1}$. $\square$

Figure 4 illustrates the collapsing procedure $\mathsf{WR}_0 \searrow_{S_{[n]}} \mathsf{WR}_1$ for $n = 2$. In this case consider the simplexes $\sigma_1 = \{(1, \{1,2\}), (2, \{0,2\}), (0, [2])\}$ and $\sigma_2 = \{(0, \{0,1\}), (1, [2]), (2, [2])\}$ then $\mathsf{WR}_0^1 \searrow_{S_{[n]}(\sigma_1^-)} \mathsf{WR}_0^2 \searrow_{S_{[n]}(\sigma_2^-)} \mathsf{WR}_0^3 = \mathsf{WR}_1$. And we have the following consequence.

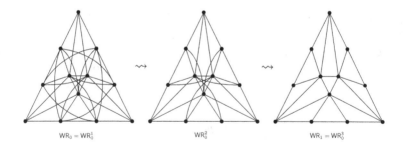

$\mathsf{WR}_0 = \mathsf{WR}_0^1$ WR_0^2 $\mathsf{WR}_1 = \mathsf{WR}_0^3$

Fig. 4. $S_{[n]}$-collapse.

Corollary 2. *For every natural number n, the simplicial complex $\mathsf{WR}(\Delta^n)$ is $S_{[n]}$-collapsible to $\chi(\Delta^n)$.*

Multi-round Protocol Complex. A *carrier map* Φ from complex $\mathcal{C}$ to complex $\mathcal{D}$ assigns to each simplex σ a subcomplex $\Phi(\sigma)$ of $\mathcal{D}$ such that $\Phi(\tau) \subseteq \Phi(\sigma)$

if $\tau \subseteq \sigma$. The protocol complex of the iterated write/read model (see Fig. 5), $k \geq 0$, is $\mathsf{WR}^{(k+1)}(\Delta^n) = \bigcup_{\sigma \in \mathsf{WR}^{(k)}(\Delta^n)} \mathsf{WR}(\sigma)$.

Corollary 3. *For all $k \geq 1$, $\mathsf{WR}^{(k)}(\Delta^n) \searrow \chi^{(k)}(\Delta^n)$.*

The collapsing procedure consists first in collapsing, in parallel, each subcomplex $\mathsf{WR}(\sigma)$ where σ is a maximal simplex of $\mathsf{WR}^{(k-1)}(\Delta^n)$ as in Theorem 2. An illustration is in Fig. 6, applied to the simplexes $\sigma_1 = \{(0, \{0,1\}), (1, \{0,1\}), (2, [2])\}$ and $\sigma_2 = \{(0, \{0,1\}), (1, [2]), (2, [2])\}$ of $\mathsf{WR}(\Delta^2)$. Second, we collapse $\chi(\mathsf{WR}^{(k-1)}(\Delta^n))$ to $\chi^{(k)}(\Delta^n)$.

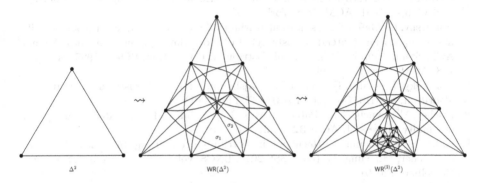

Fig. 5. Complexes of the iterated model; in $\mathsf{WR}^{(2)}(\Delta^2)$ only $\mathsf{WR}(\sigma_1)$ is depicted.

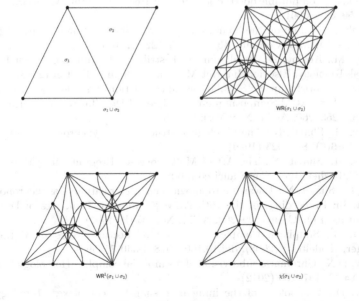

Fig. 6. First collapsing.

References

1. Afek, Y., Attiya, H., Dolev, D., Gafni, E., Merritt, M., Shavit, N.: Atomic snapshots of shared memory. J. ACM **40**(4), 873–890 (1993)
2. Attiya, H., Rajsbaum, S.: The combinatorial structure of wait-free solvable tasks. SIAM J. Comput. **31**(4), 1286–1313 (2002)
3. Biran, O., Moran, S., Zaks, S.: A combinatorial characterization of the distributed 1-solvable tasks. J. Algorithms **11**(3), 420–440 (1990)
4. Borowsky, E., Gafni, E.: Generalized FLP impossibility result for t-resilient asynchronous computations. In: Proceedings of the 25th Annual ACM Symposium on Theory of Computing, STOC, pp. 91–100. ACM, New York (1993)
5. Borowsky, E., Gafni, E.: Immediate atomic snapshots and fast renaming. In: Proceedings of the 12th ACM Symposium on Principles of Distributed Computing, PODC, pp. 41–51. ACM, New York (1993)
6. Borowsky, E., Gafni, E.: A simple algorithmically reasoned characterization of wait-free computation (extended abstract). In: Proceedings of the Sixteenth Annual ACM Symposium on Principles of Distributed Computing, PODC 1997, pp. 189–198. ACM, New York (1997)
7. Borowsky, E., Gafni, E., Lynch, N., Rajsbaum, S.: The BG distributed simulation algorithm. Distrib. Comput. **14**(3), 127–146 (2001)
8. Fischer, M., Lynch, N.A., Paterson, M.S.: Impossibility of distributed commit with one faulty process. J. ACM **32**(2), 374–382 (1985)
9. Gafni, E., Rajsbaum, S.: Recursion in distributed computing. In: Dolev, S., Cobb, J., Fischer, M., Yung, M. (eds.) SSS 2010. LNCS, vol. 6366, pp. 362–376. Springer, Heidelberg (2010)
10. Goubault, E., Mimram, S., Tasson, C.: Iterated chromatic subdivisions are coll apsible. Appl. Categorical Struct. **23**(6), 777–818 (2015)
11. Havlicek, J.: Computable obstructions to wait-free computability. Distrib. Comput. **13**(2), 59–83 (2000)
12. Havlicek, J.: A note on the homotopy type of wait-free atomic snapshot protocol complexes. SIAM J. Comput. **33**(5), 1215–1222 (2004)
13. Herlihy, M., Kozlov, D., Rajsbaum, S.: Distributed Computing Through Combinatorial Topology. Elsevier, Imprint Morgan Kaufmann, Boston (2013)
14. Herlihy, M., Rajsbaum, S.: Simulations and reductions for colorless tasks. In: Proceedings of the ACM Symposium on Principles of Distributed Computing, PODC 2012, pp. 253–260. ACM, New York (2012)
15. Herlihy, M., Shavit, N.: The topological structure of asynchronous computability. J. ACM **46**(6), 858–923 (1999)
16. Herlihy, M., Shavit, N.: The Art of Multiprocessor Programming. Morgan Kaufmann Publishers Inc., San Francisco (2008)
17. Hoest, G., Shavit, N.: Towards a topological characterization of asynchronous complexity. In: Proceedings of the 16th ACM Symposium Principles of Distributed Computing, PODC, pp. 199–208. ACM, New York (1997)
18. Jonsson, J.: Simplicial Complexes of Graphs. Lecture Notes in Mathematics. Springer, Heidelberg (2008). doi:10.1007/978-3-540-75859-4
19. Kozlov, D.N.: Chromatic subdivision of a simplicial complex. Homology Homotopy Appl. **14**(2), 197–209 (2012)
20. Kozlov, D.N.: Topology of the immediate snapshot complexes. Topology Appl. **178**, 160–184 (2014)

21. Kozlov, D.N.: Topology of the view complex. Homology Homotopy Appl. **17**(1), 307–319 (2015)
22. Loui, M.C., Abu-Amara, H.H.: Memory requirements for agreement among unreliable asynchronous processes **4**, 163–183 (1987). JAI Press
23. Rajsbaum, S., Raynal, M., Travers, C.: The iterated restricted immediate snapshot model. In: Hu, X., Wang, J. (eds.) COCOON 2008. LNCS, vol. 5092, pp. 487–497. Springer, Heidelberg (2008)
24. Saks, M., Zaharoglou, F.: Wait-free k-set agreement is impossible: the topology of public knowledge. SIAM J. Comput. **29**(5), 1449–1483 (2000)

The I/O Complexity of Computing Prime Tables

Michael A. Bender[1], Rezaul Chowdhury[1], Alexander Conway[2(⊠)],
Martín Farach-Colton[2], Pramod Ganapathi[1], Rob Johnson[1],
Samuel McCauley[1], Bertrand Simon[3], and Shikha Singh[1]

[1] Stony Brook University, Stony Brook, NY 11794-2424, USA
{bender,rezaul,pganapathi,rob,smccauley,shiksingh}@cs.stonybrook.edu
[2] Rutgers University, Piscataway, NJ 08854, USA
{alexander.conway,farach}@cs.rutgers.edu
[3] LIP, ENS de Lyon, 46 allée d'Italie, Lyon, France
bertrand.simon@ens-lyon.fr

Abstract. We revisit classical sieves for computing primes and analyze
their performance in the external-memory model. Most prior sieves are
analyzed in the RAM model, where the focus is on minimizing both the
total number of operations and the size of the working set. The hope is
that if the working set fits in RAM, then the sieve will have good I/O
performance, though such an outcome is by no means guaranteed by a
small working-set size.

We analyze our algorithms directly in terms of I/Os and operations.
In the external-memory model, permutation can be the most expensive
aspect of sieving, in contrast to the RAM model, where permutations
are trivial. We show how to implement classical sieves so that they have
both good I/O performance and good RAM performance, even when
the problem size N becomes huge—even superpolynomially larger than
RAM. Towards this goal, we give two I/O-efficient priority queues that
are optimized for the operations incurred by these sieves.

Keywords: External-memory algorithms · Prime tables · Sorting ·
Priority queues

1 Introduction

According to Fox News [21], "Prime numbers, which are divisible only by themselves and one, have little mathematical importance. Yet the oddities have long
fascinated amateur and professional mathematicians." Indeed, finding prime
numbers has been the subject of intensive study for millennia.

Prime-number-computation problems come in many forms, and in this paper
we revisit the classical (and Classical) problem of computing prime tables: how
efficiently can we compute the table $P[a, b]$ of all primes from a to b and the table

This research was supported by NSF grants CCF 1217708, IIS 1247726, IIS 1251137,
CNS 1408695, CCF 1439084, CNS-1408782, IIS-1247750, and Sandia National
Laboratories.

© Springer-Verlag Berlin Heidelberg 2016
E. Kranakis et al. (Eds.): LATIN 2016, LNCS 9644, pp. 192–206, 2016.
DOI: 10.1007/978-3-662-49529-2_15

$P[N] = P[2, N]$. Such prime-table-computation problems have a rich history, dating back 23 centuries to the sieve of Eratosthenes [17, 30].

Until recently, all efficient prime-table algorithms were *sieves*, which use a partial (and expanding) list of primes to find and disqualify composites [6,7,15,30]. For example, the sieve of Eratosthenes maintains an array representing $2, \ldots, N$ and works by crossing off all multiples of each prime up to $\sqrt{N}$ starting with 2. The surviving numbers, those that have not been crossed off, comprise the prime numbers up to N.

Polynomial-time primality testing [2,18] makes another approach possible: independently test each $i \in \{2, \ldots, N\}$ (or any subrange $\{a, \ldots, b\}$) for primality. The approaches can be combined; sieving steps can be used to eliminate many candidates cheaply before relatively expensive primality tests are performed. This is a feature of the sieve of Sorenson [31] (discussed in Sect. 6) and can also be used to improve the efficiency of AKS [2] when implemented over a range.

Prime-table algorithms are generally compared according to two criteria [6,25,27,30,31]. One is the standard run-time complexity, that is, the number of RAM operations. However, when computing very large prime tables that do not fit in RAM, such a measure may be a poor predictor of performance. Therefore, there has been a push to reduce the **working-set size**, that is, the size of memory used other than the output itself [6,11,31].[1] The hope is that if the working-set size is small enough to fit in memory for larger N, larger prime tables will be computable efficiently, though there is no direct connection between working-set size and input-output (I/O) efficiency.

Sieves and primality testing offer a trade-off between the number of operations and the working-set size of prime-table algorithms. For example, the sieve of Eratosthenes performs $O(N \log \log N)$ operations on a RAM but has a working-set size of $O(N)$. The fastest primality tests take polylogarithmic time in N, and so run in $O(N \text{polylog} N)$ time for a table but enjoy polylogarithmic working space.[2] This run-time versus working-set-size analysis has lead to a proliferation of prime-table algorithms that are hard to compare.

A small working set does not guarantee a fast algorithm for two reasons. First, eventually even slowly growing working sets will be too big for RAM. But more importantly, even if a working set is small, an algorithm can still be slow if the output table is accessed with little locality of reference.

In this paper, we analyze a variety of sieving algorithms in terms of the number of **block transfers** they induce, in addition to the number of operations. For out-of-core computations, these block transfers are page faults, and for smaller computations, they are cache misses. Directly counting such I/Os are often more predictive of the efficiency of an algorithm than the working set size or the instruction count.

---

[1] In our analyses, we model each sieving algorithm as if it writes the list of primes to an append-only output tape (i.e., the algorithm cannot read from this tape). All other memory used by the algorithm counts towards its working set size.

[2] Sieves are also less effective at computing $P[a, b]$. For primality-test algorithms, one simply checks the $b - a + 1$ candidate primes, whereas sieves generally require computing many primes smaller than a.

1.1   Computational Model

In this paper, we are interested in both the I/O complexity $\mathcal{C}_{I/O}$ and the RAM complexity $\mathcal{C}_{RAM}$. We indicate an algorithm's performance using the notation $\langle \mathcal{C}_{I/O}, \mathcal{C}_{RAM} \rangle$.

We use the standard *external memory* or *disk-access machine* (*DAM*) model of Aggarwal and Vitter [1] to analyze the I/O complexity. The DAM model allows block transfers between any two levels of the memory hierarchy. In this paper, we denote the smaller level by *RAM* or *main memory* and the larger level by *disk* or *external memory*.

In the DAM model, main memory is divided into M words, and the disk is modeled as arbitrarily large. Data is transferred between RAM and disk in blocks of B words. The I/O cost of an algorithm is the number of block transfers it induces [1,33].

We use the RAM model for counting operations. It costs $O(1)$ to compare, multiply, or add machine words. As in the standard RAM, a machine word has $\Omega(\log N)$ bits.

The prime table $P[N]$ is represented as a bit array that is stored on disk. We set $P[i] = 1$ when we determine that i is prime and set $P[i] = 0$ when we determine that i is composite. The prime table fills $O(N/\log N)$ words.[3] We are interested in values of N such that $P[N]$ is too large to fit in RAM.

1.2   Sieving to Optimize both I/Os and Operations

Let's begin by analyzing the sieve of Eratosthenes. Each prime is used in turn to eliminate composites, so the ith prime p_i touches all multiples of p_i in the array. If $p_i < B$, every block is touched. As p_i gets larger, every $\lceil p_i/B \rceil$th block is touched. We bound the I/Os by $\sum_{i=2}^{\sqrt{N}} N/(B \lceil p_i/B \rceil) \leq N \log \log N$. In short, this algorithm exhibits essentially no locality of reference, and for large N, most instructions induce I/Os. Thus, the naïve implementation of the sieve of Eratosthenes runs in $\langle \Theta(N \log \log N), \Theta(N \log \log N) \rangle$.

Section 2 gives descriptions of other sieves. For large N (e.g., $N = \Omega(M^2)$), most of these sieves also have poor I/O performance. For example, the segmented sieve of Eratosthenes [7] also requires $\langle \Theta(N \log \log N), \Theta(N \log \log N) \rangle$. The sieve of Atkin [6] requires $\langle O(N/\log \log N), O(N/\log \log N) \rangle$. On the other hand, the primality-checking sieve based on AKS has good I/O performance but worse RAM performance, running in $\langle \Theta(N/(B \log N)), \Theta(N \log^c N) \rangle$, as long as $M = \Omega(\log^c N)$.[4]

---

[3] It is possible to compress this table using known prime-density theorems, decreasing the space usage further.

[4] Here the representation of $P[N]$ matters most, because the I/O complexity depends on the size (and cost to scan) $P[N]$. For most other sieves in this paper, $P[N]$ is represented as a bit array and the I/O cost to scan $P[N]$ is a lower-order term.

As a lead-in to our approach given in Sect. 3, we show how to improve the I/O complexity of the naïve sieve of Eratosthenes (based on Schöhage et al.'s algorithm on Turing Machines [12,28]) as follows. Compute the primes up to $\sqrt{N}$ recursively. Then for each prime, make a list of all its multiples. The total number of elements in all lists is $O(N \log \log N)$. Sort using an I/O-optimal sorting algorithm, and remove duplicates: this is the list of all composites. Take the complement of this list. The total I/O-complexity is dominated by the sorting step, that is, $O(\frac{N}{B}(\log \log N)(\log_{M/B} \frac{N}{B}))$. Although this is a considerable improvement in the number of I/Os, the number of operations grows by a log factor to $O(N \log N \log \log N)$. Thus, this implementation of the sieve of Eratosthenes runs in $\left\langle O(\frac{N}{B}(\log \log N)(\log_{M/B} \frac{N}{B})), \ O(N \log N \log \log N) \right\rangle$.

In our analysis of the I/O complexity of diverse prime-table algorithms, one thing becomes clear. All known fast algorithms that produce prime numbers, or equivalently composite numbers, do so out of order. Indeed, sublinear sieves seem to require the careful representation of integers according to some order other than by value.

Consequently, the resulting primes or composites need to be permuted. In RAM, permuting values (or equivalently, sorting small integers) is trivial. In external memory, permuting values is essentially as slow as sorting [1]. Therefore, our results will involve sorting bounds. Until an in-order sieve is produced, all fast external-memory algorithms are likely to involve sorting.

1.3 Our Contributions

The results in this paper comprise a collection of data structures based on buffer trees [3] and external-memory priority queues [3–5] that allow prime tables to be computed quickly, with less computation than sorting implies.

We present data structures for efficient implementation of the sieve of Eratosthenes [17], the linear sieve of Gries and Misra [15] (henceforth called the GM linear sieve), the sieve of Atkin [6], and the sieve of Sorenson [31]. Our algorithms work even when $N \gg M$.

Table 1 summarizes our main results. Throughout, we use the notation $\text{SORT}(x) = O(\frac{x}{B} \log_{M/B} \frac{x}{B})$. Thus, the I/O lower bound of permuting x elements can be written as $\min(\text{SORT}(x), x)$ [1].

The GM linear sieve and the sieve of Atkin both slightly outperform the classical sieve of Eratosthenes. The sieve of Sorenson on the other hand induces far fewer I/O operations, but the RAM complexity is dependent on some number-theoretic unknowns, and may be far higher.

Note that the sieves of Eratosthenes and Atkins use $O(\sqrt{N})$ working space, whereas the GM Linear sieve and the sieve of Sorenson use $O(N)$ working space, which is consistent with our observation that working space is not predictive of the I/O complexity of an algorithm.

Table 1. Complexities of the main results of the paper, simplified under the assumption that N is large relative to M and B (see the corresponding theorems for the full complexities and exact requirements on N, M, and B). Note that $\text{SORT}(x) = O(\frac{x}{B}\log_{M/B}\frac{x}{B})$ is used as a unitless function, when specifying the number of I/Os in the I/O column and the number of operations in the RAM column. It is denoted by "SORT" because it matches the number of I/Os necessary for sorting in the DAM model. Here $p(N)$ is the smallest prime such that the pseudosquare $L_{p(N)} > N/(\pi(p)\log^2 N)$, and π is the prime counting function (see Sect. 6). Sorensen [31] conjectures, and the extended Riemann hypothesis implies, that $\pi(p(N))$ is polylogarithmic in N.

Sieve	I/O operations	RAM operations
Eratosthenes Sect. 3	$\text{SORT}(N)$	$B\text{SORT}(N)$
GM Linear Sect. 4	$\text{SORT}\left(\frac{N}{\log\log N}\right)$	$B\text{SORT}\left(\frac{N}{\log\log N}\right)$
Atkin Sect. 5	$\text{SORT}\left(\frac{N}{\log\log N}\right)$	$B\text{SORT}\left(\frac{N}{\log\log N}\right)$
Sorenson Sect. 6	$O(N/B)$	$O(N\pi(p(N)))$ s

2   Background and Related Work

In this Section we discuss some previous work on prime sieves. For a more extensive survey on prime sieves, we refer readers to [30].

Much of the previous work on sieving has focused on optimizing the sieve of Eratosthenes. Recall that the original sieve has an $O(N)$ working set size and performs $O(N\log\log N)$ operations. The notion of chopping up the input into intervals and sieving on each of them, referred to as the **segmented sieve of Eratosthenes** [7], is used frequently [6,9,11,29,30]. Segmenting results in the same number of operations as the original but with only $O(N^{1/2})$ working space. On the other hand, linear variants of the sieve [8,15,19,27] improve the operation count by a $\Theta(\log\log N)$ factor to $O(N)$, but also require a working set size of about $\Theta(N)$; see Sect. 4.

Recent advances in sieving achieve better performance. The sieve of Atkin [6] improves the operation count by an additional $\Theta(\log\log N)$ factor to $\Theta(N/\log\log N)$, with a working set of $N^{1/2}$ words [6] or even $N^{1/3}$ [6,14]; see Sect. 5.

Alternatively, a primality testing algorithm such as AKS [2] can be used to test the primality of each number directly. Using AKS leads to a very small working set size but a large RAM complexity. The sieve of Sorenson uses a hybrid sieving approach, combining both sieving and direct primality testing. This results in polylogarithmic working space, but a smaller RAM complexity if certain number-theoretic conjectures hold; see Sect. 6.

A common technique to increase sieve efficiency is preprocessing by a **wheel sieve**, which was introduced by Pritchard [25,26]. A wheel sieve preprocesses a large set of potential primes, quickly eliminating composites with small divisors. Specifically, a wheel sieve begins with a number $W = \prod_{i=1}^{\ell}p_i$, the product of the first ℓ primes (for some ℓ). It then marks all $x < W$ that have at least one p_i as a factor by simply testing x for divisibility by each p_i. This requires $O(\ell W)$ operations and $O(W/B \log N)$ I/Os, because marks are stored in a bit vector and the machine has a word size of $\Omega(\log N)$. The wheel sieve then uses the observation that a composite $x > W$ has a prime divisor among the first ℓ primes iff $x \bmod W$ is also divisible by that prime. Thus, the wheel iterates through each interval of W consecutive potential primes, marking off a number x iff $x \bmod W$ is marked off. When using a bit vector to store these marks, this can be accomplished by copying the first W bits into each subsequent chunk of W bits. On a machine with word size $\Omega(\log N)$, the total operations for these copies is $O(N/\log N)$, and the I/O complexity is $O(N/B \log N)$, so these costs will not affect the overall complexities of our algorithms. Typically, $\ell = \sqrt{\log N}$, so $W = N^{o(1)}$. Thus, marking off the composites less than W can be done in $N^{o(1)}$ time and $N^{o(1)}/B$ I/Os using $O(\sqrt{\log N})$ space, which will not contribute to the overall complexity of the main sieving algorithm. By Mertens' Theorem [20,32], there will be $\Theta(N/\log\log N)$ potential composites left after this pre-sieving step, which can often translate into a $\Theta(\log\log n)$ speedup to the remaining steps in the sieving algorithm.

An important component of some of the data structures presented in this paper is the priority queue of Arge and Thorup [5], which is simultaneously efficient in RAM and in external memory. In particular, their priority queue can handle inserts with $O(\frac{1}{B} \log_{M/B} N/B)$ amortized I/Os and $O(\log_{M/B} N/B)$ amortized RAM operations. Delete-min requires $O(\frac{1}{B} \log_{M/B} N/B)$ amortized I/Os and $O(\log_{M/B} N/B + \log\log M)$ amortized RAM operations. They assume that each element fits in a machine word and use integer sorting techniques to achieve this low RAM cost while retaining optimal I/O complexity.

3   Sieve of Eratosthenes

In the introduction we showed that due to the lack of locality of reference, the naïve implementation of the sieve of Eratosthenes used $\langle O(N \log\log N), O(N \log\log N) \rangle$. A more sophisticated approach—creating lists of the multiples of each prime, and then sorting them together—improved the locality at the cost of additional computation, leading to a cost of $\langle \text{SORT}(N \log\log N), O(N \log N \log\log N) \rangle$. We can sharpen this approach by using a (general) efficient data structure instead of the sorting step, and then further by introducing a data structure designed specifically for this problem.

Using Priority Queues. The sieve of Eratosthenes can be implemented using only priority-queue operations: *insert* and *delete-min*. In this version, instead of crossing off all multiples of a discovered prime consecutively, we perform lazy inserts of these multiples into the priority queue.

The priority queue Q stores $\langle k, v \rangle$ pairs, where v is a prime and k is a multiple of v. That is, the composites are the **keys** in the priority queue and the corresponding prime-factor is its **value**.[5] We start off by inserting the first pair

---

[5] Note that the delete-min operations of the priority queue are on the keys, i.e., the composites.

$\langle 4, 2 \rangle$ into Q, and at each step, we extract (and delete) the minimum composite $\langle k, v \rangle$ pair in Q. Any number less than k which has never been inserted into Q must be prime. We keep track of the last deleted composite k', and check if $k > k' + 1$. If so, we declare $p = k' + 1$ as prime, and insert $\langle p^2, p \rangle$ into Q. In each of these iterations, we always insert the next multiple $\langle k + v, v \rangle$ into Q.

We implement this algorithm using the RAM-efficient priority queue of Arge and Thorup [5].

Lemma 1. *The sieve of Eratosthenes implemented using a RAM-efficient external-memory priority queue [5] has complexity* $\langle O(\text{SORT}(N \log \log N)), O\left(N \log \log N \left(\log_{M/B} N + \log \log M \right)\right)\rangle$ *and uses* $O\left(\sqrt{N}\right)$ *space for sieving primes in* $[1, N]$.

Proof. This follows from the observation that the sieve performs $\Theta\left(\sum_{\text{prime} p \in [1, \sqrt{N}]} \frac{N}{p} \right) = \Theta\left(N \log \log N \right)$ operations on Q costing $\langle O\left(\frac{1}{B} \log_{M/B} N \right), O\left(\log_{M/B} N + \log \log M \right)\rangle$ each. □

Using a Value-sensitive Priority Queue. In the above algorithm, the key-value pairs corresponding to smaller values are accessed more frequently because smaller primes have more multiples in a given range. Therefore, a structure that prioritizes the efficiency of operations on smaller primes (values) outperforms a generic priority queue. We introduce a *value-sensitive priority queue*, in which the amortized access cost of an operation with value v depends on v instead of the size of the data structure.

A value-sensitive priority queue Q has two parts—the *top part* consisting of a single internal-memory priority queue Q' and the *bottom part* consisting of $\lceil \log \log N \rceil$ external-memory priority queues $Q_1, Q_2, \ldots, Q_{\lceil \log \log N \rceil}$.

Each Q_i in the bottom-part of Q is a RAM-efficient external-memory priority queue [5] that stores $\langle k, v \rangle$ pairs, for $v \in [2^{2^i}, 2^{2^{i+1}})$. Hence, each Q_i contains fewer than $N_i = 2^{2^{i+1}}$ items. With a cache of size M, Q_i supports insert and delete-min operations in $\langle O((\log_{M/B} N_i)/B), O(\log_{M/B} N_i + \log \log M) \rangle$ amortized cost [5]. Moreover, in each Q_i we have $\log v = \Theta(\log N_i)$. Thus, the cost reduces to $\langle O((\log_{M/B} v)/B), O(\log_{M/B} v + \log \log M) \rangle$ for an item with value v. Though we divide the cache equally among all Q_i's, the asymptotic cost per operation remains unchanged assuming $M > B(\log \log N)^{1+\varepsilon}$ for some constant $\varepsilon > 0$.

The queue Q' in the top part only contains the minimum composite (key) item from each Q_i, and so the size of Q' will be $\Theta(\log \log N)$. We use the dynamic integer set data structure [22] to implement Q' which supports insert and delete-min operations on Q' in $O(1)$ time using only $O(\log n)$ space. We also maintain an array $A[1 : \lceil \log \log N \rceil]$ such that $A[i]$ stores Q_i's contributed item to Q'; thus we can access it in constant time.

To perform a delete-min, we extract the minimum key item from Q', check its value to find the Q_i it came from, extract the minimum key item from that

Q_i and insert it into Q'. To insert an item , we first check its value to determine the destination Q_i, compare it with the item in $A[i]$, and depending on the result of the comparison we either insert the new item directly into Q_i or move Q_i's current item in Q' to Q_i and insert the new item into Q'. The following lemma summarizes the performance of these operations.

Lemma 2. *Using a value-sensitive priority queue Q as defined above, insert-ing an item with value v takes $\left\langle O((\log_{M/B} v)/B), O(\log_{M/B} v) \right\rangle$, and a delete-min that returns an item with value v takes $\left\langle O((\log_{M/B} v)/B), O(\log_{M/B} v + \log \log M) \right\rangle$, assuming $M > \log N + B(\log \log N)^{1+\varepsilon}$ for some constant $\varepsilon > 0$.*

We now use this value-sensitive priority queue to efficiently implement the sieve of Eratosthenes. Each prime p is involved in $\Theta(N/p)$ priority queue oper-ations, and by the Prime Number Theorem [16], there are $O(\sqrt{N}/\log N)$ prime numbers in $[1, \sqrt{N}]$, and the ith prime number is approximately $i \ln i$. Theorem 1 now follows.

Theorem 1. *Using a value-sensitive priority queue, the sieve of Eratosthenes runs in $\left\langle \text{SORT}(N), O(N(\log_{M/B} N + \log \log M \log \log N)) \right\rangle$ and uses $O(\sqrt{N})$ space, provided $M > \log N + B(\log \log N)^{1+\varepsilon}$ for some constant $\varepsilon > 0$. We can simplify this to $\langle \text{SORT}(N), \text{BSORT}(N) \rangle$ if $\log N / \log \log N = \Omega(\log(M/B) \log \log M)$ and $\log(N/B) = \Omega(\log N)$.*

4   Linear Sieve of Gries and Misra

There are several variants of the sieve of Eratosthenes [8,13,15,19] that perform $O(N)$ operations by only marking each composite exactly once; see [27] for a survey. We will focus on one of the linear variants, the GM linear sieve [15]. Other linear-sieve variants, such as [8,13,19] share the same underlying data-structural operations, and much of the basic analysis below carries over.

The GM linear sieve is based on the following basic property of composite numbers: each composite C can be represented uniquely as $C = p^r q$ where p is the smallest prime factor of C, and either $q = p$ or p does not divide q [15].

Thus, each composite has a unique normal form based on p, q and r. Crossing off the composites in a lexicographical order based on these (p, q, r) ensures that each composite is marked exactly once. Thus the RAM complexity is $O(N)$.

Algorithm 1 describes the linear sieve in terms of subroutines. It builds a set C of composite numbers, then returns its complement.

The subroutine Insert (x, C) inserts x in C. Inverse successor (InvSucc(x, C)) returns the smallest element larger than x that is not in C.

```
C ← {1};  p ← 1;
while p ≤ √N do
    p ← InvSucc(p,C );  q ← p;
    while q ≤ N/p do
        for r = 1, 2, . . . , log_p(N/q)
        do
            Insert(p^r q, C );
            q ← InvSucc(q, C );
    return [1; N] \ C
```

Algorithm 1. GM Linear Sieve

While the RAM complexity is an improvement by a factor of $\log \log N$ over the classic sieve of Eratosthenes, the algorithm (thematically) performs poorly in the DAM model. Even though each composite is marked exactly once, resulting in $O(N)$ operations, the overall complexity of this algorithm is $\langle O(N), O(N) \rangle$, as a result poor data locality. In the rest of the section we improve the locality using a "buffer-tree-like" data structure, while also taking advantage of the bit-complexity of words to improve the performance further.

Using a Buffer Tree. We first introduce the classical buffer tree of Arge [3], and then modify the structure to improve the bounds of the GM linear sieve. We give a high-level overview of the data structure here.

The classical buffer tree has branching factor M/B, with a buffer of size M at each node. We assume a complete tree for simplicity, so its height is $\lceil \log_{M/B} N/M \rceil = O(\log_{M/B} N/B)$. Newly-inserted elements are placed into the root buffer. If the root buffer is full, all of its elements are flushed: first sorted, and then placed in their respective children. This takes $\langle O(M/B), O(M \log M) \rangle$. This process is then repeated recursively as necessary for the buffer of each child. Since each element is only flushed to one node at each level, and the amortized cost of a flush is $\langle O(1/B), O(\log M) \rangle$, the cost to flush all elements is $\left\langle O(N/B \log_{M/B} N/B), O(N \log N) \right\rangle$.

Inverse successor can be performed by searching within the tree. However, these searches are very expensive, as we must search every level of the tree—it may be that a recently-inserted element changed the inverse successor. Thus it costs at least $\left\langle O(M/B \log_{M/B} N/B), O(M \log_{M/B} N/B) \right\rangle$ for a single inverse successor query.

Using a Buffer-tree-like Structure. In order to achieve better bounds, we will need to improve the inverse successor time to match the insert time. It turns out that this will also improve the computation time considerably; we will only do $O(B)$ computations per I/O, the best possible for a given I/O bound.

As an initial optimization, we perform a wheel sieve using the primes up to $\sqrt{\log N}$. By an analogue of Merten's Theorem, this leaves only $N/\log \log N$ candidate primes. This reduces the number of insertions into the buffer tree.

To avoid the I/Os along the search path for the inverse successor queries, we adjust the branching factor to $\sqrt{M/B}$ rather than M/B, which doubles the height, and partition each buffer into $\sqrt{M/B}$ subarrays of size $\sqrt{MB}$: one for each child. Then as we scan the array, we can store the path from the root to the current leaf in $\sqrt{MB} \log_{M/B} N/B$ words. If $\sqrt{M/B} > \log_{M/B} N/B$ this path fits in memory. Thus, the inverse successor queries can avoid the path-searching I/O cost without affecting the amortized insert cost.

Next, since the elements of the leaves are consecutive integers, each can be encoded using a single bit, rather than an entire word. Recall that we can read $\Omega(B \log N)$ of these bits in a single block transfer. This could potentially speed up queries, but only if we can guarantee that the inverse successor can always be found by scanning *only* the bit array. However, during an inverse successor scan, we already maintain the path in memory; thus, we can flush all elements along the path without any I/O cost. Therefore we can in fact get the correct inverse successor by scanning the array.

As an bonus, we can improve the RAM complexity during a flush. Since our array is static and the leaves divide the array evenly, we can calculate the child being flushed to using modular arithmetic.

In total, we insert $N/\log\log N$ elements into the buffer tree. Each must be flushed through $O(\log_{M/B} N/B)$ levels, where a flush takes $\langle O\left(1/B\right), O\left(1\right)\rangle$ amortized. The inverse successor queries must scan through $N\log\log N$ elements (by the analysis of the sieve of Eratosthenes), but due to our bit array representation this only takes $\langle O(N\log\log N/B\log N), O(N\log\log N/\log N)\rangle$, a lower-order term.

Theorem 2. *The GM linear sieve implemented using our modified buffer tree structure, assuming $M > B^2$, $\sqrt{M/B} > \log_{M/B}(N/B)$, and $\sqrt{M/B} > \log^2_{M/B}(N/B)/\log\log N$, uses $O(N)$ space and has a complexity of $\langle\text{SORT}\left(N/\log\log N\right), B\text{SORT}\left(N/\log\log N\right)\rangle$.*

Using Priority Queues. The GM linear sieve can also be implemented using a standard priority queue API. While any priority-queue of choice can be used, the RAM- and I/O-efficient priority queue of Arge and Thorup [5] in particular achieves the same bounds as the modified buffer tree implementation.

The two data structures presented to implement the GM linear sieve offer a nice contrast. The buffer tree approach is self-contained and designed specifically for sieving, while the PQ based approach offers flexibility to use a PQ of your choice. The RAM-efficient PQ [5], in particular, is based on integer sorting techniques, while the buffer tree avoids such heavy machinery. We sketch the PQ-based version here for completeness.

The basic algorithm is the same (Algorithm 1), that is, enumerate composites in their unique normal form $p^r q$. However, in this variant, InvSucc is implemented using only insert and delete-min operations.

In contrast to the buffer tree approach where we build the entire set of composites $\mathcal{C}$ and eventually return its complement, we maintain a running list of potential primes as a priority queue $\mathcal{P}$. As the primes are discovered, we extract them from $\mathcal{P}$ and output. The composites $p^r q$ generated by the GM linear sieve algorithm are temporarily stored in another priority queue $\mathcal{C}$. We ensure locality of reference by lazily deleting the discovered composites in $\mathcal{C}$ from $\mathcal{P}$. In particular, we update $\mathcal{P}$ every time InvSucc is called, just as much as is required to find the next candidate for p or q, by using delete-min operations on $\mathcal{P}$ and $\mathcal{C}$.

Theorem 3. *The GM linear sieve implemented using RAM-efficient priority queues [5], assuming $N > 2M$ and $M > 2B$, uses $O(N)$ space and has a complexity of $\left\langle\text{SORT}\left(\frac{N}{\log\log N}\right), \frac{N}{\log\log N}\left(\log_{\frac{M}{B}}\frac{N}{B} + \log\log M\right)\right\rangle$.*

We can simplify this to $\left\langle\text{SORT}\left(\frac{N}{\log\log N}\right), B\text{ SORT}\left(\frac{N}{\log\log N}\right)\right\rangle$ if $\log N > \log M\log\log M$.

5   Sieve of Atkin

The sieve of Atkin [6,12] is one of the most efficient known sieves in terms of RAM computations. It can compute all the primes up to N in $O(N/\log\log N)$ time using $O(\sqrt{N})$ memory. We first describe the original algorithm from [6] and then use various priority queues to improve its I/O efficiency.

The algorithm works by exploiting the following characterization of primes using binary quadratic forms. Note that every number that is not trivially composite (divisible by 2 or 3) must satisfy one of the three congruences. For an excellent introduction to the underlying number theoretic concepts, see [10].

Theorem 4 [6]. *Let k be a square-free integer with $k \equiv 1 \pmod 4$ (resp. $k \equiv 1 \pmod 6$, $k \equiv 11 \pmod{12}$). Then k is prime if and only if the number of positive solutions to $x^2 + 4y^2 = k$ (resp. $3x^2 + y^2 = k$, $3x^2 - y^2 = k$ $(x > y)$) is odd.*

For each quadratic form $f(x,y)$, the number of solutions can be computed by brute force in $O(N)$ operations by iterating over the set $L = \{(x,y) \mid 0 < f(x,y) \le N\}$. This can be done with a working set size of $O(\sqrt{N})$ by "tracing" the level curves of f. Then, the number of solutions that occur an even number of times are removed, and by precomputing the primes less than $\sqrt{N}$, the numbers that are not square-free can be sieved out leaving only the primes as a result of Theorem 4.

The algorithm as described above requires $O(N)$ operations, as it must iterate through the entire domain L. This can be made more efficient by first performing a wheel sieve. If we choose $W = 12 \cdot \prod_{p^2 \le \log N} p$, then by an analog of Mertens' theorem, the proportion of (x,y) pairs with $0 \le x,y < W$ such that $f(x,y)$ is a unit mod W is $1/\log\log N$. By only considering the W-translations of these pairs we obtain $L' \subseteq L$, with $|L'| = O(N/\log\log N)$ and $f(x,y)$ composite on $L \setminus L'$. The algorithm can then proceed as above.

Using Priority Queues. The above algorithm and its variants require that $M = \Omega(\sqrt{N})$. By utilizing a priority queue to store the multiplicities of the values of f over L, as well as one to implement the square-free sieve, we can trade this memory requirement for I/O operations. In what follows we use an analog of the wheel sieve optimization described above, however we note that the algorithm and analysis can be adapted to omit this.

Having performed the wheel sieve as described above, we insert the values of each quadratic form f over each domain L into an I/O- and RAM-efficient priority queue Q [5]. This requires $|L|$ such operations (and their subsequent extractions), and so this takes $\langle \text{SORT}(|L|), O(|L|\log_{M/B}|L| + |L|\log\log M/\log\log N)\rangle$. Because we have used a wheel sieve, $|L| = O(N/\log\log N)$, and so this reduces to

$$\left\langle \text{SORT}\left(\frac{N}{\log\log N}\right),\ O\left(\frac{N\log_{M/B}N}{\log\log N} + \frac{N\log\log M}{\log\log N}\right)\right\rangle. \tag{1}$$

The remaining entries in Q are now either primes or squareful numbers. In order to remove the squareful numbers, we sieve the numbers in Q as follows.

We maintain a separate I/O- and RAM-efficient priority queue Q' of pairs $\langle v, p \rangle$, where $p \leq \sqrt{N}$ is a previously discovered prime and v is a multiple of p^2. For each value v we pull from Q, we repeatedly extract the min value $\langle w, p \rangle$ from Q' and insert $\langle w + p^2, p \rangle$ until either v is found, in which case v is not square-free and thus not a prime, or exceeded, in which case v is prime. If v is a prime, then we insert $\langle v^2, v \rangle$ into Q'.

Each prime $p \leq \sqrt{N}$ will be involved in at most N/p^2 operations on Q', and so will contribute $\left\langle O(\frac{N \log_{M/B} N}{p^2 B}), \ O(\frac{N}{p^2}(\log_{M/B} N + \log \log M)) \right\rangle$ operations. Summing over p, the total number of operations in this phase of the algorithm is less than $\langle O(\text{SORT}(N)/(B \log N)), \ O((\text{SORT}(N) + \log \log M)/\log N) \rangle$.

As described above, the priority queue Q may contain up to N items. We can reduce the max size of Q to $O(\sqrt{N})$ by tracing the level curves much like the sieve of Atkin.

Theorem 5. *The sieve of Atkin implemented with a wheel sieve, as well as I/O and RAM efficient priority queues runs in* $\langle \text{SORT}(N/\log \log N),$
$O((N \log_{M/B} N)/\log \log N + N \log \log M/\log \log N) \rangle$, *using* $O(\sqrt{N})$ *space.*

We can simplify this to $\langle \text{SORT}(N/\log \log N), \ B \ \text{SORT}(N/\log \log N) \rangle$ *if* $\log N = \Omega(\log(M/B) \log \log M)$ *and* $\log N/B = \Omega(\log N)$.

6   Sieve of Sorenson

The sieve of Sorenson [31] uses a hybrid approach. It first uses a wheel sieve to remove multiples of small primes. Then, it eliminates non-primes using a test based on so called pseudosquares. Finally it removes composite prime powers with another sieve.

The **pseudosquare** L_p is the smallest non-square integer with $L_p \equiv 1$ (mod 8) that is a quadratic residue modulo every odd prime $q \leq p$. The sieve of Sorenson is based on the following theorem in that its steps satisfy each requirement of the theorem explicitly.

Theorem 6 [31]. *Let x and s be positive integers. If the following hold:*

(i) *All prime divisors of x exceed s,*
(ii) *$x/s < L_p$, the p-th pseudosquare for some prime p,*
(iii) *$p_i^{(x-1)/2} \equiv \pm 1 \pmod{x}$ for all primes $p_i \leq p$,*
(iv) *$2^{(x-1)/2} \equiv -1 \pmod{x}$ when $x \equiv 5 \pmod 8$,*
(v) *$p_i^{(x-1)/2} \equiv -1 \pmod{x}$ for some prime $p_i \leq p$ when $x \equiv 1 \pmod 8$,*

then x is a prime or a prime power.

The algorithm first sets $s = \lceil \sqrt{\log N} \rceil$. It then chooses $p(N)$ so that $L_{p(N)}$ is the smallest pseudosquare satisfying $L_{p(N)} > N/s$. Thus, the algorithm must calculate $L_{p(N)}$. We omit this calculation; see [31] for an $o(N)$ algorithm to do so.

A table of the first 73 pseudosquares is sufficient for any $N < 2.9 \times 10^{24}$.[6] Next, the algorithm calculates the first s primes. We assume that $M \gg \pi(p(N))$.

The algorithm proceeds in three phases. Sorenson's original algorithm segments the range in order to fit in cache, but this step is omitted here:

1. Perform a (linear) wheel sieve to eliminate multiples of the first s primes.[7] All remaining numbers satisfy the first requirement of Theorem 6.
2. For each remaining k:
   - It verifies that $2^{(k-1)/2} \equiv \pm 1 \pmod{k}$ and is -1 if $k \equiv 5 \bmod 8$.
   - If k passes the above test, then it verifies that $p_i^{(k-1)/2} \equiv \pm 1 \pmod{k}$ for all odd primes $p_i \leq p(N)$, and that $p_i^{(k-1)/2} \equiv -1 \pmod{k}$ for at least one p_i if $k \equiv 1 \pmod 8$.

   Note that this second test determines if the remaining requirements of Theorem 6 are met.
3. Remove all prime powers, as follows. If $N \leq 6.4 \times 10^{37}$, only primes remain and this phase is unnecessary [31,34]. Otherwise construct a list of all the perfect powers less than N by repeatedly exponentiating every element of the set $\{2, \ldots, \lfloor \sqrt{N} \rfloor\}$ until it is greater than N. Sort these $O(\sqrt{N} \log N)$ elements and remove them from the prime candidate list.

The complexity of this algorithm is dominated by step 2. To analyze the RAM complexity, first note that only $O(N/\log \log N)$ elements remain after the wheel sieve. Performing each base 2 pseudoprime test takes $O(\log N)$ time, so the cumulative total is $O(N \log N/\log \log N)$. Now, only $O(N/\log N)$ numbers up to N pass the base-2 pseudoprime test (see e.g. [23,31]). For each of the remaining integers, we must do $\pi(p(N))$ modular exponentiations (to a power less than N), which requires a total of $O(N\pi(p(N)))$ operations. Thus we get a total cost of $O(N\pi(p(N)) + N \log N/\log \log N)$

We can remove the second term using recent bounds on pseudoprimes. Pomerance and Shparlinski [24] have shown that $L_p(N) \leq \exp(3p(N)/\log \log p(N))$. Thus, $N \log N/\log \log N = O(N\pi(p(N))/\log \log p(N))$, and so the running time simplifies to $O(N\pi(p(N)))$.

Theorem 7. *The sieve of Sorenson runs in* $\langle O\left(\frac{N}{B}\right), O\left(N\pi(p(N))\right) \rangle$.

We can phrase the complexity in terms of N alone by bounding p. The best known bound for p leads to a running time of roughly $O(N^{1.1516})$. On the other hand, the Extended Riemann Hypothesis implies $p < 2\log^2 N$, and Sorenson conjectures that $p \sim \frac{1}{\log 2} \log N \log \log N$ [31]; under these conjectures the RAM complexity is $O(N \log^2 N/\log \log N)$ and $O(N \log N)$ respectively.

Sieving an Interval. Note that a similar analysis shows we can efficiently sieve an interval with the sieve of Sorenson as well.

---

[6] These tables are available online. For example, see https://oeis.org/A002189/b002189.txt.

[7] Sorenson's exposition removes multiples of the small primes one by one on each segment in order to retain small working space. From an external memory point of view, building the whole wheel of size $N^{o(1)}$ is also effective.

Acknowledgments. We thank Oleksii Starov for suggesting this problem to us.

References

1. Aggarwal, A., Vitter, S.: Jeffrey: the input/output complexity of sorting and related problems. Commun. ACM **31**(9), 1116–1127 (1988)
2. Agrawal, M., Kayal, N., Saxena, N.: Primes is in P. Ann. Math. **50**, 781–793 (2004)
3. Arge, L.: The buffer tree: a technique for designing batched external data structures. Algorithmica **37**(1), 1–24 (2003)
4. Arge, L., Bender, M.A., Demaine, E.D., Holland-Minkley, B., Munro, J.I.: Cache-oblivious priority queue and graph algorithm applications. In: Proceedings of the 34th Annual Symposium on Theory of Computing, pp. 268–276 (2002)
5. Arge, L., Thorup, M.: RAM-efficient external memory sorting. In: Cai, L., Cheng, S.-W., Lam, T.-W. (eds.) Algorithms and Computation. LNCS, vol. 8283, pp. 491–501. Springer, Heidelberg (2013)
6. Atkin, A., Bernstein, D.: Prime sieves using binary quadratic forms. Math. Comput. **73**(246), 1023–1030 (2004)
7. Bays, C., Hudson, R.H.: The segmented sieve of Eratosthenes and primes in arithmetic progressions to 1012. BIT Numer. Math. **17**(2), 121–127 (1977)
8. Bengelloun, S.: An incremental primal sieve. Acta Informatica **23**(2), 119–125 (1986)
9. Brent, R.P.: The first occurrence of large gaps between successive primes. Math. Comput. **27**(124), 959–963 (1973)
10. Cox, D.A.: Primes of the Form $x^2 + ny^2$: Fermat, Class Field Theory, and Complex Multiplication. Wiley, New York (1989)
11. Dunten, B., Jones, J., Sorenson, J.: A space-efficient fast prime number sieve. IPL **59**(2), 79–84 (1996)
12. Farach-Colton, M., Tsai, M.-T.: On the complexity of computing prime tables. In: Elbassioni, K., Makino, K. (eds.) ISAAC 2015. LNCS, vol. 9472, pp. 677–688. Springer, Heidelberg (2015). doi:10.1007/978-3-662-48971-0_57
13. Gale, R., Pratt, V.: CGOL-an Algebraic Notation for MACLISP Users. MIT Artificial Intelligence Library, Cambridge (1977)
14. Galway, W.F.: Dissecting a sieve to cut its need for space. In: Bosma, W. (ed.) ANTS-IV. LNCS, vol. 1838, pp. 297–312. Springer, Heidelberg (2000)
15. Gries, D., Misra, J.: A linear sieve algorithm for finding prime numbers. Commun. ACM **21**(12), 999–1003 (1978)
16. Hardy, G.H., Wright, E.M.: An Introduction to the Theory of Numbers. Oxford University Press, Oxford (1979)
17. Horsley, S.: ΚΟΣΚΙΝΟΝ ΕΡΑΤΟΣΘΕΝΟΥΣ. or, the sieve of eratosthenes. being an account of his method of finding all the prime numbers, by the Rev. Samuel Horsley, FRS. Philos. Trans. **62**, 327–347 (1772)
18. Lenstra Jr., H.W.: Primality testing with gaussian periods. In: Agrawal, M., Seth, A.K. (eds.) FSTTCS 2002. LNCS, vol. 2556, pp. 1–1. Springer, Heidelberg (2002)
19. Mairson, H.G.: Some new upper bounds on the generation of prime numbers. Commun. ACM **20**(9), 664–669 (1977)
20. Mertens, F.: Ein beitrag zur analytischen zahlentheorie. J. fr die reine und angewandte Mathematik **78**, 46–62 (1874)
21. News, F.: World's largest prime number discovered - all 17 million digits, February 2013. https://web.archive.org/web/20130205223234/, http://www.foxnews.com/science/2013/02/05/worlds-largest-prime-number-discovered/

22. Patrascu, M., Thorup, M., Dynamic integer sets with optimal rank, select, predecessor search. In: FOCS, pp. 166–175 (2014)
23. Pomerance, C., Selfridge, J.L., Wagstaff, S.S.: The pseudoprimes to $25 \cdot 10^9$. Math. Comput. $35(151)$, 1003–1026 (1980)
24. Pomerance, C., Shparlinski, I.E.: On pseudosquares and pseudopowers. Comb. Number Theor., 171–184 (2009)
25. Pritchard, P.: A sublinear additive sieve for finding prime number. Commun. ACM $24(1)$, 18–23 (1981)
26. Pritchard, P.: Explaining the wheel sieve. Acta Informatica $17(4)$, 477–485 (1982)
27. Pritchard, P.: Linear prime-number sieves: a family tree. Sci. Comput. Program. $9(1)$, 17–35 (1987)
28. Schönhage, A., Grotefeld, A., Vetter, E.: Fast algorithms: a multitape turing machine implementation. Wissenschaftsverlag, B.I (1994)
29. Singleton, R.C.: Algorithm 357: an efficient prime number generator. Commun. ACM 12, 563–564 (1969)
30. Sorenson, J.: An introduction to prime number sieves. Technical report 909, Computer Sciences Department, University of Wisconsin-Madison (1990)
31. Sorenson, J.P.: The pseudosquares prime sieve. In: Hess, F., Pauli, S., Pohst, M. (eds.) ANTS 2006. LNCS, vol. 4076, pp. 193–207. Springer, Heidelberg (2006)
32. Villarino, M.B.: Mertens' proof of mertens' theorem. arXiv:math/0504289 (2005)
33. Vitter, J.S.: External memory algorithms and data structures: dealing with massive data. ACM Comput. Surv. (CsUR) $33(2)$, 209–271 (2001)
34. Williams, H.C.: Edouard Lucas and primality testing. Canadian Mathematics Society Series of Monographs and Advanced Texts, 22 (1998)

Increasing Diamonds

Olivier Bodini[1], Matthieu Dien[2], Xavier Fontaine[1],
Antoine Genitrini[2]([✉]), and Hsien-Kuei Hwang[3]

[1] Laboratoire d'Informatique de Paris-Nord, CNRS UMR 7030 - Institut Galilée -
Université Paris-Nord, 99, Avenue Jean-Baptiste Clément, 93430 Villetaneuse, France
Olivier.Bodini@lipn.univ-paris13.fr, Xavier.Fontaine@polytechnique.edu
[2] Sorbonne Universités, UPMC Univ Paris 06, CNRS, LIP6 UMR 7606,
4 Place Jussieu, 75005 Paris, France
{Matthieu.Dien,Antoine.Genitrini}@lip6.fr
[3] Institute of Statistical Science, Academia Sinica, Taipei 115, Taiwan
hkhwang@stat.sinica.edu.tw

Abstract. A class of diamond-shaped combinatorial structures is studied whose enumerating generating functions satisfy differential equations
of the form $f'' = G(f)$, for some function G. In addition to their own
interests and being natural extensions of increasing trees, the study of
such DAG-structures was motivated by modelling executions of series-
parallel concurrent processes; they may also be used in other digraph
contexts having simultaneously a source and a sink, and are closely connected to a few other known combinatorial structures such as trees, cacti
and permutations. We explore in this extended abstract the analytic-
combinatorial aspect of these structures, as well as the algorithmic issues
for efficiently generating random instances.

1 Introduction

Simple combinatorial structures that are both mathematically tractable and
physically useful in different modeling purposes have received much attention
in the literature. Typical representative examples include the simply-generated
family of trees characterized by the functional equation (see [13])

$$f = zG(f),$$

and the varieties of increasing trees by the differential equation (see [3])

$$f' = G(f).$$

Due to their simplicity, these tree models also appeared naturally under various
guises in many areas. Three simple prototypical cases are given in the following
table.

This research was partially supported by the ANR MetACOnc project ANR-15-
CE40-0014.

E. Kranakis et al. (Eds.): LATIN 2016, LNCS 9644, pp. 207–219, 2016.
DOI: 10.1007/978-3-662-49529-2_16

G	$f = zG(f)$	$f' = G(f)$
$1 + z^2$	Binary tree	Binary increasing tree
$(1 + z)^2$	(Catalan tree)	(Binary search tree)
$\exp(z)$	Cayley tree	Recursive tree
$\frac{1}{1-z}$	Planted (ordered) tree	Plane-oriented recursive tree

In particular, binary trees have long been studied in the computer science literature (see Knuth's book [9]) and a compilation of 214 combinatorial objects leading to the same enumerating Catalan numbers can be found in Stanley's recent book [14]. On the other hand, binary increasing trees are isomorphic to binary search trees, which represent another class of fundamental data structures with a huge number of variants; they are also closely related to Quicksort in Algorithms, to Yule-Harding models in Phylogenetics, to random permutations in Combinatorics, Rényi's car-parking problem in Applied Probability, and to Eden model in Statistical Physics, to name just a few; see [6,8] for more information.

We explore in this paper another class of combinatorial structures, which we call *increasing diamonds*: they are *labelled, directed acyclic graphs (DAGs) with a source and a sink such that the labels along any path are increasing*; see Fig. 1 for an illustration of two different diamonds. In standard symbolic notation (see [8]), increasing diamonds can be described as

$$\mathcal{F} = \mathcal{Z}^\square + \mathcal{Z}^\square \star \mathcal{G}(\mathcal{F}) \star \mathcal{Z}^\blacksquare. \tag{1}$$

where $\mathcal{G}$ is some functional operation specifying possible degrees and construction rules, and the two symbols $\square$ and $\blacksquare$ represent the smallest and the largest labels, respectively. This equation then translates into the differential equation satisfied by the enumerating generating function[1]

$$f''(z) = G(f(z)), \quad \text{with } f(0) = 0 \text{ and } f'(0) = 1. \tag{2}$$

Here $f(z) = \sum_{n \geq 1} a_n z^n / n!$, where a_n enumerates the number of increasing diamonds with n labels.

We study in this paper three simple representative cases, and focus on asymptotic enumeration and random generation. The following table lists the dominant term in the corresponding asymptotic approximation in each case. Here OEIS stands for Sloane's Online Encyclopedia of Integer Sequences, C_m is a constant (see Theorem 3), and ρ_{binary}, $\rho_{m\text{-ary}}$ and ρ_{plane} are three constants given in (9), (12) and (14), respectively. While most properties are expected to be similar to those of increasing trees (see [3]), the higher order derivative introduces more technical difficulties, as visible from the less common asymptotic order produced when $G = 1/(1 - z)$.

---

[1] We limit our discussion in this paper to the situation when $f'(0) = 1$ for simplicity.

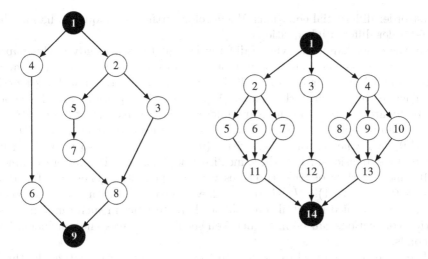

Fig. 1. A binary (left) and a ternary (right) increasing diamonds of size 9 and 14, respectively.

G	OEIS	Other description	$\frac{a_n}{n!} \sim$	Error term of order
$\exp(z)$	A000111	Euler or up/down numbers	$2n(\frac{2}{\pi})^n$	Exponential
$(1+z)^2$	A007558	Shifts 2 places left after squared	$6n\rho_{\text{binary}}^{-n-2}$	Exponential
$1+z^m$ $(m \geq 3)$	–	–	$C_m n^{-\frac{m-3}{m-1}} \rho_{m\text{-ary}}^{-n-\frac{2}{m-1}}$	Polynomial
$\frac{1}{1-z}$	A032035	Triangular cacti with bridges	$\frac{\rho_{\text{plane}}^{1-n}}{n^2\sqrt{2\log n}}$	Logarithmic
$\frac{1}{(1-z)^3}$	A001147	Double factorial	$\frac{2^n}{\sqrt{\pi n}}$	Polynomial

Structurally, increasing diamonds are bipolar digraphs with a downward planarity; they are also special cases of series-parallel graphs and are more expressive than quasi-trees in [2]. Since DAGs with a unique source and a unique sink appear naturally in many concrete applications, our increasing diamonds may be of potential use in modelling structural parameters or problem complexity in these contexts. Typical examples include: partial orders and their linear extensions, computational processes and their executions in parallel computing, network or data flows, food-webs, register sharing, machine learning, streaming analysis, grid computing, etc.

To be useful for modelling concrete structures in applications, we need either more precise statistical properties or more efficient generation algorithms for random increasing diamonds. The former will be addressed elsewhere, and for the latter, we will focus on the by now popular Boltzmann sampling algorithm proposed in [7], which was recently extended in [5] to deal with the situation of first-order differential equations. We develop further techniques to handle the second-order differential equations.

On the other hand, the type of differential equations we study in this paper $(f'' = G(f))$ also emerges naturally in other contexts, notably in a recent paper by Kuba and Panholzer [12] on multi-labeled increasing trees and hook-length formulae; see also their earlier paper [11]. While the equations are the same, our combinatorial structures here are different and to some extent more natural, and such a difference is reflected by the initial conditions: we focus on $f(0) = 0$ and $f'(0) = 1$ whereas they deal with $f(0) = f'(0) = 0$. Also we will derive asymptotic expansions. Along the same direction, Kuba and Panholzer examined in [10] another class of tree structures whose exponential generating function satisfies $f^{(m)} = (m-1)! e^{mf}$, where $m \geq 2$, which coincides with our model when $m = 2$. They studied in detail some shape characteristics in such random trees. Further connections can be made between such bucket trees and our increasing diamonds.

The paper is organized as follows. In the next section, we analyze the three classes of increasing diamonds in detail. Then Sect. 3 is devoted to the development of algorithmic tools for generating efficiently random diamonds that rely on the notion of uniform Boltzmann sampling.

2   Exact Enumeration and Asymptotics

In this section, we first discuss the general solution of the differential equation $f'' = G(f)$ subject to the initial conditions $f(0) = 0$ and $f'(0) = 1$ (other initial conditions can be dealt with in a similar manner), and then concentrate our discussion on a few special cases for which we will derive more precise asymptotic approximations.

2.1   General Solution of $f'' = G(f)$

Multiplying both sides of (2) by $2f'$, we obtain $2f''f' = 2G(f)f'$, which implies that

$$f'(z)^2 = f'(0)^2 + \int_0^z 2G(f(t))f'(t)\mathrm{d}t = 1 + 2\mathscr{G}(f(z)),$$

where $\mathscr{G}(z) := \int_0^z G(t)\mathrm{d}t$. Thus $f'(z) = \pm\sqrt{1 + 2\mathscr{G}(f(z))}$, or

$$\pm \int_0^{f(z)} \frac{1}{\sqrt{1 + 2\mathscr{G}(t)}} \, \mathrm{d}t = z. \tag{3}$$

Lemma 1. *The solution to the differential equation $f'' = G(f)$ with $f(0) = 0$ and $f'(0) = 1$ is given by*

$$\int_0^{f(z)} \frac{1}{\sqrt{1 + 2\mathscr{G}(t)}} \, \mathrm{d}t = z. \tag{4}$$

Proof. By expanding the left-hand side of the second equation in (3) as a Taylor series, we exclude the negative solution and conclude (4).

The solution (4) is, although implicit, useful in our asymptotic analysis even when no further simplification is possible. First recall a useful property when f blows up near the dominant singularity, which is readily modified from Lemma 1 of [3].

Lemma 2. *Given an entire function G, the dominant real positive singularity of the function $f(z)$, solution to $Y'' = G(Y)$ with $Y(0) = 0$ and $Y'(0) = 1$, is given by*

$$\rho = \int_0^\infty \frac{dt}{\sqrt{1 + 2\int_0^t G(v)dv}},$$

provided that the integral converges.

From the brief discussion in Introduction, we see that the coefficient a_n of f is well-approximated by $a_n \sim Cn!\rho^{-n}n^\alpha(\log n)^\beta$, and from this observation we expect that the singularity analysis of Flajolet and Odlyzko (see [8]) will be useful in such an analysis, as in [3].

2.2   Non-plane (Unordered) Increasing Diamonds

We discuss in detail the class of non-plane increasing diamonds, which can be decomposed as sets of increasing diamonds:

$$\mathcal{F} = \mathcal{Z}^\square + \mathcal{Z}^\square \star \text{SET}\,(\mathcal{F}) \star \mathcal{Z}^\blacksquare,$$

so that the corresponding exponential generating function satisfies $f'' = e^f$. The two diamonds in Fig. 1 may be regarded, neglecting the order of subtrees, as two instances of non-plane increasing diamonds.

By (4) with $G(z) = e^z$ and $\mathscr{G}(z) = e^z - 1$, we see that the exponential generating function f of a_n has the solution

$$f(z) = -\log(1 - \sin z), \tag{5}$$

and the number a_n of such increasing diamonds with n labels starts with

$$\{a_n\}_{n\geq 1} = \{1, 1, 1, 2, 5, 16, 61, 272, 1385, 7936, 50521, 353792, \dots\},$$

which coincides with A000111 in Sloane's OEIS, where many other structures with identical enumerating sequence are also given (alternating permutations, zig-zag posets, some increasing trees, etc.). This shows the richness and usefulness of the equation $f'' = e^f$ in combinatorial objects.

Note that, by the differential equation $f''' = f'f''$ (obtained by the differentiation of $f'' = e^f$), we have the recurrence relation

$$a_n = \sum_{2\leq k<n} \binom{n-3}{k-2} a_k a_{n-k} \qquad (n \geq 3),$$

which is useful for numerical purposes.

Theorem 1. *The number a_n of non-plane increasing diamonds with n labels satisfies*

$$a_n = \frac{2^{n+1}\,(n-1)!}{\pi^n} \sum_{j=-\infty}^{+\infty} \frac{1}{(1+4j)^n}. \tag{6}$$

It is less obvious that the right-hand side represents an integer.

Proof. By (5), we have $f'(z) = \tan z + \sec z$, which has only simple poles at $z = (2k+\frac{1}{2})\pi$. By standard expansion for meromorphic functions ([8, Ch. IV]), we obtain the expansion (6), which is not only an asymptotic expansion (expressible as Hurwitz's zeta function)

$$a_n = \frac{2^{n+1}\,(n-1)!}{\pi^n} \sum_{j\geq 0} \left(\frac{1}{(1+4j)^n} + \frac{(-1)^n}{(4j+3)^n} \right),$$

but also an identity for $n \geq 1$.

Another exactly solvable case is when $G(z) = (1-z)^{-3}$. In this case, we have the surprisingly simple solution (cf. [12])

$$f(z) = 1 - \sqrt{1-2z}, \tag{7}$$

leading to the simple expression for the total number of size-n diamonds

$$a_n = (2n-3)!! = \frac{(2n-2)!}{2^{n-1}(n-1)!} \qquad (n \geq 1).$$

However, exact solutions as (5) and (7) are exceptional rather than commonplace, and different techniques are needed in most cases as we will see below.

2.3   *m*-ary Increasing Diamonds

Consider now increasing diamonds in which the degrees of nodes are limited to $m \geq 2$; see Fig. 1 for a binary and a ternary diamond. In this case, we have the specification

$$\mathcal{F} = \mathcal{Z}^{\square} + \mathcal{Z}^{\square} \star \mathcal{F}^m \star \mathcal{Z}^{\blacksquare}, \tag{8}$$

which leads to the differential equation $f'' = 1 + f^m$ with $f(0) = 0$ and $f'(0) = 1$. Closed-form solutions are possible when $m = 2$ and $m = 3$ (in terms of elliptic integrals), but they are not simple. So we present only the solution for $m = 2$ and derive asymptotic approximation for $m \geq 3$ (in a slightly more general formulation).

Binary increasing diamonds and Weierstrass's $\wp$-function. From (4), we see that f satisfies the equation

$$\int_0^{f(z)} \frac{1}{\sqrt{1 + 2t + \frac{2}{m+1}t^{m+1}}}\, dt = z.$$

When $m = 2$, we can express the solution in terms of Weierstrass's elliptic function $\wp$ (see [1]), which is defined periodically over a lattice that contains one double pole in a corner of each cell. Thus, by construction,

$$\wp(z; \omega_1, \omega_2) = \frac{1}{z^2} + \sum_{(k,l) \in \mathbb{Z}^2 \setminus \{(0,0)\}} \left(\frac{1}{(z + k\omega_1 + l\omega_2)^2} - \frac{1}{(k\omega_1 + l\omega_2)^2} \right),$$

where ω_1 and ω_2 are the periods of $\wp$.

Theorem 2. *The exponential generating function of the number of binary increasing diamonds can be expressed as*

$$f(z) = 6\wp \left(z - \rho; -\tfrac{1}{3}, -\tfrac{1}{36} \right) \quad \text{where } \rho := \int_0^\infty \frac{dt}{\sqrt{1 + 2t + \tfrac{2}{3}t^3}}, \tag{9}$$

and the number of size-n binary increasing diamonds is given by

$$a_n = 6 \frac{(n+1)!}{\rho^{n+2}} \sum_{(k,l) \in \mathbb{Z}^2} \frac{1}{\left(1 + \frac{k\omega_1}{\rho} + \frac{l\omega_2}{\rho} \right)^{n+2}}, \tag{10}$$

where ω_1 and ω_2 are computable constants.

Asymptotically, $a_n \sim 6(n+1)! \rho^{-n-2}$, with an exponentially small error. Note that, by starting with the initial conditions $f(0) = f'(0) = 0$, we then obtain the bi-labelled increasing trees defined in [12], which corresponds to the sequence A144849 in OEIS.

Proof. (Sketch) The $\wp$-function satisfies the differential equation

$$\wp'^2(z) = 4\wp^3(z) - g_2\wp(z) - g_3,$$

and we need only to identify the corresponding parameters.

By Lemma 2, we first determine the dominant singularity ρ; then from the series expansion of $\wp$, we deduce (10) by a direct application of singularity analysis (see [8]). □

Although few cases lead to closed-form expressions in terms of known functions, it is not difficult to derive asymptotic approximations based on complex analysis and singularity analysis, as already highlighted in the classical paper [3].

Polynomial varieties of increasing diamonds. As in [3], we consider the polynomial varieties of increasing diamonds, which are characterized by $G(z)$ being a polynomial, say

$$G(z) = \sum_{0 \leq j \leq m} b_j z^j, \tag{11}$$

where $m \geq 2$ and $b_m > 0$. For simplicity, we may assume that $G(z) \not\equiv z^\ell H(z^k)$, for some $k \geq 2$ and $\ell \geq 0$, namely, G is aperiodic.

Then by (2), the dominant singularity is given by

$$\rho = \int_0^\infty \frac{1}{\sqrt{1 + 2\sum_{0 \leq j \leq m} \frac{b_j}{j+1} t^{j+1}}} \, dt, \tag{12}$$

which is absolutely convergent since $m \geq 2$. Then we apply the same idea used in [3], and obtain

$$\rho - z = \int_{f(z)}^\infty \frac{1}{\sqrt{1 + 2\sum_{0 \leq j \leq m} \frac{b_j}{j+1} t^{j+1}}} \, dt$$

$$= \frac{\sqrt{m+1}}{(m-1)\sqrt{2b_m}} f^{-\frac{m-1}{2}} - \frac{b_{m-1}\sqrt{m+1}}{mb_m^{3/2}} f^{-\frac{m+1}{2}} + \cdots,$$

as $z \to \infty$. Then by inverting, we get

$$f(z) = \left(\frac{(m-1)\sqrt{b_m}}{\sqrt{2(m+1)}} \right)^{-\frac{2}{m-1}} (\rho - z)^{-\frac{2}{m-1}} \left(1 + O\left(|\rho - z|^{\frac{2}{m-1}} \right) \right),$$

as $z \sim \rho$, the justification following also standard line. We then deduce by the singularity analysis the following asymptotic approximation.

Theorem 3. *Assume that G is a polynomial given in (11) and $a_n > 0$ for $n \geq n_0$ for some $n_0 > 0$. Then the number of increasing diamonds with n labels in a polynomial variety satisfies*

$$a_n = \left(\frac{\sqrt{2(m+1)}}{(m-1)\sqrt{b_m}} \right)^{\frac{2}{m-1}} \frac{n^{-\frac{m-3}{m-1}}}{\Gamma(\frac{2}{m-1})} \rho^{-n-\frac{2}{m-1}} \left(1 + O\left(n^{-\frac{4}{m-1}} \right) \right),$$

for $m \geq 2$, where ρ is given in (12).

Note that the asymptotic estimates here are independent of the initial conditions.

In the special case when $m = 3$, it is possible to express f in terms of Jacobi elliptic functions, but the expression is messy.

2.4   Plane Increasing Diamonds

We now focus on plane (ordered) increasing diamonds, which are described by

$$\mathcal{F} = \mathcal{Z}^\square + \mathcal{Z}^\square \star \text{SEQ}(\mathcal{F}) \star \mathcal{Z}^\blacksquare, \tag{13}$$

leading to the differential equation $f'' = \frac{1}{1-f}$ with the initial conditions $f(0) = 0$ and $f'(0) = 1$.

The analysis of such diamonds is more involved and the asymptotic expansion we obtain has a much poorer convergence rate: instead of exponential or polynomial, the terms are now in decreasing powers of $\log n$.

Theorem 4. *The number of plane increasing diamonds with n labels satisfies*

$$a_n = \frac{n!\rho^{1-n}}{n^2\sqrt{2\log n}} \left(\sum_{0 \le k < K} \frac{P_k(\log\log n)}{(\log n)^k} + \mathcal{O}\left(\frac{(\log\log n)^K}{(\log n)^K}\right)\right),$$

where

$$\rho := \int_0^\infty \frac{1}{\sqrt{1 - 2\log(1-t)}}\, dt = \frac{\sqrt{e}}{2}\int_0^\infty v^{-\frac{1}{2}}e^{-v}dv \approx 0.65567\,95424\ldots, \quad (14)$$

and the P_k's are computable polynomials (of degree k).

In particular, $P_0(x) = 1$ and $P_1(x) = \frac{1}{8}(x - 3 - 2\gamma + \log 2 + 2\log\rho)$.
  The method of proof is the same as above, details being omitted here. The first few terms of a_n are

$$\{1, 1, 1, 3, 13, 77, 573, 5143, 54025, 650121, 8817001, 133049339, \ldots\},$$

and corresponds to sequence A032035 in OEIS, which also enumerates increasing rooted $(2,3)$-cacti with $n-1$ nodes. Note that $f_1 = f' - 1$ satisfies the differential equation $f_1' = e^{f_1 + f_1^2/2}$.

3 Random Generation via Boltzmann Samplers

3.1 Boltzmann Samplers for the Differential Classes

The Boltzmann sampling technique was first proposed in the seminal paper [7], and has been widely developed and extended since then. It captures the features any successful algorithm must have: *simple, efficient and easily extensible.*
  In this subsection, we briefly recall this technique for labeled structures.

Definition 1. *A Boltzmann sampler of parameter $x > 0$ is an algorithm that draws an object α of size $|\alpha|$ in a given combinatorial class $\mathcal{A}$ with the probability*
$$\mathbb{P}_x(\alpha) = \frac{x^{|\alpha|}}{|\alpha|!A(x)}.$$

Note that the output size N of a Boltzmann sampler is a random variable with the law $\mathbb{P}_x(N = n) = a_n x^n/(n!A(x))$, and the expectation of N is $\mathbb{E}_x(N) = xA'(x)/A(x)$. Here x is a free variable. To generate an object of size n, one can choose the parameter x to be the solution of the saddle point equation $\mathbb{E}_x(N) = n$. With this choice, it is possible to devise a linear-time algorithm to generate a random instance by repeated use of trial-and-rejection until reaching an output of size in $[(1 - \varepsilon)n, (1 + \varepsilon)n]$) (referred to as an approximate-size algorithm).

This universal method is not only very efficient but also fully automatizable. What we need is a complete symbolic (recursive or not) description of the class in order to construct a sampler. Indeed, Boltzmann samplers for the neutral and atomic classes $\mathcal{E}$ and $\mathcal{Z}$ are trivial, and from there general procedures exist for constructing more complex samplers through elementary operations such as addition, multiplication, cycle, set, etc. We refer the reader to the original paper [7] for more details. On the other hand, the Boltzmann sampler for the box-operator of two classes was addressed in [5].

Note that Boltzmann samplers does not return a labeled object, but only the unlabeled skeleton. To complete the process, a labeling algorithm is needed.

3.2   Boltzmann Samplers for Second-Order Differential Classes

It is natural to divide the problem into two cases, one in which the differential equation is induced by the general shape specification $\mathcal{F}'' = \phi(\mathcal{Z}, \mathcal{F})$, where $\mathcal{F}''$ denotes the class of objects of $\mathcal{F}$ in which two nodes are pointed, and the other by $\mathcal{F}'' = \phi(\mathcal{F})$.

Before considering these two issues, we recall some basic and classical properties. First, the box product and the derivative operator are linked together by the fact that $\mathcal{C} = \mathcal{A} \,^\square\! \star\, \mathcal{B}$ entails that $\mathcal{C}' = \mathcal{A}' \times \mathcal{B}$. Secondly, we know how to get a sampler of parameter x for $\mathcal{F}$ by just using a sampler of $\mathcal{F}'$. This surprising result is obtained by multiplying the Boltzmann parameter x by a suitable continuous random variable u in $[0, 1]$. Indeed, this yields the following algorithm described in [4], which can also be derived from results in [5].

Algorithm 1. $\Gamma_x \mathcal{F}$ from $\Gamma \mathcal{F}'$

1: **if** Bernoulli$(f(0)/f(x))$ **then**
2:     **return** an object of size 0
3: **else**
4:     Draw $U \in [0, 1]$ with the density $\delta_x(u) = f'(ux)x/(f(x) - f(0)) \cdot \mathbf{1}_{[0,1]}(u)$
5:     Draw $\gamma' = \Gamma_{Ux} \mathcal{F}'$
6:     **return** γ' where the bud is replaced by an atom.
7: **end if**

In line 5, the object contains what is called a bud in Species Theory. It can be seen as a hole, that is the reason why it is replaced by an atom (in line 6).

General Case $\mathcal{F}'' = \phi(\mathcal{Z}, \mathcal{F})$. We consider now the case $\mathcal{F}'' = \phi(\mathcal{Z}, \mathcal{F})$, which can be dealt with by applying twice Algorithm 1. But this requires to draw two continuous random variables U and V, and use only their product UV. Clearly, this can be factored by calculating directly the random variable $S = UV$. This gives the following algorithm for which the proof is similar to that of $\mathcal{F}'' = \phi(\mathcal{Z}, \mathcal{F})$ in [5].

Algorithm 2. $\Gamma_x \mathcal{F}$ generates an object in $\mathcal{F}$ from a sampling in $\mathcal{F}''$

1: Draw $W \in [0,1]$ uniformly
2: **if** $W < \dfrac{f(0)}{f(x)}$ **then**
3:     **return** an object of size 0
4: **else if** $W < \dfrac{f(0) + xf'(0)}{f(x)}$ **then**
5:     **return** an object of size 1
6: **else**
7:     Draw $S \in [0,1]$ according to the density $\delta_x(s) = \dfrac{x^2(1-s)f''(sx)}{f(x) - xf'(0) - f(0)} \mathbf{1}_{[0,1]}(s)$
8:     Draw γ'' using $\Gamma_{Sx}\mathcal{F}''$
9:     **return** replace the buds in γ'' by two atoms.
10: **end if**

Particular Case $\mathcal{F}'' = \phi(\mathcal{F})$. We consider here the special case where ϕ does not explicitly depend on $\mathcal{Z}$. The Algorithm from [4] can be amended to deal with uniform continuous random variables rather than non-uniform random variables that are hard to simulate.

Classical Boltzmann samplers Γ are parametrized by x, so the sampler draws an object α in $\mathcal{A}$ with probability $\mathbb{P}_x(\alpha) = x^{|\alpha|}/(|\alpha|!A(x))$. But in the case of functional equations where x is not explicit (such as $\mathcal{F}' = \phi(\mathcal{F})$), it has been observed in [4] that it is preferable to deal with another parameter $\tau = f(x)$. In this case, the output is distributed as $\mathbb{P}(N = n) = a_n f^{-1}(\tau)^n/(n!\tau)$. It is nevertheless always a Boltzmann sampler but with a different parametrization. To avoid confusion, we then indicate $\Gamma_{[\tau]}\mathcal{F}$ instead of $\Gamma_x\mathcal{F}$. Thus we can now give an algorithm similar to Algorithm 1 that uses only uniform random variables.

Algorithm 3. $\Gamma_{[\tau]}\mathcal{F}$ generates an object in $\mathcal{F}$ from a sampling in $\mathcal{F}'$

1: **if** Bernoulli$(f(0)/\tau)$ **then**
2:     **return** an object of size 0
3: **else**
4:     Draw U uniformly $\in [0,1]$
5:     $\tau_{new} \leftarrow U\tau + (1-U)f(0)$
6:     Draw γ' using $\Gamma_{[\tau_{new}]}\mathcal{F}'$
7:     **return** γ' where we replace the bud by an atom.
8: **end if**

In order to apply twice this procedure (because $\mathcal{F}'' = (\mathcal{F}')'$), we need to obtain $\Gamma_{[\tau_{new}]}\mathcal{F}'$ by using the Boltzmann sampler of $\mathcal{F}''$. For this, let $y = f^{-1}(\tau_{new})$. We have $f'(vy) = Vf'(y) + (1-V)f'(0)$ where V is a uniform random variable on $[0,1]$. Since we are looking for an algorithm $\Gamma_{[\tau_{new}]}\mathcal{F}'$ where $\tau_{new} = f(y)$, we need an expression of $f(vy)$ in function of τ_{new}. But since the differential equation $f''(z) = \phi(f(z))$ can be integrated (by multiplying both sides by $f'(z)$), we then get $f'(z) = g(f(z))$, where $\frac{1}{2}g^2$ is the primitive of ϕ such that $f(0) = \dfrac{f'(0)^2}{2}$. Then we get the expression $f(vy) = g^{-1}(Vg(f(y)) + (1-V)f'(0))$. Finally, we obtain the following algorithm.

Algorithm 4. $\Gamma_{[\tau_0]}\mathcal{F}$ generates an object of $\mathcal{F}$ following the Boltzmann distribution of parameter $x = f^{-1}(\tau_0)$, from a sampler of $\mathcal{F}'' = \Phi(\mathcal{F})$

1: **if** Bernoulli($f(0)/\tau_0$) **then**
2:     **return** an object of size 0
3: **else**
4:     Draw U uniformly on $[0, 1]$
5:     $\sigma \leftarrow U\tau_0 + (1 - U)f(0)$
6:     **if** Bernoulli($f'(0)/g(\sigma)$) **then**
7:         **return** an object of size 1
8:     **else**
9:         Draw V uniformly on $[0, 1]$
10:        $\tau \leftarrow g^{-1}(Vg(\sigma) + (1 - V)f'(0))$
11:        Draw γ'' using $\Gamma_{[\tau]}\mathcal{F}'' = \Gamma_{[\tau]}\Phi(\mathcal{F})$
12:        **return** γ'' where the buds are replaced by two atoms.
13:    **end if**
14: **end if**

In contrast to the previous algorithm, we do not need here to draw random variables with complicated laws. This very simple sampler is easily implemented for testing purposes. It remains to analyze its complexity. As already discussed above, the dominant singularity ρ of f is of the form $(1 - z/\rho)^{-\alpha}$ for some $\alpha > 0$. This ensures the following theorem.

Theorem 5. *Algorithm 4 provides a Boltzmann sampler, and its approximate-size version gives a linear time algorithm for drawing uniformly at random a diamond of type $\mathcal{F}'' = \Phi(\mathcal{F})$, where Φ is a polynomial.*

We implemented this algorithm in Java, and obtained the following table, which synthesizes benchmarks computed on a laptop (1.5 GHz CPU and 4G RAM). The examples we tested consist of ternary diamonds $f'' = 1 + f^3$ with initial conditions $f(0) = 0$ and $f'(0) = 1$, and with size tolerance set at 10 percent. One of such diamonds is depicted in Fig. 2. We observe that the timing results are consistent with our analysis.

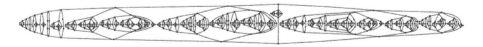

Fig. 2. A random diamond of size 591 satisfying $f'' = 1 + f^3$, with $f(0) = 0$ et $f'(0) = 1$.

Size n	10	100	1000	5000	10000	50000	100000	150000
τ_0	8.73	80.44	794	3972	7941	39752	79559	119086
Time (ms)	1	7	66	322	668	3887	7098	9812

References

1. Abramowitz, M., Stegun, I.: Handbook of Mathematical Functions: with Formulas, Graphs, and Mathematical Tables. Dover Publications, New York (2012)
2. Ando, E., Nakata, T., Yamashita, M.: Approximating the longest path length of a stochastic DAG by a normal distribution in linear time. J. Discrete Algorithms **7**(4), 420–438 (2009)
3. Bergeron, F., Flajolet, P., Salvy, B.: Varieties of increasing trees. In: Raoult, J.-C. (ed.) CAAP '92. LNCS, vol. 581, pp. 24–48. Springer, Heidelberg (1992)
4. Bodini, O.: Autour de la génération aléatoire sous modèle de Boltzmann. Habilitation thesis, UPMC (2010)
5. Bodini, O., Roussel, O., Soria, M.: Boltzmann samplers for first-order differential specifications. Discrete Appl. Math. **160**(18), 2563–2572 (2012)
6. Chern, H.-H., Fernández-Camacho, M.-I., Hwang, H.-K., Martínez, C.: Psi-series method for equality of random trees and quadratic convolution recurrences. Random Struct. Algorithms **44**(1), 67–108 (2014)
7. Duchon, P., Flajolet, P., Louchard, G., Schaeffer, G.: Boltzmann samplers for the random generation of combinatorial structures. Comb. Prob. Comput. **13**(4–5), 577–625 (2004)
8. Flajolet, P., Sedgewick, R.: Analytic Combinatorics. Cambridge University Press, Cambridge (2009)
9. Knuth, D.E.: The Art of Computer Programming, volume 1 (3rd ed.): Fundamental Algorithms, Addison Wesley Longman Publishing Co., Inc., Redwood City, CA, USA (1997)
10. Kuba, M., Panholzer, A.: A combinatorial approach to the analysis of bucket recursive trees. Theor. Comput. Sci. **411**(34–36), 3255–3273 (2010)
11. Kuba, M., Panholzer, A.: Bilabelled increasing trees and hook-length formulae. Eur. J. Combin. **33**(2), 248–258 (2012)
12. Kuba, M., Panholzer, A.: Combinatorial families of multilabelled increasing trees and hook-length formulas. Discrete Math. **339**, 227–254 (2016)
13. Meir, A., Moon, J.W.: On the altitude of nodes in random trees. Can. J. Math. **30**(5), 997–1015 (1978)
14. Stanley, R.: Catalan Numbers. Cambridge University Press, Cambridge (2015)

Scheduling Transfers of Resources over Time: Towards Car-Sharing with Flexible Drop-Offs

Kateřina Böhmová[1]([✉]), Yann Disser[2], Matúš Mihalák[1,3],
and Rastislav Šrámek[4]

[1] Department of Computer Science, ETH Zürich, Zürich, Switzerland
katerina.boehmova@inf.ethz.ch
[2] Department of Mathematics, TU Berlin, Berlin, Germany
disser@math.tu-berlin.de
[3] Department of Knowledge Engineering, Maastricht University,
Maastricht, The Netherlands
matus.mihalak@maastrichtuniversity.nl
[4] Google Zürich, Zürich, Switzerland
sramek@google.com

Abstract. We consider an offline car-sharing assignment problem with flexible drop-offs, in which n users (customers) present their driving demands, and the system aims to assign the cars, initially located at given locations, to maximize the number of satisfied users. Each driving demand specifies the pick-up location and the drop-off location, as well as the time interval in which the car will be used. If a user requests several driving demands, then she is satisfied only if *all* her demands are fulfilled. We show that minimizing the number of vehicles that are needed to fulfill *all* demands is solvable in polynomial time. If every user has exactly one demand, we show that for given number of cars at locations, maximizing the number of satisfied users is also solvable in polynomial time. We then study the problem with two locations A and B, and where every user has two demands: one demand for transfer from A to B, and one demand for transfer from B to A, not necessarily in this order. We show that maximizing the number of satisfied users is NP-hard, and even APX-hard, even if all the transfers take exactly the same (non-zero) time. On the other hand, if all the transfers are instantaneous, the problem is again solvable in polynomial time.

Keywords: Interval scheduling · Complexity · Algorithms · Transfer · Resources

1 Introduction

In car sharing services, a company manages a fleet of cars that are offered to customers for rent for a short period of time. Every car is stationed at a fixed parking location, and a customer who wishes to rent the car is usually required to return the car back to the very same location. This is a constraint that many

© Springer-Verlag Berlin Heidelberg 2016
E. Kranakis et al. (Eds.): LATIN 2016, LNCS 9644, pp. 220–234, 2016.
DOI: 10.1007/978-3-662-49529-2_17

customers would like to soften. It is thus a natural question to find alternatives allowing the customers a *flexible* drop-off possibility. We investigate this "flexible drop-off" idea in the case where the demands for driving (pick-up at location A at time t_A and drop-off at location B at time t_B) are known in advance, and we study the problem of finding a maximum number of demands that can be realized by the existing fleet of cars and parking locations.

We show that the problem can be solved in polynomial time by a reduction to the minimum-cost maximum-flow problem in a dedicated auxiliary graph. We further consider the problem when every user (customer) has multiple driving demands. A user is satisfied if all her demands are fulfilled (the user needs to get a car for all the requested drivings, and has no interest in partial rentals). We show that satisfying the maximum number of users is an APX-hard problem already when there are only two locations, every user has two demands, the time for driving is the same for every demand, and there is only one car. An exemplary problem that falls into this setting is the situation where a single car is used to commute between two popular, but (by public transportation) badly connected locations. The users want to use this car for their daily travel: Every user wants to get from one location to the other one, and later in the day also from the other location back to her original one. Interestingly, the hardness holds only whenever the travelling takes non-zero time, as we also show that for an instantaneous travel (that takes zero time), the problem becomes solvable in polynomial time.

1.1   Formal Problem Description and Outline of the Paper

We define formally only the setting with two locations, as this setting forms the base of our main results. The problem definition for more locations is straightforward. User that rents a car at location A effectively blocks the car for a fixed time interval, and makes it available at the drop-off location B. The usage and trajectory of the car in the rental period is irrelevant for our scheduling problem, and we can simply model the renting as a *transfer* of the car from location A to location B at the given time interval. We abstract from our car-sharing motivation, and refer to the cars as resources.

We consider two locations, A and B, with an initial distribution of indistinguishable resources within these two locations, say there are a resources at location A and b resources at location B in the beginning. A transfer from A to B is a movement of a resource from A to B. A transfer is possible only if there is an available resource. There are n users and each of them has one or more demands: A demand d is specified by a direction $X \rightarrow Y$ (either $A \rightarrow B$ or $B \rightarrow A$) and time interval (t_d^s, t_d^e), and represents a request for a resource transfer from location X to location Y, leaving the origin X at time t_d^s and arriving at the destination Y at time t_d^e. The demand d is *fulfilled* by moving one resource from X to Y. In this case, the resource is blocked (i.e., cannot be transferred further) for the time period (t_d^s, t_d^e). The goal is to select a feasible set of demands that maximizes the number of satisfied users. Here, a set of demands is *feasible* if: (i) whenever a demand of a user is selected, then all demands of the user are selected, and (ii) all selected demands can be fulfilled, i.e., we can move the resources as suggested by the demands.

We first considered the simplest questions: (1) Decide whether all users can be satisfied (or equivalently, decide whether all the demands can be fulfilled); (2) Compute the minimum number of resources initially needed at each location to satisfy all the users. We observed that straightforward "simulation-like" algorithms can answer these questions for any number of users, demands per user, locations, and resources.

In Sect. 2, we study the problem where each user has only one demand. We show that the problem of maximizing the number of satisfied users for given number of resources at locations (i.e., in this case, the number of fulfilled demands) is polynomially solvable, by reducing it to the minimum-cost maximum-flow problem. This approach works even if there are multiple locations and multiple resources in the system.

In Sect. 3, we study the variant where every user has exactly two demands: One transfer from A to B and one transfer from B to A, but not necessarily in this order. Recall that a user is satisfied only if both the demands are fulfilled. We show that in this setting, it is APX-hard to maximize the number of satisfied users even if (i) there is only one resource in the system, initially placed at location A, and (ii) all transfers take the same *non-zero* time (independently of the user and the direction). On the other hand, if the transfer time is always 0 (i.e., $t_d^s = t_d^e$ for every demand d), we show that this problem is polynomially solvable even if there are many resources in the system.

1.2   Related Work

Our problem lies in the area of interval scheduling (for recent surveys see [10, 11]), where, in the simplest case, one asks for a maximum non-intersecting subset of a given set of intervals. This simplest case would correspond to our setting with only one location A and every request of type "pick-up at A and drop-off at A".

In our problem with several locations, the transfers of a resource correspond to non-intersecting intervals (demands), with the following additional requirement: we label the interval with the corresponding pick-up and drop-off locations, and any two consecutive intervals for the same resource need to be compatible, i.e., the drop-off location of the first interval needs to be identical to the pick-up location of the second interval. For the setting with one resource and one demand per user, we ask for a maximum set of non-intersecting intervals with exactly this compatibility condition. With k resources (and one demand per user), we ask for k "chains" of such compatible solutions that together contain the maximum number of intervals (demands).

If every user has two or more demands, our problem relates to results on split intervals. A t-split interval is simply a union of t disjoint intervals. A t-interval graph is a conflict graph of n t-split intervals. Bar-Yehuda et al. [2] study the problem of finding the maximum number of non-intersecting t-split intervals, and show that it is APX-hard even when $t = 2$, and present a $2t$-approximation algorithm. Neither the approximation algorithm (or its techniques) nor the hardness

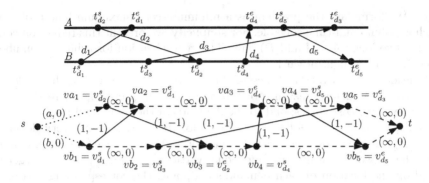

Fig. 1. An example of TRANSFERSONEDEMAND transformed into minimum-cost maximum-flow problem. The labels on the edges in the second figure specify the capacity and the cost of the edges.

result carries over to our problem with one resource and two locations. The main reason is that in our problem we require neighboring intervals in the solution to be compatible. These local compatibility requirements that we impose on the solution is what also makes our problem hard. As we will see, the hardness of our problem arises even in some configurations of intervals that would be trivially polynomially solvable under the split intervals setting (no local compatibility requirement). In particular, if every split interval intersects at most one other split interval, then the conflict graph forms a matching, and finding a maximum independent set becomes trivial. In our hardness result, in the reduction we use we obtain exactly such instances. The hardness arises due to the compatibility requirements.

Finding the maximum number of non-intersecting split intervals with certain additional pattern requirement has been studied before with the relation to problems in RNA secondary structure prediction. The 2-interval pattern problem (see e.g., [3,7]) asks for a non-intersecting subset of 2-split intervals such that every pair of selected split intervals are in one of the prescribed relations $\mathcal{R} \subseteq \{<, \sqsubset, \between\}$ (with $<$ meaning preceding, $\sqsubset$ nested, and $\between$ crossed split intervals). The complexity as well as (approximation) algorithms for different subsets of $\{<, \sqsubset, \between\}$ were studied.

2 Resource Transfers with One Demand per User

If every user has only one demand (either of the form $A \to B$ or $B \to A$), and there are, initially, a resources at location A and b resources at location B, we show that TRANSFERSONEDEMAND, the problem of maximizing the number of satisfied users (which is in this case equal to the number of fulfilled demands), is solvable in polynomial time.

Theorem 1. TRANSFERSONEDEMAND *is solvable in polynomial time, for any number of locations.*

Proof. We formulate the problem as a minimum-cost maximum-flow problem, which is polynomial-time solvable. For simplicity, we present this reduction considering two locations (A and B) only. The generalization for arbitrary number of locations is straightforward.

Consider an arbitrary instance of TRANSFERSONEDEMAND with two locations. We construct an instance of the network flow problem, where the only arcs of non-zero cost correspond to the demands of the users, and have cost -1. Formally, we proceed as follows (see Fig. 1 for illustration). For every demand $d = (t_d^s, t_d^e)$ there are two vertices in the network, v_d^s and v_d^e, one for each endpoint of d. The network contains two additional vertices – a source s, and a target t. Based on the location of each demand's endpoint, the corresponding vertex is either of type A, or B. Let $\langle va_1, \dots, va_n \rangle$ be the vertices of type A ordered by the time of the corresponding demands' endpoints, and let $\langle vb_1, \dots, vb_n \rangle$ be the vertices of type B ordered by the time of the corresponding demands' endpoints. The network contains edges of three types. For every demand $d = (t_d^s, t_d^e)$, there is a directed *demand edge* (v_d^s, v_d^e) from the vertex corresponding to the start-point of d to the vertex corresponding to its endpoint. All demand edges have capacity 1 and cost -1. For every two consecutive vertices va_i, va_{i+1} of type A, there is a directed *connecting edge* (va_i, va_{i+1}). Similarly, there is a connecting edge for every two consecutive vertices of B. There are also connecting edges (va_n, t) and (vb_n, t), from the last vertices of each type to the target vertex. For all the connecting edges, the capacity is set to ∞ and the cost is set to 0. Finally, there is an edge from s to vertex va_1 with capacity a and cost 0, and there is an edge from s to vb_1 with capacity b and cost 0. Observe that from any vertex other than s, there is unlimited capacity for a flow to t, using the connecting edges. Clearly, any st-flow has to pass via va_1 or vb_1, and the sum of capacities of (s, va_1) and (s, vb_1) is $a + b$. Thus, the maximum st-flow is of size $a + b$.

We now determine the cost of an optimum minimum-cost maximum st-flow. Since all the costs and capacities are integral, then, thanks to the integrality theorem [5], there exists an integral optimum solution (which can be found in polynomial time). From the above it follows that there is a maximum st-flow of cost 0 that does not use any demand edge. Since the network is acyclic, the capacity of a demand edge is 1, and its cost is -1, it follows that a minimum-cost st-flow aims at using as many demand edges as possible (with unit flow on each edge). We can see the integral flow as the course of the $a + b$ resources between the locations – one st-flow of size one per resource. Obviously, an integral st-flow of cost $-C$ satisfies C demands (users), and every schedule satisfying C users gives an integral flow of cost C.   □

3   Resource Transfers with Two Demands per User

If the users have more than one demand, the problem of maximizing the number of satisfied users becomes NP-hard, even APX-hard. We will show the hardness even for the special case of "using the car to commute between two badly connected locations", i.e., for the setting with two locations A and B, where every

Gx_1	Gx_2	$\cdots$	Gx_s	HFG_1	FGc_1	Gc_1	FGc_2	Gc_2	$\cdots$	FGc_r	Gc_r	HFG_2	DGc_1	DGc_2	$\cdots$	DGc_r

Fig. 2. Placement of the gadgets for the given instance Φ of MAX-3-SAT(3) with clauses $C = \{c_1, c_2, \ldots, c_r\}$ over a set of Boolean variables $X = \{x_1, x_2, \ldots, x_s\}$.

user has exactly two transfer demands, one per each direction $A \to B$, $B \to A$. We refer to this optimization problem as TRANSFERSFORCOMMUTING. We actually show that the problem is APX-hard even if there is only a single resource. We prove the hardness of TRANSFERSFORCOMMUTING by an L-reduction (see Proposition 1) from a MAX-3-SAT(3), which is an APX-hard variant [1] of the maximum satisfiability problem with at most 3 literals per clause and with each variable appearing in the formula at most twice as a positive, and (exactly) once as a negative literal.

Theorem 2. TRANSFERSFORCOMMUTING *is APX-hard even if there is only one resource, originally placed at location A, and all the transfer times are equal, but positive.*

First we describe a construction used to prove Theorem 2 and its properties.

Construction. Let Φ be an instance of MAX-3-SAT(3) given by a set of clauses $C = \{c_1, c_2, \ldots, c_r\}$ over a set of Boolean variables $X = \{x_1, x_2, \ldots, x_s\}$. We construct from Φ the following instance I of TRANSFERSFORCOMMUTING. There is a single resource in the system, initially located at A. There are two users for each occurrence of a variable in a clause and there are 26 users for each of the r clauses, in total there are at most $32r$ users. In the following we describe how the demands of the users are organized into gadgets, how they are placed, and how they interact.

For every variable $x_i \in X$ there is a *variable gadget* Gx_i. For each clause $c_j \in C$, there is a *clause gadget* Gc_j, a *dummy gadget* DGc_j, and a *light forcing gadget* FGc_j. Finally, there are two *heavy forcing gadgets* HFG_1 and HFG_2. The gadgets are placed as follows (see Fig. 2). First all the variable gadgets are placed (one per variable in Φ). After that, the heavy forcing gadget HFG_1 is placed. Then, all the clause gadgets are placed (one per clause in Φ), each preceded by a light forcing gadget. Then the heavy forcing gadget HFG_2 is placed. Finally, all the dummy gadgets are placed (again, one per clause in Φ).

For each occurrence of a variable x_i in a clause c_j there is one variable user, demanding a transfer $B \to A$ first and $A \to B$ later, and a dummy variable user demanding a transfer $A \to B$ first and $B \to A$ later. Their outbound demands are placed in Gx_i. The return demand of the variable user is placed in Gc_j, and the return demand of the dummy variable user is placed in DGc_j. For each clause c_j, there is one clause user, demanding an outbound transfer $B \to A$, placed in Gc_j, and a return transfer $A \to B$, placed in DGc_j. Finally, there is a large number of forcing users, each demanding a transfer $A \to B$ and an immediate return $B \to A$, both placed in one of the forcing gadgets FGc_j, HFG_1, or HFG_1.

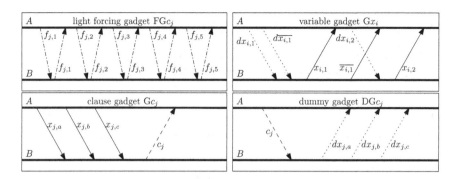

Fig. 3. Placement of the demanded transfers within the gadgets (building blocks of the hardness construction). The demands of variable users are displayed as full arrows, the demands of clause users are dashed, those of dummy variable users are dotted, and those of forcing users are dash-dotted. The heavy forcing gadget is not illustrated in the figure, since it is similar to the light forcing gadget, but consists of $10r$ users instead of 5.

We now describe the placement of the transfers within each gadget in more detail, see Fig. 3 for the exact configurations. Each light forcing gadget FGc_j, consists of five light forcing users, each demanding a transfer $A \to B$ and an immediate return $B \to A$. These demands are placed in such a way that all the users of a light forcing gadget can be satisfied together. Both heavy forcing gadgets HFG_1 and HFG_2 are similar to light forcing gadgets, but instead of 5 users, each HFG consists of $10r$ heavy forcing users (again demanding transfer $A \to B$ and an immediate return $B \to A$, placed in such a way that all can be satisfied together). The purpose of the light/heavy forcing gadgets is to ensure that at a certain moment the resource is located at A. Each forcing gadget consists of a significant number of users, such that any schedule can be transformed into the same or a larger schedule, with all forcing users satisfied.

For a variable x_i, the variable gadget Gx_i consists of the outbound demands of up to six users (two for each occurrence of x_i in Φ). We describe the case when x_i appears three times in Φ, other cases are similar. The gadget Gx_i contains 3 variable users—two *positive users* $x_{i,1}$, $x_{i,2}$ corresponding to the positive literals of x_i, and one *negative user* $\overline{x_{i,1}}$ corresponding to the negative literal of x_i. Each of these users demands in Gx_i an outbound transfer $B \to A$. Additionally, the gadget contains 3 dummy variable users $dx_{i,1}$, $dx_{i,2}$, and $d\overline{x_{i,1}}$ (complementing the variable users). Again, only their outbound demands, in direction $A \to B$, are part of Gx_i. The construction of Gx_i ensures that positive and negative variable users can never be satisfied together (the demand $\overline{x_{i,1}}$ can only be fulfilled when $x_{i,1}$ and $x_{i,2}$ are not, and vice versa). Moreover, the demands of each variable user and the demands of the corresponding dummy variable user can always be fulfilled together. These gadgets relate satisfying of positive/negative variable users of I with the true/false assignment of the corresponding variables in Φ.

For a clause c_j, the clause gadget Gc_j contains the return demands of up to 3 variable users (in the direction $A \to B$) that correspond to the variables appearing in c_j, and then an outbound demand $B \to A$ of a clause user c_j. (To simplify the notation we use c_j to denote both the clause of Φ and the corresponding user.) Each $\mathrm{G}c_j$ is preceded by $\mathrm{FG}c_j$ enforcing that whenever the demand of c_j is fulfilled in $\mathrm{G}c_j$, also a variable demand corresponding to a literal of c_j is fulfilled there, and vice versa. Thus, these gadgets bind together clause and variable users: User c_j is satisfied if and only if the variable user of a variable satisfying c_j in Φ is satisfied.

A dummy gadget $\mathrm{DG}c_j$ consists of the return demand of the clause user c_j, in the direction $A \to B$, and the return demands of up to 3 dummy variable users (again, based on the literals appearing in c_j) in the direction $B \to A$. Gadget $\mathrm{DG}c_j$ in a sense mirrors $\mathrm{G}c_j$ and allows fulfilling the return demand of c_j whenever its outbound demand is fulfilled. In a schedule where all the forcing users are satisfied, for every satisfied c_j, also a variable user and a dummy variable user (both corresponding to the same literal of c_j) will be satisfied.

Let I be an instance of TRANSFERSFORCOMMUTING constructed as above.

Lemma 1. *Given a schedule S of I, we can construct a schedule of size at least $|S|$ where all the users of heavy forcing gadgets* HFG_1 *and* HFG_2 *are satisfied.*

Proof. If $|S| < 25r$, we can construct a schedule where all the $25r$ users of heavy and light forcing gadgets are satisfied. Thus, assume that $|S| \geq 25r$. Since the total number of users of I is at most $32r$ and each HFG consists of $10r$ users, at least one user of each heavy forcing gadget HFG is satisfied. Clearly, whenever a user of a HFG is satisfied in S, all users of that HFG can be added to S. □

Lemma 2. *Given a schedule S of I, we can construct a schedule of size at least $|S|$ where all the users of light forcing gadgets are satisfied, and whenever a clause user c_j is satisfied, also a variable user corresponding to an occurrence of a literal in c_j is satisfied, as well as the corresponding dummy variable user.*

Proof. Using an iterative transformation of the given schedule S into a new schedule with the required parameters. The details are omitted due to space constraints. □

To prove APX-hardness in Theorem 2, we use the following proposition.

Proposition 1 (L-reduction [6]). *Consider two optimization problems H and P, and let H be APX-hard. Assume that for each instance Φ of H, we can construct an instance I of P in polynomial time. Also, assume that for each solution S of I, we can construct a solution ϕ of Φ in polynomial time. Let $\mathrm{OPT}(I)$, and $\mathrm{OPT}(\Phi)$ denote the size of the optimum solution of I, and Φ, respectively. Finally, assume that there exist positive constants α and β (independent on S) such that the following two conditions are met.*

(A) $\mathrm{OPT}(I) \leq \alpha \, \mathrm{OPT}(\Phi)$
(B) $|\mathrm{OPT}(\Phi) - |\phi|| \leq \beta |\mathrm{OPT}(I) - |S||$

Then, we have an L-reduction from H to P, and P is also APX-hard.

Proof (Of Theorem 2). We show the APX-hardness by an L-reduction from
MAX-3-SAT(3). Given an instance Φ of MAX-3-SAT(3) with r clauses over s
variables, construct an instance I of TRANSFERSFORCOMMUTING as above.
First we show the following: ($\Rightarrow$) For every solution ϕ of Φ of size $|\phi|$, we con-
struct a solution of I of size at least $25r + 3|\phi|$. ($\Leftarrow$) For every solution S of I
of size $|S|$, we construct a solution ϕ of Φ of size at least $(|S| - 25r)/3$. Thus, in
particular, we get $\mathrm{OPT}(I) = 25r + 3\,\mathrm{OPT}(\Phi)$.

($\Rightarrow$) Given an assignment ϕ satisfying $|\phi|$ clauses of the given instance Φ
of the MAX-3-SAT(3) problem, we construct a schedule where all $25r$ forcing
users together with exactly $3|\phi|$ other users are satisfied as follows. We schedule
$|\phi|$ clause users corresponding to the $|\phi|$ satisfied clauses. We select a subset
of (dummy) variable users: For each clause c_j, we select exactly one literal of
those that satisfy c_j in ϕ and we schedule both the corresponding variable user
and dummy variable user. We schedule all the users of the forcing gadgets.
Let us now observe that the created schedule is feasible. Clearly, the transfers
of the satisfied users do not overlap: The only overlapping transfers are those
of positive/negative literals in variable gadgets and those are never satisfied
together, since every variable is set either to TRUE, or to FALSE in ϕ. We now
observe that the movement of the resource induced by the selected transfers is
feasible. In each variable gadget the resource is moved $A \to B \to A$ for every
picked literal: $A \to B$ by a dummy variable user and then $B \to A$ by the variable
user. After all the variable gadgets, the resource is moved $10r$ times $A \to B \to A$
by the users of the forcing gadget HFG_1. In each clause gadget $\mathrm{G}c_j$ the resource
is moved $A \to B$ by a variable user and then $B \to A$ by the clause user. Before
each variable gadget, the resource is moved five times $A \to B \to A$ by the users
of the forcing gadget $\mathrm{FG}c_j$. After the last clause gadget, the resource is moved
$10r$ times $A \to B \to A$ by the users of HFG_2. Finally, in each dummy gadget
the resource moves $A \to B \to A$.

($\Leftarrow$) Now assume that we have a schedule S with $|S|$ satisfied users. It follows
from Lemmas 1 and 2 that there is a schedule S' of size at least $|S|$, where all
$25r$ forcing users are satisfied. Moreover, it also follows that at least $(|S| - 25r)/3$ clause users are satisfied, such that for each of them also a variable
user corresponding to an occurrence of a literal in c_j is satisfied, as well as
the corresponding dummy variable user are satisfied. Since the variable gadgets
ensure that for each variable, either the users corresponding to the positive
literals can be satisfied, or only the user corresponding to the negative literal
can be satisfied, we can directly construct an assignment for Φ that satisfies at
least $(|S| - 25r)/3$ clauses.

To show that the above reduction is an L-reduction, we need to prove that
conditions (A) and (B) of Proposition 1 are met. First note that $\mathrm{OPT}(\Phi) \geq r/2$
(either all-TRUE or all-FALSE assignment satisfies at least $1/2$ of all the clauses).
Recall that $\mathrm{OPT}(I) = 25r + 3\,\mathrm{OPT}(\Phi)$. Also recall that for any solution S of I,
we can construct a solution ϕ of Φ of size $|\phi| \geq (|S| - 25r)/3$. Thus we get:

(A) $\mathrm{OPT}(I) = 25r + 3\,\mathrm{OPT}(\Phi) \leq 53\,\mathrm{OPT}(\Phi)$,
(B) $(|\mathrm{OPT}(I)| - |S|) \geq 25r + 3|\mathrm{OPT}(\Phi)| - 25r - 3|\phi| = 3(|\mathrm{OPT}(\Phi)| - |\phi|)$.

It follows that the presented construction is an L-reduction from MAX-3-SAT(3) to TRANSFERSFORCOMMUTING with $\alpha = 53$ and $\beta = 1/3$.   □

3.1 Resource Transfers with Two Demands and Zero Transfer Times

In the hardness result, we used "crossing" arrows (demands) to exclude scheduling both demands. This is only possible if the transfer time is non-zero. In this section we show that whenever all transfer times are zero, i.e., when they are instantaneous, TRANSFERSFORCOMMUTING becomes tractable, even if there is more than one resource (initially, a resources at location A, and b resources at location B). As a corollary, we obtain that TRANSFERSFORCOMMUTING with non-zero transfer times is polynomially-time solvable, whenever no two demand arrows cross.

Depending on the direction of the first demanded transfer of a user, we distinguish two types of users: an ABA-type demands to transfer in direction $A \rightarrow B$ first, and in direction $B \rightarrow A$ second, whereas user of type BAB demands first the transfer in direction $B \rightarrow A$, and later in direction $A \rightarrow B$. We can equivalently specify the problem with zero transfer times as follows (see an example in Fig. 4). We represent the two demands of a user i (demanding instantaneous transfers at times $t_{i,1}$ and $t_{i,2}$) by a time interval $(t_{i,1}, t_{i,2})$, with a value $v(i) := -1$ if user i is of type ABA and $v(i) := 1$ if i is of type BAB. Each such time interval indicates the induced change in the number of available resources present at location A. That is, if user i of type ABA transfers a resource from A to B at time $t_{i,1}$ and back to A at time $t_{i,2}$, it implies that during the time $(t_{i,1}, t_{i,2})$ there is one less resource item at A (and one more at B).

Clearly, satisfying a user i in the original problem corresponds to selecting the corresponding interval i in the modified problem. In the original problem, there must be a resource available for each selected transfer, but since the transfers are instantaneous, we only need to ensure that, at any time, both locations have a non-negative amount of resources. In particular, at A there can never be more than $a + b$ items and less than 0 items. Therefore, the original goal translates to choosing a maximum subset I of the intervals (representing the users) such that at any time point t we have

$$-a \leq \sum_{t \in i \in I} v(i) \leq b.$$

In the following, we show that the problem (in the equivalent alternative formulation) is polynomially solvable by formulating it as an integral linear program (i.e., linear program that has an optimum solution which is integral). To prove that the constructed linear program $Ax \leq b$ is integral, we will show that A is totally unimodular (a matrix A is *totally unimodular* if the determinant of every square submatrix of A has value -1, 0 or 1).

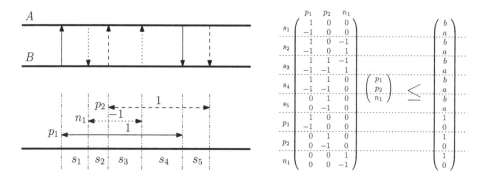

Fig. 4. An example of TRANSFERSFORCOMMUTING with zero transfer time, with 3 users; the transformation to an equivalent problem; and the corresponding system of linear inequalities $Ax \leq b$.

Theorem 3 ([12]). *Every linear program in variables x with a totally unimodular constraint matrix A is integral.*

To show that A is totally unimodular, we use the following theorem [9].

Theorem 4 (Ghouila-Houri). *A matrix is totally unimodular if and only if for every subset of rows R, there exists a function $f : R \to \{-1, +1\}$ such that $\sum_{r \in R} f(r) \cdot r \in \{-1, 0, 1\}^n$.*

Theorem 5. *The problem TRANSFERSFORCOMMUTING with zero transfer time and multiple resources is solvable in polynomial time.*

Proof. Consider the following integer linear program formulation of the problem. Let n^+ be the number of positive intervals (i.e., intervals of value 1) and let n^- be the number of negative intervals (then $n = n^+ + n^-$). For each interval we define one variable indicating whether this interval was chosen into the optimum solution or not. In particular, for each positive interval i we define a binary variable $p_i \in \{0, 1\}$ and for each negative interval j we define a binary variable $n_j \in \{0, 1\}$. The goal is to maximize

$$\sum_{i=1}^{n^+} p_i + \sum_{j=1}^{n^-} n_j$$

subject to the following constraints. We divide the time axis into N segments, defined by the endpoints of the n given intervals. Note that the number of resources present at A may change from segment to segment, but within each segment it does not change. For each segment s we have the following two constraints, based on the number of intervals that overlap s. We abuse the notation here, and use p_i and n_j both as a variable and as the corresponding interval.

$$+ \sum_{i\in[n^+],\ p_i\cap s\neq 0} p_i \quad - \sum_{j\in[n^-],\ n_j\cap s\neq 0} n_j \leq b \tag{1}$$

$$- \sum_{i\in[n^+],\ p_i\cap s\neq 0} p_i \quad + \sum_{j\in[n^-],\ n_j\cap s\neq 0} n_j \leq a \tag{2}$$

We now consider the linear relaxation of the ILP, i.e., we additionally have the linear constraints, for every $i \in [n^+]$ and $j \in [n^-]$,

$$0 \leq p_i, n_j \leq 1. \tag{3}$$

We can write the constraints as a linear system $Ax \leq b$ (see Fig. 4 for an example). To show that this linear program is integral, we show that the matrix A is totally unimodular. For that let us first dwell into the structure of the matrix. Matrix A contains one column for each variable (p_i or n_j) and one row for each constraint. We first discuss the first $2N$ rows corresponding to the constrains (1) and (2). For each segment s, the matrix A contains 2 rows. If the segment s coincides with an interval p_i, then the submatrix $A(s, p_i)$ is $(1, -1)^T$, otherwise, $A(s, p_i) = (0,0)^T$. Similarly, if s coincides with an interval n_j, then $A(s, n_j) = (-1, 1)^T$, otherwise, $A(s, n_j) = (0,0)^T$. Since all p_i and n_j are intervals, each of them spans only consecutive segments. Thus, each column (restricted to the first $2N$ rows) contains exactly one contiguous block of non-zero entries (alternating 1s and -1s). We now look at the remaining $2n = 2(n^+ + n^-)$ rows of A corresponding to the constraints (3). Clearly, each of these rows contains exactly one non-zero symbol per row and it is either 1 or -1.

Using Theorem 4, we show that A is totally unimodular as follows. We first observe that if the first $2N$ rows of A form a totally unimodular matrix A', then the whole A is totally unimodular. Each of the last $2n$ rows contains exactly one non-zero element that is either 1 or -1. Thus, for every subset R'' of these rows, we can easily find a function $f : R'' \to \{-1, +1\}$ so that each component of the vector $v'' = \sum_{r\in R''} f(r)\cdot r$ is either 0, or we can choose between 1 and -1. Then, for any vector $v' = \{-1, 0, 1\}^n$ (any vector obtained from A' due to Theorem 4) we can choose the components of $v'' = \{0, -1/1\}^n$ so that $v'+v'' = \{-1, 0, 1\}^n$. It remains to be shown that the submatrix A' corresponding to the first $2N$ rows of A is totally unimodular. Let R' be a subset of the $2N$ rows. As a preparatory step we multiply every second row of A' by -1 and obtain a matrix where each column contains a single nonzero block of either 1s (if it corresponds to p_i) or -1s (if it corresponds to n_j). We set the function $g : R' \to \{-1, +1\}$ to be alternating 1 and -1 for the $r \in R'$ ordered by row number. Since each column c contains only one block of consecutive 1s or -1s, we get $\sum_{r\in R'} g(c_r)\cdot c_r \in \{-1, 1\}$. Now we can combine the function g with the preparatory step and obtain $f : R' \to \{-1, +1\}$ such that $\sum_{r\in R'} f(r) \cdot r \in \{-1, 0, 1\}^n$.

Thus, the matrix A is totally unimodular, the constructed linear program is integral and the considered problem is polynomially solvable. □

Corollary 1. *If all the demands do not cross in their arrow representation, the problem* TRANSFERSFORCOMMUTING *is solvable in polynomial time.*

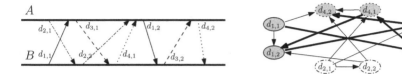

Fig. 5. An example of TRANSFERSFORCOMMUTING with 1 resource in the system, modeled as longest path problem with prescribed pairs of vertices. The edges in bold indicate the longest path.

Proof. By shrinking the given instance so that all the intervals are of length 0, we obtain an equivalent, polynomially solvable problem.   □

4   Further Notes

Longest Path Containing Subset of Prescribed Pairs of Vertices. By proving hardness in Theorem 2, we prove also the following problem to be APX-hard. Given a directed graph and a set of pairs of its vertices, the goal is to find a longest path such that for each of the given pairs it either contain both vertices or none of them. This problem is APX-hard even in directed acyclic graphs, since it can be reduced from TRANSFERSFORCOMMUTING with 1 resource as follows (see Fig. 5). Every demand is modeled as one vertex, every user defines one prescribed pair of vertices, and there is one directed edge for every pair of demands that can be consecutively fulfilled. The constructed graph is acyclic. Clearly, any path that uses from each pair either none or both vertices corresponds to a feasible schedule for TRANSFERSFORCOMMUTING and the length of the path corresponds to twice the number of satisfied users.

We haven't found this exact problem to be studied in the literature, but we link to a similar problem that received a lot of attention. Given a directed graph and a set of vertex pairs, the goal of the *longest antisymmetric path problem* is to find a longest path that does not simultaneously contain both vertices of any of the prescribed forbidden pairs. This problem arises in the area of automatic software testing and validation, and protein identification in bioinformatics. Gabow et al. [8] showed that deciding whether there is an antisymmetric *st*-path is NP-complete even if the given directed graph is acyclic and all the in- and out-degrees are at most 2. Song et al. [13] showed that the longest antisymmetric path problem cannot be approximated within $(n - 2)/2$ in polynomial time unless $P = NP$, even in directed acyclic graphs of degree at most 6.

Any of Multiple Demands Satisfies User. Consider a different variant of the problem, where each user has multiple demands, but is satisfied if any of her demands is fulfilled. This problem, in general form where also a transfer from a location L back to location L is allowed, is NP-hard. We consider a degenerated problem as follows. There are two locations A and B and exactly one unit of resource placed at each of them. There are n users and each user demands exactly one $A \to A$ transfer and one $B \to B$ transfer. Thus, each user specifies exactly one interval on A and one on B, we satisfy the user by selecting either of her intervals, and the goal is to maximize the number of satisfied users. This problem is also known as INTERVALSELECTION with 2 machines, and is known to be NP-hard [4].

Open Problems. We already know (Theorem 2) that the problem TRANSFERS-FORCOMMUTING where each user has exactly one demand in each of the two directions is APX-hard. Thus, a natural question is to seek an approximation algorithm for the problem. However, it is not clear whether it has a decent approximation, since simple approaches fail drastically. Also, given the motivation, it would be interesting to explore the problems we studied under online setting.

Acknowledgements. The authors wish to thank Peter Widmayer for many useful discussions and helpful comments, as well as Andreas Bärtschi, Barbara Geissmann, Sandro Montanari, Tobias Pröger, and Thomas Tschager for their ideas during early stage discussions on the topic. Kateřina Böhmová is supported by a Google Europe Fellowship in Optimization Algorithms. The project has been partially supported by the Swiss National Science Foundation (SNF) under the grant number 200021_156620.

References

1. Ausiello, G., Crescenzi, P., Gambosi, G., Kann, V., Marchetti-Spaccamela, A., Protasi, M.: Complexity and Approximation: Combinatorial Optimization Problems and Their Approximability Properties. Springer Science & Business Media, Heidelberg (2012)
2. Bar-Yehuda, R., Halldórsson, M.M., Naor, J., Shachnai, H., Shapira, I.: Scheduling split intervals. SIAM J. Comput. **36**(1), 1–15 (2006)
3. Blin, G., Fertin, G., Vialette, S.: New results for the 2-interval pattern problem. In: Sahinalp, S.C., Muthukrishnan, S.M., Dogrusoz, U. (eds.) CPM 2004. LNCS, vol. 3109, pp. 311–322. Springer, Heidelberg (2004)
4. Böhmová, K., Disser, Y., Mihalák, M., Widmayer, P.: Interval selection with machine-dependent intervals. In: Dehne, F., Solis-Oba, R., Sack, J.-R. (eds.) WADS 2013. LNCS, vol. 8037, pp. 170–181. Springer, Heidelberg (2013)
5. Cormen, T., Leiserson, C., Rivest, R., Stein, C.: Introduction to Algorithms, vol. 3. MIT Press, Cambridge (2001)
6. Crescenzi, P.: A short guide to approximation preserving reductions. In: 12th IEEE Conference on Computational Complexity, pp. 262–273. IEEE (1997)
7. Crochemore, M., Hermelin, D., Landau, G.M., Vialette, S.: Approximating the 2-interval pattern problem. In: Brodal, G.S., Leonardi, S. (eds.) ESA 2005. LNCS, vol. 3669, pp. 426–437. Springer, Heidelberg (2005)
8. Gabow, H.N., Maheshwari, S.N., Osterweil, L.J.: On two problems in the generation of program test paths. IEEE Trans. Softw. Eng. **2**(3), 227–231 (1976)
9. Ghouila-Houri, A.: Caracterisation des matrices totalement unimodulaires. CR Acad. Sci. Paris **254**, 1192–1194 (1962)

10. Kolen, A.W.J., Lenstra, J.K., Papadimitriou, C.H., Spieksma, F.C.R.: Interval scheduling: a survey. Naval Res. Logistics (NRL) **54**(5), 530–543 (2007)
11. Kovalyov, M.Y., Ng, C., Cheng, T.E.: Fixed interval scheduling: models, applications, computational complexity and algorithms. Eur. J. Oper. Res. **178**(2), 331–342 (2007)
12. Schrijver, A.: Theory of Linear and Integer Programming. John Wiley & Sons, Chichester (1998)
13. Song, Y., Yu, M.: On finding the longest antisymmetric path in directed acyclic graphs. Inf. Process. Lett. **115**(2), 377–381 (2015)

A 0.821-Ratio Purely Combinatorial Algorithm for Maximum k-vertex Cover in Bipartite Graphs

Édouard Bonnet[1], Bruno Escoffier[2], Vangelis Th. Paschos[3,4(✉)], and Georgios Stamoulis[3,4]

[1] Institute for Computer Science and Control,
Hungarian Academy of Sciences (MTA SZTAKI), Budapest, Hungary
bonnet.edouard@sztaki.mta.hu

[2] Sorbonne Universités,
UPMC Universite Paris 06, CNRS, LIP6 UMR 7606, Paris, France
bruno.escoffier@lip6.fr

[3] PSL* Research University, Université Paris-Dauphine, LAMSADE, Paris, France
[4] CNRS UMR 7243, Paris, France
paschos@lamsade.dauphine.fr, georgios.stamoulis@dauphine.fr

Abstract. We study the polynomial time approximation of the MAX k-VERTEX COVER problem in bipartite graphs and propose a purely *combinatorial algorithm* that beats the only such known algorithm, namely the greedy approach. We present a computer-assisted analysis of our algorithm, establishing that the worst case approximation guarantee is bounded below by 0.821.

1 Introduction

In MAX k-VERTEX COVER, a graph $G = (V, E)$ with $|V| = n$ and $|E| = m$ is given together with an integer $k \leqslant n$. The goal is to find a subset $K \subseteq V$ with k vertices such that the total number of edges covered by K is maximized. We say that an edge $e = \{u, v\}$ is covered by a subset of vertices K if $K \cap e \neq \emptyset$. MAX k-VERTEX COVER is **NP**-hard in general graphs (as a generalization of MIN VERTEX COVER) and it remains so in bipartite graphs [1,2].

The approximation of MAX k-VERTEX COVER has been originally studied in [3], where ratio $1 - (1/e)$ (≈ 0.632) is achieved by the natural greedy algorithm. This ratio is tight even in bipartite graphs [4]. Using a sophisticated linear programming method, the approximation ratio for MAX k-VERTEX COVER was improved to $3/4$ [5], which, until very recently, was the best known ratio even in bipartite graphs. The best approximation ratio in bipartite graphs is now $8/9$ and is still based on linear programming [2]. A direct reduction from MIN VER-TEX COVER shows that MAX k-VERTEX COVER can not admit a polynomial time approximation schema (PTAS), unless $\mathbf{P} = \mathbf{NP}$ [6].

Finally, we may observe that MAX k-VERTEX COVER is easy in semiregular bipartite graphs (where all the vertices of each color class have the same degree).

© Springer-Verlag Berlin Heidelberg 2016
E. Kranakis et al. (Eds.): LATIN 2016, LNCS 9644, pp. 235–248, 2016.
DOI: 10.1007/978-3-662-49529-2_18

Indeed, any k vertices in the color class of maximum degree yield an optimal solution. Obviously, if this color class contains less than k vertices, then one can cover *all* the edges.

Our Contribution. The principal motivation of this paper is to determine *to what extent combinatorial methods compete with linear programming* for MAX k-VERTEX COVER. In other words, *what ratio can a purely combinatorial algorithm guarantee?* To this purpose, we first devise a very simple algorithm that guarantees approximation ratio $2/3$, improving so the ratio of the greedy algorithm in bipartite graphs. The proof is given in [7]. Our main contribution consists of an approximation algorithm which computes six distinct solutions and returns the best among them.

Analyzing the performance guarantee of such an algorithm is a challenging task. Indeed, there is no obvious way to compare the different solutions and argue globally over a lower bound on the maximum value taken by the six solutions. The large number of variables (in all, 48) used to express the many solution values participates in the difficulty of the analysis.

Similar situation was faced, for example, in [8] where the authors gave a 0.921 approximation guarantee for MAX CUT in graphs of maximum degree 3 (and an improved 0.924 for 3-regular graphs) by a computer-assisted analysis of the quantities generated by theoretically analyzing a particular semi-definite relaxation of the problem at hand. Similarly, by setting up a suitable non-linear program and solving it, we give a computer-assisted analysis of a 0.821-approximation guarantee for MAX k-VERTEX COVER in bipartite graphs. We give all the details of the implementation in [7].

2 Preliminaries

Consider a bipartite graph $B = (V_1, V_2, E)$, instance of MAX k-VERTEX COVER and fix an optimal solution O (i.e., a set of k vertices covering a maximum number of edges in E) as well as its parts O_1 and O_2 lying in color classes V_1 and V_2, respectively.

The algorithm (called k-VC_ALGORITHM) proposed for solving MAX k-VERTEX COVER can be sketched as follows:
1. guess the cardinality k_1 and therefore $k_2 = k - k_1$ of its subsets O_1 and O_2 lying in the color classes V_1 and V_2, respectively;
2. compute the sets S_i of k_i vertices in V_i, $i = 1, 2$ that cover the most of edges; obviously S_i is a set of the k_i largest degree vertices in V_i (breaking ties arbitrarily);
3. guess the cardinalities k_i' of the intersections $S_i \cap O_i$, $i = 1, 2$;
4. compute the sets X_i of the $k_i - k_i'$ vertices from V_i, $i = 1, 2$, that cover the most of edges in graphs $B[(V \setminus S_1) \cup V_2]$ and $B[V_1 \cup (V_2 \setminus S_2)]$, respectively;
5. choose the best among six solutions built as described in Sect. 3.

Let us note that our $2/3$-approximation algorithm in [7] guarantees ratio $4/5$, when both $k_i' = 0$, $i = 1, 2$.

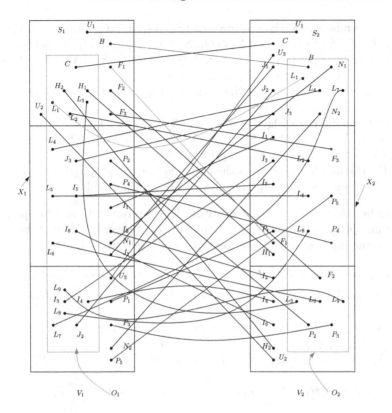

Fig. 1. Sets S_i, O_i, X_i $i = 1, 2$ and cuts between them.

Sets S_i, X_i and O_i separate each color class in 6 regions, namely, $S_i \cap O_i$, $S_i \setminus O_i$, $X_i \cap O_i$, $X_i \setminus O_i$, $O_i \setminus (S_i \cup X_i)$ (denoted by $\bar{O}_i$, in what follows) and $V_i \setminus (S_i \cup X_i \cup O_i)$. In total, there exist 36 groups of edges (cuts) among them, the group $(V_1 \setminus (S_1 \cup X_1 \cup O_1), V_2 \setminus (S_2 \cup X_2 \cup O_2))$ being irrelevant as it will become clear in the sequel. We will use the following notations to refer to the values of the 35 relevant cuts (illustrated in Fig. 1.):

B: the number of edges in the cut $(S_1 \setminus O_1, S_2 \cap O_2)$;

C: the number of edges in the cut $(S_2 \setminus O_2, S_1 \cap O_1)$;

F_1, F_2, F_3: the number of edges in the cuts $(S_1 \setminus O_1, X_2 \setminus O_2)$, $(S_1 \setminus O_1, O_2 \setminus (X_2 \cup S_2))$ and $(S_1 \setminus O_1, O_2 \cap X_2)$, respectively;

H_1, H_2: the number of edges in the cuts $(S_1 \cap O_1, X_2 \setminus O_2)$ and $(S_1 \cap O_1, V_2 \setminus (S_2 \cup X_2 \cup O_2))$, respectively;

$\{I_i\}_{i \in [6]}$: the number of edges in the cuts $(X_1 \setminus O_1, X_2 \setminus O_2)$, $(X_1 \setminus O_1, V_2 \setminus (S_2 \cup X_2 \cup O_2))$, $(O_1 \setminus (S_1 \cup X_1), X_2 \setminus O_2)$, $(O_1 \setminus (S_1 \cup X_1), V_2 \setminus (S_2 \cup X_2 \cup O_2))$, $(X_1 \cap O_1, X_2 \setminus O_2)$ and $(X_1 \cap O_1, V_2 \setminus (S_2 \cup X_2 \cup O_2))$, respectively;

J_1, J_2, J_3: the number of edges in the cuts $(S_2 \setminus O_2, X_1 \setminus O_1)$, $(S_2 \setminus O_2, O_1 \setminus (S_1 \cup X_1))$ and $(S_2 \setminus O_2, O_1 \cap X_1)$, respectively;

$\{L_i\}_{i \in [9]}$: the number of edges in the cuts $(S_1 \cap O_1, S_2 \cap O_2)$, $(S_1 \cap O_1, X_2 \cap O_2)$, $(S_1 \cap O_1, O_2 \setminus (S_2 \cup X_2))$, $(X_1 \cap O_1, S_2 \cap O_2)$, $(X_1 \cap O_1, X_2 \cap O_2)$, $(X_1 \cap O_1, O_2 \setminus (S_2 \cup X_2))$, $(O_1 \setminus (S_1 \cup X_1), S_2 \cap O_2)$, $(O_1 \setminus (S_1 \cup X_1), X_2 \cap O_2)$, and $(O_1 \setminus (S_1 \cup X_1), O_2 \setminus (S_2 \cup X_2))$, respectively;

N_1, N_2: the number of edges in the cuts $(S_2 \cap O_2, X_1 \setminus O_1)$ and $(S_2 \cap O_2, V_1 \setminus (S_1 \cup X_1 \cup O_1))$, respectively;

$\{P_i\}_{i \in [5]}$: the number of edges in the cuts $(X_2 \setminus O_2, V_1 \setminus (S_1 \cup X_1 \cup O_1))$, $(O_2 \setminus (S_2 \cup X_2), X_1 \setminus O_1)$, $(O_2 \setminus (S_2 \cup X_2), V_1 \setminus (S_1 \cup X_1 \cup O_1))$, $(X_2 \cap O_2, X_1 \setminus O_1)$, and $(X_2 \cap O_2, V_1 \setminus (S_1 \cup X_1 \cup O_1))$, respectively;

U_1, U_2, U_3: the number of edges is the cuts, $(S_1 \setminus O_1, S_2 \setminus O_2)$, $(S_1 \setminus O_1, V_2 \setminus (S_2 \cup X_2 \cup O_2))$ and $(S_2 \setminus O_2, V_1 \setminus (S_1 \cup X_1 \cup O_1))$, respectively.

Denoting by $\delta(V')$, $V' \subseteq V$, the number of edges covered by V' and by $\mathrm{opt}(B)$ the value of an optimal solution (i.e., the number edges covered) for MAX k-VERTEX COVER in the input graph B, the following holds (see also Fig. 1):

$$\delta(S_1) = B + C + F_1 + F_2 + F_3 + H_1 + H_2 + L_1 + L_2 + L_3 + U_1 + U_2 \tag{1}$$

$$\delta(S_2) = B + C + J_1 + J_2 + J_3 + L_1 + L_4 + L_7 + N_1 + N_2 + U_1 + U_3 \tag{2}$$

$$\delta(X_1) = I_1 + I_2 + I_5 + I_6 + J_1 + J_3 + \sum_{i=4}^{6} L_i + N_1 + P_2 + P_4 \tag{3}$$

$$\delta(X_2) = F_1 + F_3 + H_1 + I_1 + I_3 + I_5 + L_2 + L_5 + L_8 + P_1 + P_4 + P_5 \tag{4}$$

$$\delta(O_1) = C + H_1 + H_2 + I_3 + I_4 + I_5 + I_6 + J_2 + J_3 + \sum_{i=1}^{9} L_i \tag{5}$$

$$\delta(O_2) = B + F_2 + F_3 + \sum_{i=1}^{9} L_i + N_1 + N_2 + \sum_{i=2}^{5} P_i \tag{6}$$

$$\mathrm{opt}(B) = B + C + \sum_{i=2}^{3} F_i + \sum_{i=1}^{2} H_i + \sum_{i=3}^{6} I_i + \sum_{i=2}^{3} J_i + \sum_{i=1}^{9} L_i$$
$$+ \sum_{i=1}^{2} N_i + \sum_{i=2}^{5} P_i \tag{7}$$

Without loss of generality, we assume $k_1 \leqslant k_2$ and we set: $k_1 = \mu k_2$ $(\mu \leqslant 1)$, $k'_1 = |S_1 \cap O_1| = \nu k_1$ $(0 \leqslant \nu \leqslant 1)$ and $k'_2 = |S_2 \cap O_2| = \xi k_2$ $(0 \leqslant \xi \leqslant 1)$. Let us note that, since k'_i vertices lie in the intersections $S_i \cap O_i$, the following hold for $\bar{O}_i = O_i \setminus (S_i \cup X_i)$, $i = 1, 2$: $|\bar{O}_1| = |O_1 \setminus (S_1 \cup X_1)| \leqslant (1 - \nu)k_1 = \mu(1 - \nu)k_2$ and $|\bar{O}_2| = |O_2 \setminus (S_2 \cup X_2)| \leqslant (1 - \xi)k_2$. From the definitions of the cuts and using (1) to (6) and the expressions for $|\bar{O}_1|$ and $|\bar{O}_2|$, simple average arguments and the assumptions for k_1, k_2, k'_1 and k'_2 just above, the following holds:

$$
\begin{aligned}
\delta\left(S_1\right) &\geq \delta\left(O_1\right) \\
\delta\left(S_2\right) &\geq \delta\left(O_2\right) \\
\delta\left(X_1\right)+C+H_1+H_2+L_1+L_2+L_3 &\geq \delta\left(O_1\right) \\
\delta\left(X_2\right)+B+N_1+N_2+L_1+L_4+L_7 &\geq \delta\left(O_2\right) \\
\delta\left(S_1\right) &\geq 1/(1-\nu)\cdot\delta\left(X_1\right) \\
\delta\left(S_2\right) &\geq 1/(1-\xi)\cdot\delta\left(X_2\right) \\
\delta\left(S_1\right)+\delta\left(X_1\right) &\geq (2-\nu)/(1-\nu)\cdot\left(I_3+I_4+J_2+L_7+L_8+L_9\right) \\
\delta\left(S_2\right)+\delta\left(X_2\right) &\geq (2-\xi)/(1-\xi)\cdot\left(F_2+L_3+L_6+L_9+P_2+P_3\right) \\
B+F_1+F_2+F_3+U_1+U_2 &\geq \delta\left(X_1\right) \\
C+J_1+J_2+J_3+U_1+U_3 &\geq \delta\left(X_2\right)
\end{aligned}
\tag{8}
$$

For $i = 1, 2$, the two first inequalities in (8) hold because S_i is the set of k_i highest-degree vertices in V_i; the third and fourth ones because the lefthand side quantities are the number of edges covered by $X_i \cup (S_i \cap O_i)$; each of these sets has cardinality k_i and obviously covers more edges than O_i; the fifth and sixth inequalities because the average degree of S_i is at least the average degree of X_i and $|X_1| = (1 - \nu)k_1$ and $|X_2| = (1 - \xi)k_2$; seventh and eighth ones because the average degree of vertices in $S_i \cup X_i$ is at least the average degree of vertices in $O_i \setminus (S_i \cup X_i)$; finally, for the last two inequalities the sum of degrees of the $k_i - k_i'$ vertices in $S_i \setminus O_i$ is at least the sum of degrees of the $k_i - k_i'$ vertices of X_i.

In Sect. 3, we specify the approximation algorithm sketched above. In [7] a computer assisted analysis of its approximation-performance is presented. The non-linear program that we set up, not only computes the approximation ratio of our algorithm but it also provides an experimental study over families of graphs. Indeed, a particular configuration on the variables (i.e., a feasible value assignments on the variables that represent the set of edges $B, C, \ldots$) corresponds to a particular family of bipartite graphs with similar structural properties (characterized by the number of edges belonging to the several cut considered). Given such a configuration, it is immediate to find the ratio of the algorithm, because we can simply substitute the values of the variables in the corresponding ratios and output the largest one. We can view our program as an *experimental analysis* over all families of bipartite graphs, trying to find the particular family that implements the worst case for the approximation ratio of the algorithm. Our program not only finds such a configuration, but also provides data about the range of approximation factor on other families of bipartite graphs. Experimental results show that the approximation factor for the *absolute majority* of the instances is very close to 1, i.e., ≥ 0.95. Moreover, our program is *independent* on the size of the instance. We just need a particular configuration on the relative value of the variables $B, C, \ldots$, thus providing a compact way of representing families of bipartite graphs sharing common structural properties.

For the rest of the paper, we call "best" vertices a set of vertices that cover the most of *uncovered* edges[1] in B. Given a solution $\mathrm{SOL}_k(B)$, we denote by $\mathrm{sol}_k(B)$ its value. For the quantities implied in the ratios corresponding to these solutions, one can be referred to Fig. 1 and to expressions (1) to (7).

---

[1] For instance, "we take S_1 plus the k_2 best vertices in V_2" means that we take S_1 and then k_2 vertices of highest degree in $B[(V_1 \setminus S_1), V_2]$.

Let us note that the algorithm above, since it runs for any value of k_1 and k_2, it will run for $(k_1, k_2) = (k, 0)$ and $(0, k)$. So, it will compute the optimum for the instances of [4], where the greedy algorithm attains the ratio $(e-1)/e$. Observe finally that, when $k \geqslant \min\{|V_1|, |V_2|\}$, then $\min\{|V_1|, |V_2|\}$ is an optimal solution since it covers the whole of E. This remark will be useful for some solutions in the sequel, for example in the completion of solution $\text{SOL}_5(B)$.

3   A 0.821-Approximation for the Bipartite Max k-vertex Cover

Algorithm k-VC_ALGORITHM builds the following MAX k-VERTEX COVER-solutions:

$\text{SOL}_1(\mathbf{B})$ and $\text{SOL}_2(\mathbf{B})$, take, respectively, S_1 plus the k_2 remaining best vertices from V_2, and S_2 plus the k_1 remaining best vertices from V_1;

$\text{SOL}_3(\mathbf{B})$ takes first $S_1 \cup X_1$ in the solution and completes it with the $(1 - \mu(1 - \nu))k_2$ best vertices from V_2;

$\text{SOL}_4(\mathbf{B})$ takes S_2 and completes it either with vertices from V_2, or with vertices from both V_2 and V_1 (as specified in the next page);

$\text{SOL}_5(\mathbf{B})$ takes a π-fraction of the best vertices in S_1 and X_1, $\pi \in (0, 1/2]$; then, solution is completed with the $k_1 + k_2 - \pi(2k_1 - k_1')$ best vertices in V_2;

$\text{SOL}_6(\mathbf{B})$ takes a λ-fraction of the best vertices in S_2 and X_2, $\lambda \in (0, (1+\mu)/(2-\xi)]$; then solution is completed with the $k_1 + k_2 - \lambda(2k_2 - k_2')$ best vertices in V_1.

Let us note that the values of λ and π are parameters that we can fix. In what follows, we analyze solutions $\text{SOL}_1(B) \ldots \text{SOL}_6(B)$ computed by k-VC_ALGORITHM and give analytical expressions for their ratios. A fully detailed analysis of all these solutions is given in [7].

Solution $\text{SOL}_1(\mathbf{B})$. The best k_2 vertices in V_2, provided that S_1 has already been chosen, cover at least the maximum of the following quantities:

$$\mathcal{A}_1 = J_1 + J_2 + J_3 + L_4 + L_7 + N_1 + N_2 + U_3 \qquad \text{by } S_2$$
$$\mathcal{A}_2 = I_1 + I_3 + I_5 + L_5 + L_8 + P_1 + P_4 + P_5 \qquad \text{by } X_2$$
$$\mathcal{A}_3 = L_4 + L_5 + L_6 + L_7 + L_8 + L_9 + N_1 + N_2 + P_2 + P_3 + P_4 + P_5 \text{ by } O_2$$

So, the approximation ratio for $\text{SOL}_1(B)$ satisfies:

$$r_1 \geqslant \frac{\delta(S_1) + \max\left\{\mathcal{A}_1, \mathcal{A}_2, \mathcal{A}_3\right\}}{\text{opt}(B)}$$

Solution $\text{SOL}_2(\mathbf{B})$. The best k_1 vertices in V_1, provided that S_2 has already been chosen, cover at least the maximum of the following quantities:

$$\mathcal{B}_1 = H_1 + H_2 + F_1 + F_2 + F_3 + L_2 + L_3 + U_2 \qquad \text{by } S_1$$
$$\mathcal{B}_2 = I_1 + I_2 + I_5 + I_6 + L_5 + L_6 + P_2 + P_4 \qquad \text{by } X_1$$
$$\mathcal{B}_3 = H_1 + H_2 + I_3 + I_4 + I_5 + I_6 + L_2 + L_3 + L_5 + L_6 + L_8 + L_9 \text{ by } O_1$$

So, the approximation ratio for $\mathrm{SOL}_2(B)$ satisfies:

$$r_2 \geqslant \frac{\delta\left(S_2\right) + \max\left\{\mathcal{B}_1, \mathcal{B}_2, \mathcal{B}_3\right\}}{\mathrm{opt}(B)}$$

Solution $\mathrm{SOL}_3(\mathbf{B})$. Taking first $S_1 \cup X_1$ in the solution, $k - (k_1 + k_1 - k_1') = k_1 + k_2 - 2k_1 + k_1' = k_2 - (k_1 - k_1') = (1 - \mu(1-\nu))k_2$ vertices remain to be taken in V_2. The best such vertices will cover at least the maximum of the following quantities:

$$\mathcal{C}_1 = (1 - \mu(1-\nu))\left(J_2 + N_2 + L_7 + U_3\right) \tag{9}$$

$$\mathcal{C}_2 = \frac{1 - \mu(1-\nu)}{2 - \xi}\left(I_3 + J_2 + L_7 + L_8 + N_2 + P_1 + P_5 + U_3\right) \tag{10}$$

$$\mathcal{C}_3 = \frac{1 - \mu(1-\nu)}{3 - 2\xi}\left(I_3 + J_2 + L_7 + L_8 + L_9 + N_2 + P_1 + P_3 + P_5 + U_3\right) \tag{11}$$

where (9) corresponds to a completion by the $(1 - \mu(1-\nu))k_2$ best vertices of S_2, (10) corresponds to a completion by the $(1 - \mu(1-\nu))k_2$ best vertices of $S_2 \cup X_2$, while (11) corresponds to a completion by the $(1 - \mu(1-\nu))k_2$ best vertices of $S_2 \cup X_2 \cup \bar{O}_2$. The denominator $3 - 2\xi$ in (11) is due to the fact that, using the expression for $\bar{O}_2$, $|S_2 \cup X_2 \cup (O_2 \setminus (S_2 \cup X_2))| \leqslant (3 - 2\xi)k_2$. So, the approximation ratio for $\mathrm{SOL}_3(B)$ is:

$$r_3 \geqslant \frac{\delta\left(S_1\right) + \delta\left(X_1\right) + \max\left\{\mathcal{C}_1, \mathcal{C}_2, \mathcal{C}_3\right\}}{\mathrm{opt}(B)} \tag{12}$$

Solution $\mathrm{SOL}_4(\mathbf{B})$. Once S_2 is taken in the solution, $k_1 = \mu k_2$ are still to be taken. Completion can be done in the following ways:

1. if $k_1 \leqslant k_2 - k_2'$, i.e., $\mu \leqslant 1 - \xi$, completion can be done by vertices taken either from X_2, or from $X_2 \cup \bar{O}_2$; in the first case, the best vertices taken for completion will cover at least either a $\mu/(1-\xi)$ fraction of edges incident to X_2; in the second case, they will cover at least a $\mu/2(1-\xi)$ fraction of edges incident to $X_2 \cup \bar{O}_2$, i.e., at least $\mathcal{M}_1$ edges, where $\mathcal{M}_1$ is given by:

$$\max\left\{\frac{\mu}{1-\xi}\delta\left(X_2\right), \frac{\mu}{2(1-\xi)}\left(\delta\left(X_2\right) + F_2 + L_3 + L_6 + L_9 + P_2 + P_3\right)\right\} \tag{13}$$

2. **else**, completion can be done by taking the whole set X_2 and then the additional vertices taken:
   (a) either within the rest of V_2 covering, in particular, a $\min\{1, (\mu-1+\xi)/|\bar{O}_2|\} \geqslant \min\{1, (\mu-1+\xi)/(1-\xi)\}$ fraction of edges incident to $\bar{O}_2$ (quantity $\mathcal{M}_2$ in (14)),
   (b) or in S_1 covering, in particular, a $(\mu-1+\xi)/\mu$ fraction of uncovered edges incident to S_1 (quantity $\mathcal{M}_3$ in (14)),
   (c) or in $S_1 \cup X_1$ covering, in particular, a $(\mu-1+\xi)/\mu(2-\nu)$ fraction of uncovered edges incident to $S_1 \cup X_1$ (quantity $\mathcal{M}_4$ in (14)),

(d) or, finally, in $S_1 \cup X_1 \cup \bar{O}_1$ covering, in particular, a $^{(\mu-1+\xi)}/_{\mu(3-2\nu)}$ fraction of uncovered edges incident to this vertex-set (quantity $\mathcal{M}_5$ in (14)); in any case such a completion will cover a number of edges that is at least the maximum of the following quantities:

$$
\begin{aligned}
\mathcal{M}_2 &= \min\left\{1, \tfrac{\mu-1+\xi}{1-\xi}\right\}(F_2 + L_3 + L_6 + L_9 + P_2 + P_3) \\
\mathcal{M}_3 &= \tfrac{\mu-1+\xi}{\mu}(F_2 + H_2 + L_3 + U_2) \\
\mathcal{M}_4 &= \tfrac{\mu-1+\xi}{\mu(2-\nu)}(F_2 + H_2 + I_2 + I_6 + L_3 + L_6 + P_2 + U_2) \\
\mathcal{M}_5 &= \tfrac{\mu-1+\xi}{\mu(3-2\nu)}(F_2 + H_2 + I_2 + I_4 + I_6 + L_3 + L_6 + L_9 + P_2 + U_2)
\end{aligned}
\tag{14}
$$

Using (13) and (14), the following holds for the approximation ratio of $\mathrm{SOL}_4(B)$:

$$
r_4 \geqslant \frac{\delta(S_2) + \begin{cases} \mathcal{M}_1 & \mu \leq 1 - \xi \\ \delta(X_2) + \max\{\mathcal{M}_2, \mathcal{M}_3, \mathcal{M}_4, \mathcal{M}_5\} & \mu \geq 1 - \xi \end{cases}}{\mathrm{opt}(B)}
\tag{15}
$$

Vertical Separations – Solutions $\mathbf{SOL_5(B)}$ and $\mathbf{SOL_6(B)}$. For $i = 1, 2$, given a vertex subset $V' \subseteq V_i$, we call *vertical separation of V' with parameter $c \in (0, 1/2]$*, a partition of V' into two subsets such that one of them contains a c-fraction of the best (highest degree) vertices of V' (i.e., contains the $c|V'|$ best vertices of V'). Then, the following easy claim holds for a vertical separation of $V' \cup V''$ with parameter c.

Claim. Let $A(V')$ be a c-fraction of the best vertices in V' and $A(V'')$ the same in V''. Then $\delta(A(V')) + \delta(A(V'')) \geq c\delta(V' \cup V'')$.

Proof. Assume that in V' we have n' vertices. To form $A(V')$ we take the cn' vertices of V' with highest degree. The average degree of V' is $\delta(V')/n'$. The average degree of $A(V')$ is $\delta(A(V'))/(cn')$. But, from the selection of $A(V')$ as the cn' vertices with highest degree, we have that $\delta(A(V'))/(cn') \geq \delta(V')/n' \Rightarrow \delta(A(V')) \geq c\delta(V')$. Similarly for V'', i.e., $\delta(A(V'')) \geq c\delta(V'')$.

Solutions $\mathrm{SOL}_5(B)$ and $\mathrm{SOL}_6(B)$ are based upon vertical separations of $S_i \cup X_i$, $i = 1, 2$, with parameters π and λ, called π- and λ-vertical separations, respectively.

The idea behind vertical separation, is to handle the scenario when there is a "tiny" part of the solution (i.e. few in comparison to, let's say, k_1 vertices) that covers a large part of the solution and the "completion" of the solution done by the previous cases does not contribute more than a small fraction to the final solution. The vertical separation indeed tries to identify such a small part, and then continues the completion on the other side of the bipartition.

Solution $\mathbf{SOL_5(B)}$. It consists of *separating $S_1 \cup X_1$ with parameter $\pi \in (0, 1/2]$, of taking a π-fraction of the best vertices of S_1 and a π-fraction of the best vertices of X_1 in the solution and of completing it with the adequate vertices from V_2.* A π-vertical separation of $S_1 \cup X_1$ introduces in the solution $\pi(2k_1 - k_1') = \pi(2 - \nu)\mu k_2$ vertices of V_1, which are to be completed with $k - \pi(2 - \nu)\mu k_2 =$

$(1 + \mu)k_2 - \pi(2 - \nu)\mu k_2 = (1 - \mu(2\pi - 1) + \mu\nu\pi)k_2$ vertices from V_2. Observe that such a separation implies the cuts with corresponding cardinalities B, C, F_i, $i = 1, 2, 3$, H_1, H_2, I_1, I_2, I_5, I_6, J_1, J_3, L_j, $j = 1, \ldots, 6$, N_1, P_2, P_4, U_1 and U_2. Let us group these cuts in the following way:

$$
\begin{aligned}
\Pi_1 &= C + J_1 + J_3 + U_1 \\
\Pi_2 &= B + L_1 + L_4 + N_1 \\
\Pi_3 &= F_3 + L_2 + L_5 + P_4 \\
\Pi_4 &= I_1 + I_5 + F_1 + H_1 \\
\Pi_5 &= F_2 + L_3 + L_6 + P_2 \\
\Pi_6 &= I_2 + I_6 + H_2 + U_2
\end{aligned}
\tag{16}
$$

We may also notice that group Π_1 refers to $S_2 \setminus O_2$, Π_2 refers to $S_2 \cap O_2$, Π_3 to $X_2 \cap O_2$, Π_5 to $\bar{O}_2$ and Π_4 to $X_2 \setminus O_2$. Assume that a $\pi_i < 1$ fraction of each group Π_i, $i = 1, \ldots 6$ contributes in the π vertical separation of $S_1 \cup X_1$. Then, a π-vertical separation of $S_1 \cup X_1$ will contribute with a value:

$$
\sum_{i=1}^{6} \pi_i \Pi_i \geqslant \pi \sum_{i=1}^{6} \Pi_i
\tag{17}
$$

to $\mathrm{sol}_5(B)$. We now distinguish two cases.

Case 1: $(1 - \mu(2\pi - 1) + \mu\nu\pi)k_2 \geqslant k_2$, i.e., $1 - \mu(2\pi - 1) + \mu\nu\pi \geqslant 1$. Then we further distinguish the following two subcases *1.* and *2.*:
1. $\mu(1 - 2\pi) + \mu\nu\pi \leq 1 - \xi$; then, the partial solution induced by the π-vertical separation will be completed in such a way that the contribution of the completion is at least equal to $\max\{Z_i, i = 1, \ldots, 5\}$, where:
Z_1 refers to S_2 plus the best $(1 - \mu(2\pi - 1) + \mu\nu\pi)k_2 - k_2 = (\mu(1 - 2\pi) + \mu\nu\pi)k_2$ vertices of O_2 having a contribution of:

$$
\begin{aligned}
Z_1 = \sum_{i=1}^{2} (1 - \pi_i)\,\Pi_i &+ (J_2 + L_7 + N_2 + U_3) + \frac{\mu(1 - 2\pi) + \mu\nu\pi}{1 - \xi}\,[(1 - \pi_3)\,\Pi_3 \\
&+ (1 - \pi_5)\,\Pi_5 + (L_8 + L_9 + P_3 + P_5)]
\end{aligned}
\tag{18}
$$

Z_2 refers to S_2 plus the best $(\mu(1 - 2\pi) + \mu\nu\pi)k_2$ vertices of X_2 having a contribution of:

$$
\begin{aligned}
Z_2 = \sum_{i=1}^{2} (1 - \pi_i)\,\Pi_i &+ (J_2 + L_7 + N_2 + U_3) \\
&+ \frac{\mu(1 - 2\pi) + \mu\nu\pi}{1 - \xi}\left[\sum_{j=3}^{4} (1 - \pi_i)\,\Pi_i + (I_3 + L_8 + P_1 + P_5)\right]
\end{aligned}
$$

Z_3 and Z_4 refer to the best $(1 - \mu(2\pi - 1) + \mu\nu\pi)k_2$ vertices of $S_2 \cup X_2$ and of $S_2 \cup O_2$ having, respectively, contributions:

$$Z_3 = \frac{1 - \mu(2\pi - 1) + \mu\nu\pi}{2 - \xi} \left[\sum_{i=1}^{4} (1 - \pi_i)\,\Pi_i \right.$$
$$\left. + (I_3 + J_2 + L_7 + L_8 + N_2 + P_1 + P_5 + U_3) \right]$$

$$Z_4 = \frac{1 - \mu(2\pi - 1) + \mu\nu\pi}{2 - \xi} \left[\sum_{i=1}^{3} (1 - \pi_i)\,\Pi_i + (1 - \pi_5)\,\Pi_5 \right.$$
$$\left. + (J_2 + L_7 + L_8 + L_9 + N_2 + P_3 + P_5 + U_3) \right]$$

Z_5 refers to the best $(1 - \mu(2\pi - 1) + \mu\nu\pi)k_2$ vertices of $S_2 \cup X_2 \cup \bar{O}_2$ having a contribution of:

$$Z_5 = \frac{1 - \mu(2\pi - 1) + \mu\nu\pi}{3 - 2\xi} \left[\sum_{i=1}^{5} (1 - \pi_i)\,\Pi_i \right.$$
$$\left. + (I_3 + J_2 + L_7 + L_8 + L_9 + N_2 + P_1 + P_3 + P_5 + U_3) \right]$$

2. $\mu(1 - 2\pi) + \mu\nu\pi \geq 1 - \xi$; in this case, the partial solution induced by the π-vertical separation will be completed in such a way that the contribution of the completion is at least $\max\{\Theta_i, i = 1, \ldots, 3\}$, where:
Θ_1 refers to $S_2 \cup X_2$ plus the best $(\mu(1 - 2\pi) + \mu\nu\pi - (1 - \xi))k_2$ vertices of $\bar{O}_2$, all this having a contribution of:

$$\Theta_1 = \sum_{i=1}^{4} (1 - \pi_i)\,\Pi_i + (I_3 + J_2 + L_7 + L_8 + N_2 + P_1 + P_5 + U_3)$$
$$+ \frac{\mu(1 - 2\pi) + \mu\nu\pi - (1 - \xi)}{1 - \xi} [(1 - \pi_5)\,\Pi_5 + L_9 + P_3]$$

Θ_2 refers to $S_2 \cup O_2$ plus the best $(\mu(1 - 2\pi) + \mu\nu\pi - (1 - \xi))k_2$ vertices of $X_2 \setminus O_2$, all this having a contribution of:

$$\Theta_2 = \sum_{i=1}^{3} (1 - \pi_i)\,\Pi_i$$
$$+ (1 - \pi_5)\,\Pi_5 + (J_2 + L_7 + L_8 + L_9 + N_2 + P_3 + P_5 + U_3)$$
$$+ \frac{\mu(1 - 2\pi) + \mu\nu\pi - (1 - \xi)}{1 - \xi} [(1 - \pi_4)\,\Pi_4 + I_3 + P_1]$$

Θ_3 refers to the best $(1 - \mu(2\pi - 1) + \mu\nu\pi)k_2$ vertices of $S_2 \cup X_2 \cup \bar{O}_2$ having contribution of:

$$\Theta_3 = \frac{1 - \mu(2\pi - 1) + \mu\nu\pi}{3 - 2\xi} \left[\sum_{i=1}^{5} (1 - \pi_i)\,\Pi_i \right.$$
$$\left. + (I_3 + J_2 + L_7 + L_8 + L_9 + N_2 + P_1 + P_3 + P_5 + U_3) \right]$$

Case 2: $1 - \mu(2\pi - 1) + \mu\nu\pi < 1$. The partial solution induced by the π-vertical separation will be completed in such a way that the contribution of the completion is at least equal to $\max\{\Phi_i, i = 1, \ldots, 5\}$, where:
Φ_1 refers to the best $(1 - \mu(2\pi - 1) + \mu\nu\pi)k_2$ vertices in S_2 with a contribution:

$$\Phi_1 = (1 - \mu(2\pi - 1) + \mu\nu\pi)\left[\sum_{i=1}^{2}(1 - \pi_i)\,\Pi_i + (J_2 + L_7 + N_2 + U_3)\right]$$

Φ_2 refers to the best $(1 - \mu(2\pi - 1) + \mu\nu\pi)k_2$ vertices in X_2 with a contribution:

$$\Phi_2 = \frac{1 - \mu(2\pi - 1) + \mu\nu\pi}{1 - \xi}\left[\sum_{i=3}^{4}(1 - \pi_i)\,\Pi_i + (I_3 + L_8 + P_1 + P_5)\right]$$

Φ_3 refers to the best $(1 - \mu(2\pi - 1) + \mu\nu\pi)k_2$ vertices in O_2 with a contribution:

$$\Phi_3 = (1 - \mu(2\pi - 1) + \mu\nu\pi)\left[\sum_{i=2}^{3}(1 - \pi_i)\,\Pi_i + (1 - \pi_5)\,\Pi_5\right.$$
$$\left. + (L_7 + L_8 + L_9 + N_2 + P_3 + P_5)\right]$$

Φ_4 refers to the best $(1 - \mu(2\pi - 1) + \mu\nu\pi)k_2$ vertices in $S_2 \cup X_2$ with a contribution:

$$\Phi_4 = \frac{1 - \mu(2\pi - 1) + \mu\nu\pi}{2 - \xi}\left[\sum_{j=1}^{4}(1 - \pi_j)\,\Pi_j\right.$$
$$\left. + (I_3 + J_2 + L_7 + L_8 + N_2 + P_1 + P_5 + U_3)\right]$$

Φ_5 refers to the best $(1 - \mu(2\pi - 1) + \mu\nu\pi)k_2$ vertices in $S_2 \cup X_2 \cup \bar{O}_2$ with a contribution:

$$\Phi_5 = \frac{1 - \mu(2\pi - 1) + \mu\nu\pi}{3 - 2\xi}\left[\sum_{j=1}^{5}(1 - \pi_j)\,\Pi_j\right.$$
$$\left. + (I_3 + J_2 + L_7 + L_8 + L_9 + N_2 + P_1 + P_3 + P_5 + U_3)\right] \qquad (19)$$

Setting $Z^* = \max\{Z_i : i = 1, \ldots 5\}$, $\Theta^* = \max\{\Theta_i : i = 1, 2, 3\}$ and $\Phi^* = \max\{\Phi_i : i = 1, \ldots 5\}$, and putting (16) and (17) together with expressions (18) to (19), we get the following lower bound for ratio r_5:

$$\frac{\sum_{i=1}^{6} \pi_i \Pi_i + \begin{cases} \begin{cases} Z^* \text{ if } \mu(1 - 2\pi) + \mu\nu\pi \leq 1 - \xi \\ \Theta^* \text{ if } \mu(1 - 2\pi) + \mu\nu\pi \geq 1 - \xi \end{cases} & \text{case: } 1 - \mu(2\pi - 1) + \mu\nu\pi \geq 1 \\ \Phi^* & \text{case: } 1 - \mu(2\pi - 1) + \mu\nu\pi < 1 \end{cases}}{\text{opt}(B)}$$
$$(20)$$

Solution $\mathbf{SOL_6(B)}$. In a complete analogy with SOL_5, solution $SOL_6(B)$ consists of *separating $S_2 \cup X_2$ with parameter* $\lambda \in (0, 1/2]$. It consists of *separating*

$S_2 \cup X_2$ with parameter λ, of taking a λ fraction of the best vertices of S_2 and X_2 in the solution and of completing it with the adequate vertices from V_1. Here, we need that $\lambda(k_2 + k_2 - k_2') \leqslant k \Rightarrow \lambda(2-\xi)k_2 \leqslant (1+\mu)k_2 \Rightarrow \lambda \leqslant (1+\mu)/(2-\xi) \Rightarrow \lambda \in (0, (1+\mu)/(2-\xi)]$.

A λ-vertical separation of $S_2 \cup X_2$ introduces in the solution $\lambda(2-\xi)k_2$ vertices of V_2, which are to be completed with $k - \lambda(2-\xi)k_2 = (1+\mu)k_2 - \lambda(2-\xi)k_2 = (1 + \mu - \lambda(2-\xi))k_2$ vertices from V_1.

Observe that such a separation implies the cuts with corresponding cardinalities B, C, F_1, F_3, H_1, I_1, I_3, I_5, J_i, $i = 1, 2, 3$, L_1, L_2, L_4, L_5, L_7, L_8, N_1, N_2, P_1, P_4, P_5, U_1 and U_3. We group these cuts in the following way:

$$
\begin{aligned}
\Lambda_1 &= B + F_1 + F_3 + U_1 \\
\Lambda_2 &= C + H_1 + L_1 + L_2 \\
\Lambda_3 &= J_3 + I_5 + L_4 + L_5 \\
\Lambda_4 &= I_1 + J_1 + N_1 + P_4 \\
\Lambda_5 &= I_3 + J_2 + L_7 + L_8 \\
\Lambda_6 &= N_2 + P_1 + P_5 + U_3
\end{aligned}
\tag{21}
$$

Group Λ_1 refers to $S_1 \setminus O_1$, Λ_2 to $S_1 \cap O_1$, Λ_3 to $X_1 \cap O_1$, Λ_5 to $\bar{O}_1$ and Λ_4 to $X_1 \setminus O_1$. Assume, as previously, that a $\lambda_i < 1$ fraction of each group Λ_i, $i = 1, \ldots 6$ contributes in the λ vertical separation of $S_2 \cup X_2$. Then, a λ-vertical separation of $S_2 \cup X_2$ will contribute with a value:

$$
\sum_{i=1}^{6} \lambda_i \Lambda_i \geqslant \lambda \sum_{i=1}^{6} \Lambda_i
\tag{22}
$$

to $\mathrm{sol}_6(B)$. We again distinguish two cases.

Case 1. $(1 + \mu - \lambda(2-\xi))k_2 \geqslant \mu k_2$, i.e., $1 + \mu - \lambda(2-\xi) \geqslant \mu$. Here we have the two following subcases.

(a) $1 - \lambda(2-\xi) \leq (1-\nu)\mu$; then, the partial solution induced by the λ-vertical separation will be completed in such a way that the contribution of the completion is at least equal to $\Upsilon^* = \max\{\Upsilon_i, i = 1, \ldots, 5\}$, where: Υ_1 refers to S_1 plus the best $(1 - \lambda(2-\xi))k_2$ vertices of X_1;
Υ_2 refers to S_1 plus the best $(1 - \lambda(2-\xi))k_2$ vertices of O_1;
Υ_3 and Υ_4 refer to the best $(1 + \mu - \lambda(2-\xi))k_2$ vertices of $S_1 \cup X_1$ and $S_1 \cup O_1$;
Υ_5 refers to the best $(1 + \mu - \lambda(2-\xi))k_2$ vertices of $S_1 \cup X_1 \cup \bar{O}_1$. *(b)* $1 - \lambda(2-\xi) \geq (1-\nu)\mu$; in this case, the partial solution induced by the λ-vertical separation will be completed in such a way that the contribution of the completion is at least $\Psi^* = \max\{\Psi_i, i = 1, \ldots, 3\}$, where:
Ψ_1 refers to $S_1 \cup X_1$ plus the best $(1 - \lambda(2-\xi) - (1-\nu))k_2$ vertices of $\bar{O}_1$;
Ψ_2 refers to $S_1 \cup O_1$ plus the best $(1 - \lambda(2-\xi) - (1-\nu))k_2$ vertices of $X_1 \setminus O_1$;
Ψ_3 refers to the best $(\mu + 1 - \lambda(2-\xi))k_2$ vertices of $S_1 \cup X_1 \cup \bar{O}_1$.

Case 2. $1 + \mu - \lambda(2-\xi) \leqslant \mu$. The partial solution induced by the λ-vertical separation will be completed in such a way that the contribution of the completion is at least equal to $\Omega^* = \max\{\Omega_i, i = 1, \ldots, 5\}$, where:

Ω_1 refers to the best $(1 + \mu - \lambda(2 - \xi))k_2$ vertices in S_1;
Ω_2 refers to the best $(1 + \mu - \lambda(2 - \xi))k_2$ vertices in X_1;
Ω_3 refers to the best $(1 + \mu - \lambda(2 - \xi))k_2$ vertices in O_1;
Ω_4 refers to the best $(1 + \mu - \lambda(2 - \xi))k_2$ vertices in $S_1 \cup X_1$;
Ω_5 refers to the best $(1 + \mu - \lambda(2 - \xi))k_2$ vertices in $S_1 \cup X_1 \cup \bar{O}_1$.

Putting all this together we get:

$$
r_6 \geqslant \frac{\sum_{i=1}^{6} \lambda_i \Lambda_i + \begin{cases} \begin{cases} \Upsilon^* & \text{if } 1 - \lambda(2 - \xi) \leq (1 - \nu)\mu \\ \Psi^* & \text{if } 1 - \lambda(2 - \xi) > (1 - \nu)\mu \end{cases} \text{ case: } \mu + 1 - \lambda(2 - \xi) \geq \mu \\ \Omega^* \hspace{3.2cm} \text{case: } \mu + 1 - \lambda(2 - \xi) < \mu \end{cases}}{\text{opt}(B)}
$$

$$(23)$$

The complete study of solution $\text{SOL}_6(B)$ is deferred to [7].

4   Results and Discussion

To analyze the performance guarantee of k-VC_ALGORITHM, we set up a non-linear program and solved it to optimality. Here, we interpret the cardinalities of the edge-sets $B, C, F_i, \ldots$, as *variables*, the expressions in (8) as *constraints* and the *objective function* is $\min Z(\equiv \max_{j=1}^{6} r_j)$. In other words, we try to find a value assignments to the set of variables such that the maximum among all the six ratios defined is minimized. This value would give us the desired approximation guarantee of k-VC_ALGORITHM.

Towards this goal, we set up a GRG (Generalized Reduced Gradient [9]) program. The reasons this method is selected are presented in [7], as well as a more detailed description of the implementation. GRG is a generalization of the classical *Reduced Gradient* method [10] for solving (concave) quadratic problems so that it can handle higher degree polynomials and incorporate non-linear constraints.

As the values of parameters π and λ decrease, the approximation guarantee increases. The maximum of these ratios is attained for $\pi = \lambda = 10^{-5}$. For these values, the corresponding values of ratios $r_1 \div r_6$ computed for them are the following:

$$r_1 = 0.81806$$
$$r_2 = 0.81797$$
$$r_3 = 0.79280$$
$$r_4 = 0.79657$$
$$\mathbf{r_5 = 0.82104}$$
$$r_6 = 0.82103$$

These results correspond to the cycle that outputs the *minimum* value for the approximation factor and this is 0.821, given by solution SOL_5.

To conclude, let us note that the formulation of the non-linear program we developed for bounding the ratio below, could provide useful insights for problem's understanding and could be applied for solving the problem on other graph-classes. Finally, since the overall algorithm chooses the best among a certain number of solutions it is easily parallelizable.

Remark. As we note in [7], the GRG solver does not guarantee the global optimal solution. The 0.821 guarantee is the minimum value that the solver returns after several runs from different initial starting points. However, successive re-executions of the algorithm, starting from this minimum value, were unable to find another point with smaller value. In each one of these successive re-runs, we tested the algorithm on 1000 random different starting points (which is greater than the estimation of the number of local minima) and the solver did not find value worse that the reported one.

Acknowledgement. The work of the last author was supported by the Swiss National Research Foundation Early Post-Doc mobility grant P1TIP2_152282.

References

1. Apollonio, N., Simeone, B.: The maximum vertex coverage problem on bipartite graphs. Discrete Appl. Math. **165**, 37–48 (2014)
2. Caskurlu, B., Mkrtchyan, V., Parekh, O., Subramani, K.: On partial vertex cover and budgeted maximum coverage problems in bipartite graphs. In: Diaz, J., Lanese, I., Sangiorgi, D. (eds.) TCS 2014. LNCS, vol. 8705, pp. 13–26. Springer, Heidelberg (2014)
3. Hochbaum, D.S., Pathria, A.: Analysis of the greedy approach in problems of maximum k-coverage. Naval Res. Logistics **45**, 615–627 (1998)
4. Badanidiyuru, A., Kleinberg, R., Lee, H.: Approximating low-dimensional coverage problems. In: Dey, T.K., Whitesides, S. (eds.) SoCG 2012, pp. 161–170. ACM, Chapel Hill (2012)
5. Ageev, A.A., Sviridenko, M.I.: Approximation algorithms for maximum coverage and max cut with given sizes of parts. In: Cornuéjols, G., Burkard, R.E., Woeginger, G.J. (eds.) IPCO 1999. LNCS, vol. 1610, p. 17. Springer, Heidelberg (1999)
6. Petrank, E.: The hardness of approximation: gap location. Comput. Complex. **4**, 133–157 (1994)
7. Bonnet, E., Escoffier, B., Paschos, V.T., Stamoulis, G.: A 0.821-ratio purely combinatorial algorithm for maximum k-vertex cover in bipartite graphs. CoRR arXiv:1409.6952v2 (2015)
8. Feige, U., Karpinski, M., Langberg, M.: Improved approximation of max-cut on graphs of bounded degree. J. Algorithms **43**, 201–219 (2002)
9. Abadie, J., Carpentier, J.: Generalization of the wolfe reduced gradient method to the case of non-linear constraints. In: Abadie, J., Carpentier, J. (eds.) Optimization. Academic Publishers (1969)
10. Frank, M., Wolfe, P.: An algorithm for quadratic programming. Naval Res. Logistics Q. **3**, 95–110 (1956)

Improved Spanning Ratio
for Low Degree Plane Spanners

Prosenjit Bose, Darryl Hill$^{(\boxtimes)}$, and Michiel Smid

School of Computer Science, Carleton University, Ottawa, Canada
`darrylhill@email.carleton.ca`

Abstract. We describe an algorithm that builds a plane spanner with a maximum degree of 8 and a spanning ratio of ≈4.414 with respect to the complete graph. This is the best currently known spanning ratio for a plane spanner with a maximum degree of less than 14.

1 Introduction

Let P be a set of n points in the plane. Let G be a weighted geometric graph on vertex set P, where edges are straight line segments and are weighted according the Euclidean distance between their endpoints. Let $\delta_G(p, q)$ be the sum of the weights of the edges on the shortest path from p to q in G. If, for graphs G and H on the point set P, where G is a subgraph of H, for every pair of points p and q in P, $\delta_G(p, q) \leq t \cdot \delta_H(p, q)$ for some real number $t > 1$, then G is a t-spanner of H, and t is called the *spanning ratio*. H is called the *underlying* graph of G. In this paper the underlying graph is the Delaunay triangulation or the complete graph.

The L_1-Delaunay triangulation was first proven to be a $\sqrt{10}$-spanner of the complete graph by Chew [1]. Dobkin *et al.* [2] proved that the L_2-Delaunay triangulation is a $\frac{1+\sqrt{5}}{2}\pi$-spanner. This was improved by Keil and Gutwin [3] to $\frac{2\pi}{3\cos(\frac{\pi}{6})}$, and finally taken to its currently best known spanning ratio of 1.998 by Xia [4].

The Delaunay triangulation may have an unbounded degree. High degree nodes can be detrimental to real world applications of graphs. Thus there has been research into bounded degree plane spanners. We present a brief overview of some of the results in Table 1.

Bounded degree plane spanners are often obtained by taking a subset of edges of an existing plane spanner and ensuring that it has bounded degree, while maintaining spanning properties. We note how in Table 1 that all of the results are subgraphs of some variant of the Delaunay triangulation. All of the algorithms for building the graphs mentioned above run in $O(n \log n)$ time, including ours.

As we look down the column of results, we see a steady march towards lower degrees. Indeed, Bonichon *et al.* [11] have impressively found a degree 4 plane spanning graph. However, we believe that optimizing both degree and spanning

This work was partially supported by the Natural Sciences and Engineering Research Council of Cananda (NSERC) and by the Ontario Graduate Scholarship (OGS).

© Springer-Verlag Berlin Heidelberg 2016
E. Kranakis et al. (Eds.): LATIN 2016, LNCS 9644, pp. 249–262, 2016.
DOI: 10.1007/978-3-662-49529-2_19

Table 1. Known results for bounded degree plane spanners.

Paper	Degree	Stretch factor
Bose *et al.* [5]	27	$(\pi + 1)C_{DT} \approx 8.27$
Li and Wang [6]	23	$(1 + \pi\sin(\pi/4))C_{DT} \approx 6.44$
Bose *et al.* [7]	17	$(((1 + \sqrt{3} + 3\pi)/2) + 2\pi\sin(\pi/12))C_{DT} \approx 23.58$
Kanj *et al.* [8]	14	$(1 + (2\pi/14\cos(\pi/14))))C_{DT} \approx 2.92$
Bose *et al.* [9]	7	$(1/(1 - 2\tan(\pi/8)))C_{DT} \approx 11.65$
Bose *et al.* [9]	6	$(1/(1 - \tan(\pi/7)(1 + 1/\cos(\pi/14))))C_{DT} \approx 81.66$
Bonichon *et al.* [10]	6	6
Bonichon *et al.* [11]	4	$\sqrt{4 + 2\sqrt{2}}(19 + 29\sqrt{2}) \approx 156.82$
This paper	8	$(1 + (2\pi/(6\cos(\pi/6))))C_{DT} \approx 4.41$

C_{DT} is the spanning ratio of the Delaunay Triangulation, currently <1.998 [4]

ratio is of theoretical interest. Certainly in any practical setting both spanning ratio and degree bound are important. Thus, looking at the above results we see an opportunity for improvement in this field.

With this in mind, we present our contribution, an algorithm to construct a plane spanner of maximum degree 8 with a spanning ratio of ≈4.41. This is the lowest spanning ratio of any graph of degree less than 14.

The rest of the paper is organized as follows. In Sect. 2 we describe how to select a subset of the edges of the Delaunay triangulation $DT(P)$ to form the graph $D8(P)$. In Sect. 3 we prove that $D8(P)$ has a maximum degree of 8. In Sect. 4 we bound the spanning ratio of $D8(P)$ with respect to $DT(P)$. Since $DT(P)$ is a spanner of the complete Euclidean graph, this makes $D8(P)$ a spanner of the complete Euclidean graph as well.

2   Building D8(P)

Given as input a set P of n points in the plane, we present an algorithm for building a bounded degree plane graph with maximum degree 8 and spanning ratio bounded by a constant, which we denote as $D8(P)$. The graph denoted $D8(P)$ is constructed by taking a subset of the edges of the Delaunay triangulation of P, denoted $DT(P)$.

We assume general position of P; i.e., no three points are on a line, no four points are on a circle, and no two points form a line with slope 0, $\sqrt{3}$ or $-\sqrt{3}$.

The space around each vertex p is partitioned by *cones* consisting of 6 equally spaced rays from p. Thus each cone has an angle of $\pi/3$. See Fig. 1. We number the cones starting with the topmost cone as C_0, then number in the clockwise direction. Cone arithmetic is modulo 6. By our general position assumption we note that no point of P lies on the boundary of a cone.

We denote the Euclidean distance between two vertices p and q by $|pq|$. We also introduce a distance function known as the *bisector distance*, which is the distance from p to the orthogonal projection of q onto the bisector of C_i^p, where $q \in C_i^p$. See Fig. 1a. We denote this length $[pq]$.

Definition 1. *Let* $\{q_0, q_1, ..., q_{d-1}\}$ *be the sequence of all neighbours of* p *in* $DT(P)$ *in consecutive clockwise order. The neighbourhood* N_p, *with* apex p, *is the graph with the vertex set* $\{p, q_0, q_1, ..., q_{d-1}\}$ *and the edge set* $\{(p, q_j)\} \cup \{(q_j, q_{j+1})\}, 0 \leq j \leq d-1$, *with all values mod* d. *The edges* $\{(q_j, q_{j+1})\}$ *are called* canonical edges. N_i^p *is the subgraph of* N_p *induced by all the vertices of* N_p *in* C_i^p, *including* p. *This is called the* **cone neighbourhood** *of* p. *See Fig. 1b. If we refer to the induced subgraph of* p *and a consecutive subsequence of neighbours* $\{q_k, ..., q_l\}, 0 \leq k \leq l \leq d-1$, *we refer to this as a* **restricted neighbourhood**, *which we denote as* $N_p^{q_k, q_l}$.

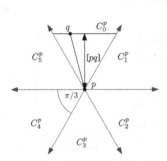 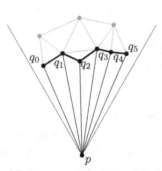

(a) Cones and bisector distance.

(b) The cone neighbourhood N_i^p in black. The thick black lines are canonical edges.

Fig. 1. Preliminaries.

The algorithm $ConstructD8(P)$ takes as input a point set P and returns the bounded degree graph $D8(P)$, with vertex set P and edge set E. The algorithm calls two subroutines. $AddIncident()$ selects a set of edges E_A. For each edge (p, r) of E_A, we call $AddCanonical(p, r)$ and $AddCanonical(r, p)$ which add edges to the set E_{CAN}. Both E_A and E_{CAN} are a subset of the edges in $DT(P)$. The final graph $D8(P)$ consists of the vertex set P and the union of edge sets E_A and E_{CAN}.

We present the algorithm here;

Algorithm: ConstructD8(P)

INPUT: Set P of n points in the plane.

OUTPUT: $D8(P)$: spanning subgraph of $DT(P)$.

Step 1: Compute the Delaunay triangulation $DT(P)$ of the point set P.

Step 2: Sort all the edges of $DT(P)$ by their bisector distance, into an ordered set L, in non-decreasing order.

Step 3: Call the function $AddIncident(L)$ with L as the argument. $AddIncident()$ selects and returns the subset E_A of the edges of L.

Step 4: For each edge (p, r) in E_A in sorted order call $AddCanonical(p, r)$ and $AddCanonical(r, p)$, which add edges to the set E_{CAN}.

Step 5: Return $D8(P) = (P, E_A \cup E_{CAN})$.

Algorithm: AddIncident(L)
INPUT: L: ordered set of edges of $DT(P)$ sorted by bisector distance.
OUTPUT: E_A: a subset of edges of $DT(P)$.

Step 1: Initialize the set $E_A = \emptyset$.
Step 2: For each $(p, q) \in L$, in non-decreasing order, do:
(a) Let i be the number of the cone of p containing q. If E_A has no edges with endpoint p in N_i^p, and if E_A has no edges with endpoint q in N_{i+3}^q, then we add (p, q) to E_A.
Step 3: Return E_A.

The next algorithm requires the following definition:

Definition 2. *Let $Can_i^{(p,r)}$ be the subgraph of $DT(P)$ consisting of the ordered subsequence of canonical edges (s, t) of N_i^p in clockwise order around apex p such that $[ps] \geq [pr]$ and $[pt] \geq [pr]$. We call $Can_i^{(p,r)}$ a* canonical subgraph. *A vertex that is the first or last vertex of $Can_i^{(p,r)}$ is called an* end vertex *of $Can_i^{(p,r)}$. A vertex that is not the first or last vertex in $Can_i^{(p,r)}$ is called an* inner vertex *of $Can_i^{(p,r)}$. Vertex r is called the* anchor *of $Can_i^{(p,r)}$. See Fig. 2.*

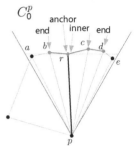

C_0^p

Fig. 2. The graph $Can_0^{(p,r)}$, based on $(p, r) \in E_A$, in red. Vertex r is the anchor, d and b are end vertices, and c and r are inner vertices (Color figure online).

Algorithm: AddCanonical(p, r)
INPUT: (p, r), an edge of E_A.
OUTPUT: A set of edges that are a subset of the edges of $DT(P)$. All edges generated by calls to $AddCanonical()$ form the set E_{CAN}.

Step 1: Without loss of generality, let $r \in C_0^p$.
Step 2: If there are at least three edges in $Can_0^{(p,r)}$, then for every canonical edge (s, t) in $Can_0^{(p,r)}$ that is not the first or last edge in the ordered subsequence of canonical edges $Can_0^{(p,r)}$, we add (s, t) to E_{CAN}.
Step 3: If the anchor r is the first or last vertex in $Can_0^{(p,r)}$, and there is more than one edge in $Can_0^{(p,r)}$, then add the edge of $Can_0^{(p,r)}$ with endpoint r to E_{CAN}. See Fig. 3b.
Step 4: Consider the first and last canonical edge in $Can_0^{(p,r)}$. Since the conditions for the first and last canonical edge are symmetric, we only describe how to process the last canonical edge (y, z). There are three possibilities.
(a) If $(y, z) \in N_5^z$ we add (y, z) to E_{CAN}. See Fig. 3c.
(b) If $(y, z) \in N_4^z$ and N_4^z does not have an edge with endpoint z in E_A, then we add (y, z) to E_{CAN}. See Fig. 3d
(c) If $(y, z) \in N_4^z$ and there is an edge with endpoint z in $E_A \cap N_4^z \setminus (y, z)$, then there is exactly one canonical edge of z with endpoint y in N_4^z. We label this edge (w, y) and add it to E_{CAN}. See Fig. 3e.

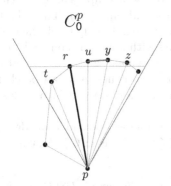

(a) Edges added to E_{CAN} in Step 2.

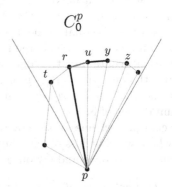

(b) Edge added to E_{CAN} in Step 3.

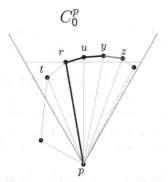

(c) Edge added to E_{CAN} in Step 4a.

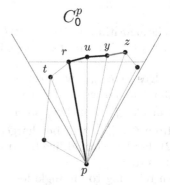

(d) Edge added to E_{CAN} in Step 4b.

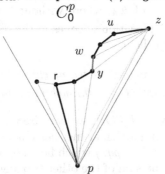

(e) Edge added to E_{CAN} in Step 4c.

Fig. 3. AddCanonical(p,r)

3   D8(P) has Maximum Degree 8

To prove $D8(P)$ has a maximum degree of 8 we use a simple charging scheme. We charge each edge (p, q) of $D8(P)$ once to p and once to q. Thus the total charge on a vertex is equal to the degree of that vertex. To help track the number of charges on a vertex, each charge is associated with a specific cone, which may not be the cone containing the edge. We show that a cone can be charged at most twice, and that for any vertex p of P, at most two cones of p can be charged twice, while the remaining cones are charged at most once, which yields our maximum degree of 8.

The charging scheme for the edges of E_A is as follows. Consider an edge (p, r) of E_A, where without loss of generality r is in C_0^p and p is in C_3^r. An edge (p, r) of E_A charges C_0^p once and C_3^r once.

Lemma 1. *Each cone of an arbitrary vertex p of the graph $D8(P)$ is charged at most once by an edge of E_A (thus yielding a maximum degree for the graph $G = (P, E_A)$ of 6).*

Proof. The algorithm $AddIncident(L)$, specifies in Step 2a that, for a vertex p and cone $C_i^p, 0 \leq i \leq 5$, at most one edge in C_i^p with endpoint p is added to E_A.   □

For edges in E_{CAN} we consider an arbitrary canonical subgraph $Can_i^{(p,r)}$, and without loss of generality let $i = 0$. We note that there are three types of vertices in $Can_0^{(p,r)}$: anchor, inner and end vertices. Thus any edge added to E_{CAN} from $Can_0^{(p,r)}$ will be charged to an inner, end or anchor vertex (refer to Fig. 2). First we would like to establish that there are empty cones to charge these edges to.

When referring to an angle formed by three points, we refer to the smaller of the two angles (that is, the angle that is $<\pi$) unless otherwise stated.

We consider the edge (p, r) of E_A, where without loss of generality, r is in C_0^p. In this section we show the location of cones in the region of $Can_0^{(p,r)}$, so we may charge edges of E_{CAN} to them.

Lemma 2. *Consider the arbitrary restricted neighbourhood $N_p^{(r,q)}$. Each vertex $x \in N_p^{(r,q)}\backslash\{p, r, q\}$ is in the circle $O_{p,r,q}$ through p, r, and q.*

Proof. Since (p, x) is an edge in $DT(P)$, we can draw a disk through p and x that is empty of points of P. In particular, neither r nor q is in this disk. Hence the sum of the angles $\angle(prx)$ and $\angle(pqx)$ which lie on opposite sides of the same chord is smaller than π, and the sum of the other two angles $\angle(rxq)$ and $\angle(rpq)$ in the quadrilateral $(prxq)$ is greater then π. That implies x is inside $O_{p,r,q}$.

Lemma 3. *Consider the restricted neighbourhood $N_p^{(r,q)}$ in cone C_i^p. Let (p, x) be an edge in $N_p^{(r,q)}$ where $x \neq r$ and $x \neq q$. Then angle $\angle(qxr) \geq \pi - \angle(qpr)$. Since the cone angle is $\pi/3$, we have that $\angle(qxr) > 2\pi/3$.*

Proof. We know by Lemma 2 that x lies inside the circle through p, r and q, which we label $O_{p,r,q}$. The angle $\angle(qxr)$ is minimized when x is on $O_{p,r,q}$. When x is on $O_{p,r,q}$, $\angle rxq = \pi - \angle(qpr)$, since the two angles lie on the same chord (r, q). Therefore $\angle(rxq) \geq \pi - \angle(qpr)$. Since both q and r are in the same cone C_i^p, and the cone angle is $\pi/3$, the $\angle(qxr) > 2\pi/3$.

Which leads to the corollary:

Corollary 1. *Let s be an inner vertex of $Can_i^{(p,r)}$ that is not the anchor. Then there is at least one empty cone of s in $Can_i^{(p,r)}$.*

Proof. Since s is not the anchor, any internal cone of $Can_i^{(p,r)}$ on vertex s is empty, and by Lemma 3, there is at least one internal cone of $Can_i^{(p,r)}$ on vertex s. Therefore there is at least one empty internal cone on s in $Can_i^{(p,r)}$.   □

Other empty cones follow a similar analysis. We outline the charging scheme below by referencing the steps of $AddCanonical(p, r)$ where edges were added to E_{CAN}.

Step 2, Step 3: **Charge vertex s:**
  i If s is the anchor (thus $s = r$), then C_2^r and C_4^r are empty cones inside $Can_0^{(p,r)}$ (since (p, r) is the shortest edge incident to p, in bisector distance, in N_0^p). If t is left of directed line segment pr, charge (r, t) to C_4^r. If t is right of pr, charge (r, t) to C_2^r. See Fig. 4a.
  ii If $s \neq r$ then by s has an empty cone C_j^s inside $Can_0^{(p,r)}$ (Corollary 1). Charge (s, t) once to C_j^s. See Fig. 4a.
Step 4a, Step 4b: **Charge vertex y:**
  i If y is the anchor, then C_2^y is empty and inside $Can_0^{(p,r)}$. Charge (y, z) to C_2^y. See Fig. 4a.
  ii Otherwise y is not the first or last vertex in $Can_0^{(p,r)}$, and has an empty cone C_j^y inside $Can_0^{(p,r)}$ (Corollary 1). Charge (y, z) to C_j^y. Figure 4b.
Charge vertex z:
  iii Step 4a : (y, z) is in C_5^z. Then C_4^z is empty and inside $Can_0^{(p,r)}$. Charge (y, z) to C_4^z. See Fig. 4c.
  iv Step 4b : (y, z) is in C_4^z, and C_4^z does not contain an edge of E_A with endpoint z. Charge (y, z) to C_4^z. See Fig. 4d.
Step 4c: **Charge vertex y:**
  i If $y = r$, then C_2^y is empty and inside $Can_0^{(p,r)}$. Charge (w, y) to C_2^y.
  ii Otherwise y is not the first or last vertex in $Can_0^{(p,r)}$, and has an empty cone C_j^y inside $Can_0^{(p,r)}$ (Corollary 1). Charge (w, y) to C_j^y. See Fig. 4e.
Charge vertex w:
  iii If $w = u$ (where (z, u) in E_A), then C_2^w is empty and inside $Can_4^{(z,w)}$ (since (z, w) is the shortest edge incident to z, in bisector distance, in N_4^z). Charge (w, y) to C_2^w.
  iv If $w \neq u$, then w is not the first or last vertex in $Can_4^{(z,u)}$, and has an empty cone C_j^w inside $Can_4^{(z,u)}$ (Corollary 1). Charge (w, y) to C_j^w. See Fig. 4f.

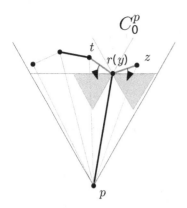

(a) Step 2, Step 3: Charge i. Step 4a, Step 4b: Charge i.

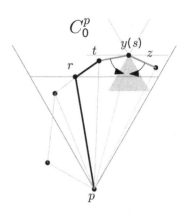

(b) Step 2, Step 3: Charge ii. Step 4a, Step 4b: Charge ii.

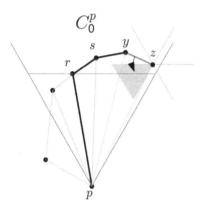

(c) Step 4a: Charge iii.

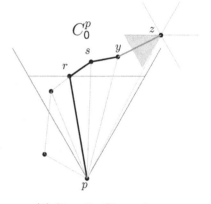

(d) Step 4b: Charge iv.

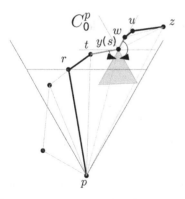

(e) Step 4c: Charge ii. (w, y) is charged to y in place of (y, z). (s, t) is charged normally.

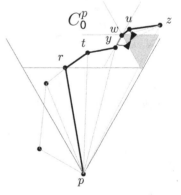

(f) Step 4c: Charge iv. (w, y) is charged to the empty cone of w.

Fig. 4. Charging scheme for edges of E_{CAN}.

An analysis of the charging scheme yields the following results.

Lemma 4. *Cones of an end vertex or anchor of a canonical subgraph are charged at most once by edges of E_{CAN}.*

Lemma 5. *Cones on an inner vertex of a canonical subgraph are charged at most twice by edges of E_{CAN}.*

Lemma 6. *The edges of E_A and E_{CAN} are never charged to the same cone.*

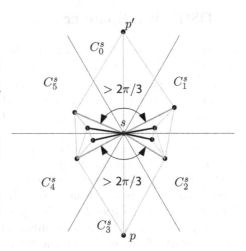

Fig. 5. A degree 8 vertex in $D8(P)$. The red edges belong to E_{CAN}, while the black edges belong to E_A. The light edges are edges of $DT(P)$ that may or may not be in $D8(P)$ (Color figure online).

Proof. The edges of E_A are charged directly to the cone they occupy on each endpoint. We know from the charging scheme above that the edges of E_{CAN} are charged to either empty cones, or to a cone that does not contain an edge of E_A. Thus the edges of E_{CAN} and E_A are never charged to the same cone. □

Lemma 7. *Consider a cone C_i^s of a vertex s in $D8(P)$ that is charged twice by edges of E_{CAN}. Then the two neighbouring cones C_{i-1}^s and C_{i+1}^s are charged at most once by edges of $D8(P)$.*

Theorem 1. *The maximum degree of $D8(P)$ is at most 8.*

Proof. Each edge (p, r) of E_A is charged once to the cone of p containing r and once to the cone of r containing p. By Lemma 1, no cone is charged more than once by edges of E_A.

No edge of E_{CAN} is charged to a cone that is charged by an edge of E_A by Lemma 6.

By Lemma 7, if a cone of a vertex s of $D8(P)$ is charged twice, then its neighbouring cones are charged at most once. This implies that there are at most 3 double charged cones on any vertex s in $D8(P)$.

Assume that we have a vertex s with 3 cones that have been charged twice. A cone of s that is charged twice is an internal cone of some cone neighbourhood N_i^p by our charging argument. Thus s is endpoint to two canonical edges (q, s) and (s, t) in N_i^p. Note that $\angle(qst) > 2\pi/3$, and this angle contains the cone of s that is charged twice. Thus to have 3 cones charged twice, the total angle around s would need to be $> 2\pi$, which is impossible. Thus there are at most two double charged cones on s, which gives us a maximum degree of 8. See Fig. 5 for an example of a degree 8 vertex. □

4   D8(P) is a Spanner

We will prove that $D8(P)$ is a spanner of $DT(P)$ with a spanning ratio of $(1 + \frac{2\pi}{3\sqrt{3}}) \approx 2.21$, thus making it a $(1 + \frac{2\pi}{3\sqrt{3}}) \cdot C_{DT}$-spanner of the complete geometric graph, where C_{DT} is the spanning ratio of the Delaunay triangulation. As of this writing, the current best bound of the spanning ratio of the Delaunay triangulation is 1.998 [4], which makes $D8(P)$ approximately a 4.42-spanner of the complete graph.

Suppose that (p, q) is in $DT(P)$ but not in $D8(P)$. We will show the existence of a short path between p and q in $D8(P)$. If the short path from p to q consists of the ideal situation of an edge (p, r) of E_A in the same cone of p as q, plus every canonical edge of p from r to q, then we have what we call the *ideal path*. We give a spanning ratio of the ideal path with respect to the *canonical triangle* T_{pq}. If $q \in C_i^p$, then T_{pq} is an equilateral triangle with vertex p, two edges on the boundary of C_i^p, with q on the third edge. Notice that in our construction, when adding canonical edges to E_{CAN} on an edge (p, r) of E_A, there are times where the first or last edges of $Can_i^{(p,r)}$ are not added to E_{CAN}. In these cases we prove the existence of alternate paths from p to q that still have the same spanning ratio. Finally we prove that the spanning ratio given in terms of the canonical triangle T_{pq} has an upper bound of $(1 + \theta/\sin\theta)|pq|$, where $\theta = \pi/3$ is the cone angle.

Ideal Paths. First we establish that a canonical subgraph is a path.

Lemma 8. *Let (p, r) be an edge in E_A in the cone C_i^p. Then $Can_i^{(p,r)}$ forms a path.*

This allows us to define the ideal path.

Definition 3. *Consider an edge (p, r) in C_i^p in E_A, and the graph $Can_i^{(p,r)}$. An ideal path is a simple path from p to any vertex in $Can_i^{(p,r)}$ using the edges of $(p, r) \cup Can_i^{(p,r)}$. See Fig. 6.*

We will prove that the length of the ideal path from p to q is not greater than $|pa| + \frac{\theta}{\sin\theta}|aq|$, where a is the corner of the canonical triangle T_{pq} to the side of (p, q) that has r, and $\theta = \pi/3$ is the cone angle.

We then use ideal paths to prove there exists a path with bounded spanning ratio between any two vertices p and q in $D8(P)$, where (p, q) is an edge in $DT(P)$. We prove a bound on the length of the path from p to q of $|pa| + \frac{\theta}{\sin\theta}|aq|$.

We note that the distance $|pa| + \frac{\theta}{\sin\theta}|aq|$ is with respect to the canonical triangle T_{pq} rather than the Euclidean distance $|pq|$. To finish the proof we show that $|pa| + \frac{\theta}{\sin\theta}|aq| \leq (1 + \frac{\theta}{\sin\theta})|pq|$.

One of the main parts of the proof comes from a paper by Bose and Keil [12], where they prove the Delaunay triangulation has a spanning ratio of ≈ 2.42 on a point set with constraints. The main lemma of that prove puts a bound on the length of the path from vertex r to q, given that there is a circle through r and

q that is empty of vertices below the line through r and q. If we take the circle $O_{(r,q)}$ that is empty of vertices below (r,q), then the path from r to q is not greater than the upper arc of $O_{(r,q)}$ from r to q. This lemma does not provide a construction, however Kanj and Perkovic [8], while working on a degree 14 spanning subgraph of the Delaunay triangulation, came up with a construction for this proof that removes the constraint that $O_{(r,q)}$ is empty below (r,q), and thus matches our ideal path.

A slightly modified version of the main lemma from the Bose and Keil [12] result is Lemma 9.

Lemma 9. *Consider the restricted neighbourhood $N_p^{(r,q)}$ in $DT(P)$ in the cone C_i^p. Let $\alpha = \angle(rpq) < \pi/3$. If no point of P lies in the triangle $\triangle(prq)$ then there is a path from r to q in $DT(P)$, using canonical edges of p, whose length satisfies:*

$$\delta(r,q) \le |rq| \frac{\alpha}{\sin \alpha}$$

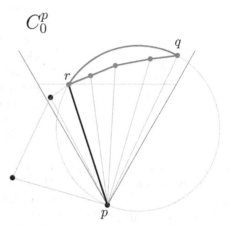

Fig. 6. The ideal path is the simple path from p to any vertex of $Can_0^{(p,r)}$ (seen in red) using the edges of $(p,r) \cup Can_0^{(p,r)}$ (Color figure online).

Our construction is slightly different from Kanj and Perkovic [8], in that we use bisector distance instead of Euclidean distance, but the construction and subsequent analysis is nonetheless largely the same. They are summed up in the following two lemmas:

Lemma 10. *Consider the restricted neighbourhood $N_p^{(r,q)}$ and without loss of generality let $N_p^{(r,q)}$ be in C_i^p. Let $\alpha = \angle(rpq)$. Let $r_q \ne p$ be the point where the line through p and r intersects the canonical triangle T_{pq}. Let $q_r \ne p$ be the point where the edge (p,q) intersects T_{pr}. If $[pr]$ is the shortest edge of all edges in $N_p^{(r,q)}$ with endpoint p, then the distance from r to q using the canonical edges of p in $N_p^{(r,q)}$ is at most $\max\{|rr_q|, |q_r q|\} + |r_q q|\frac{\theta}{\sin \theta}$. See Fig. 7.*

Using Lemma 10 we can prove the main lemma of this section:

Lemma 11. *Consider the edge (p,r) in E_A, located in Can_i^p, and the associated canonical subgraph $Can_i^{(p,r)}$. Without loss of generality, assume that $i = 0$. The length of the ideal path from p to any vertex q in $Can_0^{(p,r)}$ satisfies $\delta(p,q) \le |pa| + \frac{\theta}{\sin \theta}|aq|$, where a is the corner of T_{pq} such that $r \in \triangle(pqa)$, and $\theta = \pi/3$ is the angle of the cones.*

Proof. (Refer to Fig. 7.) By Lemma 10 the path from r to q is no greater than $\max\{|rr_q|, |q_r q|\} + |r_q q|\frac{\theta}{\sin \theta}$.

Since $|pr| + \max\{|rr_q|, |q_r q|\} \le |pa|$ and $|aq| \ge |r_q q|$ we have $\delta(p,q) \le |pr| + \max\{|rr_q|, |q_r q|\} + |r_q q|\frac{\theta}{\sin \theta} \le |pa| + |aq|\frac{\theta}{\sin \theta}$. $\square$

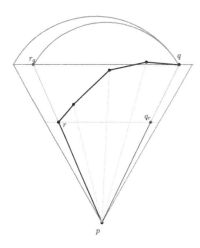

Fig. 7. Lemmas 10, 11.

A path in $D8(P)$ that approximates an edge (p, q) of $DT(P)$ may consist of the edge (p, q), or the ideal path from p to q, or the concatenation of two ideal paths from p to q, or some combination of the above. We prove that for any of these scenarios, the length of the path in $D8(P)$ from p to q, denoted $\delta(p, q)$, is not longer than $\max\{|pa| + \frac{\theta}{\sin\theta}|aq|, |pb| + \frac{\theta}{\sin\theta}|bq|\}$. Points a and b are the top left and right corners of canonical triangle T_{pq} respectively.

At this point we have a spanning ratio given in terms of T_{pq}. We use this result to prove that $D8(P)$ is a spanner with respect to the Euclidean distance $|pq|$.

We consider an edge $(p, q) \in DT(P)$. If $(p, q) \in D8(P)$ then the length of the path from p to q in $D8(P)$ is $|pq| \leq \left(1 + \frac{\theta}{\sin\theta}\right)|pq|$, as required.

Thus we assume $(p, q) \notin D8(P)$. Without loss of generality we assume q is in C_0^p. Since $(p, q) \notin D8(P)$, there is an edge (p, r) of E_A in Can_0^p or (q, u) in Can_3^q (or both), where $[pr] \leq [pq]$ and $[qu] \leq [pq]$, otherwise (p, q) would have been added to E_A in $AddIncident(L)$. Without loss of generality assume there is the edge $(p, r) \in E_A, [pr] \leq [pq]$, and that (p, q) is clockwise from (p, r) around p.

Let a be the upper left corner of T_{pq}, and b be the upper right corner. Let $\alpha = \angle(rpq)$ and $\theta = \pi/3$ be the angle of the cones.

Lemma 12. *Let $(p, r) \in E_A$, where $r \in C_i^p$. Then there is an ideal path from p to any vertex q in $Can_i^{(p,r)}$, where q is not an end vertex of Can_i^p.*

Proof. In the algorithm $AddCanonical(p, r)$, we add every canonical edge of p in $Can_i^{(p,r)}$ that is not the first or last edge. By Lemma 8, the edges of $Can_i^{(p,r)}$ form a path. Thus there is the ideal path from p to any vertex q in $Can_i^{(p,r)}$ that is not the first or last vertex. □

The next lemma establishes there is a cone that the last edge in $Can_i^{(p,r)}$ cannot be in.

Lemma 13. *Let z be the first or last vertex of $Can_i^{(p,r)}$, and assume that (p, z) is not in E_A. Let (y, z) be the first or last edge in $Can_i^{(p,r)}$. Then (y, z) is not in C_i^z.*

Let (p, r) be an edge in E_A in the graph $D8(P)$. Without loss of generality, assume that r is in C_0^p. We now turn our attention to the first or last vertex in $Can_0^{(p,r)}$. Because the cases are symmetric, we focus on the last vertex, which we designate z. If $z = r$, the path from p to z is trivial, thus we assume $z \neq r$. Let y be the neighbour of z in $Can_0^{(p,r)}$. By Lemma 13, (y, z) cannot be in C_0^z. Thus (y, z) can be in C_5^z, C_4^z, or C_3^z.

Case 1: Edge (y, z) is in C_5^z. Then (y, z) was added to E_{CAN} in $AddCanonical(p, r)$, Step 4a, and there is an ideal path from p to z.

Case 2: Edge (y, z) is in C_4^z. There are three possibilities.

    (a) If (y, z) is an edge of E_A, then there is an ideal path from p to z.

    (b) If there is no edge in E_A with endpoint z in C_4^z, then (y, z) was added to E_{CAN} in $AddCanonical(p, r)$, Step 4b, and there is an ideal path from p to z.

    (c) If there is an edge of E_A in C_4^z with endpoint z that is not (y, z), then we have added the canonical edge of z in C_4^z with endpoint y to E_{CAN} in $AddCanonical(p, r)$, Step 4c. Therefore by Lemma 12 there is an ideal path from z to y, and also an ideal path from p to y.

Case 3: Edge (y, z) is in C_3^z. Then (y, z) was not added to E_{CAN}.

In Case 1, Case 2a, and Case 2b there is an ideal path from p to q. Thus Lemma 11 tells us there is a path from p to q not longer than $|pa| + \frac{\theta}{\sin \theta} |aq|$.

In Case 2c, we have two ideal paths that meet at y, one starting at p and one starting at z. As in the case of a single ideal path, the sum of the lengths of these two paths is not more than $|pa| + \frac{\theta}{\sin \theta} |aq|$. The following lemma proves this claim:

Lemma 14. *Consider the edge (p, r) in E_A in the graph $D8(P)$, r in C_0^p. Let (y, z) be the last edge in $Can_0^{(p,r)}$, and let (y, z) be in C_4^z. Let (z, u) be an edge in E_A in C_4^z. Assume there is an ideal path from p to y in C_0^p, and an ideal path from z to y in C_4^z. Let a be the top left corner of T_{pz}. We prove an upper bound on the length $\delta(p, z)$ of $|pa| + \frac{\theta}{\sin \theta} |az|$.*

In Case 3 there is no edge from y to z. We prove the length of the path from p to z in Case 3 by induction, as part of the main lemma of this section:

Lemma 15. *Consider the edge (p, r) in E_A in the graph $D8(P)$. Without loss of generality, let r be in C_0^p. Let a and b be the top left corner and top right corner respectively of T_{pq}. For any edge $(p, q) \in DT(P)$, there exists a path from p to q in $D8(P)$ that is not longer than $\max\{|pa| + \frac{\theta}{\sin \theta} |aq|, |pb| + \frac{\theta}{\sin \theta} |bq|\}$.*

For an edge (p, q) in $DT(P)$, we have a bound on the length of the path in $D8(P)$. However, this bound is terms of the size of the canonical triangle T_{pq}, which is not the same as the Euclidean distance $|pq|$. We prove that $\max\{|pa| + \frac{\theta}{\sin \theta} |aq|, |pb| + \frac{\theta}{\sin \theta} |bq|\} \leq \left(1 + \frac{\theta}{\sin \theta}\right) |pq|$. However, due to space constraints, the proof has been omitted. Using this inequality and Lemma 15, the main theorem now follows:

Theorem 2. *For any edge $(p, q) \in DT(P)$, there is a path in $D8(P)$ from p to q with length at most $\left(1 + \frac{\theta}{\sin \theta}\right) |pq|$, where $\theta = \pi/3$ is the cone width. Thus $D8(P)$ is a $(1 + \frac{\theta}{\sin \theta}) D_T$-spanner of the complete graph, where D_T is the spanning ratio of the Delaunay triangulation (currently 1.998 [4]).*

References

1. Chew, P.: There is a planar graph almost as good as the complete graph. In: Proceedings of the Second Annual Symposium on Computational Geometry, SCG 1986, pp.169–177. ACM, New York (1986)
2. Dobkin, D., Friedman, S., Supowit, K.: Delaunay graphs are almost as good as complete graphs. Discrete & Comput. Geom. **5**, 399–407 (1990)
3. Keil, J., Gutwin, C.: Classes of graphs which approximate the complete euclidean graph. Discrete & Comput. Geom. **7**, 13–28 (1992)
4. Xia, G.: Improved upper bound on the stretch factor of Delaunay triangulations. In: Proceedings of the Twenty-Seventh Annual Symposium on Computational Geometry, SoCG 2011, pp. 264–273. ACM, New York (2011)
5. Bose, P., Gudmundsson, J., Smid, M.: Constructing plane spanners of bounded degree and low weight. In: Möhring, R.H., Raman, R. (eds.) ESA 2002. LNCS, vol. 2461, pp. 234–246. Springer, Heidelberg (2002)
6. Li, X.Y., Wang, Y.: Efficient construction of low weight bounded degree planar spanner. In: Warnow, T., Zhu, B. (eds.) Comput. Comb. Lecture Notes in Computer Science, vol. 2697, pp. 374–384. Springer, Berlin Heidelberg (2003)
7. Bose, P., Smid, M.H.M., Xu, D.: Delaunay and diamond triangulations contain spanners of bounded degree. Int. J. Comput. Geom. Appl. **19**, 119–140 (2009)
8. Kanj, I.A., Perković, L., Xia, G.: On spanners and lightweight spanners of geometric graphs. SIAM J. Comput. **39**, 2132–2161 (2010)
9. Bose, P., Carmi, P., Chaitman-Yerushalmi, L.: On bounded degree plane strong geometric spanners. J. Discrete Algorithms **15**, 16–31 (2012)
10. Bonichon, N., Gavoille, C., Hanusse, N., Perković, L.: Plane spanners of maximum degree six. In: Abramsky, S., Gavoille, C., Kirchner, C., Meyer auf der Heide, F., Spirakis, P.G. (eds.) ICALP 2010. LNCS, vol. 6198, pp. 19–30. Springer, Heidelberg (2010)
11. Bonichon, N., Kanj, I., Perković, L., Xia, G.: There are plane spanners of degree 4 and moderate stretch factor. Discrete & Comput. Geom. **53**, 514–546 (2015)
12. Bose, P., Keil, J.M.: On the stretch factor of the constrained Delaunay triangulation. In: 3rd International Symposium on Voronoi Diagrams in Science and Engineering, ISVD 2006, Banff, Alberta, Canada, pp. 25–31, 2–5 July 2006. IEEE Computer Society (2006)
13. Benson, R.: Euclidean Geometry and Convexity. McGraw-Hill, New York (1966)

Constructing Consistent Digital Line Segments

Iffat Chowdhury$^{(\boxtimes)}$ and Matt Gibson

Department of Computer Science, University of Texas at San Antonio,
San Antonio, TX, USA
iffat.chowdhury@utsa.edu, gibson@cs.utsa.edu

Abstract. Our concern is the digitalization of line segments in the unit grid as considered by Chun et al. [Discrete Comput. Geom., 2009], Christ et al. [Discrete Comput. Geom., 2012], and Chowdhury and Gibson [ESA, 2015]. In this setting, digital segments are defined so that they satisfy a set of axioms also satisfied by Euclidean line segments. The key property that differentiates this research from other research in digital line segments is that the intersection of any two segments must be connected. A system of digital line segments that satisfies these desired axioms is called a consistent digital line segments system (CDS). Our main contribution of this paper is to show that any collection of digital segments that satisfy the CDS properties in a finite $n \times n$ grid graph can be extended to a full CDS (with a segment for every pair of grid points). Moreover, we show that this extension can be computed with a polynomial-time algorithm. The algorithm is such that one can manually define the segments for some subset of the grid. For example, suppose one wants to precisely define the boundary of a digital polygon. Then we would only be interested in CDSes such that the digital line segments connecting the vertices of the polygon fit this desired boundary definition. Our algorithm allows one to manually specify the definitions of these desired segments. For any such definition that satisfies all CDS properties, our algorithm will return in polynomial time a CDS that "fits" with these manually chosen segments.

1 Introduction

This paper explores families of digital line segments as considered by Chun et al. [3], Christ et al. [2], and Chowdhury and Gibson [1]. Consider the unit grid $\mathbb{Z}^2$, and in particular the unit grid graph: for any two points $p = (p^x, p^y)$ and $q = (q^x, q^y)$ in $\mathbb{Z}^2$, p and q are neighbors if and only if $|p^x - q^x| + |p^y - q^y| = 1$. For any pair of grid vertices p and q, we'd like to define a digital line segment $R_p(q)$ from p to q. The collection of digital segments must satisfy the following five properties.

(S1) *Grid path property:* For all $p, q \in \mathbb{Z}^2$, $R_p(q)$ is the points of a path from p to q in the grid topology.

(S2) *Symmetry property:* For all $p, q \in \mathbb{Z}^2$, we have $R_p(q) = R_q(p)$.

(S3) *Subsegment property:* For all $p, q \in \mathbb{Z}^2$ and every $r, s \in R_p(q)$, we have $R_r(s) \subseteq R_p(q)$.

© Springer-Verlag Berlin Heidelberg 2016
E. Kranakis et al. (Eds.): LATIN 2016, LNCS 9644, pp. 263–274, 2016.
DOI: 10.1007/978-3-662-49529-2_20

Properties $(S2)$ and $(S3)$ are quite natural to ask for; the subsegment property $(S3)$ is motivated by the fact that the intersection of any two Euclidean line segments is connected. See Fig. 1(a) for an illustration of a violation of (S3). Note that a simple "rounding" scheme of a Euclidean segment commonly used in computer vision produces a good digitalization in isolation, but unfortunately it will not satisfy (S3) when combined with other digital segments, see Fig. 1(b) and (c).

$(S4)$ *Prolongation property:* For all $p, q \in \mathbb{Z}^2$, there exists $r \in \mathbb{Z}^2$, such that $r \notin R_p(q)$ and $R_p(q) \subseteq R_p(r)$.

The prolongation property $(S4)$ is also a quite natural property to desire with respect to Euclidean line segments. Any Euclidean line segment can be extended to an infinite line, and we would like a similar property to hold for our digital line segments. While (S1)–(S4) form a natural set of axioms for digital segments, there are pathological examples of segments that satisfy these properties which we would like to rule out. For example, Christ et al. [2] describe a CDS where a double spiral is centered at some point in $\mathbb{Z}^2$, traversing all points of $\mathbb{Z}^2$. A CDS is obtained by defining $R_p(q)$ to be the subsegment of this spiral connecting p and q. To rule out these CDSes, the following property was added.

$(S5)$ *Monotonicity property:* For all $p, q \in \mathbb{Z}^2$, if $p^x = q^x = c_1$ for any c_1 (resp. $p^y = q^y = c_2$ for any c_2), then every point $r \in R_p(q)$ has $r^x = c_1$ (resp. $r^y = c_2$).

If a system of digital line segments satisfies the axioms $(S1) - (S5)$, then it is called a *consistent digital line segments system* (CDS).

Previous Works. Unknown to Chun et al. and Christ et al. when publishing their papers, Luby [8] considers *grid geometries* which are equivalent to systems of digital line segments satisfying (S1), (S2), (S5) described in this paper. Chun et al. [3] give an $\Omega(\log n)$ lower bound on the Hausdorff Distance of a CDS where n is the number of points in the segment, and the result even applies to *consistent digital rays* or CDRs. A CDR for any point $p \in \mathbb{Z}^2$ is the set of segments $R_p(q)$

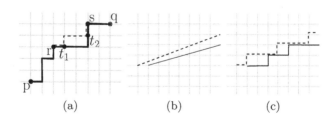

(a)   (b)   (c)

Fig. 1. (a) An illustration of the violation of (S3). The solid segment is $R_p(q)$, and the dashed segment is $R_r(s)$. (b) The dashed line and the solid line denote two different Euclidean line segments. (c) The corresponding digital line segments via a rounding approach.

for every $q \in \mathbb{Z}^2$ (i.e., all segments with p as one of the endpoints). Note that this lower bound is due to property (S3), as it is easy to see that if the requirement of (S3) is removed then digital segments with $O(1)$ Hausdorff distance are easily obtained, for example the trivial "rounding" scheme used in Fig. 1(c). Chun et al. give a construction of CDRs that satisfy the desired properties (S1)–(S5) with a tight upper bound of $O(\log n)$ on the Hausdorff distance. Christ et al. [2] extend the result to get an optimal $O(\log n)$ upper bound on Hausdorff distance for a CDS in $\mathbb{Z}^2$.

After giving the optimal CDS in $\mathbb{Z}^2$, Christ et al. [2] investigate common patterns in CDSes in an effort to obtain a characterization of CDSes. As a starting point, they are able to give a characterization of CDRs. In their effort to give a characterization, they proved a sufficient condition on the construction of the CDSes but then they give an example of a CDS that demonstrates that their sufficient condition is not necessary. They ask if there are any other interesting examples of CDSes that do not follow their sufficient condition and left open the question on how to characterize the CDSes in $\mathbb{Z}^2$. Recently, Chowdhury and Gibson [1] give this characterization by giving a set of necessary and sufficient conditions that CDSes must satisfy. Both the characterization of CDRs given in [2] and the characterization of CDS given in [1] will be considered in more detail in Sect. 2.

Our Contributions. Our main contribution of this paper is to show that any collection of digital segments that satisfy the CDS properties in a finite $n \times n$ grid graph can be extended to a to a full CDS (with a segment for every pair of grid points). Moreover, we show that this extension can be computed with a polynomial-time algorithm. The algorithm is such that one can manually define the segments for some subset of the grid. For example, suppose one wants to precisely define the boundary of a digital polygon. Then we would only be interested in CDSes such that the digital line segments connecting the vertices of the polygon fit this desired boundary definition. Also perhaps one might want to define a collection of digital line segments so that the digital arrangement of these line segments is precisely defined. Our algorithm allows one to manually specify the definitions of these desired segments. For any such definition that satisfies all CDS properties, our algorithm will return in polynomial time a CDS that "fits" with these manually chosen segments. Before giving this algorithm, we give an alternative characterization of CDS that more naturally lends itself to a polynomial-time construction algorithm. In our final result, we consider the question of Christ et al. [2] regarding the existence of "interesting" CDSes. We use our polynomial-time construction algorithm to construct a CDS in a finite grid graph that has some interesting properties that did not appear in any previously-given CDS.

Motivation and Related Work. Digital geometry plays a fundamental and substantial role in many computer vision applications, for example image segmentation, image processing, facial recognition, fingerprint recognition, and some medical applications. One of the key challenges in digital geometry is to represent Euclidean objects in a digital space so that the digital objects have a similar visual appearance as their Euclidean counterparts. Representation of Euclidean objects in a digital space has been a focus in research for over 25 years, see for example [5,7]. Digital line segments are particularly important to model accurately, as other digital objects depend on them for their own definitions (e.g. convex and star-shaped objects). In 1986, Greene and Yao [7] gave an interface between continuous domain of Euclidean line segments and discrete domain of digital line segments. Goodrich et al. [6] focused on rounding the Euclidean geometric objects to a specific resolution for better computer representation. They gave an efficient algorithm for $\mathbb{R}^2$ and $\mathbb{R}^3$ in the "snap rounding paradigm" where the endpoints or the intersection points of several different line segments are the main concerns.

2 Preliminaries

A characterization of CDRs. Before we describe our results, we first need to give some details of the Christ et al. characterization of CDRs [2]. For any point $p \in \mathbb{Z}^2$, let $Q_p^1, Q_p^2, Q_p^3, Q_p^4$ denote the first, second, third, and fourth quadrants of p respectively. Christ et al. show how to construct $R_p(q)$ for $q \in Q_p^1$ from any total order of $\mathbb{Z}$, which we denote $\prec_p^1$. We describe $R_p(q)$ by "walking" from p to q. Starting from p, the segment will move either "up" or "right" until it reaches q. Suppose on the walk we are currently at a point $r = (r^x, r^y)$. Then it needs to move to either $(r^x + 1, r^y)$ or $(r^x, r^y + 1)$. Either way, the sum of the two coordinates of the current point is increased by 1 in each step. To move from p to q, the segment will move up $q^y - p^y$ times and will move right $q^x - p^x$ times. If the line segment is at a point r for which $r^x + r^y$ is among the $q^y - p^y$ greatest values in the interval $I(p, q) := [p^x + p^y, q^x + q^y - 1]$ according to $\prec_p^1$, the line segment will move up. Otherwise, it will move right. See Fig. 2 for an example.

Property (S3) is generally the most difficult property to deal with, and we will argue that the segments $R_p(q)$ and $R_p(q')$ will not violate (S3) for any points q and q' in the first quadrant of p. As shown in [2], (S3) is violated if and only if two segments intersect at a point t_1, one segment moves vertically from t_1 while the other moves horizontally from t_1, and the segments later intersect again. Consider two digital segments that "break apart" at some point t_1

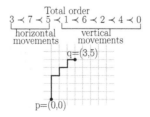

Fig. 2. The digital line segment between $p = (0,0)$ and $q = (3,5)$. According to $\prec_p$, the $q^x - p^x = 3$ smallest integers in $[0,7]$ correspond to the horizontal movements, and the $q^y - p^y = 5$ largest integers in $[0,7]$ correspond to the vertical movements.

in this manner, and suppose they do intersect again. Let t_2 be the first point at which they intersect after "splitting apart". Then we say that (t_1, t_2) is *witness to the violation of (S3)* or a *witness* for short. Therefore, one can show that any two segments satisfy (S3) by showing that they do not have witnesses, and this is how we will prove the segments satisfy (S3) now (and also in our characterization). Consider the segments $R_p(q)$ and $R_p(q')$ generated according to the Christ et al. definition, and suppose for the sake of contradiction that they have a witness (t_1, t_2) as in Fig. 1(a). One segment moves up at point t_1 and moves right into the point t_2 which implies $t_2^x + t_2^y - 1 \prec_p^1 t_1^x + t_1^y$, and the other segment moves right at point t_1 and moves up into the point t_2 which implies $t_1^x + t_1^y \prec_p^1 t_2^x + t_2^y - 1$, a contradiction. Therefore $R_p(q)$ and $R_p(q')$ do not have any witnesses and therefore satisfy (S3). Christ et al. [2] show that digital segments in quadrants Q_p^2, Q_p^3, and Q_p^4 can also be generated with total orders $\prec_p^2$, $\prec_p^3$, and $\prec_p^4$, and moreover they establish a one-to-one correspondence between CDRs and total orders. That is, (1) given any total order of $\mathbb{Z}$, one can generate all digital rays in any quadrant of p, and (2) for any set of digital rays R in some quadrant of p, there is a total order that will generate R. This provides a characterization of CDRs.

A Characterization of CDSes. In a full CDS, we have segments connecting every pair of points of $\mathbb{Z}^2$. Note that for any point p, the segments that have p as an endpoint can be viewed as a system of CDRs, and therefore they can be generated by a total order according to the CDR characterization described above. The question is now to determine the properties that the total orders must satisfy so that a CDS is obtained when each point generates its adjacent segments by its total order. Chowdhury and Gibson [1] give a set of necessary and sufficient conditions that the total orders must satisfy to obtain a CDS in $\mathbb{Z}^2$.

To understand the characterization, we need some preliminaries. It is known [1,2] that any monotone segment with non-negative slope and a monotone segment with negative slope will always have a connected intersection. Therefore it suffices to only consider segments with non-negative slope. Moreover, it suffices to only consider segments $R_p(q)$ where $q \in Q_p^1$ (i.e. "first quadrant segments"); given $R_p(q)$, the "third quadrant segment" $R_q(p)$ can simply be obtained by following $R_p(q)$ "backwards". Therefore throughout the rest of this paper, we will assume without loss of generality that our concern is first quadrant segments and from now on, we refer to a quadrant of a point p as Q_p and total order as $\prec_p$.

Consider two such segments $R_{p_1}(p_3)$ and $R_{p_2}(p_3)$ for a point $p_3 \in Q_{p_1} \cap Q_{p_2}$. A key definition in the characterization of [1] is the *layout view* of the integers in the intervals $I(p_1, p_3)$ and $I(p_2, p_3)$ that are used to define the segments $R_{p_1}(p_3)$ and $R_{p_2}(p_3)$ respectively. Without loss of generality assume $p_1^x \leq p_2^x$. The intervals are written in increasing order according to their total orders in a matrix with two rows with $I(p_1, p_3)$ in the top row and $I(p_2, p_3)$ in the bottom row. The first element of $I(p_2, p_3)$ is "shifted" to the right $(p_2^x - p_1^x)$ positions after the first element of $I(p_1, p_3)$. Note that the integers in $I(p_1, p_3)$ and $I(p_2, p_3)$ are determined by the natural total order on the integers, but then are sorted by the total orders $\prec_{p_1}$ and $\prec_{p_2}$ respectively. The advantage of the layout view is

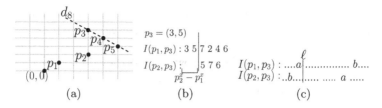

(a)   (b)   (c)

Fig. 3. An Illustration of layout view and bad pairs. (a) The points in the grid. (b) The layout view of the intervals with $p_1 = (1,1)$ and $p_2 = (3,2)$. The dividing line shown splits the intervals into the horizontal and vertical movements for $R_{p_1}(p_3)$ and $R_{p_2}(p_3)$ respectively. (c) An illustration of a bad pair.

that a single vertical line can break both of the intervals into the horizontal movements portion and the vertical movements portion. Such a line is called a *dividing line*. See Fig. 3(a) and (b). Now, let a and b be two integers in $I(p_1, p_3) \cap I(p_2, p_3)$, which are in layout view. Suppose there exists some dividing line ℓ such that in $I(p_1, p_3)$ a is on the left side of ℓ and b is on the right side of ℓ, and simultaneously in $I(p_2, p_3)$ b is on the left side of ℓ and a is on the right side of ℓ. Then $\{a, b\}$ is called a *bad pair* and ℓ *splits* the bad pair. See Fig. 3(c). Total orders $\prec_{p_1}$ and $\prec_{p_2}$ are said to not have a bad pair if for every point $p_3 \in Q_{p_1} \cap Q_{p_2}$, the intervals $I(p_1, p_3)$ and $I(p_2, p_3)$ do not have a bad pair. They complete the characterization of CDSes in $\mathbb{Z}^2$ by proving that a system of non-negative sloped line segments in $\mathbb{Z}^2$ is a CDS if and only if every pair of total orders does not have a bad pair.

3   Constructing Segments

An Alternative Characterization. In this section, we provide an alternative characterization of CDSes for a finite $\{0, 1, \ldots, n\} \times \{0, 1, \ldots, n\}$ grid that more naturally lends itself to a polynomial-time algorithm than the characterization of [1]. This characterization will use the following ordering $\mathcal{O}$ of the points in the grid. If for points p and q we have $p^x + p^y < q^x + q^y$ then p comes before q in $\mathcal{O}$. If $p^x + p^y = q^x + q^y$, then the point that has smaller x-coordinate will come first in $\mathcal{O}$.

Note that because the total orders are now finite, we can consider total orders in layout view (in $\mathbb{Z}^2$ we could only use intervals in layout view) as there is now a well-defined smallest element in a total order in this finite setting which is not the case when the total orders are infinite. Consider two points p_i and p_j; we will now define an indexing of the elements in their corresponding total orders in layout view. Let M_j denote the number of positions in $\prec_{p_j}$, and let q^* denote the point (n, n). We index the positions of the elements in $\prec_{p_j}$ from 1 to M_j. We index the positions of $\prec_{p_i}$ relative to the indexing of $\prec_{p_j}$. See Fig. 4(a). Note that if p_i comes before p_j in $\mathcal{O}$, we have $I(p_j, q^*) \subseteq I(p_i, q^*)$. For any point p, we let $\prec_p (k)$ denote the element in $\prec_p$ at index k, we let $\prec_p^{\leftarrow} (k)$ be the set of elements in $\prec_p$ which have index at most k, and we let $\prec_p^{\rightarrow} (k)$ be the set

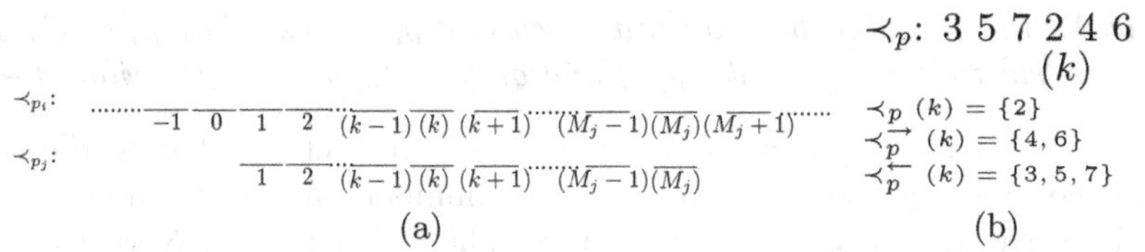

$$\prec_p\colon 3\ 5\ 7\ 2\ 4\ 6$$
$$(k)$$

$\prec_{p_i}\colon$

$$\cdots\cdots \overline{-1}\ \overline{0}\ \overline{1}\ \overline{2}\ \overline{(k-1)}\ \overline{(k)}\ \overline{(k+1)}\ \cdots\overline{(M_j-1)}\overline{(M_j)}\overline{(M_j+1)}\ \cdots\cdots$$

$\prec_{p_j}\colon$

$$\overline{1}\ \overline{2}\ \overline{(k-1)}\ \overline{(k)}\ \overline{(k+1)}\ \cdots\overline{(M_j-1)}\overline{(M_j)}$$

$\prec_p (k) = \{2\}$
$\prec_p^{\rightarrow} (k) = \{4,6\}$
$\prec_p^{\leftarrow} (k) = \{3,5,7\}$

(a)   (b)

Fig. 4. (a) The indexing of $\prec_{p_j}$ and $\prec_{p_i}$. (b) Illustration of $\prec_p (k)$, $\prec_p^{\leftarrow} (k)$, and $\prec_p^{\rightarrow} (k)$.

of elements that are in $\prec_p$ which have index at least k. See Fig. 4(b). Now, we define an important concept of the characterization.

Consider a vertical line drawn in the layout between indices k and $k + 1$. Here, we allow k to be the rightmost index (i.e., the line is just to the right of the last element) or $k + 1$ to be the leftmost index (i.e., the line is just to the left of the first element). We call this line a *contracting line* with respect to p_i if $|\prec_{p_i}^{\leftarrow} (k) \cap I(p_j, q^*)| = k$. We use α_{ij} to denote a contracting line for p_j with respect to p_i. Note that if α_{ij} is a contracting line between indices k and $k + 1$ then $|\prec_{p_i}^{\rightarrow} (k+1) \cap I(p_j, q^*)| = M_j - k$. See Fig. 5(a) and (b) for an illustration. Note there is always at least one α_{ij}, as we always have $|\prec_{p_i}^{\leftarrow} (0) \cap I(p_j, q^*)| \geq 0$, $|\prec_{p_i}^{\leftarrow} (M_j) \cap I(p_j, q^*)| \leq M_j$, and the value changes by at most one when the line is shifted one position. This implies that there must be some k where the condition for a contracting line holds.

Suppose α_{ij} is a contracting line (if there are multiple α_{ij}, we choose any one arbitrarily) between indices k and $k + 1$. Now, consider some element $a \in \prec_{p_i}^{\leftarrow} (k) \cap I(p_j, q^*)$. Let $k' \leq k$ denote the index of a in $\prec_{p_i}$, and suppose a is at index k'' in $\prec_{p_j}$. We say that k'' is *valid* if $k' \leq k'' \leq k$. Intuitively, it is valid if a only gets closer to α_{ij} without crossing over it. Similarly, if $k' \geq k + 1$ is the index of $b \in \prec_{p_i}^{\leftarrow} (k+1) \cap I(p_j, q^*)$, then k'' is a valid index for b in $\prec_{p_j}$ if $k+1 \leq k'' \leq k'$. See Fig. 5(c) and (d) where the gray colored boxes are the possible spots for 5 in $\prec_{p_j}$. If every element in $I(p_j, q^*)$ is in a valid index, then we say that $\prec_{p_j}$ is a *contraction* of $\prec_{p_i}$. If $\prec_{p_j}$ is a contraction of $\prec_{p_i}$ then we say that $\prec_{p_i}$ is an *expansion* of $\prec_{p_j}$. We have the following lemma.

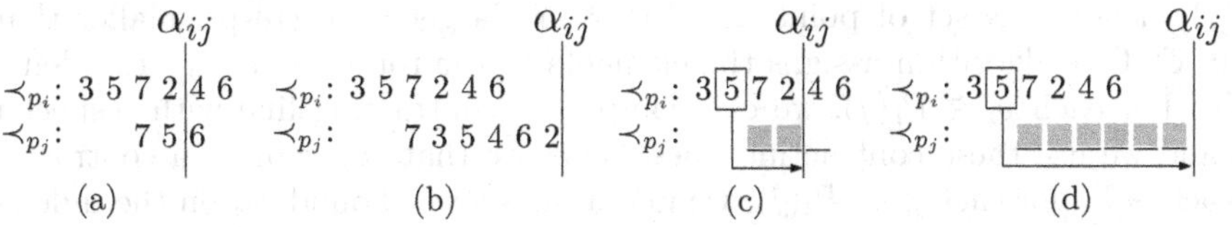

α_{ij}	α_{ij}	α_{ij}	α_{ij}

$\prec_{p_i}\colon 3\ 5\ 7\ 2\ |4\ 6$ $\prec_{p_i}\colon 3\ 5\ 7\ 2\ 4\ 6$ $\prec_{p_i}\colon 3\ |5|\ 7\ 2\ |4\ 6$ $\prec_{p_i}\colon 3\ |5|\ 7\ 2\ 4\ 6$

$\prec_{p_j}\colon \quad\ \ 7\ 5\ |6$ $\prec_{p_j}\colon \quad\ \ 7\ 3\ 5\ 4\ 6\ 2$ $\prec_{p_j}\colon$ $\prec_{p_j}\colon$

(a)   (b)   (c)   (d)

Fig. 5. Suppose $q^* = (4, 4)$. (a) α_{ij} is a contracting line for $p_i = (1,1)$ and $p_j = (3,2)$. (b) α_{ij} is a contracting line for $p_i = (1,1)$ and $p_j = (3,-1)$. (c) p_i is at $(1,1)$ and p_j is at $(3,2)$. (d) p_i is at $(1,1)$ and p_j is at $(3,-1)$.

Lemma 1. *Let p_i and p_j be two points such that p_i comes before p_j in $\mathcal{O}$. Then there is no bad pair in $\prec_{p_i}$ and $\prec_{p_j}$ if and only if $\prec_{p_j}$ is a contraction of $\prec_{p_i}$.*

Proof. Assume that $\prec_{p_j}$ is a contraction of $\prec_{p_i}$ and let α_{ij} denote the corresponding contracting line. Let a and b be two numbers in both total orders such that a is to the left of α_{ij} and b is to the right of α_{ij} in $\prec_{p_i}$. Now, to have a bad pair $\{a, b\}$, we should have that b is less than a in $\prec_{p_j}$. Then at least one of them will cross α_{ij} in $\prec_{p_j}$ because α_{ij} is between a and b. But a and b cannot cross α_{ij} because $\prec_{p_j}$ is a contraction of $\prec_{p_i}$. So, there is no bad pair $\{a, b\}$. Now assume that a and b both are on the same side of α_{ij}. Without loss of generality, assume that a and b are to the right of α_{ij} and a is closer to α_{ij} than b in $\prec_{p_i}$. Consider any dividing line ℓ between a and b in $\prec_{p_i}$ (note a is to the left of ℓ and b is to the right of ℓ). As a can only get closer to α_{ij}, a will remain on the left side of ℓ in $\prec_{p_j}$ (with b is on the right side). Therefore, we do not have a bad pair $\{a, b\}$.

Suppose, $\prec_{p_i}$ and $\prec_{p_j}$ have a bad pair $\{a, b\}$, and let ℓ be the dividing line that splits them. We will show that $\prec_{p_j}$ is not a contraction of $\prec_{p_i}$. Without loss of generality, assume that a is less than b in $\prec_{p_i}$ and b is less than a in $\prec_{p_j}$. Now, suppose there is a contracting line α_{ij} to the left of a in $\prec_{p_i}$. Then a is not in a valid position in $\prec_{p_j}$ as a moves away from α_{ij}. Similarly, b is not in a valid position if α_{ij} is to the right of b in $\prec_{p_i}$. If α_{ij} is between a and b in $\prec_{p_i}$, then clearly at least one of a or b must have crossed over α_{ij}. We conclude that $\prec_{p_j}$ is not a contraction of $\prec_{p_i}$.   □

The following Theorem 1 immediately follows from Lemma 1 and [1].

Theorem 1. *A system of digital line segments is a CDS if and only if the segments can be obtained from total orders such that each total order $\prec_{p_j}$ is a contraction of $\prec_{p_i}$ for all p_i that come before p_j in $\mathcal{O}$.*

The Algorithm. We will now show that a partial CDS can be extended to a full CDS with a polynomial-time algorithm. We are given a subset P of the grid such that each point in P has been assigned total orders that satisfy the necessary and sufficient conditions for being a CDS. We assume $P \neq \emptyset$. If no point has had its total order previously defined, then we can arbitrarily choose a point and set its total order to be whatever we wish. We let N denote the number of points in the grid. When defining the total order $\prec_{p_j}$, let $P_1(j)$ (resp. $P_2(j)$) denote the set of points in P that are larger than (resp. smaller than) p_j in $\mathcal{O}$. Our algorithm assigns the elements to the total order $\prec_{p_j}$ from left to right. For each $p_i \in P_1(j)$, we can compute a contracting line with respect to p_j, and we use these contracting lines to ensure that $\prec_{p_j}$ will be a contraction of each $\prec_{p_i}$. For each $p_l \in P_2(j)$, we maintain a lower bound β_{lj} on the index of a contracting line with respect to p_j. See Algorithm 1. We show if we construct a total order $\prec_{p_j}$ using Algorithm 1, then $\prec_{p_j}$ will be a contraction of $\prec_{p_i}$ for each $p_i \in P_1(j)$ and $\prec_{p_j}$ will be an expansion of $\prec_{p_l}$ for all $p_l \in P_2(j)$.

In the for loop at step 9, we compute a set L_i for each point $p_i \in P_1(j)$. Intuitively, this set L_i is the set of all integers that we can assign to $\prec_{p_j}(k)$ without

Algorithm 1. Construction of Total Orders

1: **for all** points $p_j \notin P$ **do**
2:    Let $P_1(j)$ be the subset of P that is larger than p_j in $\mathcal{O}$.
3:    Let $P_2(j)$ be the subset of P that is smaller than p_j in $\mathcal{O}$.
4:    Find contracting lines α_{ij} for each of the $\prec_{p_i}$ for $\prec_{p_j}$ such that $p_i \in P_1(j)$.
5:    For each $p_l \in P_2(j)$, set β_{lj} to be the index of the first position of $\prec_{p_l}$.
6:    Let D be the set of all integers in $I(p_j, q^*)$.
7:    **for all** $k = 1$ to M_j **do**

8:       If $P_1(j) = \emptyset$, let $L = D$ and go to step 18.
9:       **for all** $p_i \in P_1(j)$ **do**
10:          **if** index k is to the left of α_{ij} **then**
11:            $L_i \leftarrow D \cap \prec_{p_i}^{\leftarrow}(k)$
12:          **if** index k is to the right of α_{ij} **then**
13:            **if** $\prec_{p_i}(k) \in D$ **then**
14:               $L_i \leftarrow \{\prec_{p_i}(k)\}$
15:            **else**
16:               $L_i \leftarrow \prec_{p_i}^{\rightarrow}(k)$
17:       $L \leftarrow \cap_{i:p_i \in P_1(j)} L_i$.

18:       **for all** $p_l \in P_2(j)$ **do**
19:          **if** index k is to the left of β_{lj} **then**
20:            **if** $\prec_{p_l}(k)$ exists and $\prec_{p_l}(k) \in D$ **then**
21:               Remove every number from L not equal to $\prec_{p_l}(k)$.
22:            **else**
23:               Remove from L every number α that doesn't pass the count test.

24:       Let a be an arbitrary value in L, set $\prec_{p_j}(k) \leftarrow a$, and remove a from D.

25:       **for all** $p_l \in P_2(j)$ **do**
26:          **if** a is in $\prec_{p_l}$ and $k \leq \beta_{lj}$ **then**
27:            Let x denote the index of a in $\prec_{p_l}$.
28:            **if** $x > \beta_{lj}$ **then**
29:               $\beta_{lj} \leftarrow x + 1$
30:   $P \leftarrow P \cup \{p_j\}$

---

violating the contracting line α_{ij}. Then L is the set of integers that satisfies contracting lines for all $p_i \in P_1(j)$. In the for loop at step 18, we remove from L all integers whose placement would violate the contraction constraint with respect to our current bound β_{lj} on a contracting line for p_j and p_l. Therefore after this for loop, L is the set of all integers that can be assigned to $\prec_{p_j}(k)$ and will satisfy the conditions for all points in P. We arbitrarily choose an element from L and assign it to position k in the total order $\prec_{p_j}$. In the loop in step 25, we update the bounds on the contracting lines for the points in $P_2(j)$ if necessary. If we assign an integer a whose index x in $\prec_{p_l}$ is to the right of our current bound to a index k which is to the left of our bound, then any contracting line for p_l and p_j must be to the right of x. This follows because $\prec_{p_l}$ must be a contraction of $\prec_{p_j}$, and the integer a is sliding from position k in $\prec_{p_j}$ to some position x in $\prec_{p_l}$. It must slide closer to the contracting line without crossing it, and therefore the contracting line must be at position $x + 1$ or greater.

In step 23, we refer to a procedure called the *count test* which we will now define. In this scenario, we are considering a point $p_l \in P_2(j)$ such that position k is to the left of our current bound β_{lj}. Moreover, either k is not a position in $\prec_{p_l}$ or the element in this position was previously placed. See Fig. 6. Consider an element $\lambda \in L$ that is in $\prec_{p_l}$. Since λ is in L, it has not yet been placed and therefore must have an index in $\prec_{p_l}$ of $k + a$ for some $a > 1$. Now consider every point in P that has λ at an index greater than k in its total order. Let $D(k, k + a)$ denote the set of all integers in D with index at least k and at most $k + a$ in these points' total orders (including λ). We say that λ passes the count test if and only if $|D(k, k + a)| \le a$. Intuitively, if we were to place λ at index k of $\prec_{p_j}$, then every number of $D(k, k + a)$ must have index between $k + 1$ and $k + a$ to satisfy the contraction property, and if there are more points than there are indices then we cannot fit them.

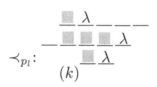

$\prec_{p_l}:$

(k)

Fig. 6. The shaded spots are the ones considered in the count test for this k and λ.

The correctness of Algorithm 1 follows from the fact that $L \ne \emptyset$ in every iteration of the algorithm. This implies that we will always be able to assign an element to $\prec_{p_j}$ (k) that satisfies the constraints for every point in P.

Lemma 2. *Algorithm 1 computes a total order $\prec_{p_j}$ that is a contraction of $\prec_{p_i}$ for all $p_i \in P_1(j)$ and is an expansion of $\prec_{p_l}$ for all $p_l \in P_2(j)$.*

This lemma gives us our main result.

Theorem 2. *Given a set of total orders for any subset P of a finite grid such that satisfies the necessary and sufficient conditions for being a CDS, there is a polynomial-time algorithm that computes total orders for the remaining grid points that satisfy the necessary and sufficient conditions for being a CDS.*

Corbett's Rotator Example. Christ et al. [2] give an example of a CDS, named the *waterline example*, where the points in $\mathbb{Z}^2$ do not all use the same total order to generate their segments. They asked if there were any other examples of interesting CDSes that use many different total orders. In this section, we give such an example of a CDS. Consider a subset S of $\mathbb{Z}$ of cardinality k, and note that there are $k!$ total permutations of the integers in S. If we sort S according to each of the total orders used in the waterline example, then we will obtain k different permutations of S. We show that for any k, there is a set S of k consecutive integers and a CDS in a $k! \times k!$ grid such that when we sort S with each of the total orders in our CDS, we obtain all $k!$ permutations of S.

We call this CDS the *Corbett's rotator example* because we make use of Corbett's rotator [4,9] to compute the total orders used by the points in the CDS. Corbett's rotator is a method of systematically generating all permutations $P_1, P_2, \ldots P_{k!}$ of a set of k elements. Corbett's rotator starts with an arbitrary permutation, and then transforms P_j to P_{j+1} using a "rotation" of the following form: the first element of P_j becomes the ith element of P_{j+1} for some integer i, all elements that were in the positions 2 to i in P_j move one position to the left in P_{j+1}, and every element with position at least $i + 1$ in P_j stays in the same position in P_{j+1}. Each rotation may use a different choice of i. The choices of i need to be carefully chosen to ensure that all permutations are generated (see [9] for the details), but for this paper it is sufficient to understand this general form of the rotations.

Consider the grid $\{1, \ldots, k!\} \times \{1, \ldots, k!\}$ and let $S = \{1, \ldots, k\}$. We now define the total orders used in our CDS. Let $\bar{S}$ be the natural ordering of the integers in $\mathbb{Z} \setminus S$, and let $P_1, P_2, \ldots P_{k!}$ be each of the permutations of S as computed by Corbett's rotator. Our CDS uses $k!$ total orders $\prec_1, \prec_2, \ldots, \prec_{k!}$, where the smallest k elements of $\prec_i$ is P_i in order followed by $\bar{S}$ in order. Now consider all points p such that $p^x + p^y = k! + 1$. Note that there are $k!$ such points in the grid. We assign $\prec_i$ to the point (p^x, p^y) such that $p^x = i$ and $p^y = k! + 1 - i$ for each $i \in \{1, 2, 3, \ldots, k!\}$. Then we can use Algorithm 1 to generate the remaining segments of the CDS.

Theorem 3. *For each positive integer k, there is a CDS and a subset S of k integers such that all $k!$ permutations are obtained when sorting S with the total orders used in the CDS.*

References

1. Chowdhury, I., Gibson, M.: A characterization of consistent digital linesegments in $\mathbb{Z}^2$. In: Algorithms - ESA 2015 - 23rd Annual European Symposium,Patras, Greece, 14–16 September 2015, Proceedings, pp. 337–348 (2015)
2. Christ, T., Pálvölgyi, D., Stojakovic, M.: Consistent digital line segments. Discrete Comput. Geom. **47**(4), 691–710 (2012)
3. Chun, J., Korman, M., Nöllenburg, M., Tokuyama, T.: Consistent digital rays. Discrete Comput. Geom. **42**(3), 359–378 (2009)
4. Corbett, P.: Rotator graphs: an efficient topology for point-to-point multiprocessor networks. IEEE Trans. Parallel Distrib. Syst. **3**, 622–626 (1992)
5. Wm. Randolph Franklin: Problems with raster graphics algorithm. In: Peters, F.J., Kessener, L.R.A., van Lierop, M.L.P. (eds.) Data Structures for Raster Graphics, Steensel, Netherlands. Springer, Heidelberg (1985)
6. Goodrich, M.T., Guibas, L.J., Hershberger, J., Tanenbaum, P.J.: Snap rounding line segments efficiently in two and three dimensions. In: Symposium on Computational Geometry, pp. 284–293 (1997)

7. Greene, D.H., Yao, F.F.: Finite-resolution computational geometry. In: 27th Annual Symposium on Foundations of Computer Science, Toronto, Canada, 27–29 October, pp. 143–152. IEEE Computer Society (1986)
8. Luby, M.G.: Grid geometries which preserve properties of euclidean geometry: a study of graphics line drawing algorithms. In: Earnshaw, R.A. (ed.) Theoretical Foundations of Computer Graphics and CAD, vol. 40, pp. 397–432. Springer, Heidelberg (1988)
9. Williams, A.: The greedy Gray code algorithm. In: Dehne, F., Solis-Oba, R., Sack, J.-R. (eds.) WADS 2013. LNCS, vol. 8037, pp. 525–536. Springer, Heidelberg (2013)

Faster Information Gathering in Ad-Hoc Radio Tree Networks

Marek Chrobak[1] and Kevin P. Costello[2]([✉])

[1] Department of Computer Science, University of California, Riverside, USA
marek@cs.ucr.edu
[2] Department of Mathematics, University of California, Riverside, USA
costello@math.ucr.edu

Abstract. We study information gathering in ad-hoc radio networks. Initially, each node of the network has a piece of information called a *rumor*, and the overall objective is to gather all these rumors in the designated target node. The ad-hoc property refers to the fact that the topology of the network is unknown when the computation starts. Aggregation of rumors is not allowed, which means that each node may transmit at most one rumor in one step.

We focus on networks with tree topologies, that is we assume that the network is a tree with all edges directed towards the root, but, being ad-hoc, its actual topology is not known. We provide two deterministic algorithms for this problem. For the model that does not assume any collision detection nor acknowledgement mechanisms, we give an $O(n \log \log n)$-time algorithm, improving the previous upper bound of $O(n \log n)$. We also show that this running time can be further reduced to $O(n)$ if the model allows for acknowledgements of successful transmissions.

1 Introduction

We study the problem of *information gathering* in ad-hoc radio networks. Initially, each node of the network has a piece of information called a *rumor*, and the objective is to gather all these rumors, as quickly as possible, in the designated target node. The nodes communicate by sending messages via radio transmissions. At any time step, several nodes in the network may transmit. When a node transmits a message, this message is sent immediately to all nodes within its range. When two nodes send their messages to the same node at the same time, a *collision* occurs. Aggregation of rumors is not allowed, which means that each node may transmit at most one rumor in one step.

The network can be naturally modeled by a directed graph, where an edge (u, v) indicates that v is in the range of u. The ad-hoc property refers to the fact that the actual topology of the network is unknown when the computation starts. We assume that nodes are labeled by integers $0, 1, ..., n-1$. An information gathering protocol determines a sequence of transmissions of a node, based on its label and on the previously received messages.

Research supported by NSF grants CCF-1217314, CCF-1536026, and NSA grant H98230-13-1-0228.

© Springer-Verlag Berlin Heidelberg 2016
E. Kranakis et al. (Eds.): LATIN 2016, LNCS 9644, pp. 275–289, 2016.
DOI: 10.1007/978-3-662-49529-2_21

Our Results. In this paper, we focus on ad-hoc networks with tree topologies, that is the underlying ad-hoc network is assumed to be a tree with all edges directed towards the root, although the actual topology of this tree is unknown. We consider two variants of the problem. In the first one, we do not assume any collision detection or acknowledgment mechanisms, so none of the nodes (in particular neither the sender nor the intended recipient) are notified about a collision after it occurred. In this model, we give a deterministic algorithm that completes information gathering in time $O(n \log \log n)$. Our result significantly improves the previous upper bound of $O(n \log n)$ from [7]. To our knowledge, no lower bound for this problem is known, except for the trivial bound of $\Omega(n)$ (since each rumor must be received by the root in a different time step).

In the second part of the paper, we also consider a variant where acknowledgments of successful transmissions are provided to the sender. All the remaining nodes, though, including the intended recipient, cannot distinguish between collisions and absence of transmissions. Under this assumption, we show that the running time can be improved to $O(n)$, which is again optimal for trivial reasons, up to the implicit constant.

While we assume that all nodes are labelled $0, 1, ..., n - 1$ (where n is the number of nodes), our algorithms' asymptotic running times remain the same if the labels are chosen from a larger range $0, 1, ..., N - 1$, as long as $N = O(n)$.

Related Work. The problem of information gathering for trees was introduced in [7], where the model without any collision detection was studied. In addition to the $O(n \log n)$-time algorithm without aggregation – that we improve in this paper – [7] develops an $O(n)$-time algorithm for the model with aggregation, where a message can include any number of rumors. Another model studied in [7], called *fire-and-forward*, requires that a node cannot store any rumors; a rumor received by a node has to be either discarded or immediately forwarded. For fire-and-forward protocols, a tight bound of $\Theta(n^{1.5})$ is given in [7].

The information gathering problem is closely related to two other information dissemination primitives that have been well studied in the literature on ad-hoc radio networks: broadcasting and gossiping. All the work discussed below is for ad-hoc radio networks modeled by arbitrary directed graphs, and without any collision detection capability.

In *broadcasting*, a single rumor from a specified source node has to be delivered to all other nodes in the network. The naïve ROUNDROBIN algorithm (see the next section) completes broadcasting in time $O(n^2)$. Following a sequence of papers [3,4,8,16] where this naïve bound was gradually improved, it is now known that broadcasting can be solved in time $O(n \log n \log \log n)$ [19] or $O(n \log^2 D)$ [11], where D is the diameter of G. This nearly matches the lower bound of $\Omega(n \log D)$ from [10]. Randomized algorithms for broadcasting have also been well studied [1,11,17].

The *gossiping* problem is an extension of broadcasting, where each node starts with its own rumor, and all rumors need to be delivered to all nodes in the network. The time complexity of deterministic algorithms for gossiping is a major open problem in the theory of ad-hoc radio networks. Obviously, the lower bound of $\Omega(n \log D)$ for broadcasting [10] applies to gossiping as well, but no better lower bound is known. It is also not known whether gossiping can be solved in time $O(n\,\mathrm{polylog}(n))$ with a deterministic algorithm, even if message aggregation is allowed. The best currently known upper bound is $O(n^{4/3} \log^4 n)$ [14] (see [8,24] for some earlier work). The case when no aggregation is allowed (or with limited aggregation) was studied in [5]. Randomized algorithms for gossiping have also been well studied [9,11,18]. Interested readers can find more information about gossiping in the survey paper [13].

Connections to Other Problems. For arbitrary graphs, assuming aggregation, one can solve the gossiping problem by running an algorithm for information gathering and then broadcasting all rumors (as one message) to all nodes in the network. Thus an $O(n\,\mathrm{polylog}(n))$-time algorithm for information gathering would resolve in positive the earlier-discussed open question about the complexity of gossiping. Due to this connection, developing an $O(n\,\mathrm{polylog}(n))$-time algorithm for information gathering on arbitrary graphs is likely to be very difficult, if possible at all.

This research, as well as the earlier work in [7], was motivated mainly by this connection to gossiping. One can think of information gathering for trees as a simple variant of the gossiping problem. We hope that developing efficient algorithms for trees, or for some other natural special cases, will ultimately lead to some insights helpful in resolving the complexity of the gossiping problem in arbitrary graphs.

Some algorithms for ad-hoc radio networks (see [5,15], for example) involve constructing a spanning subtree of the network and disseminating information along this subtree. Better algorithms for information gathering on trees may thus be useful in addressing problems for arbitrary graphs.

The problem of information gathering for trees is also related to the *contention resolution problem* in multiple-access channels (MAC). There is a myriad of variants of MAC contention resolution in the literature. (See, for example, [2,20,21].) Generally, the instance of the problem involves a collection of transmitters connected to a shared channel, like Ethernet, for example. Some of these transmitters need to send their messages across the channel, and the objective is to design a distributed protocol that will allow them to do that. The information gathering problem for trees is in essence an extension of MAC contention resolution to multi-level hierarchies of channels, where transmitters have unique identifiers, and the structure of this hierarchy is not known.

Note: Due to lack of space, in this extended abstract we only outline the main ideas behind most proofs. Complete proofs can be found in the full version of this paper [6].

2 Preliminaries

We now provide a formal definition of our model and introduce notation, terminology, and some basic properties used throughout the paper.

Radio Networks with Tree Topology. In the paper we focus exclusively on radio networks with tree topologies. Such a network will be represented by a tree $\mathcal{T}$ with root r and with $n = |\mathcal{T}|$ nodes. The edges in $\mathcal{T}$ are directed towards the root, representing the direction of information flow: a node can send messages to its parent, but not to its children. We assume that each node $v \in \mathcal{T}$ is assigned a unique label from $[n] = \{0, 1, ..., n - 1\}$, and we denote this label by label(v).

For a node v, by deg(v) we denote the *degree* of v, which is the number of v's children. For any subtree X of $\mathcal{T}$ and a node $v \in X$, X_v denotes the subtree of X rooted at v that consists of all descendants of v in X.

For any integer $\gamma = 1, 2, ..., n - 1$ and any node v of $\mathcal{T}$ define the γ-*height* of v as follows. If v is a leaf then the γ-height of v is 0. If v is an internal node then let g be the maximum γ-height of a child of v. If v has fewer than γ children of γ-height equal g then the γ-height of v is g. Otherwise, the γ-height of v is $g + 1$. The γ-height of v will be denoted by $height_\gamma(v)$. In case when more than one tree are under consideration, to resolve potential ambiguity we will write $height_\gamma(v, \mathcal{T})$ for the γ-height of v in $\mathcal{T}$. The γ-height of a tree $\mathcal{T}$, denoted $height_\gamma(\mathcal{T})$, is defined as $height_\gamma(r)$, that is the γ-height of its root.

Its name notwithstanding, the definition of γ-height is meant to capture the "bushiness" of a tree. For example, if $\mathcal{T}$ is a path then its γ-height equals 0 for each $\gamma > 1$. The concept of γ-height generalizes Strahler numbers [22,23], introduced in hydrology to measure the size of streams in terms of the complexity of their tributaries. Figure 1 gives an example of a tree and values of 3-heights for all its nodes. The lemma below is a slight refinement of an analogous lemma in [7], and it will play a critical role in our algorithms.

Lemma 1. *If $\mathcal{T}$ has q leaves, and $2 \leq \gamma \leq q$, then $height_\gamma(\mathcal{T}) \leq \log_\gamma q$.*

Equivalently, any tree having γ-height j must have at least γ^j leaves. This can be seen by induction on j – if v is a node which is furthest from the root among all nodes of γ-height j, then v by definition has γ descendants of γ-height $j - 1$, each of which has γ^{j-1} leaf descendants by inductive hypothesis.

Information Gathering Protocols. Each node v of $\mathcal{T}$ has a label (or an identifier) associated with it, and denoted label(v). When the computation is about to start, each node v has also a piece of information, ρ_v, that we call a *rumor*. The computation proceeds in discrete, synchronized time steps, numbered $0, 1, 2, ...$. At any step, v can either be in the *receiving state*, when it listens to radio transmissions from other nodes, or in the *transmitting state*, when it

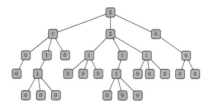

Fig. 1. An example showing a tree and the values of 3-heights for all its nodes.

is allowed to transmit. When v transmits at a time t, the message from v is sent immediately to its parent in $\mathcal{T}$. As we do not allow rumor aggregation, this message may contain at most one rumor, plus possibly $O(\log n)$ bits of other information. If w is v's parent, w will receive v's message if and only if w is in the receiving state and no other child of w transmitted at time t. In Sects. 3 and 4 we do not assume any collision detection nor acknowledgement mechanisms, so if v's message collides with one from a sibling, neither v nor w receive any notification. We relax this requirement in Sect. 5, by assuming that v (and only v) will obtain an acknowledgment from w after each successful transmission.

The objective of an information gathering protocol is to deliver all rumors from $\mathcal{T}$ to its root r, as quickly as possible. Such a protocol needs to achieve its goal even without the knowledge of the topology of $\mathcal{T}$. More formally, a gathering protocol $\mathcal{A}$ can be defined as a function that, at each time t, and for each given node v, determines the action of v at time t based only on v's label and the information received by v up to time t. The action of v at each time step t involves choosing its state (either receiving or transmitting) and, if it is in the transmitting state, choosing which rumor to transmit.

We will say that $\mathcal{A}$ runs in time $T(n)$ if, for any tree $\mathcal{T}$ and any assignment of labels to its nodes, after at most $T(n)$ steps all rumors are delivered to r.

In a simple information gathering protocol called ROUNDROBIN, nodes transmit one at a time, in n rounds, where in each round they transmit in the order $0, 1, ..., n-1$ of their labels. For any node v, when it is its turn to transmit, v transmits any rumor from the set of rumors that have been received so far (including its own rumor) but not yet transmitted. In each round, each rumor that is still not in r will get closer to r, so after n^2 steps all rumors will reach r.

Strong k-selectors. Let $\bar{S} = (S_0, S_1, ..., S_{m-1})$ be a family of subsets of $\{0, 1, ..., n-1\}$. $\bar{S}$ is called a *strong k-selector* if, for each k-element set $A \subseteq \{0, 1, ..., n-1\}$ and each $a \in A$, there is a set S_i such that $S_i \cap A = \{a\}$. As shown in [10,12], for each k there exists a strong k-selector $\bar{S} = (S_0, S_1, ..., S_{m-1})$ with $m = O(k^2 \log n)$. We will make extensive use of strong k-selectors in our algorithms. At a certain time in the computation our protocols will "run" $\bar{S}$, for an appropriate choice of k, by which we mean that it will execute a sequence of m consecutive steps, such that in the jth step the nodes from S_j will transmit, while those not in S_j will stay quiet. This will guarantee that, for any node v with at most $k-1$ siblings, there will be at least one step in the execution of $\bar{S}$ where v will transmit but none of its siblings will. Therefore at least one of v's transmissions will be successful.

3   An $O(n\sqrt{\log n})$-Time Protocol

We first give a gathering protocol SIMPLEGATHER for trees with running time $O(n\sqrt{\log n})$. Our faster protocol will be presented in the next section.

We fix three parameters $K = 2^{\lfloor \sqrt{\log n} \rfloor}$, $D = \lceil \log_K n \rceil = O(\sqrt{\log n})$ and $D' = \lceil \log K^3 \rceil = O(\sqrt{\log n})$. We also fix a strong K-selector $\bar{S} = (S_0, S_1, ..., S_{m-1})$, where $m \leq CK^2 \log n$, for some integer constant C.

By Lemma 1, we have that $height_K(\mathcal{T}) \leq D$. A node v of $\mathcal{T}$ is called *light* if $|\mathcal{T}_v| \leq n/K^3$; otherwise we say that v is *heavy*. Let $\mathcal{T}'$ be the subtree of $\mathcal{T}$ induced by the heavy nodes. By the definition of heavy nodes, $\mathcal{T}'$ has at most K^3 leaves, so $height_2(\mathcal{T}') \leq D'$. Also, obviously, $r \in \mathcal{T}'$.

To streamline the description of our algorithm we will allow each node to receive and transmit messages at the same time. We will assume a preprocessing step allowing each v to know both the size of its subtree $\mathcal{T}_v$ and its K-height. In particular, v knows whether it is in $\mathcal{T}'$ or not. We will assume that each node $v \in \mathcal{T}'$, also knows its 2-height in the subtree $\mathcal{T}'$. The full version of this paper [6] contains the details of the preprocessing and the proof of the validity of the simultaneous receive/transmit assumption.

A detailed description of Algorithm SIMPLEGATHER is given in Pseudocode 1. To distinguish between computation steps (which do not consume time) and communication steps, we use command "**at time** t". When the algorithm reaches this command it waits until time step t to continue processing. Each message transmission takes one time step. For each node v we maintain a set B_v of rumors received by v, including its own rumor ρ_v. The algorithm consists of two epochs, and we describe the computation in each epoch separately.

Epoch 1: Light Nodes. In Epoch 1, only the light nodes participate, and the goal is to move all rumors to $\mathcal{T}'$. This epoch has $D+1$ stages (numbered $h = 0, 1, \ldots, D$), with stage h beginning at time $\alpha_h = (C+1)hn$. A light node v with K-height h is only active during stage h.

Each stage has two parts. In the first part of stage h, v will transmit according to the strong K-selector $\bar{S}$. Specifically, this part has n/K^3 iterations, each corresponding to a complete execution of $\bar{S}$. During each iteration, v chooses a single rumor $\rho_z \in B_v$ that it has not yet marked, and transmits ρ_v in each time step corresponding to a set S_i containing the label of v. This ρ_z is then marked, and not chosen again during the first part. Note that if the parent u of v has degree at most K, the definition of strong K-selectors guarantees that ρ_z will be received by u, but if u's degree is larger it is possible for all transmissions of ρ_z during this stage to be blocked.

Note that the total number of steps required for this part of stage h is $(n/K^3) \cdot m \leq Cn$, so these steps will be completed before the second part of stage h starts.

In the second part, (beginning at time $\alpha_h + Cn$), we simply run the ROUNDROBIN protocol: in the l-th step of this part, if v has the rumor of the node with label l, then it transmits that rumor.

Epoch 2: Heavy Nodes. This epoch has $D' + 1$ stages, with only heavy nodes in $\mathcal{T}'$ participating. When the epoch starts, all rumors are assumed to already be in $\mathcal{T}'$. In stage g the nodes in $\mathcal{T}'$ whose 2-height is equal g will participate. Similar to the stages of epoch 1, each stage runs in time $O(n)$ and has two parts. In the first part, during each time step *every* heavy node holding a rumor it has not yet marked chooses such a rumor, marks it, and transmits it (instead of using a strong K-selector). The second part executes ROUNDROBIN, as before.

The high-level overview of the analysis of this algorithm is that rumors maintain a steady rate of progress towards the root – during stage h of Epoch 1, each

---

Pseudocode 1. SIMPLEGATHER(v)

1: $K = 2^{\lfloor \sqrt{\log n} \rfloor}$, $D = \lceil \log_K n \rceil$
2: $B_v \leftarrow \{\rho_v\}$ ▷ Initially v has only ρ_v
3: **Throughout:** all rumors received by v are automatically added to B_v
4: **if** $|T_v| \le n/K^3$ **then**   ▷ v is light (epoch 1)
5:      $h \leftarrow height_K(v, T)$; $\alpha_h \leftarrow (C+1)nh$ ▷ v participates in stage h
6:      **for** $i = 0, 1, ..., n/K^3 - 1$ **do**   ▷ iteration i
7:          **at time** $\alpha_h + im$
8:          **if** B_v has an unmarked rumor **then** ▷ Part 1: strong K-selector
9:              choose any unmarked $\rho_z \in B_v$ and mark it
10:              **for** $j = 0, 1, ..., m - 1$ **do**
11:                  **at time** $\alpha_h + im + j$
12:                      **if** label$(v) \in S_j$ **then** TRANSMIT(ρ_z)
13:      **for** $l = 0, 1, ..., n - 1$ **do**   ▷ Part 2: RoundRobin
14:          **at time** $\alpha_h + Cn + l$
15:          $z \leftarrow$ node with label$(z) = l$
16:          **if** $\rho_z \in B_v$ **then** TRANSMIT(ρ_z)
17: **else**   ▷ v is heavy (epoch 2)
18:      $g \leftarrow height_2(v, T')$; $\alpha'_g \leftarrow \alpha_{D+1} + 2ng$ ▷ v participates in stage g
19:      **for** $i = 0, 1, ..., n - 1$ **do**   ▷ Part 1: all nodes transmit
20:          **at time** $\alpha'_g + i$
21:          **if** B_v contains an unmarked rumor **then**
22:              choose any unmarked $\rho_z \in B_v$ and mark it
23:              TRANSMIT(ρ_z)
24:      **for** $l = 0, 1, ..., n - 1$ **do**   ▷ Part 2: RoundRobin
25:          **at time** $\alpha'_g + 2n + l$
26:          $z \leftarrow$ node with label$(z) = l$
27:          **if** $\rho_z \in B_v$ **then** TRANSMIT(ρ_z)

---

rumor either reaches the heavy tree or a node of K-height $h + 1$; during stage g of Epoch 2 each rumor either reaches the root or a vertex of 2 height $g + 1$ in T'. Since the heights are bounded (by Lemma 1), the algorithm will complete in the required time of $O(n\sqrt{\log n})$. The details of the analysis follow.

Analysis of Epoch 1 (The Light Nodes). We claim that the following invariant holds for all $h = 0, 1, ..., D$:

(I_h) Let $w \in T$ and let u be a light child of w with $height_K(u) \le h - 1$. Then at time α_h node w has received all rumors from T_u.

To prove this invariant we proceed by induction on h. If $h = 0$ the invariant **(I_0)** holds vacuously. So suppose that invariant **(I_h)** holds for some value of h. We want to prove that **(I_{h+1})** is true when stage $h + 1$ starts. We thus need to prove the following claim: if u is a light child of w with $height_K(u) \le h$ then at time α_{h+1} all rumors from T_u will arrive in w.

If $height_K(u) \le h - 1$ then the claim holds, immediately from the inductive assumption **(I_h)**. So assume that $height_K(u) = h$. Consider the subtree H rooted at u and containing all descendants of u whose K-height is equal to h. By the inductive assumption, at time α_h any $w' \in H$ has all rumors from the subtrees rooted at its descendants of K-height smaller than h, in addition to its own rumor $\rho_{w'}$. Therefore all rumors from T_u are already in H and each of them has exactly one copy in H, because all nodes in H were idle before time α_h.

When the algorithm executes the first part of stage h on H, then each node v in H whose parent is also in H will successfully transmit an unmarked rumor during each pass through the K selector – indeed, our definition of H guarantees that v has at most $K - 1$ siblings in H, so by the definition of strong selector it must succeed at least once. The following claim follows by simple induction:

Claim 1: At all times during stage h, the collection of nodes in H still holding unmarked rumors forms an induced tree of H.

In particular, node u will receive a new rumor during every run through the selector until it has received all rumors from its subtree. Since the tree originally held at most $|T_u| \leq n/K^3$ rumors originally, u must have received all rumors from its subtree after at most n/K^3 runs through the selector.

Note that, as $height_K(u) = h$, u will also attempt to transmit its rumors to w during this part, but, since we are not making any assumptions about the degree of w, there is no guarantee that w will receive them. This is where the second part of this stage is needed. Since in the second part each rumor is transmitted without collisions, all rumors from u will reach w before time α_{h+1}, completing the inductive step and the proof that $(\mathbf{I}_{h+1})$ holds.

In particular, using Invariant $(\mathbf{I}_h)$ for $h = D$, we obtain that after Epoch 1 each heavy node w will have received rumors from the subtrees rooted at all its light children. Therefore at that time all rumors from T will be already in T', with each rumor having exactly one copy in T'.

Analysis of Epoch 2 (The Heavy Nodes). The argument for the heavy nodes is similar as for the light nodes, but with a twist, since we do not use selectors now. In essence, we show that each stage reduces by at least one the 2-depth of the minimum subtree of T' that contains all rumors. Specifically, we show that the following invariant holds for all $g = 0, 1, ..., D'$:

$(\mathbf{J}_g)$ Let $w \in T'$ and let $u \in T'$ be a child of w with $height_2(u, T') \leq g - 1$. Then at time α'_g node w has received all rumors from T_u.

We prove invariant $(\mathbf{J}_g)$ by induction. For $g = 0$, $(\mathbf{J}_0)$ holds vacuously. Assume that $(\mathbf{J}_g)$ holds for some g. We claim that $(\mathbf{J}_{g+1})$ holds right after stage g.

Choose any child u of w with $height_2(u, T') \leq g$. If $height_2(u, T') \leq g - 1$, we are done, by the inductive assumption. So we can assume that $height_2(u, T') = g$. Let P be the subtree of T' rooted at u and consisting of all descendants of u whose 2-height in T' is equal g. Then P is simply a path. By the inductive assumption, for each $w' \in P$, all rumors from the subtrees of w' rooted at its children of 2-height at most $g - 1$ are in w'. Thus all rumors from T_u are already in P. All nodes in P participate in stage g, but their children outside P do not transmit. Therefore each transmission from any node $x \in P - \{u\}$ during stage g will be successful. Due to pipelining, all rumors from P will reach u after the first part of stage g. In the second part, all rumors from u will be successfully sent to w. So after stage g all rumors from T_u will be in w, completing the proof.

4   A Protocol with Running Time $O(n \log \log n)$

In this section we present our first main result:

Theorem 1. *The problem of information gathering on trees, without rumor aggregation, can be solved in time $O(n \log \log n)$.*

The protocol can be thought of as an iterative application of the idea behind Algorithm SIMPLEGATHER from Sect. 3. We assume that the reader is familiar with Algorithm SIMPLEGATHER and its analysis, and in our presentation we will focus on the high level ideas behind Algorithm FASTGATHER, deferring the implementation of some details to Sect. 3 and the full version of this paper [6].

As before, $\mathcal{T}$ is the input tree with n vertices. We fix some arbitrary integer constant $\beta \geq 2$. For $\ell = 1, 2, ...$, let $K_\ell = \lceil n^{\beta^{-\ell}} \rceil$. So $K_1 = \lceil n^{1/\beta} \rceil$, the sequence $(K_\ell)_\ell$ is non-increasing, and $\lim_{\ell \to \infty} K_\ell = 2$. Let L be the largest value of ℓ for which $n^{\beta^{-\ell}} \geq \log n$. (Note that L is well defined for sufficiently large n, since β is fixed). Thus $L \leq \log_\beta(\log n / \log \log n)$, $L = \Theta(\log \log n)$, and $\log n \leq K_L = K_{L+1}^\beta < (\log n)^\beta$.

For $\ell = 1, 2, ..., L$, by $\bar{S}^\ell = (S_1^\ell, S_2^\ell, ..., S_{m_\ell}^\ell)$ we denote a strong K_ℓ-selector of size $m_\ell \leq C K_\ell^2 \log n$, for some integer constant C. As discussed in Sect. 2, such selectors $\bar{S}^\ell$ exist.

Let $\mathcal{T}^{(0)} = \mathcal{T}$, and for each $\ell = 1, 2, ..., L$, let $\mathcal{T}^{(\ell)}$ be the subtree of $\mathcal{T}$ induced by the nodes v with $|\mathcal{T}_v| \geq n/K_\ell^3$. Each tree $\mathcal{T}^{(\ell)}$ is rooted at r, and $\mathcal{T}^{(\ell)} \subseteq \mathcal{T}^{(\ell-1)}$ for $\ell \geq 1$. For $\ell \neq 0$, the definition of $\mathcal{T}^{(\ell)}$ implies also that it has at most K_ℓ^3 leaves, so, by Lemma 1, its $K_{\ell+1}$-height is at most $\log_{K_{\ell+1}}(K_\ell^3)$. Since $K_\ell \leq 2n^{\beta^{-\ell}}$ and $K_{\ell+1} \geq n^{\beta^{-(\ell+1)}}$, direct calculation gives $\log_{K_{\ell+1}}(K_\ell^3) \leq 3\beta + 1$ for sufficiently large n. In particular, the $K_{\ell+1}$-height of $\mathcal{T}^{(\ell)}$ is at most $D = 3\beta + 1 = O(1)$.

Similar to the previous section we will make some simplifying assumptions. First, we will assume that all nodes can receive and transmit messages at the same time. Second, we will also assume that each node v knows the size of its subtree $|\mathcal{T}_v|$ and its K_ℓ-heights, for each $\ell \leq L$. We describe how to remove these assumptions in the full version of this paper [6].

Algorithm FASTGATHER consists of $L + 1$ epochs, numbered $1, 2, ..., L + 1$. For $\ell \leq L$, the goal of epoch L is to move all rumors from $\mathcal{T}^{(L-1)}$ to $\mathcal{T}^{(L)}$ Each of these L epochs will run in time $O(n)$, so their total running time will be $O(nL) = O(n \log \log n)$. The final epoch will move all rumors from $\mathcal{T}^{(L)}$ to the root, also in time $O(n \log \log n)$. We now provide the details.

Epochs $\ell = 1, 2, ..., L$. In epoch ℓ, only the nodes in $\mathcal{T}^{(\ell-1)} - \mathcal{T}^{(\ell)}$ participate. The computation in this epoch is very similar to the computation of light nodes (in epoch 1) in Algorithm SIMPLEGATHER. Epoch ℓ starts at time $\gamma_\ell = (D + 1)(C + 1)(\ell - 1)n$ and lasts $(D + 1)(C + 1)n$ steps.

Let $v \in \mathcal{T}^{(\ell-1)} - \mathcal{T}^{(\ell)}$. The computation of v in epoch ℓ consists of $D + 1$ identical stages. Each stage $h = 0, 1, ..., D$ starts at time step $\alpha_{\ell,h} = \gamma_\ell + (C + 1)hn$ and lasts $(C + 1)n$ steps.

Stage h has two parts. The first part starts at time $\alpha_{\ell,h}$ and lasts time Cn. During this part we execute $\lceil n/K_\ell^3 \rceil$ iterations, each iteration consisting of running the strong $K_{\ell+1}$-selector $\bar{S}^\ell$. The time needed to execute these iterations is at most $\lceil n/K_\ell^3 \rceil (CK_{\ell+1}^2 \log n)$, which can be seen by direct calculation to be at most Cn.

Thus all iterations executing the strong selector will complete before time $\alpha_{\ell,h} + Cn$. Then v stays idle until time $\alpha_{\ell,h} + Cn$, which is when the second part starts. In the second part we run the ROUNDROBIN protocol, which takes n steps. So stage h will complete right before step $\alpha_{\ell,h} + (C+1)n = \alpha_{\ell,h+1}$.

Epoch $L+1$. Due to the definition of L, we have that $\mathcal{T}^{(L)}$ contains at most $K_L^3 = O(\log^{3\beta} n)$ leaves, so its 2-depth is $D' = O(\log \log n)$, by Lemma 1. The computation in this epoch is similar to epoch 2 from Algorithm SIMPLEGATHER. As before, this epoch consists of $D' + 1$ stages, where each stage $g = 0, 1, ..., D'$ has two parts. In the first part, we have n steps in which each node transmits. In the second part, also of length n, we run one iteration of ROUNDROBIN.

The high-level analysis of Algorithm FASTGATHER is similar to that in Sect. 3: During each stage a rumor's K-height increases until it reaches the next level (tree $\mathcal{T}^{(\ell)}$), and in the last epoch its 2-height increases until it reaches the root. The full analysis is given in the full version of this paper [6].

5   An $O(n)$-time Protocol with Acknowledgments

In this section we consider a network model where acknowledgments of successful transmissions are provided to the sender. All the remaining nodes, including the intended recipient, cannot distinguish between collisions and absence of transmissions. In this section we present our second main result:

Theorem 2. *The problem of information gathering on trees without rumor aggregation can be solved in time $O(n)$ if acknowledgments are provided.*

As before, $\mathcal{T}$ is the input tree with n nodes. We will recycle the notions of light and heavy nodes from Sect. 3, although now we will use slightly different parameters. Let $\delta > 0$ be a small constant, and let $K = \lceil n^\delta \rceil$. We say that $v \in \mathcal{T}$ is *light* if $|T_v| \leq n/K^3$ and we call v *heavy* otherwise. By $\mathcal{T}'$ we denote the subtree of $\mathcal{T}$ induced by the heavy nodes.

Algorithm LinGather. Our algorithm will consist of two epochs. The first epoch is essentially identical to Epoch 1 in Algorithm SIMPLEGATHER, except for a different choice of the parameters. The objective of this epoch is to collect all rumors in $\mathcal{T}'$ in time $O(n)$. In the second epoch, only the heavy nodes in $\mathcal{T}'$ will participate in the computation, and the objective of this epoch is to gather all rumors from $\mathcal{T}'$ in the root r. This epoch is quite different from our earlier algorithms and it will use some novel combinatorial structures (obtained via probabilistic constructions) to move all rumors from $\mathcal{T}'$ to r in time $O(n)$.

Epoch 1: In this epoch only light nodes will participate, and the objective of Epoch 1 is to move all rumors into $\mathcal{T}'$. In this epoch we will not be taking advantage of the acknowledgement mechanism. As mentioned earlier, except for different choices of parameters, this epoch is essentially identical to Epoch 1 of Algorithm SIMPLEGATHER, so we only give a very brief overview here. We use a strong K-selector $\bar{S}$ of size $m \leq CK^2 \log n$.

Let $D = \lceil \log_K n \rceil \leq 1/\delta = O(1)$. By Lemma 1, the K-depth of T is at most D. Epoch 1 consists of $D + 1$ stages, where in each stage $h = 0, 1, ..., D$, nodes of K-depth h participate. Stage h consists of n/K^3 executions of $\bar{S}$, followed by an execution of ROUNDROBIN, taking total time $n/K^3 \cdot m + n = O(n)$. So the entire epoch takes time $(D + 1) \cdot O(n) = O(n)$ as well. The proof of correctness (namely that after this epoch all rumors are in T') is identical as for Algorithm SIMPLEGATHER.

Epoch 2: When this epoch starts, all rumors are already gathered in T', and the objective is to push them further to the root. The key obstacle to be overcome in this epoch is congestion stemming from the fact that nodes have many rumors to transmit. This congestion means that simply repeatedly applying k-selectors is no longer enough. For example, if the root has k children, each with n/k rumors, then repeating a k-selector n/k times would get all the rumors to the root, but take total time roughly $nk \log n$, which would be too long.

To overcome this obstacle, we introduce two novel tools that will play a critical role in our algorithm. The first tool is a so-called *amortizing selector family*. Since a parent receives at most one rumor per round, if it has k children it clearly cannot simultaneously be receiving rumors at a rate greater than $\frac{1}{k}$ from each child individually. With the amortizing family, we will be able to achieve within a constant fraction of this bound over long time intervals, so long as each child knows (approximately) how many siblings it is competing with.

Similarly to a strong selector, this amortizing family will be a collection of subsets of the underlying label set $[n]$, though now it will be doubly indexed: There will be sets S_{ij} for each $1 \leq i \leq s$ and each $j \in \{1, 2, 4, 8, \ldots, k\}$ for some parameters s and k. We say the family succeeds at cumulative rate q if the following statement is true:

For each $j \in \{1, 2, 4, \ldots, \frac{k}{2}\}$, each subset $A \subseteq \{1, \ldots, n\}$ satisfying $j/2 \leq |A| \leq 2j$, and each element $v \in A$ there are at least $\frac{q}{|A|}s$ distinct i for which

$$v \in S_{ij} \text{ and } A \cap (S_{i(j/2)} \cup S_{ij} \cup S_{i(2j)}) = \{v\}.$$

In the case $j = 1$ the set $S_{i(j/2)}$ is defined to be empty. Here s can be thought of as the total running time of the selector, and j as a node's estimate of its parent's degree, and k as some bound on the maximum degree handled by the selector. A node fires at time step i if and only if its index is contained in the set S_{ij}. What the above statement is then saying that for any subset A of siblings, if $|A|$ is at most $k/2$ and each child estimates $|A|$ within a factor of 2 then each child will transmit at rate at least $\frac{q}{|A|}$.

Theorem 3. *There are fixed constants $c, C > 0$ such that the following is true: For any k and n and any $s \geq Ck^2 \log n$, there is an amortizing selector with parameters n, k, s succeeding with cumulative rate c.*

The proof of Theorem 3 (through a probabilistic construction) appears in the full version of this paper [6].

Of course, such a family will not be useful unless a node can obtain an accurate estimate of its parent's degree, which will be the focus of our second tool, k-distinguishers. As with the amortizing selector, this will be a collection of subsets of the label set $[n]$. Let $\bar{S} = (S_1, S_2, ..., S_m)$, where each $S_j \subseteq [n]$ for each j. For $A \subseteq [n]$ and $a \in A$, define $Hits_{a,A}(\bar{S}) = \{j : S_j \cap A = \{a\}\}$, that is $Hits_{a,A}(\bar{S})$ is the collection of indices j for which S_j intersects A exactly on a. Note that, using this terminology, $\bar{S}$ is a strong k-selector if and only if $Hits_{a,A}(\bar{S}) \neq \emptyset$ for all sets $A \subseteq [n]$ of cardinality at most k and all $a \in A$.

We say that $\bar{S}$ is a k-*distinguisher* if there is a *threshold* value ξ (dependent on k) such that, for any $A \subseteq [n]$ and $a \in A$, the following conditions hold:

$$\text{if } |A| \leq k \text{ then } |Hits_{a,A}(\bar{S})| > \xi, \text{ and if } |A| \geq 2k \text{ then } |Hits_{a,A}(\bar{S})| < \xi.$$

We make no assumptions on what happens for $|A| \in \{k+1, k+2, ..., 2k-1\}$.

The idea is this: consider a fixed a, and imagine that we have some set A that contains a, but its other elements are not known. Suppose that we also have an access to a *hit oracle* that for any set S will tell us whether $S \cap A = \{a\}$ or not. With this oracle, we can then use a k-distinguisher $\bar{S}$ to extract some information about the cardinality of A by calculating the cardinality of $Hits_{a,A}(\bar{S})$. If $|Hits_{a,A}(\bar{S})| \leq \xi$ then we know that $|A| > k$, and if $|Hits_{a,A}(\bar{S})| \geq \xi$ then we know that $|A| < 2k$.

The idea of the theorem below, again by a probabilistic argument, appears in the full version of this paper [6].

Theorem 4. *For any $n \geq 2$ and $1 \leq k \leq n/2$ there exists a k-distinguisher of size $m = O(k^2 \log n)$.*

In our framework, the acknowledgement of a message received from a parent corresponds exactly to such a hit oracle. So if all nodes fire according to such a k-distinguisher, each node can determine in time $O(k^2 \log n)$ either that its parent has at least k children or that it has at most $2k$ children.

Now let λ be a fixed parameter between 0 and 1. For each $i = 0, 1, ..., \lceil \lambda \log n \rceil$, let $\bar{S}^i$ be a 2^i-distinguisher of size $O(2^{2i} \log n)$ and with threshold value ξ_i. We can then concatenate these k-distinguishers to obtain a sequence $\tilde{S}$ of size $\sum_{i=0}^{\lceil \lambda \log n \rceil} O(2^{2i} \log n) = O(n^{2\lambda} \log n)$.

We will refer to $\tilde{S}$ as a *cardinality estimator*, because applying our hit oracle to $\tilde{S}$ we can estimate a cardinality of an unknown set within a factor of 4, making $O(n^{2\lambda} \log n)$ hit queries. More specifically, consider again a scenario where we have a fixed a and some unknown set A containing a, where $|A| \leq n^\lambda$. Using the hit oracle, compute the values $h_i = |Hits_{a,A}(\bar{S}^i)|$, for all i. If i_0 is the smallest i

for which $h_i > \xi_i$, then by definition of our distinguisher we must have $2^{i_0-1} < |A| < 2(2^{i_0})$. In our gathering framework, this corresponds to each node in the tree being able to determine in time $O(n^{2\lambda} \log n)$ a value of j (specifically, $i_0 - 1$) such that the number of children of its parent is between 2^j and 2^{j+2}, which is exactly what we need to be able to run the amortizing selector.

For the remainder of this section we will assume the existence of Amortizing Selector Families and Distinguishers, and use them to construct an algorithm which completes Epoch 2 in time $O(n)$.

The Algorithm for Epoch 2: For the second epoch, we restrict our attention to the tree $\mathcal{T}'$ of heavy nodes. As before, no parent in this tree can have more than $K^3 = n^{3\delta}$ children, since each child is itself the ancestor of a subtree of size n/K^3. We will further assume the existence of a fixed amortizing selector family with parameters $k = 2K^3$ and $s = K^8$, as well as a fixed cardinality estimator with parameter $\lambda = 3\delta$ running in time $D_1 = O\left(n^{6\delta} \log n\right) = O\left(K^6 \log n\right)$.

Our protocol will be divided into stages, each consisting of $2(D_1 + K^8)$ steps. A node will be *active* in a given stage if at the beginning of the stage it has already received all of its rumors, but still has at least one rumor left to transmit (it is possible for a node to never become active, if it receives its last rumor and then finishes transmitting before the beginning of the next stage).

During each odd-numbered time step of a stage, all nodes (active or not) holding at least one rumor they have not yet successfully passed on transmit such a rumor. The even-numbered time steps are themselves divided into two parts. In the first D_1 even steps, all active nodes participate in the aforementioned cardinality estimator. At the conclusion of the estimator, each node knows a j such that their parent has between 2^j and 2^{j+2} active children. Note that active siblings do not necessarily have the same estimate for their family size. For the remainder of the even steps, each active node fires using the corresponding 2^{j+1}-selector from the amortizing family.

The stages repeat until all rumors have reached the root. Our key claim (which is proven in the full version of this paper [6]) is that the rumors aggregate at least at a steady rate over time – each node with subtree size m in the original tree $\mathcal{T}$ will receive all m rumors within $O(m)$ steps of the start of the epoch.

Acknowledgements. We thank the anonymous reviewers for constructive comments that helped us improve the presentation.

References

1. Alon, N., Bar-Noy, A., Linial, N., Peleg, D.: A lower bound for radio broadcast. J. Comput. Syst. Sci. **43**(2), 290–298 (1991)
2. Anta, A.F., Mosteiro, M.A., Munoz, J.R.: Unbounded contention resolution in multiple-access channels. Algorithmica **67**(3), 295–314 (2013)

3. Bruschi, D., Del Pinto, M.: Lower bounds for the broadcast problem in mobile radio networks. Distrib. Comput. **10**(3), 129–135 (1997)
4. Chlebus, B.S., Gasieniec, L., Gibbons, A., Pelc, A., Rytter, W.: Deterministic broadcasting in ad hoc radio networks. Distrib. Comput. **15**(1), 27–38 (2002)
5. Christersson, M., Gasieniec, L., Lingas, A.: Gossiping with bounded size messages in ad hoc radio networks. In: Widmayer, P., Triguero, F., Morales, R., Hennessy, M., Eidenbenz, S., Conejo, R. (eds.) ICALP 2002. LNCS, vol. 2380, pp. 377–389. Springer, Heidelberg (2002)
6. Chrobak, M., Costello, K.: Faster information gathering in ad-hoc radio tree networks (2015). arXiv: 1512.02179
7. Chrobak, M., Costello, K., Gasieniec, L., Kowalski, D.R.: Information gathering in ad-hoc radio networks with tree topology. In: Zhang, Z., Wu, L., Xu, W., Du, D.-Z. (eds.) COCOA 2014. LNCS, vol. 8881, pp. 129–145. Springer, Heidelberg (2014)
8. Chrobak, M., Gasieniec, L., Rytter, W.: Fast broadcasting and gossiping in radio networks. J. Algorithms **43**(2), 177–189 (2002)
9. Chrobak, M., Gasieniec, L., Rytter, W.: A randomized algorithm for gossiping in radio networks. Networks **43**(2), 119–124 (2004)
10. Clementi, A.E.F., Monti, A., Silvestri, R.: Distributed broadcast in radio networks of unknown topology. Theor. Comput. Sci. **302**(1–3), 337–364 (2003)
11. Czumaj, A., Rytter, W.: Broadcasting algorithms in radio networks with unknown topology. J. Algorithms **60**(2), 115–143 (2006)
12. Erdős, P., Frankl, P., Füredi, Z.: Families of finite sets in which no set is covered by the union of r others. Israel J. Math. **51**(1–2), 79–89 (1985)
13. Gasieniec, L.: On efficient gossiping in radio networks. In: Kutten, S., Žerovnik, J. (eds.) SIROCCO 2009. LNCS, vol. 5869, pp. 2–14. Springer, Heidelberg (2010)
14. Gasieniec, L., Radzik, T., Xin, Q.: Faster deterministic gossiping in directed ad hoc radio networks. In: Hagerup, T., Katajainen, J. (eds.) SWAT 2004. LNCS, vol. 3111, pp. 397–407. Springer, Heidelberg (2004)
15. Kowalski, D.R.: On selection problem in radio networks. In: Proceedings of the Twenty-fourth Annual ACM Symposium on Principles of Distributed Computing, PODC 2005, pp. 158–166 (2005)
16. Kowalski, D.R., Pelc, A.: Faster deterministic broadcasting in ad hoc radio networks. SIAM J. Discrete Math. **18**(2), 332–346 (2004)
17. Kushilevitz, E., Mansour, Y.: An $\Omega(D \log(N/D))$ lower bound for broadcast in radio networks. SIAM J. Comput. **27**(3), 702–712 (1998)
18. Liu, D., Prabhakaran, M.: On randomized broadcasting and gossiping in radio networks. In: Ibarra, O.H., Zhang, L. (eds.) COCOON 2002. LNCS, vol. 2387, pp. 340–349. Springer, Heidelberg (2002)
19. De Marco, G.: Distributed broadcast in unknown radio networks. In: Proceedings of the 19th Annual ACM-SIAM Symposium on Discrete Algorithms (SODA 2008), pp. 208–217 (2008)
20. De Marco, G., Kowalski, D.R.: Fast nonadaptive deterministic algorithm for conflict resolution in a dynamic multiple-access channel. SIAM J. Comput. **44**(3), 868–888 (2015)
21. De Marco, G., Kowalski, D.R.: Contention resolution in a non-synchronized multiple access channel. In: IEEE 27th International Symposium on Parallel Distributed Processing (IPDPS), pp. 525–533 (2013)

22. Strahler, A.N.: Hypsometric (area-altitude) analysis of erosional topology. Bull. Geol. Soc. Amer. **63**, 1117–1142 (1952)
23. Viennot, X.G.: A Strahler bijection between Dyck paths and planar trees. Discrete Math. **246**, 317–329 (2003)
24. Xu, Y.: An $O(n^{1.5})$ deterministic gossiping algorithm for radio networks. Algorithmica **36**(1), 93–96 (2003)

Stabbing Circles for Sets of Segments
in the Plane

Mercè Claverol[1], Elena Khramtcova[2], Evanthia Papadopoulou[2(✉)],
Maria Saumell[3], and Carlos Seara[1]

[1] Universitat Politècnica de Catalunya, Barcelona, Spain
[2] Faculty of Informatics, Università della Svizzera italiana (USI),
Lugano, Switzerland
evanthia.papadopoulou@usi.ch
[3] Department of Mathematics and European Centre of Excellence NTIS,
University of West Bohemia, Pilsen, Czech Republic

Abstract. Stabbing a set S of n segments in the plane by a line is a well-known problem. In this paper we consider the variation where the stabbing object is a circle instead of a line. We show that the problem is tightly connected to cluster Voronoi diagrams, in particular, the Hausdorff and the farthest-color Voronoi diagram. Based on these diagrams, we provide a method to compute all the combinatorially different stabbing circles for S, and the stabbing circles with maximum and minimum radius. We give conditions under which our method is fast. These conditions are satisfied if the segments in S are parallel, resulting in a $O(n \log^2 n)$ time algorithm. We also observe that the stabbing circle problem for S can be solved in optimal $O(n^2)$ time and space by reducing the problem to computing the stabbing planes for a set of segments in 3D.

1 Introduction

Let S be a set of n line segments (segments for short) in $\mathbb{R}^2$. We say that a region $\mathcal{R} \subseteq \mathbb{R}^2$ is a *stabbing region* for S if *exactly* one endpoint of each segment of S lies in the exterior of $\mathcal{R}$. The boundary of $\mathcal{R}$ (also known as a *stabber* for S) intersects all the segments in S and separates/classifies their endpoints into two classes, depending on whether or not they lie in the exterior of $\mathcal{R}$. Two stabbing regions $\mathcal{R}_1$ and $\mathcal{R}_2$ for S are *combinatorially different* if they classify the endpoints of S differently.

A natural problem is to determine the existence and compute (when possible) a representation of all combinatorially different stabbing regions for S. We are interested in stabbing regions whose boundary has constant complexity. Perhaps the simplest such region is a halfplane bounded by a line that intersects or *stabs* all the segments. Edelsbrunner et al. [11] presented an optimal $\Theta(n \log n)$ time algorithm to compute a representation of all the $O(n)$ combinatorially different

---

A preliminary version of this paper appeared in Abstracts XVI Spanish Meeting on Computational Geometry (EGC'15), pp. 112–115.

© Springer-Verlag Berlin Heidelberg 2016
E. Kranakis et al. (Eds.): LATIN 2016, LNCS 9644, pp. 290–305, 2016.
DOI: 10.1007/978-3-662-49529-2_22

Fig. 1. Left: Segment set with a stabbing circle. Right: Segment set with no stabbing circle.

stabbing lines for S. An $\Omega(n \log n)$ lower bound for the decision problem was later presented by Avis et al. [4]. For parallel segments the problem can be solved in $O(n)$ time by linear programming. If no stabbing halfplane exists for S, it is natural to ask for other types of stabbers. Computing all the combinatorially different stabbing wedges (regions defined by the intersection of two halfplanes) can be carried out in $O(n^3 \log n)$ time and $O(n^2)$ space [8]. The same question with isothetic stabbers can be answered in $O(n \log n)$ time and $O(n)$ space for stabbing strips, quadrants and 3-sided rectangles; and in $O(n^2 \log n)$ time and $O(n^2)$ space for stabbing rectangles [9].

In this paper, we focus on the *stabbing circle problem* formulated as follows. Let S be a set of n segments in the plane in general position (segments have non-zero length, no three endpoints are collinear, and no four of them are cocircular). A circle c is a stabbing circle for S if exactly one endpoint of each segment of S is contained in the exterior of the closed disk (region) induced by c; see Fig. 1. The stabbing circle problem for S consists on (1) answering whether a stabbing circle for S exists; (2) reporting a representation (for the centers) of all the combinatorially different stabbing circles for S; and (3) finding stabbing circles with the minimum and maximum radius. Note that our stabbing criterion uses only the segment endpoints, thus, S can be seen as a set of pairs of points, where a segment is simply a convenient representation for such a pair.

Other works with similar criteria are as follows: Rappaport [19] considers the problem of computing the stabbing simple polygon of minimum perimeter for a set S of general segments, where a simple polygon stabs S if *at least* one point (which is not necessarily an endpoint) of each segment is in the polygon; this minimum stabbing polygon is always a convex polygon. Díaz-Báñez et al. [10] focus on computing the stabbing simple polygons of minimum perimeter or area with a distinct criterion, specifically, that at least one endpoint of each segment is required to be in the polygon. Arkin et al. [2] consider, given a collection of compact sets, whether there exists a *convex* body $\mathcal{R}$ whose boundary intersects every set in the collection. They show that, for segment sets, deciding the existence of a convex stabber is NP-hard.

Our Results. First, we point out a connection between the stabbing circle problem and two *cluster Voronoi diagrams*, the *Hausdorff* and the *farthest-color Voronoi diagram*. This connection is interesting in its own right and it forms the base of our method to solve the stabbing circle problem. For a family of clusters (sets) of points in $\mathbb{R}^2$, the Hausdorff Voronoi diagram (HVD) is a subdivision of $\mathbb{R}^2$ into regions such that every point within one region has the same nearest cluster, where the distance between a point $p \in \mathbb{R}^2$ and a cluster C is the maximum distance between p and all points in C. The farthest-color Voronoi diagram (FCVD) is the reverse: it reveals the farthest cluster according to the minimum distance between a point and a cluster. Both diagrams have quadratic structural complexity in the worst case [1,12,14]. However, for some classes of input sites the diagrams are linear and can be constructed efficiently, see e.g. [5,18] for the HVD. Here, clusters are the pairs of segment endpoints, and S is a family of such pairs of points.

Our central object is $\mathsf{FCVD}^*(S)$, defined as the locus of points whose farthest-color neighbor (i.e., their *owner* in the farthest-color Voronoi diagram) is closer than their nearest cluster (i.e., their *owner* in the Hausdorff Voronoi diagram). We observe that any point $p \in \mathbb{R}^2$ is the center of a stabbing circle for S if and only if p lies in the interior of $\mathsf{FCVD}^*(S)$. Thus, $\mathsf{FCVD}^*(S)$ provides all the information relevant to stabbing circles: whether such circles exist, a list of all combinatorially different stabbing circles, and the stabbing circles with minimum and maximum radius. We identify sufficient conditions for fast algorithms to construct $\mathsf{FCVD}^*(S)$, and thus, to solve the stabbing circle problem. These conditions are: (1) the Hausdorff Voronoi diagram and the farthest-color Voronoi diagram have linear structural complexity and can be constructed fast; (2) any segment in S can "spoil" at most a constant number of edges of the Hausdorff Voronoi diagram, where by "spoiling" an edge e we mean a technical condition necessary to cause $e \cap \mathsf{FCVD}^*(S)$ to be disconnected. If the segments in S are parallel, conditions (1) and (2) are satisfied, thus, we obtain that the stabbing circle problem for S can be solved in $O(n \log^2 n)$ time and $O(n)$ space. As a byproduct, we establish that the farthest-color Voronoi diagram for such a set S has structural complexity $O(n)$ and can be constructed in $O(n \log n)$ time, which was not previously known.

Summary. In Sect. 2 we give the necessary definitions; in addition, we observe that, using a known technique, the stabbing circle problem for arbitrary segments can be solved in $O(n^2)$ time and space. In Sect. 3 we show the connection of $\mathsf{FCVD}^*(S)$ with the problem, and we give useful properties of $\mathsf{FCVD}^*(S)$. In Sect. 4 we present an algorithm to compute $\mathsf{FCVD}^*(S)$. Finally, in Sect. 5, we show that the stabbing circle problem for parallel segments can be solved in $O(n \log^2 n)$ time and $O(n)$ space.

2   Preliminaries and Definitions

In what follows, xx' denotes either a pair of points or a segment as convenient.

Definition 1. [12,18] *The* Hausdorff Voronoi diagram *of S is a partitioning of $\mathbb{R}^2$ into regions defined as follows:*

$\mathsf{hreg}(aa') = \{p \in \mathbb{R}^2 \mid \forall bb' \in S \setminus \{aa'\} \colon \max\{d(p,a), d(p,a')\} < \max\{d(p,b), d(p,b')\}\};$

$\quad \mathsf{hreg}(a) = \{p \in \mathsf{hreg}(aa') \mid d(p,a) > d(p,a')\}.$

Note that $\mathsf{hreg}(a)$ and $\mathsf{hreg}(a')$ are subregions of $\mathsf{hreg}(aa')$ (see Fig. 2a). The *graph structure* of this diagram is $\mathsf{HVD}(S) = \mathbb{R}^2 \setminus \bigcup_{aa' \in S} (\mathsf{hreg}(a) \cup \mathsf{hreg}(a'))$. An edge of $\mathsf{HVD}(S)$ is called *pure* if it is incident to regions of two distinct segments; and it is called *internal* if it separates the subregions of the same segment. A vertex of $\mathsf{HVD}(S)$ is called *pure* if it is incident to three pure edges, and it is called *mixed* if it is incident to an internal edge. The pure vertices are defined by three distinct sites, and the mixed vertices by two distinct sites.

Definition 2. [1,14] *The* farthest-color Voronoi diagram *is a partitioning of* $\mathbb{R}^2$ *into regions defined as follows:*

$\mathsf{fcreg}(aa') = \{p \in \mathbb{R}^2 \mid \forall bb' \in S \setminus \{aa'\} \colon \min\{d(p,a), d(p,a')\} > \min\{d(p,b), d(p,b')\}\};$

$\mathsf{fcreg}(a) = \{p \in \mathsf{fcreg}(aa') \mid d(p,a) < d(p,a')\}.$

Its graph structure is $\mathsf{FCVD}(S) = \mathbb{R}^2 \setminus \bigcup_{aa' \in S} (\mathsf{fcreg}(a) \cup \mathsf{fcreg}(a'))$. The edges and vertices of $\mathsf{FCVD}(S)$ are characterized as *pure* or *internal* (resp., *pure* or *mixed*) similarly to those of $\mathsf{HVD}(S)$ (see Fig. 2b).

When the segments in S are pairwise disjoint, the structural complexity of $\mathsf{HVD}(S)$ is $O(n)$ [12]. This is not necessarily the case for $\mathsf{FCVD}(S)$. For arbitrary segments, the complexity of both diagrams is $O(n^2)$ [1,17].

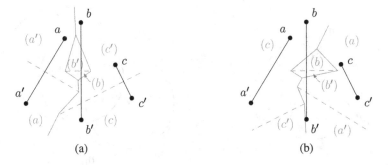

(a)   (b)

Fig. 2. (a) $\mathsf{HVD}(S)$, (b) $\mathsf{FCVD}(S)$. Pure and internal edges are represented in solid and dashed, respectively. The gray letters in parentheses label respective regions.

Let $\overline{\mathsf{hreg}}(\cdot)$ and $\overline{\mathsf{fcreg}}(\cdot)$ denote the closures of the respective regions.

Definition 3. *Given a point p, the* Hausdorff disk *(resp.,* farthest-color disk*) of p, denoted $D_h(p)$ (resp., $D_f(p)$), is the closed disk of radius $d(p,a)$, where $p \in \overline{\mathsf{hreg}}(a)$ (resp., $p \in \overline{\mathsf{fcreg}}(a)$). Its radius is called the* Hausdorff radius *(resp.,* farthest-color radius*) of p, and is denoted as $r_h(p)$ (resp., $r_f(p)$).*

Now we are ready to define $\mathsf{FCVD}^*(S)$, which satisfies that the points in its interior are exactly the centers of all stabbing circles for S (see Lemma 1).

Definition 4. *The $\mathsf{FCVD}^*(S)$ is the locus of points in $\mathbb{R}^2$ for which the farthest-color radius is less than or equal to the Hausdorff radius.*

Both HVD(S) and FCVD(S) can be viewed as envelopes of *wedges* in 3D [12, 14]: Lift up the pairs of endpoints of the segments in S onto the unit paraboloid, and join the lifted endpoints obtaining a set S' of segments in 3D. Using the transformation described in [12], each pair of endpoints of S' is transformed into a pair of planes in 3D, where the lower (resp., upper) envelope of such a pair forms a *lower* (resp., *upper*) *wedge*. The HVD(S) and FCVD(S) correspond to the upper envelope of all lower wedges and the lower envelope of all upper wedges, respectively. Thus, FCVD*(S) corresponds to the locus of points below HVD(S) and above FCVD(S). This is a set of $O(n^2)$ convex cells in 3D with $O(n^2)$ total complexity, and can be computed in $O(n^2)$ time and space [12]. Moreover, this set is a representation of all combinatorially different stabbing planes of S' (if one exists) [12]. We observe that a stabbing circle for S can be transformed into a stabbing plane for S' and vice versa. Thus, we obtain:

Observation 1. *The stabbing circle problem for a set S of n arbitrary segments can be solved in $O(n^2)$ time and space.*

Claverol [7] showed that the set S might have $\Theta(n^2)$ combinatorially different stabbing circles; see Fig. 3a. In the construction, each pair $\{a_i, a_j\}$ of points in the upper arc defines a combinatorially different stabbing circle.

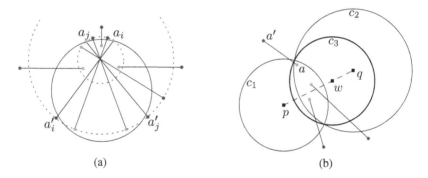

(a)   (b)

Fig. 3. (a) A set with $\Theta(n^2)$ combinatorially different stabbing circles, and the stabbing circle defined by $\{a_i, a_j\}$. (b) Illustration for the proof of Lemma 2.

3   Properties of FCVD*(S)

For brevity, we refer to the connected components of FCVD*(S) as *components* of FCVD*(S). We also let $bis(a, b)$ denote the bisector of two points $a, b \in \mathbb{R}^2$, and ∂f denote the boundary of a region f. Lemmas 1 and 2 are key facts that connect the Hausdorff and farthest-color diagrams and the stabbing circle problem.

Lemma 1. *For a point $p \in \mathbb{R}^2$, there exists a stabbing circle for S with center at p if and only if p lies in the interior of FCVD*(S).*

Proof. Let C be a circle centered at p with radius r, and let D be the closed disk induced by C. Recall that C is a stabbing circle if and only if (1) each segment in S has an endpoint outside D; and (2) each segment in S has an endpoint inside D. Condition (1) is equivalent to $r < r_h(p)$. Condition (2) is equivalent to $r \geq r_f(p)$.

Lemma 2. *For any two points p, q lying in two distinct components of $\mathsf{FCVD}^*(S)$, the stabbing circles for S with centers at p and at q are combinatorially different.*

Proof. Assume for the sake of contradiction that two points p, q lying in different components of $\mathsf{FCVD}^*(S)$ correspond to combinatorially equivalent stabbing circles c_1 and c_2. We assume that c_1 and c_2 intersect at two points, see Fig. 3b; the case when one of the circles encloses the other is similar. For any point $w \in pq$, the circle centered at w and passing through the two points of intersection between c_1 and c_2 is a stabbing circle for S, and is combinatorially equivalent to c_1 and c_2. Therefore the whole segment pq lies in $\mathsf{FCVD}^*(S)$, and p, q lie in the same component. We obtain a contradiction.

In the rest of this paper we focus on computing $\mathsf{FCVD}^*(S)$. It is not hard to see that a component of $\mathsf{FCVD}^*(S)$ is unbounded in a direction ϕ if and only if there exists a stabbing line for S that is orthogonal to ϕ. All the stabbing lines and corresponding components of $\mathsf{FCVD}^*(S)$ can be found in $O(n \log n)$ time [11]. Thus from now on we assume that S has no stabbing line.

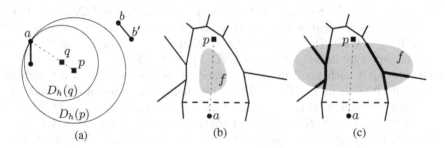

Fig. 4. Illustration for the proof of Lemma 3a: (a) visibility property, item (i); (b) $\mathsf{HVD}(S) \cap f$ cannot be empty; (c) $\mathsf{HVD}(S) \cap f$ cannot be disconnected.

Lemma 3. *Let f be a component of $\mathsf{FCVD}(S)$. Then:*

(a) Each of $\mathsf{HVD}(S) \cap f$ and $\mathsf{FCVD}(S) \cap f$ is one non-empty connected component.
(b) If f intersects an internal edge of $\mathsf{HVD}(S)$, then f contains a mixed vertex of $\mathsf{HVD}(S)$ or $\mathsf{FCVD}(S)$.
(c) ∂f is a closed polygonal line with $O(1 + H + F)$ edges, where H and F denote the number of vertices of $\mathsf{HVD}(S) \cap f$ and $\mathsf{FCVD}(S) \cap f$ respectively.

Proof. (a) The proof is based on the following visibility property of $\mathsf{FCVD}^*(S)$:
  (i) Let p be a point outside $\mathsf{FCVD}^*(S)$, and let aa' be a segment in S such that $p \in \overline{\mathsf{hreg}}(a)$. Then the entire segment $(pa \cap \overline{\mathsf{hreg}}(a))$ is outside $\mathsf{FCVD}^*(S)$.
  (ii) Let p' be a point in $\mathsf{FCVD}^*(S)$, and let aa' be a segment in S such that $p' \in \overline{\mathsf{fcreg}}(a)$. Then the entire segment $(p'a \cap \overline{\mathsf{fcreg}}(a))$ is in $\mathsf{FCVD}^*(S)$.

To prove item (i), let q be any point in $pa \cap \overline{\mathsf{hreg}}(a)$; see Fig. 4a. Clearly, $D_h(q) \subseteq D_h(p)$. Since $p \notin \mathsf{FCVD}^*(S)$, there is a segment $bb' \in S$ such that $b, b' \notin D_h(p)$. Since $D_h(q) \subseteq D_h(p)$, both b, b' are also outside $D_h(q)$. Thus $q \notin \mathsf{FCVD}^*(S)$. Item (ii) can be shown analogously using the disks $D_f(p)$ and $D_f(q)$.

Using item (i), we prove that $f \cap \mathsf{HVD}(S)$ is non-empty and connected. The same fact about $f \cap \mathsf{FCVD}(S)$ can be shown similarly, using item (ii) instead.

Suppose that $f \cap \mathsf{HVD}(S)$ is empty. Then f does not intersect any edge of $\mathsf{HVD}(S)$. Thus $f \subset \mathsf{hreg}(a)$, for some segment aa' in S; see Fig. 4b. Since f is bounded, there is a point $p \in \mathsf{hreg}(a) \setminus f$ such that segment pa intersects f; a contradiction to (i).

Suppose now that $f \cap \mathsf{HVD}(S)$ is disconnected, see Fig. 4c showing $f \cap \mathsf{HVD}(S)$ in bold. Then there is $aa' \in S$ such that $f \cap \partial\mathsf{hreg}(a)$ is disconnected, thus, $\mathsf{hreg}(a) \setminus f$ is not connected. By [17, Property 2], for any point $q \in \mathsf{hreg}(a)$, segment qa intersects $\partial\mathsf{hreg}(a)$ only once in the internal edge of $\mathsf{HVD}(S)$ (shown dashed in Fig. 4c). Then there is a point $p \in \mathsf{hreg}(a) \setminus f$ such that pa intersects f; a contradiction to (i).

(b) Let f intersect an internal edge e of $\mathsf{HVD}(S)$, and let $aa' \in S$ be such that $e \subset bis(a, a')$. By Lemma 3a, $e \cap f$ is a bounded connected component of $bis(a, a') \cap \mathsf{FCVD}^*(S)$. Note that a point $p \in bis(a, a')$ is in $\mathsf{FCVD}^*(S)$ if and only if $r_h(p) = r_f(p) = d(p, a)$. Thus, $bis(a, a') \cap \mathsf{FCVD}^*(S) = bis(a, a') \cap (\mathsf{HVD}(S) \cap \mathsf{FCVD}(S))$. All the bounded connected components of the latter are bounded by mixed vertices of $\mathsf{HVD}(S)$ or $\mathsf{FCVD}(S)$.

(c) Observe that each bend in ∂f is caused by an intersection of ∂f with an edge of $\mathsf{HVD}(S)$ or $\mathsf{FCVD}(S)$. Thus the number of edges in ∂f equals the number of edges of $\mathsf{HVD}(S) \cap f$ and $\mathsf{FCVD}(S) \cap f$ intersected by ∂f. Each of $\mathsf{HVD}(S) \cap f$ and $\mathsf{FCVD}(S) \cap f$ is a connected graph (by Lemma 3a) whose vertices have degree 3 (by the general position assumption); the claim follows from Euler's formula.   □

Lemma 3a, b guarantees that, in order to identify the components of $\mathsf{FCVD}^*(S)$, it is enough to search among the vertices of $\mathsf{HVD}(S)$ and $\mathsf{FCVD}(S)$, and among the pure edges of $\mathsf{HVD}(S)$. Lemma 3c ensures that tracing the boundary of $\mathsf{FCVD}^*(S)$ requires time proportional to the sum of the structural complexities of $\mathsf{HVD}(S)$ and $\mathsf{FCVD}(S)$. This is the base of the algorithm to compute $\mathsf{FCVD}^*(S)$ given in the next section.

4 Computing FCVD*(S)

4.1 General Algorithm

The algorithm, described in Fig. 5, has two main parts. In the first part (lines 2–9), using the characterization of Lemma 3a, b we create a list L such that, for every component f of FCVD*(S), L contains a point u that belongs to f. In the second part (lines 10–11), we use each such point u to trace the boundary of each component f.

Algorithm COMPUTING FCVD*(S)
1.   compute HVD(S) and FCVD(S);
2. $L = \emptyset$;
3.   **for** all vertices v of HVD(S) and FCVD(S) **do**
4.   **if** $v \in$ FCVD*(S) **then**
5. $L \leftarrow L \cup \{v\}$;
6.   **for** all pure edges e of HVD(S) **do**
7.   **for** all components f of FCVD*(S) such that $e \cap f \neq \emptyset$ **do**
8.   find a point w in $e \cap f$;
9. $L \leftarrow L \cup \{w\}$;
10. **for** all points $u \in L$ **do**
11.   trace the component of FCVD*(S) containing u;

Fig. 5. Algorithm to compute FCVD*(S).

Let $\mathcal{T}_{\text{HVD}(S)}$ and $\mathcal{T}_{\text{FCVD}(S)}$ denote the time to compute HVD(S) and FCVD(S) respectively. In general, both $\mathcal{T}_{\text{HVD}(S)}$ and $\mathcal{T}_{\text{FCVD}(S)}$ are $O(n^2)$ [12], but in special cases it is possible to achieve better running times.

With some abuse of notation, we denote by $|\text{HVD}(S)|$, $|\text{FCVD}(S)|$ and $|\text{FCVD}^*(S)|$ respectively the number of edges of HVD(S), FCVD(S), and ∂FCVD*(S). In time $O(|\text{HVD}(S)| \log n)$ and $O(|\text{FCVD}(S)| \log n)$ respectively we can preprocess HVD(S) and FCVD(S) to answer point-location queries in $O(\log n)$ time. Then Step 4 of the algorithm becomes simple: If v is a vertex of say HVD(S), we first locate it in FCVD(S). Once we obtain a segment $aa' \in S$ such that $v \in \overline{\text{fcreg}}(aa')$, we can compare the Hausdorff and farthest-color radii of v in constant time.

Step 11 is easy: Given a point $u \in L$ contained in a component f of FCVD*(S), we first locate u in both HVD(S) and FCVD(S). Then ∂f can be computed in time proportional to its complexity by standard tracing. By Lemma 3c, the total time of Steps 10–11 is $O(|\text{FCVD}^*(S)| \log n + |\text{HVD}(S)| + |\text{FCVD}(S)|)$. Steps 7–9 are discussed next.

4.2 Searching in a Pure Edge of HVD(S)

Let e be a pure edge of HVD(S). We compute one point for each component of FCVD*(S) that intersects e but does not contain any endpoint of e, and add it to L. We first crop e by removing any portions that contain one of its endpoints and belong to $\mathsf{FCVD}^*(S)$. If e is an infinite ray or a line, due to the assumption that S does not have any stabbing line, only a bounded portion of e may be contained in $\mathsf{FCVD}^*(S)$. Thus, we crop e in such a way that the discarded infinite portions have empty intersection with $\mathsf{FCVD}^*(S)$.

Let e be a portion of the border between $\mathsf{hreg}(a)$ and $\mathsf{hreg}(b)$, for two segments $aa', bb' \in S$. Then $e \subseteq bis(a, b)$. If the segment ab intersects the interior of e, this intersection divides e into two portions, which we process separately.

For the rest of this subsection, it is convenient to perform a rotation of the coordinate system so that e is horizontal. Let u and respectively v be the left and right endpoints of e. By hypothesis, neither of u or v belong to $\mathsf{FCVD}^*(S)$. If u is a mixed vertex of $\mathsf{HVD}(S)$, we redefine u as a point on e infinitesimally to the right, so that u is only in the boundary of $\mathsf{hreg}(a)$ and $\mathsf{hreg}(b)$. We proceed analogously with v.

The Hausdorff disks $D_h(u)$ and $D_h(v)$ contain aa', bb', and no other segment of S. Hence, every segment $cc' \in S \setminus \{aa', bb'\}$ can be classified as follows (see Fig. 6):

- cc' is of type *out* if both c and c' are outside $D_h(u) \cup D_h(v)$;
- cc' is of type *in* if either c or c' is contained in $D_h(u) \cap D_h(v)$ and the other endpoint is outside $D_h(u) \cup D_h(v)$;
- cc' is of type *left* if either c or c' is contained in $D_h(u) \setminus D_h(v)$ and the other endpoint is outside $D_h(u) \cup D_h(v)$;
- cc' is of type *right* if either c or c' is contained in $D_h(v) \setminus D_h(u)$ and the other endpoint is outside $D_h(u) \cup D_h(v)$;
- cc' is of type *middle* if either c or c' is contained in $D_h(u) \setminus D_h(v)$ and the other endpoint in $D_h(v) \setminus D_h(u)$.

Let w be any point in e. We define $type(w)$ as a set containing one element per each $cc' \in S$ such that $w \in \overline{\mathsf{fcreg}}(cc')$. The elements of $type(w)$ are defined as follows: Let cc' be a segment in S such that $w \in \overline{\mathsf{fcreg}}(cc')$. If cc' is not of type *middle*, then we add the type of cc' to $type(w)$. If cc' is of type *middle*, then either c or c' (say, c) is contained in $D_h(u) \setminus D_h(v)$, and the other endpoint (c') is contained in $D_h(v) \setminus D_h(u)$. We further differentiate the classification *middle* as follows: If w lies on $bis(c, c')$, then $mm \in type(w)$. Otherwise, if $w \in \overline{\mathsf{fcreg}}(c)$, then $ml \in type(w)$; if $w \in \overline{\mathsf{fcreg}}(c')$, then $mr \in type(w)$. When we need to specify

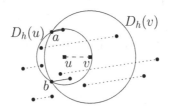

Fig. 6. From top to bottom, the types of the dotted segments are *middle*, *left*, *in*, *right*, and *out*.

cc', we do as follows: Imagine that $w \in \overline{\mathsf{fcreg}}(cc')$ and cc' is of type in. Then we say $in \in type(w)$ *caused by cc'*.

Further, we use $\tilde{l}$ to denote types *left* and *ml*; and $\tilde{r}$ to denote *right* and *mr*.

Notice that if a point w belongs to $\mathsf{FCVD}^*(S)$, then $D_h(w)$ contains at least one endpoint of every segment in S. Using this observation, we make the following analysis. Since $u \notin \mathsf{FCVD}^*(S)$, the segments $cc' \in S$ such that $u \in \overline{\mathsf{fcreg}}(cc')$ have both endpoints outside $D_h(u)$. So $in \notin type(u)$, $\tilde{l} \notin type(u)$, and $mm \notin type(u)$. On the other hand, observe that the Hausdorff disk centered at any point in e is contained in $D_h(u) \cup D_h(v)$. So if $out \in type(u)$, then no point in e lies in $\mathsf{FCVD}^*(S)$ and we can stop the search. The analysis for $type(v)$ is analogous. Therefore, it only remains to consider the case where $\tilde{r}$ (possibly with some multiplicity) is the only element in $type(u)$, and $\tilde{l}$ (possibly with some multiplicity) is the only element in $type(v)$. From now on, we assume that we are in this situation.

For any point $p \in e$, we use p_ℓ and p_r to denote two points in e infinitesimally close to p and lying to the left and right of p, respectively. When dealing with segments ts of e, or pairs (t, s) of points in e, we write the left-most point first.

Definition 5. *A point w in e is a* changing point *if $\{\tilde{r}, \tilde{l}\} \subseteq type(w)$.*

Note that a changing point w is an intersection point between a pure edge of $\mathsf{HVD}(S)$ and a pure edge of $\mathsf{FCVD}(S)$. Intuitively, at w the point giving the farthest-color radius changes from being in $D_h(v) \setminus D_h(u)$ to being in $D_h(u) \setminus D_h(v)$, i.e., $\tilde{r} \in type(w_\ell)$ and $\tilde{l} \in type(w_r)$ (see Fig. 7, left). It is easy to see that a point w' where a change in the other direction happens must be of type mm, i.e., w' is in the intersection between a pure edge of $\mathsf{HVD}(S)$ and an internal edge of $\mathsf{FCVD}(S)$. Then we have:

Observation 2. *Let f be a component of $\mathsf{FCVD}^*(S)$ such that $f \cap e \neq \emptyset$. Then there exists a point w in $f \cap e$ such that $in \in type(w)$ or w is a changing point.*

Thus, it is enough to examine the changing points of e, and the points w such that $in \in type(w)$. To find such points, we use the *find-change query* subroutine:

Definition 6 (Find-change query). *The input of the query is a pair (t, s) of points in edge e, such that $type(t)$ contains $\tilde{r}$ but not $\tilde{l}$, and $type(s)$ contains $\tilde{l}$*

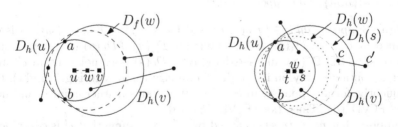

Fig. 7. Left: w is a changing point. Right: $right \in type(s)$ caused by cc' and $w \notin \mathsf{FCVD}^*(S)$.

but not $\tilde{r}$. *The query returns a point* w *in the segment* ts *such that one of the following holds: (i)* w *is a changing point; (ii) in* $\in type(w)$; *(iii) out* $\in type(w)$.

Given a pair (t, s) of points in edge e such that $type(t)$ contains $\tilde{r}$ but not $\tilde{l}$, $type(s)$ contains $\tilde{r}$, and $s_\ell \notin \mathsf{FCVD}^*(S)$, our algorithm uses a subroutine, called CHOPRIGHT, that finds the right-most point s' (if it exists) on the segment ts such that $type(s')$ contains $\tilde{l}$ but not $\tilde{r}$. Additionally, it holds that the interior of the segment $s's$ does not intersect $\mathsf{FCVD}^*(S)$. Notice that it is not possible to perform a find-change query on (t, s), but it is possible on (t, s').

Algorithm CHOPRIGHT
Input: Pair (t, s) of points in e s.t. $\tilde{r} \in type(t)$, $\tilde{l} \notin type(t)$,
  $\tilde{r} \in type(s)$, and $s_\ell \notin \mathsf{FCVD*}(S)$
1. **if** $mr \in type(s)$ caused by cc' **and** $bis(c, c') \cap ts \neq \emptyset$
    then
2. $w \leftarrow bis(c, c') \cap ts$;
3.   **if** $right \notin type(w)$ **and** $mr \in type(w)$ caused by dd'
        and $bis(d, d') \cap tw \neq \emptyset$ **then**
4.   go to line 2 (replacing cc' by dd', and s by w);
5.   **else if** $\tilde{r} \notin type(w)$ **and** $\tilde{l} \in type(w)$ **then**
6.   **return** tw;
7.   **else if** $\tilde{r} \notin type(w)$ **and** $mm \in type(w)$ **then**
8.   **return** tw_ℓ;
9.   **else return** $\emptyset$;
10. **else return** $\emptyset$;

Fig. 8. Algorithm to chop a right portion not intersecting $\mathsf{FCVD}^*(S)$ of a segment.

The pseudocode of CHOPRIGHT is in Fig. 8. It is based on the following:

Lemma 4. *Let* ts *be a segment in* e *such that* $type(t)$ *contains* $\tilde{r}$ *but not* $\tilde{l}$. *If* $type(s)$ *contains* $\tilde{r}$ *and* $s_\ell \notin \mathsf{FCVD}^*(S)$, *then:*

(a) If $right \in type(s)$, *then* $ts_\ell \cap \mathsf{FCVD}^*(S) = \emptyset$.
(b) If $mr \in type(s)$ *caused by* cc', *then* $ts_\ell \cap ws_\ell \cap \mathsf{FCVD}^*(S) = \emptyset$, *where* w *is the intersection between* $bis(c, c')$ *and the supporting line of* e.

If $type(s) = \{mm\}$, *then* $type(s_\ell) = \{\tilde{l}\}$.

Proof.(a) Suppose $right \in type(s)$ caused by cc'. Then c or c' (say, c) is contained in $D_h(v) \setminus D_h(u)$ and c' is outside $D_h(u) \cup D_h(v)$ (see Fig. 7, right). Since $s_\ell \notin \mathsf{FCVD}^*(S)$ and $s \in \overline{\mathsf{fcreg}}(cc')$, $D_h(s)$ does not contain c' and it might contain c only on its "right" boundary (in gray in Fig. 7, right). Consequently, for any point w in ts_ℓ, $D_h(w)$ contains neither c nor c' and hence $w \notin \mathsf{FCVD}^*(S)$. Thus, $ts_\ell \cap \mathsf{FCVD}^*(S) = \emptyset$.
(b) Suppose that $mr \in type(s)$ caused by cc'; we assume that c' is contained in $D_h(u) \setminus D_h(v)$ and c in $D_h(v) \setminus D_h(u)$. Then we know that $s \in \overline{\mathsf{fcreg}}(c)$. We again have that $D_h(s)$ does not contain c' and it might contain c only on its

```
Algorithm SEARCHING IN e
 1.   Q = {e};
 2.   L = ∅;
 3.   while Q ≠ ∅
 4.        u'v' ← element in Q;
 5.        Q ← Q \ u'v';
 6.        perform a find-change query on (u', v');
 7.        let w be the point returned by the query;
 8.        if out ∈ type(w) then
 9.             exit and return NULL;
10.        if w ∈ FCVD*(S) then
11.             L ← L ∪ {w};
12.             find the subsegment qq' of u'v' s.t. w ∈ qq' ⊆ FCVD*(S);
13.             Q ← Q ∪ {CHOPRIGHT(u', q), CHOPLEFT(q', v')};
14.        else (∗ w is a changing point in uv and w ∉ FCVD*(S) ∗)
15.             Q ← Q ∪ {CHOPRIGHT(u', w), CHOPLEFT(w, v')};
16.   return L;
```

Fig. 9. Algorithm to obtain a list L with a point in every component of $\mathsf{FCVD}^*(S)$ intersecting e.

"right" boundary. Furthermore, for all points w' on ts_ℓ, c lies outside $D_h(w')$. Thus w' can lie in $\mathsf{FCVD}^*(S)$ only if c' is contained in $D_h(w')$. This is only possible if c' is closer to w' than c, that is, after we cross the intersection between $bis(c, c')$ and the supporting line of e. We obtain that $ts_\ell \cap ws_\ell \cap \mathsf{FCVD}^*(S) = \emptyset$.

The last statement can be proved similarly. □

CHOPLEFT is a symmetric subroutine, whose details are analogous to the above.

Using all these subroutines, our algorithm to list a point in every component of $\mathsf{FCVD}^*(S)$ intersecting e starts by performing a find-change query on (u, v). It is sketched in Fig. 9. The correctness is based on Observation 2 and Lemma 4, which guarantees that the portions of $u'v'$ that are chopped and discarded during the subroutines CHOPRIGHT and CHOPLEFT have empty intersection with $\mathsf{FCVD}^*(S)$.

4.3   Running Time

Lemma 5. *A find-change query can be performed in $O(\log^2 n)$ time.*

Proof. For a pair (t, s) of points, we perform the find-change query as follows. We use a point-location data structure for $\mathsf{FCVD}(S)$, such that the point location for a query point q is performed by a sequence of $O(\log n)$ atomic questions of the form "is q above or below (resp., to the left or right of) a line ℓ?" (e.g., [13,15]). Notice that in our case instead of a fixed point q, we only have a pair (t, s) such that the segment ts contains desired point(s) (a changing point, a point of type *in* or *out*). An atomic question is processed as follows. If $ts \cap \ell = \emptyset$, the answer is the same for any point in ts, and we continue with the pair (t, s). Otherwise, let point p be $ts \cap \ell$. First, the normal point location for p gives us $type(p)$. If $\tilde{r} \in type(p)$ and $\tilde{l} \notin type(p)$, we continue with the pair (p, s). Symmetrically, if $\tilde{l} \in type(p)$ and $\tilde{r} \notin type(p)$, we continue with the pair (t, p). If $type(p) = \{mm\}$, then we continue with either (p, s) or (t, p). Otherwise, we stop the procedure, and return p. Clearly this happens in one of the following cases: (i) $\{\tilde{l}, \tilde{r}\} \subseteq type(p)$; (ii) $in \in type(p)$; or (iii) $out \in type(p)$.

Answering one atomic question within the procedure takes $O(\log n)$ time, and the whole find-change query takes $O(\log^2 n)$ time.

A similar trick of simulating a point location is used in [6], and in [5].   □

Let m denote the number of pairs formed by a segment $aa' \in S$ and a pure edge e of $\mathsf{HVD}(S)$ such that aa' is of type $middle$ for e. The main result of this subsection is:

Theorem 1. *Let S be a set of n segments in $\mathbb{R}^2$. Then: (a) $\mathsf{FCVD}^*(S)$ can be computed in $O(\mathcal{T}_{\mathsf{HVD}(S)} + \mathcal{T}_{\mathsf{FCVD}(S)} + |\mathsf{HVD}(S)| \log^2 n + |\mathsf{FCVD}(S)| \log n + m \log^2 n)$; (b) $|\mathsf{FCVD}^*(S)| = O(|\mathsf{HVD}(S)| + |\mathsf{FCVD}(S)| + m)$.*

Proof (sketch). In Sect. 4.1 we have argued that the cost of the algorithm Computing $\mathsf{FCVD}^*(S)$, except for Steps 7–9, is $O(\mathcal{T}_{\mathsf{HVD}(S)} + \mathcal{T}_{\mathsf{FCVD}(S)} + |\mathsf{HVD}(S)| \log n + |\mathsf{FCVD}(S)| \log n + |\mathsf{FCVD}^*(S)| \log n)$.

Regarding Steps 7–9, we first show the following: Given a pure edge e of $\mathsf{HVD}(S)$, the number of iterations of Searching in e is $O(m_e)$, where m_e is the number of segments of type $middle$ for e. Roughly speaking, this is done by showing that each iteration can be charged to a segment of type $middle$, and that each of these segments is charged at most two iterations. Indeed, new portions of uv are added to Q only when the algorithm visits line 1 of ChopRight (or ChopLeft), and in this case the bisector of a segment of type $middle$ intersects a segment of Q. Since this segment is cropped precisely at the intersection point, the same bisector does not intersect the interior of other segments of Q in future iterations. This additionally shows that $|L| = O(|\mathsf{HVD}(S)| + |\mathsf{FCVD}(S)| + m)$, where L is the list of the algorithm Computing $\mathsf{FCVD}^*(S)$. Hence, the number of components of $\mathsf{FCVD}^*(S)$ is $O(|\mathsf{HVD}(S)| + |\mathsf{FCVD}(S)| + m)$, which, together with Lemma 3c, yields (b).

Except for line 12, the running time of an iteration of Searching in e is dominated by the find-change query. In total, this amounts to $O((|\mathsf{HVD}(S)| + m) \log^2 n)$ time. Regarding line 12, in order to obtain the points q and q' we might have to cross several edges of $|\mathsf{FCVD}(S)|$. It is possible to show that, if the number of components of $\mathsf{FCVD}^*(S)$ that do not contain a vertex of $\mathsf{HVD}(S)$ is k, then the total number of edges of $\mathsf{FCVD}(S)$ intersected by these components is $O(k + |\mathsf{FCVD}(S)|)$. Thus, in total, line 12 takes $O((|\mathsf{FCVD}^*(S)| + |\mathsf{FCVD}(S)|) \log n)$ time.   □

By Lemma 2, distinct components of $\mathsf{FCVD}^*(S)$ correspond to combinatorially different stabbing circles. There might be combinatorially different stabbing circles inside a single component f of $\mathsf{FCVD}^*(S)$, but this can be easily detected while tracing ∂f. Furthermore, it is easy to see that stabbing circles of minimum and maximum radii are centered on $\partial \mathsf{FCVD}^*(S)$. We obtain:

Corollary 1. *All the combinatorially different stabbing circles for a set S of n segments and the ones with minimum and maximum radius can be computed in $O(\mathcal{T}_{\mathsf{HVD}(S)} + \mathcal{T}_{\mathsf{FCVD}(S)} + |\mathsf{HVD}(S)|\log^2 n + |\mathsf{FCVD}(S)|\log n + m\log^2 n)$ time.*

5   Parallel Segments

Let S be a set of parallel segments. The goal of this section is to prove the following.

Theorem 2. *The stabbing circle problem for a set S of n parallel segments can be solved in $O(n\log^2 n)$ time and $O(n)$ space.*

To prove the above theorem we exploit Corollary 1. First note that $\mathsf{HVD}(S)$ is an instance of *abstract Voronoi diagrams* [16] and thus $|\mathsf{HVD}(S)| = O(n)$ and $\mathcal{T}_{\mathsf{HVD}(S)} = O(n\log n)$. What remains is to show that (1) $m = O(n)$, (2) $|\mathsf{FCVD}(S)| = O(n)$, and (3) $\mathcal{T}_{\mathsf{FCVD}(S)} = O(n\log n)$. Item (1) is immediately implied by Lemma 6. Items (2) and (3) are proved in Theorem 3, which is interesting on its own right.

Lemma 6. *A segment $gg' \in S$ is of type middle for at most one pure edge of $\mathsf{HVD}(S)$.*

Proof (sketch). The proof of this lemma is very technical. Suppose that all segments in S are vertical. Assume for contradiction that segment $gg' \in S$ is of type middle for e_1 and e_2, where e_1 and e_2 are pure edges of $\mathsf{HVD}(S)$ that separate $\mathsf{hreg}(a)$ from $\mathsf{hreg}(b)$, and $\mathsf{hreg}(c)$ from $\mathsf{hreg}(d)$, respectively. We first show that $\min\{x(a), x(b)\} < x(g) < \max\{x(a), x(b)\}$ and $\min\{x(c), x(d)\} < x(g) < \max\{x(c), x(d)\}$. Then, without loss of generality, we can assume that $x(a) \le x(c) < x(b)$. The disk having a, b and g on the boundary corresponds to a disk $D_h(w)$, for some point w on the edge e_1, and does not contain both c and c'. We suppose that c is outside $D_h(w)$. Then we observe that c might lie in two distinct regions of the plane. If c is in one of the regions, we show that every disk containing c and g contains also aa' or bb'. This yields a contradiction with the fact that the Hausdorff disk centered at one of the endpoints of e_2 contains c and g but contains neither aa' nor bb' The other case is similar. □

Theorem 3. *If S is a set of parallel segments, then: (a) The combinatorial complexity of $\mathsf{FCVD}(S)$ is $O(n)$; (b) $\mathsf{FCVD}(S)$ can be constructed in $O(n\log n)$ time.*

Proof (sketch).(a) All unbounded faces of $\mathsf{FCVD}(S)$ coincide at infinity with the faces of the *farthest-segment Voronoi diagram* of S, whose total number is $O(n)$ [3]. Further, $\mathsf{FCVD}(S)$ has at most one bounded face per segment of S.

(b) We use the divide-and-conquer technique. Assuming the segments in S to be vertical, we divide S into two halves by a vertical line. We observe that the merge chain is y-monotone. Thus the merging can be done in a standard way in linear time [3]. □

Future Work. The connection between the stabbing circle problem and the cluster Voronoi diagrams allows to solve the stabbing circle problem in an efficient way under certain conditions on S. The further open question is to investigate for which other segment sets the conditions are satisfied.

Acknowledgments. M. C. and C. S. were supported by projects MTM2012-30951 and Gen.Cat. DGR2014SGR46. E. K. and E. P. were supported by SNF project 20GG21-134355, under the ESF EUROCORES, EuroGIGA/VORONOI program. M. S. was supported by project LO1506 of the Czech Ministry of Education, Youth and Sports, and by project NEXLIZ CZ.1.07/2.3.00/30.0038, co-financed by the European Social Fund and the state budget of the Czech Republic.

References

1. Abellanas, M., Hurtado, F., Icking, C., Klein, R., Langetepe, E., Ma, L.,Palop, B., Sacristán, V.: The farthest color Voronoi diagram and related problems. In: 17th European Workshop on Computational Geometry, pp. 113–116 (2001)
2. Arkin, E.M., Dieckmann, C., Knauer, C., Mitchell, J.S., Polishchuk, V., Schlipf, L., Yang, S.: Convex transversals. Comput. Geom. **47**(2), 224–239 (2014)
3. Aurenhammer, F., Klein, R., Lee, D.T.: Voronoi Diagrams and Delaunay Triangulations. World Scientific, Singapore (2013)
4. Avis, D., Robert, J., Wenger, R.: Lower bounds for line stabbing. Inform. Process. Lett. **33**(2), 59–62 (1989)
5. Cheilaris, P., Khramtcova, E., Langerman, S., Papadopoulou, E.: A randomized incremental approach for the Hausdorff Voronoi diagram of non-crossing clusters. In: Pardo, A., Viola, A. (eds.) LATIN 2014. LNCS, vol. 8392, pp. 96–107. Springer, Heidelberg (2014)
6. Cheong, O., Everett, H., Glisse, M., Gudmundsson, J., Hornus, S., Lazard, S., Lee, M., Na, H.: Farthest-polygon Voronoi diagrams. Comput. Geom. **44**(4), 234–247 (2011)
7. Claverol, M.: Problemas geométricos en morfología computacional. Ph.D. thesis, Universitat Politècnica de Catalunya (2004)
8. Claverol, M., Garijo, D., Grima, C.I., Márquez, A., Seara, C.: Stabbers of line segments in the plane. Comput. Geom. **44**(5), 303–318 (2011)
9. Claverol, M., Garijo, D., Korman, M., Seara, C., Silveira, R.I.: Stabbing segments with rectilinear objects. In: Kosowski, A., Walukiewicz, I. (eds.) FCT 2015. LNCS, vol. 9210, pp. 53–64. Springer, Heidelberg (2015)
10. Díaz-Báñez, J.M., Korman, M., Pérez-Lantero, P., Pilz, A., Seara, C., Silveira, R.I.: New results on stabbing segments with a polygon. Comput. Geom. **48**(1), 14–29 (2015)
11. Edelsbrunner, H., Maurer, H., Preparata, F., Rosenberg, A., Welzl, E., Wood, D.: Stabbing line segments. BIT **22**(3), 274–281 (1982)
12. Edelsbrunner, H., Guibas, L.J., Sharir, M.: The upper envelope of piecewise linear functions: algorithms and applications. Discrete Comput. Geom. **4**, 311–336 (1989)
13. Edelsbrunner, H., Guibas, L.J., Stolfi, J.: Optimal point location in a monotone subdivision. SIAM J. Comput. **15**(2), 317–340 (1986)
14. Huttenlocher, D.P., Kedem, K., Sharir, M.: The upper envelope of Voronoi surfaces and its applications. Discrete Comput. Geom. **9**(1), 267–291 (1993)

15. Kirkpatrick, D.: Optimal search in planar subdivisions. SIAM J. Comput. **12**(1), 28–35 (1983)
16. Klein, R.: Concrete and Abstract Voronoi Diagrams. LNCS, vol. 400. Springer, Heidelberg (1989)
17. Papadopoulou, E., Lee, D.T.: The Hausdorff Voronoi diagram of polygonal objects: a divide and conquer approach. Int. J. Comput. Geom. Appl. **14**(6), 421–452 (2004)
18. Papadopoulou, E.: The Hausdorff Voronoi diagram of point clusters in the plane. Algorithmica **40**(2), 63–82 (2004)
19. Rappaport, D.: Minimum polygon transversals of line segments. Int. J. Comput. Geom. Appl. **5**(3), 243–256 (1995)

Faster Algorithms to Enumerate Hypergraph Transversals

Manfred Cochefert[1], Jean-François Couturier[2], Serge Gaspers[3,4(✉)], and Dieter Kratsch[1]

[1] LITA, Université de Lorraine, Metz, France
manfred.cochefert@gmail.com, dieter.kratsch@univ-lorraine.fr
[2] CReSTIC, Université de Reims, Reims, France
jean-francois.couturier@univ-reims.fr
[3] University of New South Wales, Sydney, Australia
sergeg@cse.unsw.edu.au
[4] Data61 (formerly: NICTA), CSIRO, Sydney, Australia

Abstract. A transversal of a hypergraph is a set of vertices intersecting each hyperedge. We design and analyze new exponential-time polynomial-space algorithms to enumerate all inclusion-minimal transversals of a hypergraph. For each fixed $k \geq 3$, our algorithms for hypergraphs of rank k, where the rank is the maximum size of a hyperedge, outperform the previous best. This also implies improved upper bounds on the maximum number of minimal transversals in n-vertex hypergraphs of rank $k \geq 3$. Our main algorithm is a branching algorithm whose running time is analyzed with Measure and Conquer. It enumerates all minimal transversals of hypergraphs of rank 3 in time $O(1.6755^n)$. Our enumeration algorithms improve upon the best known algorithms for counting minimum transversals in hypergraphs of rank k for $k \geq 3$ and for computing a minimum transversal in hypergraphs of rank k for $k \geq 6$.

1 Introduction

A *hypergraph* H is a couple (V, E), where V is a set of vertices and E is a set of subsets of V called *hyperedges*. A *transversal* of H is a subset of vertices $S \subseteq V$ such that each hyperedge of H contains at least one vertex from S. A transversal of H is *minimal* if it does not contain a transversal of H as a proper subset. The *rank* of a hypergraph H is the maximum size of a hyperedge. Finding, counting and enumerating (minimal) transversals fulfilling certain constraints are fundamental problems in Theoretical Computer Science with many important applications, for example in artificial intelligence, biology, logics, relational and distributed databases, Boolean switching theory and model-based diagnosis [4, Section 3]. The notions *hitting set* and transversal are synonymous, both of them name a subset of elements (vertices) having non empty intersection with each subset (hyperedge) of a given set system (hypergraph). We shall use both

© Springer-Verlag Berlin Heidelberg 2016
E. Kranakis et al. (Eds.): LATIN 2016, LNCS 9644, pp. 306–318, 2016.
DOI: 10.1007/978-3-662-49529-2_23

notions interchangeably usually speaking of transversals in hypergraphs whenever adressing enumeration, while speaking of hitting sets when adressing optimization and counting. The MINIMUM HITTING SET problem is a well-studied problem that, like its dual MINIMUM SET COVER, belongs to the list of 21 problems shown to be NP-complete by Karp in 1972 [21]. It can be seen as an extension of the fundamental graph problems MINIMUM DOMINATING SET, MINIMUM VERTEX COVER and MAXIMUM INDEPENDENT SET; all of them also belonging to Karp's list [21]. These fundamental NP-complete problems have been studied extensively from many algorithmic viewpoints; among them exact, approximation and parameterized algorithms.

Prior to Our Work. Wahlström studied MINIMUM HITTING SET on hypergraphs of rank 3 and achieved an $O(1.6359^n)$ time polynomial-space algorithm as well as an $O(1.6278^n)$ time exponential-space algorithm [27,28]. Here, n denotes the number of vertices. In an attempt to show that iterative compression can be useful in exact exponential-time algorithms, Fomin et al. [10] studied MINIMUM HITTING SET on hypergraphs of rank at most k for any fixed $k \geq 2$ as well as the problem of counting minimum hitting sets and achieved best known running times for most of these problems without improving upon Wahlström's algorithms (see also [10, Table 1].) The MINIMUM HITTING SET problem on hypergraphs of fixed rank k has also been studied from a parameterized point of view by various authors [7–10,26,28] (see [10, Table 2].) A result of Cygan et al. [3] states that for every $c < 1$, MINIMUM HITTING SET cannot be solved by an $O(2^{cn})$ algorithm unless the Strong Exponential Time Hypothesis fails, while there is a $O(2^n)$ algorithm based on verifying all subsets of elements. The only nontrivial exponential-time algorithm to enumerate all minimal transversals prior to our work was given by Gaspers [14]. It is a branching algorithm for hypergraphs of rank k whose branching rule selects a hyperedge of size k in the worst case and recurses on instances with $n-1, n-2, \ldots,$ and $n-k$ vertices. We say that its *branching vector* is $(1, 2, \ldots, k)$ and the corresponding recurrence for $k = 3$ gives a running time of $O(1.8393^n)$.

Our Techniques and Results. We use various properties of minimal transversals to design polynomial-space branching algorithms and analyze them using an elaborate Measure & Conquer analysis. For an in-depth treatment of branching algorithms we refer to [12]. For details on our approach, see Sect. 3.1 and [15,16]. Our main result is the algorithm for hypergraphs of rank 3 in Sect. 3 which runs in time $O(1.6755^n)$. In Sect. 4 we show that the iterative compression approach from [10] can be extended to enumeration problems and obtain an algorithm of running time $O(1.8863^n)$ for hypergraphs of rank 4. In Sect. 5 we construct branching algorithms to enumerate the minimal transversals of hypergraphs of rank k, for all fixed $k \geq 5$. Our algorithmic results combined with implied upper bounds and new combinatorial lower bounds for the maximum number of minimal transversals in n-vertex hypergraphs of rank k are summarized in Table 1. As a byproduct, our enumeration algorithms can be used to improve upon the best known algorithms for counting minimum hitting sets in hypergraphs of rank $k \geq 3$ and for solving MINIMUM HITTING SET in hypergraphs of rank $k \geq 6$.

Table 1. Lower and upper bounds for the maximum number of minimal transversals in an n-vertex hypergraph of rank k.

Rank	Lower bound	Upper bound	Rank	Lower bound	Upper bound
2	1.4422^n	1.4423^n [24,25]	7	1.7734^n	$O(1.9893^n)$
3	1.5848^n	$O(1.6755^n)$	8	1.7943^n	$O(1.9947^n)$
4	1.6618^n	$O(1.8863^n)$	9	1.8112^n	$O(1.9974^n)$
5	1.7114^n	$O(1.9538^n)$	10	1.8253^n	$O(1.9987^n)$
6	1.7467^n	$O(1.9779^n)$	20	1.8962^n	$O(1.9999988^n)$

Other Related Work. Enumerating all minimal transversals of a hypergraph is probably the most studied problem in output-sensitive enumeration. It is still open whether it has an output-polynomial time algorithm, i.e. an algorithm whose running time is polynomial in the input size and the output size, despite efforts of more than thirty years including many of the leading researchers of the field. These efforts have produced many publications on the enumeration of the minimal transversals on special hypergraphs [4–6,13,22,23] and has also turned output-sensitive enumeration into an active field of research. Recent progress by Kanté et al. [18] showing that an output-polynomial time algorithm to enumerate all minimal dominating sets of a graph would imply an output-polynomial time algorithm to enumerate the minimal transversals of a hypergraph has triggered a lot of research on the enumeration of minimal dominating sets, both in output-sensitive enumeration and exact exponential enumeration [2,17,19,20].

Subsequent to our work, Fomin et al. [11] improved the combinatorial upper bounds presented in this paper to $(2 - 1/k)^n n^{O(1)}$ and gave corresponding exponential-space enumeration algorithms. Our algorithms remain the fastest known polynomial-space enumeration algorithms.

2   Preliminaries

We refer to the set of vertices and hyperedges of a hypergraph $H = (V, E)$ by $V(H)$ and $E(H)$, respectively. Throughout the paper we denote the number of vertices of a hypergraph by n. The *degree* of a vertex v in H, denoted $d_H(v)$, is the number of hyperedges in E containing v. The *neighborhood* of v in H, denoted $N_H(v)$, is the set of vertices that occur with v in some hyperedge of H. We denote by $d_{i,H}(v)$ the number of hyperedges of size i in H containing v. We denote $d_{\leq i,H}(v) := \sum_{j=0}^{i} d_{j,H}(v)$. We omit the subscript H when it is clear from the context. If the hypergraph is viewed as a set system E over a ground set V, transversals are also called *hitting sets*. We say that a vertex v *hits* a hyperedge e if $v \in e$. For a full version of this paper, see [1].

3   Hypergraphs of Rank 3

3.1   Measure

We will now introduce the measure we use to track the progress of our algorithm. A *measure* μ for a problem P is a function from the set of all instances for P to the set of non-negative reals. Our measure will take into account the degrees of the vertices and the number of hyperedges of size 2. Measures depending on vertex degrees have become relatively standard in the literature. As for hyperedges of size 2, our measure, similar to Wahlström's [27], indicates an advantage when we can branch on hyperedges of size 2 once or several times.

Let $H = (V, E)$ be a hypergraph of rank at most 3. Denote by n_k the number of vertices of degree $k \in \mathbb{N}$ and by m_k the number of hyperedges of size $k \in \{0, \ldots, 3\}$. Also, denote by $m_{\leq k} := \sum_{i=0}^{k} m_i$. We define the measure of H by

$$\mu(H) = \Psi(m_{\leq 2}) + \sum_{i=0}^{\infty} \omega_i n_i \ ,$$

where $\Psi : \mathbb{N} \to \mathbb{R}_{\geq 0}$ is a non-increasing non-negative function independent of n, and ω_i are non-negative reals. Clearly, $\mu(H) \geq 0$.

We will now make some assumptions simplifying our analysis and introduce notations easing the description of variations in measure. We constrain

$$\omega_i := \omega_5 \ , \qquad\qquad \Psi(i) := 0 \qquad\qquad \text{for each } i \geq 6, \qquad (1)$$
$$\Delta\omega_i := \omega_i - \omega_{i-1} \ , \qquad \Delta\Psi(i) := \Psi(i) - \Psi(i-1) \qquad \text{for each } i \geq 1, \qquad (2)$$
$$0 \leq \Delta\omega_{i+1} \leq \Delta\omega_i \quad \text{and } 0 \geq \Delta\Psi(i+1) \geq \Delta\Psi(i) \qquad \text{for each } i \geq 1 \ . \qquad (3)$$

Note that, by (3), we have that $\omega_i - \omega_{i-k} \geq k \cdot \Delta\omega_i$ for $0 \leq k \leq i$. In addition, our branching rules will add constraints on the measure. Note that a branching rule with one branch is a *reduction* rule and one with no branch is a *halting* rule. Denote by $T(\mu) := 2^\mu$ an upper bound on the number of leaves of the search trees modeling the recursive calls of the algorithm for all H with $\mu(H) \leq \mu$.

Suppose a branching rule makes k recursive calls on instances $B[1], \ldots, B[k]$, each $B[i]$ decreasing the measure by η_i. Then, we obtain the following constraint on the measure: $\sum_{i=1}^{k} T(\mu - \eta_i) \leq T(\mu)$. Dividing by 2^μ, the constraint becomes

$$\sum_{i=1}^{k} 2^{-\eta_i} \leq 1 \ . \qquad (4)$$

Given these constraints for all branching rules, we will determine values for $\Psi(0), \ldots, \Psi(5), \omega_0, \ldots, \omega_5$ so as to minimize the maximum value of $\mu(H)/|V(H)|$ taken over all rank-3 hypergraphs H when $|V(H)|$ is large. Since $\Psi(m_{\leq 2})$ is a constant, this part of the measure contributes only a constant factor to the running time. Given our assumptions on the weights, optimizing the measure amounts to solving a convex program [15,16] minimizing ω_5 subject to all constraints. If we make sure that the maximum recursion depth of the algorithm is polynomial, we obtain that the running time is within a polynomial factor of 2^{ω_5}. Formally, we will use the following lemma to upper bound the running time.

Lemma 1 ([16], **Lemma 2.5 in** [15]). *Let A be an algorithm for a problem P, $c \geq 0$ be a constant, and μ, η be measures for the instances of P, such that for any input instance I, Algorithm A transforms I into instances $I_1, \ldots, I_k$, solves these recursively, and combines their solutions to solve I, using time $O(\eta(I)^c)$ for the transformation and combination steps (but not the recursive solves), and*

$$(\forall i) \quad \eta(I_i) \leq \eta(I) - 1 \text{ , and} \tag{5}$$

$$\sum_{i=1}^{k} 2^{\mu(I_i)} \leq 2^{\mu(I)} \text{ .} \tag{6}$$

Then A solves any instance I in time $O(\eta(I)^{c+1})2^{\mu(I)}$.

3.2 Algorithm

An instance of a recursive call of the algorithm is a hypergraph $H = (V, E)$ with rank at most 3 and a set S, which is a partial hitting set for the original hypergraph. The hypergraph H contains all hyperedges that still need to be hit, and the vertices that are eligible to be added to S. Thus, $V \cap S = \emptyset$. Initially, $S = \emptyset$. Each branching rule has a condition which is a prerequisite for applying the rule. When the prerequisites of more than one rule hold, the first applicable rule is used. Our branching rules create subinstances where some vertices are selected and others are discarded.

– If we *select* a vertex v, we remove all hyperedges containing v from the subinstance of the branch, we add v to S, and remove v from V.
– If we *discard* a vertex v, we remove v from all hyperedges and from V.

We now come to the description of the branching rules, their correctness, and their analysis, i.e., the constraints they impose on the measure. Rules $0.x$ are halting rules, and Rules $1.x$ are reduction rules. Each rule first states the prerequisite, then describes its actions, then the soundness is proved if necessary, and the constraints on the measure are given for the analysis.

Rule 0.0 $\emptyset \in E$. Do nothing. The algorithm backtracks and enumerates no transversal in this recursive call since there is a hyperedge that cannot be hit.

Rule 0.1 $E = \emptyset$. If S is a minimal transversal for the original hypergraph, output S, otherwise do nothing. We are in a recursive call where no hyperedge remains to be hit. Thus, S is a transversal of the original input hypergraph. However, S might not be minimal, and the algorithm checks in polynomial time if S is a minimal transversal of the initial hypergraph and outputs it if so.

Rule 1.0 There is a vertex $v \in V$ with degree 0. Discard v. Indeed, since v hits no hyperedge of H, a transversal for H containing v is not minimal. The constraint on the measure is $2^{-\omega_0} \leq 1$, which is trivially satisfied since $\omega_0 \geq 0$.

Rule 1.1 There is a hyperedge $e_1 \in E$ that is a subset of another hyperedge $e_2 \in E$ of size 3. Remove e_2 from E. The rule is sound since each transversal of

$(V, E \setminus \{e_2\})$ is also a transversal of H. Since e_2 has size 3, this rule has no effect on the measure; the constraint $2^0 \leq 1$ is always satisfied.

Rule 1.2 There is a hyperedge e of size 1. Select v, where $\{v\} = e$. The rule is sound since v is the only vertex that hits e. Selecting v removes v and all hyperedges containing v from the instance. By (3), the decrease in measure is at least $\eta_1 = \omega_{d(v)} + \Psi(d_{\leq 2}(v)) - \Psi(0)$. To fulfill (4), we will need to constrain that $\eta_1 \geq 0$. Since $d_{\leq 2}(v) \leq d(v)$ and by (3), if suffices to constrain

$$\Psi(i) - \Psi(0) \geq -\omega_i \qquad \text{for } 1 \leq i \leq 6 . \tag{7}$$

Rule 2 There is a vertex $v \in V$ with degree one. Denote by e the unique hyperedge containing v and branch according to the following three subrules.

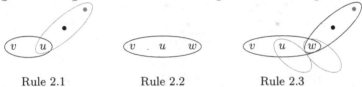

Rule 2.1   Rule 2.2   Rule 2.3

Rule 2.1 $|e| = 2$. Denote $e = \{v, u\}$. Branch into two subproblems: $B[1]$ where v is selected, and $B[2]$ where v is discarded. In $B[1]$, the vertex u is discarded due to minimality of the transversal, and the number of hyperedges of size at most 2 decreases by 1 since e is removed. By (3) the decrease in measure in $B[1]$ is at least $\eta_1 = \omega_1 + \omega_{d(u)} + \Delta\Psi(m_{\leq 2})$. In $B[2]$, the vertex u is selected by applying Rule 1.2 after having discarded v. We have that $d_{\leq 2}(u) \leq \min(d(u), m_{\leq 2})$ hyperedges of size at most 2 disappear. By (3) the decrease in measure in $B[2]$ is at least $\eta_2 = \omega_1 + \omega_{d(u)} + \Psi(m_{\leq 2}) - \Psi(\max(m_{\leq 2} - d(u), 0))$. Note that we do not take into account additional sets of size at most 2 that may be created by discarding u and degree-decreases of the other vertices in the neighborhood of u as a result of selecting u. However, these do not increase the measure due to constraints (3). Thus, we constrain that

$$2^{-\omega_1 - \omega_{d(u)} - \Delta\Psi(m_{\leq 2})} + 2^{-\omega_1 - \omega_{d(u)} - \Psi(m_{\leq 2}) + \Psi(\max(m_{\leq 2} - d(u), 0))} \leq 1 , \tag{8}$$

for $1 \leq d(u) \leq 6$ and $1 \leq m_{\leq 2} \leq 6$. Note that the value of $d(u)$ ranges up to 6 instead of 5 although $\omega_5 = \omega_6$, so that we have a constraint modeling that the value of Ψ increases from $\Psi(6) = 0$ to $\Psi(0)$ in the second branch.

In the remaining subrules of Rule 2, the hyperedge e has size 3.

Rule 2.2 The other two vertices in e also have degree 1. Branch into 3 subproblems adding exactly one vertex from e to S and discarding the other 2. Clearly, any minimal transversal contains exactly one vertex from e. The decrease in measure is $3\omega_1$ in each branch, giving the constraint

$$3 \cdot 2^{-3\omega_1} \leq 1 . \tag{9}$$

Rule 2.3 Otherwise, let $e = \{v, u, w\}$ with $d(u) \geq 2$. We create two subproblems: in $B[1]$ we select v and in $B[2]$ we discard v. Additionally, in $B[1]$ we discard u and w due to minimality. The measure decreases by at least $\eta_1 = 2\omega_1 + \omega_2$ and $\eta_2 = \omega_1 - \max_{i \geq 1}\{\Delta\Psi(i)\} = \omega_1$; the last equality holds since $\Psi(i) = 0, i \geq 6$, and by (3). We obtain the constraint

$$2^{-2\omega_1 - \omega_2} + 2^{-\omega_1} \leq 1 \ . \tag{10}$$

Rule 3 At least one hyperedge has size 2. Let v be a vertex that has maximum degree among all vertices contained in a maximum number of hyperedges of size 2. Let $e = \{v, u_1\}$ be a hyperedge containing v. Note that, due to Rule 2, every vertex has degree at least 2. We branch according to the following subrules.

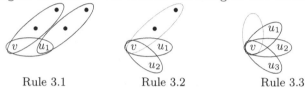

Rule 3.1   Rule 3.2   Rule 3.3

Rule 3.1 $d_2(v) = 1$. We branch into two subproblems: in $B[1]$ we select v and in $B[2]$ we discard v. Additionally, in $B[2]$, we select u_1 by Rule 1.2. Since u_1 is contained in the hyperedge e and since $d_2(u_1) \leq d_2(v)$, we have that $d_2(u_1) = 1$ and therefore $d(v) \geq d(u_1)$. In $B[1]$, we observe that the degrees of v's neighbors decrease. Also, e is a hyperedge of size 2 and it is removed; this affects the value of Ψ. The measure decrease in the first branch is therefore at least $\omega_{d(v)} + \Delta\omega_{d(u_1)} + \Delta\Psi(m_{\leq 2}) \geq \omega_{d(u_1)} + \Delta\omega_{d(u_1)} + \Delta\Psi(m_{\leq 2})$. In $B[2]$, since we select u_1, we have that $d_2(u_1) = 1$ hyperedge of size 2 disappears, and since we discard v, we have that $d(v) - 1$ sets of size 3 will become sets of size 2. Also, none of these size-2 sets already exist in E, otherwise Rule 1.1 would apply. We have a measure decrease of $\omega_{d(v)} + \omega_{d(u_1)} + \Psi(m_{\leq 2}) - \Psi(m_{\leq 2} + d(v) - 2) \geq 2\omega_{d(u_1)} + \Psi(m_{\leq 2}) - \Psi(m_{\leq 2} + d(u_1) - 2)$. We obtain the following set of constraints:

$$2^{-\omega_{d(u_1)} - \Delta\omega_{d(u_1)} - \Delta\Psi(m_{\leq 2})} + 2^{-2\omega_{d(u_1)} - \Psi(m_{\leq 2}) + \Psi(m_{\leq 2} + d(u_1) - 2)} \leq 1 \ , \tag{11}$$

for each $d(u_1)$ and $m_{\leq 2}$ with $2 \leq d(u_1) \leq 6$ and $1 \leq m_{\leq 2} \leq 6$. Here, $d(u_1)$ ranges up to 6 since $\Delta\omega_6 = 0$, whereas $\Delta\omega_5$ may be larger than 0.

Rule 3.2 $d_2(v) = 2$. We branch into two subproblems: in $B[1]$ we select v and in $B[2]$ we discard v. Denoting $\{v, u_2\}$ the second hyperedge of size 2 containing v, we additionally select u_1 and u_2 in $B[2]$. In $B[1]$, the measure decrease is at least $\eta_1 = \omega_{d(v)} + \Delta\omega_{d(u_1)} + \Delta\omega_{d(u_2)} + \Psi(m_{\leq 2}) - \Psi(m_{\leq 2} - 2)$. In $B[2]$, selecting u_1 and u_2 removes $d_2(u_1) + d_2(u_2) \leq \min(4, m_{\leq 2})$ hyperedges of size 2, and discarding v decreases the size of $d(v) - 2$ hyperedges from 3 to 2. Thus, the measure decrease is at least $\eta_2 = \omega_{d(v)} + \omega_{d(u_1)} + \omega_{d(u_2)} + \Psi(m_{\leq 2}) - \Psi(\max(m_{\leq 2} - 4, 0) + d(v) - 2)$. We obtain the following set of constraints:

$$2^{-\omega_{d(v)}-\Delta\omega_{d(u_1)}-\Delta\omega_{d(u_2)}-\Psi(m_{\leq 2})+\Psi(m_{\leq 2}-2)}$$

$$+\, 2^{-\omega_{d(v)}-\omega_{d(u_1)}-\omega_{d(u_2)}-\Psi(m_{\leq 2})+\Psi(\max(m_{\leq 2}-4,0)+d(v)-2)} \leq 1\ , \qquad (12)$$

for $2 \leq d(v), d(u_1), d(u_2) \leq 6$ and $2 \leq m_{\leq 2} \leq 6$.

Rule 3.3 $d_2(v) \geq 3$. We branch into two subproblems: in $B[1]$ we select v and in $B[2]$ we discard v. In $B[2]$ we additionally select all vertices occurring in hyperedges of size 2 with v. Denote by $\{v, u_2\}$ and $\{v, u_3\}$ a second and third hyperedge of size 2 containing v. In $B[1]$, the number of size-2 hyperedges decreases by $d_2(v)$. The measure decrease is at least $\omega_{d_2(v)} + \Delta\omega_{d(u_1)} + \Delta\omega_{d(u_2)} + \Delta\omega_{d(u_3)} + \Psi(m_{\leq 2}) - \Psi(m_{\leq 2} - d_2(v))$. In $B[2]$, the number of hyperedges of size at most 2 decreases at most by $d_2(u_1) + d_2(u_2) + d_2(u_3) \leq \min(d(u_1) + d(u_2) + d(u_3), m_{\leq 2})$. We obtain a measure decrease of at least $\omega_{d_2(v)} + \omega_{d(u_1)} + \omega_{d(u_2)} + \omega_{d(u_3)} + (d_2(v) - 3) \cdot \omega_2 + \Psi(m_{\leq 2}) - \Psi(\max(m_{\leq 2} - d(u_1) - d(u_2) - d(u_3), 0))$. The family of constraints for this branching rule is therefore

$$2^{-\omega_{d_2(v)}-\sum_{i=1}^{3}\Delta\omega_{d(u_i)}-\Psi(m_{\leq 2})+\Psi(m_{\leq 2}-d_2(v))}$$

$$+\, 2^{-\omega_{d_2(v)}-\sum_{i=1}^{3}\omega_{d(u_i)}-(d_2(v)-3)\cdot\omega_2-\Psi(m_{\leq 2})+\Psi(\max(m_{\leq 2}-\sum_{i=1}^{3}d(u_i),0))} \leq 1\ , \quad (13)$$

where $3 \leq d_2(v) \leq m_{\leq 2} \leq 6$ and $2 \leq d(u_1), d(u_2), d(u_3) \leq 6$.

Rule 4 Otherwise, all hyperedges have size 3. Choose $v \in V$ with maximum degree, and branch according to the following subrules.

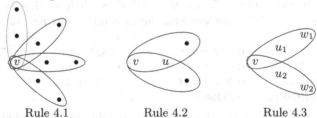

Rule 4.1   Rule 4.2   Rule 4.3

Rule 4.1 $d(v) \geq 3$. We branch into two subproblems: in $B[1]$ we select v and in $B[2]$ we discard v. In $B[1]$, the degree of each of v's neighbors decreases by the number of hyperedges it shares with v. We obtain a measure decrease of at least $\omega_{d(v)} + \sum_{u \in N(v)}(\omega_{d(u)} - \omega_{d(u)-|\{e \in E : \{u,v\} \subseteq e\}|})$, which, by (3), is at least $\omega_{d(v)} + 2 \cdot d(v) \cdot \Delta\omega_{d(v)}$. In $B[2]$, the number of hyperedges of size 2 increases from 0 to $d(v)$, for a measure decrease of at least $\omega_{d(v)} + \Psi(0) - \Psi(d(v))$. For $d(v) \in \{3, \ldots, 6\}$, this branching rules gives the constraint

$$2^{-\omega_{d(v)}-2\cdot d(v)\cdot\Delta\omega_{d(v)}} + 2^{-\omega_{d(v)}-\Psi(0)+\Psi(d(v))} \leq 1\ . \qquad (14)$$

Rule 4.2 $d(v) = 2$ and there is a vertex $u \in V \setminus \{v\}$ which shares two hyperedges with v. We branch into two subproblems: in $B[1]$ we select v and in $B[2]$ we discard v. Additionally, we discard u in $B[1]$ due to minimality. Since each vertex has degree 2, we obtain the following constraint.

$$2^{-2\omega_2 - 2\Delta\omega_2} + 2^{-\omega_2 - \Psi(0) + \Psi(2)} \leq 1 \ . \tag{15}$$

Rule 4.3 Otherwise, $d(v) = 2$ and for every two distinct $e_1, e_2 \in E$ with $v \in e_1$ and $v \in e_2$ we have that $e_1 \cap e_2 = \{v\}$. Denoting $e_1 = \{v, u_1, w_1\}$ and $e_2 = \{v, u_2, w_2\}$ the two hyperedges containing v, we branch into three subproblems: in $B[1]$ we select v and u_1; in $B[2]$ we select v and discard u_1; and in $B[3]$ we discard v. Additionally, we discard u_2 and w_2 in $B[1]$ due to minimality. Again, all vertices have degree 2 in this case. In $B[1]$, the degree of w_1 decreases to 1. Among the two hyperedges containing u_2 and w_2 besides e_2, at most one is hit by u_1, since $d(u_1) = 2$, and none of them is hit by v; thus, the branch creates at least one hyperedge of size at most 2. The measure decrease is at least $4\omega_2 + \Delta\omega_2 - \Delta\Psi(1)$. In $B[2]$, the degrees of w_1, u_2, and w_2 decrease by 1 and the size of a hyperedge containing u_1 decreases. The measure decrease is at least $2\omega_2 + 3\Delta\omega_2 - \Delta\Psi(1)$. In $B[3]$, two size-2 hyperedges are created for a measure decrease of $\omega_2 - \Delta\Psi(2)$. This gives us the following constraint:

$$2^{-4\omega_2 - \Delta\omega_2 + \Delta\Psi(1)} + 2^{-2\omega_2 - 3\Delta\omega_2 + \Delta\Psi(1)} + 2^{-\omega_2 + \Delta\Psi(2)} \leq 1 \ . \tag{16}$$

This finishes the description of the algorithm. We can now prove an upper bound on its running time, along the lines described in Sect. 3.1.

Theorem 1. *The described algorithm enumerates all minimal transversals of an n-vertex hypergraph of rank 3 in time $O(1.6755^n)$.*

Proof. Consider any input instance $H = (V, E)$ with n vertices and measure $\mu = \mu(H)$. Using the following weights, the measure μ satisfies all constraints.

i	ω_i	$\Psi(i)$	i	ω_i	$\Psi(i)$
0	0	0.566096928	4	0.742114220	0.119795899
1	0.580392137	0.436314617	5	0.744541491	0.035202514
2	0.699175718	0.306532603	6	0.744541491	0
3	0.730706814	0.211986294			

Also, $\mu \leq \omega_5 \cdot n + \Psi(0)$ by (3). Therefore, the number of times a halting rule is executed (i.e., the number of leaves of the search tree) is at most $2^{\omega_5 \cdot n + O(1)}$. Since each recursive call of the algorithm decreases the measure $\eta(H) := |V| + |E|$ by at least 1, the height of the search tree is polynomial. We conclude, by Lemma 1, that the algorithm has running time $O(1.6755^n)$ since $2^{\omega_5} = 1.6754...$ □

4   Hypergraphs of Rank 4

For hypergraphs of rank 4, we adapt an iterative compression algorithm of [10] for counting the number of minimum transversals to the enumeration setting.

Theorem 2. *Suppose there is an algorithm with running time $O^*((a_{k-1})^n)$, $1 < a_{k-1} \leq 2$, enumerating all minimal transversals in rank-$(k-1)$ hypergraphs. Then all minimal transversals in a rank-k hypergraph can be enumerated in time*

$$\min_{0.5 \leq \alpha \leq 1} \max \left\{ O^* \left(\binom{n}{\alpha n} \right), O^* \left(2^{\alpha n} (a_{k-1})^{n - \alpha n} \right) \right\} .$$

Combined with Theorem 1, the running time is minimized for $\alpha \approx 0.66938$.

Theorem 3. *The described algorithm enumerates all minimal transversals of an n-vertex hypergraph of rank 4 in time $O(1.8863^n)$.*

5   Hypergraphs of Rank at Least 5

For a hypergraph $H = (V, E)$ of rank $k \geq 5$, we use the following algorithm to enumerate all minimal transversals. As in Sect. 3.2, the instance of a recursive call of the algorithm is a hypergraph $H = (V, E)$ and set S which is a partial transversal of the original hypergraph. The hypergraph H contains all hyperedges that still need to be hit and the vertices that can be added to S. The algorithm enumerates all minimal transversals Y of the original hypergraph such that $Y \setminus S$ is a minimal transversal of H.

H1 If $E = \emptyset$, then check whether S is a minimal transversal of the original hypergraph and output S if so.
H2 If $\emptyset \in E$, then H contains an empty hyperedge, and the algorithm backtracks.
R1 If there is a vertex $v \in V$ with $d_H(v) = 0$, then discard v and recurse.
R2 If there are two hyperedges $e, e' \in E$ with $e \subseteq e'$, then remove e' and recurse.
R3 If there is a hyperedge $\{v\} \in E$, then select v and recurse.
B1 If there is a vertex $v \in V$ with $d_H(v) = 1$, then let $e \in E$ denote the hyperedge with $v \in e$. Make one recursive call where v is discarded, and one recursive call where v is selected and all vertices from $e \setminus \{v\}$ are discarded.
B2 Otherwise, select two hyperedges e, e' such that e is a smallest hyperedge and $|e \cap e'| \geq 1$. Order their vertices such that their common vertices appear first: $e = \{v_1, \ldots, v_{|e|}\}$ and $e' = \{v_1, \ldots, v_\ell, u_{\ell+1}, \ldots, u_{|e'|}\}$. Make $|e|$ recursive calls; in the ith recursive call, $v_1, \ldots, v_{i-1}$ are discarded and v_i is selected.

Theorem 4. *For any $k \geq 2$, the minimal transversals of a rank-k hypergraph can be enumerated in time $O((\beta_k)^n)$, where β_k is the positive real root of*

$$-1 + x^{-1} + \sum_{i=3}^{k}(i-2)\cdot x^{-i} + \sum_{i=k+1}^{2k-1}(2k-i)\cdot x^{-i} = 0 .$$

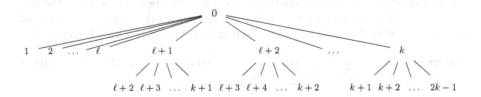

Fig. 1. Decreasing the number of vertices in recursive calls of rule B2.

Proof. The correctness of the halting and reduction rules are easy to see. For the correctness of branching rule B1, it suffices to observe that a transversal containing v and some other vertex from e is not minimal. The correctness of B2 follows since the ith recursive call enumerates the minimal transversals such that the first vertex among $v_1, \ldots, v_{|e|}$ they contain is v_i.

As for the running time, a crude analysis gives the same running time as the analysis of [14], since we can associate the branching vector $(1, |e|)$ with B1, which is $(1, 2)$ in the worst case, and the branching vector $(1, 2, \ldots, k)$ with B2.

Let us look at B2 more closely. In the worst case, $|e| = k$. Due to the reduction rules, we have that $|e \cap e'| = \ell < |e|$. We consider two cases. In the first case, $\ell = 1$. Since v_1 is discarded in branches $2, \ldots, k$, we have that the size of e' is at most $k - 1$ in each of these recursive calls, and the algorithm will either use branching rule B1 or branching rule B2 on a hyperedge of size at most $k - 1$ in branches $2, \ldots, k$. In the worst case, it uses branching rule B2 on a hyperedge of size $k - 1$ in each of these branches, leading to the branching vector $(1, 3, 4, \ldots, k + 1, 4, 5, \ldots, k + 2, 5, 6, \ldots, 2k - 1)$ whose recurrence solves to β_k. In the second case, $\ell \geq 2$, and we will show that this case is no worse than the first one. Since $v_1, \ldots, v_\ell$ are discarded in branches $\ell + 1, \ldots, k$, the size of e' is at most $k - \ell$ in each of these recursive calls, and in the worst case the algorithm will branch on a hyperedge of size $k - \ell$ in branches $\ell + 1, \ldots, k$, leading to the branching vector $(1, 2, \ldots, \ell, \ell + 2, \ell + 3, \ldots, k + 1, \ell + 3, \ell + 4, \ldots, k + 2, \ldots, k + 1, k + 2, \ldots, 2k - \ell)$. See Fig. 1. To see that this is no worse than the branching vector of the first case, follow each branch i with $2 \leq i \leq \ell$ by a k-way branching $(1, 2, \ldots, k)$, replacing the entry i in the branching vector with $1 + i, 2 + i, \ldots, k + i$. Compared with the branching vector of the first case, the only difference in branches i, $2 \leq i \leq \ell$, is that these have the additional entries $k + i$. But note that branch $\ell + 1$ has entries $k + 2, \ldots, k + \ell$ in the first case but not in the second case. We conclude that the branching vector where entries i, $2 \leq i \leq \ell$, are replaced by $1 + i, 2 + i, \ldots, k + i$ is a sub-vector of the one for $\ell = 1$.   □

Since this algorithm guarantees branching on hyperedges of size at most $k - 1$ in certain cases, its running time outperforms the one in [14] for each $k \geq 3$.

6   Lower Bounds

The graphs with a maximum number of maximal independent sets are the disjoint unions of triangles. They are hypergraphs of rank 2 with $3^{n/3}$ minimal transversals. We generalize this lower bound to hypergraphs with larger rank.

Theorem 5. *For any two integers $k, n > 0$, there is an n-vertex hypergraph of rank k with $\binom{2 \cdot k - 1}{k}^{\lfloor n/(2 \cdot k - 1) \rfloor}$ minimal transversals.*

Acknowledgments. We thank Fabrizio Grandoni for initial discussions on this research. Dieter Kratsch acknowledges support from the French Research Agency, project GraphEn (ANR-15-CE40-0009). Serge Gaspers is the recipient of an Australian

Research Council (ARC) Future Fellowship (project FT140100048) and acknowledges support under the ARC's Discovery Projects funding scheme (project DP150101134). NICTA is funded by the Australian Government through the Department of Communications and the ARC through the ICT Centre of Excellence Program.

References

1. Cochefert, M., Couturier, J.-F., Gaspers, S., Kratsch, D.: Faster algorithms to enumerate hypergraph transversals. Technical report arxiv:1510.05093 (2015)
2. Couturier, J.-F., Heggernes, P., van 't Hof, P., Kratsch, D.: Minimal dominating sets in graph classes: combinatorial bounds and enumeration. Theor. Comput. Sci. **487**(8), 2–94 (2013)
3. Cygan, M., Dell, H., Lokshtanov, D., Marx, D., Nederlof, J., Okamoto, Y., Paturi, R., Saurabh, S., Wahlström, M.: On problems as hard as CNF–SAT. In: Proceedings of CCC, pp. 74–84 (2012)
4. Eiter, T., Gottlob, G.: Identifying the minimal transversals of a hypergraph and related problems. SIAM J. Comput. **24**(6), 1278–1304 (1995)
5. Eiter, T., Gottlob, G., Makino, K.: New results on monotone dualization and generating hypergraph transversals. SIAM J. Comput. **32**(2), 514–537 (2003)
6. Elbassioni, K.M., Rauf, I.: Polynomial-time dualization of r-exact hypergraphs with applications in geometry. Discrete Math. **310**(17–18), 2356–2363 (2010)
7. Fernau, H.: Parameterized algorithmics for d-hitting set. Int. J. Comput. Math. **87**(14), 3157–3174 (2010)
8. Fernau, H.: Parameterized algorithms for d-hitting set: the weighted case. Theor. Comput. Sci. **411**(16–18), 1698–1713 (2010)
9. Fernau, H.: A top-down approach to search-trees: Improved algorithmics for 3-hitting set. Algorithmica **57**(1), 97–118 (2010)
10. Fomin, F.V., Gaspers, S., Kratsch, D., Liedloff, M., Saurabh, S.: Iterative compression and exact algorithms. Theor. Comput. Sci. **411**(7–9), 1045–1053 (2010)
11. Fomin, F.V., Gaspers, S., Lokshtanov, D., Saurabh, S.: Exact algorithms via monotone local search. Technical report arxiv:1512.01621 (2015)
12. Fomin, F.V., Kratsch, D.: Exact Exponential Algorithms. Springer, Heidelberg (2010)
13. Fredman, M.L., Khachiyan, L.: On the complexity of dualization of monotone disjunctive normal forms. J. Algorithms **21**(3), 618–628 (1996)
14. Gaspers, S.: Algorithmes exponentiels. Master's thesis, University Metz, France (2005)
15. Gaspers, S., Algorithms, E.T.: Exponential Time Algorithms: Structures, Measures, and Bounds. VDM Verlag Dr. Mueller e.K, Saarbrücken (2010)
16. Gaspers, S., Sorkin, G.B.: A universally fastest algorithm for Max 2-Sat, Max 2-CSP, and everything in between. J. Comput. Syst. Sci. **78**(1), 305–335 (2012)
17. Golovach, P.A., Heggernes, P., Kratsch, D., Villanger, Y.: An incremental polynomial time algorithm to enumerate all minimal edge dominating sets. Algorithmica **72**(3), 836–859 (2015)
18. Kanté, M.M., Limouzy, V., Mary, A., Nourine, L.: On the enumeration of minimal dominating sets and related notions. SIAM J. Discrete Math. **28**(4), 1916–1929 (2014)
19. Kanté, M.M., Limouzy, V., Mary, A., Nourine, L., Uno, T.: A polynomial delay algorithm for enumerating minimal dominating sets in chordal graphs. In: Proceedings of WG (2015)

20. Kanté, M.M., Limouzy, V., Mary, A., Nourine, L., Uno, T.: Polynomial delay algorithm for listing minimal edge dominating sets in graphs. In: Dehne, F., Sack, J.-R., Stege, U. (eds.) WADS 2015. LNCS, vol. 9214, pp. 446–457. Springer, Heidelberg (2015)
21. Karp, R.M.: Reducibility among combinatorial problems. In: Miller, R.E., Thatcher, J.W., Bohlinger, J.D. (eds.) Complexity of computer computations, pp. 85–103. Plenum Press, New York (1972)
22. Kavvadias, D.J., Stavropoulos, E.C.: An efficient algorithm for the transversal hypergraph generation. J. Graph Algorithms Appl. 9(2), 239–264 (2005)
23. Khachiyan, L., Boros, E., Elbassioni, K.M., Gurvich, V.: On the dualization of hypergraphs with bounded edge-intersections and other related classes of hypergraphs. Theor. Comput. Sci. 382(2), 139–150 (2007)
24. Miller, R.E., Muller, D.E.: A problem of maximum consistent subsets. IBM Research Report RC-240, J. T. Watson Research Center (1960)
25. Moon, J.W., Moser, L.: On cliques in graphs. Israel J. Math. 3, 23–28 (1965)
26. Niedermeier, R., Rossmanith, P.: An efficient fixed-parameter algorithm for 3-hitting set. J. Discrete Algorithms 1(1), 89–102 (2003)
27. Wahlström, M.: Exact algorithms for finding minimum transversals in rank-3 hypergraphs. J. Algorithms 51(2), 107–121 (2004)
28. Wahlström, M.: Algorithms, measures and upper bounds for satisfiability and related problems. Ph.D. thesis, Linköping University, Sweden (2007)

Listing Acyclic Orientations of Graphs with Single and Multiple Sources

Alessio Conte[1], Roberto Grossi[1], Andrea Marino[1(✉)], and Romeo Rizzi[2]

[1] Erable, Inria, Università di Pisa, Pisa, Italy
{conte,grossi,marino}@di.unipi.it
[2] Università di Verona, Verona, Italy
rizzi@di.univr.it

Abstract. We study enumeration problems for the acyclic orientations of an undirected graph with n nodes and m edges, where each edge must be assigned a direction so that the resulting directed graph is acyclic. When the acyclic orientations have single or multiple sources specified as input along with the graph, our algorithm is the first one to provide guaranteed bounds, giving new bounds with a delay of $O(m \cdot n)$ time per solution and $O(n^2)$ working space. When no sources are specified, our algorithm improves over previous work by reducing the delay to $O(m)$, and is the first one with linear delay.

1 Introduction

Acyclic orientations of graphs are related to several basic problems in graph theory. An *orientation* of an undirected graph G is the directed graph $\vec{G}$ whose arcs are obtained assigning a direction to each edge in G. The orientation $\vec{G}$ is *acyclic* when it does not contain cycles, and a node s is a *source* in $\vec{G}$ if it has indegree zero. For instance, consider the graphs in Fig. 1. The directed graph in (b) is an acyclic orientation for the undirected graph in (a). In particular, since the orientation has only one source (v_9), it is called a *single source acyclic orientation*.

Starting from the observation that each acyclic orientation corresponds to a partial order for the underlying graph, Iriarte [7] investigates which orientations maximize the number of linear extensions of the corresponding poset. Alon and Tarsi [1] look for special orientations to give bounds on the size of the maximum independent set or the chromatic number. Gallai, Roy, and Vitaver independently describe a well-known result stating that every orientation of a graph with chromatic number k contains a simple directed path with k vertices [6,12,17]. There are further problems that can be addressed by looking at acyclic orientations. For instance, Benson et al. [4] show that there exists a bijection between the set of the so-called superstable configurations of a graph and the set of its acyclic orientations with a unique source.

This work has been partially supported by the Italian Ministry of Education, University, and Research (MIUR) under PRIN 2012C4E3KT national research project AMANDA — Algorithmics for MAssive and Networked DAta.

© Springer-Verlag Berlin Heidelberg 2016
E. Kranakis et al. (Eds.): LATIN 2016, LNCS 9644, pp. 319–333, 2016.
DOI: 10.1007/978-3-662-49529-2_24

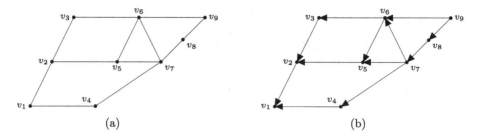

Fig. 1. An undirected connected graph without self-loops (a) and one of its acyclic orientations (b).

Counting how many acyclic orientations can be found in a graph is a fundamental problem in combinatorics, dating back to the 70 s or earlier [16]. Linial [10] proves that this problem is #P-complete. Stanley [15] shows how the number of acyclic orientations can be computed by using the chromatic polynomial (a special case of Tutte's polynomial). Another approach that concerns the number of acyclic orientations is the acyclic orientation game. Alon and Tuza [2] inquire about the amount of oriented edges needed to define a unique orientation of G, and find this number to be almost surely $\Theta(|V| \log |V|)$ in Erdős-Rényi random graphs. Pikhurko [11] shows that the number of these edges in the worst case is no greater than $(\frac{1}{4} + o(1))|V|^2$ for general graphs.

Problems Addressed. Our paper investigates new algorithms for enumerating patterns for this interesting problem in graph theory, given an undirected connected graph $G(V, E)$ with n nodes and m edges.

single source acyclic orientations (ssao): Given a node $s \in V$, enumerate all the acyclic orientations $\overrightarrow{G}$ of G, such that s is the only source.
single source acyclic orientations (weak ssao): Given a set of nodes $S \subseteq V$, enumerate all the acyclic orientations $\overrightarrow{G}$ of G such that there is exactly one source x and $x \in S$.
multiple source acyclic orientations (strong msao): Given a set of nodes $S \subseteq V$, enumerate all the acyclic orientations $\overrightarrow{G}$ of G such that *all* the nodes in S are the only sources.[1]
multiple source acyclic orientations (weak msao): Given a set of nodes $S \subseteq V$, enumerate all the acyclic orientations $\overrightarrow{G}$ of G such that if x is a source then $x \in S$.[2]
acyclic orientations (ao): Enumerate all the acyclic orientations $\overrightarrow{G}$ of G.

We will show that these problems can be reduced to SSAO, with a one-to-one correspondence between their solutions. Many other variants with constraints on the number or choice of sources can be reduced to SSAO as well: the ones

---

[1] These orientations are possible if and only if S is an independent set.
[2] Not all nodes in S must be sources, but there cannot be sources in $V \setminus S$.

we present are some of the most representative ones. We analyze the cost of an enumeration algorithm for SSAO, weak SSAO, strong MSAO, weak MSAO, and AO in terms of its *delay* cost, which is a well-known measure of performance for enumeration algorithms corresponding to the worst-case time between any two consecutively enumerated solutions (e.g. [8]). We are interested in algorithms with guaranteed delay and space.

Previous Work. We are not aware of any provably good bounds for problems SSAO, weak SSAO, strong MSAO and weak MSAO. Johnson's backtracking algorithm [9] for SSAO has been presented over 30 years ago to solve problems on network reliability. However its complexity is not given and is hard to estimate, as it is based on a backtracking approach with dead ends.

In his paper presenting an algorithm for AO, Squire [14] writes that he has been unable to efficiently implement Johnson's approach because of its dead ends. Squire's algorithm for AO uses Gray codes and has an amortized cost of $O(n)$ per solution, but its delay can be $O(n^3)$ time for a solution. The algorithm by Barbosa and Szwarcfiter [3] solves AO with an amortized time complexity of $O(n + m)$ per solution, delay $O(n \cdot m)$. The algorithm builds the oriented graph incrementally by iteratively adding the nodes to an empty directed graph.

It is worth observing that by replacing each edge with a double arc and applying any algorithm for maximal feedback arc set enumeration, one can obtain all the acyclic orientations of G. State of the art approaches for the latter problem guarantee a delay $\Omega(n^3)$ as shown by Schwikowski and Speckenmeyer [13].

All the techniques above for AO, including the one in [14], do not to extend smoothly to SSAO, weak SSAO, strong MSAO, and weak MSAO. In a previous work [5], we studied the related problem of enumerating the cyclic orientations of an undirected graph, but the proposed techniques cannot be reused for the problems in this paper.

Our Results. Our contribution is the design of the first enumeration algorithms with guaranteed bounds for SSAO, weak SSAO, strong MSAO, and weak MSAO: the complexity is $O(m \cdot n)$ delay per solution using $O(n^2)$ space. For AO, we also show how to obtain $O(m)$ delay, improving the delay of [3,13,14], but we do not improve the amortized cost of $O(n)$ in [14].

We therefore focus on SSAO in the first part of the paper, and then show an optimization that holds for the case of AO. We guarantee that, at any given partial solution, the extensions of the partial solution will enumerate new acyclic orientations. To this aim, we solve several non-trivial issues.

- We use a recursive approach where each call surely leads to a solution. For SSAO this is achieved also by using a suitable ordering of the nodes.
- We quickly identify the next recursion calls within the claimed time delay.
- We do a careful analysis of the recursion tree, and show how to check efficiently for node reachability during recursion.
- In the case of AO we exploit the fact that the recursion tree does not contain unary nodes.

The paper is organized as follows. We give the necessary definitions and terminology in Sect. 2. We then discuss how to solve SSAO in Sect. 3 and further reduce the delay for AO in Sect. 4. In Sect. 5 we show how weak SSAO, strong MSAO, weak MSAO, and AO reduce to SSAO. We draw some conclusions in Sect. 6.

2   Preliminaries

Given an undirected graph $G(V, E)$ with n nodes and m edges, an orientation of G is the directed graph $\overrightarrow{G}(V, \overrightarrow{E})$ where for any pair $\{u, v\} \in E$ either $(u, v) \in \overrightarrow{E}$ or $(v, u) \in \overrightarrow{E}$. We call $\overrightarrow{E}$ an orientation of E. We say that the orientation $\overrightarrow{G}$ is *acyclic* when it does not contain cycles. For the sake of clarity, in the following we will call *edges* the unordered pairs $\{x, y\}$ (undirected graph), while we will call *arcs* the two possible orientations (x, y) and (y, x) (directed graphs). We assume wlog that G is connected and does not contain self-loops.

Given an undirected graph $G(V, E)$, let $v_1, \ldots, v_n \in V$ be an ordering of the nodes of G. We define $V_{\leq i}$ as the set $\{v_1, \ldots, v_i\}$, $N(v_i) = \{x : \{v_i, x\} \in E\}$ as the set of neighbors of the node v_i, and $N_{\leq i}(v)$ as the set $N(v) \cap V_{\leq i}$. For brevity, $N_<(v_j)$ means $N_{\leq j-1}(v_j)$. Clearly we have $\sum_{j=1}^{n} |N_<(v_j)| = m$.

Starting from an empty directed graph, for increasing values of $i = 1, 2, \ldots, n$, our algorithms add v_i to the current graph and recursively exploit all the possible ways of directing the edges $\{v_i, x\}$ with $x \in N_<(v_i)$, called direction assignments: a *direction assignment* $\overrightarrow{Z}$ for v_i is an orientation of the set of edges $\{\{v_i, x\} : x \in N_<(v_i)\}$. We refer to the following special assignments as

$$X_i = \{(x, v_i) : x \in N_<(v_i)\}$$

$$Y_i = \{(v_i, x) : x \in N_<(v_i)\}$$

We denote by $\overrightarrow{G}_0$ the starting empty directed graph, and by $\overrightarrow{G}_i$ the graph whose last added node is v_i, with $1 \leq i \leq n$.

3   Single Source Acyclic Orientations (SSAO)

Given a graph G and a node s, this section describes how to enumerate its acyclic orientations $\overrightarrow{G}$ such that s is the unique source in $\overrightarrow{G}$. Starting from an empty graph $\overrightarrow{G}_0$, our algorithm adds v_i to $\overrightarrow{G}_{i-1}$ for $i = 1, 2, \ldots, n$. For the edges in $N_<(v_i)$, it exploits all the suitable direction assignments $\overrightarrow{Z}$: each assignment gives a certain $\overrightarrow{G}_i$, on which it recurses. Every time it adds the last vertex v_n in a recursive call, it outputs the corresponding $\overrightarrow{G}_n$ as a new solution $\overrightarrow{G}$.

The above simple scheme can lead to dead ends in its recursive calls, where partial orientations $\overrightarrow{G}_i$ cannot be extended to reach $\overrightarrow{G}_n$, i.e. an acyclic $\overrightarrow{G}$ whose only source is s. We prevent this situation by examining the nodes of G in a suitable order that allows us to exploit the following notions.

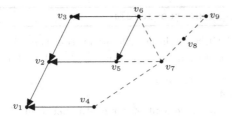

Fig. 2. A partial acyclic orientation (thick edges). To add v_7 to the partial orientation we exploit valid direction assignments for undirected edges $\{v_4, v_7\}, \{v_5, v_7\}, \{v_6, v_7\}$.

Definition 1 (full node). *Given an ordering* $v_1, \ldots, v_n$ *of the nodes in* G*, a node* v_j *($1 \leq j \leq i$) is full in* $\overrightarrow{G}_i$ *if* $N_{\leq i}(v_j) = N(v_j)$.

Definition 2 (valid direction assignment). *Given* $\overrightarrow{G}_{i-1}(V_{\leq i-1}, \overrightarrow{E})$*, the direction assignment* $\overrightarrow{Z}$ *is valid if*

- $\overrightarrow{G}_i(V_{\leq i}, \overrightarrow{E} \cup \overrightarrow{Z})$ *is acyclic, and*
- *any* $v_j \neq s$ *that is full in* $\overrightarrow{G}_i(V_{\leq i}, \overrightarrow{E} \cup \overrightarrow{Z})$ *is not a source, for* $1 \leq j \leq i$.

The rationale is the following. When dealing with $\overrightarrow{G}_i$, full nodes are the ones whose edges in G have been all already assigned in $\overrightarrow{G}_i$. This means that if a full node is a source in $\overrightarrow{G}_i$ it will be a source also in any extension of $\overrightarrow{G}_i$, i.e. in the final orientation $\overrightarrow{G}$. A valid direction assignment imposes that we do not create cycles and each full node (except s) is not a source. We will deal with orientations of $\overrightarrow{G}_i$ that are the outcome of a sequence of valid direction assignments: such orientations will be referred to as *partial acyclic orientations* of G_i.

Consider the graph in Fig. 1 (a). We want acyclic orientations whose only source is $s = v_9$ processing the nodes in the order $v_1, \ldots, v_9$. In particular, consider the situation in Fig. 2. We have the graph $\overrightarrow{G}_6$ (thick edges) and we have to add the vertex v_7, deciding how to orient the edges to the vertices in $N_{<}(v_i)$, namely $\{v_4, v_7\}, \{v_5, v_7\}, \{v_6, v_7\}$, to obtain all the possible $\overrightarrow{G}_7$. Notice that v_1, v_2, and v_3 are full in $\overrightarrow{G}_6$ since the edges in their whole neighborhood have been already oriented: hence since they are not source in $\overrightarrow{G}_6$, they will be not sources in the final orientation. The direction assignment $\{(v_7, v_4), (v_5, v_7), (v_6, v_7)\}$ is valid since the full nodes in the corresponding $\overrightarrow{G}_7$, namely v_1, v_2, v_3, v_4 are not sources, and $\overrightarrow{G}_7$ is acyclic. On the other hand, $\{(v_4, v_7), (v_5, v_7), (v_6, v_7)\}$ is not a valid direction assignment since v_4 is a source full in the corresponding $\overrightarrow{G}_7$ and it will be source in any final orientation extending $\overrightarrow{G}_7$. Notice that, since v_7, and v_9 are not in $N_{<}(v_6)$, v_6 can be source in $\overrightarrow{G}_6$ and can remain source also in $\overrightarrow{G}_7$, while it should be not a source in $\overrightarrow{G}_9$. Hence, exploiting just the valid direction assignments means taking care that no cycle is created and no full node is source.

Algorithm 1. single-source-acyclic

Input: Graph $G(V, E)$, a partial acyclic orientation $\overrightarrow{G}_{i-1}(V_{\leq i-1}, \overrightarrow{E})$, integer i
Output: Acyclic orientations of G containing $\overrightarrow{G}_{i-1}(V_{\leq i}, \overrightarrow{E})$ with source s

if $i > n$ **then** output $\overrightarrow{G}$; **return** ;
· Execute Algorithm 4;
for any valid direction assignment $\overrightarrow{Z}$ for v_i starting from $\overrightarrow{Z} = Y_i$ **do**
⌊ single-source-acyclic$(G, \overrightarrow{G}_i(V_{\leq i}, \overrightarrow{E} \cup \overrightarrow{Z}), i+1)$;

In order to efficiently find valid direction assignments, our algorithm uses an ordering of the nodes $v_1, \ldots, v_n$ that satisfies the conditions below.

Definition 3. *An ordering of the nodes* $v_1, \ldots, v_n$ *is* good *if*

– $v_n = s$, *and*
– $N_<(v_i) \subsetneq N(v_i)$, *for* $1 \leq i < n$.

The first condition in Definition 3 says that s should be the last node as it is the only source. The second condition says that there is at least one unassigned incident edge for each v_i, when adding the latter to $\overrightarrow{G}_{i-1}$. Dead ends can be avoided in this way: when adding v_i to $\overrightarrow{G}_{i-1}$, we have at least one solution extending $\overrightarrow{G}_{i-1}$ in which v_i is not a source. Indeed the following property holds.

Property 1. For any partial acyclic orientation $\overrightarrow{G}_i$, there is always an acyclic orientation $\overrightarrow{G}$ for G that has unique source s and includes $\overrightarrow{G}_i$.

Proof. For any $j > i$, consider the valid direction assignment Y_j, i.e. $\{(v_j, x) : x \in N_<(v_j)\}$ obtaining $\overrightarrow{G}$. These direction assignments cannot create cycles in $\overrightarrow{G}_j$ and the only final source is s. □

Referring to the graph in Fig. 1 (a), if $s = v_9$, the order induced by $v_1, \ldots, v_9$ is a good order. Considering $\overrightarrow{G}_6$ in Fig. 2, according to Property 1, Y_7 corresponds to $\{(v_7, v_6), (v_7, v_5), (v_7, v_4)\}$, Y_8 corresponds to $\{(v_8, v_7)\}$, while Y_9 corresponds to $\{(v_9, v_8), (v_9, v_6)\}$. These direction assignments are valid and lead to the acyclic orientation in Fig. 1 (b) whose only source is indeed v_9.

A good ordering for G and s can be found in linear time by performing a DFS from s and considering its nodes in postorder. Observe that this is a good order according to our definition: node s is the last node and, for each node, its parent in the DFS tree appears after it in the order.

Our recursive algorithm is shown in Algorithm 1, where the good ordering of the nodes is employed. The initial call is single-source-acyclic$(G, \overrightarrow{G}_0, 1)$. The algorithm recursively exploits all the possible ways of expanding the current partial solution $\overrightarrow{G}_{i-1}$ by iterating over all the valid direction assignments, starting from Y_i, which is surely valid. The latter assignments are generated by a recursive computation. The general picture of our solution can be seen as

---

Algorithm 2. Returning valid direction assignments

Input: Graph $G(V, E)$, a partial acyclic orientation $\overrightarrow{G}_{i-1}(V_{\leq i-1}, \overrightarrow{E})$, node v_i
Output: Valid direction assignments $\overrightarrow{Z}$

$F_i \leftarrow$ set of full nodes in $\overrightarrow{G}_i$ that are sources and not full in $\overrightarrow{G}_{i-1}$;
$\overrightarrow{Z} \leftarrow \{(v_i, y) : y \in F_i\}$;
Let $x_1, \ldots, x_k$ be the nodes in $N_<(v_i) \setminus F_i$;
Execute Generate $(G, \overrightarrow{G}, v_i, \overrightarrow{Z}, 1, \emptyset, \emptyset)$.

Procedure Generate $(G(V, E), \overrightarrow{G}_{i-1}(V_{\leq i-1}, \overrightarrow{E}), v_i, \overrightarrow{W}, j, R, B)$
  | if $j > k$ then add $\overrightarrow{W}$ to the output list; **return** ;
  | Update B as the set of nodes leading to v_i in $\overrightarrow{G}_i(V_{\leq i}, \overrightarrow{E} \cup \overrightarrow{W})$;
  | if $x_j \notin B$ then Generate $(G, \overrightarrow{G}_{i-1}, v_i, \overrightarrow{W} \cup \{(v_i, x_j)\}, j+1, R, B)$;
  | Update R as the set of nodes reachable from v_i in $\overrightarrow{G}_i(V_{\leq i}, \overrightarrow{E} \cup \overrightarrow{W})$;
  | if $x_j \notin R$ then Generate $(G, \overrightarrow{G}_{i-1}, v_i, \overrightarrow{W} \cup \{(x_j, v_i)\}, j+1, R, B)$;

---

follows: we have the primary recursion tree to generate all the wanted cyclic orientations (Algorithm 1), where each node has associated a secondary recursion tree to generate locally all the valid direction assignments (Algorithm 2). A naive implementation would simply consider each of the $2^{|N_<(v_i)|}$ direction assignments checking whether it is valid or not (e.g. using a DFS in $O(m)$ time). Instead, the following section introduces an efficient method that allows us to iterate through valid direction assignments only.

3.1   Iterating Over Valid Direction Assignments

Given $\overrightarrow{G}_{i-1}$ and node v_i, we show how to iterate over valid direction assignments: our approach is shown in Algorithm 2. By definition, each valid direction assignment $\overrightarrow{W}$ of the edges in $\{\{v_i, x\} : x \in N_<(v_i)\}$ should guarantee that we are not creating a cycle and no new full node becomes a source.

Let F_i be the set of nodes which are: (a) full in $\overrightarrow{G}_i$, (b) sources in $\overrightarrow{G}_{i-1}$, (c) not full in $\overrightarrow{G}_{i-1}$. Note that $F_i \subseteq N_<(v_i)$. All the valid direction assignments should guarantee that nodes in F_i are not sources in $\overrightarrow{G}_i$: this can be easily done by adding the arcs $\{(v_i, x) : x \in F_i\}$ to $\overrightarrow{W}$. Observe that this is mandatory for the orientations of the corresponding edges because otherwise a node in F_i would become a source in the final orientation $\overrightarrow{G}$. Also, $\overrightarrow{G}_i(V_{\leq i}, \overrightarrow{E} \cup \overrightarrow{W})$ is acyclic.

After that, we have to decide the orientation of the remaining edges. This part relies on procedure Generate in Algorithm 2. In particular, we have to assign a direction to the edges $\{v_i, u\}$ for each node u in $N_<(v_i) \setminus F_i$ and check if they do not create cycles. We take these nodes in arbitrary order as $x_1, \ldots, x_k$ and, for increasing values of $j = 1, 2, \ldots, k$, we do the following: if the arc $e \in \{(v_i, x_j), (x_j, v_i)\}$ does not create a cycle in $\overrightarrow{G}_i(V_{\leq i}, \overrightarrow{E} \cup \overrightarrow{W})$, proceed recursively with $\overrightarrow{W} = \overrightarrow{W} \cup \{e\}$.

For a given v_i, the reachability tests above for x_j $(j = 1, 2, \ldots, k)$ can be performed with $O(k)$ (forward and backward) DFS traversals, requiring overall $O(k \cdot m)$ time. Since k is bounded by the degree of v_i and can be $O(n)$ in the worst case, we propose a solution that reduces the cost from $O(k \cdot m)$ to $O(m)$ time. It truncates the DFSes using sets B and R to avoid visiting the nodes in these sets. Since the partially built graph is acyclic, B and R are disjoint, and a node can belong to either one of them or none of them. Below we provide an analysis based on coloring the nodes of B and R showing that the overall time required by these tests for each valid direction assignment is $O(m)$.

Lemma 1. *Algorithm 2 returns valid direction assignments with delay $O(m)$.*

Proof. The arcs directed to nodes in F_i are unchanged for all the valid direction assignments for v_i and can be computed in $O(m)$ time at the beginning.

When exploring the possible orientations of edges $\{v_i, x_j\}$, for $j = 1, 2, \ldots, k$, each time we have to decide whether (v_i, x_j) or (x_j, v_i) creates a cycle or not when added to $\overrightarrow{G}_i(V_{\leq i}, \overrightarrow{E} \cup \overrightarrow{W})$. To this aim we color incrementally the nodes: all the nodes R reachable from v_i are *red*; all the nodes B that can lead to v_i are *black*; the remaining nodes are *uncolored*. Initially, all the nodes are uncolored, are R and B are empty. Since $\overrightarrow{G}_i(V_{\leq i}, \overrightarrow{E} \cup \overrightarrow{W})$ is acyclic any node has just one color or is uncolored.

We now show that the sum of the costs to update the colors to produce a solution (valid direction assignment) is $O(m)$. Since each leaf in the secondary recursion tree induced by Generate corresponds to a distinct solution, we should bound the sum of the costs along the $k + 1$ nodes from the root to that leaf. Specifically, the delay is upper bounded by the sum of the costs along two paths: the leaf-to-root path of the current solution and the root-to-next-leaf path for the next solution (actually only the latter for the first solution). Observe that the former cost is always $O(|N_<(v_i)|)$. We prove that the sum of the costs from the root to a leaf in the secondary recursion tree induced by Generate is bounded by $O(m)$. When $j = 1$, the red colors are assigned with a forward traversal and the black colors are assigned with a backward traversal in the graph $\overrightarrow{G}_i(V_{\leq i}, \overrightarrow{E} \cup \overrightarrow{W})$. When $j > 1$, while adding the arc (v_i, x_j) to $\overrightarrow{W}$ we have only to make the traversed uncolored nodes red: since the forward traversal is rooted at v_i, we continue the traversal avoiding to visit red nodes. (No black node can be reached, otherwise x_j would be black also and thus $\overrightarrow{G}_i$ cyclic). On the other hand, when adding the arc (x_j, v_i) to $\overrightarrow{W}$ we have only to make the traversed uncolored nodes black: once again, this corresponds to continue the backward traversal rooted in v_i avoiding to visit black nodes (no red node can be reached). Since this process traverses each arc at most once for any $1 \leq j \leq k$, the sum of the costs of a root to leaf path in the secondary recursion tree induced by the Generate procedure is $O(m)$. □

Remark 1. After the last valid direction assignment has been returned, Algorithm 2 recognizes that there are no more valid direction assignments, using time $O(m)$.

Lemma 2. *Referring to Algorithm 1, the following holds.*

1. *All the acyclic orientations of G whose unique source is s are output.*
2. *Only the acyclic orientations of G whose unique source is s are output.*
3. *There are no duplicates.*

Proof. We prove the three statements separately.

1. Given a good order of the nodes, we show that any single source acyclic orientation $\overrightarrow{G}$ can be expressed as a sequence of direction assignments $\overrightarrow{Z}_i$ for v_i, for increasing values of i. Consider the following process: for decreasing values of j remove v_j from $\overrightarrow{G}$, and set $\overrightarrow{Z}_j$ equal to the current outgoing arcs from u in $\overrightarrow{G}$. The sequence of sets $\overrightarrow{Z}_1, \ldots, \overrightarrow{Z}_n$ will lead the algorithm to the discovery of $\overrightarrow{G}$. Note that each direction assignment $\overrightarrow{Z}_j$ is valid otherwise we have a cycle or a source different from s in $\overrightarrow{G}$.
2. Each solution is acyclic, since each time we add a node v_i and a valid direction assignment we do not introduce a cycle by definition of valid direction assignment. We have to show that s is the unique source; any v_j full in $\overrightarrow{G}_i$, with $j \leq i < n$, is not a source in $\overrightarrow{G}_i$: hence it is not a source in $\overrightarrow{G}_n = \overrightarrow{G}$. Indeed, for the good ordering definition, each node v_j not full in $\overrightarrow{G}_i$, with $j \leq i$, has a neighbor in $V \setminus V_{\leq i}$: if it is a source in $\overrightarrow{G}_i$, when considering its last neighbor v_z in the good order, it will not be a source anymore in $\overrightarrow{G}_z$.
3. Given any two solutions, looking at the primary recursion tree induced by Algorithm 1, they differ at least for the valid direction assignments branching in their least common ancestor. □

Theorem 1. SSAO *can be solved with delay $O(n \cdot m)$ and space $O(n^2)$.*

Proof. We exploit the properties of the primary recursion tree induced by Algorithm 1. First of all, notice that each internal node has at least one child because of Property 1. This means that all the leaves correspond to a solution. Moreover, observe that all the leaves are at the same depth n, which is the height of the recursion tree. The first solution is clearly returned in time $O(n \cdot m)$, which is the height times the cost to get the first valid direction assignment for v_i. This is bounded by $O(m)$ time by applying Lemma 1. For any two consecutive solutions, the delay is bounded by the sum of the costs along a leaf-to-root path and the root-to-next-leaf path. The former is bounded by $O(n \cdot m)$: indeed the height of the tree is $O(n)$ and each time we return we spend $O(m)$ to recognize that no more valid direction assignments are possible, as highlighted by Remark 1. The latter is still bounded by $O(n \cdot m)$, applying n times Lemma 1.

The space is bounded by $O(n^2)$, that is the space occupancy of a root-to-leaf path in the primary recursion. Indeed, in this path we have to maintain $O(n)$ times the status of the Generate iterator, whose total space is bounded by $O(n)$. □

4   Acyclic Orientations (AO)

This section deals with the problem of enumerating all the acyclic orientations of an undirected graph G. The general scheme remains the same as discussed in Sect. 3. Differently from before, we have no restriction about the possible sources when adding v_i. Hence we redefine the concept of valid direction assignment as follows.

Definition 4 (valid direction assignment). *Given $\overrightarrow{G}_{i-1}(V_{\leq i-1}, \overrightarrow{E})$, a direction assignment $\overrightarrow{Z}$ is valid if $\overrightarrow{G}_i(V_{\leq i}, \overrightarrow{E} \cup \overrightarrow{Z})$ is acyclic.*

Referring to Fig. 2, when adding v_7 to $\overrightarrow{G}_6$, this means that also the direction assignment $(v_4, v_7), (v_5, v_7), (v_6, v_7)$ is valid.

Algorithm 3. acyclic

Input: Graph $G(V, E)$, partial acyclic orientation $\overrightarrow{G}_{i-1}(V_{\leq i-1}, \overrightarrow{E})$, integer i
Output: Acyclic orientations of G containing $\overrightarrow{G}_{i-1}(V_{\leq i-1}, \overrightarrow{E})$

if $i > n$ **then** output $\overrightarrow{G}_n$; **return** ;
Execute Algorithm 2;
for any valid direction assignment $\overrightarrow{Z}$ for v_i starting from $\overrightarrow{Z} = X_i$ to $\overrightarrow{Z} = Y_i$ **do**
  $\lfloor$ acyclic$(G, \overrightarrow{G}_i(V_{\leq i}, \overrightarrow{E} \cup \overrightarrow{Z}), i+1)$;

Moreover, another difference from the previous section is that we do not need the good order (Definition 3). Namely for any order of the nodes we can prove that the following property holds.

Property 2. For any v_i there are always at least two valid direction assignments, i.e. $X_i = \{(x, v_i) : x \in N_<(v_i)\}$ and $Y_i = \{(v_i, x) : x \in N_<(v_i)\}$.

Proof. While adding v_i to $\overrightarrow{G}_{i-1}$, adding the arcs of X_i to $\overrightarrow{G}_{i-1}$ or adding the arcs of Y_i does not create a cycle. Indeed, inductively the following facts hold. $\overrightarrow{G}_0$ does not contains a cycle. Assuming that $\overrightarrow{G}_{i-1}$ does not contain a cycle, any cycle should involve v_i. Since in the two orientations above v_i is source or target, adopting one of these direction assignments cannot make $\overrightarrow{G}_i$ cyclic.   □

Property 2 simply states that in Fig. 2, both $\{(v_4, v_7), (v_5, v_7), (v_6, v_7)\}$ and $\{(v_7, v_4), (v_7, v_5), (v_7, v_6)\}$ are valid direction assignments when adding v_7 to $\overrightarrow{G}_6$. This is because $\overrightarrow{G}_6$ is acyclic and both of the new direction assignments cannot create cycles in the corresponding $\overrightarrow{G}_7$.

The actual scheme is summarized in Algorithm 3, whose starting call is acyclic$(G, \overrightarrow{G}_0, 1)$. At step i, given the acyclic orientation $\overrightarrow{G}_{i-1}(V_{\leq i-1}, \overrightarrow{E})$, each recursive call is of the kind acyclic$(G, \overrightarrow{G}_i, i+1)$ where $\overrightarrow{G}_i(V_{\leq i}, \overrightarrow{E} \cup \overrightarrow{Z})$ is obtained by adding a valid direction assignment $\overrightarrow{Z}$. By Property 2, $\overrightarrow{Z} = X_i$ and $\overrightarrow{Z} = Y_i$ are always taken. The corresponding two recursive calls are done respectively at the beginning and at the end of the procedure. All the other possible valid direction assignments (if any) are explored in the other calls of

---

Algorithm 4. Returning valid direction assignments

Input: Graph $G(V, E)$, partial acyclic orientation $\overrightarrow{G}_{i-1}(V_{\leq i-1}, \overrightarrow{E})$, node v_i
Output: Valid direction assignments $\overrightarrow{Z}$

Let $x_1, \ldots, x_k$ be the nodes of $N_<(v_i)$;
Execute Generate $(G, \overrightarrow{G}, v_i, \emptyset, 1, \emptyset, \emptyset)$.

Procedure Generate $(G(V, E), \overrightarrow{G}_{i-1}(V_{\leq i-1}, \overrightarrow{E}), v_i, \overrightarrow{W}, j, R, B)$

   | if $j > k$ **then** add $\overrightarrow{W}$ to the output list; **return** ;
   | if $\overrightarrow{W} \cap Y_i \neq \emptyset$ **then**
   | | update R as the nodes reachable from v_i in $\overrightarrow{G}_i(V_{\leq i}, \overrightarrow{E} \cup \overrightarrow{W})$
   | if $x_j \notin R$ **then** Generate $(G, \overrightarrow{G}_{i-1}, v_i, \overrightarrow{W} \cup \{(x_j, v_i)\}, j + 1, R, B)$;
   | if $\overrightarrow{W} \cap X_i \neq \emptyset$ **then**
   | | update B as the nodes leading to v_i in $\overrightarrow{G}_i(V_{\leq i}, \overrightarrow{E} \cup \overrightarrow{W})$
   | if $x_j \notin B$ **then** Generate $(G, \overrightarrow{G}_{i-1}, v_i, \overrightarrow{W} \cup \{(v_i, x_j)\}, j + 1, R, B)$;

---

the for cycle. When $i = n + 1$, all the nodes have been added and all the edges have been assigned a direction. In this case $\overrightarrow{G}_n$ is an acyclic orientation to be output.

Lemma 3. *Referring to Algorithm 3, the following holds.*

1. *All the acyclic orientations of G are output.*
2. *Just the acyclic orientations of G are output.*
3. *There are no duplicates.*

Proof. Similar to the proof of Lemma 2. □

We introduce a method that, at running time, allows us to iterate just on valid direction assignments: in particular, this method gets the first valid direction assignment X_i in $O(|N_<(v_i)|)$ time and the remaining ones with delay $O(m)$, one after the other. The point is that we spend $O(m)$ time and get a new solution. Here it is crucial that Y_i is the last valid direction assignment for this purpose as it indicates when stopping the search for valid direction assignments. The motivation for this choice is given by the following lemma.

Lemma 4. *When iterating over valid direction assignments $\overrightarrow{Z}$, suppose that the first assignment X_i is returned in $O(|N_<(v_i)|)$ time and the delay between any two consecutive valid direction assignments is $O(m)$. Then Algorithm 3 has delay $O(m)$.*

Proof. In the primary recursion tree, the internal nodes have at least two children, the leaves are all the solutions, and their depth is n. The solution in the first leaf is obtained by using always X_i sets, each one corresponding to a node v_i, and the cost is $\sum_{i=1}^{n} O(|N_<(v_i)|) = O(m)$, using the hypothesis of the lemma.

The delay between two consecutive solutions is upper bounded by the sum of the costs of the recursive calls to go from a leaf to the next leaf in preorder. Now let S and T be the solutions in two consecutive leaves. The paths from the root to S and to T share the prefix until a recursive call R_j, corresponding to the recursion while adding v_j to $\overrightarrow{G}_{j-1}$, for some $1 \leq j \leq n-1$. Let S' and T' be R_j's children that are respectively the ancestors of S and T. Note that the path from S' to S is made of Y_i branches while the path from T' to T is made of X_i branches $(j + 1 \leq i \leq n)$. The cost from S to S' is $O(n-j)$ as we do not need to check for further valid direction assignments after each Y_i. The cost from S' to T' is $O(m)$ by hypothesis as they are two consecutive valid direction assignments for v_j. The cost from T' to T is $O(\sum_{i=j+1}^{n} |N_<(v_i)|) = O(m)$ by hypothesis on the costs to get each X_i. $\square$

In the next section we will provide a way of iterating over valid direction assignments fitting the hypothesis of Lemma 4.

4.1   Iterating Over Valid Direction Assignments

Given the node v_i and the current acyclic directed graph $\overrightarrow{G}_{i-1}$, this section describes how to get all the valid direction assignments (i.e. such that adding one of them and v_i to $\overrightarrow{G}_{i-1}$, we obtain a $\overrightarrow{G}_i$ which is still acyclic).

Algorithm 4 extends the current partial valid direction assignment $\overrightarrow{W}$ for v_i: for each edge in $\{v_i, x\}$ such that $x \in N_<(v_i)$, it adds the arc (v_i, x) or the arc (x, v_i) to $\overrightarrow{W}$ whether the partial direction assignment is still valid, i.e. no cycles are created. It explores all of these extensions.

More formally, given $\overrightarrow{G}_{i-1}(V_{\leq i}, \overrightarrow{E})$ and v_i, we consider the nodes $x_1, \ldots, x_k$ in $N_<(v_i)$ (where $|N_<(v_i)| = k$) one after the other. Initially, let $\overrightarrow{W}$ be an empty set. For increasing values of j, with $1 \leq j \leq k$, we do the following. If the arc $e \in \{(v_i, x_j), (x_j, v_i)\}$ does not create a cycle in $\overrightarrow{G}_{i-1}(V_{\leq i}, \overrightarrow{E} \cup \overrightarrow{W})$, this arc can be added to the ongoing solution $\overrightarrow{W}$, exploring recursively the case $\overrightarrow{W} = \overrightarrow{W} \cup \{e\}$.

Let R and B be respectively the set of nodes reachable from v_i in $\overrightarrow{G}_i(V_{\leq i}, \overrightarrow{E} \cup \overrightarrow{W})$ and the set of nodes leading to v_i in $\overrightarrow{G}_i(V_{\leq i}, \overrightarrow{E} \cup \overrightarrow{W})$. Adding the arc (v_i, x_j) to $\overrightarrow{W}$ creates a cycle if and only if $x_j \notin B$. Analogously, adding the arc (x_j, v_i) to $\overrightarrow{W}$ creates a cycle if and only if $x_j \notin R$. The scheme of the iterator is summarized by Algorithm 4. Notice that the first valid direction assignment produced is X_i and the last valid direction assignment is Y_i, as required by Algorithm 3. Moreover observe that the update of R and B is respectively not required when $\overrightarrow{W} \cap Y_i = \emptyset$ and $\overrightarrow{W} \cap X_i = \emptyset$, since these conditions means respectively that the outdegree and the indegree of v_i in $\overrightarrow{G}_i(V_{\leq i}, \overrightarrow{E} \cup \overrightarrow{W})$ is zero.

We remark that there are no dead ends. Indeed, if $x_j \in R$ then $x_j \notin B$, meaning that even if the first call is skipped the second one is performed. Similarly $x_j \in B$ implies $x_j \notin R$. This means that each call produces at least another call unless the direction assignment is completed, that is, each call returns at least one solution.

Lemma 5. *Algorithm 4 returns the first valid direction assignment in time* $O(|N_<(v_i)|)$, *and the remaining ones with delay* $O(m)$.

Proof. Recall that the direction assignment X_i is always returned first while Y_i is returned last. For increasing values of i, adding the arcs of X_i (respectively Y_i) does not create cycles: in this case no check is needed. In particular X_i is returned in time $O(|N_<(v_i)|)$ since the update of R is not needed and never performed in Algorithm 4 (because $\overrightarrow{W} \cap Y_i = \emptyset$, i.e. the outdegree of v_i in $\overrightarrow{G}_i(V_{\leq i}, \overrightarrow{E} \cup \overrightarrow{W})$ is zero). Checking $\overrightarrow{W} \cap Y_i \neq \emptyset$ and $\overrightarrow{W} \cap X_i \neq \emptyset$ can be done in constant time: these are the out- and in-degree of x_j in $\overrightarrow{G}(V_{\leq i}, \overrightarrow{E} \cup \overrightarrow{W})$ that can be updated while updating $\overrightarrow{W}$. The time is hence dominated by the cost to update R and B. As in Lemma 1, this can be done by growing R and B continuing the same forward and backward traversals from v_i: since each arc is traversed at most once, the overall time to update R and B is $O(m)$. □

By combining Lemmas 4 and 5, we obtain the following result.

Theorem 2. *Problem* AO *can be solved with delay* $O(m)$ *and space* $O(n^2)$.

5 Reducing to Single Source Acyclic Orientations

We show that the problems mentioned in Sect. 1 can be reduced to SSAO. It is easy to see that weak SSAO can be solved simply by enumerating all the SSAOs in G with source s for each $s \in S$. It is worth observing that for each s there is at least a solution, meaning that the size of S does not influence the delay of weak SSAO.

Let us consider weak MSAO. To solve it, we create a dummy node s, and connect it to every node in S. More formally, we build $G'(V \cup \{s\}, E \cup E_s)$, where $E_s = \{\{s, x\} : x \in S\}$. Any weak MSAO of G can be transformed into a SSAO of G' if we add s and all edges in E_s (oriented away from s): s is a source and all nodes in S are no longer sources since they can be reached from s, hence s is the single source. Note that the orientation is still acyclic as s is a source and cannot be part of a cycle. The opposite is true as well: any SSAO of G' can be transformed into a weak MSAO of G by removing s and the edges in E_s. This process only removes edges incident to nodes in S, hence only nodes in S possibly become sources. Clearly the orientation is still acyclic as removing nodes and edges cannot create cycles.

Finally, consider strong MSAO. To solve it, we simply collapse all nodes of S into one node s. More formally, we generate $G''(V \cup \{s\} \setminus S, E \cup E_s)$, where $E_s = \{\{s, x\} : \exists\, y \in S \text{ with } \{y, x\} \in E\}$. As s and all nodes in S must be sources, all of their incident edges must be oriented away from them in all acyclic orientations, while the rest of the graph is exactly the same for both cases. Clearly, any SSAO for G'' induces a strong MSAO of G that can be obtained by removing s and E_s and re-integrating S and the edges between S and $V \setminus S$ (oriented away from S). Similarly, removing S (and the edges between S and

$V \setminus S$) and integrating s and E_s (with edges oriented away from s), creates a SSAO for G'': there is an edge from s to any node in $V \setminus S$ that was previously connected with S, hence these nodes cannot be sources; all other nodes in $V \setminus S$ were not connected to S and hence their in-degrees and out-degrees are unchanged.

As for AO, it can be obtained from weak MSAO by setting $S = V$. A direct reduction to SSAO is described in [14], although the paper does not provide an algorithm for SSAO.

By Theorem 1, observing that the above transformations requires $O(m)$ time, we can conclude the following result.

Theorem 3. *Problems* SSAO, *weak* SSAO, *strong* MSAO, *and weak* MSAO *can be solved with delay* $O(m \cdot n)$ *and space* $O(n^2)$.

6 Conclusions

In this paper we have shown the first enumeration algorithms with guaranteed bounds for SSAO, weak SSAO, strong MSAO, and weak MSAO, whose delay is $O(m \cdot n)$ time and $O(n^2)$ space. The delay reduces to $O(m)$ in the case of AO, improving prior work. It would be interesting to reduce the delay of the former problems to $O(m)$ as well.

References

1. Alon, N., Tarsi, M.: Colorings and orientations of graphs. Combinatorica **12**(2), 125–134 (1992)
2. Alon, N., Tuza, Z.: The acyclic orientation game on random graphs. Random Struct. Algorithms **6**(2–3), 261–268 (1995)
3. Barbosa, V.C., Szwarcfiter, J.L.: Generating all the acyclic orientations of an undirected graph. Inf. Process. Lett. **72**(1), 71–74 (1999)
4. Benson, B., Chakrabarty, D., Tetali, P.: G-parking functions, acyclic orientations and spanning trees. Discrete Math. **310**(8), 1340–1353 (2010)
5. Conte, A., Grossi, R., Marino, A., Rizzi, R.: Enumerating cyclic orientations of a graph. In: IWOCA, 26th International Workshop on Combinatorial Algorithms (2015, to appear)
6. Erdős, P., Katona, G., Társulat, B.J.M.: Theory of Graphs: Proceedings of the Colloquium Held at Tihany, Hungary, September 1966. Academic Press, New York (1968)
7. Iriarte, B.: Graph orientations and linear extensions. In: DMTCS Proceedings, pp. 945–956 (2014)
8. Johnson, D.S., Papadimitriou, C.H., Yannakakis, M.: On generating all maximal independent sets. Inf. Process. Lett. **27**(3), 119–123 (1988)
9. Johnson, R.: Network reliability and acyclic orientations. Networks **14**(4), 489–505 (1984)
10. Linial, N.: Hard enumeration problems in geometry and combinatorics. SIAM J. Algebraic Discrete Methods **7**(2), 331–335 (1986)
11. Pikhurko, O.: Finding an unknown acyclic orientation of a given graph. Comb. Probab. Comput. **19**, 121–131 (2010)

12. Roy, B.: Nombre chromatique et plus longs chemins d'un graphe. Rev. Fr. D'informatique Rech. Opérationnelle **1**(5), 129–132 (1967)
13. Schwikowski, B., Speckenmeyer, E.: On enumerating all minimal solutions of feedback problems. Discrete Appl. Math. **117**(1), 253–265 (2002)
14. Squire, M.B.: Generating the acyclic orientations of a graph. J. Algorithms **26**(2), 275–290 (1998)
15. Stanley, R.: Acyclic orientations of graphs. In: Gessel, I., Rota, G.-C. (eds.) Classic Papers in Combinatorics. Modern Birkhäuser Classics, pp. 453–460. Birkhäuser, Boston (1987)
16. Stanley, R.P.: What Is Enumerative Combinatorics?. Springer, New York (1986)
17. Vitaver, L.M.: Determination of minimal coloring of vertices of a graph by means of boolean powers of the incidence matrix. Dokl. Akad. Nauk SSSR **147**, 728 (1962)

Linear-Time Sequence Comparison Using Minimal Absent Words & Applications

Maxime Crochemore[1], Gabriele Fici[2], Robert Mercaş[1,3],
and Solon P. Pissis[1(✉)]

[1] Department of Informatics, King's College London, London, UK
{maxime.crochemore,solon.pissis}@kcl.ac.uk
[2] Dipartimento di Matematica e Informatica, Università di Palermo, Palermo, Italy
gabriele.fici@unipa.it
[3] Department of Computer Science, Kiel University, Kiel, Germany
rgm@informatik.uni-kiel.de

Abstract. Sequence comparison is a prerequisite to virtually all comparative genomic analyses. It is often realized by sequence alignment techniques, which are computationally expensive. This has led to increased research into alignment-free techniques, which are based on measures referring to the composition of sequences in terms of their constituent patterns. These measures, such as q-gram distance, are usually computed in time linear with respect to the length of the sequences. In this article, we focus on the complementary idea: how two sequences can be efficiently compared based on information that does not occur in the sequences. A word is an *absent word* of some sequence if it does not occur in the sequence. An absent word is *minimal* if all its proper factors occur in the sequence. Here we present the first linear-time and linear-space algorithm to compare two sequences by considering *all* their minimal absent words. In the process, we present results of combinatorial interest, and also extend the proposed techniques to compare circular sequences.

Keywords: Algorithms on strings · Sequence comparison · Alignment-free comparison · Absent words · Forbidden words · Circular words

1 Introduction

Sequence comparison is an important step in many basic tasks in bioinformatics, from phylogenies reconstruction to genomes assembly. It is often realized by sequence alignment techniques, which are computationally expensive, requiring quadratic time in the length of the sequences. This has led to increased research into *alignment-free* techniques. Hence standard notions for sequence comparison are gradually being complemented and in some cases replaced by alternative ones [10]. One such notion is based on comparing the words that are absent in each sequence [1]. A word is an *absent word* (or a forbidden word) of some sequence if it does not occur in the sequence. Absent words represent a type of *negative information*: information about what does not occur in the sequence.

© Springer-Verlag Berlin Heidelberg 2016
E. Kranakis et al. (Eds.): LATIN 2016, LNCS 9644, pp. 334–346, 2016.
DOI: 10.1007/978-3-662-49529-2_25

Given a sequence of length n, the number of absent words of length at most n is exponential in n. However, the number of certain classes of absent words is only linear in n. This is the case for *minimal absent words*, that is, absent words in the sequence whose all proper factors occur in the sequence [5]. An upper bound on the number of minimal absent words is known to be $\mathcal{O}(\sigma n)$ [9,23], where σ is the size of the alphabet Σ. Hence it may be possible to compare sequences in time proportional to their lengths, for a fixed-sized alphabet, instead of proportional to the product of their lengths. In what follows, we consider sequences on a *fixed-sized alphabet* since the most commonly studied alphabet is $\Sigma = \{\texttt{A},\texttt{C},\texttt{G},\texttt{T}\}$.

An $\mathcal{O}(n)$-time and $\mathcal{O}(n)$-space algorithm for computing all minimal absent words on a fixed-sized alphabet based on the construction of suffix automata was presented in [9]. The computation of minimal absent words based on the construction of suffix arrays was considered in [28]; although this algorithm has a linear-time performance in practice, the worst-case time complexity is $\mathcal{O}(n^2)$. New $\mathcal{O}(n)$-time and $\mathcal{O}(n)$-space suffix-array-based algorithms were presented in [2,3,15] to bridge this unpleasant gap. An implementation of the algorithm presented in [2] is currently, and to the best of our knowledge, the fastest available for the computation of minimal absent words. A more space-efficient solution to compute all minimal absent words in time $\mathcal{O}(n)$ was also presented in [6].

In this article, we consider the problem of comparing two sequences x and y of respective lengths m and n, using their sets of minimal absent words. In [7], Chairungsee and Crochemore introduced a measure of similarity between two sequences based on the notion of minimal absent words. They made use of a length-weighted index to provide a measure of similarity between two sequences, using sample sets of their minimal absent words, by considering the length of each member in the symmetric difference of these sample sets. This measure can be trivially computed in time and space $\mathcal{O}(m + n)$ provided that these sample sets contain minimal absent words of some bounded length ℓ. For unbounded length, the same measure can be trivially computed in time $\mathcal{O}(m^2 + n^2)$: for a given sequence, the cumulative length of all its minimal absent words can grow *quadratically* with respect to the length of the sequence.

The same problem can be considered for two *circular* sequences. The measure of similarity of Chairungsee and Crochemore can be used in this setting provided that one extends the definition of minimal absent words to circular sequences. In Sect. 4, we give a definition of minimal absent words for a circular sequence from the Formal Language Theory point of view. We believe that this definition may also be of interest from the point of view of Symbolic Dynamics, which is the original context in which minimal absent words have been introduced [5].

Our Contribution. Here we make the following threefold contribution:

(a) We present an $\mathcal{O}(m + n)$-time and $\mathcal{O}(m + n)$-space algorithm to compute the similarity measure introduced by Chairungsee and Crochemore by considering *all* minimal absent words of two sequences x and y of lengths m and n, respectively; thereby showing that it is indeed possible to compare two sequences in time proportional to their lengths (Sect. 3).

(b) We show how this algorithm can be applied to compute this similarity measure for two circular sequences x and y of lengths m and n, respectively, in the same time and space complexity as a result of the extension of the definition of minimal absent words to circular sequences (Sect. 4).

(c) We provide an open-source code implementation of our algorithms and investigate potential applications of our theoretical findings (Sect. 5).

2   Definitions and Notation

We begin with basic definitions and notation. Let $y = y[0]y[1] .. y[n-1]$ be a *word* of *length* $n = |y|$ over a finite ordered *alphabet* Σ of size $\sigma = |\Sigma| = \mathcal{O}(1)$. For two positions i and j on y, we denote by $y[i .. j] = y[i] .. y[j]$ the *factor* (sometimes called *substring*) of y that starts at position i and ends at position j (it is empty if $j < i$), and by ε the *empty word*, word of length 0. We recall that a prefix of y is a factor that starts at position 0 ($y[0 .. j]$) and a suffix is a factor that ends at position $n-1$ ($y[i .. n-1]$), and that a factor of y is a *proper* factor if it is not y itself. The set of all the factors of the word y is denoted by $\mathcal{F}_y$.

Let x be a word of length $0 < m \leq n$. We say that there exists an *occurrence* of x in y, or, more simply, that x *occurs in* y, when x is a factor of y. Every occurrence of x can be characterised by a starting position in y. Thus we say that x occurs at the *starting position* i in y when $x = y[i .. i + m - 1]$. Opposingly, we say that the word x is an *absent word* of y if it does not occur in y. The absent word x of y is *minimal* if and only if all its proper factors occur in y. The set of all minimal absent words for a word y is denoted by $\mathcal{M}_y$. For example, if $y = abaab$, then $\mathcal{M}_y = \{aaa, aaba, bab, bb\}$. In general, if we suppose that all the letters of the alphabet appear in y of length n, the length of a minimal absent word of y lies between 2 and $n + 1$. It is equal to $n + 1$ if and only if y is the catenation of n copies of the same letter. So, if y contains occurrences of at least two different letters, the length of a minimal absent word for y is bounded from above by n.

A *language* over the alphabet Σ is a set of finite words over Σ. A language is *regular* if it is recognized by a finite state automaton. A language is *factorial* if it contains all the factors of its words. A language is *antifactorial* if no word in the language is a proper factor of another word in the language. Given a word x, the language *generated* by x is the language $x^* = \{x^k \mid k \geq 0\} = \{\varepsilon, x, xx, xxx, \dots\}$. The *factorial closure* of a language L is the language $\mathcal{F}_L = \{\mathcal{F}_y \mid y \in L\}$. Given a factorial language L, one can define the (antifactorial) language of minimal absent words for L as $\mathcal{M}_L = \{aub \mid aub \notin L, au, ub \in L\}$. Notice that $\mathcal{M}_L$ is not the same language as the union of $\mathcal{M}_x$ for $x \in L$.

We denote by SA the *suffix array* of y of length n, that is, an integer array of size n storing the starting positions of all (lexicographically) sorted suffixes of y, i.e. for all $1 \leq r < n$ we have $y[\mathsf{SA}[r-1] .. n-1] < y[\mathsf{SA}[r] .. n-1]$ [22]. Let $\mathsf{lcp}(r, s)$ denote the length of the longest common prefix between $y[\mathsf{SA}[r] .. n-1]$ and $y[\mathsf{SA}[s] .. n-1]$ for all positions r, s on y, and 0 otherwise. We denote by LCP the *longest common prefix* array of y defined by $\mathsf{LCP}[r] = \mathsf{lcp}(r-1, r)$ for all $1 \leq r < n$, and $\mathsf{LCP}[0] = 0$. The inverse iSA of the array SA is defined by $\mathsf{iSA}[\mathsf{SA}[r]] = r$, for all $0 \leq r < n$. It is known that SA [25], iSA, and LCP [12] of a word of length n can be computed in time and space $\mathcal{O}(n)$.

In what follows, as already proposed in [2], for every word y, the set of minimal words associated with y, denoted by $\mathcal{M}_y$, is represented as a set of tuples $\langle a, i, j \rangle$, where the corresponding minimal absent word x of y is defined by $x[0] = a$, $a \in \Sigma$, and $x[1 .. m - 1] = y[i .. j]$, where $j - i + 1 = m \geq 2$. It is known that if $|y| = n$ and $|\Sigma| = \sigma$, then $|\mathcal{M}_y| \leq \sigma n$ [23].

In [7], Chairungsee and Crochemore introduced a measure of similarity between two words x and y based on the notion of minimal absent words. Let $\mathcal{M}_x^\ell$ (resp. $\mathcal{M}_y^\ell$) denote the set of minimal absent words of length at most ℓ of x (resp. y). The authors made use of a length-weighted index to provide a measure of the similarity between x and y, using their sample sets $\mathcal{M}_x^\ell$ and $\mathcal{M}_y^\ell$, by considering the length of each member in the symmetric difference $(\mathcal{M}_x^\ell \triangle \mathcal{M}_y^\ell)$ of the sample sets. For sample sets $\mathcal{M}_x^\ell$ and $\mathcal{M}_y^\ell$, they defined this index to be

$$\mathsf{LW}(\mathcal{M}_x^\ell, \mathcal{M}_y^\ell) = \sum_{w \in \mathcal{M}_x^\ell \triangle \mathcal{M}_y^\ell} \frac{1}{|w|^2}.$$

This work considers the following generalized version of the same problem.

MAW-SequenceComparison
Input: a word x of length m and a word y of length n
Output: $\mathsf{LW}(\mathcal{M}_x, \mathcal{M}_y)$, where $\mathcal{M}_x$ and $\mathcal{M}_y$ denote the sets of minimal absent words of x and y, respectively.

We also consider the aforementioned problem for two circular words. A circular word of length m can be viewed as a traditional linear word which has the left- and right-most letters wrapped around and stuck together in some way. Under this notion, the same circular word can be seen as m different linear words, which would all be considered equivalent. More formally, given a word x of length m, we denote by $x^{\langle i \rangle} = x[i .. m - 1]x[0 .. i - 1]$, $0 \leq i < m$, the i-th *rotation* of x, where $x^{\langle 0 \rangle} = x$. Given two words x and y, we define $x \sim y$ if and only if there exist i, $0 \leq i < |x|$, such that $y = x^{\langle i \rangle}$. A *circular word* $\tilde{x}$ is a conjugacy class of the equivalence relation $\sim$. Given a circular word $\tilde{x}$, any (linear) word x in the equivalence class $\tilde{x}$ is called a *linearization* of the circular word $\tilde{x}$. Conversely, given a linear word x, we say that $\tilde{x}$ is a *circularization* of x if and only if x is a linearization of $\tilde{x}$. The set $\mathcal{F}_{\tilde{x}}$ of factors of the circular word $\tilde{x}$ is equal to the set $\mathcal{F}_{xx} \cap \Sigma^{\leq |x|}$ of factors of xx whose length is at most $|x|$, where x is any linearization of $\tilde{x}$.

Note that if $x^{\langle i \rangle}$ and $x^{\langle j \rangle}$ are two rotations of the same word, then the factorial languages $\mathcal{F}_{(x^{\langle i \rangle})^*}$ and $\mathcal{F}_{(x^{\langle j \rangle})^*}$ coincide, so one can unambiguously define the (infinite) language $\mathcal{F}_{\tilde{x}^*}$ as the language $\mathcal{F}_{x^*}$, where x is any linearization of $\tilde{x}$.

In Sect. 4, we give the definition of the set $\mathcal{M}_{\tilde{x}}$ of minimal absent words for a circular word $\tilde{x}$. We will prove that the following problem can be solved with the same time and space complexity as its counterpart in the linear case.

MAW-CIRCULARSEQUENCECOMPARISON
Input: a word x of length m and a word y of length n
Output: $\mathsf{LW}(\mathcal{M}_{\tilde{x}}, \mathcal{M}_{\tilde{y}})$, where $\mathcal{M}_{\tilde{x}}$ and $\mathcal{M}_{\tilde{y}}$ denote the sets of minimal absent words of the circularizations $\tilde{x}$ of x and $\tilde{y}$ of y, respectively.

3   Sequence Comparison

The goal of this section is to provide the first linear-time and linear-space algorithm for computing the similarity measure (see Sect. 2) between two words defined over a fixed-sized alphabet. To this end, we consider two words x and y of lengths m and n, respectively, and their associated sets of minimal absent words, $\mathcal{M}_x$ and $\mathcal{M}_y$, respectively. Next, we give a linear-time and linear-space solution for the MAW-SEQUENCECOMPARISON problem. It is known from [9] and [2] that we can compute the sets $\mathcal{M}_x$ and $\mathcal{M}_y$ in linear time and space with respect to the two lengths m and n, respectively. The idea of our strategy consists of a merge sort on the sets $\mathcal{M}_x$ and $\mathcal{M}_y$, after they have been ordered with the help of suffix arrays.

To this end, we construct the suffix array associated to the word $w = xy$, together with the implicit LCP array corresponding to it. All of these structures can be constructed in time and space $\mathcal{O}(m + n)$, as mentioned earlier. Furthermore, we can preprocess the array LCP for range minimum queries, which we denote by $\mathsf{RMQ}_{\mathsf{LCP}}$ [13]. With the preprocessing complete, the longest common prefix LCE of two suffixes of w starting at positions p and q can be computed in constant time [19], using the formula $\mathsf{LCE}(w, p, q) = \mathsf{LCP}[\mathsf{RMQ}_{\mathsf{LCP}}(\mathsf{iSA}[p] + 1, \mathsf{iSA}[q])]$.

Using these data structures, it is straightforward to sort the tuples in the sets $\mathcal{M}_x$ and $\mathcal{M}_y$ lexicographically. That is, two tuples $x_1, x_2 \in \mathcal{M}_x$, are ordered such that the one being the prefix of the other one comes first, or according to the letter following their longest common prefix, when the former is not the case. To do this, we simply go once through the suffix array associated to w and assign to each tuple in $\mathcal{M}_x$, respectively $\mathcal{M}_y$, the rank of the suffix starting at the position indicated by its second component, in the suffix array. Since sorting an array of n distinct integers, such that each is in $[0, n - 1]$, can be done in time $\mathcal{O}(n)$ (using bucket sort, for example), we can sort now each of the sets of minimal absent words, taking into consideration the letter on the first position and these ranks. Thus, from now on, we assume that $\mathcal{M}_x = (x_0, x_1, \ldots, x_k)$ where x_i is lexicographically smaller than x_{i+1}, for $0 \leq i < k \leq \sigma m$, and $\mathcal{M}_y = (y_0, y_1, \ldots, y_\ell)$, where y_j is lexicographically smaller than y_{j+1}, for $0 \leq j < \ell \leq \sigma n$.

Provided these tools, we now proceed to do the merge. Thus, considering that we are analysing the $(i + 1)$th tuple in $\mathcal{M}_x$ and the $(j + 1)$th tuple in $\mathcal{M}_y$, we note that the two are equal if and only if $x_i[0] = y_j[0]$ and

$$\mathsf{LCE}(w, x_i[1], |x| + y_j[1]) \geq \ell, \text{ where } \ell = x_i[2] - x_i[1] = y_j[2] - y_j[1].$$

In other words, the two minimal absent words are equal if and only if their first letters coincide, they have equal length $\ell + 1$, and the longest common prefix of the suffixes of w starting at the positions indicated by the second components of the tuples has length at least ℓ.

Such a strategy will empower us with the means for constructing a new set $\mathcal{M}_{xy} = \mathcal{M}_x \cup \mathcal{M}_y$. At each step, when analysing tuples x_i and y_j we proceed as following:

$$\mathcal{M}_{xy} = \begin{cases} \mathcal{M}_{xy} \cup \{x_i\}, & \text{and increment } i, & \text{if } x_i < y_j; \\ \mathcal{M}_{xy} \cup \{y_j\}, & \text{and increment } j, & \text{if } x_i > y_j; \\ \mathcal{M}_{xy} \cup \{x_i = y_j\}, & \text{and increment both } i \text{ and } j, & \text{if } x_i = y_j. \end{cases}$$

Observe that the last condition is saying that basically each common tuple is added only once to their union.

Furthermore, simultaneously with this construction we can also calculate the similarity between the words, given by $\mathsf{LW}(\mathcal{M}_x, \mathcal{M}_y)$, which is initially set to 0. Thus, at each step, when comparing the tuples x_i and y_j, we update

$$\mathsf{LW}(\mathcal{M}_x, \mathcal{M}_y) = \begin{cases} \mathsf{LW}(\mathcal{M}_x, \mathcal{M}_y) + \frac{1}{|x_i|^2}, & \text{and increment } i, & \text{if } x_i < y_j; \\ \mathsf{LW}(\mathcal{M}_x, \mathcal{M}_y) + \frac{1}{|y_j|^2}, & \text{and increment } j, & \text{if } x_i > y_j; \\ \mathsf{LW}(\mathcal{M}_x, \mathcal{M}_y), & \text{and increment both } i \text{ and } j, & \text{if } x_i = y_j. \end{cases}$$

We impose the increment of both i and j in the case of equality as in this case we only look at the symmetric difference between the sets of minimal absent words.

As all these operations take constant time, once per each tuple in $\mathcal{M}_x$ and $\mathcal{M}_y$, it is easily concluded that the whole operation takes in the case of a fixed-sized alphabet time and space $\mathcal{O}(m + n)$. Thus, we can compute the symmetric difference between the *complete* sets of minimal absent words, as opposed to [7], of two words defined over a fixed-sized alphabet, in linear time and space with respect to the lengths of the two words. We obtain the following result.

Theorem 1. *Problem* MAW-SEQUENCECOMPARISON *can be solved in time and space* $\mathcal{O}(m + n)$.

4   Circular Sequence Comparison

Next, we discuss two possible definitions for the minimal absent words of a circular word, and highlight the differences between them.

We start by recalling some basic facts about minimal absent words. For further details and references the reader is recommended [11]. Every factorial language L is uniquely determined by its (antifactorial) language of minimal absent words $\mathcal{M}_L$, through the equation $L = \Sigma^* \setminus \Sigma^* \mathcal{M}_L \Sigma^*$. The converse is also true, since by the definition of a minimal absent word we have $\mathcal{M}_L = \Sigma L \cap L \Sigma \cap (\Sigma^* \setminus L)$. The previous equations define a bijection between factorial and antifactorial languages. Moreover, this bijection preserves regularity. In the case of a single (linear) word x, the set of minimal absent words for x is indeed the antifactorial language $\mathcal{M}_{\mathcal{F}_x}$. Furthermore, we can retrieve x from its set of minimal absent words in linear time and space [9].

Recall that given a circular word $\tilde{x}$, the set $\mathcal{F}_{\tilde{x}}$ of factors of $\tilde{x}$ is equal to the set $\mathcal{F}_{xx} \cap \Sigma^{\leq |x|}$ of factors of xx whose lengths are at most $|x|$, where x is any linearization of $\tilde{x}$. Since a circular word $\tilde{x}$ is a conjugacy class containing all the rotations of a linear word x, the language $\mathcal{F}_{\tilde{x}}$ can be seen as the factorial closure of the set $\{x^{\langle i \rangle} \mid i = 0, \ldots, |x| - 1\}$. This leads to the first definition of the set of minimal absent words for $\tilde{x}$, that is the set $\mathcal{M}_{\mathcal{F}_{\tilde{x}}} = \{aub \mid a, b \in \Sigma, aub \notin \mathcal{F}_{\tilde{x}}, au, ub \in \mathcal{F}_{\tilde{x}}\}$. For instance, if $x = abaab$, we have

$$\mathcal{M}_{\mathcal{F}_{\tilde{x}}} = \{aaa, aabaa, aababa, abaaba, ababaa, baabab, babaab, babab, bb\}.$$

The advantage of this definition is that we can retrieve uniquely $\tilde{x}$ from $\mathcal{M}_{\mathcal{F}_{\tilde{x}}}$. However, the total size of $\mathcal{M}_{\mathcal{F}_{\tilde{x}}}$ (that is, the sum of the lengths of its elements) can be very large, as the following lemma suggests.

Lemma 2. *Let $\tilde{x}$ be a circular word of length $m > 0$. The set $\mathcal{M}_{\mathcal{F}_{\tilde{x}}}$ contains precisely ℓ words of maximal length $m+1$, where ℓ is the number of distinct rotations of any linearization x of $\tilde{x}$, that is, the cardinality of $\{x^{\langle i \rangle} \mid i = 0, \ldots, |x| - 1\}$.*

Proof. Let $x = x[0]x[1] \ldots x[m-1]$ be a linearization of $\tilde{x}$. The word obtained by appending to x its first letter, $x[0]x[1] \ldots x[m-1]x[0]$, belongs to $\mathcal{M}_{\mathcal{F}_{\tilde{x}}}$, since it has length $m + 1$, hence it cannot belong to $\mathcal{F}_{\tilde{x}}$, but its maximal proper prefix $x = x^{\langle 0 \rangle}$ and its maximal proper suffix $x^{\langle 1 \rangle} = x[1] \ldots x[m-1]x[0]$ belong to $\mathcal{F}_{\tilde{x}}$.

The same argument shows that for any rotation $x^{\langle i \rangle} = x[i]x[i+1] \ldots x[m-1]x[0] \ldots x[i-1]$ of x, the word $x[i]x[i+1] \ldots x[m-1]x[0] \ldots x[i-1]x[i]$, obtained by appending to $x^{\langle i \rangle}$ its first letter, belongs to $\mathcal{M}_{\mathcal{F}_{\tilde{x}}}$.

Conversely, if a word of maximal length $m + 1$ is in $\mathcal{M}_{\mathcal{F}_{\tilde{x}}}$, then its maximal proper prefix and its maximal proper suffix are words of length m in $\mathcal{F}_{\tilde{x}}$, so they must be consecutive rotations of x.

Therefore, the number of words of maximal length $m + 1$ in $\mathcal{M}_{\mathcal{F}_{\tilde{x}}}$ equals the number of distinct rotations of x, hence the statement follows. $\square$

This is in sharp contrast with the situation for linear words, where the set of minimal absent words can be represented on a trie having size linear in the length of the word. Indeed, the algorithm MF-TRIE, introduced in [9], builds the tree-like deterministic automaton accepting the set of minimal absent words for a word x taking as input the factor automaton of x, that is the minimal deterministic automaton recognizing the set of factors of x. The leaves of the trie correspond to the minimal absent words for x, while the internal states are those of the factor automaton. Since the factor automaton of a word x has less than $2|x|$ states (for details, see [8]), this provides a representation of the minimal absent words of a word of length n in space $O(\sigma n)$.

This algorithmic drawback leads us to the second definition. This second definition of minimal absent words for circular strings has been already introduced in [26, 27]. First, we give a combinatorial result which shows that when considering circular words it does not make sense to look at absent words obtained from more than two rotations.

Lemma 3. *For any positive integer k and any word u, the set $V = \{v \mid k|u|+1 < |v| \leq (k+1)|u|\} \cap (\mathcal{M}_{u^{k+1}} \setminus \mathcal{M}_{u^k})$ is empty.*

Proof. This obviously holds for all words u of length 1. Assume towards a contradiction that this is not the case in general. Hence, there must exist a word v of length m that fulfills the conditions in the lemma, thus $v \in V$ and $m > 2$. Furthermore, since the length $m - 1$ prefix and the length $m - 1$ suffix of every minimal absent word occur in the main word at non-consecutive positions, there must exist positions $i < j \leq n = |u|$ such that

$$v[1 \mathinner{.\,.} m - 2] = u^{k+1}[i + 1 \mathinner{.\,.} i + m - 2] = u^{k+1}[j + 1 \mathinner{.\,.} j + m - 2]. \tag{1}$$

Obviously, following Eq. (1), since $m - 2 \geq kn$, we have that $v[1 \mathinner{.\,.} m - 2]$ is $(j-i)$-periodic. But, we know that $v[1 \mathinner{.\,.} m-2]$ is also n-periodic. Thus, following a direct application of the periodicity lemma we have that $v[1 \mathinner{.\,.} m - 2]$ is $p = \gcd(j-i, n)$-periodic. But, in this case we have that u is p-periodic, and, therefore, $u[i] = u[j]$, which leads to a contradiction with the fact that v is a minimal absent word, whenever i is defined. Thus, it must be the case that $i = -1$. Using the same strategy and looking at positions $u[i+m-2]$ and $u[j+m-2]$, we conclude that $j + m - 2 = (k+1)n$. Therefore, in this case, we have that $m = kn + 1$, which is a contradiction with the fact that the word v fulfills the conditions of the lemma. This concludes the proof. □

Observe now that the set V consists in fact of all extra minimal absent words generated whenever we look at more than one rotation, that do not include the length arguments. That is, V does not include the words bounding the maximum length that a word is allowed, nor the words created, or lost, during a further concatenation of an image of u. However, when considering an iterative concatenation of the word, these extra elements determined by the length constrain cancel each other.

As observed in Sect. 2, two rotations of the same word x generate two languages that have the same set of factors. So, we can unambiguously associate to a circular word $\tilde{x}$ the (infinite) factorial language $\mathcal{F}_{\tilde{x}^*}$. It is therefore natural to define the set of minimal absent words for the circular word $\tilde{x}$ as the set $\mathcal{M}_{\mathcal{F}_{\tilde{x}^*}}$. For instance, if $\tilde{x} = abaab$, then we have

$$\mathcal{M}_{\mathcal{F}_{\tilde{x}^*}} = \{aaa, aabaa, babab, bb\}.$$

This second definition is much more efficient in terms of space, as we show below. In particular, the length of the words in $\mathcal{M}_{\mathcal{F}_{\tilde{x}^*}}$ is bounded from above by $|x|$, hence $\mathcal{M}_{\mathcal{F}_{\tilde{x}^*}}$ is a finite set.

Recall that a word x is *a power* of a word y if there exists a positive integer $k > 1$ such that x is expressed as k consecutive concatenations of y, denoted by $x = y^k$. Conversely, a word x is *primitive* if $x = y^k$ implies $k = 1$. Notice that a word is primitive if and only if any of its rotation is. We can therefore extend the definition of primitivity to circular words. The definition of $\mathcal{M}_{\mathcal{F}_{\tilde{x}^*}}$ does not allow one to uniquely reconstruct $\tilde{x}$ from $\mathcal{M}_{\mathcal{F}_{\tilde{x}^*}}$, unless $\tilde{x}$ is known to be primitive, since it is readily verified that $\mathcal{F}_{\tilde{x}^*} = \mathcal{F}_{\widetilde{xx}^*}$ and therefore also the minimal absent words of these two languages coincide. However, from the algorithmic point of view, this issue can be easily managed by storing the length $|x|$ of a linearization x of $\tilde{x}$ together with the set $\mathcal{M}_{\mathcal{F}_{\tilde{x}^*}}$. Moreover, in most practical cases, for example when dealing with biological sequences, it is highly unlikely that the circular word considered is not primitive.

The difference between the two definitions above is presented in the next lemma.

Lemma 4. $\mathcal{M}_{\mathcal{F}_{\tilde{x}^*}} = \mathcal{M}_{\mathcal{F}_{\tilde{x}}} \cap \Sigma^{\leq |x|}$.

Proof. Clearly, $\mathcal{F}_{\tilde{x}^*} \cap \Sigma^{\leq |x|} = \mathcal{F}_{\tilde{x}}$. The statement then follows from the definition of minimal absent words.   □

Based on the previous discussion, we set $\mathcal{M}_{\tilde{x}} = \mathcal{M}_{\mathcal{F}_{\tilde{x}^*}}$, while the following corollary comes straightforwardly as a consequence of Lemma 3.

Corollary 5. *Let $\tilde{x}$ be a circular word. Then $\mathcal{M}_{\tilde{x}} = \mathcal{M}_{xx}^{|x|}$.*

Corollary 5 was first introduced as a definition for the set of minimal absent words of a circular word in [26]. Using the result of Corollary 5, we can easily extend the algorithm described in the previous section to the case of circular words. That is, given two circular words $\tilde{x}$ of length m and $\tilde{y}$ of length n, we can compute in time and space $\mathcal{O}(m+n)$ the quantity $\mathsf{LW}(\mathcal{M}_{\tilde{x}}, \mathcal{M}_{\tilde{y}})$. We obtain the following result.

Theorem 6. *Problem* MAW-CIRCULARSEQUENCECOMPARISON *can be solved in time and space $\mathcal{O}(m+n)$.*

5   Implementation and Applications

We implemented the presented algorithms as programme scMAW to perform pairwise sequence comparison for a set of sequences using minimal absent words. scMAW uses programme MAW [2] for linear-time and linear-space computation of minimal absent words using suffix array. scMAW was implemented in the C programming language and developed under GNU/Linux operating system. It takes, as input argument, a file in MultiFASTA format with the input sequences, and then any of the two methods, for *linear* or *circular* sequence comparison, can be applied. It then produces a file in PHYLIP format with the distance matrix as output. Cell $[x, y]$ of the matrix stores $\mathsf{LW}(\mathcal{M}_x, \mathcal{M}_y)$ (or $\mathsf{LW}(\mathcal{M}_{\tilde{x}}, \mathcal{M}_{\tilde{y}})$ for the circular case). The implementation is distributed under the GNU General Public License (GPL), and it is available at http://github.com/solonas13/maw, which is set up for maintaining the source code and the man-page documentation. Notice that *all* input datasets and the produced outputs referred to in this section are publicly maintained at the same web-site.

An important feature of the proposed algorithms is that they require space linear in the length of the sequences (see Theorems 1 and 6). Hence, we were also able to implement scMAW using the Open Multi-Processing (OpenMP) PI for shared memory multiprocessing programming to distribute the workload across the available processing threads without a large memory footprint.

Application. Recently, there has been a number of studies on the biological significance of absent words in various species [1,16,31]. In [16], the authors presented dendrograms from dinucleotide relative abundances in sets of minimal absent words for prokaryotes and eukaryotic genomes. The analyses support the hypothesis that minimal absent words are inherited through a common ancestor, in addition to lineage-specific inheritance, only in vertebrates. Very recently, in [31], it was shown that there exist three minimal words in the Ebola virus genomes which are absent from human genome. The authors suggest that the identification of such species-specific sequences may prove to be useful for the development of both diagnosis and therapeutics.

In this section, we show a potential application of our results for the construction of dendrograms for DNA sequences with circular structure. Circular DNA sequences can be found in viruses, as plasmids in archaea and bacteria, and in the mitochondria and plastids of eukaryotic cells. Circular sequence comparison thus finds applications in several contexts such as reconstructing phylogenies using viroids RNA [24] or Mitochondrial DNA (MtDNA) [17]. Conventional tools to align circular sequences could yield an incorrectly high genetic distance between closely-related species. Indeed, when sequencing molecules, the position where a circular sequence starts can be totally arbitrary. Due to this *arbitrariness*, a suitable rotation of one sequence would give much better results for a pairwise alignment [4,18]. In what follows, we demonstrate the power of minimal absent words to pave a path to resolve this issue by applying Corollary 5 and Theorem 6. Next we do not claim that a solid phylogenetic analysis is presented but rather an investigation for potential applications of our theoretical findings.

We performed the following experiment with synthetic data. First, we simulated a basic dataset of DNA sequences using INDELible [14]. The number of taxa, denoted by α, was set to 12; the length of the sequence generated at the root of the tree, denoted by β, was set to 2500bp; and the substitution rate, denoted by γ, was set to 0.05. We also used the following parameters: a deletion rate, denoted by δ, of 0.06 *relative* to substitution rate of 1; and an insertion rate, denoted by ϵ, of 0.04 *relative* to substitution rate of 1. The parameters were chosen based on the genetic diversity standard measures observed for sets of MtDNA sequences from primates and mammals [4]. We generated another instance of the basic dataset, containing one *arbitrary* rotation of each of the α sequences from the basic dataset. We then used this randomized dataset as input to scMAW by considering $\mathsf{LW}(\mathcal{M}_{\tilde{x}}, \mathcal{M}_{\tilde{y}})$ as the distance metric. The output of scMAW was passed as input to NINJA [33], an efficient implementation of neighbor-joining [30], a well-established hierarchical clustering algorithm for inferring dendrograms (trees). We thus used NINJA to infer the respective tree T_1 under the neighbor-joining criterion. We also inferred the tree T_2 by following the same pipeline, but by considering $\mathsf{LW}(\mathcal{M}_x, \mathcal{M}_y)$ as distance metric, as well as the tree T_3 by using the *basic* dataset as input of this pipeline and $\mathsf{LW}(\mathcal{M}_{\tilde{x}}, \mathcal{M}_{\tilde{y}})$

Table 1. Accuracy measurements based on relative pairwise RF distance

Dataset $< \alpha, \beta, \gamma, \delta, \epsilon >$	T_1 vs. T_3	T_2 vs. T_3
$< 12, 2500, 0.05, 0.06, 0.04 >$	100 %	100 %
$< 12, 2500, 0.20, 0.06, 0.04 >$	100 %	88,88 %
$< 12, 2500, 0.35, 0.06, 0.04 >$	100 %	100 %
$< 25, 2500, 0.05, 0.06, 0.04 >$	100 %	100 %
$< 25, 2500, 0.20, 0.06, 0.04 >$	100 %	100 %
$< 25, 2500, 0.35, 0.06, 0.04 >$	100 %	100 %
$< 50, 2500, 0.05, 0.06, 0.04 >$	100 %	97,87 %
$< 50, 2500, 0.20, 0.06, 0.04 >$	100 %	97,87 %
$< 50, 2500, 0.35, 0.06, 0.04 >$	100 %	100 %

as distance metric. Hence, notice that T_3 represents the original tree. Finally, we computed the pairwise Robinson-Foulds (RF) distance [29] between: T_1 and T_3; and T_2 and T_3.

Let us define *accuracy* as the difference between 1 and the relative pairwise RF distance. We repeated this experiment by simulating different datasets $< \alpha, \beta, \gamma, \delta, \epsilon >$ and measured the corresponding accuracy. The results in Table 1 (see T_1 vs. T_3) suggest that by considering $\mathsf{LW}(\mathcal{M}_{\tilde{x}}, \mathcal{M}_{\tilde{y}})$ we can always reconstruct the original tree even if the sequences have been first arbitrarily rotated (Corollary 5). This is not the case (see T_2 vs. T_3) if we consider $\mathsf{LW}(\mathcal{M}_x, \mathcal{M}_y)$. Notice that 100 % accuracy denotes a (relative) pairwise RF distance of 0.

6   Final Remarks

In this article, complementary to measures that refer to the composition of sequences in terms of their constituent patterns, we considered sequence comparison using minimal absent words, information about what does not occur in the sequences. We presented the first linear-time and linear-space algorithm to compare two sequences by considering *all* their minimal absent words (Theorem 1). In the process, we presented some results of combinatorial interest, and also extended the proposed techniques to circular sequences. The power of minimal absent words is highlighted by the fact that they provide a tool for sequence comparison that is as efficient for circular as it is for linear sequences (Corollary 5 and Theorem 6); whereas, this is not the case, for instance, using the general edit distance model [21]. Finally, a preliminary experimental study shows the potential of our theoretical findings.

Our immediate target is to consider the following *incremental* version of the same problem: given an appropriate encoding of a comparison between sequences x and y, can one incrementally compute the answer for x and ay, and the answer for x and ya, efficiently, where a is an additional letter? Incremental sequence comparison, under the edit distance model, has already been considered in [20].

In [18], the authors considered a more powerful generalization of the q-gram distance (see [32] for definition) to compare x and y. This generalization comprises partitioning x and y in β blocks each, as evenly as possible, computing the q-gram distance between the corresponding block pairs, and then summing up the distances computed blockwise to obtain the new measure. We are also planning to apply this generalization to the similarity measure studied here and evaluate it using real and synthetic data.

Acknowledgements. We warmly thank Alice Heliou for her inestimable code contribution and Antonio Restivo for useful discussions. Gabriele Fici's work was supported by the PRIN 2010/2011 project "Automi e Linguaggi Formali: Aspetti Matematici e Applicativi" of the Italian Ministry of Education (MIUR) and by the "National Group for Algebraic and Geometric Structures, and their Applications" (GNSAGA – INdAM). Robert Mercaş's work was supported by the P.R.I.M.E. programme of DAAD co-funded by BMBF and EU's 7th Framework Programme (grant 605728). Solon P. Pissis's work was supported by a Research Grant (#RG130720) awarded by the Royal Society.

References

1. Acquisti, C., Poste, G., Curtiss, D., Kumar, S.: Nullomers: really a matter of natural selection? PLoS ONE **2**(10), e1022 (2007)
2. Barton, C., Heliou, A., Mouchard, L., Pissis, S.P.: Linear-time computation of minimal absent words using suffix array. BMC Bioinform. **15**, 388 (2014)
3. Barton, C., Heliou, A., Mouchard, L., Pissis, S.P.: Parallelising the computation of minimal absent words. In: PPAM, LNCS. Springer, Heidelberg (2015)
4. Barton, C., Iliopoulos, C.S., Kundu, R., Pissis, S.P., Retha, A., Vayani, F.: Accurate and efficient methods to improve multiple circular sequence alignment. In: Bampis, E. (ed.) SEA 2015. LNCS, vol. 9125, pp. 247–258. Springer, Heidelberg (2015)
5. Béal, M., Mignosi, F., Restivo, A., Sciortino, M.: Forbidden words in symbolic dynamics. Adv. Appl. Math. **25**(2), 163–193 (2000)
6. Belazzougui, D., Cunial, F., Kärkkäinen, J., Mäkinen, V.: Versatile succinct representations of the bidirectional burrows-wheeler transform. In: Bodlaender, H.L., Italiano, G.F. (eds.) ESA 2013. LNCS, vol. 8125, pp. 133–144. Springer, Heidelberg (2013)
7. Chairungsee, S., Crochemore, M.: Using minimal absent words to build phylogeny. Theor. Comput. Sci. **450**, 109–116 (2012)
8. Crochemore, M., Hancart, C., Lecroq, T.: Algorithms on Strings. Cambridge University Press, New York, NY, USA (2007)
9. Crochemore, M., Mignosi, F., Restivo, A.: Automata and forbidden words. Inf. Process. Lett. **67**, 111–117 (1998)
10. Domazet-Lošo, M., Haubold, B.: Efficient estimation of pairwise distances between genomes. Bioinformatics **25**(24), 3221–3227 (2009)
11. Fici, G.: Minimal Forbidden Words and Applications. Ph.D. thesis, Université de Marne-la-Vallée (2006)
12. Fischer, J.: Inducing the LCP-array. In: Dehne, F., Iacono, J., Sack, J.-R. (eds.) WADS 2011. LNCS, vol. 6844, pp. 374–385. Springer, Heidelberg (2011)

13. Fischer, J., Heun, V.: Space-efficient preprocessing schemes for range minimum queries on static arrays. SIAM J. Comput. **40**(2), 465–492 (2011)
14. Fletcher, W., Yang, Z.: INDELible: a flexible simulator of biological sequence evolution. Mol. Biol. Evol. **26**(8), 1879–1888 (2009)
15. Fukae, H., Ota, T., Morita, H.: On fast and memory-efficient construction of an antidictionary array. In: ISIT, pp. 1092–1096. IEEE (2012)
16. Garcia, S.P., Pinho, A.J., Rodrigues, J.M.O.S., Bastos, C.A.C., Ferreira, P.J.S.G.: Minimal absent words in prokaryotic and eukaryotic genomes. PLoS ONE **6**(1), e16065 (2011)
17. Goios, A., Pereira, L., Bogue, M., Macaulay, V., Amorim, A.: mtDNA phylogeny and evolution of laboratory mouse strains. Genome Res. **17**(3), 293–298 (2007)
18. Grossi, R., Iliopoulos, C.S., Mercaş, R., Pisanti, N., Pissis, S.P., Retha, A., Vayani, F.: Circular sequence comparison with q-grams. In: Pop, M., Touzet, H. (eds.) WABI 2015. LNCS, vol. 9289, pp. 203–216. Springer, Heidelberg (2015)
19. Ilie, L., Navarro, G., Tinta, L.: The longest common extension problem revisited and applications to approximate string searching. J. Discrete Algorithms **8**(4), 418–428 (2010)
20. Landau, G.M., Myers, E.W., Schmidt, J.P.: Incremental string comparison. SIAM J. Comput. **27**(2), 557–582 (1998)
21. Maes, M.: On a cyclic string-to-string correction problem. Inf. Process. Lett. **35**(2), 73–78 (1990)
22. Manber, U., Myers, E.W.: Suffix arrays: a new method for on-line string searches. SIAM J. Comput. **22**(5), 935–948 (1993)
23. Mignosi, F., Restivo, A., Sciortino, M.: Words and forbidden factors. Theor. Comput. Sci. **273**(1–2), 99–117 (2002)
24. Mosig, A., Hofacker, I.L., Stadler, P.F.: Comparative analysis of cyclic sequences: viroids and other small circular RNAs. GCB, LNI **83**, 93–102 (2006)
25. Nong, G., Zhang, S., Chan, W.H.: Linear suffix array construction by almost pure induced-sorting. In: DCC, pp. 193–202. IEEE (2009)
26. Ota, T., Morita, H.: On a universal antidictionary coding for stationary ergodic sources with finite alphabet. In: ISITA, pp. 294–298. IEEE (2014)
27. Ota, T., Morita, H.: On antidictionary coding based on compacted substring automaton. In: ISIT, pp. 1754–1758. IEEE (2013)
28. Pinho, A.J., Ferreira, P.J.S.G., Garcia, S.P., Rodrigues, J.M.O.S.: On finding minimal absent words. BMC Bioinform. **10**(1), 1 (2009)
29. Robinson, D., Fould, L.: Comparison of phylogenetic trees. Math. Biosci. **53**(1–2), 131–147 (1981)
30. Saitou, N., Nei, M.: The neighbor-joining method: a new method for reconstructing phylogenetic trees. Mol. Biol. Evol. **4**(4), 406–425 (1987)
31. Silva, R.M., Pratas, D., Castro, L., Pinho, A.J., Ferreira, P.J.S.G.: Three minimal sequences found in Ebola virus genomes and absent from human DNA. Bioinformatics **31**(15), 2421–2425 (2015)
32. Ukkonen, E.: Approximate string-matching with q-grams and maximal matches. Theor. Comput. Sci. **92**(1), 191–211 (1992)
33. Wheeler, T.J.: Large-scale neighbor-joining with NINJA. In: Salzberg, S.L., Warnow, T. (eds.) WABI 2009. LNCS, vol. 5724, pp. 375–389. Springer, Heidelberg (2009)

The Grandmama de Bruijn Sequence for Binary Strings

Patrick Baxter Dragon, Oscar I. Hernandez, and Aaron Williams[⊠]

Division of Science, Math and Computing, Bard College at Simon's Rock,
Great Barrington, USA
{pdragon,ohernandez13}@simons-rock.edu, haron@uvic.ca

Abstract. A de Bruijn sequence is a binary string of length 2^n which, when viewed cyclically, contains every binary string of length n exactly once as a substring. Knuth refers to the lexicographically least de Bruijn sequence for each n as the "granddaddy" and Fredricksen et al. showed that it can be constructed by concatenating the aperiodic prefixes of the binary necklaces of length n in lexicographic order. In this paper we prove that the granddaddy has a lexicographic partner. The "grandmama" sequence is constructed by instead concatenating the aperiodic prefixes in co-lexicographic order. We explain how our sequence differs from the previous sequence and why it had not previously been discovered.

Keywords: de Bruijn sequence · Lexicographic order · Necklace · Lyndon word · FKM construction · Ford sequence

1 Introduction

Let $B(n)$ be the set of binary strings of length n. A *de Bruijn sequence* is a binary string of length 2^n that contains each element of $B(n)$ as a substring that is allowed to wrap-around from the end to the beginning. For example,

$$D = 0000100110101111 \tag{1}$$

is a de Bruijn sequence for $n = 4$ since its substrings are

$$0000, 0001, 0010, 0100, 1001, 0011, 0110, 1101,$$
$$1010, 0101, 1011, 0111, 1111, 1110, 1100, 1000,$$

where the last three substrings wrap-around. Another example for $n = 4$ is

$$M = 0000101001101111. \tag{2}$$

Although D and M are not equal, they are *equivalent*, meaning that they are in the same equivalence class generated by the following operations:

© Springer-Verlag Berlin Heidelberg 2016
E. Kranakis et al. (Eds.): LATIN 2016, LNCS 9644, pp. 347–361, 2016.
DOI: 10.1007/978-3-662-49529-2_26

- *rotation* which maps $b_1 b_2 \cdots b_{2^n}$ to $b_2 \cdots b_{2^n} b_1$;
- *complementation* which maps $b_1 b_2 \cdots b_{2^n}$ to $\bar{b}_1 \bar{b}_2 \cdots \bar{b}_{2^n}$ where $\bar{b} = 1 - b$;
- *reversal* which maps $b_1 b_2 \cdots b_{2^n}$ to $b_{2^n} \cdots b_2 b_1$.

In particular, the sequences in (1) and (2) are complemented reversals of each other.

De Bruijn sequences are often used in the education of discrete mathematics and theoretical computer science (see *Concrete Mathematics* by Knuth, Patashnik, and Graham [7]). Historically, they have been used in many interesting applications (see the 'Memory Wheels' chapter in *Mathematics: The Man-Made Universe* [18]). More recently they have been used in genome assembly [2]. For these reasons it is helpful to have simple constructions and algorithms for generating specific de Bruijn sequences.

The most well-known de Bruijn sequence for each n is the *lexicographically-least*, which means the first in lexicographic order when viewed as a binary string of length 2^n. In particular, the lexicographically-least for $n = 4$ is D from (1).

Lexicographically-least de Bruijn sequences were first constructed greedily for all n by Martin in 1934 [9], and Knuth refers to Martin's construction as the "granddaddy of all de Bruijn sequence constructions" [8]. We follow Knuth's playful nomenclature, although we use *granddaddy* to refer to the lexicographically-least de Bruijn sequence, instead of Martin's construction of it. (The term *Ford sequence* also describes these sequences due to Ford's independent work [4].)

The granddaddy can be efficiently generated using the *FKM construction*, which is named after the work of Fredricksen, Kessler, and Maiorana [5,6]. The construction uses the concepts of necklaces, Lyndon words, and aperiodic prefixes. For the purposes of this introduction it is sufficient to know the following:

- The set of necklaces for $n = 4$ is $\{0000, 0001, 0011, 0101, 0111, 1111\}$.
- The Lyndon words whose length divides $n = 4$ are $\{0, 0001, 0011, 01, 0111, 1\}$.
- The aperiodic prefixes of 0000, 0101, and 1111 are respectively 0, 01, and 1.

The FKM construction is defined in one of two ways:

1. Concatenate the Lyndon words whose length divides n in lexicographic order.
2. Concatenate the aperiodic prefixes of the necklaces of length n in lexicographic order.

The two definitions always give identical concatenations, although the distinction will prove to be important. The result for $n = 4$ matches D from (1) as follows:

$$D = 0 \cdot 0001 \cdot 0011 \cdot 01 \cdot 0111 \cdot 1. \tag{3}$$

We provide a new de Bruijn sequence that is closely-related to the granddaddy, yet non-equivalent for $n \geq 6$. The *grandmama* is obtained by changing lexicographic order to co-lexicographic order in the necklace-based definition of the FKM construction[1]. The result for $n = 4$ matches M from (2) as follows:

$$M = 0 \cdot 0001 \cdot 01 \cdot 0011 \cdot 0111 \cdot 1. \tag{4}$$

---

[1] Note: In the grandmama construction the necklaces are still the lexicographically least representatives for their rotational equivalence class, as clarified in Sect. 4.1.

Co-lexicographic order is the same as lexicographic order, except that strings are compared symbol-by-symbol from right-to-left instead of left-to-right. For example, if $\alpha = 0011$ and $\beta = 0101$, then α comes before β in lexicographic order, but α comes after β in co-lexicographic order. This change is the only difference between the granddaddy and grandmama when $n = 4$, as seen by the aperiodic prefixes 0011 and 01 swapping places in (3) and (4). Surprisingly, the Lyndon word definition of the FKM construction does not create a de Bruijn sequence when using co-lexicographic order. In particular, when $n = 4$ the concatenation gives

$$0 \cdot 1 \cdot 01 \cdot 0001 \cdot 0011 \cdot 0111 = 0101000100110111, \qquad (5)$$

which is not a de Bruijn sequence because it does not contain 0000 or 1111, and instead has duplicates of 1010 and 1101. Although this modification of the FKM construction does not work, our positive result suggests that the granddaddy and grandmama sequences are part of a larger family of de Bruijn sequences that can be generated using generalizations of the FKM construction. Also, the specific definition of the FKM construction is crucial when considering generalizations.

The remainder of the paper is organized as follows:

- Section 2 defines the FKM construction and granddaddy de Bruijn sequence.
- Section 3 focuses on necklaces in co-lexicographic order.
- Section 4 gives our new construction and verifies its correctness.
- Section 5 investigates the originality of our new sequence, including its non-equivalence to the granddaddy sequence for $n \geq 6$, and the necessity of using necklaces instead of Lyndon words with its FKM construction.
- Section 6 concludes with additional observations and open problems.

2   The Granddaddy de Bruijn Sequence

2.1   Necklaces and Lyndon Words

We define an equivalence relation on the set of binary strings of length n as follows: two strings are equivalent if one is a rotation of the other. The equivalence classes with respect to this relation are called *necklace classes*. For example, the necklace class for $n = 4$ containing the string 1000 is $\{0001, 0010, 0100, 1000\}$. The preferred representatives for necklace classes are the lexicographically-least, and are called *necklaces representatives* or often, simply *necklaces*. For example, the necklace for the necklace class given above is 0001. Let $N(n)$ denote the set of necklaces of length n. For example,

$$N(4) = \{0000, 0001, 0011, 0101, 0111, 1111\}.$$

A string α is called *periodic* if it can be written as $\alpha = \beta^k$, for some integer $k > 1$. Otherwise, the string is called *aperiodic*. For example, $01010101 = (01)^4$, and so 01010101 is periodic. In contrast, the string 00101101 is aperiodic.

Given a string α, its *aperiodic prefix*, denoted $ap(\alpha)$, is its shortest prefix β such that $\alpha = \beta^k$ for some integer $k \geq 1$. We define the *period* of α to be $|ap(\alpha)|$.

For example, if $\alpha = 01010101$ then $ap(\alpha) = 01$ and α has period two. Note, if α is aperiodic, then $ap(\alpha) = \alpha$, and the period of α is its length $|\alpha|$.

An aperiodic necklace is called a *Lyndon word*. Let $L(n)$ denote the set of Lyndon words of length n. For example, $L(4) = \{0001, 0011, 0111\}$.

2.2   FKM Construction

The FKM construction of the granddaddy de Bruijn sequence $G_d(n)$ can be defined in two ways:

Definition 1. $G_d(n) = L_1 L_2 \ldots L_m$, *where* $L_1, L_2, \ldots, L_m$ *lists the Lyndon words whose lengths divide* n, *in lexicographic order.*

Definition 2. $G_d(n) = ap(\alpha_1) \cdot ap(\alpha_2) \cdots ap(\alpha_k)$, *where* $\alpha_1, \alpha_2, \ldots, \alpha_k$ *lists the necklaces of length* n, *in lexicographic order.*

For example, choose $n = 4$. The divisors of 4 are 1, 2, and 4 and the corresponding Lyndon words are

$$L(1) = \{0, 1\} \text{ and } L(2) = \{01\} \text{ and } L(4) = \{0001, 0011, 0111\}.$$

In lexicographic order these Lyndon words are $0, 0001, 0011, 01, 0111, 1$. Concatenating them gives the sequence $G_d(4) = 0000100110101111$ which matches D from (1) and (3). On the other hand, the necklaces for $n = 4$ were listed in $N(4)$ above. Their aperiodic prefixes are $0, 0001, 0011, 01, 0111, 1$. We see that $G_d(4)$ coincides with the Lyndon word definition.

Notice that Definition 1 requires lexicographic order to be defined on strings of different lengths, whereas Definition 2 does not. We give a precise definition of lexicographic order in Sect. 3. Regardless of which definition is used, $G_d(n)$ is the lexicographically least de Bruijn sequence. This property has been proven in more general settings by Sawada, Williams, and Wong for binary [16] and k-ary strings [17], and also for k-ary strings by Moreno and Perrin [12] based on [10].

Theorem 1 ([12,16]). $G_d(n)$ *is the lexicographically least de Bruijn sequence.*

The distinction between Definitions 1 and 2 was discussed in Ruskey, Sawada, and Williams [14] who suggested that Definition 2 lends itself more readily to generalizations; the results of Sect. 5.2 further justify that opinion. This subtlety, along with the literature's predominant use of Definition 1 helps explain why our otherwise natural result had not previously been discovered.

We also mention that there does not appear to be a direct analogue of Theorem 1 for the grandmama sequence. In particular, the co-lexicographically least de Bruijn sequence is simply the granddaddy written in reverse, and hence is not equivalent to the grandmama.

3   Co-lexicographic Order of Necklaces

In this section we define co-lexicographic order and lexicographic order for arbitrary sets of binary strings. Then we specifically look at co-lexicographic order for necklaces, and prove several small results that will be helpful in Sect. 4. For convenience, we often use the term *co-lex* instead of *co-lexicographic* and *lex* instead of *lexicographic*.

3.1   Co-lex vs. Lex

Suppose that L is an arbitrary set of binary strings. Let x/L denote the set of strings of L with prefix $x \in \{0,1\}$ removed. More specifically,

$$x/L = \{b_2 b_3 \ldots b_n \mid b_1 b_2 \ldots b_n \in L \text{ and } b_1 = x\}.$$

The *lexicographic order* of L is defined recursively as follows:

$$\text{lex}(L) = \begin{cases} \varepsilon, \; 0 \cdot \text{lex}(0/L), \; 1 \cdot \text{lex}(1/L) & \text{if } \varepsilon \in L \\ 0 \cdot \text{lex}(0/L), \; 1 \cdot \text{lex}(1/L) & \text{if } \varepsilon \notin L \end{cases}$$

where ε denotes the empty string. Notice that this definition orders symbols as $\varepsilon < 0 < 1$ and applies the order from left-to-right. Thus, prefixes of a given string are ordered before the string itself. For example, 0 is ordered before 0001.

Co-lexicographic order maintains the symbol ordering as $\varepsilon < 0 < 1$ but it applies this order from right-to-left. To formalize this idea, let $L \backslash x$ denote set of strings of L with suffix $x \in \{0,1\}$ removed. That is,

$$L \backslash x = \{b_1 b_2 \ldots b_{n-1} \mid b_1 b_2 \ldots b_{n-1} b_n \in L \text{ and } b_n = x\}$$

The *co-lexicographic order* of L is then defined recursively as follows:

$$\text{colex}(L) = \begin{cases} \varepsilon, \; \text{colex}(L \backslash 0) \cdot 0, \; \text{colex}(L \backslash 1) \cdot 1 & \text{if } \varepsilon \in L \\ \text{colex}(L \backslash 0) \cdot 0, \; \text{colex}(L \backslash 1) \cdot 1 & \text{if } \varepsilon \notin L \end{cases}$$

Notice that in this definition the suffixes of a given string are ordered before the string itself, so 01 is ordered before 0001.

3.2   Necklaces in Co-lex Order

Now we consider the necklaces of length n in co-lex order. We focus on the *co-lexicographic successor* of a given necklace, which is the necklace that immediately follows it in co-lex order. (The successor will be undefined for $\alpha = 1^n$ since this is the last necklace in co-lex order.)

We begin with the following lemma, which proves that necklaces are closed under replacing any prefix of length k by k copies of 0. Before proving this lemma, it is worth noting that necklaces are <u>not</u> closed under replacing any prefix of length k by a lexicographically smaller prefix of length k. For example, consider $\alpha = 0110111 \in N(7)$. If we replace α's prefix 011 with the lexicographically smaller 010, then the result is $0100111 \notin N(7)$. On the other hand, if we replace the prefix with 000 then the result is $0000111 \in N(7)$.

Lemma 1. *If $b_1 b_2 \ldots b_n$ is a necklace, then $0^k b_{k+1} b_{k+2} \ldots b_n$ is a necklace for each integer k such that $1 \leq k \leq n$.*

Proof. The proof is trivial when $b_1 \ldots b_n = 0^n$ and when $k = n$, so we ignore these cases. Thus, $b_n = 1$ since 0^n is the only necklace ending in 0. Without loss of generality we can also assume that $b_k = 1$ since it is sufficient to prove the lemma for prefixes that end in 1. Therefore, $b_k = b_n = 1$ and $k < n$.

Let $\chi = x_1 \ldots x_n = 0^k b_{k+1} \ldots b_n$. From our earlier choices, $x_n = b_n = 1$. For the sake of contradiction, suppose that χ is not a necklace. Therefore, χ's lexicographically smallest rotation is $x_i \ldots x_n x_1 \ldots x_{i-1}$ for some $i > 1$. Now we place two bounds on the value of i:

- $i + k - 1 < n$. Since χ contains 0^k, its smallest rotation's prefix must also begin with 0^k. That is, $x_i \ldots x_{i+k-1} = 0^k$. The inequality follows from the fact that $x_n = 1$.
- $i > k + 1$. Recall that $x_1 \ldots x_k = 0^k$. Therefore, if $i \leq k + 1$ then the last symbol in χ's lexicographically smallest rotation is $x_{i-1} = 0$, which is not possible since χ is not equal to 0^n and so its smallest rotation cannot end in 0.

These two bounds imply that $x_i \ldots x_{i+k-1} = b_i \ldots b_{i+k-1}$. Therefore, $b_1 \ldots b_{i+k-1}$ contains a substring equal to 0^k. This contradicts that $b_1 \ldots b_n$ is a necklace because $b_k = 1$, and hence its prefix of length k is strictly larger than 0^k.   □

Increment Index. Now we define a specific index for every necklace that has a successor. We explain why the index is well-defined and provide an example after the definition.

Definition 3. *The* increment index *of necklace $x_1 x_2 \ldots x_n \neq 1^n$ is the smallest index k such that $x_k = 0$ and $0^{k-1} 1 x_{k+1} x_{k+2} \ldots x_n$ is a necklace.*

To see why the increment index is well-defined, let d be the largest index such that $x_d = 0$. Substituting $x_1 \ldots x_d$ with $0^{d-1} 1$ creates the string $0^{d-1} 1^{n-d+1}$, which is a necklace for all $1 \leq d \leq n$.

Remark 1. The increment index is well-defined for all necklaces $x_1 \ldots x_n \neq 1^n$.

The increment index is the smallest index in which this substitution creates a necklace. For example, consider the necklace $\alpha = 00010010001011 \in N(14)$. We consider the indices of the 0-bits from left-to-right:

- $10010010001011 \notin N(14)$;
- $01010010001011 \notin N(14)$;
- $00110010001011 \notin N(14)$;
- $00001010001011 \in N(14)$.

Therefore, α's increment index is $k = 5$.

Co-lex Successor. Now we prove that the successor in co-lex order is precisely the necklace defined in Definition 3.

Lemma 2. *The successor of necklace* $\chi = x_1 \ldots x_n \neq 1^n$ *in co-lex order is* $0^{k-1}1x_{k+1} \ldots x_n$, *where* k *is its increment index.*

Proof. From Remark 1 we know that $\psi = 0^{k-1}1x_{k+1} \ldots x_n$ is well-defined. Furthermore, it is a necklace that appears after χ in co-lex order. Therefore, in order to prove that $0^{k-1}1x_{k+1} \ldots x_n$ is the successor, we only need to prove that there are no necklaces between χ and ψ in co-lex order.

For the sake of contradiction suppose that there is another necklace between χ and ψ in co-lex order. This necklace must have suffix $1x_{j+1} \ldots x_n$ for some j that satisfies $x_j = 0$ and $j < k$. By Lemma 1 this implies that $0^{j-1}1x_{j+1} \ldots x_n$ is a necklace. However, this contradicts the definition of χ's increment index k. □

Properties. When proving the correctness of the grandmama construction, we will need to consider common substrings between successive necklaces, including those that are periodic. We now prove three Lemmata and one remark related to this goal.

Lemma 3. *Suppose* $\alpha = a_1a_2 \ldots a_n$ *and* $\beta = b_1b_2 \ldots b_n$ *are consecutive necklaces in co-lex order, and* i *is the minimum value such that* $a_{i+1} \ldots a_n = b_{i+1} \ldots b_n$. *Then* $b_1b_2 \ldots b_i = 0^{i-1}1$.

Proof. By Lemma 2, β is obtained from α by replacing its prefix $a_1 \ldots a_k$ by $0^{k-1}1$. Therefore, the result follows by taking $i = k$. □

Lemma 4. *Suppose that* $\beta = b_1 \ldots b_n$ *and* $\gamma = c_1 \ldots c_n$ *are consecutive necklaces in co-lex order, and* j *is the maximum value such that* $b_1 \ldots b_j = c_1 \ldots c_j$. *Then* $b_1 \ldots b_j = 0^j$.

Proof. By Lemma 2, γ's prefix of the form $0^{k-1}1$ differs from β's prefix of length k. Therefore, any common prefix is of the form 0^j for some $j \leq k - 1$. □

Next we consider periodic necklaces and the necklaces that come immediately before and after. An example of Lemma 5 appears below, where underlines illustrate the common suffix and prefix lengths with β:

$$\alpha = 000\underline{011}\ \underline{001011}\ \underline{001011}\ \underline{001011}$$
$$\beta = \underline{001011}\ \underline{001011}\ \underline{001011}\ \underline{001011}$$
$$\gamma = \underline{000}111\ 001011\ 001011\ 001011$$

Lemma 5. *Suppose that* β *is a periodic necklace with* $\beta \notin \{0^n, 1^n\}$. *Let* $ap(\beta) = \zeta = z_1 \ldots z_j$ *with* $z_1 \ldots z_{i+1} = 0^i1$ *and* $k = \frac{n}{j}$. *The necklace before* β *is aperiodic and shares the same suffix of length* $n-i-1$, *and the necklace after* β *is aperiodic and shares the same prefix of length* i.

Proof. Consider the string $\alpha = 0^{i+1}z_{i+2}\cdots z_j\zeta^{k-1}$. Note that α is a necklace since 0^{i+1} is lexicographically least substring of length $i+1$ in α. Furthermore, α is aperiodic. The increment index of α is at least i due to the prefix 0^i in ζ, and it is at most i due to the necklace β. Therefore, the increment index of alpha is equal to i, and hence α is followed by β in co-lex order by Lemma 2. Thus, α shares the suffix $z_{i+2}z_{i+3}\ldots z_j\zeta^{k-1}$ with β. Next notice that the increment index of β is at least $i+2$. Therefore, by Lemma 2, γ must begin with at least $i+1$ copies of 0. Hence γ shares the prefix 0^i with β.   □

The following simplification of Lemma 5 will be helpful in Sect. 4.

Remark 2. If χ and ψ are consecutive necklaces in co-lex order, and ψ is periodic, then $ap(\chi)ap(\psi)$ has ψ as a suffix.

4   The Grandmama de Bruijn Sequence

4.1   Definition

The grandmama sequence $G_m(n)$ is constructed by concatenating the aperiodic prefixes of necklaces in co-lex order as in Definition 2. It is important to note that we still use the lexicographically least representative for each necklace. In other words, co-lex order replaces lex order only in the ordering of the necklaces, and not in the representatives of each necklace.

Definition 4. *Let* $G_m(n) = ap(\alpha_1)\cdot ap(\alpha_2)\cdots ap(\alpha_k)$ *where* $\alpha_1,\cdots,\alpha_k$ *lists the necklaces of length n in co-lex order.*

For example, the necklaces of length 4 in co-lex order are given by

$$\mathsf{colex}(N(4)) = 0000, 0001, 0101, 0011, 0111, 1111.$$

Therefore,

$$G_m(4) = ap(0000) \cdot ap(0001) \cdot ap(0101) \cdot ap(0011) \cdot ap(0111) \cdot ap(1111)$$
$$= 0 \cdot 0001 \cdot 01 \cdot 0011 \cdot 0111 \cdot 1$$
$$= 0000101001101111$$

which is M in (2). Recall that D and M in (1) and (2) are equivalent, so when $n = 4$ the granddaddy sequence $G_d(4)$ and the grandmama sequence $G_m(4)$ are equivalent. In Sect. 5 we will see that this equivalence does not hold for $n \geq 6$.

4.2   Verification

Now we verify that $G_m(n)$ is a de Bruijn sequence.

Theorem 2. $G_m(n)$ *is a de Bruijn sequence.*

Proof. Observe that $|G_m(n)| = 2^n$ since it is a reordering of the bits in $G_d(n)$. Thus, we can prove that $G_m(n)$ is a de Bruijn sequence by showing that it contains every n-bit binary string when $G_m(n)$ is viewed cyclically. To prove this we consider an arbitrary necklace β, and we prove that each rotation of β appears as a substring in $G_m(n)$ when $G_m(n)$ is viewed cyclically. For clarity we often underline the specific rotations.

<u>Case One:</u> $\beta = 0^m 1^{n-m}$ for some $0 \le m \le n$. This case covers the first two necklaces in co-lex order, namely 0^n and $0^{n-1}1$, and the last two necklaces in co-lex order, namely 01^{n-1} and 1^n. Also, note that $G_m(n)$ has suffix 1^n and prefix 0^n, from $ap(01^{n-1}) \cdot ap(1^n) = 01^n$ and $ap(0^n) \cdot ap(01^{n-1}) = 0^n 1$, respectively. Therefore, all of the substrings that "wrap-around" in $G_m(n)$ have the form 1^*0^* and hence belong to a necklace of the form $0^m 1^{n-m}$. Therefore, this case also covers all necklaces that have a rotation in the "wrap-around".

If $m = n$, then the only rotation of $\beta = 0^n$ is in $ap(0^n) \cdot ap(0^{n-1}1) = \underline{0 \cdot 0^{n-1}1}$. If $m = 0$, then the only rotation of $\beta = 1^m$ is in $ap(01^{n-1}) \cdot ap(1^n) = 0\underline{1^{n-1}1}$. Otherwise, β has n distinct rotations, which we consider in the following cases:

- The rotation $0^m 1^{n-m}$ is in $ap(0^m 1^{n-m}) = \underline{0^m 1^{n-m}}$.
- The rotation $1^{n-m}0^m$ is in $ap(01^{n-1}) \cdot ap(1^n) \cdot ap(0^n) \cdot ap(0^{n-1}1)$ since this is equal to $01^{n-1}100^{n-1}1 = 01^m\underline{1^{n-m}0^m}0^{n-m}1$.
- Consider a rotation $0^x 1^{n-m}0^{m-x}$ with $0 < x < m$. Let β' be the last necklace in co-lex order with suffix $0^x 1^{n-m}$ and let γ' be the necklace after β' in co-lex order. (The choice of β' is well-defined since β has this suffix.) Therefore, the increment index of β' is greater than $m - x$, and by Lemma 2 γ' has prefix 0^{m-x}. Therefore, $ap(\beta') \cdot ap(\gamma')$ contains $\underline{0^x 1^{n-m} \cdot 0^{m-x}}$.
- Consider a rotation $1^x 0^m 1^{n-m-x}$ with $0 < x < n - m$. Let α be the last necklace in co-lex order with suffix 01^{n-m-1}. Notice that $ap(\alpha)$ must also have suffix 01^{n-m-1}. Furthermore, the necklace following α in co-lex order will be the first necklace with suffix 1^{n-m}, which is β. Therefore, $ap(\alpha) \cdot ap(\beta)$ contains $01^{n-m-1} \cdot 0^m 1^{n-m}$ which itself contains $\underline{1^x \cdot 0^m 1^{n-m-x}}$.

<u>Case Two:</u> β is periodic and $\beta \notin \{0^n, 1^n\}$. Let $|ap(\beta)| = j$ and so β has j distinct rotations. By our choice of β, it must have a prefix of the form $0^i 1$ for some positive integer $i < n$. Let α and γ be the necklaces immediately before and after β, respectively. By Lemma 5, α shares a suffix of length $n - (i+1)$ with β, and γ shares a prefix of length i with β. Furthermore, both α and γ are aperiodic. Thus, $ap(\alpha) \cdot ap(\beta) \cdot ap(\gamma)$ has a substring of length $n - (i+1) + j + i = n + j - 1$ which contains all j rotations of β.

<u>Case Three:</u> $\beta = b_1 b_2 \cdots b_n$ is aperiodic and $\beta \neq 0^m 1^{n-m}$ for any $0 \le m \le n$. Let $\alpha = a_1 a_2 \ldots a_n$ and $\gamma = c_1 c_2 \ldots c_n$ be the necklaces before and after β respectively. By Remark 2, $G_m(n)$ must contain the following substring,

$$\alpha \cdot \beta \cdot ap(\gamma).$$

Now consider the values of i and j from Lemmas 3 and 4 respectively. Thus,

1. $a_{i+1} \ldots a_n = b_{i+1} \ldots b_n$
2. $b_1 \ldots b_j = c_1 \ldots c_j = 0^j$
3. $b_1 \ldots b_i = 0^{i-1}1$

From items 2 and 3, $j \leq i - 1$. Notice that the 0^j prefix in γ is also in $ap(\gamma)$. (This is due to the fact that $\beta \neq 0^{n-1}1$ and so the necklace following it in co-lex order is not 0^n.) Therefore, $G_m(n)$ has the following substring:

$$a_{i+1} \ldots a_n b_1 \ldots b_n c_1 \ldots c_j = b_{i+1} \ldots b_n b_1 \ldots b_n b_1 \ldots b_j.$$

Notice that $n-i+j+1$ distinct rotations of β are included in this substring. The 'missing' rotations of β are those starting from the symbols $b_{j+2}, b_{j+3}, \ldots, b_i$. (If $j = i - 1$, then there are no 'missing' rotations of β.) Consider an arbitrary one of these rotations $b_x \ldots b_n b_1 \ldots b_{x-1}$ where $j + 2 \leq x \leq i$. Notice the following:

- $b_1 \ldots b_{x-1} = 0^{x-1}$ since $x \leq i$.
- $b_x \ldots b_n$ is a suffix of β;
- $b_x \ldots b_n \neq 1^{n-x+1}$ since otherwise $\beta = 0^m 1^{n-m}$ for some $0 \leq m \leq n$.

Therefore, there exists a last necklace β' in co-lex order whose suffix is $b_x \ldots b_n$. The next necklace γ' in co-lex order must not have this suffix. Therefore, by Lemma 2, γ' has prefix 0^{x-1}. Therefore, $G_m(n)$ contains the following substring

$$b_x \ldots b_n 0^{x-1} = b_x \ldots b_n b_1 \ldots b_{x-1}$$

where periodicity is addressed as above. Hence, this 'missing' rotation is contained in $G_m(n)$ and similarly all 'missing' rotations can be found.   □

5   Originality of the Grandmama Sequence

In this section we give two results on the originality of the grandmama sequence:

5.1   Distinctness

In this subsection we prove that the granddaddy and grandmama sequences are not equivalent for sufficiently large n. Recall that two de Bruijn sequences are equivalent if one can be obtained from the other by rotations, complementations, and reversals. For example, $G_d(4)$ and $G_m(4)$ are equivalent by complementation and reversal. The same operations also show equivalence when $n = 5$. That is,

$$\overline{G_d(5)}^R = \overline{00000100011001010011101011011111}^R$$
$$= 00000100101000110101100111011111$$
$$\text{and } G_m(5) = 00000100101000110101100111011111$$

where overline represents bitwise complement and ^{R} denotes reversal. However, these operations do not show equivalence when $n = 6$ as shown below,

$$\overline{G_d(6)}^R = \overline{0000001000011000101000110010010110011010011110101011011011111}^R$$
$$= 0000001001000101010000110100110010110110001110101110011110111111$$
$$\text{but } G_m(6) = 0000001001000101010011010000110010110110001110101110011110111111$$

where the non-equal bits appear in bold. In fact, no series of operations make the sequences equal, so they are not equivalent. Although the sequences for $n = 6$ are not equivalent, this example shows that they can have long prefixes and suffixes in common up to reversal and complement. Thus, we will need to venture further into the sequences to prove that they are distinct.

To formalize our results we use two equivalence relations on binary strings. *Rotational equivalence* $\equiv_r$ is simply equivalence under rotation. We previously used this equivalence when defining necklaces, however it is helpful in this section to have explicit notation. Its equivalence classes are defined as follows:

$$[b_1 b_2 \cdots b_n]_r = \{b_i b_{i+1} \cdots b_n b_1 b_2 \cdots b_{i-1} \mid 1 \le i \le n\}.$$

For example, the rotational equivalence class for the granddaddy $G_d(3)$ is

[00010111]$_r$ = {00010111, 00101110, 01011100, 10111000, 01110001, 11100010, 11000101, 10001011}.

De Bruijn equivalence $\equiv_{dB}$ matches our notion of equivalent de Bruijn sequences, meaning equivalence under rotation, complementation, and reversal. We define its equivalence classes below as the union of four rotational equivalence classes:

$$[b_1 b_2 \cdots b_n]_{dB} = [b_1 b_2 \cdots b_n]_r \cup [\overline{b_1 b_2 \cdots b_n}]_r \cup [b_n \cdots b_2 b_1]_r \cup [\overline{b_n \cdots b_2 b_1}]_r. \quad (6)$$

For example, the de Bruijn equivalence class for $G_d(3)$ is

$[00010111]_r = [00010111]_r \cup [11101000]_r \cup [11101000]_r \cup [00010111]_r = [00010111]_r \cup [11101000]_r$

$= \{00010111, 00101110, 01011100, 10111000, 01110001, 11100010, 11000101, 10001011\} \cup$

$\{11101000, 11010001, 10100011, 01000111, 10001110, 00011101, 00111010, 01110100\}.$

We now proceed with two lemmas.

Lemma 6. *When viewed cyclically, every de Bruijn sequence for the binary strings of length n contains exactly one copy of $10^n 1$ and $01^n 0$ as substrings.*

Proof. A de Bruijn sequence has exactly one copy of the substring 0^n. Furthermore, this substring must be flanked on the left and right by 1s, since otherwise it would contain 0^{n+1} and hence two copies of 0^n. Therefore, it must contain exactly one copy of $10^n 1$. A similar argument works for $01^n 0$. □

The next lemma points out that the substrings discussed in Lemma 6 overlap each other in both the granddaddy and grandmama.

Lemma 7. *Both $G_d(n)$ and $G_m(n)$ contain exactly one copy of $01^n 0^n 1$.*

Proof. Both sequences contain $ap(01^{n-1}) \cdot ap(1^n) \cdot ap(0^n) \cdot ap(0^{n-1}1) = 01^n 0^n 1$, and by Lemma 6 it is not possible for either to contain more than one copy. □

The string in Lemma 7 has two properties: Its reverse equals its complement, and its complemented reversal equals itself. That is,

$$\overline{01^n 0^n 1} = 10^n 1^n 0 = 01^n 0^n 1^R \quad \text{and} \quad \overline{01^n 0^n 1}^R = 01^n 0^n 1. \tag{7}$$

Now we prove the main result of this subsection.

Theorem 3. *The de Bruijn sequences $G_d(n)$ and $G_m(n)$ are not equivalent under de Bruijn equivalence $\equiv_{dB}$ for all $n \geq 6$.*

Proof. The definition of de Bruijn equivalence in (6) uses four rotational equivalences. We consider each equivalence in turn and prove that it does not hold.

Case One: $G_d(n) \equiv_r G_m(n)$.
The first three necklaces in lex and co-lex order are given below:

$$\text{lex begins } 0^n, \ 0^{n-1}1, \ 0^{n-2}11$$

$$\text{co-lex begins } 0^n, \ 0^{n-1}1, \ 0^{\lceil \frac{n-2}{2} \rceil} 10^{\lfloor \frac{n-2}{2} \rfloor} 1$$

Therefore, $G_d(n)$ contains $0^n 10^{n-2}11$ while $G_m(n)$ contains $0^n 10^{\lceil \frac{n-2}{2} \rceil} 10^{\lfloor \frac{n-2}{2} \rfloor} 1$. These substrings are different for $n \geq 4$. Furthermore, they both contain 0^n. Therefore, $G_d(n)$ and $G_m(n)$ are not rotationally equivalent by Lemma 6.

Case Two: $G_d(n) \equiv_r \overline{G_m(n)}$.
By Lemma 7, $G_d(n)$ contains $01^n 0^n 1$ while $\overline{G_m(n)}$ contains $10^n 1^n 0$. These substrings reverse the order of 0^n and 1^n, so $G_d(n)$ and $\overline{G_m(n)}$ are not rotationally equivalent by Lemma 6.

Case Three: $G_d(n) \equiv_r G_m(n)^R$.
This case is identical to Case Two by (7).

Case Four: $G_d(n) \equiv_r \overline{G_m(n)}^R$.
In this case the string in Lemma 7 is not sufficient for distinguishing the two sequences by itself due to (7), so we must consider more bits in the sequences. The first eight necklaces in lex order and the last eight necklaces in co-lex order are given below for any $n \geq 6$:

$$\text{lex begins } 0^n, \ 0^{n-1}1, \ 0^{n-2}11, \ 0^{n-3}101, \ 0^{n-3}111, \ 0^{n-4}1001, \ 0^{n-4}1011, \ 0^{n-4}1101$$

$$\text{co-lex ends } 00001^{n-4}, \ 00101^{n-4}, \ 01101^{n-4}, \ 0001^{n-3}, \ 0101^{n-3}, \ 001^{n-2}, \ 01^{n-1}, \ 1^n$$

When $n \geq 7$ all of these necklaces are aperiodic except for 0^n and 1^n. Therefore, when $n \geq 7$ the granddaddy $G_d(n)$ contains

$$00^{n-1}10^{n-2}110^{n-3}1010^{n-3}1110^{n-4}10010^{n-4}10110^{n-4}1101 \tag{8}$$

and the grandmama $G_m(n)$ contains

$$00001^{n-4}00101^{n-4}01101^{n-4}0001^{n-3}0101^{n-3}001^{n-2}01^{n-1}1. \tag{9}$$

Thus, the complemented and reversed grandmama will contain the following

$$00^{n-1}10^{n-2}110^{n-3}1010^{n-3}1110^{n-4}10010^{n-4}10110^{n-4}1111. \tag{10}$$

Both (8) and (10) contain 0^n, but they are not equal. Therefore, $G_d(n)$ and $\overline{G_m(n)}^R$ are not rotationally equivalent for $n \geq 7$ by Lemma 6. The non-equivalence for $n = 6$ can be similarly derived with the periodicity of $0^{n-4}1001 = 001001$ and $01101^{n-4} = 011011$ changing the above concatenations slightly. $\square$

5.2 Lyndon Words in Co-lex Order

Now we return to the distinction between Definitions 1 and 2. While both definitions create the granddaddy sequence when using lex order, we will now show that using co-lex order with Definition 1 does not create a de Bruijn sequence. In particular, (5) showed that Definition 1 does not work for $n = 4$. More specifically, the divisors of $n = 4$ are $d = 1, 2, 4$ and hence the concatenation gives

$$\mathsf{colex}(\{0,1\} \cup \{01\} \cup \{0001, 0011, 0111\}) = 0 \cdot 1 \cdot 01 \cdot 0001 \cdot 0011 \cdot 0111.$$

As previously mentioned, this is not a de Bruijn sequence since it does not contain 0000 and 1111 as a substring, and instead contains 1010 and 1101 twice. More generally, Theorems 2 and 4 imply that only the FKM construction specified by Definition 2 provides a de Bruijn sequence when using co-lex order.

Theorem 4. *The Lyndon word definition of the FKM construction does not create a de Bruijn sequence when using co-lex order for all $n \geq 4$.*

Proof. When $n \geq 4$ the last two Lyndon words in co-lex order are 001^{n-2} and 01^{n-1}, and the first two Lyndon words are 0 and 1. Thus, Definition 1 includes

$$001^{n-2} \cdot 01^{n-1} \cdot 0 \cdot 1,$$

which is not part of a de Bruijn sequence since it has two copies of $1^{n-2}01$. $\square$

6 Open Problems and Additional Results

6.1 Efficient Generation

The FKM construction allows the granddaddy sequence to be generated in amortized $O(1)$-time per bit without creating the underlying de Bruijn graph or storing previously generated bits. This important result was obtained by generating necklaces in lex order in amortized $O(1)$-time (see Ruskey, Savage, and Wang [13]). If necklaces can be generated in amortized $O(1)$-time in co-lex order, then that would lead to an amortized $O(1)$-time per bit algorithm for the grandmama sequence. Fixed-weight binary necklaces (and Lyndon words) have already been generated in amortized $O(1)$-time in co-lex order by Sawada and Williams [15].

6.2   Properties of the Grandmama Sequence

The 'discrepancy' of the granddaddy sequence has been analyzed [3]. Also, many of its subsequences have been shown to provide de Bruijn sequences for certain subsets of binary strings. For example, see Au [1], Moreno [11], and Sawada, Williams, and Wong [16,17]. It will be interesting to further analyze the grandmama sequence for properties like these and others.

6.3   Generalization to Larger Alphabets

The authors have observed that Theorem 2 holds for k-ary strings, and will prove this generalization in an upcoming paper.

The authors would like to thank Joe Sawada and the anonymous first reviewer for many helpful comments.

References

1. Au, Y.H.: Shortest sequences containing primitive words and powers. Discrete Math. **338**(12), 2320–2331 (2015)
2. Compeau, P.E.C., Pevzner, P.A., Tesler, G.: How to apply de Bruijn graphs to genome assembly. Nat. Biotechnol. **29**, 987–991 (2011)
3. Cooper, J., Heitsch, C.: The discrepancy of the lex-least de Bruijn sequence. Discrete Math. **310**, 1152–1159 (2014)
4. Ford, L.R.: A cyclic arrangement of m-tuples. Report No. P-1071, RAND Corp., Santa Monica (1957)
5. Fredricksen, H., Kessler, I.J.: An algorithm for generating necklaces of beads in two colors. Discrete Math. **61**, 181–188 (1986)
6. Fredricksen, H., Maiorana, J.: Necklaces of beads in k colors and k-ary de Bruijn sequences. Discrete Math. **23**, 207–210 (1978)
7. Graham, R.L., Knuth, D.E., Patashnik, O., Mathematics, C.: A Foundation for Computer Science, 2nd edn. Addison-Wesley Professional, Reading (1994)
8. Knuth, D.E.: The Art of Computer Programming. Combinatorial Algorithms, vol. 4A. Addison-Wesley Professional, Boston (2011)
9. Martin, M.H.: A problem in arrangements. Bull. Am. Math. Soc. **40**, 859–864 (1934)
10. Moreno, E.: On the theorem of Fredricksen and Maiorana about de Bruijn sequences. Adv. Appl. Math. **33**, 413–415 (2004)
11. Moreno, E.: On the theorem of Fredricksen and Maiorana about de Bruijn sequences. Adv. Appl. Math. **33**(2), 413–415 (2004)
12. Moreno, E., Perrin, D.: Corrigendum to "On the theorem of Fredricksen and Maiorana about de Bruijn sequences". Adv. Appl. Math. **62**, 184–187 (2015)
13. Ruskey, F., Savage, C.D., Wang, T.M.Y.: Generating necklaces. J. Algorithms **13**(3), 414–430 (1992)
14. Ruskey, F., Sawada, J., Williams, A.: De Bruijn sequences for fixed-weight binary strings. SIAM J. Discrete Math. **26**(2), 605–617 (2012)
15. Sawada, J., Williams, A.: A Gray code for fixed-density necklaces and Lyndon words in constant amortized time. Theoret. Comput. Sci. **502**, 46–54 (2013)

16. Sawada, J., Williams, A., Wong, D.: The lexicographically smallest universal cycle for binary strings with minimum specified weight. J. Discrete Algorithms **28**, 31–40 (2014). StringMasters 2012 & 2013 Special Issue
17. Sawada, J., Williams, A., Wong, D., Generalizing the classic greedy, necklace constructions for de Bruijn sequences, universal cycles. Electron. J. Comb., **23**(1) (2016). Paper #1.24
18. Stein, S.K.: Mathematics: The Man-Made Universe, 3rd edn. W. H. Freeman and Company, San Francisco (1994)

Compressing Bounded Degree Graphs

Pål Grønås Drange[1], Markus Dregi[1(✉)], and R.B. Sandeep[2]

[1] Department of Informatics, University of Bergen, Bergen, Norway
{pal.drange,markus.dregi}@ii.uib.no
[2] Department of CSE, IIT Hyderabad, Medak, India
cs12p0001@iith.ac.in

Abstract. Recently, Aravind et al. (IPEC 2014) showed that for any finite set of connected graphs $\mathcal{H}$, the problem $\mathcal{H}$-FREE EDGE DELETION admits a polynomial kernelization on bounded degree input graphs. We generalize this theorem by no longer requiring the graphs in $\mathcal{H}$ to be connected. Furthermore, we complement this result by showing that also $\mathcal{H}$-FREE EDGE EDITING admits a polynomial kernelization on bounded degree input graphs.

We show that there exists a finite set $\mathcal{H}$ of connected graphs such that $\mathcal{H}$-FREE EDGE COMPLETION is incompressible even on input graphs of maximum degree 5, unless the polynomial hierarchy collapses to the third level. Under the same assumption, we show that C_{11}-FREE EDGE DELETION—as well as $\mathcal{H}$-FREE EDGE EDITING—is incompressible on 2-degenerate graphs.

1 Introduction

Graph modification problems have been a fundamental part of computational graph theory throughout its history [11, A1. GraphTheory]. In these classical problems you are to apply at most k modifications to an input graph G to make it adhere to a specific set of properties, where both the modifying operations and the target properties are problem specific. Unfortunately, even when considering vertex deletion to hereditary graph classes, the modification problems often regarded as the most tractable, almost all of them are NP-complete [17]. A similar dichotomy is yet to appear for edge modification problems and hence the classical complexity landscape seems far more involved. However, various results display the NP-hardness of the edge variants as well [3,7,19]. Due to this inherent intractability we need to find other ways of coping. A well-established tool for tackling hard problems, in practice as well as in theory, is preprocessing of data. In theoretical computer science, preprocessing is best described within the framework of parameterized complexity as kernelization. For our purposes a problem admits a kernel of size $f(k)$ if given a graph G and an integer k as

The research leading to these results has received funding from the TCS Research Scholarship, Bergen Research Foundation under the project Beating Hardness by Preprocessing and the European Research Council under the European Union's Seventh Framework Programme (FP/2007-2013) / ERC Grant Agreement no. 267959.

© Springer-Verlag Berlin Heidelberg 2016
E. Kranakis et al. (Eds.): LATIN 2016, LNCS 9644, pp. 362–375, 2016.
DOI: 10.1007/978-3-662-49529-2_27

input, one can in polynomial time output an equivalent instance (G', k') such that both the size of G' and the value k' is bounded by $f(k)$. If f is a polynomial we say that the problem admits a polynomial kernelization.

In this paper we will restrict our attention to hereditary graph classes characterized by finite sets of forbidden induced subgraphs. Hence, for every graph class studied there is a set of finite graphs $\mathcal{H}$ such that a graph G is in the graph class if and only if no graph in $\mathcal{H}$ is an induced subgraph of G. In this situation Cai's theorem [4] shows that all $\mathcal{H}$-free modification problems are *fixed parameter tractable*, that is, they are all solvable in time $f(k) \cdot \mathrm{poly}(n)$. And furthermore, every vertex deletion problem admits a classic $O(k^d)$ polynomial kernel, based on the sunflower lemma [1,10]. However, for edge modification problems the landscape is much less understood. In particular, P_4-free edge deletion admits a polynomial kernel, C_4-free edge deletion does not and for S_4 and the claw $(K_{1,3})$, nobody knows.

The edge modification problems characterized by a finite set of forbidden induced subgraphs $\mathcal{H}$ are often referred to as $\mathcal{H}$-Free Edge Completion, $\mathcal{H}$-Free Edge Deletion and $\mathcal{H}$-Free Edge Editing, where one is to add, remove or both add and remove k edges to make the graph $\mathcal{H}$-free. In dealing with the inherent intractability of graph modification problems Natanzon, Shamir, and Sharan [18] suggested to study $\mathcal{H}$-Free Edge Deletion on bounded degree input graphs. Recently, following this direction of research, Aravind, Sandeep and Sivadasan [2] were able to show that as long as every graph $H \in \mathcal{H}$ is connected, the problem $\mathcal{H}$-Free Edge Deletion admits a polynomial kernel of size

$$O\left(\Delta^c \cdot k^d\right),$$

where c is depending only on $\mathcal{H}$ and d on $\mathcal{H}$ and Δ. In particular, this yields a polynomial kernel for every fixed maximum degree Δ.

The first result of the paper is several, simultaneously applicable improvements upon the above mentioned result. First, we are able to remove the condition requiring all graphs of $\mathcal{H}$ to be connected. As many interesting graph classes (threshold graphs, split graphs e.g.) are described by disconnected forbidden subgraphs, this is a major extension. Second, we complement it by proving that the same kernels can be obtained when considering $\mathcal{H}$-Free Edge Editing. And third, we improve the kernel dependency on Δ. The novelty of our approach lies within a better understanding of how forbidden subgraphs are introduced when edges are modified in the input graph. Due to this, we can localize the crucial part of the instance even when both forbidden subgraphs and modifications are spread throughout the graph.

We continue by providing several hardness results. First, we prove that somewhat surprisingly the positive result does not extend to the completion variant. Due to page restrictions, we have deferred some of the proofs from the kernelization section to the full version. The statements to which these proofs belong have been marked with a ♠.

Table 1. Overview of polynomial kernelization complexity for graph modification on bounded degree and degenerate input graphs. The table shows that there is no distinction between disconnected graphs, and that the completion variant is notoriously incompressible—bounded degree does not help compressing completion problems.

	Deletion	Completion	Editing	Vertex deletion
bounded degree	Yes ([2], Theorem 4)	No (Theorem 1)	Yes (Theorem 4)	Yes
2-degenerate	No (Theorem 3)	No (Theorem 1)	No (Theorem 2)	Yes

Theorem 1 (♠). *There exists a finite set* $\mathcal{H}$ *such that* $\mathcal{H}$-FREE EDGE COMPLETION *does not admit a polynomial kernel, even on input graphs of maximum degree 5, unless* NP $\subseteq$ coNP/poly.

Furthermore, we prove that for both $\mathcal{H}$-FREE EDGE EDITING and $\mathcal{H}$-FREE EDGE DELETION there is no hope for polynomial kernels, even when restricted to 2-degenerate graphs. It can easily be observed that the same proofs can be applied to generalize the results to K_9-minor free graphs.

Theorem 2 (♠). *There is a finite set of connected graphs* $\mathcal{H}$ *such that* $\mathcal{H}$-FREE EDGE EDITING *does not admit a polynomial kernel, even on 2-degenerate graphs, unless* NP $\subseteq$ coNP/poly.

Theorem 3 (♠). *There is a finite set of connected graphs* $\mathcal{H}$ *such that* $\mathcal{H}$-FREE EDGE DELETION *does not admit a polynomial kernel, even on 2-degenerate graphs, unless* NP $\subseteq$ coNP/poly.

We now have complete information on the kernelization complexity of edge and vertex modification problems when the target graph class is characterized by a finite set of forbidden induced subgraphs, on bounded degree and 2-degenerate input graphs. Recall that the yes answer for the vertex deletion version on general graphs is obtained by a simple reduction from the $\mathcal{H}$-FREE VERTEX DELETION problem to the d-HITTING SET problem, which, using the sunflower lemma [8], can be shown to admit a polynomial kernel [1].

Related work. One should note that many modification problems remains NP-complete for bounded degree graphs. Komusiewicz and Uhlmann showed [15] that even for simple cases like $\mathcal{H} = \{P_3\}$, the path on three vertices, $\mathcal{H}$-FREE EDGE DELETION—also known as CLUSTER DELETION—is NP-complete, even on graphs of maximum degree 6. Later, it was also shown that P_4-FREE EDGE DELETION and EDITING (COGRAPH EDITING) and $\{C_4, P_4\}$-FREE EDGE DELETION and EDITING (TRIVIALLY PERFECT EDITING) [6] had similar properties; NP-complete, even on graphs of maximum degree 4.

Gramm et al. [12], and Guo [14] showed kernels for several graph modification problems to graph classes characterized by a finite set of forbidden induced

subgraphs. Several positive results followed, which led Fellows, Langston, Rosamond, and Shaw to ask whether all $\mathcal{H}$-free modification problems admit polynomial kernels [9].

This was refuted by Kratsch and Wahlström [16] who showed that for $\mathcal{H} = \{H\}$
where H is a certain graph on seven vertices, $\mathcal{H}$-FREE EDGE DELETION, as
well as $\mathcal{H}$-FREE EDGE EDITING, does not admit a polynomial kernel unless
$NP \subseteq coNP/poly$.[1] Without stating it explicitly, but revealed by a more careful
analysis of the inner workings of their proofs, Kratsch and Wahlström actually
showed something even stronger; namely that the result holds when restricted
to 6-degenerate graphs, both for the deletion and for the editing version.

This line of research was followed up by Guillemot, Havet, Paul, and Perez
[13] showing large classes of simple graphs for which $\mathcal{H}$-FREE EDGE DELETION
is incompressible, which was further developed by Cai and Cai [5]; Combining
these results, we now know that when H is a path or a cycle, $\mathcal{H}$-FREE EDGE
DELETION, EDITING *and* COMPLETION is compressible if and only if H has at
most three edges, that is, only for the simplest graphs.

Notation

We consider only simple finite undirected graphs. Let $G = (V, E)$ be a graph
on n vertices with $v \in V$. When $X \subseteq V(G)$, we write $G - X$ to denote the
graph $(V \setminus X, E)$. Similarly, when $F \subseteq [V]^2$, we write $G - F$ to denote the graph
$(V, E \setminus F)$ and $G \triangle F$ to mean $(V, E \triangle F)$ where $\triangle$ is the *symmetric difference
operator*, i.e., $A \triangle B = (A \setminus B) \cup (B \setminus A)$.

We say that a graph H is a *subgraph* of G if $V(H) \subseteq V(G)$ and $E(H) \subseteq
E(G) \cap [V(H)]^2$. Furthermore, we say that H is an *induced subgraph* of G if
H is a subgraph of G and $E(H) = E(G) \cap [V(H)]^2$. For a set $X \subseteq V(G)$
we denote the induced subgraph of G with X as its vertices by $G[X]$. We lift
the notion of neighborhoods to subgraphs by letting $N(H) = N_G(V(H))$ and
$N[H] = N_G[V(H)]$. In addition, if H is a subgraph of G and $F \subseteq E(G)$ we denote
by $H \triangle F$ the graph $H \triangle (F \cap [V(H)]^2)$. The *diameter* of a connected graph G,
denoted $\mathrm{diam}(G)$, is defined as the number of edges in a longest shortest path of
G, $\mathrm{diam}(G) = \max_{u,v \in V(G)} \mathrm{dist}_G(u, v)$. If G is disconnected, we define $\mathrm{diam}(G)$
to be $\max_C \mathrm{diam}(C)$, over all connected components C of G. For a graph G, a
vertex $v \in V(G)$ and a set of vertices $X \subseteq V(G)$ we define the *distance from v to
X*, denoted $\mathrm{dist}(v, X)$ as $\min_{u \in X} \mathrm{dist}(v, u)$. When provided with a non-negative
integer r in addition, we define the *ball around X of radius r*, denoted $B(X, r)$,
as the set $\{v \in V(G)$ such that $\mathrm{dist}(v, X) \leq r\}$.

Obstructions. An *obstruction set* $\mathcal{H}$ is a finite set of graphs. Given an obstruction
set $\mathcal{H}$, a graph G and an induced subgraph H of G we say that H is an *obstruction*
in G if H is isomorphic to some element of $\mathcal{H}$. If there is no obstruction H in
G we say that G is $\mathcal{H}$-free. The size of the largest graph in $\mathcal{H}$ we denote by

---

[1] $NP \subseteq coNP/poly$ implies that PH is contained in Σ_3^p. It is widely believed that
PH does not collapse, and hence it is also believed that $NP \not\subseteq coNP/poly$. We will
throughout this section assume that $NP \not\subseteq coNP/poly$.

$n_\mathcal{H} = \max\{|V(H)|$ for $H \in \mathcal{H}\}$. In addition, we lift the notation of diameter to account for a finite set of graphs $\mathcal{H}$, denoted $\text{diam}(\mathcal{H})$, being the maximum of $\text{diam } G$ for $G \in \mathcal{H}$.

Given a graph G and an integer k the problem $\mathcal{H}$-Free Edge Deletion asks whether there is a set $F \subseteq E(G)$ with $|F| \leq k$ such that $G - F$ is $\mathcal{H}$-free. And similarly, $\mathcal{H}$-Free Edge Editing asks whether there is a set $F \subseteq E(G)$ with $|F| \leq k$ such that $G \triangle F$ is $\mathcal{H}$-free. We say that a set of edges F is an $\mathcal{H}$-*solution* if $G \triangle F$ is $\mathcal{H}$-free. When $\mathcal{H}$ is clear from context, we will refer to F simply as a solution. When the problem at hand is the deletion problem, we furthermore assume $F \subseteq E(G)$, and when the problem at hand is the completion problem, we assume $F \cap E(G) = \emptyset$, as is expected.

Definition 1 (H-packing). *Given a graph G and an obstruction H we say that $\mathcal{X} \subseteq 2^{V(G)}$ forms an H-packing in G if*

(i) $G[X]$ and H are isomorphic for every $X \in \mathcal{X}$, and
(ii) X and Y are disjoint for every $X, Y \in \mathcal{X}$.

Observation 1. Given a graph G and an obstruction H we can obtain a maximal H-packing $\mathcal{X}$ in $O(n^{|V(H)|+1})$ time.

The problem we are dealing with in this article is the following, where we may replace *editing* with *deletion* or *completion*, by simply putting restrictions on where we chose F from.

$\mathcal{H}$-Free Edge Editing

Input:	A graph G and an integer k
Parameter:	k
Question:	Is there a set F of at most k edges s.t. $G \triangle F$ is $\mathcal{H}$-free?

2 Graph Modification on Bounded Degree Graphs

In this section we prove that for any finite set of obstructions $\mathcal{H}$, the problem of deleting or editing at most k edges to make an input graph of bounded degree $\mathcal{H}$-free admits polynomial kernels. More precisely, both $\mathcal{H}$-Free Edge Editing and $\mathcal{H}$-Free Edge Deletion admits polynomial kernels on bounded degree graphs.

The argument consists of two parts. First, we identify a set of critical vertices in the input graph G, called the obstruction core Z. Based on this set we can decompose any set of modifications F in G. The decomposition leads to the construction of a set of vertices in the graph, called the extended obstruction core Z^+. The first crucial property of Z^+ is that if F modifies $G[Z^+]$ into an $\mathcal{H}$-free graph, then F also modifies G into an $\mathcal{H}$-free graph. In other words, the obstructions you want to eliminate in your graph, should be eliminated within the extended obstruction core. The second crucial property is that the extended obstruction core can be proved to live within a ball around the obstruction core, were the radius depends on how well the solution decomposes. This ball will in the end constitute the kernel.

In the second part of the argument we prove that every minimal solution decomposes well. Hence we can bound the size of the ball containing the extended obstruction core and obtain a kernel.

We point out that we have considered the editing variant of the problem where you are allowed to surpass the original maximum degree in the graph by adding edges. However, it is the case that there is always a solution that at most doubles the maximum degree of the graph since if more edges are added one might as well remove all edges incident to the vertex. The validity of this is proved in Lemma 9. It can furthermore be argued that this version of the problem is the most general one. This is due to the fact that adding every supergraph of the star with $\Delta(G) + 1$ leaves to the obstruction set ensures that any solution respects the current maximum degree.

2.1 Cores and Layers

In this section we introduce the concepts of obstruction cores and extended obstruction cores. They are heavily based on the notion of shattered obstructions; the set of obstructions you get from $\mathcal{H}$ if you take every connected component as an obstruction. It follows immediately that every shattered obstruction is connected.

Definition 2. (Shattered obstructions). *Given a set of obstructions $\mathcal{H}$ we define the* shattered obstructions, *denoted $\mathcal{H}^{\blacktriangledown}$ as the set of all connected components of graphs of $\mathcal{H}$.*

Based on shattered obstructions we now define an obstruction core and explain how such a set of not too large size can be obtained.

Definition 3. (Obstruction core). *Let (G, k) be an instance of $\mathcal{H}$-FREE EDGE EDITING ($\mathcal{H}$-FREE EDGE DELETION). We then say that a set $Z \subseteq V(G)$ is an* obstruction core *in G if for every shattered obstruction H in G it holds that either:*

(i) $V(H) \subseteq Z$ or
(ii) there is an H-packing in $G[Z]$ of size at least $(\Delta(G) + 1) \cdot n_{\mathcal{H}} + 2k + 1$.

Observe that for every H satisfying *(ii)* it holds that even if you discard an arbitrary obstruction in G and its entire neighborhood, together with all vertices touched by a set of at most k edges, it still holds that H occurs in $G[Z]$. This is very useful if you want to replace some part of an obstruction.

Observation 2. Given an instance (G, k) of $\mathcal{H}$-FREE EDGE EDITING ($\mathcal{H}$-FREE EDGE DELETION) we can in $O(|\mathcal{H}^{\blacktriangledown}|n^{n_{\mathcal{H}}+1})$ time obtain an obstruction core Z in G of size at most $|\mathcal{H}^{\blacktriangledown}|((\Delta(G) + 1) \cdot n_{\mathcal{H}} + 2k + 1)$.

Proof. Let Z be the empty set initially. Then for every shattered obstruction H we find a maximal H-packing $\mathcal{X} = X_1, \ldots, X_t$ and add the following set $\bigcup_{i=1}^{p} X_i$ to Z, where $p = \min(t, (\Delta(G) + 1) \cdot n_{\mathcal{H}} + 2k + 1)$. The time complexity follows by Observation 1.

The next definitions are the ones of layer decompositions and core extensions, arguably the most central definitions of the kernelization algorithm. They are both with respect to a fixed obstruction core Z and set of edges F. The solution is decomposed into several layers such that the first layer consists of the edges of F that are contained in Z. The second layer consists of the edges of F that are contained in scattered obstructions created when the modifications in Z was done, and so forth. The extended core is a set of vertices encapsulating all scattered obstructions either in $G[Z]$ or created in G when doing the modifications of the layers. It should be observed by the reader that the consider solution F is not constructed, but analyzed implicitly with the intention to locate a part of the input graph that encapsulates all the crucial information of the instance.

Layer decompositions and core extensions. Let (G, k) be an instance of $\mathcal{H}$-FREE EDGE EDITING ($\mathcal{H}$-FREE EDGE DELETION), $F \subseteq [V(G)]^2$ and Z an obstruction core. We construct the *layer decomposition* $F_1, \ldots, F_\ell$ of F as follows: Let $G_1 = G$, $R_1 = F$ and $Z_1 = Z$. Then, inductively we construct the set $X = R_i \cap [Z_i]^2$. If X is empty we stop the process, otherwise we let $F_i = X$, $G_{i+1} = G_i \triangle F_i$ and $R_{i+1} = R_i \setminus F_i$. Furthermore, we let

$$W_{i+1} = \{v \in H : H \text{ is a shattered obstruction in } G_{i+1} \text{ with } [V(H)]^2 \cap F_i \neq \emptyset\},$$

and based on this we let $Z_{i+1} = Z_i \cup W_{i+1}$.

With the construction above in mind, we will refer to G_i as the *ith intermediate graph*, R_i as the *ith remainder*, Z_i as the *ith core extension* and ℓ as the *solution depth* (all with respect to G, Z and F). Furthermore, we will refer to $G^+ = G_{\ell+1}$ as the *resulting graph* and $Z^+ = Z_{\ell+1}$ as the *extended core*.

The next lemma says that if there is an obstruction in some intermediate graph such that every connected component of the obstruction is either inside the corresponding core extension or not modified at all so far by the layers, then there is an isomorphic obstruction contained entirely within the core extension. The intuition is that any untouched connected component has a large packing in Z and hence it can be replaced by an isomorphic subgraph inside Z that both avoids the modifications and the neighborhood of the rest of the obstruction.

Lemma 3. *Let (G, k) be an instance of $\mathcal{H}$-FREE EDGE EDITING ($\mathcal{H}$-FREE EDGE DELETION), Z an obstruction core of G, and $F \subseteq [V(G)]^2$ with $|F| \leq k$ and $F_1, \ldots, F_\ell$ a layer decomposition of F. For an integer $j \in [1, \ell+1]$ let G_j be the intermediate graph and Z_j the core extension with respect to G, Z and F. Let H be an obstruction in G_j with connected components $H_1, \ldots, H_t$ such that every H_i satisfies either: (i) $V(H_i) \subseteq Z_j$ or (ii) $H_i = G[V(H_i)]$. Then there is an obstruction H' in G_j isomorphic to H with $V(H') \subseteq Z_j$ and $V(H') \setminus V(H) \subseteq Z$.*

Proof. For convenience we denote neighborhoods in G_j by N_j. Let H' be the disjoint union of every H_i such that $V(H_i) \subseteq Z_j$ and $\mathcal{L}$ the list containing every H_i not added to H'. Let H_i be an element of $\mathcal{L}$. We will now prove that there is an H_i' in $G_j[Z_j \setminus N_j[H']]$ such that H_i and H_i' are isomorphic. Let $\mathcal{X}_i$ be the maximal H_i-packing obtained when constructing Z. Since $V(H_i)$ is not contained in Z_j (and hence Z) and H_i's edges are as in G it holds that $|\mathcal{X}_i| \geq (\Delta(G)+1) \cdot n_{\mathcal{H}} + 2k + 1$ by the definition of obstruction cores. This yields that $(\Delta(G) + 1) \cdot n_{\mathcal{H}} + 2k + 1$ of the elements of the packing was added to Z. Furthermore, we observe that $|V(H')| \leq n_{\mathcal{H}}$ and hence that $|N_G(H')| \leq \Delta \cdot n_{\mathcal{H}}$. It follows immediately that $|N_j(H')| \leq \Delta \cdot n_{\mathcal{H}} + k$ and hence that $|N_j[H']| \leq (\Delta+1) \cdot n_{\mathcal{H}} + k$. By the previous arguments it follows that there is an H_i-packing in $G_j[Z \setminus N_j[H']]$ of size at least $k + 1$. And hence, by the pigeon hole principle there is an H'_i isomorphic to H_i in $G_j[Z_j \setminus N_j[H']]$ such that $[V(H'_i)]^2$ and F' are disjoint.

To complete the proof we do the following for every H_i in $\mathcal{L}$. We find an H'_i as described above, add H'_i to H' and remove H_i from $\mathcal{L}$. Since $H_1, \ldots, H_t$ are the connected components of H it follows that H and H' are isomorphic. Furthermore, $V(H')$ is clearly contained in Z_j and $V(H') \setminus V(H)$ in Z.

This possibility of moving obstructions to the inside of core extensions immediately yields several very useful lemmata.

Lemma 4. *Let (G, k) be an instance of $\mathcal{H}$-FREE EDGE EDITING($\mathcal{H}$-FREE EDGE DELETION), Z an obstruction core of G, and $F \subseteq [V(G)]^2$. Construct the layer decomposition $F_1, \ldots, F_\ell$ of F with respect to Z, let $F' = \cup_{i=1}^{\ell} F_i$ and let Z^+ be the extended core with respect to Z and F. It then holds that: $(G \triangle F')[Z^+]$ is $\mathcal{H}$-free if and only if $G \triangle F'$ is $\mathcal{H}$-free.*

Proof. Recall that $G^+ = G \triangle F'$. It is trivial that if there is an obstruction H in $G^+[Z^+]$ then H is also an obstruction in G^+. For the other direction, let H be an obstruction in G^+ and $H_1, \ldots, H_t$ the connected components of H. Observe that by the definition of Z^+ it holds that every H_i satisfies either *(i)* or *(ii)* of Lemma 3 with $j = \ell + 1$. It follows that there is an obstruction H' in G^+ with $V(H') \subseteq Z^+$. Hence H' is an obstruction in $G^+[Z^+]$, which completes the argument.

Lemma 5. *Let (G, k) be an instance of $\mathcal{H}$-FREE EDGE EDITING($\mathcal{H}$-FREE EDGE DELETION), Z an obstruction core of G, F a minimal solution and $F_1, \ldots, F_\ell$ the layer decomposition of F with respect to Z. It then holds that $F_1, \ldots, F_\ell$ forms a partition of F.*

Proof. Let F_i and F_j be two layers with $i < j$. It follows immediately from the definition of layer decomposition that $F_j \subseteq R_j \subseteq R_i \setminus F_i$ and hence F_i and F_j are disjoint. For convenience we let $F' = \cup_{i \in [1,\ell]} F_i$. We now prove that $F' = F$. It follows from the definition of layer decomposition that $F' \subseteq F$. Assume for a contradiction that $F' \subsetneq F$. Consider the final graph $G^+ = G \triangle F'$. If G^+ is $\mathcal{H}$-free it follows that F is not a minimal solution, yielding a contradiction.

Hence, G^+ is not $\mathcal{H}$-free. It follows immediately from Lemma 4 that $G^+[Z^+]$ is also not $\mathcal{H}$-free. Furthermore, we know by the definition of layer decompositions that $G^+[Z^+] = (G \triangle F)[Z^+]$. And hence $G \triangle F$ is not $\mathcal{H}$-free, contradicting that F is a solution.

We finish the section by stating two important properties of the core; The first one gives the true power of an extended core, namely that if a set of edges is a solution for the graph induced on its extended core it also is a solution for the entire graph. The second lemma gives us a partial tool for encapsulating an extended core without knowing the solution beforehand. The next section is dedicated to turning this partial tool into a true hammer.

Lemma 6. *Let (G, k) be an instance of $\mathcal{H}$-FREE EDGE EDITING ($\mathcal{H}$-FREE EDGE DELETION), Z an obstruction core of G, $F \subseteq [V(G)]^2$ and Z^+ the extended core with respect to Z and F. If $F \subseteq [Z^+]^2$ then $(G \triangle F)[Z^+]$ is $\mathcal{H}$-free if and only if $G \triangle F$ is $\mathcal{H}$-free.*

Proof. Since $F \subseteq [Z^+]^2$ it holds that $G^+ = G \triangle F$. It trivially holds that if G^+ is $\mathcal{H}$-free, then so is $G^+[Z^+]$. Let H be an obstruction in G^+ with connected components $H_1, \ldots, H_t$. Observe that if H_i contains an edge of F then $V(H_i) \subseteq Z^+$ due to the definition of Z^+ and the assumption that $F \subseteq [Z^+]$. Apply Lemma 3 with $j = \ell + 1$ to obtain an obstruction H' in Z^+.

Lemma 7. *Let (G, k) be an instance of $\mathcal{H}$-FREE EDGE EDITING ($\mathcal{H}$-FREE EDGE DELETION), Z an obstruction core of G, $F \subseteq [V(G)]^2$ and Z^+ the extended core with respect to Z and F. It then holds that: $Z^+ \subseteq B(Z, \ell \cdot \text{diam}(\mathcal{H}))$.*

Proof. Let $Z_1, \ldots, Z_{\ell+1}$ be the extended cores. Instead of proving the lemma directly we prove the following, stronger claim:

$(\star)$ For every Z_i it holds that $Z_i \subseteq B(Z, (i - 1) \cdot \text{diam}(\mathcal{H}))$.

Since $Z^+ = Z_{\ell+1}$, it is clear that $(\star)$ implies the lemma. The proof of $(\star)$ is by induction. First, we observe that $(\star)$ holds for $i = 1$ by the definition of balls, since $Z = Z_1$. Assume for the induction step that $(\star)$ holds for i. Let v be a vertex in Z_{i+1}. If v is also in Z_i we are done by assumption. Hence, we assume v to be a vertex in $Z_{i+1} \setminus Z_i$. Or in other words, v is in W_{i+1}. By definition there is a scattered obstruction H in G_{i+1} and an edge uw in F_i such that both u, v and w are in H. Observe that the distance between u and v is at most $\text{diam}(H)$ and recall that u is in $Z_i \subseteq B(Z, (i - 1) \cdot \text{diam}(\mathcal{H}))$. It follows immediately that v is in $B(Z, i \cdot \text{diam}(\mathcal{H}))$ and hence the proof is complete.

2.2 Solutions are Shallow

In this section we prove that the depth of any solution is bounded logarithmically by the size of the solution. This, combined with Lemma 7 gives that linearly in k many balls of logarithmic radius is sufficient to encapsulate an extended core. To motivate that we obtain a polynomial kernel, observe that a ball of logarithmic radius in a bounded degree graph is of polynomial size.

First, we prove that when considering any layer we can always find a set of vertices of the same size, the removal of which would result in an $\mathcal{H}$-free graph. Next we prove that as long as the graph is not very small, removing a set of vertices from the graph has the same effect as modifying the graph such that the set becomes a set of isolates.

Lemma 8. *Let (G, k) be an instance of $\mathcal{H}$-FREE EDGE EDITING ($\mathcal{H}$-FREE EDGE DELETION), Z an obstruction core of G, F a minimal solution of the instance and $F_1, \ldots, F_\ell$ the layer partition of F with respect to Z. For every $i \in [1, \ell]$ there exist a set Y with $Y \leq |F_i|$ such that $G_i - Y$ is $\mathcal{H}$-free.*

Proof. We construct Y as follows: For every edge uv in F_i we add to Z the endpoint furthest away from Z. If it is a tie, we choose an arbitrary endpoint. Assume for a contradiction that $G_i - Y$ is not $\mathcal{H}$-free. Let H be an obstruction in $G_i - Y$ and $H_1, \ldots, H_t$ the connected components of H.

First, we consider the case when $i = 1$. We then apply a modification of the proof of Lemma 3. The idea is as follows: Let H' be the disjoint union of the components of H contained in Z and H_x a component not in Z. Then there is a H_x-packing of size $k+1$ in Z avoiding the closed neighborhood of H'. We observe that Y intersects with at most k of the elements of the packing and hence we can find a subgraph H'_x in $G[Z]$ not intersecting with Y such that H_x and H'_x are isomorphic. Add H'_x to H' and continue with the next component not contained in Z. It follows immediately that H' is also an obstruction in G_2. By definition $G_2[Z] = G^+[Z]$ and hence H' is an obstruction in G^+. This contradicts F being a solution.

If $i \geq 2$ it holds that Y and Z are disjoint. This is true since if both endpoints of an edge are included in Z, the edge would be in F_1 and not F_i. It holds by the definition of Y that $[V(H)]^2 \cap F_i$ is empty. Furthermore, by the definition of layer decompositions it holds that if some H_x intersects with some F_j with $j < i$ then $V(H_x) \subseteq Z_{j+1} \subseteq Z_i$. Hence we can apply Lemma 3 to obtain an obstruction H' in G_i with $V(H') \subseteq Z_i$. Since $V(H) \subseteq V(G) \setminus Y$ and $V(H') \setminus V(H) \subseteq Z$ it follows that H' is an obstruction in $G_i \setminus Y$. It follows immediately that H' is also an obstruction in G_{i+1}. By definition $G_{i+1}[Z_i] = G^+[Z_i]$ and hence H' is an obstruction in G^+. This contradicts F being a solution and completes the proof. □

Lemma 9. *Let (G, k) be an instance of $\mathcal{H}$-FREE EDGE EDITING ($\mathcal{H}$-FREE EDGE DELETION), X a set of vertices in G and E_X the set of edges incident to vertices in X. It then holds that either*

(i) $|V(G)| < |X| + k + 2(\Delta(G) + 1)n_{\mathcal{H}}$ or
(ii) the instances $(G - X, k')$ and $(G - E_X, k')$ are equivalent for every k'.

Proof. We assume that (i) does not apply and prove that this implies (ii). It is trivial that if $(G - E_X, k')$ is a yes-instance then $(G - X, k')$ is also a yes-instance. For the other direction, assume for a contradiction that $(G - X, k')$ is a yes-instance and that $(G - E_X, k')$ is a no-instance. Let F be a solution of $(G - X, k')$. For convenience we define $G_V = (G - X) \triangle F$ and $G_E = (G - E_X) \triangle F$. Let H an obstruction in G_E and B the set of vertices $V(H) \setminus X$. Observe that $G_V[B] = G_E[B]$ and that $|N_{G_E}(V(H))| \leq \Delta(G) \cdot n_{\mathcal{H}} + k$. It follows immediately that

$$|V(G_E) \setminus (X \cup N_{G_E}[V(H)])|$$
$$\geq |V(G_E)| - |X| - |N_{G_E}[V(H)]|$$
$$\geq |X| + k + 2(\Delta(G) + 1)n_{\mathcal{H}} - |X| - n_{\mathcal{H}} - \Delta(G) \cdot LH - k$$
$$= 2(\Delta(G) + 1)n_{\mathcal{H}} - n_{\mathcal{H}} - \Delta(G) \cdot n_{\mathcal{H}}$$
$$= (\Delta(G) + 1)n_{\mathcal{H}}.$$

Hence we can obtain an independent set I of size $X \cap V(H)$ that is contained entirely outside of both X and $N_{G_E}[V(H)]$. Let $H' = G_V[I \cup B]$ and observe that H' is isomorphic to H, contradicting G_V being $\mathcal{H}$-free.

With the two previous lemmata in mind we present the main intuition of the shallowness of a solution. Basically, if for any level of a decomposed solution you do a factor $\Delta(G)$ more modifications in the future than you do in this particular level you could instead remove a set of edges related to this layer and stop any further propagation. This ensures that in any optimal solution the size of the union of the remaining layers are bounded by a layer and the maximum degree of the graph.

Lemma 10. *Given an instance (G, k) of $\mathcal{H}$-FREE EDGE EDITING($\mathcal{H}$-FREE EDGE DELETION), an obstruction core Z, an optimal solution F and its layer decomposition $F_1, \ldots, F_\ell$ it holds that either*

(i) $|V(G)| \leq k + 2(\Delta(G) + 1) \cdot n_{\mathcal{H}}$ or
(ii) $\Delta(G) \cdot |F_i| \geq |R_{i+1}|$ for every $i \in [1, \ell]$.

Proof. We assume that *(i)* does not apply and hence that $|V(G)| > k + (\Delta(G) + 2) \cdot n_{\mathcal{H}}$. Assume for a contradiction that there is an $i \in [1, \ell]$ such that *(ii)* does not hold. Specifically, i is so that $\Delta(G) \cdot |F_i| < |R_{i+1}|$. By Lemma 8 there is a set of vertices Y with $|Y| \leq |F_i|$ such that $G_i - Y$ is $\mathcal{H}$-free. It follows by Lemma 9 with $k' = 0$ that $G_i - E_X$ is also $\mathcal{H}$-free. Let $F' = (\cup_{j \in [1, i-1]} F_j) \cup E_X$ and observe that $G \triangle F'$ is $\mathcal{H}$-free. By the following calculations;

$$|F'| \leq |\cup_{j \in [1, i-1]} F_j| + |E_X| < |\cup_{j \in [1, i-1]} F_j| + |R_{i+1}| = |F|,$$

we conclude that $|F'| < |F|$. This contradicts the optimality of $|F|$ and hence our proof is complete.

Lemma 11. *Given a instance (G, k) of $\mathcal{H}$-FREE EDGE EDITING($\mathcal{H}$-FREE EDGE DELETION), an optimal solution F and its layer decomposition $F_1, \ldots, F_\ell$ it holds that either*

(i) $|V(G)| \leq k + 2(\Delta(G) + 1) \cdot n_{\mathcal{H}}$ or
(ii) $\ell \leq 1 + \log_{\frac{\Delta(G)+1}{\Delta(G)}} |F|$.

Proof. Assume that *(i)* does not hold and hence that $|V(G)| > k + 2(\Delta(G) + 1) \cdot n_{\mathcal{H}}$. It follows immediately that *(ii)* in Lemma 10 applies.

$$
\begin{aligned}
|F| = |R_1| = |F_1| + |R_2| \\
\geq \frac{|R_2|}{\Delta(G)} + |R_2| = \frac{\Delta(G) + 1}{\Delta(G)} \cdot |R_2| = \frac{\Delta(G) + 1}{\Delta(G)} \cdot (|F_2| + |R_3|) \\
\geq \cdots \geq \left(\frac{\Delta(G) + 1}{\Delta(G)}\right)^{\ell - 1} \cdot |R_\ell| \\
= \left(\frac{\Delta(G) + 1}{\Delta(G)}\right)^{\ell - 1} \cdot |F_\ell| \\
\geq \left(\frac{\Delta(G) + 1}{\Delta(G)}\right)^{\ell - 1}
\end{aligned}
$$

This gives that $\ell \leq 1 + \log_{\frac{\Delta(G)+1}{\Delta(G)}} |F|$ and hence the argument is complete.

2.3   Obtaining the Kernels

We now have all the necessary tools for providing the kernels. We reduce the graph to a ball of small radius around any obstruction core Z and by this obtain a kernelized instance. Both the size bounds and the correctness of the reduction rule follows by combining the tools developed during the section.

Rule 1. *Given an instance* (G, k) $\mathcal{H}$-FREE EDGE EDITING *(*$\mathcal{H}$-FREE EDGE DELETION) such that $|V(G)| > k + 2(\Delta(G) + 1) \cdot n_{\mathcal{H}}$, we find an obstruction core Z in G and return $(G[B(Z, r)], k)$ where $r = \operatorname{diam}(\mathcal{H}) \cdot (1 + \log_{\frac{\Delta(G)+1}{\Delta(G)}} k)$.*

Lemma 12. *Let* (G, k) *be an instance of* $\mathcal{H}$-FREE EDGE EDITING *(*$\mathcal{H}$-FREE EDGE DELETION) *and* (G', k) *the instance obtained when applying Rule 1 to* (G, k). *Then* (G, k) *is a yes-instance if and only if* (G', k) *is a yes-instance.*

Proof. It follows immediately from G' being an induced subgraph of G that if (G, k) is a yes-instance, then so is (G', k). For the other direction, let (G', k) be a yes-instance and let Z be the obstruction core of G obtained when applying Rule 1. Clearly, Z is also an obstruction core of G'. Let F be an optimal solution of (G', k) and construct the layer decomposition $F'_1, \ldots, F'_{\ell'}$ and the core extensions Z'_i with respect to Z and F in G'. Now we construct the layer decomposition $F_1, \ldots, F_\ell$ and the core extensions Z_i with respect to Z and F in G. By the definition core extensions it holds that $Z'_i \subseteq Z_i$ and hence $\ell \leq \ell'$. By Lemma 6 it holds that $Z_G^+ = Z_{\ell+1} \subseteq B_G(Z, \ell \cdot \operatorname{diam}(\mathcal{H})) \subseteq B_G(Z, \ell' \cdot \operatorname{diam}(\mathcal{H}))$. By Lemma 11 applied to F in G' it holds that $\ell \leq 1 + \log_{\frac{\Delta(G)+1}{\Delta(G)}} |F| \leq 1 + \log_{\frac{\Delta(G)+1}{\Delta(G)}} k$ and hence $Z_G^+ \subseteq V(G')$. It follows immediately that $(G \triangle F)[Z_G^+]$ is $\mathcal{H}$-free. By Lemma 5 it holds that $F \subseteq [Z'_{\ell'+1}]^2$ and hence $F \subseteq [Z_{\ell+1}]^2$. It follows immediately that Lemma 6 applies and hence $G \triangle F$ is $\mathcal{H}$-free. Hence (G, k) is a yes-instance and the proof is complete.

For ease of readability, we denote diam($\mathcal{H}$) simply by D and $\Delta(G)$ by Δ.

Theorem 4 (♠). *Both $\mathcal{H}$-FREE EDGE DELETION and $\mathcal{H}$-FREE EDGE EDITING admit kernels with at most $2n_{\mathcal{H}}|\mathcal{H}^{\blacktriangledown}|\Delta^{D+1}k^{1+D(\Delta \log \Delta)}$ vertices. For fixed $\mathcal{H}$ and Δ this is a kernel with $k^{O(1)}$ vertices.*

3 Conclusion

We showed that for any finite set $\mathcal{H}$ of forbidden induced subgraphs, both $\mathcal{H}$-FREE EDGE EDITING and $\mathcal{H}$-FREE EDGE DELETION admit polynomial kernelizations on bounded degree input graphs. This extendes and generalizes the result of Aravind et al. [2], who showed that $\mathcal{H}$-FREE EDGE DELETION admits kernel when $\mathcal{H}$ is connected on bounded degree input. We not only extend their kernel, but also improve on the size of their kernel.

We showed two lower bounds: (1) for a finite set $\mathcal{H}$ of connected graphs, $\mathcal{H}$-FREE EDGE COMPLETION does not admit a polynomial kernel on bounded degree input graphs, unless NP $\subseteq$ coNP/poly. (2) Under the same assumption, C_{11}-FREE EDGE DELETION does not have a polynomial kernel on 2-degenerate graphs, nor does $\mathcal{H}$-FREE EDGE EDITING.

Since there is a finite set $\mathcal{H}$ of connected graphs such $\mathcal{H}$-FREE EDGE COMPLETION does not admit a polynomial kernel, we encourage a further study of these problems. We leave it as an open problem whether there is a dichotomy for when $\mathcal{H}$-FREE EDGE COMPLETION admits a polynomial kernel, restricted to bounded degree graphs and connected, forbidden induced subgraphs.

References

1. Abu-Khzam, F.N.: A kernelization algorithm for d-hitting set. J. Comput. Syst. Sci. **76**(7), 524–531 (2010)
2. Aravind, N.R., Sandeep, R.B., Sivadasan, N.: On polynomial kernelization of $\mathcal{H}$-free edge deletion. In: IPEC (2014)
3. Burzyn, P., Bonomo, F., Durán, G.: NP-completeness results for edge modification problems. Discrete Appl. Math. **154**(13), 1824–1844 (2006)
4. Cai, L.: Fixed-parameter tractability of graph modification problems for hereditary properties. Inf. Proc. Lett. **58**(4), 171–176 (1996)
5. Cai, L., Cai, Y.: Incompressibility of H-free edge modification problems. Algorithmica **71**(3), 731–757 (2015)
6. Drange, P.G., Pilipczuk, M.: A polynomial kernel for trivially perfect editing. In: ESA (2015)
7. El-Mallah, E.S., Colbourn, C.J.: The complexity of some edge deletion problems. IEEE Trans. Circ. Syst. **35**(3), 354–362 (1988)
8. Erdős, P., Rado, R.: Intersection theorems for systems of sets. J. Lond. Math. Soc. **1**(1), 85–90 (1960)
9. Fellows, M.R., Langston, M.A., Rosamond, F.A., Shaw, P.: Efficient parameterized preprocessing for cluster editing. In: Csuhaj-Varjú, E., Ésik, Z. (eds.) FCT 2007. LNCS, vol. 4639, pp. 312–321. Springer, Heidelberg (2007)
10. Flum, J., Grohe, M.: Parameterized Complexity Theory. Springer, New York (2006)

11. Garey, M.R., Johnson, D.S.: Computers and Intractability: a Guide to the Theory of NP-Completeness. W. H. Freeman & Co., New York (1979)
12. Gramm, J., Guo, F., Hüffner, J., Niedermeier, R.: Data reduction and exact algorithms for clique cover. ACM J. Exp. Algorithmics **13**, 1–14 (2008)
13. Guillemot, S., Havet, F., Paul, C., Perez, A.: On the (non-)existence of polynomial kernels for P_l-free edge modification problems. Algorithmica **65**(4), 900–926 (2013)
14. Guo, J.: Problem kernels for NP-complete edge deletion problems: split and related graphs. In: Tokuyama, T. (ed.) ISAAC 2007. LNCS, vol. 4835, pp. 915–926. Springer, Heidelberg (2007)
15. Komusiewicz, C., Uhlmann, J.: Cluster editing with locally bounded modifications. Discrete Appl. Math. **160**(15), 2259–2270 (2012)
16. Kratsch, S., Wahlström, M.: Two edge modification problems without polynomial kernels. Discrete Optim. **10**(3), 193–199 (2013)
17. Lewis, J.M., Yannakakis, M.: The node-deletion problem for hereditary properties is np-complete. J. Comput. Syst. Sci. **20**(2), 219–230 (1980)
18. Natanzon, A., Shamir, R., Sharan, R.: Complexity classification of some edge modification problems. Discrete Appl. Math. **113**(1), 109–128 (2001)
19. Yannakakis, M.: Edge-deletion problems. SIAM J. Comput. **10**(2), 297–309 (1981)

Random Partial Match in Quad-K-d Trees

A. Duch[(✉)], G. Lau, and C. Martínez

Computer Science Department, Technical University of Catalonia – Barcelona Tech,
Barcelona, Catalonia, Spain
{duch,glau,conrado}@cs.upc.edu

Abstract. Quad-K-d trees were introduced by Bereczky et al. [3] as a generalization of several well-known hierarchical multidimensional data structures such as K-d trees and quad trees. One of the interesting features of quad-K-d trees is that they provide a unified framework for the analysis of associative queries in hierarchical multidimensional data structures. In this paper we consider partial match, one of the fundamental associative queries, and prove that the expected cost of a random partial match in a random quad-K-d tree of size n is of the form $\Theta(n^\alpha)$, with $0 < \alpha < 1$, for several families of quad-K-d trees including, among others, K-d trees and quad trees. We actually give a general result that applies to any family of quad-K-d trees where each node has a *type* that is independent of the type of other nodes. We derive, exploiting Roura's Continuous Master Theorem, the general equation satisfied by α, in terms of the dimension K, the number of specified coordinates s in the partial match query, and also the additional parameters that characterize each of the families of quad-K-d trees considered in the paper. We also conduct an experimental study whose results match our theoretical findings; as a by-product we propose an implementation of the partial match search in quad-K-d trees in full generality.

1 Introduction

Partial match queries have been widely studied in the literature since they are a fundamental associative query [13, 16]. Given a collection (or file) $\mathcal{F}$ of n records, in which each *record* in $\mathcal{F}$ is an ordered K-tuple of values (the attributes or coordinates of the record), a *query* of $\mathcal{F}$ is a retrieval of the records whose attributes satisfy certain given conditions. The query is considered *associative* when the imposed conditions deal with more than one of the attributes.

In particular, *partial match* (PM hereinafter) queries consist of retrieving all the records in $\mathcal{F}$ whose attributes match the attributes of the query record that are specified. Other examples of associative queries are *nearest-neighbor* queries whose aim is to retrieve a record in $\mathcal{F}$ that is closest to a given record under a

This work has been partially supported by funds from the Spanish Ministry for Economy and Competitiveness (MINECO) and the European Union (FEDER funds) under grant COMMAS (ref. TIN2013-46181-C2-1-R) and AGAUR grant SGR 2014:1034 (ALBCOM).

E. Kranakis et al. (Eds.): LATIN 2016, LNCS 9644, pp. 376–389, 2016.
DOI: 10.1007/978-3-662-49529-2_28

given distance metric, or *orthogonal range* queries, in which we want to retrieve all the records in $\mathcal{F}$ that fall inside a given hyper-rectangle whose sides are parallel to the coordinate axes.

The study of PM queries is fundamental because of their intrinsic interest but also because its analysis is the basis of the analysis of other associative queries (in particular, of orthogonal range and nearest neighbor queries). Indeed, several papers address the analysis of PM queries (either random or fixed) in a wide variety of general purpose hierarchical multidimensional data structures [4,5,7, 8,10,11,14]. Undoubtedly, a unified analysis of PM queries and other associative queries has great interest and might give us a better insight on their performance. The work we present here is a step in that direction.

To do so, we use the general framework of quad-K-d trees introduced in [3] and follow most of the definitions, terminology and conventions therein to prove that the expected cost of a random PM query in a random quad-K-d tree of size n is of the form $\Theta(n^\alpha)$, with $0 < \alpha < 1$. This result applies to any family of quad-K-d trees when each node has a *type* that is independent of the type of the other nodes. This includes random relaxed K-d trees and quad trees.

Standard K-d trees were introduced by Bentley [1]. They are binary search trees such that, at every level of the tree, one of the K coordinates is used to discriminate: at level 0 is coordinate 0, at level 1 is coordinate 1 and, in general, at level i is coordinate i mod K. In order to facilitate their randomization, relaxed K-d trees were proposed in [9]. They are K-d trees in which the coordinate to discriminate at each individual node is uniformly chosen among the K possibilities. When all the K coordinates are used to discriminate at each node, the tree obtained is a quad tree [2]. A quad-K-d tree is a multidimensional tree in which each node discriminates with respect to some number m (potentially distinct for each node), $1 \leq m \leq K$, of coordinates (and thus it has 2^m subtrees).

The paper is organized as follows. In Sect. 2 we give some preliminaries and introduce some families of quad-K-d trees. We then derive, in Sect. 3, using Roura's Continuous Master Theorem [15] (Roura's CMT hereinafter), the expected cost of a random PM query in a random quad-K-d tree (Subsect. 3.1). We also propose an implementation of the PM search algorithm in quad-K-d trees in full generality and conduct an experimental study whose results match our theoretical findings (Subsect. 3.2). We finish with conclusions and possibilities of further work in Sect. 4.

2   Random Quad-K-d Trees

The formal definition of quad-K-d trees as given in [3] is the following:

Definition 1 (Bereczky et al. [3]). *A quad-K-d search tree T of size $n \geq 0$ stores a set of n K-dimensional records, each holding a key $\mathbf{x} = (x_0, \ldots, x_{K-1}) \in D$ and a coordinate split Boolean vector $\delta = (\delta_0, \ldots, \delta_{K-1})$, where $D = D_0 \times \cdots \times D_{K-1}$ and each D_j, $0 \leq j < K$, is a totally ordered domain. The quad-K-d tree T is such that*

- *either it is empty when $n = 0$, or*
- *its root stores a record with key* $\mathbf{x}$, *a bit-vector* δ *that contains exactly m ones (i.e., it is of order m), with $1 \le m \le K$, and pointers to its 2^m subtrees that store the $n - 1$ remaining records as follows: each subtree, let us call it T_w, is associated with a string $w = w_0 w_1 \ldots w_{K-1} \in \{0, 1, \#\}^K$, such that $\forall w \in \{0, 1, \#\}^K$, T_w is a quad-K-d tree and, for any key $\mathbf{y} \in T_w$ and $0 \le j < K$, it holds that*
  - *if $\delta_j = 0$ then $w_j = \#$*
  - *if $\delta_j = 1$ and $w_j = 0$, then $y_j < x_j$*
  - *if $\delta_j = 1$ and $w_j = 1$, then $y_j > x_j$.*

We assume, without loss of generality, that $D_j = [0, 1]$, $0 \le j < K$.

It is worth noting that with this definition, as is the case for binary search trees, a quad-K-d tree can not handle keys that have some coordinates that are equal, to do so the definition as well as the algorithms supported by the data structure should be slightly modified. However, because of our assumptions in their analysis, we can safely disregard this situation since, as we will see, the probability that two records share one coordinate value is 0.

A 3-dimensional quad-K-d tree is shown in Fig. 1. Inside each node appears its label and in the table therein is the 3-dimensional key with which the node is associated together with its split vector. Next to every edge appears the label (the string w) of the subtree it points to.

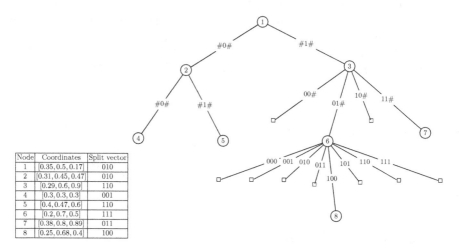

Node	Coordinates	Split vector
1	[0.35, 0.5, 0.17]	010
2	[0.31, 0.45, 0.47]	010
3	[0.29, 0.6, 0.9]	110
4	[0.3, 0.3, 0.3]	001
5	[0.4, 0.47, 0.6]	110
6	[0.2, 0.7, 0.5]	111
7	[0.38, 0.8, 0.89]	011
8	[0.25, 0.68, 0.4]	100

Fig. 1. An example of a 3-dimensional quad-K-d tree, omitting some empty subtrees

As we have already mentioned, both K-d trees and quad trees are special cases of quad-K-d trees. In fact, for any quad-K-d tree T of size n, if the split vector δ associated with every node of T contains all the K coordinates ($\delta_j = 1$ for all j) then T is a quad tree, and if it contains exactly one 1 for every node then T is a relaxed K-d tree.

A node with key **x** and split vector δ in a quad-K-d tree is *of type m* if and only if δ is of order m, $1 \leq m \leq K$. A direct consequence of this definition is that nodes of type m have 2^m children each. A *m-regular* (or *m-ary*) quad-K-d tree is a quad-K-d tree that has all its internal nodes of type m.

Quad trees of dimension K are K-regular quad-K-d trees. All variants of K-d trees are 1-regular quad-K-d trees.

In order to proceed to the average-case analysis of random PM queries in quad-K-d trees we need to introduce the random model that will be used; it is equivalent to the conventional random models found in the literature for the particular cases of random quad trees and random relaxed K-d trees.

Definition 2. *A random relaxed quad-K-d tree of size n is a quad-K-d tree built by n random insertions[1] into an initially empty tree, and it additionally satisfies the following two conditions:*

1. *The types of its n nodes are given by n i.i.d. random variables in the set $\{1, \ldots, K\}$. We denote τ_m as the probability that an arbitrary node is of type m.*
2. *Any subset of m coordinates out of K is equally likely to be the set of discriminating coordinates (that is, those for which $\delta_j = 1$) for a node of type m; in other words, the probability that the discriminating coordinates of a node of type m are $0 \leq i_0 < i_1 < \cdots < i_{m-1} < K$ is $1/\binom{K}{m}$ for any subset $\{i_0, \ldots, i_{m-1}\} \subseteq \{0, \ldots, K-1\}$.*

The above definition includes the following families of search trees:

- Random quad trees: these are K-regular random quad-K-d trees; here $\tau_K = 1$, and $\tau_m = 0$ if $m \neq K$.
- Random relaxed K-d trees: these are 1-regular random quad-K-d trees. We have $\tau_1 = 1$, and $\tau_m = 0$ if $m \neq 1$.
- Random Split [3] quad-K-d trees. We will use the name Pseudo-binomial Split to refer to this family hereinafter. Given some real value p, $0 < p < 1$, for $m > 1$ we have

$$\tau_m = \binom{K}{m} p^m q^{K-m}, \qquad q = 1 - p,$$

and $\tau_1 = q^K + Kpq^{K-1}$.

We now introduce the following new families of quad-K-d trees:

1. Uniform Split quad-K-d trees. We have here simply $\tau_m = 1/K$, for any m.
2. Binomial Split quad-K-d trees. Given some real value p, $0 < p < 1$, we have

$$\tau_m = \binom{K-1}{m-1} p^{m-1} q^{K-m}, \qquad q = 1 - p.$$

---

[1] There are several ways to characterize random insertions. The one we will consider here is that every coordinate of the data point to be inserted is independently drawn from some continuous distribution in $[0, 1]$.

3. Geometric Split quad-K-d trees. Given some real value p, $0 < p < 1$, we have, for $m < K$

$$\tau_m = qp^{m-1}, \qquad q = 1 - p,$$

and $\tau_K = 1 - \sum_{1 \leq m \leq K-1} \tau_m = p^{K-1}$.

4. m-regular relaxed quad-K-d trees. All types are given by the "degenerate" random variable $X \sim m$; that is, $\tau_m = 1$ and $\tau_\ell = 0$ for all $\ell \neq m$.

In the case of Pseudo-binomial Split (a.k.a. Random Split), Binomial Split and Geometric Split quad-K-d trees we have a "dial" to control the transition from random relaxed K-d trees to quad trees. When $p = 0$ all these families coincide with random relaxed K-d trees, while they become quad trees when $p = 1$. In the case of m-regular relaxed quad-K-d trees we also have the transition from relaxed K-d trees ($m = 1$) to quad trees ($m = K$).

The expected amount of memory used in a quad-K-d tree is obviously related to the arity of the different nodes. If $\overline{d}$ is the average arity of the quad-K-d tree, we might expect to need $\overline{d}n + 1$ pointers, of which approximately $(\overline{d} - 1)n$ will be null. Indeed, the average arity $\overline{d}$ of a random relaxed quad-K-d tree is given by

$$\overline{d} = \sum_{1 \leq m \leq K} \tau_m 2^m$$

Table 1. Average arity in several families of random relaxed quad-K-d trees

Family	Average arity
m-regular	2^m
Uniform Split	$\frac{2}{K}(2^K - 1)$
Pseudo-binomial Split	$(1 + p)^K + (1 - p)^K$
Binomial Split	$2(1 + p)^{K-1}$
Geometric Split	$\frac{(2p)^K + 2p - 2}{2p - 1}$ if $p \neq 1/2$, $K + 1$ if $p = 1/2$

Table 1 gives the average arity for the families of random relaxed quad-K-d trees studied in this paper. In the case of Geometric Split it is interesting to observe that there are three regimes, according to $p < 1/2$, $p = 1/2$ or $p > 1/2$. If $p < 1/2$ then the average degree is constant $(\overline{d} \approx \frac{2q}{1-2p})$, irrespective of K, it grows linearly with K when $p = 1/2$ $(\overline{d} = K + 1)$, and it grows exponentially with K when $p > 1/2$ $(\overline{d} \approx \frac{1}{2p-1}(2p)^K)$.

3 Partial Match in Quad-K-d Trees

In this section we consider the expected cost of a random PM query in a random relaxed quad-K-d tree. The cost is measured, as usual, by the number of nodes visited by the PM search algorithm.

A random PM query is a pair $\langle \mathbf{q}, \mathbf{u} \rangle$, where $\mathbf{q} = (q_0, \ldots, q_{K-1})$ is a K-dimensional point independently drawn from the same continuous distribution as the data points, and $\mathbf{u} = (u_0, \ldots, u_{K-1})$ is the *pattern* of the query; each $u_i = S$ (the i-th attribute of the query is *specified*) or $u_i = *$ (the i-th attribute is *unspecified*). The goal of the PM search is to report all data points $\mathbf{x} = (x_0, \ldots, x_{K-1})$ in the tree such that $x_i = q_i$ whenever $u_i = S$. The number of specified coordinates will be denoted by s; the interesting cases are when $0 < s < K$.

3.1   Analysis

Let P_n denote the expected cost of a random PM query in a random relaxed quad-K-d tree of size n. Since the discriminating coordinates at each node of the quad-K-d tree are randomly chosen, the pattern of the PM query is irrelevant and there is no need to make it appear as a parameter, the required parameter in this case is the value s. Let $P_n^{(m)}$ denote the same cost, conditional on the root of the quad-K-d tree being of type m; similarly, let $P_n^{(i,m)}$ denote the expected cost of a random PM query conditional on the root being of type m and that exactly i, $0 \leq i \leq \min(m, s)$ of the discriminating coordinates of the root are specified in the PM query. Then, since the probability that the root of a random quad-K-d tree is of type m is τ_m we have

$$P_n = \sum_{m=1}^{K} \tau_m P_n^{(m)}, \tag{1}$$

$$P_n^{(m)} = \sum_{0 \leq i \leq m} \mu_{i,m} P_n^{(i,m)}, \tag{2}$$

$$P_n^{(i,m)} = 1 + \sum_{0 \leq k < n} \pi_{n,k}^{(i,m)} P_k, \tag{3}$$

with $\mu_{i,m}$ the probability that exactly i of the m discriminating coordinates are specified (these quantities also depend on s and K) and $\pi_{n,k}^{(i,m)}$ the average number of recursive calls on subtrees of the root of size k when the tree is of size n, its root is of type m and exactly i discriminating coordinates out of m are specified. It is not difficult to prove that

$$\mu_{i,m} = \frac{\binom{m}{i}\binom{K-m}{s-i}}{\binom{K}{s}}$$

and

$$P_n = 1 + \sum_{0 \leq k < n} \pi_{n,k} P_k, \tag{4}$$

where

$$\pi_{n,k} = \sum_{m=1}^{K} \tau_m \sum_{0 \leq i \leq m} \mu_{i,m} \pi_{n,k}^{(i,m)}.$$

In order to apply Roura's CMT [15] to solve the recurrence, we need to find a *shape* function, that is, a normalized continuous approximation to the weights $\pi_{n,k}$, the expected number of recursive calls on subtrees of the root of size k, when the input tree is of size n.

If $\omega_{i,m}(z)$ is a shape function for the sequence of weights $\{\pi_{n,k}^{(i,m)}\}_{0 \le k < n}$ then

$$\omega_m(z) = \sum_{0 \le i \le m} \frac{\binom{m}{i}\binom{K-m}{s-i}}{\binom{K}{s}} \omega_{i,m}(z)$$

is a shape function for the sequence $\left\{\pi_{n,k}^{(m)} := \sum_{0 \le i < m} \mu_{i,m} \pi_{n,k}^{(i,m)}\right\}_{0 \le k < n}$ and

$$\omega(z) = \sum_{m=1}^{K} \tau_m \omega_m(z)$$

is a shape function for the sequence $\{\pi_{n,k}\}_{0 \le k < n}$.

The shape function $\omega_{i,m}(z)$ follows from a reasoning analogous to that used by Flajolet et al. in their paper "Analytic Variations on Quadtrees" [12] where they analyze random PM searches in quad trees when $K = m$ and $s = i$. Actually, in [12] the authors show that

$$\pi_{n,k}^{(i,m)} = \frac{2^m}{n(n+1)} \sum_{\mathcal{L}} \left[\frac{1}{(\ell_1 + 2) \dots (\ell_{i-1} + 2)}\right]\left[\frac{1}{(\ell_{i+1} + 1) \dots (\ell_{m-1} + 1)}\right],$$

where $\mathcal{L}$ is the condition $n > \ell_1 \ge \ell_2 \ge \dots \ge \ell_{m-1} \ge \ell_m = k$. An alternative way to find $\pi_{n,k}^{(i,m)}$ can be found in [6].

Lemma 1. *For $0 \le i \le m$, we have*

$$\omega_{i,m}(z) = (-1)^{i+m} 2^m \left[\sum_{k=0}^{i-1} d_k^{(i,m)} z \ln^k z - \sum_{k=0}^{m-i-1} c_k^{(i,m)} \ln^k z\right], \tag{5}$$

where

$$c_k^{(i,m)} = \binom{m-2-k}{i-1}\frac{1}{k!},$$

and

$$d_k^{(i,m)} = (-1)^k \binom{m-2-k}{m-i-1}\frac{1}{k!}.$$

Proof. The application of Roura's CMT requires finding a shape function $\omega(z)$, a continuous approximation of the discrete weights $\omega_{n,k}$ (in our instance, $\omega_{n,k} := \pi_{n,k}^{(i,m)}$) fulfilling the condition

$$\sum_{k=0}^{n-1} \left|\omega_{n,k} - \int_{k/n}^{(k+1)/n} \omega(z)\,dz\right| = O(n^{-d}),$$

for some $d > 0$. Such shape function can be easily found here replacing the sums in the definition of $\pi_{n,k}^{(i,m)}$ by integrals so that

$$\pi_{n,k}^{(i,m)} \sim \frac{2^m}{n} \int_{k/n}^1 \frac{dy_{m-1}}{y_{m-1}} \int_{y_{m-1}}^1 \frac{dy_{m-2}}{y_{m-2}} \int_{y_{m-2}}^1 \cdots \int_{y_{i+1}}^1 dy_i \int_{y_i}^1 \frac{dy_{i-1}}{y_{i-1}} \int_{y_{i-1}}^1 \cdots \int_{y_2}^1 \frac{dy_1}{y_1},$$

and letting $\omega(z) = \lim_{n\to\infty} n \cdot \pi_{n,z\cdot n}^{(i,m)}$. □

From the explicit form of $\omega_{i,m}(z)$ given above, the next lemma follows from straightforward calculations and a few easy binomial identities (a computer algebra system like Maple is of great help here).

Lemma 2. *For any $a > -1$, and for any $0 \le i \le m$,*

$$\int_0^1 z^a \omega_{i,m}(z)\, dz = \frac{2^m}{(a+1)^{m-i}(a+2)^i}.$$

Now,

$$\omega(z) = \sum_{m=1}^K \tau_m \sum_{0 \le i \le s} \frac{\binom{m}{i}\binom{K-m}{s-i}}{\binom{K}{s}} \omega_{i,m}(z)$$

and, hence, for any $a > -1$,

$$\int_0^1 z^a \omega(z)\, dz = \sum_{m=1}^K \tau_m \sum_{0 \le i \le s} \frac{\binom{m}{i}\binom{K-m}{s-i}}{\binom{K}{s}} \frac{2^m}{(a+1)^{m-i}(a+2)^i}.$$

In particular, for a random PM search the toll function is $t_n = 1$, hence, computing the constant-entropy with $a = 0$ we get

$$\mathcal{H} = 1 - \int_0^1 z^0 \omega(z)\, dz = 1 - \sum_{m=1}^K \tau_m \sum_{0 \le i \le s} \frac{\binom{m}{i}\binom{K-m}{s-i}}{\binom{K}{s}} 2^{m-i}.$$

Since $2^{m-i} \ge 1$ for all $0 \le i \le m$, it follows that $\mathcal{H} \le 0$. Furthermore, whenever $i < m$, $2^{m-i} > 0$ and since these cases have positive probability (because $s < K$) it follows that $\mathcal{H} < 0$. Then, by direct application of Roura's CMT we have the following theorem.

Theorem 1. *The expected cost P_n of a random PM in a random relaxed quad-K-d tree of size n (measured as the number of nodes visited by Program 1) is*

$$P_n = \Theta(n^\alpha)$$

where α is the unique real solution in $[0, 1]$ of

$$\sum_{m=1}^K \tau_m \sum_{0 \le i \le s} \frac{\binom{m}{i}\binom{K-m}{s-i}}{\binom{K}{s}} \frac{2^m}{(\alpha+1)^{m-i}(\alpha+2)^i} = 1. \tag{6}$$

3.2   Implementation and Experiments

In this section we give (in Program 1) a general implementation of the PM search algorithm that receives as parameters a quad-K-d tree t of any kind and a random PM query q and returns the cost of the PM search, i.e., the number of nodes of the tree visited during the execution of the program.

Program 1. Implementation in C++ of the PM search algorithm

```
// Since  the  query  is  random , we  assume  w.l.o.g.  that  the
// s specified   coordinates  are  the  first  s  coordinates
typedef  vector<int>  VI;
typedef  vector<double>  VD;
typedef  struct  QKnode*  tree;
typedef  vector<tree>  VT;
struct  QKnode  {
    VD  key;
    VI  disc;  // sparse  representation  of  the  bit  vectors
    VT  T;
    int  elems;
    int  dim;
};
//Post:  Returns  the  number  of  visited  nodes  in  t
int  partial_match(tree  t,  const  VD&  query)  {
    if  (t  ==  NULL)  return  0;
    int  res  =  1;
    int  masc  =  0;
    int  q  =  query.size();
    int  m  =  t->disc.size();
    int  j;
    for  (j  =  0;  j  <  m  and  t->disc[j]  <  q;  ++j)  {
        int  i  =  t->disc[j];
        if  (query[i]  >  t->key[i])  masc  |=  (1<<j);
    }
    for  (int  i  =  0;  i  <  m  -  j  +  1;  ++i)  {
        int  st  =  (i  <<  j)  |  masc;
        res  +=  partial_match(t->T[st],  query);
    }
    return  res;
}
```

Our theoretical analysis of the expected cost of a PM search for random queries in quad-K-d trees is only asymptotic and provides the exponent of the leading order term. We have conducted some experiments in order to investigate whether our theoretical findings are good predictions of the average behavior of PM searches and for which input sizes we can expect to start getting relatively good predictions.

As we will see, the main conclusion of our experiments is that, despite the asymptotic nature of our theoretical results, they do a reasonably good job at predicting the average cost of random PM queries, even for relatively small values of n (less than 50000).

Each run of our experiments can be described by a tuple $\langle \mathcal{T}, \mathbf{q}, n, M \rangle$ where $\mathcal{T}$ is a type of quad-K-d tree (regular, binomial or geometric), $\mathbf{q}$ is the query, n is the size of the trees in the sample and M is the size of the sample.

For each run we generate M random quad-K-d trees of type $\mathcal{T}$ and size n. In each tree we perform a PM search with the random query $\mathbf{q}$, counting the total number of visited nodes and taking the corresponding sample mean $\bar{P}_n := 1/M \sum_{i=1}^{M} P_n^{(i)}$, where the $P_n^{(i)}$ are independent identically distributed realizations of P_n.

We conjecture that the average cost of random PM searches in quad-K-d trees is of the form $P_n = \beta n^\alpha + l.o.t.$, as is the case for random PM searches in K-d trees and quad trees. The conjectured constant β will depend on s, K and the family of quad-K-d trees under consideration, just as the exponent α does.

Figures 2 and 3 show the behavior of quad-K-d trees built using Binomial Split (top), Geometric Split (middle) and m-regular (bottom). Figure 2 plots $\log \bar{P}_n / \log n$ for n in the range $[25000, 50000]$. The empirical curves should converge to the theoretical value of α given by Theorem 1 in all cases, but the convergence can be quite slow (as can be seen in most cases).

In Fig. 3 the plots depict the normalized mean cost $\bar{P}_n / n^\alpha$, where α is the appropriate exponent for each case. The plots do not contradict our conjecture that $P_n = \beta n^\alpha + o(n^\alpha)$; in fact, they can be taken as evidence to support it. If our conjecture is true then the curves in Fig. 3 should be close (closer as n grows) to the values of the conjectured constants β. For 1-regular trees it corresponds to the constant β of relaxed K-d trees while for K-regular trees it is that of quad trees.

In Table 2 we show the numerical values of α and the average arity for several families of quad-K-d trees fixing $K = 6$ and varying the probability p or the value of m, whichever is appropriate, and the ratio $\rho = s/K$. We appreciate, that for fixed values of p and ρ, the average arity of Geometric quad-K-d trees is less than that of Pseudo-binomial quad-K-d trees which in turn is less than that of Binomial quad-K-d trees. Consequently, for the values of α, the inequalities go in reverse sense, since the higher the arity the lower the value of α. A similar phenomenon occurs for m-regular quad-K-d trees; while m decreases the average arity decreases and α grows.

Both Table 2 and Fig. 3 show a fact already known from the analysis of random PM queries in random K-d trees and quad trees, namely, when the ratio $\rho = s/K$ decreases the value of α decreases as well since, specifying more coordinates in the query, fewer times the PM search algorithm will follow more than one subtree of t.

All the programs used in the experiments were written in the C++ programming language and compiled with the GNU gcc compiler version v4.4.3. The experiments were run on a Pentium Genuine Intel x86_64 64-bit dual 32 K core processor.

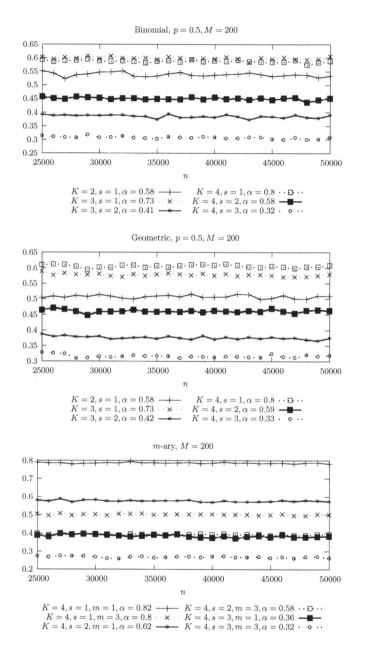

Fig. 2. Empirical estimates of α $(\log \bar{P}_n / \log n)$ as a function of n for Binomial Split (top), Geometric Split (middle) and m-regular (bottom)

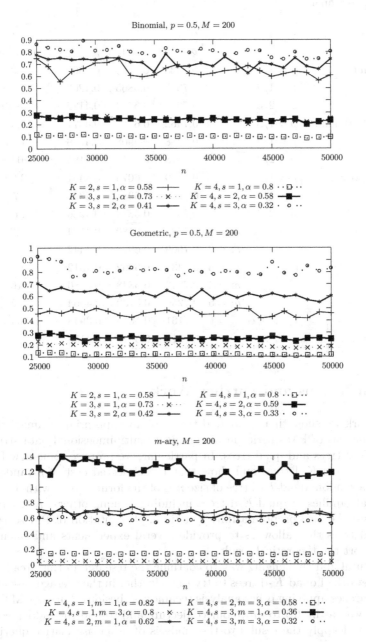

Fig. 3. Empirical normalized cost $(\bar{P}_n/n^{\alpha})$ of Binomial Split (top), Geometric Split (middle) and m-regular (bottom)

Table 2. Performance (exponent α, average degree $\overline{d}$) of Binomial Split, Geometric Split and m-regular

$K = 6$		α			$\overline{d}$
		$\rho = 1/3$	$\rho = 1/2$	$\rho = 2/3$	Average arity
Binomial	$p = 1/3$	0.7423	0.5955	0.4308	8.32
	$p = 1/2$	0.7357	0.5865	0.4209	15.18
	$p = 2/3$	0.7295	0.5784	0.4122	25.20
Pseudo Binomial	$p = 1/3$	0.7462	0.6	0.4371	5.62
	$p = 1/2$	0.7388	0.5908	0.4256	11.4
	$p = 2/3$	0.7318	0.5815	0.4155	20.92
Geometric	$p = 1/3$	0.7509	0.6076	0.4451	3.69
	$p = 1/2$	0.7442	0.5985	0.4348	7
	$p = 2/3$	0.7358	0.5874	0.4226	14.4
m-regular	K-d tree $m = 1$	0.7583	0.6180	0.4574	2
	$m = 2$	0.7509	0.6073	0.4442	4
	$m = 3$	0.743	0.5962	0.4312	8
	$m = 4$	0.7346	0.5848	0.4186	16
	$m = 5$	0.7256	0.5732	0.4064	32
	Quad tree $m = 6$	0.7161	0.5615	0.3948	64
Uniform		0.7313	0.5812	0.4156	21

4   Conclusions and Further Work

In this work we show that quad-K-d trees are an appropriate framework for a unified analysis of PM queries in hierarchical multidimensional data structures such as K-d trees and quad trees. In particular, we propose some new families of random quad-K-d trees and show that the expected cost of a random PM query in a random quad-K-d tree of size n is of the form $\Theta(n^\alpha)$, with $0 < \alpha < 1$, for several families of quad-K-d trees including, among others, K-d trees and quad trees. We also propose a fully generic implementation of PM searches in quad-K-d trees that allow us to provide several experiments and calculations that support the theoretical result.

A natural and challenging way to continue the research in this area is to use the framework of quad-K-d trees to try to go further in the average case analysis of PM searches applying other analysis techniques (besides Roura's CMT and in a similar way as in [6]) to prove our conjecture and to find the hidden constant factor β and apply the results to the analysis of other associative queries and data structures.

An additional way to continue research on quad-K-d trees is to explore their usefulness in practice by finding (or defining) benchmarks that would allow to decide, among the models tested here, which one is better in practical situations.

References

1. Bentley, J.L.: Multidimensional binary search trees used for associative searching. Commun. ACM **18**(9), 509–517 (1975)
2. Bentley, J.L., Finkel, R.A.: Quad trees: a data structure for retrieval on composite keys. Acta Informatica **4**, 1–9 (1974)
3. Bereczky, N., Duch, A., Németh, K., Roura, S.: Quad-kd trees: a general framework for kd trees and quad trees. Theor. Comput. Sci. **616**, 126–140 (2016). doi:10.1016/j.tcs.2015.12.030
4. Broutin, N., Neininger, R., Sulzbach, H.: A limit process for partial match queries in random quadtrees and 2-d trees. Ann. Appl. Probab. **23**(6), 2560–2603 (2013)
5. Chern, H.-H., Hwang, H.-K.: Partial match queries in random k-d trees. SIAM J. Comput. **35**(6), 1440–1466 (2006)
6. Chern, H.-H., Hwang, H.-K.: Partial match queries in random quadtrees. SIAM J. Comput. **32**(4), 904–915 (2003)
7. Cunto, W., Lau, G., Flajolet, P.: Analysis of kdt-trees: kd-trees improved by local reorganisations. In: Dehne, F., Sack, J.R., Santoro, N. (eds.) WADS 1989. LNCS, vol. 382, pp. 24–38. Springer, Heidelberg (1989)
8. Curien, N., Joseph, A.: Partial match queries in two-dimensional quadtrees: a probabilistic approach. Adv. Appl. Probab. **43**(1), 178–194 (2011)
9. Duch, A., Estivill-Castro, V., Martínez, C.: Randomized K-dimensional binary search trees. In: Chwa, K.-Y., Ibarra, O.H. (eds.) ISAAC 1998. LNCS, vol. 1533, pp. 199–208. Springer, Heidelberg (1998)
10. Duch, A., Lau, G., Martínez, C.: On the cost of fixed partial match queries in K-d trees. Algorithmica (2016). doi:10.1007/s00453-015-0097-4
11. Flajolet, P., Puech, C.: Partial match retrieval of multidimensional data. J. ACM **33**(2), 371–407 (1986)
12. Flajolet, P., Gonnet, G.H., Puech, C., Robson, J.M.: Analytic variations on quadtrees. Algorithmica **10**(6), 473–500 (1993)
13. Gaede, V., Günther, O.: Multidimensional access methods. ACM Comput. Surv. **30**(2), 170–231 (1998)
14. Martínez, C., Panholzer, A., Prodinger, H.: Partial match queries in relaxed multidimensional search trees. Algorithmica **29**(1–2), 181–204 (2001)
15. Roura, S.: Improved master theorems for divide-and-conquer recurrences. J. ACM **48**(2), 170–205 (2001)
16. Samet, H.: The Design and Analysis of Spatial Data Structures. Addison-Wesley, Reading (1990)

From Discrepancy to Majority

David Eppstein$^{(\boxtimes)}$ and Daniel S. Hirschberg

Department of Computer Science, University of California, Irvine, USA
david.eppstein@gmail.com

Abstract. We show how to select an item with the majority color from n two-colored items, given access to the items only through an oracle that returns the discrepancy of subsets of k items. We use $n/\lfloor \frac{k}{2} \rfloor + O(k)$ queries, improving a previous method by De Marco and Kranakis that used $n - k + k^2/2$ queries. We also prove a lower bound of $n/(k-1) - O(n^{1/3})$ on the number of queries needed, improving a lower bound of $\lfloor n/k \rfloor$ by De Marco and Kranakis.

1 Introduction

A large body of theoretical computer science research concerns problems of computing a function using a minimal number of calls to an oracle for another function on small subsets of input values. Such problems include sorting with a minimum number of comparisons, as well as *combinatorial group testing*, in which the goal is to identify the positions of a small set of true values among a larger number of false values using an oracle that returns the disjunction of an arbitrary subset of values [1,2]. Other problems with this flavor include Valiant's work on computing the majority of n values by shallow circuits of 3-input majority gates [3] and recent work by the authors using two-input disjunctions to identify a small number of slackers among a larger number of workers [4].

De Marco and Kranakis [5] provide another interesting class of such problems. Their input consists of n items, each having one of two colors. The goal is to select an item of the majority color or, if the input is equally balanced between colors, to report that fact rather than returning an item. The algorithm may only access the input by *counting* queries on k-item subsets of the input. If a subset has b black items and $w = k - b$ white items, then the result of the query is $c = \min(b, w)$, the size of the smaller of the two color classes. Equivalently, one may ask for the *discrepancy* $d = |b - w|$ of the query subset; the count can be calculated from the discrepancy or vice versa via the identity $2c + d = k$. The motivating application of De Marco and Kranakis is in fault diagnosis of distributed systems, which requires a majority of processors to be non-faulty. Their queries model tests that examine a small number of processors per test in order to determine whether the fault-free processors are indeed a majority.

The case $k = 2$ of this problem had been previously studied [6–8], and optimal bounds are known [6,8]. De Marco and Kranakis [5] provide more general

---

David Eppstein was supported in part by NSF grant CCF-1228639.

E. Kranakis et al. (Eds.): LATIN 2016, LNCS 9644, pp. 390–402, 2016.
DOI: 10.1007/978-3-662-49529-2_29

solutions that apply whenever k is sufficiently smaller than n. They show that it is possible to find a majority item for even k using only $n - k + k^2/2$ counting queries[1], and they prove a lower bound of $\lfloor n/k \rfloor$ on the number of queries that are necessary for this problem for all k.

The upper bound of De Marco and Kranakis for counting queries is greater than the lower bound by a factor of k in its leading term. In this work, we reduce this upper bound by a factor of approximately $k/2$ to $n/\lfloor \frac{k}{2} \rfloor + O(k)$, matching the lower bound to within a constant factor independent of k.

De Marco and Kranakis also considered a more powerful type of query, the *output model*, in which the answer to a query is a partition of the queried set into two monochromatic subsets (not revealing the colors of each subset). For this problem De Marco and Kranakis provided an upper bound of $\lceil (n-1)/(k-1) \rceil$ queries, and showed that the same $\lfloor n/k \rfloor$ lower bound for counting queries applies also to the output model. For odd k, we show that their upper bound is tight by proving a matching lower bound. For even k, we slightly improve their upper bound and prove a new lower bound that is within an additive $O(n^{1/3})$ lower-order term of the upper bound. Our new lower bounds apply both to counting queries and to the output model, and the $n/(k-1)$ leading terms of the new lower bounds improve the n/k leading term of the previously known bound.

Our results can also be interpreted in the framework of *discrepancy theory*, the study of how small the discrepancy of the sets in a set system can be made by choosing an appropriate 2-coloring of the set elements [9]. The first stage in our counting-query algorithm, finding an unbalanced query, is equivalent to constructing a system of k-element sets with discrepancy > 1, and our results for this stage provide examples of such unbalanced k-set systems.

1.1   Notational Conventions and Problem Statement

We use the following shorthand notation for sets:

- $[m]$ denotes the set $\{1, 2, \ldots, m\}$ of the first m positive integers.
- If S is a set, i is an element of S, and j is not an element of S, then S_i^j denotes the set $(S \setminus \{i\}) \cup \{j\}$. That is, we replace i by j in S.
- With the same conventions, if A is a subset of S and B is disjoint from S, then S_A^B denotes the set $(S \setminus A) \cup B$.
- If S and T are two sets of numbers with $|S| \geq |T|$, then $S \triangleleft T$ is the set formed from S by removing the $|T \setminus S|$ largest elements of $S \setminus T$ and replacing them by the elements of $T \setminus S$. The result is a set with the same size as S that forms a subset of $S \cup T$ and a superset of T. By abuse of notation, when t is a number, we write $S \triangleleft t$ as a shorthand for $S \triangleleft \{t\}$.

To avoid confusion with the equality predicate, we use the notation $x := y$ to indicate that a variable x of our algorithm should be assigned the new value y.

An instance of the majority problem may be parameterized by two values, n (the number of input items) and k (the size of queries), with $n > k$.

---

[1] There is a bug in their method for odd k, in Case 1 of Theorem 4.1, when $i = \lfloor k/2 \rfloor$.

We may represent an input to the problem by an n-tuple X of numbers x_i ($i \in [n]$) where each x_i is a member of the set $\{0, 1\}$. The argument to a query made by a majority-finding algorithm may be represented by a set $Q \subset [n]$ with $|Q| = k$. Then we may define the results of the input queries count and partition as

$$\text{count}(Q) = \min \left\{ \sum_{i \in Q} x_i, \sum_{i \in Q} (1 - x_i) \right\}$$
$$\text{partition}(Q) = \left\{ \{i \mid x_i = 0\}, \{i \mid x_i = 1\} \right\}.$$

By extension, we allow these functions to be applied to any set, not necessarily of cardinality k, with the same definitions.

For odd k it will be convenient to partition the set $[n]$ into two complementary subsets, M and L. M is the set of indices i whose associated values x_i equal the majority value of $[k]$. (This may differ from the majority of $[n]$.) Similarly, L is the set of indices i whose associated values x_i equal the minority value in $[k]$.

We say that a query set Q is *homogeneous* if all of its elements have the same value; that is, it is homogeneous when $\text{count}(Q) = 0$ and when $\text{partition}(Q) = \{\emptyset, Q\}$. We say that a query is *inhomogeneous* if it is not homogeneous. We say that a query set is *balanced* if it is equally partitioned between elements of the two values (or as near to equal as possible when k is odd). That is, Q is balanced when its discrepancy is at most 1 or when $\text{count}(Q) = \lfloor k/2 \rfloor$. We say that Q is *unbalanced* when it is not balanced.

1.2   New Results

We prove the following new results.

- A majority element may be found by making $n/\lfloor \frac{k}{2} \rfloor + O(k)$ count queries. The best previous bound, by De Marco and Kranakis [5], was $n - k + k^2/2$.
- When n is odd, a majority element may be found by making $\lceil (n-2)/(k-1) \rceil$ partition queries. This improves for some values of k the best previous upper bound, by De Marco and Kranakis [5], of $\lceil (n-1)/(k-1) \rceil$.
- Determining the majority element requires at least $\lceil (n-1)/(k-1) \rceil$ queries, for odd k, and $n/(k-1) - O(n^{1/3})$ queries, for even k, regardless of whether the queries are of type count or partition. The best previous lower bound for both these query types, by De Marco and Kranakis [5], was $\lfloor n/k \rfloor$.

In addition our methods prove the following discrepancy-theoretic result:

- For even k, there exists a family of at most $2 \log_2 k + 1$ sets, each having k elements, that cannot be 2-colored to make every set balanced. For odd k, there exists a family of at most $k + 3 \log_2 k + 4$ sets with the same property.

2   Upper Bounds for Counting

For our new upper bounds for counting we use an algorithm with the following four stages:

1. Find an unbalanced query U.
2. Use U to find a homogeneous query H.
3. Use H to determine count($[n]$).
4. Based on the value of count($[n]$), find the result of the majority problem.

We describe these four stages in the following four subsections.

2.1 Finding an Unbalanced Query

Throughout this section, when a subroutine discovers that a set U is unbalanced, we will abort the subroutine and its callers, and pass U on to the next stage of the algorithm. To indicate that this action is not simply returning to the subroutine's caller, we describe it using the Java-like pseudocode "throw U".

For even k, we do not need to find an unbalanced set, as our algorithm for finding a homogeneous set does not require it. However, the solution below serves as a warmup for the odd-k case. It maintains a homogeneous subset H of a balanced set B, repeatedly doubling H until it is too large to be a subset of a balanced set. To double H, we query a set B_H^Q; if it is balanced, then Q and H have the same composition and the doubled set $H \cup Q$ is homogeneous.

Subroutine 1 to find an unbalanced set when k is even:

1. Set $B := [k]$ and $H := \{1\}$.
2. Repeat the following steps:
   (a) If B is unbalanced, throw B.
   (b) Let Q be a set disjoint from B with $|Q| = |H|$.
   (c) If B_H^Q is unbalanced, throw B_H^Q.
   (d) Set $H := H \cup Q$ and then set $B := B \triangleleft H$.

Throughout the loop, H remains homogeneous, and doubles in size at each iteration. The loop terminates on or before the iteration for which $k/2 < |H| \leq k$, after at most $2 \log_2 k + 1$ queries, because substituting such a large homogeneous set into B will always produce an unbalanced set. Thus, $|H|$ cannot grow larger than k and cause B_H^Q to become undefined. For the subroutine to work correctly, we must have $n \geq 3k/2$ to ensure that a large enough subset Q disjoint from B can be chosen in step 2(b).

When k is odd we use a similar idea, doubling the size of a small unbalanced seed set until it overwhelms the whole set, but the details are more complicated. In the first place, the seed set for the doubling routine in the even case is always the set $\{1\}$, found without any queries, but in the odd case we choose our seed more carefully to have the form $\{j, j'\}$ where $\{j, j'\} \subset L$. To construct this seed, we choose j and j' to be arbitrary indexes disjoint from $[k]$ and then verify that they both belong to L by using the following subroutine:

Subroutine 2 star(j) (for $j > k$) verifies that $j \in L$ or finds an unbalanced set:

1. If $[k]$ is unbalanced, throw $[k]$.
2. For $i := 1, 2, \ldots (k+3)/2$, if $[k]_i^j$ is unbalanced, throw $[k]_i^j$.

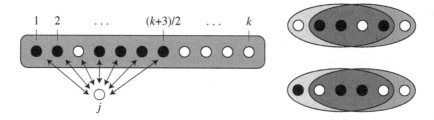

Fig. 1. Left: the arrows connect pairs of elements swapped into and out of the queries made by star(j). Right: if two overlapping queries (shown as ellipses) differ in a single element, and are both balanced, then either the two swapped elements have equal values (top) or they are unequal but both are in the majority for their query (bottom).

The subroutine name refers to the fact that the pairs (i, j) defining the queries form the edges of a star graph (Fig. 1, left).

Lemma 1. *If* star(j) *terminates without finding an unbalanced set, then* $j \in L$.

Proof. There are two different possible ways that the sets $[k]$ and $[k]_i^j$ queried by the algorithm can both be balanced (Fig. 1, right): either $x_i = x_j$ (the upper case in the figure), or $i \in M$ and $j \in L$ (the lower case). The first of these two possibilities, that $x_i = x_j$, can happen only for $\lceil k/2 \rceil$ choices of i, for otherwise too many of the members of $[k]$ would be equal to x_j (and each other) for $[k]$ to be balanced. However, star(j) tests a larger number of pairs than that. Therefore, if it tests all of these pairs and fails to find an unbalanced set, then it must be the case that $j \in L$. $\square$

We define a set S with even cardinality to be *L-heavy* if a majority of S belongs to L, and *L-balanced* if S is either balanced or L-heavy. Because we assume $|S|$ is even, an L-heavy set must contain at least $1 + |S|/2$ elements of L, and an L-balanced set must contain at least $|S|/2$ elements of L. The disjoint union of an L-heavy and an L-balanced set must itself be L-heavy, for if X and Y are disjoint with X containing at least $1 + |X|/2$ elements of L and $|Y|$ containing at least $|Y|/2$ elements of L, then $X \cup Y$ contains at least $1 + |X|/2 + |Y|/2 = 1 + |X \cup Y|/2$ elements of L. Our algorithm for the odd case of stage 1 depends on the following result, which lets us determine an L-heavy set of size double that of a previously known L-heavy set using $O(1)$ queries.

Lemma 2. *Suppose that* S *and* T *are sets disjoint from* $[k]$, *that* S *is L-heavy, that* $|S| = |T| \leq k$, *and that* $[k]$, $[k] \lhd S$, *and* $[k] \lhd T$ *are all balanced. Then* T *is necessarily L-balanced.*

Proof. Let U be the set of the largest $|S|$ elements of $[k]$; this is the subset of $[k]$ removed to make way for S in the set $[k] \lhd S$ (Fig. 2). For $[k]$ and $[k] \lhd S$ to be balanced, U can have at most one more member of M than S does; that is, U is L-balanced. Again, for $[k]$ and $[k] \lhd T$ to be balanced, T must have at least as many members of L as U does; therefore, T is also L-balanced. $\square$

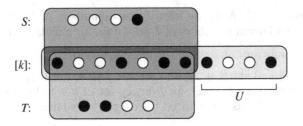

Fig. 2. The sets S (top), T (bottom), and U (middle right), and the query sets $[k]$ (yellow), $[k] \lhd S$ (red), and $[k] \lhd T$ (blue), used in the proof of Lemma 2 (Color figure online).

Based on Lemma 2, we define a second subroutine multiply(P, m) that transforms an L-heavy set P into a larger L-heavy set of cardinality $m|P|$. It takes as input an L-heavy set P, where P has even size and is disjoint from $[k]$, and a positive integer m with $m|P| \leq k$. It either finds an unbalanced set U (aborting the subroutine) or returns as output an L-heavy set of cardinality $m|P|$. We assume as a precondition for this subroutine that $[k]$ has already been determined to be balanced. The subroutine uses the binary representation of m to find its return value in a small number of doublings.

Subroutine 3 multiply(P, m) (where m and P are as described above) finds an unbalanced set or returns an L-heavy set disjoint from $[k]$ of size $m|P|$:

1. If $m = 1$, return P.
2. If $[k] \lhd P$ is unbalanced, throw $[k] \lhd P$.
3. Choose Q disjoint from both P and $[k]$ with $|Q| = |P|$.
4. If $[k] \lhd Q$ is unbalanced, throw $[k] \lhd Q$.
5. Set $R := $ multiply$(P \cup Q, \lfloor m/2 \rfloor)$.
6. If m is even, return R.
7. Choose S disjoint from R and from $[k]$ with $|S| = |P|$.
8. If $[k] \lhd S$ is unbalanced, throw $[k] \lhd S$.
9. Return $R \cup S$.

By Lemma 2, if multiply does not find an unbalanced set, then Q and S must both be L-balanced, and their disjoint union with an L-heavy set is another L-heavy set. Therefore, the set returned by this subroutine is L-heavy, and (by induction on the number of recursive calls) has the desired cardinality. The number of levels of recursion (counting only levels that can perform queries) is $\lfloor \log_2 m \rfloor$; at each level it performs either two or three queries, depending on whether m is even or odd. Therefore, in the worst case, it performs at most $3 \log_2 m$ queries.

Putting star and multiply together, we have the following algorithm to find an unbalanced set when k is odd. It uses star twice to find a two-element L-heavy set Y, then uses multiply to expand this set to an L-heavy set of $k-1$ elements. If this L-heavy set together with one element $i \in [k]$ remains unbalanced, it must be the case that $i \in M$. After we identify two members of M, we can replace them with the two known members of L to obtain an unbalanced set.

Subroutine 4 finds an unbalanced set when k is odd:

1. Call star$(k + 1)$ and star$(k + 2)$, and set $Y := \{k + 1, k + 2\}$.
2. Set $Z := \mathsf{multiply}(Y, (k - 1)/2)$, an L-heavy set of $k - 1$ elements.
3. If $Z \cup \{1\}$ or $Z \cup \{2\}$ is unbalanced, throw the unbalanced set.
4. Throw $[k]^Y_{\{1,2\}}$.

The two calls to star (after eliminating the shared query of set $[k]$) take a total of $k+4$ queries. The call to multiply takes at most $3(\log_2 k - 1)$ queries. The remaining steps of the algorithm use at most two queries. Therefore, the total number of queries made in this stage of the algorithm is at most $k + 3 \log_2 k + 3$. In order to work, this algorithm needs n to be at least $2k - 1$ so that it can find enough elements in the disjoint sets that it chooses.

For the algorithms in this stage, the sequence of queries made by the algorithm is non-adaptive: whenever a query finds an unbalanced set, the algorithm terminates, so the sequence of queries can be found by simulating the algorithm using an oracle that knows nothing about the input and always returns a balanced result. Eventually, the algorithm will determine that some particular set is unbalanced without querying it. The sequence of query sets together with the final unqueried and unbalanced set form a family of k-sets with the property that, no matter how their elements are colored, at least one set in the family will be unbalanced. This proves the following result:

Theorem 1. *When k is even, there exists a family of at most $2 \log_2 k + 1$ sets, each having k elements, that cannot be 2-colored to make every set in the family be balanced. When k is odd, there exists a family of at most $k + 3 \log_2 k + 4$ sets with the same property.*

These bounds are not tight for many values of k. When $k = 2 \pmod 4$, three k-sets with pairwise intersections of size $k/2$ cannot all be balanced. And for many odd values of k our bound can be improved by using optimal addition chains. However, such improvements would make our algorithms more complex and would affect only a low-order term of our overall analysis.

2.2   Finding a Homogeneous Query

After the previous stage of the algorithm, we have obtained an unbalanced query U. We may also assume that we know the result of the query count(U), for the algorithm of the previous stage will either query this number itself or it will find an unbalanced query U for which count(U) can be determined without making a query. Our algorithm for finding a homogeneous query is based on the principle that, for any two indices i and j with $i \in U$ and $j \notin U$, we can test whether $x_i = x_j$ in a single additional query, by testing whether count$(U^j_i) = \mathsf{count}(U)$. If $x_i = x_j$ then the count stays the same, clearly. However, with U unbalanced, it is not possible for the two indices to have different values while preserving the count.

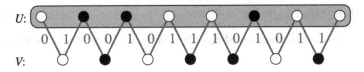

Fig. 3. Finding a homogeneous query. Given an unbalanced k-element query U (top, yellow), we find a disjoint set V of $k-1$ elements (bottom), and construct a spanning tree of the complete bipartite graph that has U and V as its two vertex sets (blue edges). We then query each set U_i^j for each spanning tree edge ij and use the result to label each edge 0 (if $x_i = x_j$) or 1 (otherwise). Any two elements of $U \cup V$ have the same value if and only if the spanning tree path connecting them has even label sum (Color figure online).

Subroutine 5 to find a homogeneous query:

1. Let V be a set of $k-1$ elements disjoint from U.
2. Construct a spanning tree T of the complete bipartite graph $K_{k,k-1}$ having U and V as the two sides of its bipartition.
3. For each edge (i,j) of T, with $i \in U$ and $j \in V$, query $\mathsf{count}(U_i^j)$. Label the edge with the number 1 if the query value is different from $\mathsf{count}(U)$ and instead label the edge with the number 0 if the two query values are equal.
4. Define two elements of $U \cup V$ to be equivalent when the path connecting them in T has even label sum, and partition $U \cup V$ into the two equivalence classes X and Y of the resulting equivalence relation.
5. Return a subset of k elements from the larger of the two equivalence classes.

The algorithm is illustrated in Fig. 3. This stage performs $2k-2$ queries and requires that $n \geq 2k-1$.

2.3   Finding the Count

We next use the known homogeneous query H to compute $\mathsf{count}([n])$.

Subroutine 6 to compute $\mathsf{count}([n])$, given a homogeneous set H:

1. Partition $[n] \setminus H$ into $(n-k)/\lfloor \frac{k}{2} \rfloor$ subsets $S_1, S_2, \ldots$, each having at most $\lfloor \frac{k}{2} \rfloor$ elements.
2. For each subset S_i of the partition, query $\mathsf{count}(H \triangleleft S_i)$. Since $S_i \leq k/2$ and the remaining elements of $H \triangleleft S_i$ are homogeneous, this query determines the number of elements of S_i that are not the same type as H.
3. Let c be the sum of the query values, and return $\min(c, n-c)$.

As well as computing $\mathsf{count}([n])$, the same algorithm can determine whether H is in the majority (according to whether c or $n-c$ is the minimum) and, if not, find an inhomogeneous query I for which $|H \cap I| \geq k/2$ (any of the queries with a nonzero query value). The number of queries it needs is

$$\frac{n-k}{\lfloor k/2 \rfloor} \leq \frac{n}{\lfloor k/2 \rfloor} - 2.$$

2.4   Finding the Majority

After the previous three stages of the algorithm, we have the following information:

– A homogeneous query H.
– The number count$([n])$.
– Whether the elements of H are in the majority.
– An inhomogeneous query I (if H is not in the majority), with $|H \cap I| \geq k/2$.

If count$([n]) = n/2$, we report that there is no majority. If H is a subset of the majority, we may return any element of H as the majority element. In the remaining case, we find an element of I that is not of the same type as the elements of H, using binary search:

Subroutine 7 uses binary search to find a majority element:

1. Let $U := I \setminus H$, a set containing an element not the same type as H.
2. Let $c := $ count(I), the number of majority elements in U already determined in stage three of the algorithm.
3. While $|U| > c$, do the following steps:
   (a) Let $V := $ any subset of $\lfloor |U|/2 \rfloor$ elements of U.
   (b) Query count$(H \triangleleft V)$.
   (c) If the result of the query is nonzero, let $U := V$ and let $c := $ the query result. Otherwise, let $U := U \setminus V$ and leave c unchanged.
4. Return any element of the remaining set U.

By induction, for a given set U, this algorithm uses at most $\lfloor 1 + \log_2(|U| - 1) \rfloor$ queries. The worst case occurs when $|U|$ is one plus a power of two and the query result is zero, resulting in a case of the same type in the next step. Since initially $|U| \leq k/2$, it follows that the total number of queries for this stage of the algorithm is less than $\log_2 k$. This bound can be improved by making a more careful choice of the set I to ensure that the initial values in the algorithm satisfy $c > |U|/2$, but this improvement is unnecessary for our results.

2.5   Counting Analysis

By adding together the numbers of queries made in the four stages of our algorithm we obtain the following result.

Theorem 2. *Let k and n be given integers with $n \geq 2k - 1$ and $k > 1$. Then it is possible to find a majority element of a set of n 2-colored elements, or to report that there is no majority, using at most $n/\lfloor \frac{k}{2} \rfloor + 3k + 4 \log_2 k$ count queries on subsets of k elements.*

In the full version of the paper we remove the constraint that $n \geq 2k - 1$ by providing substitute algorithms for the case that $k < n < 2k - 1$, using $O(k)$ queries.

3 Lower Bounds

In contrast to our upper bounds for counting queries, our lower bounds are simpler and tighter in the case that k is odd, so we begin with that case first.

Our lower bound for odd k uses partition queries, as they are the most powerful and can simulate count queries: if it is impossible to find the majority using a given number of partition queries, it is also impossible with the same number of queries of the other types. We prove our lower bound by an adversary argument: we design an algorithm for answering queries that, unless enough queries are made, will be able to force the querying algorithm into making a wrong choice of answer to the majority problem.

At any point during the interaction of the querying algorithm and adversary, we define the *query graph* to be a bipartite graph that has the n given set elements on one side of its bipartition and the queries made so far on the other side of the bipartition. We make each query be adjacent to the elements in it. As a shorthand, we use the word *component* to refer to a connected component of the query graph. The querying algorithm can be assumed to know the results of applying the partition and count functions to any subset of elements within a single component, as those results can be inferred from the queries actually performed within the component. Note also that, if any component C has discrepancy zero, the querying algorithm may safely ignore that component for the rest of the querying process, as removing its elements from the problem will not change the majority.

To simplify the task of the adversary, we restrict the querying algorithm to make only *reasonable queries*, which we define as queries that never include elements from components with zero discrepancy, and that (unless the result of the query leaves at most one nonzero-discrepancy component) never include more than one element from the same pre-query component. It follows from these properties that the querying algorithm must stop making queries, and choose an output for the majority problem, if it ever reaches a state where at most one component has nonzero discrepancy.

Lemma 3. *Any lower bound for an algorithm that makes only reasonable queries will be valid as a lower bound for all querying algorithms.*

Proof. An arbitrary querying algorithm can be transformed into one that makes only reasonable queries by skipping any query whose elements belong to one component, removing query elements that come from zero-discrepancy components or that duplicate the component of another element, and replacing the removed elements by elements from new components. This modification produces components that are supersets of the original ones, from which the results of the original queries can be inferred. □

By induction, with only reasonable queries for k odd, if more than one component remains, then all components have odd cardinality and therefore odd discrepancy. We design an adversary that maintains for each odd component a partition of its elements into two subsets (consistent with previous answers) that has discrepancy one. If a query produces a single component of even cardinality, we allow the adversary to choose any partition consistent with previous answers. If a query merges multiple discrepancy-one components, then (by choosing slightly more than half of the input components to have a majority that coincides with the majority of the merged component, and slightly fewer than half of the input components to have a majority that falls into the minority of the merged component) we can always find a consistent partition with discrepancy one. Therefore, by induction, the adversary can always achieve the goals stated above.

Lemma 4. *If a querying algorithm that makes reasonable queries does not reduce the input to a single component before producing its output, then the adversary described above can force it to compute an incorrect answer.*

Proof. Unless there is one component, more than one answer to the majority problem is consistent with the choices already made by the adversary.

In particular, if there are evenly many odd components of discrepancy one, then by choosing the majorities of all components to be the majority of the whole input, it is possible to cause the whole input to have a majority. But by choosing half of the components to have a majority of value 0 and half of the components to have a majority of value 1, it is also possible to cause the whole input to be evenly split between the two values and have no majority. Thus, regardless of whether the querying algorithm declares that there is no majority or whether it chooses a majority element, it can be made to be incorrect.

If there are an odd number of odd components, then a majority always exists. We may achieve discrepancy one for the whole input set of elements by choosing slightly more than half of the components to have majority value 1 and slightly fewer than half to have majority value 0; however, each component can be either on the majority 1 or majority 0 side, so each element can be either in the majority or in the minority. Regardless of which element the querying algorithm determines to belong to the majority, it can be made to be incorrect.   □

Theorem 3. *When k is odd, any algorithm that always correctly finds the majority of n elements by making* partition *or* count *queries must use at least $\lceil (n-1)/(k-1) \rceil$ queries.*

Proof. As above, the algorithm can be assumed to make only reasonable partition queries, and must make enough queries to reduce the query graph to a single component. This graph initially has n components, and each query reduces the number of components by at most $k-1$, from which the result follows.   □

De Marco and Kranakis showed that the majority problem on n elements may be solved using $\lceil (n-1)/(k-1) \rceil$ partition queries on subsets of k elements, matched by the lower bound of Theorem 3. For odd n, this bound may be improved to $\lceil (n-2)/(k-1) \rceil$ by applying it only to the first $n-1$ elements, and either returning the result (if it is a majority) or the final element (if the first $n-1$ elements have no majority). However, this modification to their algorithm can reduce the number of queries only when $k-1$ evenly divides

$n - 2$, which only happens when k is even. Therefore, this improvement does not contradict Theorem 3. When $k = 2$ a similar improvement can be continued recursively by pairing up elements, eliminating balanced pairs, and recursively finding the majority of a set of representative elements from each pair. The resulting algorithm uses $n - b$ queries, where b is the number of nonzero bits in the binary representation of n, and a matching lower bound is known [8]. Again, this does not contradict Theorem 3 because $k = 2$ is even. These improvements to the upper bound of De Marco and Kranakis raise the question of whether the majority can be found with significantly fewer queries whenever k is even. However, we show in the full version of the paper that the answer is no. An adversary strategy similar to the odd-k strategy but more complicated than it can be used to prove a lower bound of $n/(k - 1) - O(n^{1/3})$ on the number of queries.

4   Conclusions

We have provided new bounds for the majority problem, for count and partition queries. For partition queries with odd query size, our bounds are tight, and for even query size we achieve a matching leading term in our upper and lower bounds. However, for count queries, our upper and lower bounds are separated from each other by a factor of two. Reducing this gap remains open.

Recently, Gerbner et al. have given bounds for the majority problem for a different type of query that returns an element of the majority of a three-tuple [10]. It would be of interest to extend their results to k-tuples as well.

Our work also raises the discrepancy-theoretic question of how many sets are needed in a family of k-element sets that cannot be balanced. In this, also, our bounds are not tight and further improvement would be of interest.

References

1. Du, D.Z., Hwang, F.K.: Combinatorial Group Testing and its Applications. Ser. Appl. Math., vol. 12, 2nd edn. World Scientific, New York (2000)
2. Eppstein, D., Goodrich, M.T., Hirschberg, D.S.: Improved combinatorial group testing algorithms for real-world problem sizes. SIAM J. Comput. 36(5), 1360–1375 (2007)
3. Valiant, L.G.: Short monotone formulae for the majority function. J. Algor. 5(3), 363–366 (1984)
4. Eppstein, D., Goodrich, M.T., Hirschberg, D.S.: Combinatorial pair testing: distinguishing workers from slackers. In: Dehne, F., Solis-Oba, R., Sack, J.-R. (eds.) WADS 2013. LNCS, vol. 8037, pp. 316–327. Springer, Heidelberg (2013)
5. De Marco, G., Kranakis, E.: Searching for majority with k-tuple queries. Discrete Math. Algor. Appl. 7(2), 1550009 (2015)
6. Alonso, L., Reingold, E.M., Schott, R.: Determining the majority. Inform. Process. Lett. 47(5), 253–255 (1993)
7. Alonso, L., Reingold, E.M., Schott, R.: The average-case complexity of determining the majority. SIAM J. Comput. 26(1), 1–14 (1997)

 8. Saks, M.E., Werman, M.: On computing majority by comparisons. Combinatorica **11**(4), 383–387 (1991)
 9. Beck, J., Chen, W.W.L.: Irregularities of Distribution. Cambridge Tracts in Mathematics, vol. 89. Cambridge University Press, Cambridge (2008)
10. Gerbner, D., Keszegh, B., Pálvölgyi, D., Patkós, B., Vizer, M., Wiener, G.: Finding a majority ball with majority answers. In: Proceedings of the 8th European Conference on Combinatorics, Graph Theory, and Applications (EuroComb 2015). Elect. Notes Discrete Math., vol. 49, pp. 345–351. Elsevier (2015)

On the Planar Split Thickness of Graphs

David Eppstein[1], Philipp Kindermann[2], Stephen Kobourov[3],
Giuseppe Liotta[4], Anna Lubiw[5], Aude Maignan[6], Debajyoti Mondal[7]([✉]),
Hamideh Vosoughpour[5], Sue Whitesides[8], and Stephen Wismath[9]

[1] University of California, Irvine, USA
eppstein@uci.edu
[2] FernUniversität Hagen, Hagen, Germany
philipp.kindermann@fernuni-hagen.de
[3] University of Arizona, Tucson, USA
kobourov@cs.arizona.edu
[4] Università Degli Studi di Perugia, Perugia, Italy
giuseppe.liotta@unipg.it
[5] University of Waterloo, Waterloo, Canada
{alubiw,hvosough}@uwaterloo.ca
[6] Université Grenoble Alpes, Grenoble, France
aude.maignan@imag.fr
[7] University of Manitoba, Winnipeg, Canada
jyoti@cs.umanitoba.ca
[8] University of Victoria, Victoria, Canada
sue@uvic.ca
[9] University of Lethbridge, Lethbridge, Canada
wismath@uleth.ca

Abstract. Motivated by applications in graph drawing and information visualization, we examine the planar split thickness of a graph, that is, the smallest k such that the graph is k-splittable into a planar graph. A k-split operation substitutes a vertex v by at most k new vertices such that each neighbor of v is connected to at least one of the new vertices.

We first examine the planar split thickness of complete and complete bipartite graphs. We then prove that it is NP-hard to recognize graphs that are 2-splittable into a planar graph, and show that one can approximate the planar split thickness of a graph within a constant factor. If the treewidth is bounded, then we can even verify k-splittablity in linear time, for a constant k.

1 Introduction

Transforming one graph into another by repeatedly applying an operation such as vertex/edge deletion, edge flip or vertex split is a classic problem in graph theory [15]. In this paper, we examine graph transformations under the vertex split operation. Specifically, a *k-split operation* at some vertex v inserts at most k new vertices $v_1, v_2, \ldots, v_k$ in the graph, then, for each neighbor w of v, adds at least one edge (v_i, w) where $i \in [1, k]$, and finally deletes v along with its incident edges.

© Springer-Verlag Berlin Heidelberg 2016
E. Kranakis et al. (Eds.): LATIN 2016, LNCS 9644, pp. 403–415, 2016.
DOI: 10.1007/978-3-662-49529-2_30

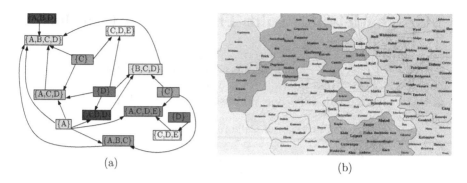

(a)   (b)

Fig. 1. (a) A 2-split visualization of subset relations among 10 sets. (b) Visualization of a social network. Note the 3 yellow clusters at the lower left of the map.

We define a *k-split* of graph G as a graph that is obtained by applying a k-split to each vertex of G at most once. We say that G is *k-splittable* into G^k. If $\mathcal{G}$ is a class of graphs, we say that G is *k-splittable* into a graph of $\mathcal{G}$ (or "*k-splittable* into $\mathcal{G}$") if there is a k-split of G that lies in $\mathcal{G}$. We introduce the $\mathcal{G}$ *split thickness* of a graph G as the minimum integer k such that G is k-splittable into a graph of $\mathcal{G}$.

Graph transformation via vertex splits is important in graph drawing and information visualization. For example, assume that we want to visualize the subset relation among a collection S of n sets. Construct an n-vertex graph G with a vertex for each set and an edge when one set is a subset of another. A planar drawing of this graph gives a nice visualization of the subset relation. Since the graph is not necessarily planar, a natural approach is to split G into a planar graph and then visualize the resulting graph, as illustrated in Fig. 1(a). Let's now consider another interesting scenario where we want to visualize a graph G of a social network, see Fig. 1(b). First, group the vertices of the graph into clusters by running a clustering algorithm. Now, consider the cluster graph: every cluster is a node and there is an edge between two cluster-nodes if there exists a pair of vertices in the corresponding clusters that are connected by an edge. In general, the cluster graph is non-planar, but we would like to draw the clusters in the plane. Thus, we may need to split a cluster into two or more sub-clusters. The resulting "cluster map" will be confusing if clusters are broken into too many disjoint pieces, which leads to the question of minimizing the planar split thickness.

Related Work. The problem of determining the planar split thickness of a graph G seems to be related to the graph thickness [1], empire-map [12] and k-splitting [15] problem. The *thickness* of a graph G is the minimum integer t such that G admits an edge-partition into t planar subgraphs. One can assume that these planar subgraphs are obtained by applying a t-split operation at each vertex. Hence, thickness is an upper bound on the planar split thickness, e.g., the thickness and thus the planar split thickness of graphs with treewidth ρ and maximum-degree-4 is at most $\lceil \rho/2 \rceil$ [5] and 2 [6], respectively. Analogously, the planar split thickness of a graph is bounded by its *arboricity*, that is, the minimum number of forests into which its edges can be partitioned. We will later show that both parameters also provide an asymptotic lower bound on the planar split thickness.

A *k-pire* map is a k-split planar graph, i.e., each *empire* consists of at most k vertices. In 1890, Heawood [11] proved that every 12 mutually adjacent empires can be drawn as a 2-pire map where each empire is assigned exactly two regions. Later, Ringel and Jackson [19] showed that for every integer $k \geq 2$ a set of $6k$ mutually adjacent empires can be drawn as a k-pire map. This implies an upper bound of $\lceil n/6 \rceil$ on the planar split thickness of a complete graph on n vertices.

A rich body of literature considers the planarization of non-planar graphs via *vertex splits* [7,10,15,16]. Here a *vertex split* is one of our 2-split operations. These results focus on minimizing the *splitting number*, i.e., the total number of vertex splits. Note that upper bounding the splitting number does not necessarily guarantee any good upper bound on the planar split thickness.

Knauer and Ueckerdt [13] studied the *folded covering number* which is equivalent to our problem and stated several results for splitting graphs into star forests, caterpillar forests, or interval graphs, e.g., planar graphs are 4-splittable into a star forest, and planar bipartite graphs as well as outerplanar graphs are 3-splittable into a star forest or a caterpillar forest. It follows from Scheinerman and West [20] that planar graphs are 3-splittable into interval graphs and 4-splittable into a caterpillar forest, while outerplanar graphs are 2-splittable into interval graphs.

Our Contribution. In this paper, we examine the planar split thickness for non-planar graphs. Initially, we focus on splitting the complete and complete bipartite graphs into planar graphs. We then prove that it is NP-hard to recognize graphs that are 2-splittable into a planar graph, while we describe a technique for approximating the planar split thickness within a constant factor. Finally, for bounded treewidth graphs, we present a technique to verify planar k-splittablity in linear time, for any constant k. Because our results are for planar k-splittability, we will drop the word "planar", and use "k-splittable" to mean "planar k-splittable".

2 Planar Split Thickness of K_n and $K_{m,n}$

In this section, we focus on the planar split thickness of K_n and $K_{m,n}$, and on graphs with maximum degree Δ.

2.1 Complete Graphs

Let $f(G)$ be the planar split thickness of the graph G. Recall that Ringel and Jackson [19] showed that $f(K_n) \leq \lceil n/6 \rceil$ for every $n \geq 12$. Since a $(n/6)$-split graph contains at most $n^2/2 - 6$ edges, and the largest complete graph with at most $n^2/2 - 6$ edges is K_n, this bound is tight. Besides, for every $n < 12$, it is straightforward to construct a 2-split graph of K_n by deleting $2(12 - n)$ vertices from the 2-split graph of K_{12}. Hence, we obtain the following theorem.

Theorem 1 (Ringel and Jackson [19]). *If $n \leq 4$, then $f(K_n) = 1$, and if $5 \leq n \leq 12$, then $f(K_n) = 2$. Otherwise, $f(K_n) = \lceil n/6 \rceil$.*

Let K_{12}^2 be any 2-split graph of K_{12}. Then, K_{12}^2 exhibits some useful structure, as stated in the following lemma.

Lemma 1. *Any planar embedding Γ of K_{12}^2 is a triangulation, where each vertex of K_{12} is split exactly once and no two vertices that correspond to the same vertex in K_{12} can appear in the same face.*

Proof. K_{12} has 66 edges. The 2-split operation produces a graph with at most twice the number of vertices and at least the original number of edges, so any graph K_{12}^2 has 24 vertices and 66 edges, since that is the largest number of edges for a 24-vertex planar graph by Eulers formula. Therefore, if K_{12}^2 is planar, it must be maximal planar, with all faces triangles. If two copies of the same vertex appear on a face, then those copies would not be adjacent and that face could not be a triangle.   □

Let H be the graph consisting of 2 copies of K_{12} attached at a common vertex v. Then, H provides an example of a graph that is not 2-splittable even though its edge count does not preclude its possibility of being 2-splittable.

Lemma 2. *The graph H is not 2-splittable.*

Proof. Consider a 2-split graph H' of one copy of K_{12}. By Lemma 1, the vertices v_1 and v_2 in H' that correspond to the same vertex in K_{12} cannot appear in the same face. Since v can be split only once, the 2-split graph H'' of the other copy of K_{12} must lie inside some face that is incident to either v_1 or v_2. Without loss of generality, assume that it is incident to some face incident to v_1. Note that both H' and H'' need a copy of v in some face which is not incident to v_1. Since both H' and H'' are triangulations, this would introduce a crossing in any 2-split graph of H.   □

2.2 Complete Bipartite Graphs

Hartsfield et al. [10] showed that the splitting number of $K_{m,n}$, where $m, n \geq 2$, is exactly $\lceil (m-2)(n-2)/2 \rceil$. However, their construction does not guarantee tight bounds on the splitting thickness of complete bipartite graphs. For example, if m is an even number, then their construction does not duplicate any vertex of the set A with m vertices, but uses $n + (m/2 - 1)(n - 2)$ vertices to represent the set B of n vertices. Therefore, at least one vertex in the set B is duplicated at least $(n + (m/2 - 1)(n - 2))/n = m/2 - m/n + 2/n \geq 3$ times, for $m \geq 6$ and $n \geq 5$. On the other hand, we show that $K_{m,n}$ is 2-splittable in some of these cases, as stated in the following theorem.

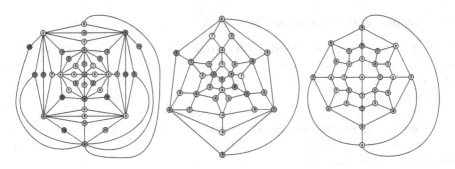

Fig. 2. The 2-split graphs of $K_{5,16}$, $K_{6,10}$ and $K_{7,8}$.

Theorem 2. *The graphs $K_{5,16}$, $K_{6,10}$, and $K_{7,8}$ are 2-splittable, and their 2-split graphs are quadrangulations, which implies that for complete bipartite graphs $K_{m,n}$, where $m = 5, 6, 7$, those are the largest graphs with planar split thickness 2.*

Proof. The sufficiency can be observed from the 2-split construction of $K_{5,16}$, $K_{6,10}$, and $K_{7,8}$, as shown in Fig. 2. A planar bipartite graph can have at most $2n - 4$ edges [10]. Since the graphs $K_{5,16}, K_{6,10}$ and $K_{7,8}$ contain exactly $4(m + n) - 4$ edges, their 2-split graphs are quadrangulations, which in turn implies that the result is tight. □

The following theorem gives a necessary condition for a complete bipartite graph to be k-splittable based on the edge count argument.

Theorem 3. *If $d \geq 4k + 4\sqrt{k^2 - 1}$ and $n > \frac{d - \sqrt{d^2 - 8kd + 16}}{2}$, then $K_{n,d-n}$ is not k-splittable.*

Proof. Note that any k-split graph H^k of $K_{n,m}$ must be a planar bipartite graph. Therefore, if p and q are the number of vertices and edges in H^k, respectively, then the inequality $q \leq 2p - 4$ holds.

Consider a complete bipartite graph $K_{n,d-n}$ that is k-splittable. The number of edges in this graph is $n \times (d - n)$. Since any k-split graph of $K_{n,d-n}$ can have at most kd vertices, we have

$$n(d - n) \leq 2kd - 4 \Leftrightarrow n^2 - nd + 2kd - 4 \geq 0 \tag{1}$$

The factorization of the previous polynomial (1) gives

$$n^2 - nd + 2kd - 4 = \left(n - \frac{d - \sqrt{d^2 - 8kd + 16}}{2} \right) \left(n - \frac{d + \sqrt{d^2 - 8kd + 16}}{2} \right),$$

when $d \geq 4k + 4\sqrt{k^2 - 1}$. Therefore, Eq. (1) holds if $n \leq \frac{d - \sqrt{d^2 - 8kd + 16}}{2}$ or $n \geq \frac{d + \sqrt{d^2 - 8kd + 16}}{2}$. □

2.3 Graphs with Maximum Degree Δ

Recall that the planar split thickness of a graph is bounded by its arboricity. By definition, any maximum-degree-Δ graph has degeneracy[1] at most Δ and, thus, arboricity at most Δ. Hence, the planar split thickness of a maximum-degree-Δ graph is bounded by Δ.

Moreover, since every 2-regular graph is planar, the planar split thickness of any graph with maximum degree Δ is bounded by $\lceil \Delta/2 \rceil$. Therefore, the planar split thickness of a maximum-degree-5 graph is at most 3. The following theorem states that this bound is tight.

Theorem 4. *For any nontrivial minor-closed property P, there exists a graph G of maximum degree five whose P split thickness is at least 3.*

Proof. This follows from a combination of the following observations:

1. There exist arbitrarily large 5-regular graphs with girth $\Omega(\log n)$ [17].
2. Splitting a graph cannot decrease its girth.
3. For every h, the K_h-minor-free n-vertex graphs all have at most $O(nh\sqrt{\log h})$ edges [21].
4. Every graph with n vertices, m edges, and girth g has a minor with $O(n/g)$ vertices and $m - n + O(n/g)$ edges [2].

Thus, let h be large enough that K_h does not have property P. If G is a sufficiently large n-vertex 5-regular graph with logarithmic girth (Observation 1), then any 2-split of G will have $2n$ vertices and $5n/2$ edges. By Observation 4, this 2-split will have a minor whose number of edges is larger by a logarithmic factor than its number of vertices, and for n sufficiently large this factor will be large enough to ensure that a K_h minor exists within the 2-split of G (by Observation 3). Thus, G cannot be 2-split into a graph with property P.   □

3 NP-Hardness and Approximation

Faria et al. [7] showed that determining the splitting number of a graph is NP-hard, even when the input is restricted to cubic graphs. Since cubic graphs are 2-splittable, their hardness proof does not readily imply the hardness of 2-splittable graph recognition. In this section, we show that it is NP-hard to recognize graphs that are 2-splittable into a planar graph. We then show that the arboricity of k-splittable graphs is bounded by $3k + 1$ and that testing k-splittability is fixed-parameter tractable in the treewidth of the given graph.

---

[1] A graph G is k-degenerate if every subgraph of G contains a vertex of degree at most k.

3.1   NP-Hardness of 2-Splittability

The reduction is from planar 3-SAT with a cycle through the clause vertices [14]. Specifically the input is an instance of 3-SAT with variables X and clauses C such that the following graph is planar: the vertex set is $X \cup C$; we add edge (x, c) if variable x appears in clause c; and we add a cycle through all the clause vertices. Kratochvíl et al. [14] showed that this version of 3-SAT remains NP-complete.

For our construction, we will need to restrict the splitting options for some vertices. For a vertex v, *attaching K_{12} to v* means inserting a new copy of K_{12} into the graph and identifying v with a vertex of this K_{12}. A vertex that has a K_{12} attached will be called a "K-vertex".

Lemma 3. *If C is a cycle of K-vertices then in any planar 2-split, the cycle C appears intact, i.e. for each edge of C there is a copy of the edge in the 2-split such that the copies are joined in a cycle.*

Proof. Let v be a vertex of cycle C. We will argue that the two edges incident to v in C are incident to the same copy of v in the planar 2-split. This implies that the cycle appears intact in the planar 2-split.

Suppose the vertices of C are $v = c_0, c_1, c_2, \ldots, c_t$ in that order, with an edge (v, c_t). As noted earlier in the paper, a planar 2-split of K_{12} must split all vertices, and no two copies of a vertex share a face in the planar 2-split. Furthermore, any planar 2-split of K_{12} is connected.

Let H_i be the induced planar 2-split of the K_{12} incident to c_i. Let v^1 and v^2 be the two copies of v in H_0. Suppose that the copy of edge (v, c_1) in the planar 2-split is incident to v^1. Our goal is to show that the copy of edge (v, c_t) in the planar 2-split is also incident to v^1. H_1 must lie in a face F of H_0 that is incident to v^1. Since there is an edge (c_1, c_2), H_2 must also lie in face F of H_0. Continuing in this way, we find that H_t must also lie in the face F. Therefore, the copy of the edge (c_t, v) must be incident to v^1 in the planar 2-split.   □

Note that the Lemma extends to any 2-connected subgraph of K-vertices.

Given an instance of planar 3-SAT with a cycle through the clause vertices, we construct a graph as follows. We will make a K-vertex c_j for each clause c_j, and join them in a cycle as given in the input instance. By the Lemma above, this "clause" cycle will appear intact in any planar 2-split of the graph.

Let T be any other cycle of K-vertices, disjoint from the clause cycle. T will also appear intact in any planar 2-split, so we can identify the "outside" of the cycle T as the side that contains the clause cycle. The other side is the "inside".

For each variable v_i, we create a vertex gadget as shown in Figs. 3(a)–(b) with six K-vertices: two special vertices v_i and $\bar{v}_i$ and four other vertices forming a "variable cycle" $v_i^1, v_i^2, v_i^3, v_i^4$ together with two paths v_i^1, v_i, v_i^3 and $v_i^2, \bar{v}_i, v_i^4$. Observe that, in an embedding of any planar 2-split, the vertex gadget will appear intact, and exactly one of v_i and $\bar{v}_i$ must lie inside the variable cycle and exactly one must lie outside the variable cycle. Our intended correspondence is that the one that lies outside is the one that is set to *true*.

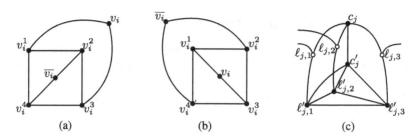

Fig. 3. (a) A variable gadget shown in the planar configuration corresponding to $v_i =$ *true* and (b) in the planar configuration corresponding to $v_i = $ *false*. (c) A clause gadget—a K_5 with added subdivision vertices $\ell_{j,1}, \ell_{j,2}, \ell_{j,3}$ corresponding to the literals in the clause. The half-edges join the corresponding variable vertices.

For each clause c_j with literals $\ell_{j,k}$, $k = 1, 2, 3$, we create a K_5 clause gadget, as shown in Fig. 3(c), with five K-vertices: two vertices c_j, c'_j and three vertices $\ell'_{j,k}$. Furthermore, we subdivide each edge $(c_j, \ell'_{j,k})$ by a vertex $\ell_{j,k}$ that is *not* a K-vertex. If literal $\ell_{j,k}$ is v_i, then we add an edge $(v_i, \ell_{j,k})$ and if literal $\ell_{j,k}$ is $\bar{v}_i$, then we add an edge $(\bar{v}_i, \ell_{j,k})$. Figure 4 shows an example of the construction.

Note that the only non-K-vertices are the $\ell_{j,k}$'s, which have degree 3 and can be split in one of three ways as shown in Figs. 5(a)–(c). In each possibility, one edge incident to $\ell_{j,k}$ is "split off" from the other two. If the edge to the variable gadget is split off from the other two, we call this the *F-split*.

Observe that if, in the clause gadget for c_j, all three of $\ell_{j,1}, \ell_{j,2}, \ell_{j,3}$ use the F-split (or no split), then we effectively have edges from c_j to each of $\ell'_{j,1}, \ell'_{j,2}, \ell'_{j,3}$, so the clause gadget is a K_5 which must remain intact after the 2-split and is not planar. This means that in any planar 2-split of the clause gadget, at least one of $\ell_{j,1}, \ell_{j,2}, \ell_{j,3}$ must be split with a non-F-split.

Lemma 4. *If the formula is satisfiable, then the graph has a planar 2-split.*

Proof. For every literal $\ell_{j,k}$ that is set to *false*, we do an F-split on the vertex $\ell_{j,k}$. For every literal $\ell_{j,k}$ that is set to *true*, we split off the edge to $\ell'_{j,k}$; see Fig. 5(b). For any K-vertex v incident to edges E_v outside its K_{12}, we split all vertices of the K_{12} as required for a planar 2-split of K_{12} but we keep the edges of E_v incident to the same copy of v, which we identify as the "real" v.

If variable v_i is set to *true*, we place (real) vertex v_i outside the variable cycle and we place vertex $\bar{v}_i$ and its dangling edges inside the variable cycle. If variable v_i is set to *false*, we place vertex $\bar{v}_i$ outside the variable cycle and we place vertex v_i and its dangling edges inside the variable cycle.

Consider a clause c_j. It has a true literal, say $\ell_{j,1}$. We have split off the edge from $\ell_{j,1}$ to $\ell'_{j,1}$ which cuts one edge of the K_5 and permits a planar drawing of the clause gadget as shown in Fig. 5(d), with $\ell'_{j,1}$ and its dangling edge inside the cycle $c', \ell'_{j,2}, \ell'_{j,3}$.

Because we started with an instance of planar 3-SAT with a cycle through the clause vertices, we know that the graph of clauses versus variables plus the

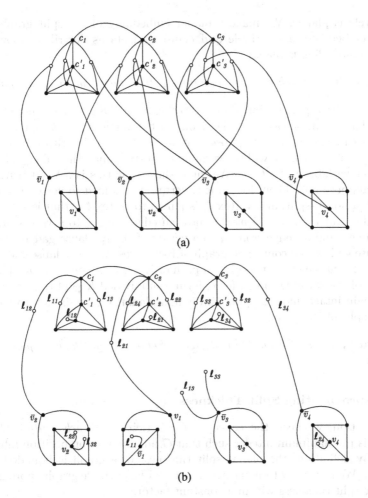

(a)

(b)

Fig. 4. (a) A graph that corresponds to the 3-SAT instance $\phi = (\bar{v}_1 \vee \bar{v}_2 \vee \bar{v}_3) \wedge (v_1 \vee v_2 \vee v_4) \wedge (v_2 \vee \bar{v}_3 \vee \bar{v}_4)$. (b) A planarization of the graph in (a) that satisfies ϕ: $v_1 = $ true, $v_2 = v_3 = v_4 = $ false

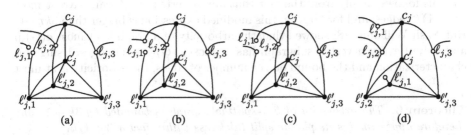

(a)   (b)   (c)   (d)

Fig. 5. (a)–(c) The three ways of splitting $\ell_{j,1}$; (a) is the F-split. (d) A planar drawing of the clause gadget when literal $\ell_{j,1}$ is set to *true* and the split of vertex $\ell_{j,1}$ results in a dangling edge to $\ell'_{j,1}$.

clause cycle is planar. We make a planar embedding of the split graph based on this, embedding the variable and clause gadgets as described above. The resulting embedding is planar.   □

Lemma 5. *If the graph has a planar 2-split, then the formula is satisfiable.*

Proof. Consider a planar embedding of a 2-split of the graph. As noted above, in each clause gadget, say c_j, at least one of the vertices $\ell_{j,k}$, $k = 1, 2, 3$, must be split with a non-F-split. Suppose that vertex $\ell_{j,k}$ is split with a non-F-split. If literal $\ell_{j,k}$ is v_i then we will set variable v_i to *true*; and if literal $\ell_{j,k}$ is $\bar{v}_i$ then we will set variable v_i to *false*. We must show that this is a valid truth-value setting. Suppose not. Then, for some i, vertex v_i is joined to vertex $\ell_{j,k}$ that is split with a non-F-split, and vertex $\bar{v}_i$ is joined to vertex $\ell_{r,s}$ that is split with a non-F-split. But then we essentially have an edge from v_i to a vertex of the c_j clause gadget and an edge from $\bar{v}_i$ to a vertex of the c_r clause gadget. Because each clause gadget is a connected graph of K-vertices, and the clause gadgets are joined by the clause cycle, this gives a path of K-vertices from v_i to $\bar{v}_i$. Then the 6 vertices of the variable gadget for v_i form a subdivided $K_{3,3}$ of K-vertices. This must remain intact under 2-splits and is non-planar. Contradiction to having a planar 2-split of the graph.   □

Theorem 5. *It is NP-hard to decide whether a graph has planar split thickness 2.*

3.2   Approximating Split Thickness

In this section, we need the concept of arboricity. The *arboricity* $a(G)$ of a graph G is the minimum integer such that G admits a decomposition into $a(G)$ forests. By definition, the planar split thickness of a graph is bounded by its arboricity. We now show that the arboricity of a k-splittable graph approximates its planar split thickness within a constant factor.

Let G be a k-splittable graph with n vertices and let G^k be a k-split graph of G. Since G^k is planar, it has at most $3kn - 6$ edges. Therefore, the number of edges in an n-vertex graph is also at most $(3k + 1)(n - 1)$: for n at most $6k$, this follows simply from the fact that any n-vertex graph can have at most $n(n-1)/2$ edges, and for larger n this modified expression is bigger than $3kn - 6$. But Nash-Williams [18] showed that the arboricity of a graph is at most $a(G)$ if and only if every n-vertex subgraph has at most $a(G)(n - 1)$ edges. Using this characterization and the bound on the number of edges, the arboricity is at most $3k + 1$.

Theorem 6. *The arboricity of a k-splittable graph is bounded by $3k + 1$, and therefore approximates its planar split thickness within factor $3 + 1/k$.*

Note that the thickness of a graph is bounded by its arboricity, and thus also approximates the planar split thickness within factor $3 + 1/k$.

3.3   Fixed-Parameter Tractability

Although k-splittability is NP-complete, we show in this section that it is solvable in polynomial time for graphs of bounded treewidth. The result applies not only to planarity, but to many other graph properties.

Theorem 7. *Let P be a graph property, such as planarity, that can be tested in monadic second-order graph logic, and let k and w be fixed constants. Then it is possible to test in linear time whether a graph of treewidth at most w is k-splittable into P in linear time.*

Proof. We use Courcelle's theorem [3], according to which any monadic second-order property can be tested for bounded-treewidth graphs in linear time. We modify the formula for P into a formula for the graphs k-splittable into P.

To do so, we need to be able to distinguish the two endpoints of each edge of our given graph G, within the modified formula. Thus, we wrap the formula in existential quantifiers for an edge set T and a vertex r, and we form the conjunction of the formula with the conditions that every partition of the vertices into two subsets is crossed by an edge, that every nonempty vertex subset includes at least one vertex with at most one neighbor in the subset, and that, for every edge e that is not part of T, there is a path in T starting from r whose vertices include the endpoints of e. These conditions ensure that T is a depth-first search tree of the given graph, in which the two endpoints of each edge of the graph are related to each other as ancestor and descendant; we can orient each edge from its ancestor to its descendant [4].

With this orientation in hand, we wrap the formula in another set of existential quantifiers, asking for k^2 edge sets, and we add conditions to the formula ensuring that these sets form a partition of the edges of the given graph. If we number the split copies of each vertex in a k-splitting of the given graph from 1 to k, then these k^2 edge sets determine, for each input edge, which copy of its ancestral endpoint and which copy of its descendant endpoint are connected in the graph resulting from the splitting.

Given these preliminary modifications, it is straightforward but tedious to modify the formula for P itself so that it applies to the graph whose splitting is described by the above variables rather than to the input graph. To do so, we need only replace every vertex set variable by k such variables (one for each copy of each vertex), expand the formula into a disjunction or conjunction of k copies of the formula for each individual vertex variable that it contains, and modify the predicates for vertex-edge incidence within the formula to take account of these multiple copies.   □

4   Conclusion

In this paper, we have explored the split thickness of graphs while transforming them to planar graphs. We have proved some tight bounds on the planar split thickness of complete and complete bipartite graphs. In general, we have proved that recognizing 2-splittable graphs is NP-hard, but it is possible to approximate the planar split thickness of a graph within a constant factor. Furthermore, if the treewidth of the input graph is bounded, then for any fixed k, one can decide k-splittability into planar graphs in linear time.

Splitting number has been examined also on the projective plane [9] and torus [8]. Hence, it is natural to study split thickness on different surfaces. We observed that any graph that can be embedded on the torus or projective plane is 2-splittable. For the projective plane, use the hemisphere model of the projective plane, in which points on the equator of the sphere are identified with the opposite point on the equator; then expand the hemisphere to a sphere with two copies of each point, and choose arbitrarily which of the two copies to use for each edge. For the torus, draw the torus as a square with periodic boundary conditions, make two copies of the square, and when an edge crosses the square boundary connect it around between the two squares.

Acknowledgments. Most of the results of this paper were obtained at the McGill-INRIA-Victoria Workshop on Computational Geometry, Barbados, February 2015. We would like to thank the organizers of these events, as well as many participants for fruitful discussions and suggestions. The first, fourth, sixth, and eighth authors acknowledge the support from NSF grant 1228639, 2012C4E3KT PRIN Italian National Research Project, PEPS egalite project, and NSERC respectively.

References

1. Beineke, L.W., Harary, F.: The thickness of the complete graph. Canad. J. Math. **14**(17), 850–859 (1965)
2. Borradaile, G., Eppstein, D., Zhu, P.: Planar induced subgraphs of sparse graphs. In: Duncan, C., Symvonis, A. (eds.) GD 2014. LNCS, vol. 8871, pp. 1–12. Springer, Heidelberg (2014)
3. Courcelle, B.: The monadic second-order logic of graphs. I. Recognizable sets of finite graphs. Inform. Comput. **85**(1), 12–75 (1990)
4. Courcelle, B.: On the expression of graph properties in some fragments of monadic second-order logic. In: Immerman, N., Kolaitis, P.G. (eds.) Proc. Descr. Complex. Finite Models. DIMACS, vol. 31, pp. 33–62. Amer. Math. Soc. (1996)
5. Dujmovic, V., Wood, D.R.: Graph treewidth and geometric thickness parameters. Discrete Comput. Geom. **37**(4), 641–670 (2007)
6. Duncan, C.A., Eppstein, D., Kobourov, S.G.: The geometric thickness of low degree graphs. In: Snoeyink, J., Boissonnat, J. (eds.) Proceedings of the 20th ACM Symposium on Computational Geometry (SOCG 2004). pp. 340–346. ACM (2004)
7. Faria, L., de Figueiredo, C.M.H., de Mendonça Neto, C.F.X.: Splitting number is NP-complete. Discrete Appl. Math. **108**(1–2), 65–83 (2001)
8. Hartsfield, N.: The toroidal splitting number of the complete graph K_n. Discrete Math. **62**, 35–47 (1986)
9. Hartsfield, N.: The splitting number of the complete graph in the projective plane. Graphs Comb. **3**(1), 349–356 (1987)
10. Hartsfield, N., Jackson, B., Ringel, G.: The splitting number of the complete graph. Graphs Comb. **1**(1), 311–329 (1985)
11. Heawood, P.J.: Map colour theorem. Quart. J. Math. **24**, 332–338 (1890)

12. Hutchinson, J.P.: Coloring ordinary maps, maps of empires, and maps of the moon. Math. Mag. **66**(4), 211–226 (1993)
13. Knauer, K., Ueckerdt, T.: Three ways to cover a graph. Arxiv report (2012). http:// arxiv.org/abs/1205.1627
14. Kratochvíl, J., Lubiw, A., Nesetril, J.: Noncrossing subgraphs in topological layouts. SIAM J. Discrete Math. **4**(2), 223–244 (1991)
15. Liebers, A.: Planarizing graphs - a survey and annotated bibliography. J. Graph Algor. Appl. **5**(1), 1–74 (2001)
16. de Mendonça Neto, C.F.X., Schaffer, K., Xavier, E.F., Stolfi, J., Faria, L., de Figueiredo, C.M.H.: The splitting number and skewness of $C_n \times C_m$. Ars Comb. 63 (2002)
17. Morgenstern, M.: Existence and explicit constructions of $q+1$ regular Ramanujan graphs for every prime power q. J. Comb. Theory, Ser. B **62**(1), 44–62 (1994)
18. Nash-Williams, C.: Decomposition of finite graphs into forests. J. London Math. Soc. **39**(1), 12 (1964)
19. Ringel, G., Jackson, B.: Solution of Heawood's empire problem in the plane. J. Reine Angew. Math. **347**, 146–153 (1984)
20. Scheinerman, E.R., West, D.B.: The interval number of a planar graph: Three intervals suffice. J. Comb. Theory, Ser. B **35**(3), 224–239 (1983)
21. Thomason, A.: The extremal function for complete minors. J. Comb. Theory, Ser. B **81**(2), 318–338 (2001)

A Bounded-Risk Mechanism for the Kidney Exchange Game

Hossein Esfandiari[1(✉)] and Guy Kortsarz[2]

[1] University of Maryland, College Park, USA
hossein@cs.umd.edu
[2] Rutgers University, Camden, USA

Abstract. In this paper we consider the *pairwise kidney exchange game*. This game naturally appears in situations that some service providers benefit from pairwise allocations on a network, such as the kidney exchanges between hospitals.

Ashlagi et al. [1] present a 2-approximation randomized truthful mechanism for this problem. This is the best known result in this setting with multiple players. However, we note that the variance of the utility of an agent in this mechanism may be as large as $\Omega(n^2)$, which is not desirable in a real application. In this paper we resolve this issue by providing a 2-approximation randomized truthful mechanism in which the variance of the utility of each agent is at most $2 + \epsilon$.

As a side result, we apply our technique to design a *deterministic* mechanism such that, if an agent deviates from the mechanism, she does not gain more than $2\lceil \log_2 m \rceil$.

1 Introduction

Kidney transplant is the only treatment for several types of kidney diseases. Since people have two kidneys and can survive with only one kidney, they can potentially donate one of their kidneys. It may be the case that a patient finds a family member or a friend willing to donate her kidney. Nevertheless, at times the kidney's donor is not compatible with the patient. These patient-donor pairs create a list of incompatible pairs. Consider two incompatible patient-donor pairs. If the donor of the first pair is compatible with the patient of the second pair and vise-versa, we can efficiently serve both patients without affecting the donors.

In this paper we consider pairwise kidney exchange, even though there can be a more complex combinations of transplantation of kidneys, that involves three or more pairs. Nevertheless, such chains are complicated to deal with in the real life applications since they need six or more simultaneous surgeries.

To make the pool of donor-patient pairs larger, hospitals combine their lists of pairs to one big pool, trying to increase the number of treated patients by exchanging pairs from different hospitals. This process is managed by some national supervisor. A centralized mechanism can look at all of the hospitals

H. Esfandiari and G. Kortsarz—Supported by NSF grant number 1218620.

E. Kranakis et al. (Eds.): LATIN 2016, LNCS 9644, pp. 416–428, 2016.
DOI: 10.1007/978-3-662-49529-2_31

together and increase the total number of kidney exchanges. The problem is that for a hospital its key interest is to increase the number of its own served patients. Thus, the hospital may not report some patient-donors pairs, namely, the hospital may report a partial list. This partial list is then matched by the national supervisors. Undisclosed set of pairs are matched by the hospitals locally, without the knowledge of the supervisor. This may have a negative effect on the number of served patients.

A challenging problem is to design a mechanism for the national supervisor, to convince the hospitals not to hide information, and report all of their pairs. In fact, if hiding any subset of vertices does not increases the utility of an agent, she has no intention to hide any vertex.

1.1   Notations and Definitions

To model this and similar situations hospitals are called *agents*, and each patient-donor pair is modeled by a vertex. Let m be the number of agents. Each agent owns a disjoint set of vertices. We denote the vertex set of the i-th agent by V_i and $\vec{V} = \{V_1, V_2, ..., V_m\}$ is called the vector of vertices of the agents. Denote an instance of the kidney exchange problem by $(G, \vec{V})$, where G is the underlying graph, $\vec{V}$ is the vector of vertices of the agents and E is the edge set. Each vertex in $G = (V, E)$ belongs to exactly one agent. Thus, $V = \cup_{i=1}^{m} V_i$ holds.

In this game, the utility of an agent i is the expected number of matched vertices in V_i and is denoted by u_i. Similarly, the utility of an agent i with respect to a matching M is the number of vertices of V_i matched by M and is denoted by $u_i(M)$. The social welfare of a mechanism is the size of the output matching.

A mechanism for the kidney-exchange game is the mechanism employed by the national supervisor to choose edges among the reported vertices. The process is a three step process. First the agent expose some of their vertices. Then the mechanism chooses a matching on the reported graph. Finally, each agent matches her unmatched vertices, including her non disclosed vertices, privately.

Formally, a kidney exchange mechanism F is a function from an instance of a kidney exchange problem $(G, \vec{V})$ to a matching M of G. The mechanism F may be randomized.

Given that some pairs are undisclosed, we say a kidney exchange mechanism F is α-approximation if for every graph G the number of matched vertices in the maximum matching of G is at most α times the expected number of matched vertices in $F(G)$. This means that for every graph G

$$\frac{|Opt(G)|}{E[|F(G)|]} \leq \alpha,$$

where $Opt(G)$ is the maximum matching in graph G, and the expectation is over the run of the mechanism F.

We define the notion of *bounded-risk* mechanisms as follow.

Definition 1. *A mechanism is a bounded-risk mechanism if the variance of the utility of each agent is is bounded by a constant.*

We say a kidney exchange mechanism is truthful if no agent has incentive to hide any vertex i.e., for each agent i, we have

$$\forall_{V_i' \in V_i} \quad u_i(F(G)) \geq u_i(F(G \backslash V_i')) + u_i(F(G \backslash V_i'), V_i')$$

where $u_i(F(G \backslash V_i'), V_i')$ is the [expected] number of vertices that agent i matches privately if she hides V_i'. We also define *almost truthful* mechanisms, to use for our side result.

Definition 2. *We say that a kidney exchange mechanism is almost truthful if by deviating from the mechanism, an agent have an additional gain of at most $O(\log m)$ vertices in the utility, where m is the number of agents.*

Indeed, in a real application finding the right subset of vertices to hide is costly. Roughly speaking, this cost involves extracting the information of m other agents, and hence, is an increasing function in m. Thus, in an almost truthful mechanism we hope that agents ignore gaining $O(\log m)$ vertices compare to the cost involved, and report the true information. Remark that, in this paper we do not consider a cost for deviating from the mechanism, and thus, we use two different definitions of truthful mechanisms and almost truthful mechanisms.

1.2   Related Work

The model considered in this paper was initiated by Sönmez and Ünver [9] and Ashlagi and Roth [2]. Sönmez and Ünver [9] show that there is no deterministic truthful mechanism that gets the maximum possible social welfare. See Fig. 1. In this example, the number of vertices is odd. Therefore, any mechanism that provide a maximum matching leaves exactly one vertex unmatched. Consider a mechanism that leaves a vertex of the first agent unmatched. In this case the utility of the first agent is 2. If this agent hides the fifth and the sixth vertices, any maximum matching matches the first vertex to the second vertex and the third vertex to the fourth vertex. Later, agent one matches the fifth and sixth vertices, privately. This increases the utility of the first agent to 3, and means that such a mechanism is not truthful. Similarly, if the mechanism leaves a vertex of the second agent unmatched, she can increases her utility by hiding the second and third vertices and matching them privately. This shows that a mechanism that always reports a maximum matching is not truthful.

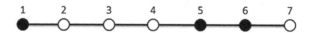

Fig. 1. Black vertices belong to the first agent and white vertices belong to the second agent.

Achieving social welfare optimal mechanisms, which are truthful, is thus not possible. However, achieving approximate truthful mechanisms may be possible. Ashlagi et al. [1] used the same example as in Fig. 1 to show that there is no deterministic truthful mechanism for the kidney-exchange game, with approximation ratio better than 2. Moreover, they show that there is no randomized truthful mechanism with an approximation ratio better than 8/7. They also introduce a deterministic 2-approximation truthful mechanism for the two player kidney exchange game and a randomize 2-approximation truthful mechanism for the multi-agent kidney exchange game. Later Caragiannis et al. [4] improved the approximation ratio for two agents to an expected 3/2-approximation truthful mechanism. It is conjectured that there is no deterministic constant-approximation truthful mechanism for the multi-agent kidney exchange game, even for three agents [1].

Almost truthful mechanisms has been widely studied (See [5–7]) with slightly different definitions. However, all use the concept that an agent should not gain more than small amount by deviating from the truthful mechanism.

1.3   Our Results

First, we show that the variance of the utility of an agent in the mechanism proposed by Ashlagi et al. [1] may be as large as $\Omega(n^2)$, where n is the number of vertices. The variance of the utility can be interpreted as the risk of the agent caused by the randomness in the mechanism. Indeed, in a real application agents prefer to take less risk for the same expected utility. In Sect. 2, we provide a tool to lower the variance of the utility of each agent in a kidney exchange mechanism while keeping the expected utility of each agent the same. The following theorem is an application of this tool to the mechanism proposed by Ashlagi et al. [1] low variance.

Theorem 1. *There exists a bounded-risk truthful 2-approximation mechanism for multi-agent kidney exchange. Specifically, in this mechanism the variance of the utility of each agent is at most $2 + \epsilon$, where ϵ is an arbitrary small constant.*

As a side result, in Sect. 3, we provide a derandomization of our mechanism. Specifically, we design an almost truthful deterministic 2-approximation mechanism for this problem. To the best of our knowledge this is the first non-trivial deterministic mechanism for the multi-agent kidney exchange game.

Theorem 2. *There exists an almost truthful deterministic 2-approximation mechanism for multi-agent kidney exchange.*

Remark that in a real application hiding vertices involves extracting the information about other agents and finding the right subset of vertices to hide, which is costly itself. Thus, in a real application, if the loss stemming from being truthful is *negligible*, it is likely that the hospital will absorb the small loss and remain truthful.

2   A Truthful Mechanism with Small Utility Variance

Ashlagi et al. in EC'10 [1] study the multi-agent kidney exchange game. They provide a polynomial time truthful 2-approximation mechanism called Mix and

Match. The Mix and Match mechanism is described as follows; independently label each agent either by 1 or 0 each with probability 0.5. Remove the edges between different agents with the same labels, i.e., for each edge $(u, v) \in E$, if u and v belongs to different agents and these agents have the same label, remove the edge (u, v) from G. Let G' be the new graph. Consider all matchings in G' that contain a maximum matching over the induced subgraph of each agent separately. Output the one with the maximum cardinality. Ties are broken serially in favor of agents with label 1 and then agents with labels 0. The following example shows that in this mechanism the variance of an agent utility may be as large as $\Omega(n^2)$.

Example 1. Consider a game with three agents. Each agent has $\frac{n}{3}$ vertices, where n is the number of vertices in the graph. There is a perfect matching with $\frac{n}{3}$ edges between vertices of agent 1 and agent 2 and there is no other edges (see Fig. 2).

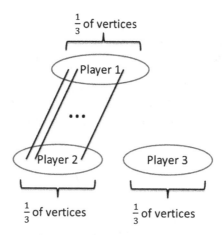

Fig. 2. In this example, the variance of the first agent utility, is $\Omega(n^2)$

In this example, with probability 0.5, agent 1 and agent 2 get the same label and all of the edges between these two agents are removed. In this case, all edges are removed and thus the utility of each agent is zero. However, with probability 0.5, agent 1 and agent 2 get different labels and we have a matching of size $\frac{n}{3}$ between the vertices of these two agents. In this case, the utility of agent 1 is $\frac{n}{3}$. Therefore, the variance of the first agent utility is

$$\sigma^2 = 0.5(0 - \frac{n}{6})^2 + 0.5(\frac{n}{3} - \frac{n}{6})^2 = \frac{n^2}{36},$$

which is $\Omega(n^2)$.

Our mechanism uses a randomized truthful mechanism as a core mechanism. We take two matchings resulting from two independent runs of the core mechanism. These two matchings are combined into a new matching. The way we choose the new matching is randomized too. The new matching preserves the expected utility of each agent, and in addition, decreases the variance of the utility of each agent by a constant factor. This gives a mechanism with a lower utility-variance. We repeat this procedure iteratively (See Fig. 3) and decrease the variance of the utilities to $O(1)$. We show that for this purpose it is enough to apply the combination of two matchings mechanism a logarithmic number of times. For the purpose of this section, we use Mix and Match as the core mechanism and show that there exists a truthful 2-approximation mechanism such that the variance of each utility is at most $2 + \epsilon$, where ϵ is an arbitrary small constant.

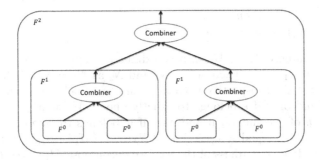

Fig. 3. Hierarchy of mechanisms

One can think of our mechanism as a multi-layered mechanism. The layer-0 mechanism is the core mechanism. In the i-th layer we combine two outputs of the layer $i - 1$ mechanism. Lemma 2 shows that we can use the lower layer mechanism and create a mechanism where the variance of the utilities is almost halved.

Note that the utility of an agent can be completely different for two matchings M_1 and M_2. Let $M_1 \oplus M_2$ denote the symmetric difference of M_1 and M_2. There may be a path from u to v in $M_1 \oplus M_2$ in which u is a vertex of agent i and v is a vertex of agent j and $i \neq j$. One of the two matchings will have a utility smaller by 1 for agent i. As the number of such paths in $M_1 \oplus M_2$ may be very large, the difference in utility of an agent with respect to M_1 and M_2 can be very large. We show how to find two matchings N_1 and N_2 such that the utility of each agent with respect to these two matchings is almost equal.

Let $(G, \vec{V})$ be an instance of the kidney exchange graph. Consider two matchings M_1 and M_2 derived by independent runs of the previous layer. Let $P = \{p_1, p_2, \ldots, p_k\}$ be a subset of distinct paths in $M_1 \oplus M_2$ (ignoring cycles).

Definition 3. *The* contraction graph $Cont((G, \vec{V}), P)$ *is defined as follows:*

- *Each vertex in $Cont((G, \vec{V}), P)$ corresponds to one agent in $(G, \vec{V})$.*
- *There is an edge in $Cont((G, \vec{V}), P)$ between agent i and j if and only if there is a path in P that begins with a vertex of agent i and ends with a vertex of agent j.*

We call this graph the contraction graph *because the paths are replaced by edges. When the instance of the kidney exchange game is clear from the context, we drop $(G, \vec{V})$ from the notation of the contraction graph.*

The following lemma proves that any two matchings can be transformed into two other matchings such that for every agent i the utility of agent i in the two new matchings has a difference of at most 2.

Lemma 1. *Let M_1 and M_2 be two matchings of graph G. There exist two matchings N_1 and N_2 such that for any agent i we have*

- $|u_i(N_1) - u_i(N_2)| \leq 2$ *and*
- $u_i(N_1) + u_i(N_2) = u_i(M_1) + u_i(M_2)$.

Proof. We decompose $M_1 \oplus M_2$ into two different matchings N_1' and N_2' such that $N_1' \cup N_2' = M_1 \oplus M_2$. Then define N_1 and N_2 as $N_1 = N_1' \cup (M_1 \cap M_2)$ and $N_2 = N_2' \cup (M_1 \cap M_2)$ respectively. Clearly, an edge e belongs to exactly one of N_1 or N_2, if and only if e belongs to exactly one of M_1 or M_2. In addition, e belongs to *both* of N_1 and N_2 if and only if it belongs to *both* of M_1 and M_2. This means that the equality $u_i(N_1) + u_i(N_2) = u_i(M_1) + u_i(M_2)$ holds for all agents. It holds true, regardless of the way we decompose $M_1 \oplus M_2$ into N_1' and N_2'. We now describe our approach to achieve the main property, namely, change $M_1 \oplus M_2$ into two matchings N_1' and N_2' such that $N_1' \cup N_2' = M_1 \oplus M_2$ and $|u_i(N_1') - u_i(N_2')| \leq 2$.

Consider the subgraph induced in G by $M_1 \oplus M_2$. The degrees of vertices in this graph are either zero or one or two. There are three types of connected components: cycles, even-length paths and odd-length paths. We explain how to decompose the different parts of $M_1 \oplus M_2$ in these three cases. Every path p decomposes into two matchings M_1^p and M_2^p. In any such decompositions, all the vertices of p except the endpoints are covered by both M_1^p and M_2^p.

- **Case 1: Components that are cycles.** Each of these cycles is the union of two matchings. It means that every other edge in each cycle belongs to one of the matchings. We add one of these matchings to N_1' and we add the other one to N_2'. Since these two matchings cover the same set of vertices, they have the same effect on the utility of agents.
- **Case 2: Edges of even length paths.** Let p be an even size path between two vertices v and u. Let M_1^p and M_2^p be a decomposition of p into matching such that M_1^p covers all vertices in p except u and M_2^p covers all vertices in p except v. For each path, we add one of M_1^p and M_2^p to N_1' and add the other one to N_2'. However, the assignment cannot be arbitrary. The assignment is derived by performing computations on the contraction graph. We represent the selection of M_1^p by directing the edge in the contraction graph from the

agent that contains u to the agent that contains v. Note that we deal with all even paths *simultaneously*. The difference of the outgoing and in-going degrees of each agent exactly equals the difference of her utilities caused by edges of even length paths in N_1' and N_2'. Thus, we just need to direct edges of the contraction graph of even length paths, so as to minimize this difference. We can direct the edges of this graph such that for each vertex the difference of outgoing and in-going edges is at most one. This is done by adding a matching between the odd degree vertices, directing the edges through an Eulerian cycle and removing the added edges. We adopt this strategy to get our almost balanced in and out degree pair of matchings, derived from all even sized paths in $M_1 \oplus M_2$.

- **Case 3: Odd length paths.** Let p be a path between vertices v and u which has an odd number of edges. We can decompose p into two matchings M_1^p and M_2^p such that M_1^p covers all vertices in p and M_2^p covers all vertices in p except v and u. Again in this case, for each path, we add one of M_1^p or M_2^p to N_1' and add the other one to N_2'.

In one of N_1' and N_2', both endpoints are matched and in the other none of the two endpoints are matched.

We represent the selection of M_1^p by coloring the edge corresponding to p blue. Otherwise, we color the edge red. Let the blue (red) degree of a vertex be the number of blue (red) edges touching the vertex. The difference between the red and blue degrees of each agent, exactly equals the difference in her utilities caused by odd length paths in $M_1 \oplus M_2$.

We can color the edges of any arbitrary graph with blue and red such that for each vertex the difference between the red and blue degree is at most 2. This again is done by adding a matching of dummy edges between the odd degree vertices, and coloring every second edge in the Eulerian cycle red and every other second edge blue. Then we remove the fake edges added at the beginning. Note that the start vertex of the cycle may be touched by two red or two blue, edges. On the other hand, the other vertices in this cycle are touched by the same number of blue and red edges. We use the following rule: if the start vertex has a dummy edge, we use this edge as the first in the Euler cycle. This is done such that the difference of blue and red degrees will not accumulate to 3 (2 may be added to the difference due to the fact that the path starts and ends in the same color, and an additional 1 can be added to the difference when we take the dummy edge out). This clearly implies a difference of 2 in the utility of any agent with respect to the new matchings.

If we combine the matchings of case 2 and case 3, the difference of the utilities for any agent with respect to N_1' and N_2' may grow up to at most 3. We want to avoid this situation. Let i be some agent for which the utility difference is 3. The difference of the utilities is derived as follows:

- Agent i has a difference of two between the number of red edges and blue edges. Note that this means that the other agents have a difference of at most 1 between the number of red and blue edges, as the agents are not start vertices of the cycle.

– Agent i has a difference of one between the number of outgoing and in-going edges. This means that the vertex is an odd degree vertex.
– The effect of these two differences accumulate and cause a difference of three between N_1' and N_2'.

In this case, we flip the color of edges in the component that contains i. This decreases *for* i the difference of N_1' and N_2' from 3 to 1. Note that any vertex that was not a start vertex of the Euler cycle has a difference of at most 1 in the edge coloring stage. The flipping of colors still implies that the maximum difference in the utility of every agent that is not a beginning of a cycle, is at most 1. Together with the difference caused by even length paths, the total difference is at most 2. Note that cycles of two different agents with difference 2 in their utility, are disjoint. Only one agent with two red or two blue edges can exist in every connected component. □

The following lemma uses Lemma 1 to combine outcomes of two independent runs of a mechanism.

Lemma 2. *Let F be a mechanism for the multi-agent kidney exchange game and let x_i be the random variable of the utility of agent i in the mechanism F. Then there exist a mechanism F^* such that for every input graph and every agent i the following holds:*

$$\mathrm{Var}(y_i) \leq \frac{\mathrm{Var}(x_i)}{2} + 1, E[y_i] = E[x_i],$$

where y_i is the random variable that indicates the utility of agent i in mechanism F^*.

Proof. We run mechanism F two times independently. Let M_1 and M_2 be the random matchings resulting from these two runs. We apply Lemma 1 on M_1 and M_2 and let N_1 and N_2 be the resulting matchings. The mechanism F^* chooses one of the two matchings N_1 and N_2 uniformly at random. Note that:

$$E[y_i] = \frac{E[u_i(N_1)] + E[u_i(N_2)]}{2} = \frac{E[u_i(M_1)] + E[u_i(M_2)]}{2} = \frac{2E[x_i]}{2} = E[x_i]$$

where the second equality is an application of Lemma 1. This means that each agent has the same expected utility in F and F^*. Now, we need to bound the variance of the utilities of the agents in F^*. Let D_i be the difference between $u_i(N_1)$ and the average of $u_i(M_1)$ and $u_i(M_2)$. Note that $u_i(N_1) = \frac{u_i(M_1)+u_i(M_2)}{2} + D_i$ and $u_i(N_2) = \frac{u_i(M_1)+u_i(M_2)}{2} - D_i$. Let I be the random variable that indicates whether F^* reports N_1 or not. Thus, we have

$$\text{Var}(y_i) \tag{1}$$

$$=\text{Var}\left(\frac{u_i(M_1) + u_i(M_2)}{2} + ID_i\right)$$

$$=\text{Var}\left(\frac{u_i(M_1) + u_i(M_2)}{2}\right) + \text{Var}\left(ID_i\right) + 2Cov\left(\frac{u_i(M_1) + u_i(M_2)}{2}, ID_i\right)$$

$$=\frac{\text{Var}(u_i(M_1))}{4} + \frac{\text{Var}(u_i(M_2))}{4} + \text{Var}(ID_i) + 2Cov\left(\frac{u_i(M_1) + u_i(M_2)}{2}, ID_i\right)$$

$$=\frac{\text{Var}(x_i)}{2} + \text{Var}(ID_i) + 2Cov\left(\frac{u_i(M_1) + u_i(M_2)}{2}, ID_i\right)$$

$$=\frac{\text{Var}(x_i)}{2} + \text{Var}(ID_i). \tag{2}$$

It remains to bound $\text{Var}(ID_i)$. From Bhatia-Davis Inequality [3] for any random variable X: $\text{Var}(X) \le (Sup(X) - \mu)(\mu - Inf(X))$, where μ is the expected value of X, $Sup(X)$ is the supremum of X and $Inf(X)$ is the infimum of X. By applying Bhatia-Davis Inequality to ID_i, we have

$$\text{Var}(ID_i) \le (Sup(ID_i) - \mu)(\mu - Inf(ID_i)) \le (1 - \mu)(\mu + 1) \le 1$$

where the second inequality is by definition of D_i and Lemma 1. Combining this with inequality (2), gives us $\text{Var}(y_i) \le \frac{\text{Var}(x_i)}{2} + 1$ as desired.   □

Lemma 2 provides a way to decrease the variance of the utilities, iteratively. New we are ready to prove Theorem 1.

Proof (of Theorem 1).

Let F^0 be the Mix and Match mechanism and let F^i be the combination of two independent runs of F^{i-1} from Lemma 2. We call the mechanism a *multi-layered* mechanism and F^i denoted i-th layer mechanism in the multi-layered mechanism. It is easy to see that F^k is a combination of 2^k independent runs of the Mix and Match mechanism which is the layer 0 mechanism. Recall that all those combinations preserve the expected utility of every agent. Thus, this process preserves the social welfare function, which is the sum of utilities of all of the agents. Thus, the assumption that F^0 is a 2-approximation mechanism immediately gives us that F^k is a 2-approximation mechanism, for any k.

Now, we show that for any k, F^k is truthful. We prove this by contradiction. Suppose that F^k is not truthful. Without loss of generality we assume that if agent 1 deviates, her expected utility increases. Since, F^k preserves the expected utility of each agent, this deviation should increase the expected utility of this agent all the way back to F^0. However, this contradicts the truthfulness of F^0. Therefore, F^k is truthful.

Now, we need to bound the variance of the utility of each agent. Without loss of generality, we fix an agent i and bound the variance of the utility of that agent. The same bound holds for all agents, by symmetry.

Let σ^2 be the variance of the utility of the agent in F^0. We prove by induction that the variance of the utility of this agent in F^k is $\frac{\sigma^2}{2^k} + 2 - \frac{2}{2^k}$. The base case is clear since $\frac{\sigma^2}{2^0} + 2 - \frac{2}{2^0} = \sigma^2$. Assume the bound for F^k and then we prove it for F^{k+1}. By applying Lemma 2, for F^{k+1} the variance of the utility of the agent is at most

$$\frac{\frac{\sigma^2}{2^k} + 2 - \frac{2}{2^k}}{2} + 1 = \frac{\sigma^2}{2^{k+1}} + 1 - \frac{1}{2^k} + 1 = \frac{\sigma^2}{2^{k+1}} + 2 - \frac{2}{2^{k+1}}.$$

This completes the induction. The utility of an agent cannot exceed the total number of vertices. Thus, we have $\sigma^2 \leq n^2$. If we set k to $2\log(n) + log(\frac{1}{\epsilon})$, then the variance of the utility of each agent in F^k is at most

$$\frac{\sigma^2}{2^k} + 2 - \frac{2}{2^k} = \frac{\sigma^2}{2^{2log(n)+log(\frac{1}{\epsilon})}} + 2 - \frac{2}{2^{2log(n)+log(\frac{1}{\epsilon})}}$$
$$= \frac{\sigma^2 \epsilon}{n^2} + 2 - \frac{2\epsilon}{n^2} \leq \epsilon + 2 - \frac{2\epsilon}{n^2} \leq 2 + \epsilon.$$

We note that the running time is polynomial. Indeed, running F^k, combines 2^{k-i} instances of F^0, mechanisms. Since we set k to $2\log n + log(\frac{1}{\epsilon})$, F^k operates on $\frac{n^2}{\epsilon}$ instances of F^0 and contains $\frac{n^2}{\epsilon} - 1$ combinations of matchings in higher levels. Since here both F^0 and the combination process runs in polynomial in n, F^k runs polynomial time as well. □

3   An Almost Truthful Deterministic Mechanism

In some applications, agents may not accept any risk. In this section, we modify our randomized mechanism to a deterministic one. This deterministic mechanism is not truthful anymore. However, it is almost truthful i.e., by deviating from the mechanism, an agent may gain an additive factor of at most $2\lceil \log_2(m) \rceil$, where m is the number of agents.

The analysis of the Mix and Match mechanism does not use the property that the labels of agents are fully independent. It just uses the fact that for every two fixed agents i and j, with probability 0.5, we assign different labels to agents i and j. In fact, this holds even if we use m *pairwise independent* random bits. We can generate m pairwise independent random bits using $\lceil \log_2(m) \rceil$ fully independent random bits [8]. We call this modified mechanism that just uses $\lceil \log_2(m) \rceil$ random bits, Modified Mix and Match.

Proof (of Theorem 2). For simplicity of notation, we replace $\log_2(.)$ by $\log(.)$. Our deterministic mechanism can be seen as a multi-layered mechanism defined as follows. In Layer 0, we run the modified Mix and Match mechanism over all possible values of the $\lceil \log(m) \rceil$ random bits. The collection of all resulted matchings is called *layer 0 matchings*. Note that the number of matchings for layer 0 is at most $2^{\lceil \log(m) \rceil} \leq 2m$. We now describe $\lceil \log(m) \rceil$ steps to combine these matchings together, into a single matching. We note that each layer will halve the number of matchings, and so $\lceil \log(m) \rceil$ applications of the mechanism give a layer with a single matching and this matching is our output. After the i-th step we inductively construct the matchings of the $i+1$-th layer as follows. We decompose the matchings in the i-th layer into arbitrary pairs of matchings. We use the procedure of Lemma 2 on every pair. Unlike the randomized version, here we always output the matching between N_1 and N_2 that has the largest number of edges. Clearly, in each step, the number of matchings in the layer is halved. Thus, after $\lceil \log(m) \rceil$ steps we have exactly one matchings. This matching is the output of our mechanism.

This mechanism contains at most $2\,m$ runs of the modified Mix and Match mechanism and at most $2\,m$ combinations of such matchings. Both the modified Mix and Match mechanism and the combination procedure run in polynomial time. Thus, this mechanism is a polynomial time mechanism.

Note that the average number of edges in the 0-th layer is exactly equal to the expected social welfare in the modified Mix and Match mechanism. This follows because every labeling among the $2^{\lceil \log(m) \rceil}$ is equally likely. In each step, we replace each pair with one of the matching obtained from Lemma 2. Selecting the matching between N_1 and N_2 that contains the largest number of edges, combined with the second property of Lemma 2, implies that the average number of edges cannot decrease when we go to the next layer. Thus, the number of edges in the last-layer matching is at least that of the modified Mix and Match mechanism. Thus, this mechanism is a 2-approximation mechanism. We now discuss individual utilities. The average utility of every agent in layer 0 exactly equals her expected utility in the modified Mix and Match mechanism. Using the first property of Lemma 2, it is clear that for a fixed agent, the difference between its modified Mix and Match utility, and the average of her utilities in the 1-th layer is at most 1. This holds true each time we go from one layer to the next. Namely, for every agent, the difference in utility between the modified Mix and Match strategy and our strategy goes up by at most 1. Thus, in the i-th layer the difference between the utility in our deterministic mechanism and the modified Mix and Match mechanism is at most i.

We now inspect how much an agent can gain by deviating from the mechanism. Since the modified Mix and Match mechanism is truthful, her utility in the modified Mix and Match mechanism does not increase with the new strategy. The difference between her expected utility in the modified Mix and Match mechanism and our mechanism is at most $\lceil \log(m) \rceil$. A non truthful strategy can increase the utility by at most an additive factor of $2\lceil \log(m) \rceil$. $\square$

References

1. Ashlagi, I., et al.: Mix and match: a strategyproof mechanism for multi-hospital kidney exchange. Games Econ. Behav. (2013)
2. Ashlagi, I., Roth, A.: Individual rationality and participation in large scale, multi-hospital kidney exchange. In: Proceedings of the 12th ACM Conference on Electronic Commerce, pp. 321–322. ACM (2011)

3. Bhatia, R., Davis, C.: A better bound on the variance. Am. Math. Mon. **107**(4), 353–357 (2000)
4. Caragiannis, I., Filos-Ratsikas, A., Procaccia, A.D.: An improved 2-agent kidney exchange mechanism. In: Chen, N., Elkind, E., Koutsoupias, E. (eds.) WINE 2011. LNCS, vol. 7090, pp. 37–48. Springer, Heidelberg (2011)
5. Dughmi, S., Roughgarden, T., Vondrák, J., Yan, Q.:An approximately truthful-in-expectation mechanism for combinatorial auctions using value queries. arXiv preprint arXiv:1109.1053 (2011)
6. Kothari, A., Parkes, D.C., Suri, S.: Approximately-strategyproof and tractable multiunit auctions. Decis. Support Syst. **39**(1), 105–121 (2005)
7. Lesca, J., Perny, P.: Almost-truthful mechanisms for fair social choice functions. In: ECAI, pp. 522–527 (2012)
8. Luby, M.: A simple parallel algorithm for the maximal independent set problem. SIAM J. Comput. **15**(4), 1036–1053 (1986)
9. Sonmez, T., Unver, M.U.: Market design for kidney exchange. In: The Handbook of Market Design, p. 93 (2013)

Tight Approximations of Degeneracy in Large Graphs

Martín Farach-Colton and Meng-Tsung Tsai[✉]

Rutgers University, New Brunswick, NJ 08901, USA
{farach,mtsung.tsai}@cs.rutgers.edu

Abstract. Given an n-node m-edge graph G, the degeneracy of graph G and the associated node ordering can be computed in linear time in the RAM model by a greedy algorithm that iteratively removes the node of min-degree [28]. In the semi-streaming model for large graphs, where memory is limited to $\mathcal{O}(n \operatorname{polylog} n)$ and edges can only be accessed in sequential passes, the greedy algorithm requires too many passes, so another approach is needed.

In the semi-streaming model, there is a deterministic log-pass algorithm for generating an ordering whose degeneracy approximates the minimum possible to within a factor of $(2+\varepsilon)$ for any constant $\varepsilon > 0$ [12]. In this paper, we propose a randomized algorithm that improves the approximation factor to $(1 + \varepsilon)$ with high probability and needs only a single pass. Our algorithm can be generalized to the model that allows edge deletions, but then it requires more computation and space usage.

The generated node ordering not only yields a $(1+\varepsilon)$-approximation for the degeneracy but gives constant-factor approximations for arboricity and thickness.

Keywords: Degeneracy · Arboricity · Thickness · Semi-streaming algorithm · Space lower bound

1 Introduction

Any ordering of the nodes of an n-node, m-edge simple undirected graph G defines an acyclic orientation of the edges in which each edge is oriented from the earlier node in the ordering to the latter. The *degeneracy* of an ordering is the maximum out-degree it induces. The *degeneracy* of G, denoted by $d(G)$, is the smallest degeneracy among all orderings[1], and an ordering whose degeneracy

---

Work supported by CNS-1408782 and IIS-1247750.

[1] The degeneracy of a graph was originally defined to be the maximum minimum degree among all subgraphs [2,5–7,14,28,34]. The definition here is a slight modification of the *coloring number* [5,6,14] of a graph, a dual definition of degeneracy. The coloring number of a graph was shown to be one larger than the degeneracy [5,6,14], and our definition yields the same value as the original definition of degeneracy.

© Springer-Verlag Berlin Heidelberg 2016
E. Kranakis et al. (Eds.): LATIN 2016, LNCS 9644, pp. 429–440, 2016.
DOI: 10.1007/978-3-662-49529-2_32

is $d(G)$ is called a **degenerate ordering**. An ordering is d-**degenerate** if it has degeneracy at most d.

Degenerate orderings have many uses. Given a degenerate ordering, one can: decompose a graph into at most twice the minimum number of disjoint forests [2,5]; decompose a graph into at most six times the minimum number of disjoint planar graphs [5,9]; speed up the counting of the number of short paths or cycles [2], for example, counting the exact number of 3-cycles in $Q(md(G))$ time; find a component of density at least half the maximum density of any subgraph, i.e. a 1/2-approximation [7]; identify a dominating set of cardinality at most $Q(d^2(G))$ times the cardinality of a minimum dominating set [26] as well as some variations of dominating sets [10], e.g. k-dominating sets; etc. Although most of these problems can be solved exactly in polynomial time [7,15,16,24], the approximation algorithms based on degenerate orderings are faster, use less space or yield better approximation factors for large graphs. For example, such orderings yield a better approximation algorithm for decomposing a graph into a minimum number of planar subgraphs than other algorithms using $O(n)$ space [22,27]. Although all of the results listed originally relied on (optimally) degenerate orderings, in [12], we show that orderings that are nearly degenerate orderings, that is, whose degeneracy approximates rather than matches the graph degeneracy also yield good approximation algorithms.

The degeneracy of graph G and the associated node ordering can be computed in linear time in the RAM model by a greedy algorithm that iteratively removes the node of min-degree, as shown by Matula and Beck [28]. However, iteratively removing a single min-degree node is inefficient when graphs are larger than memory. Thus, another approach is needed.

We consider algorithms in the semi-streaming model [30,32,33], in which we are allowed $\mathcal{O}(n \operatorname{polylog} n)$ working space, and edges can be accessed in sequential read-only passes through the graph. The goal is then to minimize the number of passes and the time complexity of the algorithm.

Some graph problems that have similar complexities in the RAM model can have quite different complexities in the semi-streaming model. Some graph problems, e.g. connectivity, minimum spanning tree, finding bridges and articulation points, can be solved optimally [11,13]. Other graph problems, e.g. counting the number of 3-cycles, maximum matching and graph degeneracy, can be approximated [1,3,12]. Some fundamental problems, such as breath-first search, depth-first search, topological sorting, and directed connectivity, are believed to be difficult to solve in a small number of passes [18,32,33].

In [12], we give a deterministic log-pass algorithm for generating a node ordering whose degeneracy approximates the minimum possible to within a factor of $(2 + \varepsilon)$ for any constant $\varepsilon > 0$. In this paper, we propose a randomized algorithm that improves the approximation factor to $(1 + \varepsilon)$ and reduces the number of pass to one but which has a small probability of failure. Theorem 1 is our main result.

Theorem 1. *In the semi-streaming model, there exists an $\mathcal{O}(m)$-time 1-pass randomized algorithm that outputs a node ordering whose degeneracy approximates the minimum possible to within a factor of $(1 + \varepsilon)$ for any constant $\varepsilon > 0$ with probability $1 - 1/n^{\Omega(1)}$ using a space of $\mathcal{O}(\varepsilon^{-2} n \log^2 n)$ bits.*

Our algorithm can be generalized to the model that allows edge deletions, known as the dynamic stream [4, 8, 17, 20, 21, 25, 29] or the turnstile model [23, 35], by appealing the Jowhari et al. [19] result on building L_0-samplers, which are data structures that can be updated in the streaming model and can generate a sampled edge once the stream has been processed, and the algorithm that deals with sets of L_0-samplers efficiently, due to McGregor et al. [29]. In our case, we need $\mathcal{O}(\varepsilon^{-2}n\log n)$ L_0-samplers to produce a sampled subgraph with $\mathcal{O}(\varepsilon^{-2}n\log n)$ edges. The generalization to the turnstile model increases the time- and space-complexity by polylog n factors. We summarize the result in Theorem 2.

Theorem 2. *In the turnstile model, there exists an $\mathcal{O}(m\,\mathrm{polylog}\,n)$-time 1-pass randomized algorithm that outputs a node ordering whose degeneracy approximates the minimum possible to within a factor of $(1+\varepsilon)$ for any constant $\varepsilon > 0$ with probability $1 - 1/n^{\Omega(1)}$ using a space of $\mathcal{O}(\varepsilon^{-2}n\log^3 n)$ bits.*

In addition to these upper bounds, we also show that computing the degeneracy in the semi-streaming model in one pass with constant success rate has a space lower bound of $\Omega(n\log n)$ bits. The space lower bound also holds in the turnstile model, because the turnstile model is a generalization of the semi-streaming model. We note that our algorithm in the semi-streaming model is optimal in both time- and pass-complexity and has a nearly-optimal space-complexity, which is no more $\log n$ times optimal.

To illustrate how to apply the low-degeneracy node ordering to other problems, we also show constant-factor approximations for arboricity and thickness.

Our Techniques. We sample a small random subgraph $H \subseteq G$ such that H fits in memory. Then, we show that the degenerate ordering of H is a low-degenerate ordering of G, as follows:

In [12], we show that iteratively removing a node of degree no more than $(1+\varepsilon)$ times the minimum degree generates a node ordering whose degeneracy is no more than $(1+\varepsilon)$ times the graph degeneracy. This fact leaves some flexibility in picking the next node to remove, rather than always having to pick the min-degree node, as required in the exact algorithm [28]. Since low degree nodes in H are likely to be low degree nodes in G, we are about to exploit this flexibility to minimize the probability of error in the final order.

Organization. We prove that the degenerate ordering of a random subgraph H is a low-degenerate ordering of graph G in Sect. 2. To obtain a random subgraph H, in Sect. 3 we devise algorithms to sample an H in the semi-streaming model and in the turnstile model. In Sect. 4, we show a lower bound on the space needed to compute a low-degenerate ordering in one pass. Lastly, in Sect. 5, we present some applications of low-degenerate orderings.

2 Degeneracy and Random Subgraphs

We revisit some properties of degeneracy and, based on those, we show that the degenerate ordering of $H = G(p)$ is also a low-degenerate ordering of G, where

$G(p)$ is a random subgraph of G such that every edge in G is included in $G(p)$ independently with probability p.

To begin, let v (resp. $\hat{v}$) be the min-degree node of G (resp. H). By $d_G(v)$ we denote the degree of v in G. Intuitively, since $H = G(p)$ is a sketch of G, the difference between $d_G(v)$ and $d_G(\hat{v})$ is likely to be small. We claim that if p is set to be $\Omega(\varepsilon^{-2} n \log n / m)$,

$$d_G(\hat{v}) \leq \max\left\{(1 + \varepsilon)d_G(v), m/n\right\}$$

with probability $1 - 1/n^{\Omega(1)}$. We prove a stronger form of this claim in Lemma 3, in which G_U denotes the subgraph of G induced by node set U, and $\delta(G_U)$ denotes the minimum node degree in G_U.

Lemma 3. *Let $H = G(p)$ be a random subgraph of an n-node m-edge graph G. For any node set U, the node $\hat{v}$ that has minimum degree in H_U has degree in G_U bounded by*

$$\max\left\{(1 + \varepsilon)\delta(G_U), m/n\right\} \text{ for any constant } \varepsilon > 0$$

with probability $1 - 1/n^{\Omega(1)}$ if $p = \Omega(\varepsilon^{-2} n \log n / m)$.

Proof. Let v be the min-degree node in G_U, and let Q be the set of bad candidates of $\hat{v}$; formally,

$$Q = \left\{x \in G_U : d_{G_U}(x) > \max\left\{(1 + \varepsilon)d_{G_U}(v), m/n\right\}\right\}.$$

We show that the degree $d_{G_U}(\hat{v})$ is bounded as required w.h.p. by considering the two probabilities

$$\Pr\left[\hat{v} \in Q \mid C_1 : d_{G_U}(v) \geq m/n\right] \text{ and } \Pr\left[\hat{v} \in Q \mid C_2 : d_{G_U}(v) < m/n\right],$$

such that $d_{G_U}(\hat{v})$ is not bounded as required. Here we bound the first probability by the Chernoff and Union bounds as follows:

$$\begin{aligned}
\Pr\left[\hat{v} \in Q \mid C_1\right] &\leq \sum_{x \in Q} \Pr\left[d_{H_U}(x) \leq d_{H_U}(v)\right] \\
&\leq \sum_{x \in Q} \Pr\left[d_{H_U}(x) \leq (1 - c)pd_{G_U}(x) \vee d_{H_U}(v) \geq (1 + c)pd_{G_U}(v)\right] \\
&\leq \sum_{x \in Q} \exp\left(-\frac{c^2}{2}pd_{G_U}(x)\right) + \exp\left(-\frac{c^2}{2 + c}pd_{G_U}(v)\right) \quad (1) \\
&\leq n \exp\left(-\Omega(\log n)\right) \\
&= 1/n^{\Omega(1)},
\end{aligned}$$

where $c = \varepsilon/(2 + \varepsilon)$, so that $(1 - c)pd_{G_U}(x) \geq (1 + c)pd_{G_U}(v)$ if $x \in Q$. The second probability has the same upper bound as well because $\Pr\left[\hat{v} \in Q \mid C_1\right] \geq \Pr\left[\hat{v} \in Q \mid C_2\right]$. $\square$

Lemma 3 implies that the degenerate ordering of $H = G(p)$ is a low-degeneracy ordering of G w.h.p. Before proceeding to the proof of our main claim, we observe the following facts about degeneracy:

1. $d(G) \geq m/n$: the degeneracy of an n-node m-edge graph is at least m/n, because the sum of the out-degrees is m for any node ordering, and therefore the maximum out-degree induced by any ordering is at least m/n;
2. $d(G) \geq d(H)$: the degeneracy of graph G is no less than the degeneracy of any subgraph $H \subseteq G$, because adding edges cannot decrease the degeneracy;
3. $d(G) \geq \delta(G)$: the minimum degree is no more than the degeneracy, because for any node ordering, the induced out-degree of the first node equals its degree and of course cannot be less than the minimum.

We are now in a position to show our main claim, as stated in Theorem 4. We prove this by construction. We obtain the degenerate ordering of $H = G(p)$ by the greedy algorithm [28] that iteratively removes the min-degree node from H until H becomes empty. Such a node removal gives a node ordering $\hat{v}_1, \hat{v}_2, \ldots, \hat{v}_n$. Let the remainder of the graph after the node removal be $G_0 = G$, $G_k = G_{k-1} \setminus \{\hat{v}_k\}$ for each $k > 0$. Note that G_k is a subgraph of G induced by the node set $\{\hat{v}_i : i > k\}$. By Lemma 3, if $p = \Omega(\varepsilon^{-2} n \log n/m)$, we have that

$$\Pr\left[d_{G_{k-1}}(\hat{v}_k) > \max\left\{(1+\varepsilon)\delta(G_{k-1}), m/n\right\}\right] < 1/n^{\Omega(1)} \text{ for each } k,$$

and therefore we have, by the Union bound,

$$\Pr\left[\bigvee_{k \in [1,n]} \left(d_{G_{k-1}}(\hat{v}_k) > \max\left\{(1+\varepsilon)\delta(G_{k-1}), m/n\right\}\right)\right] < 1/n^{\Omega(1)}.$$

In other words, we can say that with probability $1 - 1/n^{\Omega(1)}$ every $\hat{v}_k$ has degree no more than the minimum degree in G_{k-1} or the quantity m/n. Hence, $\hat{v}_1, \hat{v}_2, \ldots, \hat{v}_k$ is a low-degeneracy ordering of G whose degeneracy is bounded by

$$\max\left\{(1+\varepsilon)\delta(G_{k-1}), m/n, d(G_{k-1} \setminus \{\hat{v}_k\})\right\} \leq (1+\varepsilon)d(G)$$

with probability $1 - 1/n^{\Omega(1)}$, where the inequality immediately follows from the abovementioned three facts and by induction on k. As a result, we have:

Theorem 4. *The degenerate ordering of $H = G(p)$ is a low-degeneracy ordering of G whose degeneracy approximates the minimum possible to within a factor of $(1 + \varepsilon)$ with probability $1 - 1/n^{\Omega(1)}$ if $p = \Omega(\varepsilon^{-2} n \log n/m)$.*

3 Algorithms

We now present how to compute a node ordering in the considered models such that the computation takes a single pass, and the degeneracy of the ordering is a $(1 + \varepsilon)$-approximation of the graph degeneracy w.h.p.

To recap, we propose an algorithm that samples a random subgraph H from the entire graph G, and then outputs the degenerate ordering of the random subgraph. We have shown in Theorem 4 that the ordering is a low-degenerate ordering of the entire graph with a fairly good probability. In particular, we let $H = G(p)$ for $p = \Theta(\varepsilon^{-2} n \log n/m)$, and thus the output ordering has degeneracy that approximates the minimum possible to within a factor of $(1 + \varepsilon)$ with probability $1 - 1/n^{\Omega(1)}$.

The sampling procedure is quite different in the semi-streaming and the turnstile models. We discuss them separately. In the semi-streaming model, each edge in the data stream is contained in the final edge set of G. Thus, obtaining a sampled subgraph H is straightforward. To make the algorithm optimal in runtime, we batch the edges, as described below.

On the other hand, an edge on the data stream might be subsequently deleted in the turnstile model. This uncertainty makes sampling hard. To handle the uncertainty we appeal to the L_0-sampler construction of Jowhari et al. [19] and the algorithm for efficiently maintaining sets of such L_0-samplers due to McGregor et al. [29]. In this approach, the sampled subgraph H is an approximate of $G(p)$ that has a small distortion from $G(p)$. We adapt Lemma 3 to account for this distortion.

3.1   In the Semi-streaming Model

In this model the edge set of G is presented in a read-only stream. We assume that n, the number of nodes, is known at the beginning of data stream, but m, the number of edges, is not known until the end of data stream. The desired size of H is

$$s = \Theta(\varepsilon^{-2} n \log n).$$

We note here that s is known at the beginning because it only depends on n and ε. Our goal is to pick a p so that $s = mp$, that is, so that $H = G(p)$ has size $\mathcal{O}(s)$, w.h.p. However, m is unknown, and therefore p is also unknown, when we start sampling. To sample each edge with an unknown probability p, we guess that $m = s$ and set $p = 1$. If there are more than s edges, we adjust the guess to be $m = 2s$, set $p = 1/2$, and then kick out some sampled edges to make sure that all edges were sampled with probability $1/2$. We keep adjusting the guess for m and p and resampling, until we run out of edges. In order to implement this intuition efficiently, we use the following algorithm.

We allocate a working space of size $2s$ to hold the sampled edges, and call this space the **pool**. If we ever end up selecting more than $2s$ edges, our algorithm goes into a low-probability failure mode in which it outputs an arbitrary node ordering.

Now suppose, for ease of presentation, that m is a multiple of s. The pool is empty at the beginning. Our algorithm works in rounds: in the k-th round, the algorithm brings the k-th group of s edges into a buffer. Then, the algorithm kicks out some of the edges that are already in the pool, if any, each with probability $p_k = 1/k$, and migrates the edges that are in the buffer to the pool, each with probability $q_k = 1/k$, discarding those that fails to migrate. At the end of the (m/s)-th round, the pool contains a randomly sampled subgraph H in which each edge is sampled independently from G with probability $p = s/m$ even though m was unknown initially. See Fig. 1 for an illustration of the sampling procedure.

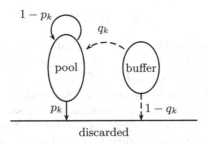

Fig. 1. Sampling procedure in the k-th round.

In the case that m is not a multiple of s, then there are αs edges for $\alpha \in (0,1)$ in the last-round buffer. For this last round, it suffices to set $p_k = \alpha/(k-1+\alpha)$ and $q_k = 1/(k-1+\alpha)$.

To quantify how often the sampling procedure has a pool of size greater than $2s$, we give a bound on the probability of such a pool overflows in a round. By the Chernoff bound, we have

$$\Pr\left[\sum_{i=[1,m]} X_i > 2mp\right] \leq \exp\left(-\Omega(\varepsilon^{-2} n \log n)\right) \leq 1/n^{\Omega(n)},$$

where X_i is an indicator variable denoting whether the i-th edge of graph G is included in the random subgraph H. Then, by the Union bound, the probability that the pool overflows at some point of the entire sampling procedure is $1/n^{\Omega(1)}$, which is dominated by the claimed failure rate.

Lastly, we analyze the time complexity of the proposed algorithm. The above sampling procedure has $\mathcal{O}(m/s)$ rounds, and in each round it deals with at most $3s$ edges. Thus, the sampling procedure takes $\mathcal{O}(m)$ time. After obtaining the random sampled subgraph H, the algorithm computes the degenerate ordering of H using the in-memory algorithm for computing degenerate ordering introduced by Matula and Beck [28], which takes $\mathcal{O}(s) = \mathcal{O}(mp)$ time. Hence, the total runtime is bounded by $\mathcal{O}(m)$.

This completes the proof of Theorem 1.

3.2 In the Turnstile Model

In this model a sequence of edge insertions and deletions is presented in a read-only stream. The procedure described in the Sect. 3.1 does not work in this model because the pool in Sect. 3.1 might end up choosing edges that are about to be deleted, making $H = \emptyset$ rather than $H = G(p)$.

Since that the working space cannot keep the entire edge set E ($|E| = m$), using an L_0-sampler [19] one can sample an edge e from E with success rate at least $1 - \delta$ using $\mathcal{O}(\log^2 n \log(1/\delta))$ bits, such that the probability that an edge $e \in E$ is picked in the sample is

$$1/m + 1/n^{\Omega(1)},$$

which is roughly the desired $1/m$ but has a small distortion $1/n^{\Omega(1)}$.

To sample a random subgraph $H = G(p)$, one can use $2s$ L_0-samplers to sample $2s = 2mp/(1 - \delta) = \Theta(\varepsilon^{-2} n \log n)$ for constant $\delta < 1$ independent edges from the final edge set E of G. These might be repeated, so pick the first t distinct edges from the $2s$ samples, where t is a random variate sampled from $Bin(m, p)$, the binomial distribution of m trials and success rate p. Note that an L_0-sampler fails to return a sample with probability $1 - \delta$, and so we set $2s = 2mp/(1 - \delta)$ to balance the failure rate. One can assert that there exist t distinct edges among the $2s$ samples with probability $1 - 1/n^{\Omega(1)}$ by the Chernoff bound. To handle the failure case, again, the algorithm outputs an arbitrary node ordering.

It follows from the above sampling procedure, for each edge $e \in G$, that the probability that H contains e is

$$\Pr[e \in H] = \sum_{k=[0,m]} \Pr[t \sim Bin(m,p) = k] \left(k/m + k/n^{\Omega(1)} \right) = p \left(1 + 1/n^{\Omega(1)} \right).$$

Hence, $H \approx G(p)$.

Since $H \neq G(p)$, to prove the correctness of Theorem 2, we adapt Lemma 3 to accommodate the small distortion. The distortion changes the expected values of $d_{H_U}(x)$ and $d_{H_U}(\hat{v})$. However the change is so small that the bounds on the tail probabilities in Eq. 1 still hold. Therefore, Lemma 3 works as well for $H \approx G(p)$.

A naïve implementation of the above requires $\mathcal{O}(\varepsilon^{-2} nm \operatorname{polylog} n)$ time. We appeal to the alternative sampling procedure devised by McGregor et al. [29] to obtain the abovementioned $2s$ samples in $\mathcal{O}(m \operatorname{polylog} n)$ time.

This establishes Theorem 2.

4   Space Lower Bounds

In this section, we show that any randomized algorithm in the semi-streaming model that can approximate degeneracy, arboricity or thickness in one sequential pass with constant success rate requires a working space of $\Omega(n \log n)$ bits. Our proofs rely on the space lower bounds of cycle-freeness [35, Theorem 7] and planarity testing [35, Corollary 12] shown recently by Sun and Woodruff.

We observe that, combining the space lower bound of cycle-freeness testing [35, Theorem 7] with Lemma 5, the space lower bound for approximating degeneracy to within a factor of $(2 - \varepsilon)$ is immediate, summarized in Theorem 6.

Lemma 5. *Graph G has degeneracy $d(G) = 1$ if and only if G is cycle-free.*

Proof. If G is cycle-free, then G is a forest and one can make every tree in G rooted. In this way, let the orientation of the edges in G from the descendant to the ancestor. Since every node except the roots in such rooted cycle-free graph has a single parent node. The out-degree of every node is either 0 or 1. In other words, the degeneracy $d(G) = 1$.

Otherwise, G contains a cycle C. Since the degeneracy $d(C) \geq \delta(C)$ (Fact 3 in Sect. 2) and $\delta(C) = 2$, then $d(C) \geq 2$. Combining that $d(G) \geq d(C)$ (Fact 2 in Sect. 2), $d(G) \geq 2$.   □

Theorem 6. *In the semi-streaming model, any randomized algorithm that can approximate the degeneracy to within a factor of $(2 - \varepsilon)$ for any constant $\varepsilon > 0$ with constant success rate has a space lower bound of $\Omega(n \log n)$ bits.*

In addition, we show similar space lower bounds for computing arboricity and thickness. The proof directly follows from the definition of arboricity and thickness. Recall that the arboricity (resp. thickness) of graph G is the minimum number of forests (resp. planar subgraphs) whose union forms G. Therefore, computing the arboricity (resp. thickness) is no easier than cycle-freeness (resp. planarity) testing. Combining the space lower bounds of cycle-freeness and planarity testing shown in [35], we have Theorem 7.

Theorem 7. *In the semi-streaming model, any randomized algorithm that can approximate arboricity or thickness to within a factor of $(2 - \varepsilon)$ for any constant $\varepsilon > 0$ with constant success rate has a space lower bound of $\Omega(n \log n)$ bits.*

We note that the above space lower bounds also hold in the turnstile model, because the turnstile model is a generalization of the semi-streaming model.

5   Applications

A low-degeneracy node ordering has many applications. Here we present how to use the ordering to partition a graph into edge-disjoint forests such that the number of forests approximates the minimum possible, i.e. the arboricity.

The Nash-William theorem [31] states that if m_S (resp. n_S) denotes the number of edges (resp. nodes) in the subgraph S, the arboricity can be stated as

$$\alpha(G) = \max_{S \subseteq G} \lceil m_S/(n_S - 1) \rceil,$$

which has a form similar to that of the density of the densest subgraph. The algorithm in [29] that approximates the density of the densest subgraph can be adapted to approximate arboricity to within a factor of $(1 + \varepsilon)$.

It is unclear how to exploit the actual value of the arboricity to actually partition the input graph into a small number of forests, however a d-degenerate node ordering has a direct application to such a partition. Considering that in the acyclic orientation induced by the ordering, the out-degree of each node is no more than d, and thus one can partition the edge set into d subgraphs, each of which contains the i-th out-going edge of each node, if any. Note that the subgraphs are forests because all of them have degeneracy 1 or, equivalently, are cycle-free due to Lemma 5.

The above procedure has a simple two-pass implementation in the semi-streaming model. We use the d-degenerate ordering obtained in the first pass to assign the orientation of the incoming edges in the second pass. Then, maintaining the out-degree of each node suffices to partition the edges as described.

Since the computed d-degenerate ordering has $d \leq (1 + \varepsilon)d(G)$, and each of the degeneracy, arboricity and thickness approximates each other, we have the result in Theorem 8. Our result improves the approximation ratio by a factor of 2, compared to that in [12].

Theorem 8. *In the semi-streaming model, there exists an $\mathcal{O}(m)$-time 2-pass randomized algorithm that partition the edges into forests (resp. planar graphs) such that the number of forests (resp. planar graphs) approximates the minimum possible to within a factor of $(2 + \varepsilon)$ (resp. $(6 + \varepsilon)$) with probability $1 - 1/n^{\Omega(1)}$ using a space of $\mathcal{O}(\varepsilon^{-2} n \log^2 n)$ bits.*

References

1. Ahn, K.J., Guha, S.: Linear programming in the semi-streaming model with application to the maximum matching problem. In: Aceto, L., Henzinger, M., Sgall, J. (eds.) ICALP 2011, Part II. LNCS, vol. 6756, pp. 526–538. Springer, Heidelberg (2011)
2. Alon, N., Yuster, R., Zwick, U.: Finding and counting given length cycles. Algorithmica **17**(3), 209–223 (1997)
3. Bar-Yossef, Z., Kumar, R., Sivakumar, D.: Reductions in streaming algorithms, with an application to counting triangles in graphs. In: Proceedings of the thirteenth annual ACM-SIAM symposium on Discrete algorithms (SODA), pp. 623–632. SIAM (2002)
4. Bhattacharya, S., Henzinger, M., Nanongkai, D., Tsourakakis, C.E.: Space- and time-efficient algorithm for maintaining dense subgraphs on one-pass dynamic streams. In: Proceedings of the Forty-Seventh Annual ACM on Symposium on Theory of Computing (STOC), pp. 173–182 (2015)
5. Bollobás, B.: Extremal Graph Theory. Academic Press, London (1978)
6. Bollobás, B.: The evolution of sparse graphs. In: Graph Theory and Combinatorics, Proceedings of the Cambridge Combinatorial Conference in honor of Paul Erdős, pp. 35–57. Academic Press (1984)
7. Charikar, M.: Greedy approximation algorithms for finding dense components in a graph. In: Jansen, K., Khuller, S. (eds.) APPROX 2000. LNCS, vol. 1913, pp. 84–95. Springer, Heidelberg (2000)
8. Chitnis, R.H., Cormode, G., Esfandiari, H., Hajiaghayi, M., McGregor, A., Monemizadeh, M., Vorotnikova, S.: Kernelization via sampling with applications to dynamic graph streams, CoRR abs/1505.01731 (2015)
9. Dean, A.M., Hutchinson, J.P., Scheinerman, E.R.: On the thickness and arboricity of a graph. J. Comb. Theor. Series B **52**(1), 147–151 (1991)

10. Dvořák, Z.: Constant-factor approximation of the domination number in sparse graphs. Eur. J. Comb. **34**(5), 833–840 (2013)
11. Farach-Colton, M., Hsu, T., Li, M., Tsai, M.-T.: Finding articulation points of large graphs in linear time. In: Dehne, F., Sack, J.-R., Stege, U. (eds.) WADS 2015. LNCS, vol. 9214, pp. 363–372. Springer, Heidelberg (2015)
12. Farach-Colton, M., Tsai, M.-T.: Computing the degeneracy of large graphs. In: Pardo, A., Viola, A. (eds.) LATIN 2014. LNCS, vol. 8392, pp. 250–260. Springer, Heidelberg (2014)
13. Feigenbaum, J., Kannan, S., McGregor, A., Suri, S., Zhang, J.: On graph problems in a semi-streaming model. Theor. Comput. Sci. **348**(2), 207–216 (2005)
14. Frank, A., Gyarfas, A.: How to orient the edges of a graph. In: Proceedings of the Fifth Hungarian Colloquium on Combinatorics. vol. I, Combinatorics, pp. 353–364 (1976)
15. Gabow, H., Westermann, H.: Forests, frames, and games: algorithms for matroid sums and applications. In: Proceedings of the twentieth annual ACM Symposium on Theory of Computing (STOC), pp. 407–421. ACM (1988)
16. Goldberg, A.V.: Finding a maximum density subgraph. Technical report (1984)
17. Guha, S., McGregor, A., Tench, D.: Vertex and hyperedge connectivity in dynamic graph streams. In: Proceedings of the 34th ACM Symposium on Principles of Database Systems (PODS), pp. 241–247 (2015)
18. Guruswami, V., Onak, K.: Superlinear lower bounds for multipass graph processing. In: 28th Conference on Computational Complexity (CCC), pp. 287–298. IEEE (2013)
19. Jowhari, H., Sağlam, M., Tardos, G.: Tight bounds for L_p samplers, finding duplicates in streams, and related problems. In: Proceedings of the 30th ACM Symposium on Principles of Database Systems (PODS), pp. 49–58. ACM (2011)
20. Kapralov, M., Lee, Y.T., Musco, C., Musco, C., Sidford, A.: Single pass spectral sparsification in dynamic streams. In: 55th IEEE Annual Symposium on Foundations of Computer Science (FOCS), pp. 561–570 (2014)
21. Kapralov, M., Woodruff, D.P.: Spanners and sparsifiers in dynamic streams. In: ACM Symposium on Principles of Distributed Computing (PODC), pp. 272–281 (2014)
22. Kawano, S., Yamazaki, K.: Worst case analysis of a greedy algorithm for graph thickness. Inf. Process. Lett. **85**(6), 333–337 (2003)
23. Konrad, C.: Maximum matching in turnstile streams. In: Bansal, N., Finocchi, I. (eds.) ESA 2015. LNCS, vol. 9294, pp. 840–852. Springer, Verlag (2015)
24. Kowalik, Ł.: Approximation scheme for lowest outdegree orientation and graph density measures. In: Asano, T. (ed.) ISAAC 2006. LNCS, vol. 4288, pp. 557–566. Springer, Heidelberg (2006)
25. Kutzkov, K., Pagh, R.: Triangle counting in dynamic graph streams. In: Ravi, R., Gørtz, I.L. (eds.) SWAT 2014. LNCS, vol. 8503, pp. 306–318. Springer, Heidelberg (2014)
26. Lenzen, C., Wattenhofer, R.: Minimum dominating set approximation in graphs of bounded arboricity. In: Lynch, N.A., Shvartsman, A.A. (eds.) DISC 2010. LNCS, vol. 6343, pp. 510–524. Springer, Heidelberg (2010)
27. Mansfield, A.: Determining the thickness of graphs is NP-hard. Math. Proc. Cambridge Philos. Soc. **93**, 9–23 (1983)
28. Matula, D.W., Beck, L.L.: Smallest-last ordering and clustering and graph coloring algorithms. J. ACM **30**(3), 417–427 (1983)

29. McGregor, A., Tench, D., Vorotnikova, S., Vu, H.T.: Densest subgraph in dynamic graph streams. In: Italiano, G.F., Pighizzini, G., Sannella, D.T. (eds.) MFCS 2015. LNCS, vol. 9235, pp. 472–482. Springer, Heidelberg (2015)
30. Muthukrishnan, S.: Data streams: algorithms and applications. Technical report (2003)
31. Nash-Williams, C.S.A.: Edge-disjoint spanning trees of finite graphs. J. Lond. Math. Soc. s1–36(1), 445–450 (1961)
32. O'Connell, T.C.: A survey of graph algorithms under extended streaming models of computation. In: Ravi, S.S., Shukla, S.K. (eds.) Fundamental Problems in Computing, pp. 455–476. Springer, The Netherlands (2009)
33. Ruhl, J.M.: Efficient algorithms for new computational models. Ph.D. thesis, Massachusetts Institute of Technology, September 2003
34. Schank, T., Wagner, D.: Finding, counting and listing all triangles in large graphs, an experimental study. In: Nikoletseas, S.E. (ed.) WEA 2005. LNCS, vol. 3503, pp. 606–609. Springer, Heidelberg (2005)
35. Sun, X., Woodruff, D.P.: Tight bounds for graph problems in insertion streams. In: Approximation, Randomization, and Combinatorial Optimization. Algorithms and Techniques, APPROX/RANDOM, pp. 435–448 (2015)

Improved Approximation Algorithms for Capacitated Fault-Tolerant k-Center

Cristina G. Fernandes[1], Samuel P. de Paula[1], and Lehilton L.C. Pedrosa[2(✉)]

[1] Department of Computer Science, University of São Paulo, São Paulo, Brazil
{cris,samuelp}@ime.usp.br
[2] Institute of Computing, University of Campinas, Campinas, Brazil
lehilton@ic.unicamp.br

Abstract. In the k-center problem, given a metric space V and a positive integer k, one wants to select k elements (centers) of V and an assignment from V to centers, minimizing the maximum distance between an element of V and its assigned center. One of the most general variants is the *capacitated α-fault-tolerant k-center*, where centers have a limit on the number of assigned elements, and, if α centers fail, there is a reassignment from V to non-faulty centers. In this paper, we present a new approach to tackle fault tolerance, by selecting and pre-opening a set of backup centers, then solving the obtained residual instance. For the $\{0, L\}$-capacitated case, we give approximations with factor 6 for the basic problem, and 7 for the so called *conservative* variant, when only clients whose centers failed may be reassigned. Our algorithms improve on the previously best known factors of 9 and 17, respectively. Moreover, we consider the case with general capacities. Assuming α is constant, our method leads to the first approximations for this case.

1 Introduction

The *k-center* is the minimax problem in which, given a metric space V and a positive integer k, we want to choose a set of k centers such that the maximum distance from an element of V to its closest center is minimum. More precisely, we want to select $S \subseteq V$ with $|S| = k$ that minimizes

$$\max_{u \in V} \min_{v \in S} d(u, v),$$

where $d(u, v)$ is the distance between u and v. The decision version of the k-center appears in Garey and Johnson's list of NP-complete problems, identified by MS9 [8]. It is well known that k-center has a 2-approximation which is best possible unless P = NP [7,9,11–13]. The elements of the set S are usually referred to as *centers*, and the elements of V as *clients*.

In a typical application of k-center, set V represents the nodes of a network, and one may want to install k routers so that the network latency is minimized.

Partially supported by CAPES, CNPq (grants 308523/2012-1, 477203/2012-4, and 456792/2014-7), FAPESP (grants 2013/03447-6 and 2014/14209-1), and MaCLinC.

E. Kranakis et al. (Eds.): LATIN 2016, LNCS 9644, pp. 441–453, 2016.
DOI: 10.1007/978-3-662-49529-2_33

Other applications have additional constraints, so variants of the k-center have been considered as well. For example, the number of nodes that a router may serve might be limited. In the *capacitated k-center*, in addition to the set of selected centers, we also want to obtain an assignment from the set of clients to centers such that at most a number L_u of clients are assigned to each center u. The value L_u is called the *capacity* of u. The first approximation for this version of the problem is due to Bar-Ilan et al. [2], who gave a 10-approximation for the particular case of uniform capacities, where there is a number L such that $L_u = L$ for every u in V. This was improved by Khuller and Sussmann [14], who obtained a 6-approximation, and also considered the *soft* capacitated case, in which multiple centers may be opened at the same location, obtaining a 5-approximation, both for uniform capacities.

Despite the progress in the approximation algorithms for related problems, such as the metric facility location problem, the first constant approximation for the (non-uniform) capacitated k-center was obtained only in 2012, by Cygan et al. [5]. Differently from algorithms for the uniform case, the algorithm of Cygan et al. is based on the relaxation of a linear programming (LP) formulation. Since the natural formulation for the k-center has unbounded integrality gap, a preprocessing is used, what allows considering only instances whose LP has bounded gap. They also presented an 11-approximation for the soft capacitated case. Later, An et al. [1] presented a cleaner rounding algorithm, and obtained an improved approximation with factor 9 (while the previous approximation had a large constant factor, not explicitly calculated). As for negative results, it has been shown that the capacitated k-center has no approximation with factor better than 3 unless P = NP [5].

Another natural variant of the k-center comprises the possibility that centers may fail during operation. This was first discussed by Krumke [16], who considered the version in which clients must be connected to a minimum number of centers. In the *fault-tolerant k-center*, for a given number α, we consider the possibility that any subset of centers of size at most α may fail. The objective is to minimize the maximum distance from a client to its $\alpha + 1$ nearest centers. For the variant in which selected centers do not need to be served, Krumke [16] gave a 4-approximation, later improved to a (best possible) 2-approximation by Chaudhuri et al. [3], and Khuller et al. [15]. For the standard version, when a client must be served even if a center is installed at client's location, there is a 3-approximation by Khuller et al. [15], that also gave a 2-approximation for the particular case of $\alpha \leq 2$.

Chechik and Peleg [4] considered a common generalization of the capacitated k-center and the fault-tolerant k-center, where centers have limited capacity and may fail during operation. They defined only the uniform capacitated version, presenting a 9-approximation. Also, they considered the case in which, after failures, only clients that were assigned to faulty centers may be reassigned. For this variant, called the conservative fault-tolerant k-center, a 17-approximation was obtained for the uniform capacitated case. For the special case in which $\alpha < L$, the so called *large capacities* case, they obtained a 13-approximation.

1.1   Our Contributions and Techniques

We consider the *capacitated α-fault-tolerant k-center* problem. Formally, an instance for this problem consists of a metric space V with corresponding distance function $d : V \times V \to \mathbb{Q}_+$, non-negative integers k and α, with $\alpha < k$, and a non-negative integer L_v for each v in V. A solution is a subset S of V with $|S| = k$, such that, for each $F \subseteq S$ with $|F| \leq \alpha$, there exists an assignment $\phi_F : V \to S \setminus F$ with $|\phi_F^{-1}(v)| \leq L_v$ for each v in $S \setminus F$. For a given F, we denote by ϕ_F^* an assignment ϕ_F with minimum $\max_{u \in V} d(u, \phi_F(u))$. The problem's objective is to find a solution that minimizes

$$\max_{u \in V, F \subseteq V : |F| \leq \alpha} d(u, \phi_F^*(u)).$$

We also consider the *capacitated conservative α-fault-tolerant k-center*. In this variant, in addition to the set S, a solution comprises an initial assignment ϕ_0. We require that an assignment ϕ_F for a failure scenario F differs from ϕ_0 only for vertices assigned by ϕ_0 to centers in F. Precisely, given $F \subseteq S$ with $|F| \leq \alpha$, we say that an assignment ϕ_F is *conservative* (with respect to ϕ_0) if $\phi_F(u) = \phi_0(u)$ for every $u \in V$ with $\phi_0(u) \notin F$. A solution for the problem is a pair (S, ϕ_0) such that, for each $F \subseteq S$ with $|F| \leq \alpha$, there exists a conservative assignment ϕ_F. The objective function is defined analogously.

Our major technical contribution is a new strategy to deal with the fault-tolerant and capacitated problems. Namely, we solve the considered problems in two phases. In the first phase, we identify clusters of vertices where an optimal solution must install a minimum of α centers. For each cluster, we carefully select α of its vertices, and pre-open them as centers. These α centers will have enough *backup* capacity so that, in the case of failure events, the unused capacity of all pre-opened centers will be sufficient to obtain a reassignment for all clients. While the α guessed centers of a cluster may not correspond to centers in an optimal solution, we may meticulously modify the instance if necessary, so that our choice always leads to a good solution. In the second phase, we are left with a residual instance, where part of a solution is already known. Depending on the problem, obtaining the remaining centers of a solution may be reduced to the non-fault-tolerant variant. Otherwise, we can make stronger assumptions over the input and the solution, so that the task of obtaining a fault-tolerant solution is simplified.

A good feature of our approach is that it can be used in combination with different techniques and algorithms (be they well known algorithms, or newly proposed ones). Indeed, we apply it to obtain approximations for both the conservative and non-conservative variants of the capacitated fault-tolerant k-center. Moreover, each of the algorithms uses novel and specific techniques that are of particular interest. For the conservative variant, we present elegant combinatorial algorithms that reduce the problem to the non-fault-tolerant case. For the non-conservative variant, our algorithms are based on the rounding of a new LP formulation for the problem. Interestingly, we use the set of pre-opened centers to obtain a partial solution for the LP variables with integral values. We hope that other problems can benefit from similar techniques.

1.2   Obtained Approximations and Paper Organization

In Sect. 3, we present a 7-approximation for the $\{0, L\}$-capacitated conservative α-fault-tolerant k-center. This is the subset of the problem where the capacities are either 0 or L, for some L. Notice that this generalizes the uniform capacitated case, when all capacities are equal to L. This result improves on the previously known factors of 17 and 13 by Chechik and Peleg [4], that apply to particular cases with uniform capacities, and uniform large capacities, respectively. In Sect. 4, we study the case of general capacities, and present a $(9 + 6\alpha)$-approximation when α is constant. To the best of our knowledge, this is the first approximation for the problem with arbitrary capacities. For the non-conservative variant, our algorithms are based on the rounding of a new LP formulation. We present the LP formulation, and give an overview of the algorithms in Section 5. First we consider the case of arbitrary capacities, and present a 10-approximation when α is constant. Once again, this is the first approximation for the problem with arbitrary capacities. The rounding algorithm is adapted for the $\{0, L\}$-capacitated fault-tolerant k-center, for which we obtain a 6-approximation with α being part of the input. We remark that this factor matches the best known factor for the problem without fault tolerance [1,14], and improves on the best previously known algorithm for the fault-tolerant version, which achieves factor 9 for the uniform capacitated case [4].

For the problems with arbitrary capacities, we require that α is constant. We remark that, when α is part of the input, even deciding whether a given set of vertices S is a feasible solution with a given cost is not trivial. Indeed, while the capacitated α-fault-tolerant k-center is NP-hard, we do not know whether its decision version is in NP. Nevertheless, our assumption for this case is consistent with most of the problem's applications, as α corresponds to an upper bound on the number of failing centers. A summary of the results is given in Table 1.

Table 1. Summary of the obtained approximation factors.

Version	Capacities	Value of α	Previous	This paper
Conservative	uniform	given in the input	17 [4]	7
Conservative	arbitrary	fixed	–	$9 + 6\alpha$
Non-conservative	uniform	given in the input	9 [4]	6
Non-conservative	arbitrary	fixed	–	10

2   Preliminaries

Let $G = (V, E)$ be an undirected and unweighted graph. We denote by d_G the metric induced by G, that is, for u and v in V, let $d_G(u, v)$ be the length of a shortest path between u and v in G. For given nonempty sets A, $B \subseteq V$, we define $d_G(A, B) = \min_{a \in A, b \in B} d_G(a, b)$.

For an integer ℓ, we let $N_G^\ell(u) = \{v \in V : d_G(v, u) \leq \ell\}$. For a subset $U \subseteq V$, let $N_G^\ell(U) = \bigcup_{u \in U} N_G^\ell(u)$. We may omit the superscript ℓ when $\ell = 1$, and the subscript G when the graph is clear from the context. For a directed graph G, we define $d_G(u, v)$ as the length of a shortest directed path from u to v in G, and define $N_G^\ell(u)$ similarly. We also define the (power) graph $G^\ell = (V, E^\ell)$, where $\{u, v\} \in E^\ell$ if $v \in N^\ell(u) \setminus \{u\}$.

2.1   Reduction to the Unweighted Case

As it is standard for the k-center problem, we will use the bottleneck method [12], so that we can consider the case in which the metric space is induced by an unweighted undirected graph. Suppose we have an algorithm that, given an unweighted graph, either gives a distance-r solution for the unweighted problem, or a certificate that no distance-1 solution exists. We may then use this algorithm to obtain an r-approximation for the general metric case.

Let V be a metric space with distance $c : V \times V \to \mathbb{R}_+$. For a certain number τ in $\mathbb{R}_+$, we consider the *threshold graph* defined as $G_{\leq\tau} = (V, E_{\leq\tau})$, where $E_{\leq\tau} = \{\{u, v\} : c(u, v) \leq \tau\}$. Next we obtain a sequence of values of $c(u, v)$ for (u, v) in V^2, in increasing order. For each τ in this ordering, we obtain $G_{\leq\tau}$, and use the algorithm for the unweighted case; we stop when the algorithm fails to provide a negative certificate, and return the obtained solution. Notice that there must be a distance-1 solution for $G_{\leq\text{OPT}}$, where OPT denotes the optimum value for the problem. Since OPT is in the considered ordering for τ, the algorithm always stops, and returns a solution for some $\tau \leq$ OPT, so we obtain a solution for the original problem of cost at most $r \cdot \tau \leq r \cdot$ OPT. So, from now on, we assume an unweighted graph $G = (V, E)$ is given, and the goal is to either obtain a certificate that no distance-1 solution exists, or return a distance-r solution for some constant r.

2.2   Preprocessing and Reduction to the Connected Case

We also may assume without loss of generality that G is connected [4,5,14]. If this is not the case, we may proceed as follows. Suppose there is an algorithm that, given a connected graph $\tilde{G}$, and an integer $\tilde{k}$, produces a distance-r solution with $\tilde{k}$ vertices, or gives a certificate that no distance-1 solution with $\tilde{k}$ vertices exists. Now, consider a given arbitrary unweighted graph G, and a given integer k. We decompose G into its connected components, say $G_1, \ldots, G_t$. For each connected component G_i, with $1 \leq i \leq t$, we run the algorithm for each $\tilde{k} = \alpha + 1, \ldots, k$, and find the minimum value k_i, if any, for which the algorithm obtains a distance-r solution. As the failure set is arbitrary, in the worst case all faulty centers might be in a component. If there is some G_i for which there is no distance-1 solution with k centers or if $k_1 + \cdots + k_t > k$, then clearly there is no distance-1 solution for G with k centers; otherwise, conjoining the solutions obtained for each component leads to a distance-r solution for G with no more than k centers, and this solution is tolerant to the failure of α centers. From now on, we will assume that G is connected.

3 $\{0, L\}$-Capacitated Conservative Fault-Tolerant k-Center

After a failure, a distance-1 conservative solution has to reassign each unserved client to an open center in its vicinity with available capacity. This requires some kind of "local available center capacity". The next definition describes a set of vertices that are nice candidates to be open as backup centers. This set can be partitioned into clusters of at most α vertices, with the clusters sufficiently apart from each other. The idea is that failures in the vicinity of one of these clusters do not affect centers in the other clusters. More precisely, the vicinities of different clusters do not intersect, therefore, in a distance-1 conservative solution, any client that is assigned to a center in a certain cluster cannot be reassigned to a center in the vicinity of any of the other clusters.

Definition 1. *Consider a graph $G = (V, E)$ and non-negative integers α and ℓ. A set W of vertices of G is (α, ℓ)-independent if it can be partitioned into sets $C_1, \ldots, C_t$, such that $|C_i| \leq \alpha$ for $1 \leq i \leq t$, and $d(C_i, C_j) > \ell$ for $1 \leq i < j \leq t$.*

In what follows, we denote by (G, k, L, α) an instance of the capacitated conservative α-fault-tolerant k-center as obtained by Sect. 2. We say that (G, k, L, α) is feasible if there exists a distance-1 solution for it.

Lemma 1. *Let (G, k, L, α) be a feasible instance for the capacitated conservative α-fault-tolerant k-center, and let (S^*, ϕ_0^*) be a corresponding distance-1 solution. If $W \subseteq S^*$ is an $(\alpha, 4)$-independent set in G, then $(G, k - |W|, L')$ is feasible for the capacitated k-center, where $L'_u = 0$ for u in W, and $L'_u = L_u$ otherwise.*

Proof. Since W is $(\alpha, 4)$-independent, there must be a partition $C_1, \ldots, C_t$ of W such that $d(C_i, C_j) > 4$ for any pair i, j, with $1 \leq i < j \leq t$. Also, each part C_i has at most α vertices, and thus there exists a conservative assignment $\phi_{C_i}^*$ with $(\phi_{C_i}^*)^{-1}(C_i) = \emptyset$. Therefore, $\phi_{C_i}^*$ is a distance-1 solution for the $(G, k - |C_i|, L^i)$ instance of the capacitated k-center problem, where $L_u^i = 0$ for u in C_i, and $L_u^i = L_u$ otherwise. Moreover, as ϕ_0^* is conservative, $\phi_{C_i}^*$ differs from ϕ_0^* only in $(\phi_0^*)^{-1}(C_i)$. So, if a center u in S^* is such that $(\phi^*)^{-1}(u) \neq (\phi_{C_i}^*)^{-1}(u)$, then $u \in N^2(C_i)$. As W is $(\alpha, 4)$-independent, $N^2(C_i) \cap N^2(C_j) = \emptyset$ for every $j \in [t] \setminus \{i\}$. Let ψ be an assignment such that, for each client v,

$$\psi(v) = \begin{cases} \phi_{C_i}^*(v) & \phi_0^*(v) \in C_i \text{ for some } i \text{ in } [t], \\ \phi_0^*(v) & \text{otherwise.} \end{cases}$$

Therefore, set $\psi^{-1}(u)$ is empty if $u \in W$; is $(\phi_{C_i}^*)^{-1}(u)$ if there exists $i \in [t]$ such that $u \in N^2(C_i) \setminus C_i$; and is $(\phi^*)^{-1}(u)$ otherwise. This means that, for L' as in the statement of the lemma, $|\psi^{-1}(u)| \leq L'_u$ for every u, and so (S^*, ψ) is a solution for the $(G, k - |W|, L')$ instance of the capacitated k-center problem. $\square$

A set of vertices $A \subseteq V$ is *7-independent* in G if every pair of vertices in A is at distance at least 7 in G. This definition was also used by Chechik and Peleg [4] and, as we will show, such a set is useful to obtain an $(\alpha, 4)$-independent set in G.

Lemma 2. *Let A be a 7-independent set in G, for each a in A, let $B(a)$ be any set of α vertices in $N(a)$, and let $B = \cup_{a \in A} B(a)$. If (G, k, L, α) is feasible for the capacitated conservative α-fault-tolerant k-center, then $(G, k - |B|, L')$ is feasible for the capacitated k-center, where $L'_u = 0$ for u in B, and $L'_u = L_u$ otherwise.*

Proof. Let (S^*, ϕ_0^*) be a solution for (G, k, L, α). For each $a \in A$, there must be at least α centers in $S^* \cap N(a)$. Let $W(a)$ be the union of $S^* \cap B(a)$ and other $\alpha - |S^* \cap B(a)|$ centers in $S^* \cap N(a)$. Let $W = \cup_{a \in A} W(a)$. Since A is 7-independent, $N^3(a)$ and $N^3(b)$ are disjoint for any two a and b in A, and so $N^2(W(a)) \cap N^2(W(b)) = \emptyset$. Thus, W is $(\alpha, 4)$-independent.

Now let L'' be such that $L''_u = 0$ if $u \notin S^*$, and $L''_u = L_u$ otherwise. Observe that (G, k, L'', α) is feasible (as we only set to zero the capacities of non-centers). By Lemma 1, the instance $(G, k - |W|, L''')$ is feasible, where $L'''_u = 0$ if $u \in W$, and $L'''_u = L''_u$ otherwise. Notice that $L'_u \geq L'''_u$ for every u, and $|B| = |W|$. Therefore, since $(G, k - |W|, L''')$ is feasible, so is $(G, k - |B|, L')$. □

Now we present a 7-approximation for the $\{0, L\}$-capacitated conservative α-fault-tolerant k-center. For this case, rather than using a capacity function, it is convenient to consider the subset V^L of vertices with capacity L. We denote by (G, k, V^L, α) and by (G, k, V^L) instances of the fault-tolerant and non-fault-tolerant versions.

In the following procedure, we denote by ALG an approximation algorithm for the $\{0, L\}$-capacitated k-center.

Algorithm 1. $\{0, L\}$-capacitated conservative α-fault-tolerant k-center.

Input: connected graph G, k, V^L, and α

1   $A \leftarrow$ a maximal 7-independent vertex set in G
2   **foreach** $a \in A$ **do**
3      $B(a) \leftarrow \alpha$ vertices chosen arbitrarily in $N(a) \cap V^L$
4   $B \leftarrow \cup_{a \in A} B(a)$
5   **if** ALG$(G, k - |B|, V^L \setminus B)$ *returns* FAILURE **then**
6   **return** FAILURE
7   **else**
8   Let (S, ϕ) be the solution returned by ALG$(G, k - |B|, V^L \setminus B)$
9   **return** $(S \cup B, \phi)$

Theorem 1. *If ALG is a β-approximation for the $\{0, L\}$-capacitated k-center, then Algorithm 1 is a $\max\{7, \beta\}$-approximation for the $\{0, L\}$-capacitated conservative α-fault-tolerant k-center.*

Proof. Consider an instance (G, k, V^L, α) of the $\{0, L\}$-capacitated conservative α-fault-tolerant k-center problem, with $G = (V, E)$. Let A, $B(a)$ for a in A, and B be as defined in Algorithm 1 with (G, k, V^L, α) as input. Assume that (G, k, V^L, α) is feasible. Since A is 7-independent, by Lemma 2, the instance $(G, k - |B|, V^L \setminus B)$, where we set to zero the capacities of all vertices in B, is also feasible for the $\{0, L\}$-capacitated k-center problem. This means that, if

Algorithm 1 executes Line 6, then the given instance is indeed infeasible. On the other hand, if ALG returns a solution (S, ϕ), then, since $|S| \leq k - |B|$, the size of $S \cup B$ is at most k, and ϕ is a valid initial center assignment. Moreover, ϕ is such that: (1) each vertex u is at distance at most β from $\phi(u)$; and (2) no vertex is assigned to B.

Let $F \subseteq S \cup B$ with $|F| = \alpha$ be a failure scenario. We describe a conservative center reassignment for $(S \cup B, \phi)$. We only need to reassign vertices initially assigned to centers in $F \setminus B$ (as no vertex was assigned to a vertex in B). Thus, at most $L|F \setminus B|$ vertices need to be reassigned. For each such u, we can choose $a \in A$ at distance at most 6 from u (as A is maximal), and let $\tilde{\phi}(u) = a$. Then, for each $a \in A$, and for each u with $\tilde{\phi}(u) = a$, reassign u to some non-full center of $B(a) \setminus F$. Notice that $B(a) \setminus F$ can absorb all reassigned vertices. Indeed, the available capacity of $B(a) \setminus F$ before the failure event is $L|B(a) \setminus F| = L|F \setminus B(a)| \geq L|F \setminus B|$, where we used $|B(a)| = |F| = \alpha$. Since for a reassigned vertex u, $d(u, \tilde{\phi}(u)) \leq 6$, and u is reassigned to some center $v \in N(\tilde{\phi}(u))$, the distance between u and v is at most 7. Also, if a vertex u was not reassigned, then the distance to its center is at most β.   □

Now, using the 6-approximation for the $\{0, L\}$-capacitated k-center by An et al. [1], we obtain the following.

Corollary 1. *Algorithm 1 using the algorithm by An et al. [1] for the $\{0, L\}$-capacitated k-center is a 7-approximation for the $\{0, L\}$-capacitated conservative α-fault-tolerant k-center.*

4   Capacitated Conservative Fault-Tolerant k-Center

In this section, we consider the capacitated conservative α-fault-tolerant k-center. Recall that this is the case in which capacities may be arbitrary. An instance for this problem is denoted by (G, k, L, α) for some $G = (V, E)$ and $L : V \to \mathbb{Z}_{\geq 0}$. Under the assumption that α is bounded by a constant, we present the first approximation for the problem.

In the $\{0, L\}$-capacitated case, each vertex assigned to a faulty center could be reassigned to a non-faulty center in $B(a)$, for an arbitrary nearby element a of a 7-independent set A. Each $B(a)$ could absorb *all* reassigned vertices. With arbitrary capacities, the set B of pre-opened centers must be obtained much more carefully, as the capacities of non-zero-capacitated vertices are not necessarily all the same. Once the set B of backup centers is selected, one needs to ensure that the residual instance for the capacitated k-center problem is feasible. In Sect. 3, an $(\alpha, 4)$-independent set is obtained from A, and Lemma 1 is used. This lemma is valid for arbitrary capacities, and so it is useful here as well. To obtain an $(\alpha, 4)$-independent set from B, we make sure that B can be partitioned in such a way that any two parts are at least at distance 7. This is done by Algorithm 2 below. In the following procedure, ALG denotes an approximation for the capacitated k-center problem.

Algorithm 2. Capacitated conservative α-fault-tolerant k-center, fixed α.

Input: connected graph $G = (V, E)$, k, and $L : V \to \mathbb{Z}_{\geq 0}$

1 **foreach** $u \in V$ **do**
2   **if** $L_u > |V|$ **then** $L_u \leftarrow |V|$
3 $B \leftarrow \emptyset$
4 **while** *there is a set* $U \subseteq V$ *with* $|U| = \alpha$ *and* $L(U) > L(B \cap N^6(U))$ **do**
5   Let $U \subseteq V$ be such that $|U| = \alpha$ and $L(U) > L(B \cap N^6(U))$
6     $B \leftarrow (B \setminus N^6(U)) \cup U$
7 **foreach** $u \in V$ **do**
8   **if** $u \in B$ **then** $L'_u \leftarrow 0$ **else** $L'_u \leftarrow L_u$
9 **if** $\text{ALG}(G, k - |B|, L')$ *returns* FAILURE **then**
10   **return** FAILURE
11 **else**
12   Let (S, ϕ) be the solution returned by $\text{ALG}(G, k - |B|, L')$
13   **return** $(S \cup B, \phi)$

Algorithm 2 is polynomial on the size of G, k, and L. The test in Line 4 is equivalent to finding a set $U \subseteq V$ with $|U| = \alpha$ that minimizes function $L(B \cap N^6(U)) - L(U)$ (note that this is a particular case of minimizing a submodular function with cardinality constraint). It would be interesting to settle the complexity of this subproblem, as an efficient algorithm for it would imply that Algorithm 2 is polynomial on α. When α is fixed, we may enumerate the sets U, and so we may obtain the next theorem (due to space restrictions, the proof is left for the full version of the paper).

Theorem 2. *Algorithm 2 using the algorithm by An et al. [1] for the capacitated k-center is a $(9 + 6\alpha)$-approximation for the capacitated conservative α-fault-tolerant k-center with fixed α.*

5 Capacitated Fault-Tolerant k-Center

5.1 An LP-Formulation

Recall that we are given a connected unweighted graph, and the objective is to decide whether there is a distance-1 solution (see Sect. 2). As in [6], we use an integer LP that formulates the problem. If, after relaxing the integrality constraints, the LP is infeasible, then we know there is no distance-1 solution, otherwise we round the solution, and obtain an approximate solution.

In the following, variable y_u indicates whether a vertex u is chosen as a center. In the fault-tolerant k-center, for each failure scenario $F \subseteq V$ with $|F| \leq \alpha$, we must have a different assignment from vertices to non-faulty centers opened by y. For the sake of simplicity, rather than using assignment variables specific for each F, we will use an equivalent formulation based on Hall's condition, which is a necessary and sufficient condition for a bipartite graph to have a perfect matching [10]. It is simple to verify that the next feasibility integer linear program, denoted by $ILP_{k,\alpha}(G)$, formulates the problem.

$$
\begin{array}{ll}
\sum_{u \in V} y_u = k & \\
|U| \leq \sum_{u \in N_G(U) \setminus F} y_u L_u & \forall\, U \subseteq V,\ F \subseteq V : |F| = \alpha \\
y_u \in \{0, 1\} & \forall\, u \in V.
\end{array}
$$

While one could try relaxing $ILP_{k,\alpha}(G)$, the induced "integrality gap" is unbounded, that is, there might be a "fractional" distance-1 solution for some graph for which the best solution has arbitrary cost s. The reason is that the opening fraction on y for a failure scenario F of α fractionally opened centers might be strictly less than α, that is, $y(F) < \alpha$. Thus, the considered constraints do not capture the real failure event.

Pre-opening Centers. Suppose that we knew a subset B of the centers of an optimal solution that might fail. Then we could set $y_u = 1$ for each u in B, that is, we pre-open u so that, for any failure scenario $F \subseteq B$, we would have $y(F) = |F|$. Obviously, we do not have such B, so we aim at more relaxed goals: (1) we pre-open centers that are known to be close to centers of an optimal solution; and (2) we consider that only this selected subset of the centers may fail, and that this comprises the worst case scenario. To achieve these goals, we first make use of a standard clustering technique [2,14,16]. In particular, Khuller and Sussmann [14] obtain:

Lemma 3 ([14])**.** *Given a connected graph $G = (V, E)$, one can obtain a set of midpoints $\Gamma \subseteq V$, and a partition of V into sets $\{C_v\}_{v \in \Gamma}$, such that*

- *there exists a rooted tree T on Γ, with $d_G(u, v) = 3$ for every edge (u, v);*
- *$N_G(v) \subseteq C_v$ for every v in Γ; and*
- *$d_G(u, v) \leq 2$ for every v in Γ and every u in C_v.*

We apply Lemma 3 to G and obtain a clustering of V. Let v in Γ be a cluster midpoint, and consider any distance-1 solution. Since up to α centers in this solution may fail, there must be at least $\alpha + 1$ centers in $N(v)$. Moreover, since sets $N(v)$ are disjoint, there are at least $\alpha+1$ centers per cluster in any distance-1 solution. Now we consider a failure event, and reason on the total capacity that may become unavailable in each cluster. In the worst case, this does not exceed the accumulated capacity of the α most capacitated vertices in the cluster. This suggests that we might pre-open such centers. For each v in Γ, let $B_v \subseteq C_v$ be a set of α elements of C_v with largest capacities. The set of all pre-opened centers is $B = \cup_{v \in \Gamma} B_v$.

When we establish a partial solution, we may turn the original linear formulation infeasible, that is, it is possible that there is no distance-1 solution whose set of centers contains B. However, since in any distance-1 solution there are at least $\alpha + 1$ centers in a given cluster, each center in such distance-1 solution is within distance 3 to a distinct element of B of non-smaller capacity. To obtain an LP relaxation that opens all centers in B, we modify the supporting graph G

as follows: we define the directed graph $G' = G'(G, \{C_v\}_{v \in \Gamma}) = (V, E')$, where E' is the set of arcs (u, w) such that $\{u, w\} \in E$, or there exist v in Γ and t in $N(v)$ such that $\{u, t\} \in E$ and $w \in B_v$ (see Fig. 1). We use a directed graph, as we only want a reassignment from a client to a center in B.

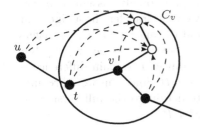

The white vertices represent B_v, the most capacitated vertices in C_v for $\alpha = 2$. The dashed lines represent the arcs added to the graph G in order to obtain G'. Solid lines represent duplicated arcs in opposite directions. Notice that u is not in C_v in this example.

Fig. 1. The modified graph.

A New Formulation. In the new formulation, we consider only scenarios $F \subseteq B$ such that we always have $y(F) = |F|$. Also, for each cluster midpoint v, we (fractionally) open at least one non-faulty center in $N(v)$ for each failure scenario. While this is implicit in $ILP_{k,\alpha}(G)$, for fractional y there might be high capacity centers that satisfy the local demand with less than one open unit. Thus, we have an additional constraint for each midpoint v. We obtain a new linear program, denoted by $LP_{k,\alpha}(G, \{C_v\}_{v \in \Gamma})$:

$$
\begin{aligned}
\sum_{u \in V} y_u &= k \\
|U| &\leq \sum_{u \in N_{G'}(U) \setminus F} y_u L_u \quad && \forall\, U \subseteq V,\ F \subseteq B : |F| = \alpha \\
1 &\leq \sum_{u \in N_G(v) \setminus B} y_u && \forall\, v \in \Gamma \\
y_u &= 1 && \forall\, u \in B \\
0 \leq y_u &\leq 1 && \forall\, u \in V.
\end{aligned}
$$

Note that the new LP depends on the clustering. The next lemma states that $LP_{k,\alpha}(G, \{C_v\}_{v \in \Gamma})$ is a "relaxation" of $ILP_{k,\alpha}(G)$.

Lemma 4. *If $ILP_{k,\alpha}(G)$ is feasible, then $LP_{k,\alpha}(G, \{C_v\}_{v \in \Gamma})$ is feasible.*

Though $LP_{k,\alpha}(G, \{C_v\}_{v \in \Gamma})$ has an exponential size, it can be separated by a min-cut max-flow algorithm, thus the LP is solvable in polynomial-time. For the $\{0, L\}$-capacitated case, we can solve the LP even if α is part of the input.

5.2 The Algorithm

Our algorithm consists of two parts. In the first, we round a fractional solution y of $LP_{k,\alpha}(G, \{C_v\}_{v \in \Gamma})$, and obtain a set R of k centers. In the second part, for each failure scenario $F \subseteq R$ with $|F| \leq \alpha$, we have to obtain an assignment from V to $R \setminus F$.

Rounding. Our rounding algorithm is based on the concept of *distance-r trans-fers* used by An et al. [1]. The main idea is that the fractional opening of centers can be transferred to other centers at distance up to r, while preserving the local capacity. If, in the modified solution, all centers are integrally open, then we obtain a set R of k centers, and may obtain an assignment from V to R. Since, by the LP constraints, each vertex of V could be (fractionally) assigned to centers at distance 1 in y, in the modified solution we may transfer the assignment as well, and obtain an assignment to centers of R at distance $r+1$. The transfers are well understood for the capacitated (*non*-fault-tolerant) k-center [1,6], but need several new ideas in the case with fault tolerance. The main difference is that we do not transfer opening from or to vertices in B. Since this would require several pages of description, we left the complete details to the full version. In particular, we obtain an integral distance-8 transfer R.

Assignment. After opening centers in R, we get an assignment for each failure scenario $F \subseteq R$ with $|F| = \alpha$. First, suppose F is a subset of the pre-opened centers B. In this case, $LP_{k,\alpha}(G, \{C_v\}_{v \in \Gamma})$ guarantees that each u can be assigned to a certain set X of (fractionally opened) centers. Consider a v in X to which u is assigned. Notice that $X \subseteq N_{G'}(u) \backslash F$, thus either $v \in N_G(u)$, or $v \in N_G^4(u) \cap B$ (the latter occurs when u is connected to v through an added arc of the modified graph). Since R is a distance-8 transfer and we do not transfer opening from B, if the opening is transfered to some vertex v', then $v, v' \notin B$ and $d_G(v, v') \leq 8$. So each vertex u can be assigned to a center in $R \setminus F$ at distance at most 9.

Now, we suppose that F is not a subset of B. Let F_v be the set of centers that failed in a cluster C_v, and $F_v' \subseteq B$ be the set of $|F_v|$ most capacitated centers in C_v. In this case, we will first obtain a distance-9 assignment ϕ' for the failure scenario $F' = \cup_{v \in \Gamma} F_v'$. Notice that ϕ' may assign clients to a faulty center $w \in F \setminus F'$. However, for each such w, by the construction of F', there must be some unused center $s(w) \in F' \setminus F$ in the same cluster with non-smaller capacity. Therefore, we reassign each client u connected to w to center $s(w)$, obtaining a feasible assignment.

A naive analysis of the preceding algorithm would yield a 13-approximation (as $d(u, s(w)) \leq d(u, w) + d(w, s(w)) \leq 9 + 4$). Let $\delta(w)$ be the midpoint of the cluster that contains w, that is, $\delta(w)$ is the midpoint v such that $w \in C_v$. To obtain a factor 10, we carefully bound $d(u, \delta(w))$ by 8, so that $d(u, s(w)) \leq d(u, \delta(w)) + d(\delta(w), s(w)) \leq 8 + 2$. This is formalized next.

Lemma 5. *Consider $F \subseteq B$ with $|F| = \alpha$ and let R be the integral transfer obtained from y. One can find, in polynomial time, an assignment $\phi : V \to R \backslash F$ such that $d_G(u, \phi(u)) \leq 9$ and $d_G(u, \delta(\phi(u))) \leq 8$ for each u in V.*

Corollary 2. *For constant α, there is a 10-approximation for the capacitated α-fault-tolerant k-center.*

5.3   The $\{0, L\}$-Capacitated Case

We apply our strategy to the $\{0, L\}$-capacitated case. As in [1], we further pre-process the input graph before solving the LP so that a better clustering may be obtained. Since, in this case, non-zero capacitated centers have the same capacities, the set of pre-opened centers B_v for a cluster C_v can be chosen in $N(v)$. Moreover, we use the uniformity of capacities of pre-opened centers to simplify the LP so that it can be solved even if α is part of the input. We obtain the following result.

Theorem 3. *There exists a 6-approximation for the $\{0, L\}$-capacitated α-fault-tolerant k-center (with α as part of the input).*

References

1. An, H.-C., Bhaskara, A., Chekuri, C., Gupta, S., Madan, V., Svensson, O.: Centrality of trees for capacitated k-center. In: Lee, J., Vygen, J. (eds.) IPCO 2014. LNCS, vol. 8494, pp. 52–63. Springer, Heidelberg (2014)
2. Bar-Ilan, J., Kortsarz, G., Peleg, D.: How to allocate network centers. J. Algorithms **15**(3), 385–415 (1993)
3. Chaudhuri, S., Garg, N., Ravi, R.: The p-neighbor k-center problem. Inf. Process. Lett. **65**(3), 131–134 (1998)
4. Chechik, S., Peleg, D.: The fault-tolerant capacitated k-center problem. Theor. Comput. Sci. **566**, 12–25 (2015)
5. Cygan, M., Hajiaghayi, M., Khuller, S.: LP rounding for k-centers with non-uniform hard capacities. In: IEEE 53rd Annual Symposium on Foundations of Computer Science (FOCS), pp. 273–282 (2012)
6. Cygan, M., Kociumaka, T.: Constant factor approximation for capacitated k-center with outliers. In: 31st International Symposium on Theoretical Aspects of ComputerScience (STACS), vol. 25, pp. 251–262 (2014)
7. Feder, T., Greene, D.: Optimal algorithms for approximate clustering. In: Proceedings of the Twentieth Annual ACM Symposium on Theory of Computing (STOC), pp. 434–444. ACM, New York (1988)
8. Garey, M.R., Johnson, D.S.: Computers and Intractability: A Guide to the Theory of NP-Completeness. Freeman, New York (1979)
9. Gonzalez, T.F.: Clustering to minimize the maximum intercluster distance. Theor. Comput. Sci. **38**, 293–306 (1985)
10. Hall, P.: On representatives of subsets. J. London Math. Soc **10**(1), 26–30 (1935)
11. Hochbaum, D.S., Shmoys, D.B.: A best possible heuristic for the k-center problem. Math. Oper. Res. **10**(2), 180–184 (1985)
12. Hochbaum, D.S., Shmoys, D.B.: A unified approach to approximation algorithms for bottleneck problems. J. ACM **33**(3), 533–550 (1986)
13. Hsu, W.L., Nemhauser, G.L.: Easy and hard bottleneck location problems. Discrete Appl. Math. **1**(3), 209–215 (1979)
14. Khuller, S., Sussmann, Y.J.: The capacitated k-center problem. SIAM J. Discrete Math. **13**(3), 403–418 (2000)
15. Khuller, S., Pless, R., Sussmann, Y.J.: Fault tolerant k-center problems. Theor. Comput. Sci. **242**(1–2), 237–245 (2000)
16. Krumke, S.: On a generalization of the p-center problem. Inf. Process. Lett. **56**(2), 67–71 (1995)

Bundled Crossings in Embedded Graphs

Martin Fink[1]($\boxtimes$), John Hershberger[2], Subhash Suri[1], and Kevin Verbeek[3]

[1] Department of Computer Science, University of California,
Santa Barbara, CA, USA
fink@cs.ucsb.edu
[2] Mentor Graphics Corporation, Wilsonville, OR, USA
[3] Department of Mathematics and Computer Science, TU Eindhoven,
Eindhoven, The Netherlands

Abstract. Edge crossings in a graph drawing are an important factor in the drawing's quality. However, it is not just the presence of crossings that determines the drawing's quality: any drawing of a nonplanar graph in the plane necessarily contains crossings, but the geometric structure of those crossings can have a significant impact on the drawing's readability. In particular, the structure of two disjoint groups of locally parallel edges (*bundles*) intersecting in a complete crossbar (a *bundled crossing*) is visually simpler—even if it involves many individual crossings—than an equal number of random crossings scattered in the plane.

In this paper, we investigate the complexity of partitioning the crossings of a given drawing of a graph into a minimum number of bundled crossings. We show that this problem is *NP*-hard, propose a constant-factor approximation scheme for the case of *circular embeddings*, where all vertices lie on the outer face, and show that the bundled crossings problem in general graphs is related to a minimum dissection problem.

1 Introduction

We introduce and investigate the problem of minimizing the number of *bundled crossings* in a given embedding of a graph, where each bundled crossing is formed by two groups of edges that cross each other in a local region. See Fig. 1 for an example. In particular, let $G = (V, E)$ be a simple graph that is drawn in the plane. The drawing defines a *combinatorial embedding* $\mathcal{E}$ of the graph: the circular order of edges around each vertex and each crossing, and the relative order of crossings along each edge. We are free to alter the drawing in any way to improve its appearance but must adhere to the prescribed combinatorial embedding $\mathcal{E}$. For ease of reference, we will identify each crossing e_1, e_2 with a *crossing vertex*, labeled with the pair (e_1, e_2) of edges involved in the crossing. A *bundled crossing* in the embedding $\mathcal{E}$ is a subset C of the crossing vertices satisfying the following two conditions:

(i) C contains exactly the crossings of the form (e_1, e_2) for all $e_1 \in E_1$, $e_2 \in E_2$, where $E_1, E_2 \subseteq E$ are two disjoint subsets of the edges. (E_1 and E_2 are the *bundles* of the *bundled crossing*.)

© Springer-Verlag Berlin Heidelberg 2016
E. Kranakis et al. (Eds.): LATIN 2016, LNCS 9644, pp. 454–468, 2016.
DOI: 10.1007/978-3-662-49529-2_34

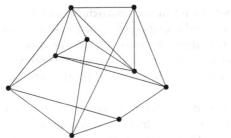 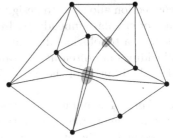

Fig. 1. Two drawings of a graph with the same embedding: a straight-line drawing with 8 crossings, and a drawing with two bundled crossings (highlighted).

(ii) C can be separated from all remaining crossings of the embedding by a pseudodisk D—a closed polygonal curve that crosses each edge at most twice—and no other edge $e \notin E_1 \cup E_2$ intersects D. This requirement (ii) ensures that the bundled crossing can be visually separated from the rest of the drawing.

The *bundled crossing number* $\mathrm{bc}(\mathcal{E})$ of the embedding $\mathcal{E}$ is the minimum number of bundled crossings into which the set of crossings can be partitioned. In this paper we study the following problem.

Bundled Crossing Number Problem. *Given a nonplanar embedding $\mathcal{E}$ of a graph $G = (V, E)$, determine* $\mathrm{bc}(\mathcal{E})$.

Motivation. We propose the bundled crossing number as a new measure for the visual quality of a nonplanar drawing. We believe that the number of bundled crossings is a better measure of visual clutter than the number of individual crossings, because bundled crossings take the structure of crossings into account. A drawing with few bundled crossings is simpler than one with individual crossings scattered over the drawing in an unstructured way.

Since minimizing the number of bundled crossings may conflict with other aesthetic criteria and application constraints, we expect bundled crossing minimization to be used as a postprocessing step for improving the output of a separate graph drawing algorithm. Once the embedding is fixed, the crossings can be partitioned into a small number of bundled crossings. Then, by appropriately routing edges, e.g., as splines, we can bring the edges of bundles close together to emphasize the simple structure of the bundled crossings (see Fig. 1).

Our Contribution. Our first result is that the problem of minimizing the number of bundled crossings in general is *NP*-hard. We further show that minimizing bundled crossings is a generalization of the problem of dissecting a rectilinear polygon-with-holes into a minimum number of rectangles. In particular, the number of bundled crossings is equal to the number of combinatorial rectangles needed to partition a special combinatorial polygon derived from the input graph embedding.

This connection allows us to design a constant-factor approximation for the simpler case of circular embeddings. In a circular embedding, all vertices lie on the boundary of a topological disk, and the edges are routed inside the disk. Circular embeddings are often used in applications with dense graphs where vertices placed in the interior of the drawing can easily be obscured by the crossings.

Related Work. The number of crossings in a graph drawing is perhaps the most obvious criterion of drawing quality. Computing a drawing with a minimum number of crossings is therefore a natural optimization problem, but unfortunately it is *NP*-hard [5].

The number of crossings is not the only determiner of a drawing's readability, and many other criteria have been explored. For instance, even if the combinatorial embeddings of two drawings are the same, one with larger crossing angles is more readable [9], since larger angles make it easier to distinguish edges. Recently, Hu and Shi [8] considered using different colors for edges; in their approach, pairs of edges with a small crossing angle should have distinct colors to make the edges easier to distinguish from each other.

Eppstein et al. [2] suggested another way to evaluate the readability of a nonplanar drawing. On each crossing they use shading to indicate which of the crossing edges is on top. They claim that for a single edge it is best to be either always above or always below other edges that are crossed; their aim is, hence, to minimize the number of *switches* of this above-below status for all edges.

Holten and van Wijk introduced *edge bundling* [6,7]. In bundled drawings, groups of edges are drawn close together for some time when being routed through the drawing. While this naturally leads to a bundled crossing when two edge bundles cross, the existing work on edge bundling does not use the number of bundled crossings as an optimization criterion. Instead, heuristics using, e.g., the similarity of edges or their closeness in an initial drawing are used to compute bundles and output a bundled drawing. In a recent survey, Schaefer [10] has asked for a crossing number variant making use of bundles, partially motivated by the work of Fink et al., who introduced a method similar to bundled crossings, called *block crossings*, for presenting the crossings of metro lines in transportation networks of cities [3]. However, the similarity between our bundled crossing problem and the block crossings in metro line drawings is only superficial—the block crossing number of an instance says nothing about its bundled crossing number because the metro lines and block crossings must be routed along edges of the given graph. While minimizing the number of block crossings of metro lines is *NP*-hard, Fink et al. developed approximation algorithms and heuristics for special networks as well as general metro networks. Due to the similarity between edge bundles and metro lines, these algorithms can also be applied for the crossings of edges within a bundle in edge bundling.

2   Preliminaries

A drawing of a graph $G = (V, E)$ in the plane imposes a combinatorial structure, namely, the circular order of edges around each vertex, the order in which crossings occur along each edge, and the order of edges around each crossing. This combinatorial embedding $\mathcal{E}$ contains all the relevant information for the bundled crossing minimization problem. Therefore, in the following, we use the combinatorial embedding instead of the actual drawing. Equivalently, $\mathcal{E}$ is the planar combinatorial embedding of the planar drawing resulting from replacing each crossing by a dummy vertex. We assume that in $\mathcal{E}$ we can distinguish between crossing vertices and vertices of G, and we can recognize the edges of the original graph.

We assume that both the graph and the drawing are simple. This means that there are no pairs of edges that touch, except in a crossing or vertex, and there is no vertex contained in the interior of an edge. Furthermore, no three edges can meet in a point that is not a vertex, no edge crosses itself, no two adjacent edges cross, and no two edges in E cross more than once. Hence, each crossing vertex has degree four and there are no faces of degree 2.

For the purpose of minimizing bundled crossings, we can treat the graph as a matching: since the order of edges around each vertex is fixed, we can remove a small ε-environment around each vertex from a drawing, that is, a disk of radius ε centered at the vertex that is so small that it contains only the vertex and parts of the incident edges, but no other vertices, crossings, or other edges. After removing the ε-environment we solve the problem, and then reinsert the removed parts. This is equivalent to replacing each vertex v by $(\deg v)$ many vertices, where each vertex is an endpoint of a single edge. It follows that the order of crossings along each edge and order of edges around each crossing are the only relevant input.

Observation. *Solving the bundled crossing number problem in general can be reduced to the special case in which the embedded graph G is a perfect matching.*

3   Bundled Crossing Minimization Is *NP*-Hard

We begin by showing that the bundled crossing number problem is *NP*-complete.

Theorem 1. *Given a nonplanar combinatorial embedding $\mathcal{E}$ of a matching $G = (V, E)$ and a number k, it is NP-complete to decide whether $\mathrm{bc}(\mathcal{E}) \leq k$.*

Proof. $\mathrm{bc}(\mathcal{E}) \leq k$ would be witnessed by a partition of the crossings into k subsets. Furthermore, if we take the smallest (combinatorial) pseudodisc around each set of crossings, it is easy to check whether a partition of the crossings is a feasible partitioning into bundled crossings. Hence, the problem is in NP.

We show *NP*-hardness by reduction from PLANAR 3SAT. Let $(X, \mathcal{C})$ be a PLANAR 3SAT instance, where X is a set of variables and $\mathcal{C}$ is a set of clauses on these variables. Each clause $C \in \mathcal{C}$ contains negated or unnegated variables, that is, $C \subseteq X \cup \{\neg x \mid x \in X\}$, and we know that $|C| \in \{2, 3\}$. The problem remains *NP*-hard even if every variable occurs in only three clauses [1]. The graph $G_{XC} = (X \cup \mathcal{C}, E_{XC})$ with $E_{XC} = \{(x, C) \mid x \in C \text{ or } \neg x \in C\}$ describing the occurrence of variables in clauses has a planar embedding $\mathcal{E}_{XC}$.

Based on the input, we create a nonplanar embedding $\mathcal{E}$ of a graph G that models the 3SAT instance. By placing variable and clause gadgets on top of their vertices in $\mathcal{E}_{XC}$ and routing edges accordingly, we ensure that $\mathcal{E}$ contains only crossings that we added intentionally as part of the gadgets.

In our construction no inner face will have degree 4. As a result, all bundled crossings must be "thin", that is, consist of a single edge that crosses one or more other edges. Each edge is part of at most two gadgets, a variable gadget and a clause gadget. On an edge connecting two gadgets, there can be a bundled crossing containing crossings of both gadgets, allowing us to save bundled crossings in some cases. Therefore, in such cases, we also say that a bundled crossing extends from the variable gadget into the clause gadget.

Our basic tool is a hexagon, that is, a hexagonal face bounded by six edges and six crossings. In any solution, the six crossings of the hexagon are part of at least three bundled crossings; there are two different solutions with only three bundled crossings for the hexagon, which we can associate with true and false.

Variable Gadget. For each variable $x \in X$, we place a variable gadget g_x consisting of four hexagons that are connected as shown in Fig. 2. In a solution with the minimum number of bundled crossings, the central hexagon needs three bundled crossings; each of the other hexagons can use one bundled crossing extending from the central hexagon and needs only two additional bundled crossings. Hence, there are only two solutions with 9 bundled crossings for the gadget, the one indicated in bold—the true state—and the symmetric one (false); all other solutions have more bundled crossings since each bundled crossing contains at most three crossings and there can be only three bundled crossings with three crossings. The marked corners can be used for connecting the variable to clauses. All three locally have the same status in an optimum solution. For a negated variable, we take the corner marked by a dashed circle (and the one right of it).

Clause Gadget. For each clause $C \in \mathcal{C}$, we build a clause gadget g_C and place it according to $\mathcal{E}_{XC}$. We connect each variable of the clause to the gadget via two edges coming from the variable gadget; see an example in Fig. 3. Which hexagon of the variable gadget is used for the connection is determined based on the clockwise order of edges in the embedding; the same holds for the order of the literals in the clause. Note that for the middle literal (y) the second used edge comes from the vertex right of the marked vertex in the literal's gadget. By routing the two edges between variable and clause gadget following the corresponding edge in the embedding $\mathcal{E}_{XC}$ we make sure that there are no crossings except the ones that are part of the gadgets.

For each variable, only one of two edges can extend a bundled crossing into the gadget—the bold one if the variable is true, the other one if it is false. By considering the eight possible cases, one can check that if at least one literal is true, one additional bundled crossing suffices for the gadget (see bold bundled crossings in Fig. 3). If all literals are false, two additional bundled crossings are necessary; otherwise, the crossings of the triangular face cannot be covered.

A clause $x \vee y$ with two literals is modeled as a single crossing between the positive edges of the literals. If a literal is true, the crossing can be part of an

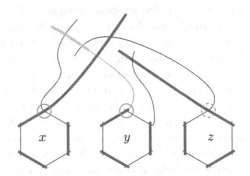

Fig. 2. Variable gadget in true-state. Vertices marked by circles are used for connecting to a clause gadget; for negated occurrences, vertices marked by dashed circles are used instead.

Fig. 3. Clause gadget for the clause $x \vee y \vee \neg z$; by rotating the respective hexagon by one corner, a variable can be replaced by its negation (here shown for z).

existing bundled crossing of the corresponding variable gadget; otherwise, an additional bundled crossing is necessary.

Let k_3 be the number of clauses with three literals and let $k = 9|X| + k_3$. We now claim that the crossings can be partitioned into at most $k = 9|X| + k_3$ bundled crossings if and only if the 3SAT formula is satisfiable, where k_3 is the number of clauses with three literals.

First assume that we have a satisfying truth assignment. For every variable $x \in X$, we partition the crossings of the corresponding variable gadget into 9 bundled crossings in one of the two possible ways, depending on the truth assignment for x. Then, since for every clause at least one variable is true, bundled crossings starting in variable gadgets can be extended in such a way that every clause gadget with three variables is covered with only one additional bundled crossing; no additional bundled crossing is necessary for clauses with two literals. Hence, we have partitioned all crossings into only $9|X| + k_3 = k$ bundled crossings.

Now, assume that we are given a solution with at most k bundled crossings. We want to construct a satisfying truth assignment. Note that no gadget contains an inner quadrilateral face bounded by four crossings. Furthermore, edges leaving a variable gadget either go to the same clause gadget, or they are separated from each other by several crossings on the boundary of the gadget. Hence, also no face bounded by edges of different gadgets can be an inner quadrilateral that may be contained in a bundled crossing. Therefore, all bundled crossings consist of a single edge that crosses one or more other edges.

We assign each bundled crossing that contains crossings of two gadgets to the variable gadget. Since each variable gadget needs at least 9 bundled crossings, there can be no more than k_3 crossings not assigned to a variable gadget. Assume that a clause gadget with variables x, y, and z (see Fig. 3) does not have any bundled crossing assigned. This is only possible if both edges coming from y's variable gadget extend a bundled crossing into the clause gadget. The connecting hexagon then has at least three bundled crossings not extending from the variable gadget's central hexagon, and the variable gadget must have at least 10 bundled crossings in total. There are several cases that locally are possible for the connecting hexagon; for each of the cases it is easy to verify that a local repartitioning is possible such that (i) the number of bundled crossings does not change, (ii) the hexagon has only two bundled crossing not extending from the central hexagon, and (iii) the clause gadget has one bundled crossing not extending from a variable gadget. If we apply this repartitioning step for all variable gadgets, if necessary, property (ii) will ensure that every variable gadget has only 9 bundled crossings.

We have seen, that 9 bundled crossings per variable gadget and one bundled crossing for every clause gadget with three variables are necessary. Therefore, our solution has exactly this number of bundled crossings for each of the gadget types. Since there are only two possible solutions with 9 crossings for a variable gadget, we automatically get a corresponding truth assignment, and we argue that this truth assignment satisfies all clauses.

For a clause $x \lor y$, the corresponding crossing must be part of a bundled crossing coming from one of the variable gadgets. Hence, the corresponding literal is **true**. On the other hand, a clause gadget for $x \lor y \lor z$ must have one bundled crossing coming from the positive edge of one of the literals, which, hence, is **true** and satisfies the clause.   □

4   Bundled Crossings via Minimum Dissection

In this section we reduce the problem of computing the bundled crossing number of a given embedding $\mathcal{E}$ to that of dissecting a combinatorial space $\mathcal{C}(\mathcal{E})$ into as few "rectangles" as possible. This latter problem is actually a generalization of the minimum dissection problem for rectilinear polygons with holes. A rectilinear polygon with holes can be dissected into the minimum possible number of rectangles in polynomial time [11]. As Theorem 1 suggests, this result cannot be extended to our generalization of the problem, but the insight gained from reducing the problem, combined with the ideas in [11], allow us to obtain approximate solutions for the bundled crossing number.

4.1   The Crossing Complex

For a given embedding $\mathcal{E}$ we can construct a combinatorial space: the *crossing complex* $\mathcal{C}(\mathcal{E})$ is a special type of cell complex obtained by gluing together quadrilateral cells according to the structure of $\mathcal{E}$. Specifically, $\mathcal{C}(\mathcal{E})$ can be obtained from $\mathcal{E}$ using the following three construction rules (see Fig. 4):

1. There is one quadrilateral cell $\mathcal{C}(v)$ in $\mathcal{C}(\mathcal{E})$ for every crossing vertex v of $\mathcal{E}$. Since v has degree four, every side of $\mathcal{C}(v)$ is associated with one of the incident edges of v in $\mathcal{E}$.

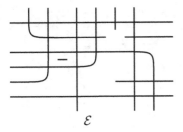
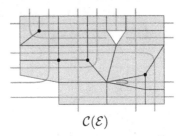

$\mathcal{E}$ $\qquad\qquad\qquad\qquad\qquad$ $\mathcal{C}(\mathcal{E})$

Fig. 4. Embedding $\mathcal{E}$ and crossing complex $\mathcal{C}(\mathcal{E})$. Black dots represent point holes.

2. If two crossing vertices u and v share an edge in $\mathcal{E}$, then the cells $\mathcal{C}(u)$ and $\mathcal{C}(v)$ are glued together along the sides corresponding to the shared edge.
3. If $\mathcal{E}$ contains an empty quadrilateral face, then the unique corner shared by the cells corresponding to the four crossing vertices of the quadrilateral face is added to $\mathcal{C}(\mathcal{E})$.

As the last construction rule already suggests, the cells of $\mathcal{C}(\mathcal{E})$ should be considered as open cells. That is, a corner of a cell is not in $\mathcal{C}(\mathcal{E})$ unless it is added due to Rule 3, and a side is not in $\mathcal{C}(\mathcal{E})$ unless it was involved in a gluing operation. Corners and sides that are in $\mathcal{C}(\mathcal{E})$ are referred to as *internal.* Other corners and sides belong to the boundary of $\mathcal{C}(\mathcal{E})$. Note that corners that correspond to nonempty quadrilateral faces or other faces of $\mathcal{E}$ (not containing vertices of G) are actually point holes in $\mathcal{C}(\mathcal{E})$—holes consisting of a single point.

4.2 Dissecting the Crossing Complex

Given a crossing complex $\mathcal{C}(\mathcal{E})$, the goal is to dissect $\mathcal{C}(\mathcal{E})$ into rectangular subcomplexes. A *rectangular subcomplex* of a crossing complex consists of a subset of the cells arranged in a $k \times \ell$ grid (for $k, \ell \geq 1$) without any holes. A *chord* of a crossing complex consists of a sequence of internal sides and internal corners connecting two boundary corners. A chord must be straight, that is, the two internal sides incident on an internal corner (which must have degree four) must be opposite to each other (see Fig. 5 left). A crossing complex can be dissected into rectangular subcomplexes by cutting the crossing complex along internally disjoint chords (see Fig. 5 right).

Lemma 1. *Partitioning the crossing vertices of an embedding $\mathcal{E}$ into bundled crossings is equivalent to dissecting the crossing complex $\mathcal{C}(\mathcal{E})$ into rectangular subcomplexes.*

Proof. We show that every bundled crossing of $\mathcal{E}$ corresponds to a rectangular subcomplex of $\mathcal{C}(\mathcal{E})$ and vice versa. Assume we have a bundled crossing of $\mathcal{E}$ consisting of all crossings of the edge bundles E_1 and E_2 (of the original graph), say with $|E_1| = k$ and $|E_2| = \ell$. Thus, the crossings among these edges

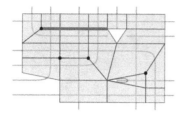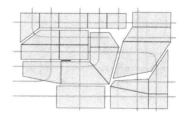

Fig. 5. Left: a chord (bold). Right: dissection into rectangular subcomplexes.

must be arranged in a $k \times \ell$ grid, which corresponds to a rectangular subcomplex of $\mathcal{C}(\mathcal{E})$. Furthermore, the internal faces inside the bundled crossing must be empty quadrilaterals, so the corresponding rectangular subcomplex does not have holes due to construction rule 3.

Now assume we have a rectangular subcomplex of $\mathcal{C}(\mathcal{E})$ of size $k \times \ell$. By construction of $\mathcal{C}(\mathcal{E})$, every column (and every row) of the rectangular subcomplex corresponds to a sequence of crossing vertices of $\mathcal{E}$ that are part of the same edge. Thus, there is a set E_1 of k edges and a set E_2 of ℓ edges, and cells of the rectangular subcomplex correspond to the crossings between all pairs $(e_1, e_2) \in E_1 \times E_2$. Since the rectangular subcomplex does not have holes, the internal quadrilateral faces among these edges in $\mathcal{E}$ must be empty, and hence the crossings between the edges in $E_1 \times E_2$ can be bundled. $\square$

Lemma 1 directly implies that the bundled crossing number of an embedding $\mathcal{E}$ is exactly the minimum number of rectangular subcomplexes into which $\mathcal{C}(\mathcal{E})$ can be dissected. Extending ideas from [11], we define a *measure* $m(\mathcal{C})$ for the "concavity" of a crossing complex $\mathcal{C}$. Let $\partial \mathcal{C}$ be the set of boundary corners of $\mathcal{C}$. For a boundary corner $c \in \partial \mathcal{C}$ we define the angle $\alpha(c)$ as the number of cells in $\mathcal{C}$ it bounds. If a cell uses a corner twice (see, e.g., the small hole in Fig. 4), then this cell must be counted twice toward the angle of this corner. The measure $m(\mathcal{C})$ can now be defined as follows.

$$m(\mathcal{C}) = \sum_{c \in \partial \mathcal{C}} \begin{cases} \lfloor \frac{\alpha(c)-1}{2} \rfloor + 1, & \text{if } c \text{ is a point hole;} \\ \lfloor \frac{\alpha(c)-1}{2} \rfloor, & \text{otherwise.} \end{cases}$$

Note that the measure of a rectangular subcomplex is zero. Furthermore, if the measure of a crossing complex is nonzero, then it contains a point hole or a boundary corner of angle at least three, which means it is not a rectangular subcomplex. Thus, the measure of a crossing complex is zero if and only if it consists of a set of rectangular subcomplexes. Next we investigate the influence of cutting along a chord on the measure of a crossing complex.

Lemma 2. *Let $\mathcal{C}'$ be the result of cutting a crossing complex $\mathcal{C}$ along a chord. Then $m(\mathcal{C}') \geq m(\mathcal{C}) - 2$.*

Proof. A chord can influence the measure at only two boundary corners of $\mathcal{C}$, namely the endpoints of the chord. If a boundary corner is a point hole, then the angle of this boundary corner is not changed, but $m(\mathcal{C})$ is reduced by one. Otherwise, a boundary corner with angle k is split into two boundary corners, one with angle k_1 and one with angle k_2, where $k_1 + k_2 = k$. If k_1 or k_2 is odd, then $\lfloor \frac{k_1-1}{2} \rfloor + \lfloor \frac{k_2-1}{2} \rfloor \geq \frac{k_1+k_2-3}{2} = \frac{k-1}{2} - 1$. Otherwise, k is even and $\lfloor \frac{k_1-1}{2} \rfloor + \lfloor \frac{k_2-1}{2} \rfloor \geq \frac{k_1+k_2-4}{2} = \lfloor \frac{k-1}{2} \rfloor - 1$. Again, $m(\mathcal{C})$ is reduced by at most one and the result follows. $\qquad\square$

Lemma 3. *If $m(\mathcal{C}) > 0$, then there exists a chord that can be used to reduce $m(\mathcal{C})$ by one.*

Proof. Since $m(\mathcal{C}) > 0$, there must be a boundary corner c with $\alpha(c) \geq 3$ (this must also hold for point holes). If c is a point hole, then we can simply choose one of the incident internal sides and construct a chord starting from there, effectively removing the point hole. This chord is uniquely defined and must reach a boundary corner eventually. This can be seen by observing that internal corners of the chord in $\mathcal{C}(\mathcal{E})$ correspond to a sequence of empty quadrilaterals in $\mathcal{E}$; if a chord loops onto itself, then some edges of the original graph must intersect themselves or make a loop, which is not allowed. If c is not a point hole, we choose a chord that splits c into two boundary corners with angles 2 and $\alpha(c) - 2$. This again reduces $m(\mathcal{C})$ by one. $\qquad\square$

We should note that the optimal dissection of a crossing complex can always be obtained by cutting along chords completely. Following the arguments of Lemma 2, a partial chord can reduce the measure by at most one, and by Lemma 3 such partial cuts are never needed to obtain an optimal solution. Lemma 3 directly suggests a simple algorithm to dissect a crossing complex into rectangular subcomplexes: keep applying Lemma 3 until the measure is zero. Next we analyze this simple algorithm, which we call REPEATEDCUT.

For the remainder of this section we assume that we are given an embedding $\mathcal{E}$ and its crossing complex $\mathcal{C} = \mathcal{C}(\mathcal{E})$, and that $\mathcal{C}$ is connected; otherwise we can run the algorithm on each connected component separately. The number of holes $h(\mathcal{C})$ of $\mathcal{C}$ is the number of connected components of the boundary of $\mathcal{C}$ minus the number of connected components of $\mathcal{C}$. When $\mathcal{C}$ is cut along a chord to obtain $\mathcal{C}'$, one of two things can happen: (1) $h(\mathcal{C}') = h(\mathcal{C}) - 1$, or (2) $\mathcal{C}'$ has exactly one more connected component than $\mathcal{C}$. In the end $m(\mathcal{C}) = h(\mathcal{C}) = 0$ must hold, and the goal is to minimize the number of connected components.

Let $\mathrm{ALG}(\mathcal{E})$ be the number of bundled crossings obtained by algorithm REPEATEDCUT, using the equivalence of Lemma 1. We obtain the following result.

Theorem 2. *For any embedding $\mathcal{E}$ we have $\mathrm{ALG}(\mathcal{E}) \leq 2\,\mathrm{bc}(\mathcal{E}) + h(\mathcal{C}(\mathcal{E})) - 1$.*

Proof. Every time we cut $\mathcal{C}(\mathcal{E})$ along a chord and reduce $m(\mathcal{C}(\mathcal{E}))$ by at least one, either the number of holes of $\mathcal{C}(\mathcal{E})$ reduces by one, or the number of connected components of $\mathcal{C}(\mathcal{E})$ increases by one (starting at one). So we obtain that $\text{ALG}(\mathcal{E}) - 1 + h(\mathcal{C}(\mathcal{E})) \leq m(\mathcal{C}(\mathcal{E}))$. Similarly, from Lemma 2 it follows that $\text{bc}(\mathcal{E}) - 1 + h(\mathcal{C}(\mathcal{E})) \geq \frac{m(\mathcal{C}(\mathcal{E}))}{2}$. Thus, we obtain:

$$\text{ALG}(\mathcal{E}) \leq m(\mathcal{C}(\mathcal{E})) + 1 - h(\mathcal{C}(\mathcal{E})) = 2\frac{m(\mathcal{C}(\mathcal{E}))}{2} + 1 - h(\mathcal{C}(\mathcal{E}))$$
$$\leq 2\,\text{bc}(\mathcal{E}) + h(\mathcal{C}(\mathcal{E})) - 1. \qquad \square$$

The algorithm REPEATEDCUT can easily be implemented to run in linear time with respect to the complexity of $\mathcal{E}$: after constructing $\mathcal{C}(\mathcal{E})$, Lemma 3 is repeatedly applied. We maintain a list of holes with positive measure that are used for starting chords. Since we cut at most once through every side of a cell, the total length of cuts is linear in the complexity of $\mathcal{E}$, and so is the runtime.

4.3 Effective Chords

Unlike the algorithm for dissecting rectilinear polygons with holes [11], our algorithm is unable to compute the minimum dissection of crossing complexes efficiently. This is unsurprising, given the hardness result (Theorem 1), but in this section we explore the reasons behind this discrepancy.

The algorithm in [11] relies on so-called *effective chords* that reduce the measure by two. Unlike our chords, these can contain boundary corners in their interior. If a chord cuts an even angle into two odd angles, then the measure does not change at the corner. In that case boundary corners can be in the interior of the chord. Note that effective chords cannot be used simultaneously if they intersect or meet at a boundary corner. As a result, the maximum independent set of effective chords must be found, but since the crossing graph is bipartite for axis-aligned chords, this can be computed efficiently.

In our case, the crossing graph of effective chords is not bipartite, since we do not have horizontal and vertical edges, and we cannot compute the maximum independent set efficiently. However, even if this were the case, there is another problem to deal with: the angle of a boundary corner may be larger than four and an effective chord passing through a boundary corner is not uniquely defined. Furthermore, multiple chords can meet at the same boundary corner, and their effectiveness at this corner may depend on each other's presence in the final solution. Unfortunately, finding the largest set of effective chords seems very hard. Finally, note that even an approximation of the maximum set of effective chords cannot guarantee an approximation of the bundled crossing number.

5 Circular Embeddings

We now consider circular embeddings, in which all vertices lie on the outer face. Such embeddings occur in several applications and have also been used in the context of edge bundling [6].

We will show that the algorithm REPEATEDCUT developed in the previous section yields a constant-factor approximation for the bundled crossing number of circular embeddings. The reason is that for circular embeddings there is a linear lower bound for the bundled crossing number in terms of the number of holes, that is, the number of inner faces that cannot be contained completely in a single bundled crossing. We first relate the number of triangular faces in circular embeddings to the number of faces of higher degree. We then prove a lower bound for the bundled crossing number in terms of the number of triangles. Using the bound and the insight on face-degrees, we can then show that the algorithm yields a constant-factor approximation.

Faces in Circular Embeddings. Let $V = \{v_1, v_2, \ldots, v_n\}$ be the clockwise order of the vertices on the outer face (i.e., on the circle). To aid our analysis, we modify the graph (and the embedding) and assume that the cycle $(v_1, v_2, \ldots, v_n, v_1)$ is contained in G; in the embedding $\mathcal{E}$ this cycle forms the outer face. Let $G = (V', E')$ be the planar graph of the embedding $\mathcal{E}$ whose vertices are vertices of G and crossing vertices.

V contains n vertices and a number n_c of crossing vertices. The vertices of V lie on the outer face; all crossing vertices are inner vertices. For $k \geq 3$ let f_k be the number of inner faces of degree k. Since there is an outer face of degree n, the total number of faces in $\mathcal{E}$ is $f = 1 + \sum_{k \geq 3} f_k$. We relate the number of faces of degree three to the total complexity of faces of degree five and more.

Lemma 4.
$$f_3 = 4 + \sum_{k \geq 5} f_k \cdot (k - 4)$$

Proof. Every vertex $v \in V$ has degree 3 and every crossing vertex has degree 4. We express the number of edges in E' in two ways: (1) $|E'| = 3n/2 + 2n_c$ by counting the vertex-degrees and (2) $|E'| = n/2 + 1/2 \sum_{k \geq 3} f_k \cdot k$ by counting the face-degrees. By Euler's formula $|V'| - |E'| + f = 2$, that is, $|E'| - |V'| + 2 = 1 + \sum_{k \geq 3} f_k$. Hence, we can determine the number of triangles in the embedding as follows.

$$f_3 = 4f_3 - 3f_3 = 4 \left(1 + \frac{1}{2}|E'| + \frac{1}{2}|E'| - |V'| - \sum_{k \geq 4} f_k \right) - 3f_3$$

$$= 4 \left(1 + \frac{1}{2} \left(\frac{3}{2}n + 2n_c \right) + \frac{1}{2} \left(\frac{n}{2} + \frac{1}{2} \sum_{k \geq 3} f_k \cdot k \right) - (n + n_c) - \sum_{k \geq 4} f_k \right) - 3f_3$$

$$= 4 \left(1 + \frac{1}{4} 3 f_3 + \frac{1}{4} \sum_{k \geq 4} f_k \cdot (k - 4) \right) - 3f_3 = 4 + \sum_{k \geq 5} f_k \cdot (k - 4) \qquad \square$$

Hence, the number of triangles is directly correlated to the additional complexity of faces more complex than quadrilaterals.

Lower Bound. We will now develop a lower bound on the bundled crossing number in terms of the numbers of triangles in the embedding $\mathcal{E}$.

Lemma 5.
$$\mathrm{bc}(\mathcal{E}) \geq \frac{1}{4} f_3$$

Proof. Assume that we are given an optimum solution that partitions all crossings into $\mathrm{bc}(\mathcal{E})$ bundled crossings. Faces for which all crossings are part of a bundled crossing must be quadrilaterals. Hence, a bundled crossing can only contain one or two crossings of a given triangle. Furthermore, for every triangle there must be a bundled crossing that contains exactly one of its crossings, since a triangle has either three crossings or—if it is adjacent to the outer cycle—one crossing.

Consider a bundled crossing C of bundles E_1 and E_2 enclosed by a pseudodisk D. If a triangle contains two edges from E_1 (or two from E_2), two crossings of the triangle are part of C. If we follow the edges of $E_1 \cup E_2$ as they cross the boundary of D in clockwise order (each edge twice), a triangle with one crossing as part of the bundled crossing must lie on the transition between E_1 and E_2. However, there are exactly four such transitions. Since every triangle must occur in at least one transition of a bundled crossing, we get $f_3 \leq 4\,\mathrm{bc}(\mathcal{E})$. $\square$

By using the relation between triangles and other faces, we get the alternative bound $\mathrm{bc}(\mathcal{E}) \geq \frac{f_3}{4} = \frac{1}{4}\left(4 + \sum_{k \geq 5} f_k \cdot (k-4)\right) \geq 1 + \sum_{k \geq 5} f_k \cdot \frac{k-4}{4}$.

We get a combined bound by taking the average of the two.

Lemma 6.
$$\mathrm{bc}(\mathcal{E}) \geq \frac{1}{2} + \frac{1}{8} f_3 + \sum_{k \geq 5} f_k \cdot \frac{k-4}{8}$$

There exist embeddings with $\Theta(m^2)$ triangles, since there are arrangements of m straight lines with $\Theta(m^2)$ triangles [4]. Such an embedding has an (optimal) bundled crossing number of $\Theta(m^2)$. The approximation algorithm described next can be used to detect such "bad" circular embeddings, i.e., ones whose bundled crossing number is large with respect to m.

Approximation. Using the lower bound of Lemma 6 we can show that for a circular embedding, the algorithm REPEATEDCUT yields a constant-factor approximation. Here, every face of degree other than four is a (point) hole unless it is adjacent to a vertex on the circle. Hence we have $h(\mathcal{C}(\mathcal{E})) \leq f_3 + \sum_{k \geq 5} f_k$. Let $\mathrm{ALG}(\mathcal{E})$ be the number of bundled crossings that the algorithm creates. By Theorem 2 we have $\mathrm{ALG}(\mathcal{E}) \leq 2\,\mathrm{bc}(\mathcal{E}) + h(\mathcal{C}(\mathcal{E})) - 1$. On the other hand, we get $h(\mathcal{C}(\mathcal{E})) \leq f_3 + \sum_{k \geq 5} f_k \leq 8 \cdot \left(1/2 + f_3/8 + \sum_{k \geq 5} f_k(k-4)/8\right) \leq 8\,\mathrm{bc}(\mathcal{E})$ using Lemma 6. Hence, $\mathrm{ALG}(\mathcal{E}) \leq 10\,\mathrm{bc}(\mathcal{E})$.

Theorem 3. *Algorithm* REPEATEDCUT *yields a 10-approximation for the bundled crossing number of circular embeddings.*

We can also formulate the algorithm REPEATEDCUT directly in terms of the embedding (with the same runtime). We consider the embedding $\mathcal{E}$ with the outer cycle added. An inner face (bounded only by crossing vertices) with degree other than four corresponds to a (concave) point hole. Similarly, a face of degree $k \geq 5$ adjacent to the outer cycle corresponds to a concave boundary corner of angle $k - 2$. Finally, cutting along a chord corresponds to breaking edges into two parts while following two parallel edges.

6   Conclusion and Open Problems

We have introduced the bundled crossing number, which takes not only the number of crossings but also their structure into account. We showed that computing the bundled crossing number of an embedding is NP-hard. Furthermore, we observed that this problem generalizes the dissection of a rectilinear polygon into rectangles. This insight allowed us to obtain an easy-to-implement algorithm that yields a constant-factor approximation for circular embeddings.

There are some interesting open questions for the bundled crossing number. Most importantly, is there an approximation for general embeddings? The related dissection problem could provide helpful ideas for this case. Furthermore, is computing the bundled crossing number NP-hard even for circular embeddings?

Another direction for future research is to find an embedding with small bundled crossing number for a given graph. We conjecture that this new version of the crossing number of a graph is NP-hard.

Acknowledgments. The research of Martin Fink was partially supported by a fellowship within the Postdoc-Program of the German Academic Exchange Service (DAAD), and by NSF grants CCF-1161495 and CCF-1525817. The research of Subhash Suri was partially supported by NSF grants CCF-1161495 and CCF-1525817.

References

1. Dahlhaus, E., Johnson, D.S., Papadimitriou, C.H., Seymour, P.D., Yannakakis, M.: The complexity of multiterminal cuts. SIAM J. Comput. **23**(4), 864–894 (1994). http://dx.doi.org/10.1137/S0097539792225297

2. Eppstein, D., van Kreveld, M.J., Mumford, E., Speckmann, B.: Edges and switches, tunnels and bridges. Comput. Geom. **42**(8), 790–802 (2009). http://dx.doi.org/10.1016/j.comgeo.2008.05.005

3. Fink, M., Pupyrev, S., Wolff, A.: Ordering metro lines by block crossings. J. Graph Algorithms Appl. **19**(1), 111–153 (2015). http://dx.doi.org/10.7155/jgaa.00351

4. Füredi, Z., Palásti, I.: Arrangements of lines with a large number of triangles. Proc. Am. Math. Soc. **92**(4), 561–566 (1984). http://dx.doi.org/10.1090/S0002-9939-1984-0760946-2

5. Garey, M.R., Johnson, D.S.: Crossing number is NP-complete. SIAM J. Algebraic Discrete Methods **4**(3), 312–316 (1983). http://epubs.siam.org/doi/abs/10.1137/0604033

6. Holten, D.: Hierarchical edge bundles: visualization of adjacency relations in hierarchical data. IEEE Trans. Vis. Comput. Graph. **12**(5), 741–748 (2006). http://doi.ieeecomputersociety.org/10.1109/TVCG.2006.147

7. Holten, D., van Wijk, J.J.: Force-directed edge bundling for graph visualization. Comput. Graph. Forum **28**(3), 983–990 (2009). http://dx.doi.org/10.1111/j.1467-8659.2009.01450.x

8. Hu, Y., Shi, L.: A coloring algorithm for disambiguating graph and map drawings. In: Duncan, C., Symvonis, A. (eds.) GD 2014. LNCS, vol. 8871, pp. 89–100. Springer, Heidelberg (2014). http://dx.doi.org/10.1007/978-3-662-45803-7_8

9. Huang, W., Hong, S.H., Eades, P.: Effects of crossing angles. In: Proceedings of 7th International IEEE Asia-Pacific Symposium Information Visualisation (PacificVIS 2008), pp. 41–46 (2008). http://dx.doi.org/10.1109/PACIFICVIS.2008.4475457

10. Schaefer, M.: The graph crossing number and its variants: asurvey. Electron. J. Comb. Dyn. Surv. **21**, 1–100 (2013). http://www.combinatorics.org/ojs/index.php/eljc/article/view/DS21

11. Soltan, V., Gorpinevich, A.: Minimum dissection of a rectilinear polygon with arbitrary holes into rectangles. Discrete Comp. Geom. **9**(1), 57–79 (1993). http://dx.doi.org/10.1007/BF02189307

Probabilistic Analysis of the Dual Next-Fit Algorithm for Bin Covering

Carsten Fischer[(✉)] and Heiko Röglin

Department of Computer Science, University of Bonn, Bonn, Germany
carsten.fischer@uni-bonn.de, roeglin@cs.uni-bonn.de

Abstract. In the bin covering problem, the goal is to fill as many bins as possible up to a certain minimal level with a given set of items of different sizes. Online variants, in which the items arrive one after another and have to be packed immediately on their arrival without knowledge about the future items, have been studied extensively in the literature. We study the simplest possible online algorithm Dual Next-Fit, which packs all arriving items into the same bin until it is filled and then proceeds with the next bin in the same manner. The competitive ratio of this and any other reasonable online algorithm is 1/2.

We study Dual Next-Fit in a probabilistic setting where the item sizes are chosen i.i.d. according to a discrete distribution and we prove that, for every distribution, its expected competitive ratio is at least $1/2 + \epsilon$ for a constant $\epsilon > 0$ independent of the distribution. We also prove an upper bound of 2/3 and better lower bounds for certain restricted classes of distributions. Finally, we prove that the expected competitive ratio equals, for a large class of distributions, the random-order ratio, which is the expected competitive ratio when adversarially chosen items arrive in uniformly random order.

1 Introduction

In the *bin covering problem* one is given a set of items with sizes $s_1, \ldots, s_n \in [0, 1]$ and the goal is to fill as many bins as possible with these items, where a bin is counted as filled if it contains items with a total size of at least 1. More precisely, we are interested in finding the maximal number ℓ of pairwise disjoint sets $X_1, \ldots, X_\ell \subseteq \{1, \ldots, n\}$ such that $\sum_{i \in X_j} s_i \geq 1$ for every j. We call the sets X_j *bins* and we say that a bin is *filled* or *covered* if the total size of the items it contains is at least 1. Variants of the bin covering problem occur frequently in industrial applications, e.g., when packing food items with different weights into boxes that each need to have at least the advertised weight.

Bin covering is a well-studied NP-hard optimization problem. A straightforward reduction from the partition problem shows that it cannot be approximated within a factor of $1/2 + \epsilon$ for any $\epsilon > 0$. On the positive side, Jansen and Solis-Oba [10] presented an asymptotic fully polynomial-time approximation scheme.

This research was supported by ERC Starting Grant 306465 (BeyondWorstCase).

E. Kranakis et al. (Eds.): LATIN 2016, LNCS 9644, pp. 469–482, 2016.
DOI: 10.1007/978-3-662-49529-2_35

In many applications, it is natural to study online variants, in which items arrive one after another and have to be packed directly into one of the bins without knowing the future items. It is also often natural to restrict the number of open bins, i.e., bins that contain at least one item but are not yet covered, that an online algorithm may use.

We study the simple online algorithm DUAL NEXT-FIT (DNF) that packs all arriving items into the same bin until it is filled. Then the next items are packed into a new bin until it is filled, and so on. The (asymptotic) competitive ratio of DNF is $1/2$ [2], which is best possible for deterministic online algorithms [7]. In fact, all deterministic online algorithms that do not add items to a bin that is already covered and have at most a constant number of open bins at any point in time have a competitive ratio of exactly $1/2$ [3]. Since competitive analysis does not yield much insight for bin covering, alternative measures have been suggested. Most notably are probabilistic models in which the item sizes are drawn at random from a fixed distribution [5] or in which the item sizes are adversarial but the items arrive in random order [3].

In this article, we study the *asymptotic average performance ratio* and the *asymptotic random-order ratio* of DNF. We give now an intuitive explanation of these measures (formal definitions are given in Sect. 1.1). In order to define the former, we allow an adversary to choose an arbitrary distribution F on $[0, 1]$ with finite support. The asymptotic average performance ratio AAPR(DNF, F) of DNF with respect to the distribution F is then defined as the expected competitive ratio of DNF on instances with $n \to \infty$ items whose sizes are independently drawn according to F. Furthermore, let AAPR(DNF) $= \inf_F$ AAPR(DNF, F). In order to define the latter, we allow an adversary to choose $n \to \infty$ item sizes $s_1, \ldots, s_n \in [0, 1]$. The asymptotic random-order ratio RR(DNF) is then defined as the expected competitive ratio of DNF on instances in which these items arrive in uniformly random order. It is assumed that the adversary chooses item sizes that minimize this expected value.

We prove several new results on the asymptotic average performance ratio and the asymptotic random-order ratio of DNF and the relation between these two measures. First of all, observe that both RR(DNF) and AAPR(DNF) lie between $1/2$ and 1 because even in the worst case DNF has a competitive ratio of $1/2$. Using ideas of Kenyon [12], it follows that AAPR(DNF) $\geq$ RR(DNF). We show that RR(DNF) $\leq$ AAPR(DNF) $\leq 2/3$, which improves a result by Christ et al. [3] who proved that RR(DNF) $\leq 4/5$. To the best of our knowledge the bound by Christ et al. is the only non-trivial result about the random-order ratio of DNF in the literature.

Csirik et al. [5] have proved that AAPR(DNF, F) $= 2/e$ if F is the uniform distribution on $[0, 1]$. We are not aware, however, of any lower bound for AAPR(DNF, F) that holds for any discrete distribution F, except for the trivial bound of $1/2$. We obtain the first such bound and prove that AAPR(DNF) $\geq 1/2 + \epsilon$ for a small but constant $\epsilon > 0$. We prove even better lower bounds for certain classes of distributions that we will describe in detail in Sect. 1.4. Finally we study the connection between the performance measures and prove that AAPR(DNF) = RR(DNF) if the adversary in the random-order model is restricted to inputs $s_1, \ldots, s_n$ with $\sum_i s_i = \omega(n^{2/3})$.

1.1 Performance Measures

Before we discuss our results in more detail and mention further related work, let us formally introduce the performance measures that we employ.

Definition 1. *A* discrete distribution *F is defined by a vector $s = (s_1, \ldots, s_m)$ of non-negative rational item sizes and an associated vector $p = (p_1, \ldots, p_m)$ of positive rational probabilities such that $\sum_{i=1}^m p_i = 1$.*

We denote by $I_n(F) = (X_1, \ldots, X_n)$ a list of n items, where the X_i are drawn i.i.d. according to F. For an algorithm A and a list of item sizes L we denote by $A(L)$ the number of bins that A fills on input L.

Definition 2. *Let A be an algorithm for the bin covering problem, and let F be a discrete distribution. We define the* asymptotic average performance ratio *as*

$$\mathrm{AAPR}(A, F) = \liminf_{n \to \infty} \mathbb{E}\left[\frac{A(I_n(F))}{\mathrm{OPT}(I_n(F))}\right]$$

and the asymptotic expected competitive ratio *as*

$$\mathrm{AECR}(A, F) = \liminf_{n \to \infty} \frac{\mathbb{E}\left[A(I_n(F))\right]}{\mathbb{E}\left[\mathrm{OPT}(I_n(F))\right]}.$$

For a set $\mathcal{F}$ of discrete distributions, we define

$$\mathrm{AAPR}(A, \mathcal{F}) = \inf_{F \in \mathcal{F}} \mathrm{AAPR}(A, F) \quad and \quad \mathrm{AECR}(A, \mathcal{F}) = \inf_{F \in \mathcal{F}} \mathrm{AECR}(A, F).$$

We denote by $\mathcal{D}$ the set of all discrete distributions and we define

$$\mathrm{AAPR}(A) = \mathrm{AAPR}(A, \mathcal{D}) \quad and \quad \mathrm{AECR}(A) = \mathrm{AECR}(A, \mathcal{D}).$$

Both the asymptotic average performance ratio and the asymptotic expected competitive ratio have been studied in the literature (sometimes under different names). We will see later that for our purposes there is no need to distinguish between them because they coincide for DNF.

Let $L = (a_1, \ldots, a_N)$ be a list of length N, and let $\sigma \in S_N$ be a permutation of N elements (S_N denotes the symmetric group of order N). Then $\sigma(L)$ denotes a permutation of L.

Definition 3. *In bin covering, the* asymptotic random-order ratio *for an algorithm A is defined as*

$$\mathrm{RR}(A) = \liminf_{\mathrm{OPT}(L) \to \infty} \frac{\mathbb{E}_\sigma[A(\sigma(L))]}{\mathrm{OPT}(L)},$$

where σ is drawn uniformly from $S_{|L|}$.

The asymptotic random-order ratio for bin covering and bin packing has been introduced in [3] and [12], respectively. All definitions above can also be adapted to the bin packing problem; we only have to replace inf and lim inf by sup and lim sup, respectively.

1.2   Related Work

Csirik et al. [6] presented an algorithm (which requires an unlimited number of open bins) whose asymptotic average performance ratio is 1 for all discrete distributions. Csirik et al. [5] have proved that the asymptotic expected competitive ratio of DNF is $2/e$ if F is the uniform distribution on $[0, 1]$. Kenyon [12] introduced the notion of asymptotic random-order ratio for bin packing and proved that the asymptotic random-order ratio of the best-fit algorithm lies between 1.08 and 1.5. Coffman et al. [11] showed that the random-order ratio of the next-fit algorithm is 2. Christ et al. [3] adapted the asymptotic random-oder ratio to bin covering and proved that RR(DNF) $\leq 4/5$. The article of Kenyon [12] contains in Sect. 3 an argument for AECR(DNF) $\geq$ RR(DNF) (even though this is not stated explicitly). Asgeirsson and Stein [1] developed a heuristic for online bin covering based on Markov chains and demonstrated its good performance in experiments.

1.3   Definitions and Notation

Let $L = (a_1, \ldots, a_N) \in [0, 1]^N$ be a list of items. We denote by $s(L) := \sum_{i=1}^N a_i$ the total size of the items in L and by $N(L) := N$ the length of L. For an algorithm A, we define $W^A(L) := s(L) - A(L)$ as the *waste of algorithm A on list L*. We denote by OPT an optimal offline algorithm. Of particular interest are distributions that an optimal offline algorithm can pack with sublinear waste.

Definition 4. *We say that a discrete distribution F is a* perfect-packing distri-bution, *if it satisfies the perfect-packing property, i.e.,*

$$\mathbb{E}\left[W^{\mathrm{OPT}}(I_n(F))\right] = o(n).$$

We denote the set of all perfect-packing distributions by $\mathcal{P}$.

Let F be a discrete distribution with associated item sizes $s = (s_1, \ldots, s_m)$ and probabilities $p = (p_1, \ldots, p_m)$. We say that $b = (b_1, \ldots, b_m) \in \mathbb{N}_0^m$ is a *perfect-packing configuration*, if $\sum_{i=1}^m b_i s_i = 1$. Let Λ_F denote the closure under convex combinations and positive scalar multiplication of the set of perfect-packing con-figurations. Courcoubetis and Weber [4] found out, that F is a perfect-packing distribution if and only if $p \in \Lambda_F$.

Let $L = (a_1, \ldots, a_N)$ be a list. We say that a discrete distribution F is induced by L, if the vector of item sizes $(s_1, \ldots, s_m)$ contains exactly all the item sizes arising in L, and the vector of probabilities p is given by $p_i := p(s_i) = |\{1 \leq j \leq N : a_j = s_i\}|/N$. Vice versa we can find for every discrete distribution F a list L, such that F is induced by L.

1.4   Outline and Our Results

In Sect. 2 we discuss how DNF can be interpreted as a Markov chain and we prove some properties using this interpretation. We investigate how the different performance measures are related and we point out that $\text{AECR}(\text{DNF}, F) = \text{AAPR}(\text{DNF}, F)$ for any discrete distribution F. Since the asymptotic expected competitive ratio and the asymptotic average performance ratio coincide for all discrete distributions, we will only consider the former in the following even though all mentioned results are also true for the latter. The main result of Sect. 3 is a proof that $\text{AECR}(\text{DNF}) = \text{RR}(\text{DNF})$ if the adversary in the random-order model is restricted to inputs $s_1, \ldots, s_n$ with $\sum_i s_i = \omega(n^{2/3})$.

We start Sect. 4 by showing that perfect-packing distributions are the worst distributions for the considered measures, i.e., $\text{AECR}(\text{DNF}, \mathcal{P}) = \text{AECR}(\text{DNF})$. Similarly we show that for proving a lower bound on the random order ratio of DNF it suffices to consider sequences of items that can be packed with waste zero. Then we show that $\text{AECR}(\text{DNF}) \leq 2/3$, which implies $\text{RR}(\text{DNF}) \leq 2/3$. The main contribution of Sect. 4 are various new lower bounds on the asymptotic expected competitive ratio of DNF. We first prove that $\text{AECR}(\text{DNF}) \geq 1/2 + \epsilon$ for a small but constant $\epsilon > 0$. Then we consider the following special cases for which we show better lower bounds.

- Let $\mathcal{P}_x$ be the set of all perfect packing distributions, where the maximum item size is bounded from above by x. For $x \in [1/2, 1]$, we prove by an application of Lorden's inequality for the overshoot of a stopped sum of random variables that $\text{AECR}(\text{DNF}, \mathcal{P}_x) \geq \frac{1}{1 + x^2 + (1-x)^2}$.
- Let F be a discrete perfect-packing distribution with associated item sizes $s = (s_1, \ldots, s_m)$ and probabilities $p = (p_1, \ldots, p_m)$. According to our discussion after Definition 4, the vector p lies in Λ_F. Hence, there exist perfect-packing configurations $b^1, \ldots, b^N$ and coefficients $\alpha_1, \ldots, \alpha_N \geq 0$ with $p = \alpha_1 b^1 + \ldots + \alpha_N b^N$. We denote the smallest N for which such b^i and α_i exist the *degree* of p. Let $\mathcal{P}^{(N)}$ denote all discrete perfect-packing distributions with degree N. We prove that

$$\frac{2}{3} \leq \text{AECR}(\text{DNF}, \mathcal{P}^{(1)}) \leq \left(\sum_{i=1}^{\infty} \frac{(i-1)!}{i^i} \right)^{-1} \approx 0.736.$$

  If the maximum item size is greater than or equal to $\frac{1}{2}$ the lower bound can be improved to $\left(1 + \sum_{i=2}^{\infty} \frac{1}{i^2} \cdot \left(1 - \frac{1}{i}\right)^{i-2} \right)^{-1} \approx 0.686$.
- Let F be a discrete perfect-packing distribution with items $s = (s_1, \ldots, s_m)$ and probabilities $p = (p_1, \ldots, p_m)$ and let $p = \alpha_1 b^1 + \ldots + \alpha_N b^N$ for perfect-packing configurations $b^1, \ldots, b^N$ and coefficients $\alpha_1, \ldots, \alpha_N \geq 0$. Let $\mathcal{P}_{\text{two}}$ denote all discrete perfect-packing distributions for which there exists such a representation in which every perfect-packing configuration b^i contains at most two non-zero entries. We show that $\text{AECR}(\text{DNF}, \mathcal{P}_{\text{two}}) = 2/3$.

In Sect. 5 we give some conclusions and present open problems. Due to space limitations, some of the proofs and basics about Markov chains, which are relevant for this paper, are deferred to the full version [9].

2   Basic Statements

Let L_1 and L_2 be two lists and let L_1L_2 denote the concatenation of them. At first, we want to point out that OPT as well as DNF are superadditive, i.e., it holds $\mathrm{OPT}(L_1) + \mathrm{OPT}(L_2) \leq \mathrm{OPT}(L_1L_2)$ and $\mathrm{DNF}(L_1) + \mathrm{DNF}(L_2) \leq \mathrm{DNF}(L_1L_2)$. Now let F be a fixed discrete distribution. The limits $\gamma(F) := \lim_{n\to\infty} \mathbb{E}\left[\mathrm{OPT}(I_n(F))\right]/n$ and $\lim_{n\to\infty} \mathbb{E}\left[\mathrm{DNF}(I_n(F))\right]/n$ exist due to Fekete's lemma. This guarantees that the lim inf in the definition of $\mathrm{AECR}(\mathrm{DNF}, F)$ is in fact a limit.

Furthermore, the performance measures mentioned in Definition 2 coincide in our case:

Lemma 5. *Let F be a discrete distribution. It holds*

$$\mathrm{AAPR}(\mathrm{DNF}, F) = \mathrm{AECR}(\mathrm{DNF}, F).$$

A proof of a similar statement can be found in the extended version of [15]. For our purposes it is easier to deal with $\mathrm{AECR}(\mathrm{DNF}, F)$, so we will only mention this measure in the following.

In order to study $\mathbb{E}\left[\mathrm{DNF}(I_n(F))\right]$, it will be useful to think of $\mathrm{DNF}(I_n(F))$ as a Markov chain. We will give a brief introduction to Markov chains in the full version of the paper. A comprehensive overview can be found in [13]. The state space is given by the possible arising bin levels, where we subsume all bin levels greater than or equal to 1 and the bin level 0 to a special state, which we call the *closed* state. This Markov chain is irreducible.

Sometimes it will be necessary that the Markov chain does not start in the closed state, but with bin level ℓ. $\mathrm{DNF}(\ell, L)$ denotes the number of bins that DNF closes on input L, starting with bin level ℓ. We set $\mathrm{DNF}(L) := \mathrm{DNF}(c, L)$, where c denotes the closed state.

A first important observation is that we can restrict ourselves to discrete distributions F, such that the Markov chain induced by F and DNF is aperiodic.

Lemma 6. *Let F be a discrete distribution and $d \in \mathbb{N}_{\geq 2}$. If the Markov chain induced by F and DNF is d-periodic then $\mathrm{AECR}(\mathrm{DNF}, F) = 1$.*

Therefore, we will assume in the following that the Markov chain induced by a discrete distribution F and DNF is aperiodic, and so it converges to a unique stationary distribution π_F. It holds $\pi_F(c) = \mathbb{E}\left[T_{\mathrm{DNF}}^F\right]^{-1}$, where T_{DNF}^F denotes the number of items we need to close a bin, starting with bin level zero.

Lemma 7. *Let F be a perfect-packing distribution and X be distributed according to F. Then*

$$\lim_{n\to\infty} \frac{\mathbb{E}\left[\mathrm{OPT}(I_n(F))\right]}{n} = \mathbb{E}\left[X\right].$$

For every discrete distribution F, it holds

$$\lim_{n\to\infty} \frac{\mathbb{E}\left[\mathrm{DNF}(I_n(F))\right]}{n} = \frac{1}{\mathbb{E}\left[T_{\mathrm{DNF}}^F\right]}.$$

So, if F is a perfect-packing distribution, it holds

$$\text{AECR}(\text{DNF}, F) = \frac{1}{\mathbb{E}[X] \cdot \mathbb{E}[T^F_{\text{DNF}}]}. \tag{1}$$

3 Connection Between Asymptotic Expected Competitive Ratio and Random-Order Ratio

In this section we want to examine the connection between the asymptotic expected competitive ratio and the random-order ratio. At first we want to mention a result, which follows from [12].

Lemma 8. *It holds*

$$\text{RR}(\text{DNF}) \leq \text{AECR}(\text{DNF}).$$

Proof. Let $L_n = \{L = (a_1, \ldots, a_n) : \mathbb{P}[I_n(F) = L] > 0\}$. Then there exists a set of lists $\mathcal{L}_n$, such that $L_n = \bigcup_{H \in \mathcal{L}_n} \{L : \exists \sigma \in S_n \text{ s.t. } L = \sigma(H)\}$. Using the inequality $(\sum_{i=1}^n b_i)/(\sum_{i=1}^n c_i) \geq \min_{1 \leq i \leq n} b_i/c_i$, it follows

$$\frac{\mathbb{E}[\text{DNF}(I_n(F))]}{\mathbb{E}[\text{OPT}(I_n(F))]} \geq \min_{H \in \mathcal{L}_n} \frac{\mathbb{E}_\sigma[\text{DNF}(\sigma(H))]}{\mathbb{E}_\sigma[\text{OPT}(\sigma(H))]} = \min_{H \in \mathcal{L}_n} \frac{\mathbb{E}_\sigma[\text{DNF}(\sigma(H))]}{\text{OPT}(H)}.$$

$\square$

We will show that the performance measures coincide if the sum of the items increases fast enough in terms of the number of items. A side product of the results of this section is the following: In [8] the authors noted that in bin packing for the algorithm Next-fit and a certain list L, the following relationship holds:

$$\lim_{j \to \infty} \frac{\mathbb{E}_\sigma[\text{NF}(\sigma(L^j))]}{\text{OPT}(L^j)} = \text{AECR}(\text{NF}, F), \tag{2}$$

where L^j denotes the concatenation of j copies of L and F is the discrete distribution induced by L. They asked if this result holds for arbitrary lists L. We can show that the answer is true in the context of DNF.

In the following let K denote a universal constant, which does not depend on the considered list L. We establish the following two bounds:

Theorem 9. *Let L be an arbitrary instance and let F be the induced discrete distribution. Then it holds*

$$|\text{OPT}(L) - N(L) \cdot \gamma(F)| \leq K \cdot N(L)^{2/3}.$$

Theorem 10. *Let L be an arbitrary instance and let F be the induced discrete distribution. We assume that the Markov chain induced by DNF and F possesses a unique stationary measure π_F. Then it holds*

$$|\mathbb{E}_\sigma[\text{DNF}(\sigma(L))] - N(L) \cdot \pi_F(c)| \leq K \cdot N(L)^{2/3}.$$

The first step of the proofs is to split up $I_{N(L)}(F)$ or $\sigma(L)$ into smaller sublists, an idea which was brought up in [8]. The following lemma shows that the difference between sampling with and without replacement can be controlled if the length of the sublists is sufficiently small compared to $N(L)$.

Lemma 11. *Let* $L = (a_1, \ldots, a_N)$, F *be the corresponding induced discrete distribution, and* $b \in \mathbb{N}$. *We set* $\sigma(L)_{[1:b]} = (a_{\sigma(1)}, \ldots, a_{\sigma(b)})$, *where* σ *is an arbitrary permutation of* L. *Then for* $A \in \{\mathrm{DNF}, \mathrm{OPT}\}$ *it holds*

$$\left| \mathbb{E}\left[A(\sigma(L)_{[1:b]}) \right] - \mathbb{E}\left[A(I_b(F)) \right] \right| \leq \frac{b^3}{N}.$$

The proof of the lemma is based on estimates of the *total variation distance*. In the remaining parts of the proofs of the theorems we have to show that we also can control other errors, e.g., stemming from beginning with different bin levels in the case of DNF.

Theorem 12. *If there exists a sequence of lists* $L^{(i)}$ *such that*

$$\lim_{i \to \infty} \frac{\mathbb{E}_\sigma \left[\mathrm{DNF}(\sigma(L^{(i)})) \right]}{\mathrm{OPT}(L^{(i)})} = \mathrm{RR}(\mathrm{DNF}),$$

and $s(L^{(i)}) \in \omega(N(L^{(i)})^{2/3})$, *then* $\mathrm{RR}(\mathrm{DNF}) = \mathrm{AECR}(\mathrm{DNF})$.

Proof. Let $\epsilon > 0$ be arbitrary. Since $s(L^{(i)}) \geq \mathrm{OPT}(L^{(i)}) \geq \mathrm{DNF}(L^{(i)}) \geq s(L^{(i)})/2$, it also holds $\mathrm{OPT}(L^{(i)}) \in \omega(N(L^{(i)})^{2/3})$. Let F^i denote the discrete distribution induced by $L^{(i)}$. Using that $\mathrm{AECR}(\mathrm{DNF}, F^i) = \pi_{F^i}(c)/\gamma(F^i)$ and the basic inequality $|a/b - a'/b'| \leq |(a - a')/b| + |a'/b'| \cdot |(b - b')/b|$, we obtain

$$\left| \frac{\mathbb{E}_\sigma \left[\mathrm{DNF}(L^{(i)}) \right]}{\mathrm{OPT}(L^{(i)})} - \mathrm{AECR}(\mathrm{DNF}, F^i) \right| \leq K \cdot \frac{N(L^{(i)})^{2/3}}{\mathrm{OPT}(L^{(i)})}.$$

Hence, if we choose i large enough, we can find a distribution F^i, such that $\mathrm{AECR}(\mathrm{DNF}, F^i) \leq \mathrm{RR}(\mathrm{DNF}) + \epsilon$. Then

$$\mathrm{RR}(\mathrm{DNF}) + \epsilon \geq \mathrm{AECR}(\mathrm{DNF}, F^i) \geq \mathrm{AECR}(\mathrm{DNF}) \geq \mathrm{RR}(\mathrm{DNF}),$$

i.e., both performance measures would coincide. $\square$

The following corollary follows from the proof of Theorem 12.

Corollary 13. *Let* $\mathrm{RR}'(\mathrm{DNF})$ *denote the random-order ratio of* DNF *restricted to instances* L *with* $s(L) \in \omega(N(L)^{2/3})$. *It holds* $\mathrm{RR}'(\mathrm{DNF}) = \mathrm{AECR}(\mathrm{DNF})$.

Using the same method as in the proof we can also show that (2) holds true for the dual next-fit algorithm.

4 Upper and Lower Bounds for Dual Next-Fit on Perfect-Packing Distributions

At first we show that we can restrict ourselves to studying perfect-packing distributions. They represent the worst-case with respect to the investigated performance measures.

Lemma 14. *Let $L = (a_1, \ldots, a_N)$ be a list for bin covering. Then there exists a list H that can be packed perfectly, i.e., $W^{\mathrm{OPT}}(H) = 0$, such that*

$$\frac{\mathbb{E}_\sigma\left[\mathrm{DNF}(\sigma(L))\right]}{\mathrm{OPT}(L)} \geq \frac{\mathbb{E}_\sigma\left[\mathrm{DNF}(\sigma(H))\right]}{\mathrm{OPT}(H)}.$$

Furthermore, for each distribution F there exists a perfect-packing distribution G such that

$$\lim_{n\to\infty} \frac{\mathbb{E}\left[\mathrm{DNF}(I_n(F))\right]}{\mathbb{E}\left[\mathrm{OPT}(I_n(F))\right]} \geq \lim_{n\to\infty} \frac{\mathbb{E}\left[\mathrm{DNF}(I_n(G))\right]}{\mathbb{E}\left[\mathrm{OPT}(I_n(G))\right]}.$$

4.1 Upper and Lower Bounds for Arbitrary Perfect-Packing Distributions

We begin presenting an upper bound for the considered performance measures, which improves a result in [3].

Theorem 15. *It holds*

$$\mathrm{RR(DNF)} \leq \mathrm{AECR(DNF)} \leq \frac{2}{3}.$$

Proof. Let $F(m, k)$ be the uniform distribution on the item sizes

$$\left(\frac{1}{k}, 1 - \frac{1}{k}, \left(\frac{1}{k}\right)^2, 1 - \left(\frac{1}{k}\right)^2, \ldots, \left(\frac{1}{k}\right)^m, 1 - \left(\frac{1}{k}\right)^m\right).$$

It is clear, that $F(m, k)$ is a perfect-packing distribution. We show that for every $\epsilon > 0$ there are parameters m and k, such that $\mathbb{E}\left[T_{\mathrm{DNF}}^{F(m,k)}\right] \geq 3 - \epsilon$.

It holds

$$\mathbb{E}\left[T_{\mathrm{DNF}}^{F(m,k)}\right] = \sum_{i=0}^\infty \mathbb{P}\left[T_{\mathrm{DNF}}^{F(m,k)} > i\right] \geq 2 + \sum_{i=2}^{k-1} \mathbb{P}\left[T_{\mathrm{DNF}}^{F(m,k)} > i\right].$$

Simple counting yields for $i \geq 2$

$$\mathbb{P}\left[T_{\mathrm{DNF}}^{F(m,k)} > i\right] = \frac{m^i}{(2m)^i} + \frac{1}{(2m)^i}\sum_{j=2}^m i \cdot (j-1)^{i-1} = \frac{1}{2^i} + \frac{i}{2^i m^i}\sum_{j=1}^{m-1} j^{i-1}$$

$$\geq \frac{1}{2^i} + \frac{i}{2^i m^i} \cdot \int_0^{m-1} x^{i-1}\, dx = \frac{1}{2^i} \cdot \left[1 + \left(1 - \frac{1}{m}\right)^i\right].$$

Therefore, if we choose at first k, and then m large enough

$$\mathbb{E}\left[T_{\text{DNF}}^{F(m,k)}\right] \geq 2 + \sum_{i=2}^{k-1} \frac{1}{2^i} \cdot \left[1 + \left(1 - \frac{1}{m}\right)^i\right] \geq 3 - \epsilon.$$

Using Lemma 7 the statement follows.   □

Now we show, that in a probabilistic setting, we behave better than in the worst-case.

Theorem 16. *There exists an $\epsilon > 0$ such that*

$$\text{AECR(DNF)} \geq \frac{1}{2} + \epsilon.$$

We want to give here the idea of the proof: Let $X_1, X_2, \ldots$ denote i.i.d. random variables distributed according to F, and $S_n = \sum_{i=1}^{n} X_i$. The waste, which occurs closing the bin, is given by $R = S_{T_{\text{DNF}}^F} - 1$. We will denote R also as *overshoot*. Due to Wald's equation it holds that $1 + \mathbb{E}[R] = \mathbb{E}\left[S_{T_{\text{DNF}}^F}\right] = \mathbb{E}\left[T_{\text{DNF}}^F\right] \cdot \mathbb{E}[X]$. From (1) it follows that $\text{AECR(DNF}, F) = (1 + \mathbb{E}[R])^{-1}$. We show that there exists an $\epsilon > 0$, independent of the perfect-packing distribution F, such that $\mathbb{E}[R] \leq 1 - \epsilon$.

We assume that F is induced by ℓ^* perfectly packed bins and a uniform distribution on the items. Otherwise we could copy bins to achieve such a setting. We call an item *large* if it is strictly larger than $1/2$. Otherwise we call the item *small*. Our goal is to show that we will close a bin with a small item with a constant probability, independent of F.

We denote by $\ell \leq \ell^*$ the number of large items and by n the number of small items. We assume that $n \geq \ell$ and that n is a multiple of ℓ. If this is not the case, we add an appropriate number of items of size 0. This will not change the probability of closing a bin with a small item.

Let $b_1, \ldots, b_\ell$ denote the large items and assume that $b_1 \geq b_2 \geq \ldots \geq b_\ell > 1/2$. Let $b^\star := b_{\lceil \ell/2 \rceil}$ and $s^\star = 1 - b^\star$.

Let T denote the time step at which we draw for the first time a large item, and let $\mathcal{E}$ denote the event $\{T \geq 15n/\ell+1\} \cap \{\sum_{i=1}^{15n/\ell} X_i \geq s^\star\}$. We can show that this event happens with positive probability, independent of the distribution F.

Lemma 17. *It holds*

$$\mathbb{P}[\mathcal{E}] \geq 1.1 \cdot 10^{-12}.$$

Proof. Using that $n \geq \ell$, and so $\ell/(n + \ell) \leq 1/2$, we obtain

$$\mathbb{P}\left[T \geq \frac{15n}{\ell} + 1\right] = \left(1 - \frac{\ell}{n+\ell}\right)^{15n/\ell} \geq \left(1 - \frac{\ell}{n+\ell}\right)^{15(n+\ell)/\ell} \geq \frac{1}{4^{15}}.$$

Let us condition in the following on the event that $T \geq \frac{15n}{\ell} + 1$. We now want to show, that the event, that the sum of the first $15n/\ell$ small items is at least s^*, occurs with positive probability independent of F. To prove this, we reduce the summation to a kind of coupon-collectors problem. We assume that the coupons have numbers from 1 to n, and we can use for a coupon with number i, also a coupon with a higher number. Let $m \geq n$ and $v \in \{1, \ldots, n\}^m$. We denote by v^* the vector with the same entries as v except that the entries are ordered in non-increasing order. We say that v covers the vector $(n, \ldots, 1)$ if $v_i^* \geq n - i + 1$ for all $i \in [n]$.

We partition the small items according to their size into n/ℓ groups with ℓ items each. The ℓ smallest items are in the first group and so on. We denote by h the total size of items in the last group and Z the total size of all small items. Let $v = (m_1, \ldots, m_{15n/\ell})$, where m_i denotes the number of the group the i-th drawn item belongs to. We say that v covers all groups three times, if there exists a permutation v^σ of v s.t. $(v_1^\sigma, \ldots, v_{5n/\ell}^\sigma)$, $(v_{5n/\ell+1}^\sigma, \ldots, v_{10n/\ell}^\sigma)$ and $(v_{10n/\ell+1}^\sigma, \ldots, v_{15n/\ell}^\sigma)$ cover $(n, \ldots, 1)$ respectively. We can show that under the condition that $T \geq 15n/\ell + 1$, v covers all groups three times with probability at least 0.956^3. The proof of this statement is deferred to the full version. For $i \in [n/\ell]$ let g_i denote the largest item in group i. If all groups are covered three times, the sum of the small items drawn is at least

$$3 \sum_{i=1}^{n/\ell-1} g_i \geq \frac{3(Z-h)}{\ell}.$$

Furthermore, the total weight of all small items is at least

$$Z \geq (\ell - \lceil \ell/2 \rceil + 1)s^* \geq \ell s^*/2.$$

For the following argument we can assume w.l.o.g. that there is no small item with size larger than s^* because we are only interested in the probability that the small items drawn add up to at least s^*. If all groups are covered three times, the sum of the small items is at least

$$3 \sum_{i=1}^{n/\ell-1} g_i \geq \frac{3(Z-h)}{\ell} \geq \frac{3(\ell s^*/2 - h)}{\ell} = 3s^*/2 - 3h/\ell.$$

If $h \leq \ell s^*/6$, then the sum of the small items drawn is at least s^*. Hence, in this case

$$\mathbb{P}[\mathcal{E}] \geq 0.873 \cdot \frac{1}{4^{15}} \geq 8 \cdot 10^{-10}.$$

If $h > \ell s^*/6$ then at least $\ell/11$ small items have size at least $s^*/12$. We can see this as follows: Let x denote the number of items in group 1, which have size at least $s^*/12$. Then h is bounded from above by $xs^* + (\ell - x)s^*/12$. Since $h > \ell s^*/6$, it follows

$$(\ell s^*)/6 < h \leq xs^* + (\ell - x)s^*/12.$$

A simple computation then yields $x > \ell/11$. The probability that exactly 12 of these items are drawn under the condition $T \geq 15n/\ell + 1$ is at least

$$\binom{15n/\ell}{12} \cdot \left(1 - \frac{\ell/11}{n}\right)^{15n/\ell - 12} \cdot \left(\frac{\ell/11}{n}\right)^{12}$$

$$\geq \left(\frac{15n}{12\ell}\right)^{12} \cdot \left(1 - \frac{\ell}{11n}\right)^{15n/\ell} \cdot \left(\frac{\ell}{11n}\right)^{12}$$

$$\geq \left(\frac{15}{11 \cdot 12}\right)^{12} \cdot 0.239 \geq 1.1 \cdot 10^{-12}.$$

$\square$

We now show that from this it follows that the overshoot is strictly less than 1. Let $A := \{\sum_{i=1}^{T-1} X_i < 1/2\}$. We set $q := \mathbb{P}[\mathcal{E}^c]$, $p_1 := \mathbb{P}[\mathcal{E} \cap A]$, and $p_2 := \mathbb{P}[\mathcal{E} \cap A^c]$. Then, the following inequalities hold:

$$\mathbb{E}[R] \leq (1/2) \cdot p_2^2 + 1 \cdot (1 - p_2^2) = 1 - p_2^2/2$$
$$\mathbb{E}[R] \leq (1/2) \cdot p_1/2 + 1 \cdot (p_1/2 + p_2 + q) = 1 - p_1/4.$$

The first inequality follows from the observation that $\mathbb{P}[A^c]^2 \geq p_2^2$ is a lower bound on the probability that the bin gets filled with only small items, in which case the waste is at most $1/2$. The second inequality follows because if the event $\mathcal{E} \cap A$ occurs then the small items that arrive before the first large item have a total size of at least $s^\star$ and at most $1/2$. If the first large item has size at least $b^\star$, which happens with probability at least $1/2$, then it closes the bin with waste at most $1/2$. Since $p_1 + p_2 \geq 1.1 \cdot 10^{-12}$, it follows that $\mathbb{E}[R] < 1$.

4.2   Improved Lower Bounds for Certain Classes of Perfect-Packing Distributions

At first we look at the case that the maximum item size in the perfect-packing distribution is bounded from above by x. Let $\mathcal{P}_x$ denote the set of all such distributions.

Theorem 18. *If* $x \geq \frac{1}{2}$, *then*

$$\mathrm{AECR}(\mathrm{DNF}, \mathcal{P}_x) \geq \frac{1}{1 + x^2 + (1 - x)^2}.$$

The given lower bound slightly improves the worst-case bound $(1 + x)^{-1}$ in the case that the maximum item size is greater than $\frac{1}{2}$. Csirik et al. pointed out in [5] that in the case of DNF there is a connection between the bin covering problem and renewal theory, and so it is obvious to use tools from this field. The proof is based on an estimate of the overshoot, given by Lorden:

Lemma 19. (Lorden's inequality,[14]). *Suppose* $X_1, X_2, \ldots$ *are non-negative i.i.d. random variables with* $\mathbb{E}[X_1] > 0$ *and* $\mathbb{E}[X_1^2] < \infty$. *Let* $S_n = X_1 + \ldots + X_n$, $T = \inf\{n \in \mathbb{N} : S_n \geq 1\}$, *and* $R = S_T - 1$. *Then*

$$\mathbb{E}[R] \leq \mathbb{E}[X_1^2]/\mathbb{E}[X_1].$$

Now we want to look at the case $F \in \mathcal{P}^{(1)}$, i.e., we have a vector of item sizes s, and the vector of probabilities p_F is given by $p_F = b/|b|_1$, where b is a perfect-packing configuration.

Theorem 20. *It holds*

$$\frac{2}{3} \leq \text{AECR}(\text{DNF}, \mathcal{P}^{(1)}) \leq \left(\sum_{i=1}^{\infty} \frac{(i-1)!}{i^i} \right)^{-1} \approx 0.736.$$

If the maximum item size is greater than or equal to $\frac{1}{2}$ the lower bound can be improved to $\left(1 + \sum_{i=2}^{\infty} \frac{1}{i^2} \cdot \left(1 - \frac{1}{i}\right)^{i-2}\right)^{-1} \approx 0.686.$

We see that even in the analysis of this simple case there is room for improvement. Based on simulations, we suppose that the upper bound represents the truth. Furthermore we were not able to improve the worst-case bound $(1+x)^{-1}$ in the case that the maximum item size is bounded from above by $\frac{1}{2}$.

Finally, let $\mathcal{P}_{\text{two}}$ denote all discrete perfect-packing distributions for which there exists a representation in which every perfect-packing configuration b^i contains at most two non-zero entries.

Theorem 21. *Let $F \in \mathcal{P}_{\text{two}}$, then*

$$\text{AECR}(\text{DNF}, F) \geq 2/3.$$

Combining this with the proof of Theorem 15, we obtain $\text{AECR}(\text{DNF}, \mathcal{P}_{\text{two}}) = 2/3$.

5   Conclusions and Further Research

We have proven the first lower bound better than $1/2$ for the asymptotic expected competitive ratio of DNF that holds for any discrete distribution. Our lower bound is only slightly better than $1/2$ and there is still a considerable gap to the best known upper bound of $2/3$, which we also proved in this article. It is an interesting problem to close the gap between the lower and the upper bound. We conjecture that the lower bound can be improved to $2/3$. Furthermore, we have shown that the asymptotic random-order ratio coincides with the asymptotic expected competitive ratio under the mild assumption that the adversary is not allowed to add too many too small items. We believe that this assumption is not needed and we conjecture that also for arbitrary inputs the asymptotic random-order ratio coincides with the asymptotic expected competitive ratio for DNF.

Our analysis in Sect. 3 that shows the connection between the asymptotic random-order ratio and the asymptotic expected competitive ratio under the previously mentioned assumption can easily be adapted to the next-fit algorithm for bin packing. We expect that it can also be generalized to more sophisticated algorithms for bin packing (e.g., to all bounded-space algorithms with only a constant number of open bins).

References

1. Asgeirsson, E.I., Stein, C.: Bounded-space online bin cover. J. Sched. **12**(5), 461–474 (2009)
2. Assmann, S.F., Johnson, D.S., Kleitman, D.J., Leung, J.Y.-T.: On a dual version of the one-dimensional bin packing problem. J. Algorithms **5**(4), 502–525 (1984)
3. Christ, M.G., Favrholdt, L.M., Larsen, K.S.: Online bin covering: expectations vs. guarantees. Theor. Comput. Sci. **556**, 71–84 (2014)
4. Courcoubetis, C., Weber, R.R.: Stability of on-line bin packing with random arrivals and long-run average constraints. Probab. Eng. Informational Sci. **4**(4), 447–460 (1990)
5. Csirik, J., Frenk, J.B.G., Galambos, G., Kan, A.H.G.R.: Probabilistic analysis of algorithms for dual bin packing problems. J. Algorithms **12**(2), 189–203 (1991)
6. Csirik, J., Johnson, D.S., Kenyon, C.: Better approximation algorithms for bin covering. In: Proceedings of the 12th ACM-SIAM Symposium on Discrete Algorithms (SODA), pp. 557–566 (2001)
7. Csirik, J., Totik, V.: Online algorithms for a dual version of bin packing. Discrete Appl. Math. **21**(2), 163–167 (1988)
8. Coffman Jr., E.G., Csirik, J., Rónyai, L., Zsbán, A.: Random-order bin packing. Discrete Appl. Math. **156**(6), 2810–2816 (2008)
9. Fischer, C., Röglin, H.: Probabilistic analysis of the dual next-fit algorithm for bin covering, December 2015. http://arxiv.org/abs/1512.04719
10. Jansen, K., Solis-Oba, R.: An asymptotic fully polynomial time approximation scheme for bin covering. Theor. Comput. Sci. **306**(1–3), 543–551 (2003)
11. Coffman Jr., E.G., Csirik, J., Rónyai, L., Zsbán, A.: Random-order bin packing. Discrete Appl. Math. **156**(14), 2810–2816 (2008)
12. Kenyon, C.: Best-fit bin-packing with random order. In: Proceedings of the 17th ACM-SIAM Symposium on Discrete Algorithms (SODA), pp. 359–364 (1996)
13. Levin, D.A., Peres, Y., Wilmer, E.L.: Markov Chains and Mixing Times. AMS (2009)
14. Lorden, G.: On excess over the boundary. Ann. Math. Stat. **41**(2), 520–527 (1970)
15. Naaman, N., Rom, R.: Average case analysis of bounded space bin packing algorithms. Algorithmica **50**, 72–97 (2008)

Deterministic Sparse Suffix Sorting on Rewritable Texts

Johannes Fischer, Tomohiro I., and Dominik Köppl[(⊠)]

Department of Computer Science, TU Dortmund, Dortmund, Germany
johannes.fischer@cs.tu-dortmund.de, tomohiro@ai.kyutech.ac.jp,
dominik.koeppl@tu-dortmund.de

Abstract. Given a rewritable text T of length n on an alphabet of size σ, we propose an online algorithm computing the sparse suffix array and the sparse longest common prefix array of T in $\mathcal{O}(c\sqrt{\lg n} + m \lg m \lg n \lg^* n)$ time by using the text space and $\mathcal{O}(m)$ additional working space, where $m \leq n$ is the number of suffixes to be sorted (provided online and arbitrarily), and $c \geq m$ is the number of characters that must be compared for distinguishing the designated suffixes.

1 Introduction

Sorting suffixes of a long text lexicographically is an important first step for many text processing algorithms [15]. The complexity of the problem is quite well understood, as for integer alphabets suffix sorting can be done in optimal linear time [10], and also almost in-place [14]. In this article, we consider a variant of the problem: instead of computing the order of *every* suffix, we address the **sparse suffix sorting problem**. Given a text $T[1..n]$ of length n and a set $\mathcal{P} \subseteq [1..n]$ of m arbitrary positions in T, the problem asks for the (lexicographic) order of the suffixes starting at the positions in $\mathcal{P}$. The answer is encoded by a permutation of $\mathcal{P}$, which is called the **sparse suffix array (SSA)** of T (with respect to $\mathcal{P}$).

Like the "full" suffix arrays, we can enhance $\mathsf{SSA}(T, \mathcal{P})$ by the length of the longest common prefix (LCP) between adjacent suffixes in $\mathsf{SSA}(T, \mathcal{P})$, which we call the **sparse longest common prefix array (SLCP)**. In combination, $\mathsf{SSA}(T, \mathcal{P})$ and $\mathsf{SLCP}(T, \mathcal{P})$ store the same information as the **sparse suffix tree**, i.e., they implicitly represent a compacted trie over all suffixes starting at the positions in $\mathcal{P}$. This allows us to use the SSA as an efficient index for pattern matching, for example.

Based on classic suffix array construction algorithms [10,14], sparse suffix sorting is easily conducted in $\mathcal{O}(n)$ time if $\mathcal{O}(n)$ additional working space is available. For $m = o(n)$, however, the needed working space may be too large, compared to the final space requirement of $\mathsf{SSA}(T)$. Although some special choices of $\mathcal{P}$ admit space-optimal $\mathcal{O}(m)$ construction algorithms (see [2]), the problem of sorting arbitrary choices of suffixes in small space seems to be much harder. We are aware of the following results: As a deterministic algorithm,

© Springer-Verlag Berlin Heidelberg 2016
E. Kranakis et al. (Eds.): LATIN 2016, LNCS 9644, pp. 483–496, 2016.
DOI: 10.1007/978-3-662-49529-2_36

Kärkkäinen et al. [10] gave a trade-off using $\mathcal{O}(\tau m + n\sqrt{\tau})$ time and $\mathcal{O}(m + n/\sqrt{\tau})$ working space with a parameter $\tau \in [1..\sqrt{n}]$. If randomization is allowed, there is a technique based on Karp-Rabin fingerprints, first proposed by Bille et al. [2] and later improved by I et al. [8]. The latest one works in $\mathcal{O}(n \lg n)$ expected time and $\mathcal{O}(m)$ additional space.

1.1   Computational Model

We assume that the text of length n is loaded into RAM. Our algorithms are allowed to overwrite parts of the text, as long as they can restore the text into its original form at termination. Apart from this space, we are only allowed to use $\mathcal{O}(m)$ additional words. The positions in $\mathcal{P}$ are assumed to arrive online, implying in particular that they need not be sorted. We aim at worst-case efficient *deterministic* algorithms.

Our computational model is the word RAM model with word size $\Omega(\lg n)$. Here, characters use $\lceil \log \sigma \rceil$ bits, where σ is the alphabet size; hence, $\lg_\sigma n$ characters can be packed into one word. Comparing two strings X and Y therefore takes $\mathcal{O}(lcp(X,Y)/\lg_\sigma n)$ time, where $lcp(X,Y)$ denotes the length of the longest common prefix of X and Y.

1.2   Algorithm Outline and Our Results

Our main algorithmic idea is to insert the suffixes starting at positions of $\mathcal{P}$ into a self-balancing binary search tree [9]; since each insertion invokes $\mathcal{O}(\lg m)$ suffix-to-suffix comparisons, the time complexity is $\mathcal{O}(t_S m \lg m)$, where t_S is the cost for each suffix-to-suffix comparison. If all suffix-to-suffix comparisons are conducted by naively comparing the characters, the resulting worst case time complexity is $\mathcal{O}(nm \lg m)$. In order to speed this up, our algorithm identifies large identical substrings at different positions during different suffix-to-suffix comparisons. Instead of performing naive comparisons on identical parts over and over again, we build a data structure (stored in redundant text space) that will be used to accelerate subsequent suffix-to-suffix comparisons. Informally, when two (possibly overlapping) substrings in the text are detected to be the same, one of them can be overwritten.

To accelerate suffix-to-suffix comparisons, we focus on a data structure called *edit sensitive parsing (ESP) tree* [5]. The ESP tree supports *longest common extension (LCE)* queries. An LCE query on an ESP tree asks for the length of the longest common prefix of two suffixes of the string on which the tree is built. Besides answering LCE queries, ESP trees are *mergeable*, allowing us to build a dynamically growing LCE index on substrings read in the process of the sparse suffix sorting. Consequently, comparing two already indexed substrings is done by a single LCE query.

In their plain form, ESP trees need more space than the text itself; to overcome this space problem, we devise a *truncated* version of the ESP tree, yielding a trade-off parameter between space consumption and LCE query time. By choosing this parameter appropriately, the truncated ESP tree fits into the text space.

However, the need for merging still causes a problem due to the fact that leaves of an ESP tree point to substrings of the text. Although we can prohibit overwriting those referred substrings, a merging may create a new leaf whose substring is already overwritten by the in-text construction of a different ESP tree. To cope with this situation, we propose a new variant of ESP, called **hierarchical stable parsing (HSP)**, allowing us to quickly find a surrogate substring. With a text space management specialized on the properties of the HSP, we achieve the result of Theorem 1 below.

We make the following definition that allows us to analyze the running time more accurately. Define $\mathcal{C} := \bigcup_{p,p' \in \mathcal{P}, p \neq p'} [p..p + lcp(T[p..], T[p'..])]$ as the set of positions that must be compared for distinguishing the suffixes from $\mathcal{P}$. Then sparse suffix sorting is trivially lower bounded by $\Omega(|\mathcal{C}| / \lg_\sigma n)$ time.

We now state the main result of this article as follows:

Theorem 1. *Given a text T of length n that is loaded into RAM, the SSA and SLCP of T for a set of m arbitrary positions can be computed deterministically in $\mathcal{O}(|\mathcal{C}| \sqrt{\lg n} + m \lg m \lg n \lg^* n)$ time, using $\mathcal{O}(m)$ additional working space.*

1.3 Relationship Between Suffix Sorting and LCE Queries

The LCE-problem is to preprocess a text T such that subsequent LCE-queries $lce(i, j) := lcp(T[i..], T[j..])$ giving the length of the longest common prefix of the suffixes starting at positions i and j can be answered efficiently. Data structures for LCE and sparse suffix sorting are closely related, as shown in the following observation:

Observation 1. *Given a data structure that computes LCE in $\mathcal{O}(\tau)$ time for $\tau > 0$, we can compute sparse suffix sorting for m positions in $\mathcal{O}(\tau m \lg m)$ time by inserting suffixes in a balanced binary search tree.*

Conversely, given an algorithm computing the SSA and the SLCP of a text T of length n for m positions in $\mathcal{O}(m)$ space and $\mathcal{O}(f(n, m))$ time for some f, we can construct a data structure in $\mathcal{O}(f(n, m))$ time and $\mathcal{O}(m)$ space, answering LCE queries on T in $\mathcal{O}(n^2/m^2)$ time [4], (using a difference cover sampling modulo n/m [10]).

The currently best deterministic data structure for LCE we are aware of is due to Bille et al. [3], using $\mathcal{O}(n/\tau)$ space and answering LCE queries in $\mathcal{O}(\tau)$ time, for any $1 \leq \tau \leq n$. However, this data structure has a preprocessing time of $\Omega(n^2)$, and is thus not helpful for sparse suffix sorting. We develop a new data structure for LCE with the following properties.

Theorem 2. *There is a data structure using $\mathcal{O}(n/\tau)$ space that answers LCE queries in $\mathcal{O}(\lg^* n \, (\lg (n/\tau) + \tau^{\lg 3}/\lg_\sigma n))$ time, where $1 \leq \tau \leq n$. We can build the data structure in $\mathcal{O}(n (\lg^* n + (\lg n)/\tau + (\lg \tau)/\lg_\sigma n))$ time with additional $\mathcal{O}(\tau^{\lg 3} \lg^* n)$ words during construction.*

An advantage of our data structure against the deterministic data structures in [3] is its faster construction time, which is upper bounded by $\mathcal{O}(n \lg n)$.

1.4   Outline of this Article

The first part of the paper (Sect. 2) is dedicated to answering LCE queries (Theorem 2) with the (truncated) ESP tree. In Sect. 3 we describe our algorithm for the sparse suffix sorting problem with the abstract data type dynLCE that supports LCE queries and a merging operation. In Sect. 4 we study how the text space can be exploited to lower the memory footprint. To this end, we develop (truncated) HSP trees. By the properties of the HSP tree, we finally solve the sparse suffix sorting problem (Theorem 1) in the claimed time and space.

1.5   Preliminaries

Let Σ be an ordered alphabet of size σ. We assume that a character in Σ is represented by an integer. For a string $X \in \Sigma^*$, let $|X|$ denote the length of X. For a position i in X, let $X[i]$ denote the i-th character of X. For positions i and j, let $X[i..j] = X[i]X[i+1]\cdots X[j]$. For $W = XYZ$ with $X, Y, Z \in \Sigma^*$, we call X, Y and Z a prefix, substring, suffix of W, respectively. In particular, the suffix beginning at position i is denoted by $W[i..]$.

   An *interval* $\mathcal{I} = [b..e]$ is the set of consecutive integers from b to e, for $b \leq e$. For an interval $\mathcal{I}$, we use the notations $\mathsf{b}(\mathcal{I})$ and $\mathsf{e}(\mathcal{I})$ to denote the beginning and end of $\mathcal{I}$; i.e., $\mathcal{I} = [\mathsf{b}(\mathcal{I})..\mathsf{e}(\mathcal{I})]$. We write $|\mathcal{I}|$ to denote the length of $\mathcal{I}$; i.e., $|\mathcal{I}| = \mathsf{e}(\mathcal{I}) - \mathsf{b}(\mathcal{I}) + 1$.

2   Answering LCE Queries with ESP Trees

Edit sensitive parsing (ESP) and ESP trees were proposed by Cormode and Muthukrishnan [5] to approximate the edit distance with moves efficiently. A similar technique is *signature encoding* [12]. Based on signature encoding, Alstrupet et al. [1] and Nishimoto et al. [13] derive new data structures for supporting LCE queries. For several reasons, these data structures cannot be used in our context; we therefore show in this section that ESP trees can also be used to answer LCE queries.

2.1   Edit Sensitive Parsing

The aim of the ESP technique is to decompose a string $Y \in \Sigma^*$ into substrings of length 2 or 3 such that each substring of this decomposition is determined uniquely by its neighboring characters. To this end, it first identifies so-called **meta-blocks** in Y, and then further refines these meta-blocks into **blocks** of length 2 or 3.

   The meta-blocks are created in the following 3-stage process:

(1) Identify maximal regions of repeated symbols (i.e., maximal substrings of the form c^ℓ for $c \in \Sigma$ and $\ell \geq 2$). Such substrings form the type 1 meta-blocks.
(2) Identify remaining substrings of length at least 2 (which must lie between two type 1 meta-blocks). Such substrings form the type 2 meta-blocks.

(3) Any substring not yet covered by a meta-block consists of a single character and cannot have **type 2** meta-blocks as its neighbors. Such characters $Y[i]$ are fused with the **type 1** meta-block to their right[1], or, if $Y[i]$ is the last character in Y, with the **type 1** meta-block to its left. The meta-blocks emerging from this are called **type M** (mixed).

Meta-blocks of **type 1** and **type M** are collectively called **repeating meta-blocks**.

Although meta-blocks are defined by the comprising characters, we treat them as intervals on the text range.

Meta-blocks are further partitioned into **blocks**, each containing two or three characters from Σ. Blocks inherit the type of the meta-block they are contained in. How the blocks are partitioned depends on the type of the meta-block:

Repeating meta-blocks. A repeating meta-block is partitioned greedily: create blocks of length three until there are at most four, but at least two characters left. If possible, create a single block of length 2 or 3; otherwise create two blocks, each containing two characters.

Type-2 meta-blocks. A **type 2** meta-block μ is processed in $\mathcal{O}(|\mu| \lg^* \sigma)$ time by a technique called *alphabet reduction* [5]. The first $\lg^* \sigma$ characters are blocked in the same way as repeating meta-blocks. Any remaining block β is formed such that β's interval boundaries are determined by $Y[\max(b(\beta) - \Delta_{\mathrm{L}}, b(\mu)) .. \min(e(\beta) + \Delta_{\mathrm{R}}, e(\mu))]$, where $\Delta_{\mathrm{L}} := \lceil \lg^* \sigma \rceil + 5$ and $\Delta_{\mathrm{R}} := 5$ (see [5, Lemma 8]).

We call the substring $Y[b(\beta) - \Delta_{\mathrm{L}} .. e(\beta) + \Delta_{\mathrm{R}}]$ the **local surrounding** of β, if it exists. Blocks whose local surroundings exist are also called **surrounded**.

Let $\tilde{\Sigma} \subseteq \Sigma^2 \cup \Sigma^3$ denote the set of blocks resulting from ESP (the "new alphabet"). We use $esp: \Sigma^* \to \tilde{\Sigma}^*$ to denote the function that parses a string by ESP and returns a string in $\tilde{\Sigma}^*$.

2.2 Edit Sensitive Parsing Trees

Applying esp recursively on its output generates a context free grammar (CFG) as follows. Let $Y_0 := Y$ be a string on an alphabet $\Sigma_0 := \Sigma$ with $\sigma_0 = |\Sigma_0|$. The output of $Y_h := esp^h(Y) = esp(esp^{h-1}(Y))$ is a sequence of blocks, which belong to a new alphabet Σ_h ($h > 0$). A block $b \in \Sigma_h$ contains a string $b \in \Sigma_{h-1}^*$ of length two or three. Since each application of esp reduces the string length by at least $1/2$, there is a $k = \mathcal{O}(\lg |Y|)$ such that $esp(Y_k)$ returns a single block τ. We write $\mathcal{V} := \bigcup_{1 \le h \le k} \Sigma_h$ for the set of all blocks in $Y_1, Y_2, \ldots, Y_k$.

We use a (deterministic) dictionary $\mathfrak{D}: \Sigma_h \to \Sigma_{h-1}^2 \cup \Sigma_{h-1}^3$ to map a block to its characters, for each $1 \le h \le k$. The dictionary entries are of the form $b \to xy$ or $b \to xyz$, where $b \in \Sigma_h$ and $x, y, z \in \Sigma_{h-1}$. The CFG for Y is represented by

---

[1] The original version prefers the left meta-block, but we change it for a more stable behavior.

the non-terminals $\mathcal{V}$, the terminals Σ_0, the dictionary $\mathfrak{D}$, and the start symbol τ. This grammar exactly derives Y.

Our representation differs from that of Cormode and Muthukrishnan [5] because it does not use hash tables.

Definition 1. *The **ESP tree** $\mathsf{ET}(Y)$ of a string Y is a slightly modified derivation tree of the CFG defined above. The internal nodes are elements of $\mathcal{V} \setminus \Sigma_1$, and the leaves are from Σ_1. Each leaf refers to a substring in Σ_0^2 or Σ_0^3. Its root node is the start symbol τ.*

For convenience, we count the height of nodes from 1, so that the sequence of nodes on height h, denoted by $\langle Y \rangle_h$, is corresponding to Y_h. The ***generated substring*** of a node $\langle Y \rangle_h[i]$ is the substring of Y generated by the symbol $Y_h[i]$ (applying $\mathfrak{D}$ recursively on $Y_h[i]$). Each node v represents a block that is contained in a meta-block μ, for which we say that μ ***builds*** v. More precisely, a node $v := \langle Y \rangle_h[i]$ is said to be built on a meta-block represented by $\langle Y \rangle_{h-1}[b..e]$ iff $\langle Y \rangle_{h-1}[b..e]$ contains the children of v. Like with blocks, nodes inherit the type of meta-block on which they are built.

Surrounded Nodes. A leaf is called surrounded iff its representing block on text-level is surrounded. Given an internal node v on height $h+1$ ($h \geq 1$) whose children are $\langle Y \rangle_h[\beta]$, we say that v is ***surrounded*** iff the nodes $\langle Y \rangle_h[\mathsf{b}(\beta) - \Delta_L..\mathsf{e}(\beta) + \Delta_R]$ are surrounded.

2.3   Tree Representation

We store the ESP tree as a CFG. Every non-terminal is represented by a ***name***. The name is a pointer to a data-field, which is composed differently for leaves and internal nodes:

Leaves. A leaf stores a position i and a length $l \in \{2, 3\}$ such that $Y[i..i+l-1]$ is the generated substring.

Internal Nodes. An internal node stores the length of its generated substring, and the names of its children. If it has only two children, we use a special, invalid name 0 for the non-existing third child such that all data fields are of the same length.

This representation allows us to navigate top-down in the ESP tree by traversing the tree from the root, in time linear in the height of the tree.

We keep the invariant that the roots of *isomorphic* subtrees have the *same* names. In other words, before creating a new name for the rule $b \to xyz$, we have to check whether there already exists a name for xyz. To perform this look-up efficiently, we need also the *reverse* dictionary of $\mathfrak{D}$, with the right hand side of the rules as search keys. We use a dictionary of size $\mathcal{O}(|Y|)$, supporting lookup and insert in $\mathcal{O}(t_\lambda)$ time.

More precisely, we assume there is a dictionary data structure, storing n elements in $\mathcal{O}(n)$ space, supporting lookup and insert in $\mathcal{O}(t_\lambda + l / \lg_\sigma n)$ time for a key of length l, where $t_\lambda = t_\lambda(n)$ depends on n. For instance, Franceschini and Grossi's data structure [7] with word-packing supports $t_\lambda = \mathcal{O}(\lg n)$.

Lemma 1. *An ESP tree of a string of length n can be built in $\mathcal{O}(n\,(\lg^* n + t_\lambda))$ time. It consumes $\mathcal{O}(n)$ space.*

2.4   LCE Queries in ESP Trees

ESP trees are fairly stable against edit operations: The number of nodes that are differently parsed after prepending or appending a string to the input is upper bounded by $\mathcal{O}(\lg n \lg^* n)$ [5, Lemma 11]. To use this property in our context of LCE queries, we consider nodes of $\mathsf{ET}(Y)$ that are still present in $\mathsf{ET}(XYZ)$; a node v in $\mathsf{ET}(Y)$ generating $Y[i_0..j_0]$ is said to be **stable** iff, for *all* strings X and Z, there exists a node in $\mathsf{ET}(XYZ)$ that has the same name as v and generates $(XYZ)[|X| + i_0..|X| + j_0]$. We also consider repeating nodes that are present with slight shifts; a non-stable repeating node v in $\mathsf{ET}(Y)$ generating $Y[i_0..j_0]$ is said to be **semi-stable** iff, for *all* strings X and Z, there exists a node in $\mathsf{ET}(XYZ)$ that has the same name as v and generates a substring intersecting with $(XYZ)[|X| + i_0..|X| + j_0]$. Then, the proof of Lemma 9 of [5] says that, for each height, $\mathsf{ET}(Y)$ contains $\mathcal{O}(\lg^* n)$ nodes that are not (semi-)stable, which we call **fragile**. Since the children of the (semi-)stable nodes are also (semi-)stable, there is a border in $\mathsf{ET}(Y)$ separating the (semi-)stable nodes from the fragile ones.

In order to use semi-stable nodes to answer LCE queries efficiently, we let each node have an additional property, called **surname**. A node $v := \langle Y \rangle_h[i]$ is said to be **repetitive** iff there exists $\langle Y \rangle_{h'}[\mathcal{I}]$ at some height $h' < h$ with $Y_{h'}[\mathcal{I}] = d^{|\mathcal{I}|}$, where $\langle Y \rangle_{h'}[\mathcal{I}]$ is the sequence of nodes on height h' in the subtree rooted at $\langle Y \rangle_h[i]$ and $d \in \Sigma_{h'}$. The surname of a repetitive node $v := \langle Y \rangle_h[i]$ is the name of a highest non-repetitive node in the subtree rooted at v. The surname of a non-repetitive node is the name of the node itself. It is easy to compute and store the surnames while constructing ESP trees.

The connection between semi-stable nodes and the surnames is based on the fact that a semi-stable node is repetitive: Let u be the node whose name is the surname of a semi-stable node v. If u is on height h, v's subtree consists of a repeat of u's on height h. A shift of v can only be caused by adding u's. So the shift is always a multiple of the length of the generated substring of u.

We now state a lemma that shows how ESP trees can be used for LCE queries; the proof (as well as all other missing proofs) can be found in the full version [6].

Lemma 2. *Let X and Y be strings with $|X| \le |Y| \le n$. Given $\mathsf{ET}(X)$ and $\mathsf{ET}(Y)$ built with the same dictionary and two text-positions $1 \le i_X \le |X|, 1 \le i_Y \le |Y|$, we can compute $l := lcp(X[i_X..], Y[i_Y..])$ in $\mathcal{O}(\lg|Y| + \lg l \lg^* n)$ time.*

2.5   Truncated ESP Trees

Building an ESP tree over a string Y requires $\mathcal{O}(|Y|)$ words of space, which might be too much in some scenarios. Our idea is to truncate the ESP tree at some fixed height, discarding the nodes in the lower part. The truncated version stores just the upper part, while its (new) leaves refer to (possibly long) substrings of Y. The resulting tree is called the ***truncated ET (tET)***. More precisely, we define a height η and delete all nodes at height less than η, which we call ***lower nodes***. A node higher than η is called an ***upper node***. The nodes at height η form the new leaves and are called η-***nodes***. Similar to the former leaves, their names are pointers to their generated substrings appearing in Y. Remembering that each internal node has two or three children, an η-node generates a string of length at least 2^η and at most 3^η. So the maximum number of nodes in a truncated ESP tree of a string of length n is $n/2^\eta$.

Similar to leaves, we use the generated substring X of an η-node v for storing and looking up v: It can be looked up or inserted in $\mathcal{O}(|X|/\lg_\sigma n + t_\lambda)$ time.

Lemma 3. *We can build a truncated ESP tree of a string Y of length n in $\mathcal{O}(n(\lg^* n + \eta/\lg_\sigma n + t_\lambda/2^\eta)$ time, using $\mathcal{O}(3^\eta \lg^* n)$ words of working space. The tree consumes $\mathcal{O}(n/2^\eta)$ space.*

Lemma 4. *Let X and Y be strings with $|X|, |Y| \leq n$. Given $\mathsf{ET}(X)$ and $\mathsf{ET}(Y)$ built with the* same *dictionary and two text-positions $1 \leq i_X \leq |X|, 1 \leq i_Y \leq |Y|$, we can compute $lcp(X[i_X..], Y[i_Y..])$ in $\mathcal{O}(\lg^* n(\lg(n/2^\eta) + 3^\eta/\lg_\sigma n))$ time.*

With $\tau := 2^\eta$ we get Theorem 2.

3 Sparse Suffix Sorting

Borrowing the technique of Irving and Love [9], an AVL tree on a set of strings $\mathcal{S}$ can be augmented with LCP values so that we can compute $l := \max\{lcp(X, Y) \mid X \in \mathcal{S}\}$ for a string Y in $\mathcal{O}(l/\lg_\sigma n + \lg|\mathcal{S}|)$ time. Inserting a new string into the tree is supported in the same time complexity. Irving and Love [9] called this data structure the ***suffix AVL tree*** on $\mathcal{S}$; we denote it by $\mathsf{SAVL}(\mathcal{S})$.

Given a text T of length n, we will use $\mathsf{SAVL}(Suf(\mathcal{P}))$ as a representation for $\mathsf{SSA}(T, \mathcal{P})$ and $\mathsf{SLCP}(T, \mathcal{P})$. Our goal is to build $\mathsf{SAVL}(Suf(\mathcal{P}))$ efficiently. However, inserting suffixes naively suffers from the lower bound $\Omega(n|\mathcal{P}|/\lg_\sigma n)$ on time. How to speed up the comparisons by exploiting a data structure for LCE queries is topic of this section.

3.1 Abstract Algorithm

Our idea is that creating a mergeable LCE data structure on the read substrings may be helpful for later queries. We call this abstract data type ***dynamic LCE (dynLCE)***; it supports the following operations:

- dynLCE(Y) constructs a dynLCE data structure M on a substring Y of T. Let $M.\mathtt{text}$ denote the string Y on which M is constructed.
- LCE(M_1, M_2, p_1, p_2) computes $lcp(M_1.\mathtt{text}[p_1..], M_2.\mathtt{text}[p_2..])$, where $p_i \in [1..|M_i.\mathtt{text}|]$ for $i = 1, 2$.
- merge(M_1, M_2) merges two dynLCEs M_1 and M_2 such that the output is a dynLCE on the concatenation of $M_1.\mathtt{text}$ and $M_2.\mathtt{text}$.

We use the expression $t_C(|Y|)$ to denote the construction time of such a data structure on a string Y. Further, $t_L(|X| + |Y|)$ and $t_M(|X| + |Y|)$ denote the LCE query time and the time for merging two such data structures on strings X and Y, respectively. Querying a dynLCE built on a string of length ℓ is faster than the word-packed character comparison iff $\ell = \Omega(t_L(\ell) \lg n / \lg \sigma)$. Hence, there is no point in building a dynLCE on a text smaller than $g := \Theta(t_L(g) \lg n / \lg \sigma)$.

We store the text intervals covered by the dynLCEs such that we know the text-positions where querying a dynLCE is possible. Such an interval is called an **LCE interval**. An LCE interval $\mathcal{I}$ stores a pointer to its dynLCE data structure M, and an integer i such that $M.\texttt{text}[i..i + |\mathcal{I}| - 1] = T[\mathcal{I}]$. The LCE intervals themselves are maintained in a self-balancing binary search tree of size $\mathcal{O}(|\mathcal{P}|)$, storing their starting positions as keys.

For a new position $1 \leq \hat{p} \leq |T|, \hat{p} \notin \mathcal{P}$, updating $\mathsf{SAVL}(Suf(\mathcal{P}))$ to $\mathsf{SAVL}(Suf(\mathcal{P} \cup \{\hat{p}\}))$ involves two parts: first locating the insertion node for $\hat{p}$ in $\mathsf{SAVL}(Suf(\mathcal{P}))$, and then updating the set of LCE intervals.

Locating. The suffix AVL tree performs an LCE computation for each node encountered while locating the insertion point of $\hat{p}$. Assume that the task is to compare the suffixes $T[i..]$ and $T[j..]$ for some $1 \leq i, j \leq |T|$. First check whether the positions i and j are contained in an LCE interval, in $\mathcal{O}(\lg m)$ time. If both positions are covered by LCE intervals, then query the respective *dynLCEs*. Otherwise, look up the position where the next LCE interval starts. Up to that position, naively compare both substrings. Finally, repeat the above check again at the new positions, until finding a mismatch. After locating the insertion point of $\hat{p}$ in $\mathsf{SAVL}(Suf(\mathcal{P}))$, we obtain $\bar{p} := \mathrm{mlcparg}_{\hat{p}}$ and $l := \mathrm{mlcp}_{\hat{p}}$ as a byproduct, where $\mathrm{mlcparg}_p := \mathrm{argmax}_{p' \in \mathcal{P}, p \neq p'} lcp(T[p..], T[p'..])$ and $\mathrm{mlcp}_p := lcp(T[p..], T[\mathrm{mlcparg}_p..])$ for $1 \leq p \leq |T|$.

Updating. The LCE intervals are updated dynamically, subject to the following constraints:

C1: The length of each LCE interval is at least g.

C2: For every $p \in \mathcal{P}$ the interval $[p..p + \mathrm{mlcp}_p - 1]$ is covered by an LCE interval *except at most g* positions at its left and right ends.

C3: There is a gap of at least g positions between every pair of LCE intervals.

These constraints guarantee that there is at most one LCE interval that intersects with $[p..p + \mathrm{mlcp}_p - 1]$ for a $p \in \mathcal{P}$.

The following instructions will satisfy the constraints: If $l < g$, we do nothing. Otherwise, we have to care about C2. Fortunately, there is at most one position in $\mathcal{P}$ that possibly invalidates C2 after adding $\hat{p}$, and this is $\bar{p}$; otherwise, by transitivity, we would have created some larger LCE interval previously. Let $U \subset [1..n]$ be the positions that belong to an LCE interval. The set $[\hat{p}..\hat{p}+l-1] \setminus U$ can be represented as a set of disjoint intervals of maximal length. For each interval $\mathcal{I} := [\hat{p} + i..\hat{p} + j] \subset [\hat{p}..\hat{p} + l - 1]$ of that set (for some $0 \leq i \leq j < l$), apply the following rules with $\mathcal{J} := [\bar{p} + i..\bar{p} + j]$ sequentially:

R1: If $\mathcal{J}$ is a sub-interval of an LCE interval, then declare $\mathcal{I}$ as an LCE interval and let it refer to the dynLCE of the larger LCE interval.

R2: If $\mathcal{J}$ intersects with an LCE interval $\mathcal{K}$, enlarge $\mathcal{K}$ to $\mathcal{K} \cup \mathcal{J}$, enlarging its corresponding dynLCE (We can enlarge an dynLCE by creating a new instance and merge both instances). Apply R1.

R3: Otherwise, create a dynLCE on $\mathcal{I}$, and make $\mathcal{I}$ to an LCE interval.

R4: If C3 is violated, then a newly created or enlarged LCE interval is adjacent to another LCE interval. Merge those LCE intervals and their dynLCEs.

We also need to satisfy C2 on $[\bar{p}..\bar{p}+l-1]$. To this end, update U, compute the set of disjoint intervals $[\bar{p}..\bar{p}+l-1] \setminus U$ and apply the same rules on it.

Although we might create some LCE intervals covering less than g characters, we will restore C1 by merging them with a larger LCE interval in R4. In fact, we introduce at most two new LCE intervals. C1 is easily maintained, since we will never shrink an LCE interval.

Lemma 5. *Given a text T of length n that is loaded into RAM, the SSA and SLCP of T for a set of m arbitrary positions can be computed deterministically in $\mathcal{O}(t_C(|\mathcal{C}|) + t_L(|\mathcal{C}|)m \lg m + m t_M(|\mathcal{C}|))$ time.*

3.2 Sparse Suffix Sorting with ESP Trees

We will show that the ESP tree is a suitable data structure for dynLCE. In order to merge two ESP trees, we use a *common* dictionary $\mathfrak{D}$ that is stored *globally*. Fortunately, it is easy to combine two ESP trees by updating just a handful of nodes, which are fragile.

Lemma 6. *Assume that we have already created $\mathsf{ET}(X)$ and $\mathsf{ET}(Y)$ on two strings $X, Y \in \Sigma^*$. Merging both trees into $\mathsf{ET}(XY)$ takes $\mathcal{O}(t_\lambda(\Delta_L \lg |Y| + \Delta_R \lg |X|))$ time.*

The following theorem combines the results of Lemmas 5 and 6.

Theorem 3. *Given a text T of length n and a set of m text positions $\mathcal{P}$, $\mathsf{SSA}(T, \mathcal{P})$ and $\mathsf{SLCP}(T, \mathcal{P})$ can be computed in $\mathcal{O}(|\mathcal{C}| (\lg^* n + t_\lambda) + m \lg m \lg n \lg^* n)$ time.*

4   Hierarchical Stable Parsing

Remembering the outline in the introduction, the key idea is to solve the limited space problem by storing dynLCEs in text space. Taking two LCE intervals on the text containing the same substring, we overwrite *one* part while marking the *other* part as a reference. By choosing a suitably large η, we can overwrite the text of one LCE interval with a truncated ESP tree (tET) whose η-nodes refer to substrings of the other LCE interval. Merging two tETs involves a reparsing of some η-nodes. Assume that we want to reparse an η-node v, and that its generated substring gets enlarged due to the parsing. We have to locate a substring in the text that contains its new generated substring X. Although we can create a suitably large string containing X by concatenating the generated substrings of its preceding and succeeding siblings, these η-nodes may point to text intervals that may not be consecutive. Since the name of an η-node is the representation of a *single* substring, we have to search for a substring equal to X in the text. Because this would be too inefficient, we will show a slight modification of the ESP technique that circumvents this problem.

4.1 Hierarchical Stable Parse Trees

Our modification, which we call *hierarchical stable parse trees* or **HSP trees**, affects only the definition of meta-blocks. The factorization of meta-blocks is done by relaxing the check whether two characters are equal; instead of comparing names we compare by surname.[2] A more detailed study of HTs can be read in the full version of the paper [6].

Lemma 7. *An HSP tree on an interval of length l can be built in $\mathcal{O}(l\,(\lg^* n + t_\lambda))$ time. It consumes $\mathcal{O}(l)$ space.*

The modified parsing allows us to claim the following lemma:

Lemma 8. *If a surrounded node is neither stable nor semi-stable, it can only be changed to a node whose generated substring is a prefix of the generated substring of an already existing node.*

4.2 Sparse Suffix Sorting in Text Space

The **truncated HSP tree (tHT)** is the truncated version of the HSP tree. It is defined analogously as the truncated ESP tree (see Sect. 2.5), with the exception of the surnames: For each repetitive node, we mark whether its surname is the name of an upper node, of an η-node, or of a lower node. Therefore, we need to save the names of certain lower nodes in the reverse dictionary of $\mathfrak{D}$. This is only necessary when an upper node or an η-node v has a surname that is the name of a lower node. If v is an upper node having a surname equal to the name of a lower node, the η-nodes in the subtree rooted at v have the same surname, too. So the number of lower node entries in the reverse dictionary is upper bounded by the number of η-nodes, and each lower node generates a substring of length less than 3^η. We conclude that the results of Lemmas 3 and 4 apply to the tHT, too.

Assume that we want to store $\mathsf{tHT}(T[\mathcal{I}])$ on some text interval $\mathcal{I}$. Since $\mathsf{tHT}(T[\mathcal{I}])$ could contain nodes with $|\mathcal{I}|$ distinct names, it requires $\mathcal{O}(|\mathcal{I}|)$ words, i.e., $\mathcal{O}(|\mathcal{I}|\lg n)$ bits of space that do not fit in the $|\mathcal{I}|\lg\sigma$ bits of $T[\mathcal{I}]$. Taking some constant α (independent of n and σ, but dependent of the size of a single node), we can solve this space issue by setting $\eta := \log_3(\alpha\lg^2 n/\lg\sigma)$:

Lemma 9. *With η as defined above, the number of nodes in a truncated HSP tree is bounded by $\mathcal{O}(l(\lg\sigma)^{0.7}/(\lg n)^{1.2})$. Further, an η-node generates a substring containing at most $\lceil \alpha(\lg n)^2/(\lg\sigma)\rceil$ characters.*

---

[2] The check is relaxed since nodes with different surnames cannot have the same name.

Applying Lemma 9 to the results elaborated in Sect. 2.5 for the tETs yields.

Corollary 1. *We can compute a tHT on a substring of length l in $\mathcal{O}(l \lg^* n + t_\lambda l/2^\eta + l \lg \lg n)$ time. The tree takes $\mathcal{O}(l/2^\eta)$ space. We need a working space of $\mathcal{O}(\lg^2 n \lg^* n / \lg \sigma)$ characters.*

Corollary 2. *An LCE query on two tHTs can be answered in $\mathcal{O}(\lg^* n \lg n)$ time.*

We analyze the merging when applied by the sparse suffix sorting algorithm in Sect. 3.1. Assume that our algorithm found two intervals $[i..i + l - 1]$ and $[j..j + l - 1]$ with $T[i..i + l - 1] = T[j..j + l - 1]$. Ideally, we want to construct $\mathsf{tHT}(T[i..i + l - 1])$ in the text space $[j..j+l-1]$, leaving $T[i..i+l-1]$ untouched so that parts of this substring can be referenced by the η-nodes. Unfortunately, there are two situations that make the life of a tHT complicated: the need for merging tHTs, and possible overlapping of the intervals $[i..i+l-1]$ and $[j..j+l-1]$.

Partitioning of LCE Intervals. In order to merge trees, we have to take special care of those η-nodes that are fragile, because their names may have to be recomputed during a merge. In order to recompute the name of an η-node v, consisting of a pointer and a length, we have to find a substring that consists of v's generated substring and some adjacent characters with respect to the original substring in the text. That is because the parsing may assign a new pointer and a new length to an η-node, possibly enlarging the generated substring, or letting the pointer refer to a different substring.

The name for a surrounded fragile η-nodes v is easily recomputable thanks to Lemma 8: Since the new generated substring of v is a prefix of the generated substring of an already existing η-node w, which is found in the reverse dictionary for η-nodes, we can create a new name for v from the generated substring of w.

Unfortunately, the same approach does not work with the non-surrounded η-nodes. Those nodes have a generated substring that is found on the border area of $T[j..j + l - 1]$. If we leave this area untouched, we can use it for creating names of a non-surrounded η-node during a reparsing. Therefore, we mark those parts of the interval $[j..j+l-1]$ as read-only. Conceptually, we partition an LCE interval into subintervals of green and red intervals; we free the text of a ***green interval*** for overwriting, while prohibiting write-access on a ***red interval***. The green intervals are managed in a dynamic, global list. We keep the invariant that

Invariant 1: $f := \lceil 2\alpha \lg^2 n \Delta_\mathrm{L} / \lg \sigma \rceil = \Theta(g)$ positions of the left and right ends of each LCE interval are *red*.

This invariant solves the problem for the non-surrounded nodes.

Allocating Space. We can store the upper part of the tHT in a green interval, since $l/2^\eta \lg n \leq l\alpha^{0.6}(\lg \sigma)^{0.7}/(\lg n)^{0.2} = \mathrm{o}(l \lg \sigma)$ holds. By choosing g and α properly, we can always leave $f \lg \sigma / \lg n = \mathcal{O}(\lg^* n \lg n)$ words on a green interval untouched, sufficiently large for the working space needed by Corollary 1. Therefore, we pre-compute α and g based on the input T, and set both as *global* constants. Since the same amount of free space is needed during a later merging when reparsing an η-node, we add the invariant that

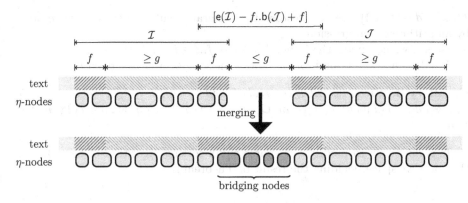

Fig. 1. The merging is performed only if the gap between both trees is less than g. The substring $T[e(\mathcal{I}) - f..b(\mathcal{J}) + f]$ is marked red for the sake of the bridging nodes (Color figure online).

Invariant 2: Each LCE interval has $f \lg \sigma / \lg n$ free space left on a green interval.

For the merging, we need a more sophisticated approach that respects both invariants:

Merging. We introduce a merge operation that allows the merge of two tHTs whose LCE intervals have a gap of less than g characters. The merge operation builds new η-nodes on the gap. The η-nodes whose generated substrings intersect with the gap are called ***bridging*** nodes. The bridging nodes have the same problem as the non-surrounded η-nodes, since the gap may be a unique substring of T.

Let $\mathcal{I}$ and $\mathcal{J}$ be two LCE intervals with $0 \leq b(\mathcal{J}) - e(\mathcal{I}) \leq g$, where on each interval a tHT has been computed. We compute $\mathsf{tHT}(T[b(\mathcal{I})..e(\mathcal{J})])$ by merging both trees. By Lemma 6, at most $\mathcal{O}(\Delta_\mathrm{L} + \Delta_\mathrm{R})$ nodes at every height on each tree have to be reprocessed, and some bridging nodes connecting both trees have to be built. Unfortunately, the text may not contain another occurrence of $T[e(\mathcal{I}) - f..b(\mathcal{J}) + f]$ such that we could overwrite $T[e(\mathcal{I}) - f..b(\mathcal{J}) + f]$. Therefore, we mark this interval as red. So we can use the characters contained in $T[e(\mathcal{I}) - f..b(\mathcal{J}) + f]$ for creating the bridging η-nodes, and for modifying the non-surrounded nodes of both trees (Fig. 1). Since the gap consists of less than g characters, the bridging nodes need at most $\mathcal{O}(\lg n \lg^* n)$ additional space. By choosing g and α sufficiently large, we can maintain Invariant 2 for the merged LCE interval.

Interval Overlapping. Assume that the LCE intervals $[i..i + l - 1]$ and $[j..j + l - 1]$ overlap, without loss of generality $j > i$. Our goal is to create $\mathsf{tHT}(T[i..i + l - 1])$. First, we compute the smallest period $d \leq j - i$ of $T[i..j + l - 1]$ in $\mathcal{O}(l)$ time [11]. The substring $T[i..i + d + f - 1]$ is used as a reference and therefore marked red. Keeping the original characters in

$T[i..i + d + f - 1]$, we can restore the generated substrings of every η-node by an arithmetic progression.

Hence, we can mark the interval $[i + d + f..j + l - 1 - f]$ *green*.

Finally, the time bound for the above merging strategy is given by

Corollary 3. *Given two LCE intervals $\mathcal{I}$ and $\mathcal{J}$ with $0 \leq \mathsf{b}(\mathcal{J}) - \mathsf{e}(\mathcal{I}) \leq g$. We can build $\mathsf{tHT}(T[\mathsf{b}(\mathcal{I})..\mathsf{e}(\mathcal{J})])$ in $\mathcal{O}(g \lg^* n + t_\lambda g/2^\eta + g\eta/\lg_\sigma n + t_\lambda \lg^* n \lg n)$ time.*

It is now easy to modify our sparse suffix sorting algorithm of Sect. 3.1 for tHT on text space, yielding the result of Theorem 1.

References

1. Alstrup, S., Brodal, G.S., Rauhe, T.: Pattern matching in dynamic texts. In: SODA, pp. 819–828 (2000)
2. Bille, P., Fischer, J., Gørtz, I.L., Kopelowitz, T., Sach, B., Vildhøj, H.W.: Sparse suffix tree construction in small space. In: Fomin, F.V., Freivalds, R., Kwiatkowska, M., Peleg, D. (eds.) ICALP 2013, Part I. LNCS, vol. 7965, pp. 148–159. Springer, Heidelberg (2013)
3. Bille, P., Gørtz, I.L., Knudsen, M.B.T., Lewenstein, M., Vildhøj, H.W.: Longest common extensions in sublinear space. In: Cicalese, F., Porat, E., Vaccaro, U. (eds.) CPM 2015. LNCS, vol. 9133, pp. 65–76. Springer, Heidelberg (2015)
4. Bille, P., Gørtz, I.L., Sach, B., Vildhøj, H.W.: Time-space trade-offs for longest common extensions. In: Kärkkäinen, J., Stoye, J. (eds.) CPM 2012. LNCS, vol. 7354, pp. 293–305. Springer, Heidelberg (2012)
5. Cormode, G., Muthukrishnan, S.: The string edit distance matching problem with moves. ACM Trans. Algorithms **3**(1), 2 (2007)
6. Fischer, J., I, T., Köppl, D.: Deterministic sparse suffix sorting on rewritable texts. arXiv:1509.07417 (2015)
7. Franceschini, G., Grossi, R.: No sorting? better searching! In: Foundations of Computer Science, pp. 491–498, October 2004
8. I, T., Kärkkäinen, J., Kempa, D.: Faster sparse suffix sorting. In: STACS, pp. 386–396 (2014)
9. Irving, R.W., Love, L.: The suffix binary search tree and suffix AVL tree. J. Discrete Algorithms **1**(5–6), 387–408 (2003)
10. Kärkkäinen, J., Sanders, P., Burkhardt, S.: Linear work suffix array construction. J. ACM **53**(6), 918–936 (2006)
11. Kolpakov, R., Kucherov, G.: Finding maximal repetitions in a word in linear time. In: Foundations of Computer Science, FOCS, pp. 596–604 (1999)
12. Mehlhorn, K., Sundar, R., Uhrig, C.: Maintaining dynamic sequences under equality-tests in polylogarithmic time. In: SODA, pp. 213–222. SIAM (1994)
13. Nishimoto, T., I, T., Inenaga, S., Bannai, H., Takeda, M.: Dynamic index, LZ factorization, and LCE queries in compressed space. arXiv:1504.06954 (2015)
14. Nong, G., Zhang, S., Chan, W.H.: Two efficient algorithms for linear time suffix array construction. IEEE Trans. Comput. **60**(10), 1471–1484 (2011)
15. Puglisi, S.J., Smyth, W.F., Turpin, A.: A taxonomy of suffix array construction algorithms. ACM Comput. Surv. **39**(2), 4 (2007)

Minimizing the Number of Opinions for Fault-Tolerant Distributed Decision Using Well-Quasi Orderings

Pierre Fraigniaud[1], Sergio Rajsbaum[2]($\boxtimes$), and Corentin Travers[3]

[1] University Paris Diderot and CRNS, Paris, France
pierre.fraigniaud@liafa.univ-paris-diderot.fr
[2] Instituto de Matemáticas, UNAM, Mexico City, Mexico
rajsbaum@im.unam.mx
[3] University of Bordeaux, Talence, France
travers@labri.fr

Abstract. The notion of deciding a *distributed language* $\mathcal{L}$ is of growing interest in various distributed computing settings. Each process p_i is given an input value x_i, and the processes should collectively decide whether their set of input values $x = (x_i)_i$ is a valid state of the system w.r.t. to some specification, i.e., if $x \in \mathcal{L}$. In *non-deterministic* distributed decision each process p_i gets a local certificate c_i in addition to its input x_i. If the input $x \in \mathcal{L}$ then there exists a certificate $c = (c_i)_i$ such that the processes collectively accept x, and if $x \notin \mathcal{L}$, then for every c, the processes should collectively reject x. The collective decision is expressed by the set of *opinions* emitted by the processes.

In this paper we study non-deterministic distributed decision in systems where asynchronous processes may crash. It is known that the number of opinions needed to deterministically decide a language can grow with n, the number of processes in the system. We prove that every distributed language $\mathcal{L}$ can be non-deterministically decided using only three opinions, with certificates of size $\lceil \log \alpha(n) \rceil + 1$ bits, where α grows at least as slowly as the inverse of the Ackerman function. The result is optimal, as we show that there are distributed languages that cannot be decided using just two opinions, even with arbitrarily large certificates.

To prove our upper bound, we introduce the notion of *distributed encoding of the integers*, that provides an explicit construction of a long *bad sequence* in the *well-quasi-ordering* $(\{0,1\}^*, \leq_*)$ controlled by the successor function. Thus, we provide a new class of applications for well-quasi-orderings that lies outside logic and complexity theory. For the lower bound we use combinatorial topology techniques.

Keywords: Runtime verification · Distributed decision · Distributed verification · Well-quasi-ordering · Wait-free computing · Combinatorial topology

Supported by ECOS-CONACYT Nord grant M12A01, ANR project DISPLEXITY, INRIA project GANG and UNAM-PAPIIT grant.

© Springer-Verlag Berlin Heidelberg 2016
E. Kranakis et al. (Eds.): LATIN 2016, LNCS 9644, pp. 497–508, 2016.
DOI: 10.1007/978-3-662-49529-2_37

1   Introduction

In *distributed decision* each process has only a local perspective of the system, and collectively the processes have to decide if some predicate about the global system state is valid. Recent work in this area includes but is not limited to, deciding locally whether the nodes of a network are properly colored, checking the results obtained from the execution of a distributed program [11,14], designing time lower bounds on the hardness of distributed approximation [7], estimating the complexity of logics required for distributed run-time verification [13], and elaborating a distributed computing complexity theory [10,15].

The predicate to be decided in a distributed decision problem is specified as the set of all valid input vectors, called a *distributed language* $\mathcal{L}$. Each process p_i is given an input value x_i, and should produce an output value $o_i \in U$, where U is the set of possible *opinions*. The processes should collectively decide whether their vector of input values $x = (x_i)_i$ represents a valid state of the system w.r.t. to the specification, i.e., if $x \in \mathcal{L}$. The collective decision is expressed by the vector of opinions $o = (o_i)_i$ emitted by the processes.

In a distributed system where n processes are unable to agree on what the global system state is (e.g. due to failures, communication delays, locality, etc.), it is unavoidable that processes have different opinions about the validity of the predicate at any given moment (a consequence of consensus impossibility [9]). Often processes emit two possible opinions, $U = \{true, false\}$, and the collective opinion is interpreted as the conjunction of the emitted values. Some languages $\mathcal{L}$ may be decided by emitting only two opinions, but not all. In fact, it is known that up to n different opinions may be necessary to decide some languages [13], irrespectively of how the opinions are interpreted. For example, for the *k-set agreement* language, specifying that at most k leaders are elected, a set U of $\min\{2k, n\} + 1$ opinions is necessary and sufficient in a system where n asynchronous processes may crash [14]. A measure of the complexity of $\mathcal{L}$ is the minimum number of opinions needed to decide it.

Non-deterministic Distributed Decision. In *non-deterministic* distributed decision, each process p_i gets a local certificate c_i in addition to its input x_i. If the input vector x is in $\mathcal{L}$ then there exists a certificate $c = (c_i)_i$ such that the processes collectively accept x, and if $x \notin \mathcal{L}$, then for every c, the processes should collectively reject x (i.e., the protocol cannot be fooled by "fake" certificates on illegal instances). Notice that as for the input, the certificate is also distributed; each process only knows its local part of the certificate. As in the deterministic case, the collective decision is expressed by the opinions emitted by the processes.

This non-deterministic framework is inspired by classical complexity theory, but it has been used before also in various distributed settings, e.g. in distributed complexity [10], in silent self-stabilization [4] (as captured by the concept of proof-labeling schemes [19]), as well as failure detectors [5] where an underlying layer produces certificates giving information about process failures — the failure detector should provide certificates giving sufficient information about process failures to solve e.g. consensus, but an incorrect certificate should not lead to an invalid consensus solution.

In several of these contexts, it is natural to seek certificates that are as small as possible, perhaps for information theoretic purposes, privacy purposes, or because certificates have to be exchanged among processes [4,19]. As we shall prove in this paper, it is possible to use small certificates to enable the number of opinions to be drastically reduced. We do so in the standard framework of asynchronous crash-prone processes communicating by writing and reading shared variables[1].

Our Contribution. We show that, for every distributed language $\mathcal{L}$, it is possible to design a non-deterministic protocol using very small certificates, while using a small set U of opinions. Our solution is based on a combinatorial construction called a *distributed encoding*.

We define a distributed encoding of the (natural) integers as a collection of code-words providing every integer n with a code $w = (w_i)_{i=1,...,n}$ in Σ^n, where Σ is a (possibly infinite) alphabet, such that, for any $k \in [1, n)$, no subwords[2] $w' \in \Sigma^k$ of w is encoding k. Trivially, every integer $n \geq 1$ can be (distributedly) encoded by the word $w = (\text{bin}(n), ..., \text{bin}(n)) \in \Sigma^n$ with $\Sigma = \{0, 1\}^*$, where $\text{bin}(n)$ is the binary representation of n. Hence, to encode the first n integers, one can use words on an alphabet with n symbols, encoded on $O(\log n)$ bits.

Our first result is a constructive proof that there is a distributed encoding of the integers which encodes the first n integers using words on an alphabet with symbols of $\lceil \log \alpha(n) \rceil + 1$ bits, where α is a function growing at least as slowly as the inverse-Ackerman function. This first result is obtained by considering the *well-quasi-ordering* $(\Lambda, =)$ where $\Lambda = \{0, 1\}$ is composed of two incomparable elements 0 and 1, and by constructing long (so-called) *bad sequences* of words over $(\Lambda^*, \leq_*)$ starting from any word $a \in \Lambda^*$, and controlled by the successor function $g(x) = x + 1$. (See Sect. 2 for the formal definitions of these concepts, and for the definition of the relation $\leq_*$ over Λ^*).

Our second result is an application of distributed encoding of the integers to distributed computing. This is a novel use of well-quasi-orderings, that lies outside the traditional applications to logic and complexity theory. Specifically, we prove that any distributed language $\mathcal{L}$ can be non-deterministically decided with certificates of $\lceil \log \alpha(n) \rceil + 1$ bits, and a set U of only three opinions. Each opinion provides an estimation of the correctness of the execution from the perspective of one process. Moreover, using arguments from combinatorial topology, we show that the result is best possible. Namely, there are distributed languages for which two opinions are insufficient, even with only three processes, and regardless of the size of the certificates.

This motivates a new line of research in distributed computing, consisting in designing distributed algorithms producing *certified* outputs, i.e., outputs

---

[1] The theory of read/write wait-free computation is of considerable significance, because results in this model can be transferred to other message-passing and f-resilient models e.g. [2,17].

[2] Such a subword is of the form $w' = (w_{i_j})_{j=1,...,k}$ with $i_j < i_{j+1}$ for $j \in [1, k)$.

that can be verified afterward by another algorithm. This can be achieved in the framework of asynchronous systems with *transient* failures [4]. Our results demonstrate that, conceptually, this can also be achieved in asynchronous systems with *crash* failures, at low costs, in term of both certificate size and number of opinions.

Due to space limitations, proofs and additional material can be found in a companion technical report [12].

Related Work. The area of *decentralized runtime verification* is concerned with a set of failure-free monitors observing the behavior of system executions with respect to some correctness property, specified in some form of temporal logic. It is known, for instance, that linear temporal logic (LTL) is not sufficient to handle all system executions, some of them requiring multi-valued logics [3]. Further references to this area appear in the recent work [22], where 3-valued semantics of LTL specifications are considered.

Deterministic distributed decision in the context of asynchronous, crash-prone distributed computing was introduced in [11] with the name *checking*, where a characterization of the tasks that are AND-checkable is provided. The results where later on extended in [14] to the set agreement task and in [13] proving nearly tight bounds on the number of opinions required to check any distributed language. In [10,15] the context of *local* distributed network computing is considered. It was shown that not all network decision tasks can be solved locally by a non-deterministic algorithm. On the other hand, every languages can be locally decided non-deterministically if one allows the verifier to err with some probability.

Our construction of distributed encoding of the integers relies very much on the notion of *well-quasi-ordering* (wqo) [20]. This important tool in logic and computability has a wide variety of applications — see [21] for a survey. One important application is providing *termination arguments* in decidability results [6]. Indeed, thirteen years after publishing his undecidability result, Turing [27] proposed the now classic method of proving program termination using so-called *bad sequences*, with respect to a wqo. In this setting, the problem of *bounding* the length of bad sequences is of utmost interest as it yields upper bounds on terminating program executions. Hence, the interest in *algorithmic aspects* of wqos has grown recently [8,23], as witnessed by the amount of work collected in [24]. Our paper is tackling the study of wqos, from a *distributed algorithm* perspective. Also, lower bounds showing Ackermanian termination growth have been identified in several applications, including lossy counter machines and reset Petri nets [24,26]. For more applications and related work on wqos, including rewriting systems, tree embeddings, lossy channel systems, and graph minors, see recent work [16,24].

2   Distributed Encoding of the Integers

Given a finite or infinite alphabet Σ, a *word* of size n on Σ is an ordered sequence $w = w_1, w_2, \ldots, w_n$ of symbols $w_i \in \Sigma$. The set of all finite words over Σ

is Σ^*, and the set of all words of size n is Σ^n. A *sub-word of w* is a word $w' \in \Sigma^*$, which is sub-sequence of w, $w' = w_{i_1}, w_{i_2}, \ldots, w_{i_k}$ with $k < n$ and $1 \le i_1 < i_2 < \cdots < i_k \le n$.

Definition 1. *A distributed encoding of the positive integers is a pair (Σ, f) where Σ is a (possibly infinite) alphabet, and $f : \Sigma^* \to \{\text{true}, \text{false}\}$ satisfying that, for every integer $n \ge 1$, there exists a word $w \in \Sigma^n$ with $f(w) = \text{true}$, such that for every sub-word w' of w, $f(w') = \text{false}$. The word w is called the distributed code of n.*

A trivial distributed encoding of the integers can be obtained using the infinite alphabet $\Sigma = \{0, 1\}^*$ (each symbol is a sequence of 0's and 1's). The distributed code of n consists in repeating n times the binary encoding of n, for each positive integer n, $w = \text{bin}(n), \ldots, \text{bin}(n)$. For every integer $n \ge 1$ and every word $w \in \Sigma^n$, we set $f(w) = \text{true}$ if and only if $w_i = \text{bin}(n)$ for every $i \in \{1, \ldots, n\}$. However, this encoding is quite redundant, and consumes an alphabet of n symbols to encode the first n positive integers (i.e., $O(\log n)$ bits per symbol).

A far more compact distributed encoding of the integers can be obtained, using a variant of the Ackermann function. Given a function $f : \mathbb{N} \to \mathbb{N}$, we denote by $f^{(n)}$ the nth iterate of f, with $f^{(0)}$ the identity function. Let $A_k : \mathbb{N} \to \mathbb{N}, k \ge 1$ be the family of functions defined recursively as follows:

$$A_k(n) = \begin{cases} 2n + 2 & \text{if } k = 1 \\ A_{k-1}(\ldots A_{k-1}(0)) = A_{k-1}^{(n+1)}(0) & \text{otherwise.} \end{cases} \tag{1}$$

Hence $A_k(0) = 2$ for every $k \ge 1$, and, for $n \ge 0$, $A_2(n) = 2^{n+2} - 2$, and $A_3(n) = 2^{2^{\cdot^{\cdot^{\cdot^2}}}} - 2$, where the tower is of height $n + 2$. (Many versions of the Ackerman function exist, and a possible definition [25] is $Ack(n) = A_n(1)$). Let $F : \mathbb{N} \to \mathbb{N}$ be the function: $F(k) = A_1(A_2(\ldots(A_{k-1}(A_k(0))))) + 1$. Finally, let $\alpha : \mathbb{N} \to \mathbb{N}$ be the function:

$$\alpha(k) = \min\{i \ge 1 : F^{(i)}(1) > k\}. \tag{2}$$

Hence, α grows extremely slowly. In addition, note that $F^{(n)}(1) > n$ for every $n \ge 1$. Hence, a crude lower bound of $F^{(n)}(1)$ is $F^{(n)}(1) \ge Ack(n-1)$. Therefore the function α grows at least as slowly as the inverse-Ackermann function.

Theorem 1. *There is a distributed encoding (Σ, f) of the positive integers which encodes the first n integers using words on an alphabet with symbols on $\lceil \log \alpha(n) \rceil + 1$ bits, where α is defined in Eq. (2).*

The proof of Theorem 1 heavily relies on the notion of *well-quasi-ordering*. Recall that a *well-quasi-ordering* (wqo) is a quasi-ordering that is well-founded and has finite antichains. That is, a wqo is a pair $(A, \le)$, where $\le$ is a reflexive and transitive relation over a set A, such that every infinite sequence of elements $a^{(0)}, a^{(1)}, a^{(2)}, \cdots$ from A contains an *increasing pair*, i.e., a pair $(a^{(i)}, a^{(j)})$ with

$i < j$ and $a^{(i)} \leq a^{(j)}$. Sequences (finite or infinite) with an increasing pair of elements are called *good* sequences. Instead, sequences where no such increasing pair can be found are called *bad*. Therefore, every infinite sequence over a wqo A is good, and, as a consequence, bad sequences over a wqo A are finite. Often, $a \in A$ is a finite word over some domain Λ, i.e., $a \in \Lambda^*$. Assuming $(\Lambda, \leq)$ itself is a wqo, then Higman's Lemma says that $(\Lambda^*, \leq_*)$ is a wqo, where $\leq_*$ is the *subword* ordering defined as follows. For any $a = a_1, a_2, \ldots, a_n \in \Lambda^*$, and any $b = b_1, b_2, \ldots, b_m \in \Lambda^*$,

$$a \leq_* b \iff \exists 1 \leq i_1 < i_2 < \cdots < i_n \leq m : (a_1 \leq b_{i_1}) \wedge \cdots \wedge (a_n \leq b_{i_n}).$$

As said before, the longest bad sequence starting on any $a \in \Lambda^*$ is of interest for practical applications (e.g., to obtain upper bounds on the termination time of a program). This length is strongly related to the *growth* of the words' length in Λ^*. More generally, let $|\cdot|$ be a norm on a wqo A that defines the *size* $|a|$ of each $a \in A$. For any $a \in A$, there is a longest bad sequence $a^{(0)}, a^{(1)}, a^{(2)}, \ldots, a^{(k)}$ starting on $a^{(0)} = a$, provided that, for every $i \geq 0$, the size of $a^{(i+1)}$ does not grow unboundedly with respect to the size of the previous element $a^{(i)}$. Given an increasing function g, the *length function* $L_g(n)$ is defined as the length of the longest sequence over all sequences *controlled* by g, starting in an element a with $|a| \leq n$. The function g controls the sequence in the sense that it bounds the growth of elements as we iterate through the sequence. That is, $L_g(n)$ is the length of the longest sequence $a^{(0)}, a^{(1)}, \ldots$ such that $|a^{(0)}| \leq n$, and, for any $i \geq 0$, $|a^{(i+1)}| \leq g(|a^{(i)}|)$. The Length Function Theorem of [23] provides an upper bound on bad sequences parametrized by a control function g and by the size $p = |\Lambda|$ of the alphabet.

Proof (Theorem 1). Consider the *well-quasi-ordering* $(\Lambda, =)$ where $\Lambda = \{0, 1\}$ is composed of two incomparable elements 0 and 1. We construct a bad sequence $B(a)$ of words over $(\Lambda^*, \leq_*)$ starting from any words $a \in \Lambda^*$, and controlled by the successor function $g(x) = x + 1$. That is, the difference between the length of two consecutive words in the bad sequence $B(a)$ must be at most 1. We obtain an infinite sequence $\mathcal{S} = \mathcal{S}^{(1)}, \mathcal{S}^{(2)}, \ldots$ of words over Λ^* by concatenating bad sequences. See Fig. 1. More specifically, $\mathcal{S} = B(\mathcal{S}^{(0)})|B(\mathcal{S}^{(t_1)})|B(\mathcal{S}^{(t_2)})|\ldots$ where "$|$" denotes the concatenation of sequences, $\mathcal{S}^{(0)} = 0$, and, for $k \geq 1$, $\mathcal{S}^{(t_k)} = (0, \ldots, 0)$, where the number of 0s is equal to the length of the last word of the bad sequence $B(\mathcal{S}^{(t_{k-1})})$, plus 1. For further references, we call these long bad *multi-diagonal* sequences. An example is in Fig. 2.

Given the infinite sequence $\mathcal{S}$, we construct our distributed encoding (Σ, f) of the integers as follows. We set $\Sigma = \{0, 1\}^* \times \Lambda$, and the distributed code of $n \geq 1$ is $w = w_1 w_2 \ldots w_n \in \Sigma^n$ with $w_i = (\text{bin}(k), \mathcal{S}_i^{(n)})$ where $k \geq 1$ is such that the nth word $\mathcal{S}^{(n)}$ in the sequence $\mathcal{S}$ belongs to the kth multi-diagonal sequence $B(\mathcal{S}^{(t_k)})$, and $\mathcal{S}_i^{(n)} \in \Lambda$ is the ith bit of $\mathcal{S}^{(n)}$, $i = 1, \ldots, n$. For each integer $n \geq 1$ and every word $w \in \Sigma^n$, we set:

$$f(w) = \text{true} \iff \forall i \in \{1, \ldots, n\}, w_i = (\text{bin}(k), \mathcal{S}_i^{(n)}) \text{ with } \mathcal{S}^{(n)} \in B(\mathcal{S}^{(t_k)}).$$

$\mathcal{S}^{(1)} = 0$ (1st bad sequence starts)

$\mathcal{S}^{(2)} = 11$ (1st bad sequence ends)

$\mathcal{S}^{(3)} = 000$ (2nd bad sequence starts)

$\mathcal{S}^{(4)} = 0110$

$\mathcal{S}^{(5)} = 11010$

$\mathcal{S}^{(6)} = 101011$

$\mathcal{S}^{(7)} = 0101111$

$\mathcal{S}^{(8)} = 11111100$

$\mathcal{S}^{(9)} = 111110011$

$\mathcal{S}^{(10)} = 1111001111$

$\mathcal{S}^{(11)} = 11100111111$

$\mathcal{S}^{(12)} = 110011111111$

$\mathcal{S}^{(13)} = 1001111111111$

$\mathcal{S}^{(14)} = 00111111111111$

$\mathcal{S}^{(15)} = 111111111111110$

$\mathcal{S}^{(16)} = 1111111111111011$

$\mathcal{S}^{(17)} = 11111111111101111$

$\mathcal{S}^{(18)} = 111111111110111111$

$\vdots \quad \vdots \qquad \vdots$

$\mathcal{S}^{(29)} = 011111111111111111111111111111$

$\mathcal{S}^{(30)} = 111111111111111111111111111111$ (2nd bad sequence ends)

$\mathcal{S}^{(31)} = 000000000000000000000000000000$ (3rd bad sequence starts)

$\mathcal{S}^{(32)} = 000000000000000000000000000110$

$\mathcal{S}^{(33)} = 000000000000000000000000011010$

$\mathcal{S}^{(34)} = 000000000000000000000001101010$

$\vdots \quad \vdots \qquad \vdots$

Fig. 1. The beginning of the infinite sequence $\mathcal{S}$.

This is a correct distributed encoding since, for every integer $n \geq 1$, there exists a word $w \in \Sigma^n$ such that $f(w) = true$, and, for every subword w' of w, $f(w') = false$. The latter holds because every subword w' must be of the form $w' = (w_{i_j})_{j=1,\dots,m}$ with $i_j < i_{j+1}$ for $j \in [1, m)$, and if the mth element $\mathcal{S}^{(m)}$ in the sequence $\mathcal{S}$ satisfies $\mathcal{S}^{(m)} \leq_* \mathcal{S}^{(n)}$, then it cannot be the case that $\mathcal{S}^{(m)} \in B(\mathcal{S}^{(t_k)})$ too. Indeed, by construction, $B(\mathcal{S}^{(t_k)})$ is a bad sequence. See [12] for a complete proof and more details on the construction of $\mathcal{S}$.

		$(x^{(i)}), \mu_i$
$M^{(1)}$	$= 0000$	$(0,0,0,0), 0$
$M^{(2)}$	$= 00110$	$(0,0,2), 0$
$M^{(3)}$	$= 011010$	$(0,2,1), 0$
$M^{(4)}$	$= 1101010$	$(2,1,1), 0$
$M^{(5)}$	$= 10101011$	$(1,1,1), 2$
$M^{(6)}$	$= 010101111$	$(0,1,1), 4$
$M^{(7)}$	$= 1111110010$	$(6,0,1), 0$
$M^{(8)}$	$= 11111001011$	$(5,0,1), 2$
$M^{(9)}$	$= 111100101111$	$(4,0,1), 4$
$M^{(10)}$	$= 1110010111111$	$(3,0,1), 6$
$M^{(11)}$	$= 11001011111111$	$(2,0,1), 8$
$M^{(12)}$	$= 100101111111111$	$(1,0,1), 10$
$M^{(13)}$	$= 0010111111111111$	$(0,0,1), 12$
$M^{(14)}$	$= 01111111111111100$	$(0,14,0), 0$
$M^{(15)}$	$= 110111111111111100$	$(2,13,0), 0$
$\vdots$	$\vdots \quad \vdots$	$\vdots$
$M^{(24)}$	$= 011111111111100111111111111$	$(0,12,0), 12$
$M^{(25)}$	$= 1111111111111101111111111100$	$(14,11,0), 0$
$\vdots$	$\vdots \quad \vdots$	$\vdots$
		$(0,0,0), A_3(2) - 2$
		$(0, A_3(2)), 0$
$\vdots$	$\vdots \quad \vdots$	$\vdots$
		$(0,0), A_2(A_3(2)) - 2$
		$(A_2(A_3(2))), 0$
$\vdots$	$\vdots \quad \vdots$	$\vdots$
$M^{(F(4)-5)}$	$= 01111111111\ldots\ldots\ldots 111111111111111111$	$(0), A_1(A_2(A_3(2))) - 2$
$M^{(F(4)-4)}$	$= 11111111111\ldots\ldots\ldots 1111111111111111111$	$(), A_1(A_2(A_3(2)))$

Fig. 2. The beginning of a long bad (multi-diagonal) sequence starting at 0000. Note that $A_4(0) = 2$, and thus $A_1(A_2(A_3(2))) = F(4) - 1$.

3   Distributed Decision

In this section, we present the application of distributed encoding of the integers to *distributed decision*. First, we describe the computational model (more details can be found in e.g. [2,17]), and then we formally define the notions of distributed languages and decision (based on the framework of [10,11,13]).

Computational Model. We consider the standard *asynchronous wait-free read/write shared memory* model. Each process runs at its own speed, that may vary along with time, and the processes may fail by *crashing* (i.e., halt and never recover). We consider the *wait-free* model [2] in which any number of processes may crash in an execution. The processes communicate through a shared memory composed of atomic registers. We associate each process p to a positive integer, its *identity* id(p), and the registers are organized as an array of single-writer/multiple-reader (SWMR) registers, one per process. A register

i supports two operations: read() that returns the value stored in the register, and can be executed by any process, and write(x) that writes the value x in the register, and can be executed only by process with ID i. For simplicity, we use a *snapshot* operation by which a process can read all registers, in such a way that a snapshot returns a copy of all the values that were simultaneously present in the shared memory at some point during the execution of the operation. We may assume snapshots are available because they can be implemented by a wait-free algorithm using only the array of SWMR registers [1].

Distributed Languages. A correctness specification that is to be monitored is stated in terms of a *distributed language*. Suppose a set of processes $\{id_1, \ldots, id_k\} \subseteq [n]$ observe the system, and get samples $\{a_1, \ldots, a_k\}$, respectively, over a domain A. A distributed language $\mathcal{L}$ specifies whether $s = \{(id_1, a_1), \ldots, (id_k, a_k)\}$ corresponds to a *legal* or an *illegal* system behavior. Such a set s consisting of pairs of processes and samples is called an *instance*, and a distributed language $\mathcal{L}$ is simply the set of all legal instances of the underlying system, over a domain A of possible samples. Given a language $\mathcal{L}$, we say that an instance s is *legal* if $s \in \mathcal{L}$ and *illegal* otherwise. Given an instance $s = \{(id_1, a_1), \ldots, (id_k, a_k)\}$ let $ID(s) = \{id_1, \ldots, id_k\}$ the set of identities in s and $val(s)$ the multiset of values in s.

Each process $i \in [n]$ has a read-only variable, $input_i$, initially equal to a symbol $\perp$ (not in A), and where the process sample a_i is deposited. We consider only the simplest scenario, where these variables change only once, from the value $\perp$, to a value in A, and this is the first thing that happens when the process starts running. The goal is for the processes to decide that, collectively, the values deposited in these variables are correct: after communicating with each other, processes output opinions. Each process i eventually deposits its opinion in its write-once variable $output_i$. Due to failures, it may be the case that only a subset of processes $P \subseteq [n]$ participate. The *instance* of such an *execution* is $s = \{(id_i, a_i) \mid id_i \in P\}$ and we consider only all executions where all processes in P run to completion (the others do not take any steps), and each one produces an opinion $u_i \in U$, where U is a set of possible opinions.

Deciding a Distributed Language. Deciding a language $\mathcal{L}$ involves two components: an *opinion-maker* M, and an *interpretation* μ. The opinion-maker is the distributed algorithm executed by the processes. Each process produces an individual *opinion* in U about the legality of the global instance. The interpretation μ specifies the way one should interpret the collection of individual opinions produced by the processes. It guarantees the minimal requirement that the opinions of the processes should be able to distinguish legal instances from illegal ones according to $\mathcal{L}$. Consider the set of all multi-sets over U, each one with at most n elements. Then $\mu = (\mathbf{Y}, \mathbf{N})$ is a partition of this set. $\mathbf{Y}$ is called the "yes" set, and $\mathbf{N}$ is called the "no" set.

For instance, when $U = \{0, 1\}$, process may produce as an opinion either 0 or 1. Together, the monitors produce a multi-set of at most n boolean values. We do not consider which process produce which opinion, but we do consider how many processes produce a given opinion. The partition produced by the

AND-operator [11] is as follows. For every multi-set of opinions S, set $S \in \mathbf{Y}$ if every opinion in S is 1, otherwise, $S \in \mathbf{N}$.

Given a language $\mathcal{L}$ over an alphabet A, a *distributed monitor for $\mathcal{L}$* is a pair (M, μ), an opinion maker M and an interpretation μ, satisfying the following, for every execution E of M starting with instance $s = \{(\mathrm{id}_i, a_i) \mid \mathrm{id}_i \in P\}$, $P \subseteq [n]$.

Definition 2. *The pair (M, μ) decides $\mathcal{L}$ with opinions U if every execution E on instance $s = \{(\mathrm{id}_i, a_i) \mid \mathrm{id}_i \in P, \ a_i \in A\}$ satisfies*

- *The input of process i is a_i, and the opinion-maker M outputs on execution E an opinion $u_i \in U$.*
- *The instance $s \in \mathcal{L}$ if and only if the processes produce a multiset of opinions $S \in \mathbf{Y}$. Given that $(\mathbf{Y}, \mathbf{N})$ is a partition of the multisets over U, $s \notin \mathcal{L}$ if and only $S \notin \mathbf{Y}$.*

Non-deterministic Distributed Decision. Similarly to the way NP extends P, we extend the notion of distributed decision to distributed *verification*. In addition to its input x_i, process id_i receives a string $c_i \in \{0, 1\}^*$. The set $c = \{(\mathrm{id}_i, c_i) \mid \mathrm{id}_i \in P\}$ is called a *distributed certificate* for processes P. The pair (M, μ) is a *distributed verifier* for $\mathcal{L}$ with opinions U if for any $s = \{(\mathrm{id}_i, a_i) \mid \mathrm{id}_i \in P, \ a_i \in A\}$, the following hold

1. For any certificate $c = \{(\mathrm{id}_i, c_i) \mid \mathrm{id}_i \in P\}$, the input of process i is the pair (a_i, c_i), and the opinion-maker M outputs on every execution E an opinion $u_i \in U$.
2. (a) If instance $s \in \mathcal{L}$ then there exists a certificate c such that in every execution the processes produce a multiset of opinions $S \in \mathbf{Y}$.
   (b) If instance $s \notin \mathcal{L}$ then for any certificate c the processes produce a multiset of opinions $S \in \mathbf{N}$.

Note that we do not enforce any constraints on the running time of the opinion maker M. Nevertheless, M must be *wait-free*, and must not be fooled by any "fake" certificate c for an instance $s \notin \mathcal{L}$.

4   Efficient Non-deterministic Decision

We show that it is possible to verify every distributed language using three opinions, with small size certificates. Then we show that with constant size certificates, almost constant size number of opinions are sufficient.

Verification with a Constant Number of Opinions. Ideally, we would like to deal with opinion-makers using very few opinions (e.g., just true or false), and with simple interpreters (e.g., the boolean AND operator). However, the following result shows that even very classical languages like consensus cannot be verified with such simple verifiers.

Theorem 2. *There are languages that cannot be verified using only two opinions, even restricted to instances of dimension at most 2 (i.e., 3 processes), and regardless of the size of the certificates.*

The proof of Theorem 2 uses arguments from combinatorial topology. Indeed, it is known (see e.g., [17]) that, roughly, a task is wait-free solvable if and only if there is a simplicial map from a subdivision of its *input complex* to its *output complex*. For instance, consensus is not *wait-free solvable* because any subdivision preserves the connectivity of the consensus input complex, while the consensus output complex is disconnected, from which it follows that a simplicial map between the two complexes cannot exist. We use a similar style argument to show that *binary* consensus among three processes is not *wait-free verifiable* with only two opinions.

On the other hand, it was proved in [18] that every distributed language can be verified using only three opinions (true, false, undetermined). However, the verifier in [18] exhibited to establish this result uses certificates of size $O(\log n)$ bits for n-dimensional instances. The following shows how to improve this bound using distributed encodings and function α (Eq. (2)).

Theorem 3. *Every distributed language can be verified using three opinions, with certificates of size $\lceil \log \alpha(n) \rceil + 1$ bits for n-process instances.*

Verification with Constant-Size Certificates. We can reduce the size of the certificates even further, at the cost of slightly increasing the number of opinions.

Theorem 4. *Every language can be verified with 1-bit certificates, using $2\,\alpha(n) + 1$ opinions for n-dimensional instances.*

Acknowledgment. The third author is thankful to Philippe Duchon and Patrick Dehornoy for fruitful discussions on wqos.

References

1. Afek, Y., Attiya, H., Dolev, D., Gafni, E., Merritt, M., Shavit, N.: Atomic snapshots of shared memory. J. ACM **40**(4), 873–890 (1993)
2. Attiya, H., Welch, J.: Distributed Computing: Fundamentals, Simulations, and Advanced Topics. Wiley, Chichester (2004)
3. Bauer, A., Leucker, M., Schallhart, C.: Comparing LTL semantics for runtime verification. J. Log. Comput. **20**(3), 651–674 (2010)
4. Blin, L., Fraigniaud, P., Patt-Shamir, B.: On proof-labeling schemes versus silent self-stabilizing algorithms. In: Felber, P., Garg, V. (eds.) SSS 2014. LNCS, vol. 8756, pp. 18–32. Springer, Heidelberg (2014)
5. Chandra, T., Toueg, S.: Unreliable failure detectors for reliable distributed systems. J. ACM **43**(2), 225–267 (1996)
6. Cook, B., Podelski, A., Rybalchenko, A.: Proving program termination. Commun. ACM **54**(5), 88–98 (2011)
7. Sarma, A., Holzer, S., Kor, L., Korman, A., Nanongkai, D., Pandurangan, G., Peleg, D., Wattenhofer, R.: Distributed verification and hardness of distributed approximation. SIAM J. Comput. **41**(5), 1235–1265 (2012)

8. Figueira, D., Figueira, S., Schmitz, S., Schnoebelen, P.: Ackermannian and primitive-recursive bounds with dickson's lemma. In: Proceedings of 26th IEEE Symposium on Logic in Computer Science (LICS), pp. 269–278 (2011)
9. Fischer, M., Lynch, N., Paterson, M.: Impossibility of distributed consensus with one faulty process. J. ACM **32**(2), 374–382 (1985)
10. Fraigniaud, P., Korman, A., Peleg, D.: Towards a complexity theory for local distributed computing. J. ACM **60**(5), 35 (2013)
11. Fraigniaud, P., Rajsbaum, S., Travers, C.: Locality and checkability in wait-free computing. Distrib. Comput. **26**(4), 223–242 (2013)
12. Fraigniaud, P., Rajsbaum, S., Travers, C.: Minimizing the Number of Opinions for Fault-Tolerant Distributed Decision Using Well-Quasi Orderings Technical report #hal-01237873 (2015). https://hal.archives-ouvertes.fr/hal-01237873v1
13. Fraigniaud, P., Rajsbaum, S., Travers, C.: On the number of opinions needed for fault-tolerant run-time monitoring in distributed systems. In: Bonakdarpour, B., Smolka, S.A. (eds.) RV 2014. LNCS, vol. 8734, pp. 92–107. Springer, Heidelberg (2014)
14. Fraigniaud, P., Rajsbaum, S., Roy, M., Travers, C.: The opinion number of set-agreement. In: Aguilera, M.K., Querzoni, L., Shapiro, M. (eds.) OPODIS 2014. LNCS, vol. 8878, pp. 155–170. Springer, Heidelberg (2014)
15. Göös, M., Suomela, J.: Locally checkable proofs. In: Proceedings of 30th ACM Symposium on Principles of Distributed Computing (PODC), pp. 159–168 (2011)
16. Haase, C., Schmitz, S., Schnoebelen, P.: The power of priority channel systems. In: D'Argenio, P.R., Melgratti, H. (eds.) CONCUR 2013 – Concurrency Theory. LNCS, vol. 8052, pp. 319–333. Springer, Heidelberg (2013)
17. Herlihy, M., Kozlov, D., Rajsbaum, S.: Distributed Computing Through Combinatorial Topology. Morgan Kaufmann (2013)
18. Jeanmougin, M.: Checkability in Asynchronous Error-Prone Distributed Computing Using Few Values. Master Thesis Report, University Paris Diderot (2013)
19. Korman, A., Kutten, S., Peleg, D.: Proof labeling schemes. Distrib. Comput. **22**(4), 215–233 (2010)
20. Kruskal, J.: The theory of well-quasi-ordering: a frequently discovered concept. J. Comb. Theor. A **13**(3), 297–305 (1972)
21. Milner, E.: Basic WQO- and BQO-theory. In: Rival, I. (ed.) The Role of Graphs in the Theory of Ordered Sets and Its Applications. NATO ASI Series, vol. 147, pp. 487–502. Springer, Netherlands (1985)
22. Mostafa, M., Bonakdarpour, B.: Decentralized runtime verification of LTL specifications in distributed systems. In: Proceedings of IEEE Parallel and Distributed Processing Symposium (IPDPS), pp. 494–503 (2015)
23. Schmitz, S., Schnoebelen, P.: Multiply-recursive upper bounds with higman's lemma. In: Aceto, L., Henzinger, M., Sgall, J. (eds.) ICALP 2011, Part II. LNCS, vol. 6756, pp. 441–452. Springer, Heidelberg (2011)
24. Schmitz, S., Schnoebelen, P.: Algorithmic Aspects of WQO Theory. Technical report Hal#cel-00727025 (2013). https://cel.archives-ouvertes.fr/cel-00727025v2
25. Schnoebelen, P.: Verifying lossy channel systems has nonprimitive recursive complexity. Inf. Process. Lett. **83**(5), 251–261 (2002)
26. Schnoebelen, P.: Revisiting ackermann-hardness for lossy counter machines and reset petri nets. In: Hliněný, P., Kučera, A. (eds.) MFCS 2010. LNCS, vol. 6281, pp. 616–628. Springer, Heidelberg (2010)
27. Turing, A.: Checking a large routine. In: Report of a Conference on High Speed Automatic Calculating Machines, pp. 67–69 (1949)

Unshuffling Permutations

Samuele Giraudo[✉] and Stéphane Vialette

Université Paris-Est, LIGM (UMR 8049), CNRS, UPEM, ESIEE Paris, ENPC,
77454 Marne-la-Vallée, France
{samuele.giraudo,vialette}@univ-mlv.fr

Abstract. A permutation is said to be a square if it can be obtained by shuffling two order-isomorphic patterns. The definition is intended to be the natural counterpart to the ordinary shuffle of words and languages. In this paper, we tackle the problem of recognizing square permutations from both the point of view of algebra and algorithms. On the one hand, we present some algebraic and combinatorial properties of the shuffle product of permutations. We follow an unusual line consisting in defining the shuffle of permutations by means of an unshuffling operator, known as a coproduct. This strategy allows to obtain easy proofs for algebraic and combinatorial properties of our shuffle product. We besides exhibit a bijection between square (213, 231)-avoiding permutations and square binary words. On the other hand, by using a pattern avoidance criterion on oriented perfect matchings, we prove that recognizing square permutations is **NP**-complete.

1 Introduction

The *shuffle product*, denoted ⧢, is a well-known operation on words first defined by Eilenberg and Mac Lane [6]. Given three words u, v_1, and v_2, u is said to be a *shuffle* of v_1 and v_2 if it can be formed by interleaving the letters from v_1 and v_2 in a way that maintains the left-to-right ordering of the letters from each word. Besides purely combinatorial questions, the shuffle product of words naturally leads to the following computational problems:

1. Given two words v_1 and v_2, compute the set $v_1 ⧢ v_2$.
2. Given three words u, v_1, and v_2, decide if u is a shuffle of v_1 and v_2.
3. Given words u, v_1, ..., v_k, decide if u is in $v_1 ⧢ \cdots ⧢ v_k$.
4. Given a word u, decide if there is a word v such that u is in $v ⧢ v$.

Even if these problems seem similar, they radically differ in terms of time complexity. Let us now review some facts about these. In what follows, n denotes the size of u and m_i denotes the size of each v_i. A solution to Problem 1 can be computed in $O\left((m_1 + m_2)\binom{m_1+m_2}{m_1}\right)$ time [14]. An improvement and a generalization of Problem 1 has been proposed in [1], where it is proved that given words v_1, ..., v_k, the iterated shuffle $v_1 ⧢ \cdots ⧢ v_k$ can be computed in $O\left(\binom{m_1+\cdots+m_k}{m_1,\ldots,m_k}\right)$ time. Problem 2 is in **P**; it is indeed a classical textbook exercise

© Springer-Verlag Berlin Heidelberg 2016
E. Kranakis et al. (Eds.): LATIN 2016, LNCS 9644, pp. 509–521, 2016.
DOI: 10.1007/978-3-662-49529-2_38

to design an efficient dynamic programming algorithm solving it. It can be tested in $O\left(n^2/\log(n)\right)$ time [15]. To the best of our knowledge, the first $O(n^2)$ time algorithm for this problem appeared in [9]. This algorithm can easily be extended to check in polynomial-time whether or not a word is in the shuffle of any fixed number of given words. Nevertheless, Problem 3 is **NP**-complete [9,17]. This remains true even if the ground alphabet has size 3 [17]. Of particular interest, it is shown in [17] that Problem 3 remains **NP**-complete even if all the words v_i, $i \in [k]$, are identical, thereby proving that, for two words u and v, it is **NP**-complete to decide whether or not u is in the iterated shuffle of v. Again, this remains true even if the ground alphabet has size 3. Let us now finally focus on Problem 4. It is shown in [3,11] that it is **NP**-complete to decide if a word u is a *square* (w.r.t. the shuffle), that is a word u with the property that there exists a word v such that u is a shuffle of v with itself. Hence, Problem 4 is **NP**-complete.

This paper is intended to study a natural generalization of ⊔⊔, denoted by •, as a shuffle of permutations. Roughly speaking, given three permutations π, σ_1, and σ_2, π is said to be a *shuffle* of σ_1 and σ_2 if π (viewed as a word) is a shuffle of two words that are order-isomorphic to σ_1 and σ_2. This shuffle product was first introduced by Vargas [16] under the name of *supershuffle*. Our intention in this paper is to study this shuffle product of permutations • both from a combinatorial and from a computational point of view by focusing on *square* permutations, that are permutations π being in the shuffle of a permutation σ with itself. Many other shuffle products on permutations appear in the literature. For instance, in [5], the authors define the *convolution product* and the *shifted shuffle product*. For this last product, π is a shuffle of σ_1 and σ_2 if π is in the shuffle, as words, of σ_1 and the word obtained by incrementing all the letters of σ_2 by the size of σ_1. It is a simple exercise to prove that, given three permutations π, σ_1, and σ_2, deciding if π is in the shifted shuffle of σ_1 and σ_2 is in **P**.

This paper is organized as follows. In Sect. 3 we provide a precise definition of •. This definition passes through the preliminary definition of an operator Δ, allowing to *unshuffle* permutations. This operator is in fact a coproduct, endowing the linear span of all permutations with a coalgebra structure (see [8] or [7] for the definition of these algebraic structures). By duality, the unshuffling operator Δ leads to the definition of our shuffle operation on permutations. This approach has many advantages. First, some combinatorial properties of • depend on properties of Δ and are more easy to prove on the coproduct side. Second, this way of doing allows to obtain a clear description of the multiplicities of the elements appearing in the shuffle of two permutations, which are worthy of interest from a combinatorial point of view. Section 4 is devoted to showing that the problems related to the shuffle of words has links with the shuffle of permutations. In particular, we show that binary words that are square are in one-to-one correspondence with square permutations avoiding some patterns (Proposition 1). Next, Sect. 5 presents some algebraic and combinatorial properties of •. We show that • is associative and commutative (Proposition 2), and that if a permutation is a square, its mirror, complement, and inverse are also squares (Proposition 3).

Finally, Sect. 6 presents the most important result of this paper: the fact that deciding if a permutation is a square is **NP**-complete (Proposition 4). This result is obtained by exhibiting a reduction from the pattern involvement problem [2] which is **NP**-complete.

2   Notations

If S is a finite set, the cardinality of S is denoted by $|S|$, and if P and Q are two disjoint sets, $P \sqcup Q$ denotes the disjoint union of P and Q. For any nonnegative integer n, $[n]$ is the set $\{1, \ldots, n\}$.

We follow the usual terminology on words [4]. Let us recall here the most important ones. Let u be a word. The length of u is denoted by $|u|$. The *empty word*, the only word of null length, is denoted by ϵ. We denote by $\tilde{u}$ the *mirror image* of u, that is the word $u_{|u|}u_{|u|-1} \ldots u_1$. If P is a subset of $[|u|]$, $u_{|P}$ is the subword of u consisting in the letters of u at the positions specified by the elements of P. If u is a word of integers and k is an integer, we denote by $u[k]$ the word obtained by incrementing by k all letters of u. The *shuffle* of two words u and v is the set recursively defined by $u \sqcup\!\sqcup \epsilon = \{u\} = \epsilon \sqcup\!\sqcup u$ and $ua \sqcup\!\sqcup vb = (u \sqcup\!\sqcup vb)a \cup (ua \sqcup\!\sqcup v)b$, were a and b are letters. A word u is a *square* if there exists a word v such that u belongs to $v \sqcup\!\sqcup v$.

We denote by S_n the set of permutations of size n and by S the set of all permutations. In this paper, permutations of a size n are specified by words of length n on the alphabet $[n]$ and without multiple occurrence of a letter, so that all above definitions about words remain valid on permutations. The only difference lies on the fact that we shall denote by $\pi(i)$ (instead of π_i) the i-th letter of any permutation π. For any nonnegative integer n, we write $\nearrow_n$ (resp. $\searrow_n$) for the permutation $12 \ldots n$ (resp. $n(n-1) \ldots 1$). If π is a permutation of S_n, we denote by $\bar{\pi}$ the *complement* of π, that is the permutation satisfying $\bar{\pi}(i) = n - \pi(i) + 1$ for all $i \in [n]$. The *inverse* of π is denoted by π^{-1}.

If u is a word of integers without multiple occurrences of a same letter, $s(u)$ is the *standardized* of u, that is the unique permutation of the same size as u such that for all $i, j \in [|u|]$, $u_i < u_j$ if and only if $s(u)(i) < s(u)(j)$. In particular, the image of the map s is the set S of all permutations. Two words u and v having the same standardized are *order-isomorphic*. If σ is a permutation, there is an *occurrence* of (the *pattern*) σ in π if there is a set P of indexes of letters of π such that σ and $\pi_{|P}$ are order-isomorphic. When π does not admit any occurrence of σ, π *avoids* σ. The set of permutations of size n avoiding σ is denoted by $S_n(\sigma)$.

Let us now provide some definitions about graphs and oriented perfect matchings that are used in the sequel. If G is an oriented graph without loops, two different edges of G are *independent* if they do not share any common vertex. We say that G is an *oriented matching* if all edges of G are pairwise independent. Moreover, G is *perfect* if any vertex of G belongs to at least one arc. For any permutation π of S_n, an *oriented perfect matching* on π is an oriented perfect matching $\mathcal{M}$ on the set of vertices $[n]$. In the sequel, we shall consider a natural notion of pattern avoidance in oriented perfect matchings on permutations.

For instance, an oriented perfect matching $\mathcal{M}$ on a permutation π *admits an occurrence* of the pattern ⁅⁀⁀⁆ if there are four positions $i < j < k < \ell$ in π such that $(\pi(k), \pi(i))$ and $(\pi(j), \pi(\ell))$ are arcs of $\mathcal{M}$. When $\mathcal{M}$ does not admit any occurrence of a pattern $\mathcal{P}$, we say that $\mathcal{M}$ *avoids* $\mathcal{P}$. The definition naturally extends to sets of patterns: $\mathcal{M}$ *avoids* $P = \{\mathcal{P}_i : 1 \le i \le k\}$ if it avoids every pattern $\mathcal{P}_i$.

3 Shuffle Product on Permutations

The purpose of this section is to define a shuffle product $\bullet$ on permutations. Recall that a first definition of this product was provided by Vargas [16]. To present an alternative definition of this product adapted to our study, we shall first define a coproduct denoted by Δ, enabling to unshuffle permutations. By duality, Δ implies the definition of $\bullet$. The reason why we need to pass by the definition of Δ to define $\bullet$ is justified by the fact that a lot of properties of $\bullet$ depend of properties of Δ, and that this strategy allows to write concise and clear proofs of them. We invite the reader unfamiliar with the concepts of coproduct and duality to consult [8] or [7].

Let us denote by $\mathbb{Q}[S]$ the linear span of all permutations. We define a linear coproduct Δ on $\mathbb{Q}[S]$ in the following way. For any permutation π, we set

$$\Delta(\pi) = \sum_{P_1 \sqcup P_2 = [|\pi|]} \mathrm{s}\left(\pi_{|P_1}\right) \otimes \mathrm{s}\left(\pi_{|P_2}\right). \tag{1}$$

We call Δ the *unshuffling coproduct of permutations*. For instance,

$$\Delta(213) = \epsilon \otimes 213 + 2 \cdot 1 \otimes 12 + 1 \otimes 21 + 2 \cdot 12 \otimes 1 + 21 \otimes 1 + 213 \otimes \epsilon, \tag{2}$$

$$\Delta(1234) = \epsilon \otimes 1234 + 4 \cdot 1 \otimes 123 + 6 \cdot 12 \otimes 12 + 4 \cdot 123 \otimes 1 + 1234 \otimes \epsilon, \tag{3}$$

$$\Delta(1432) = \epsilon \otimes 1432 + 3 \cdot \mathbf{1} \otimes \mathbf{132} + 1 \otimes 321 + 3 \cdot 12 \otimes 21 \\ + 3 \cdot 21 \otimes 12 + 3 \cdot 132 \otimes 1 + 321 \otimes 1 + 1432 \otimes \epsilon. \tag{4}$$

Observe that the coefficient of the tensor $1 \otimes 132$ is 3 in (4) because there are exactly three ways to extract from the permutation 1432 two disjoint subwords respectively order-isomorphic to the permutations 1 and 132.

As announced, let us now use Δ to define a shuffle product on permutations. As any coproduct, Δ leads to the definition of a product obtained by duality in the following way. From (1), for any permutation π, we have

$$\Delta(\pi) = \sum_{\sigma, \nu \in S} \lambda_{\sigma, \nu}^{\pi} \, \sigma \otimes \nu, \tag{5}$$

where the $\lambda_{\sigma, \nu}^{\pi}$ are nonnegative integers. Now, by definition of duality, the dual product of Δ, denoted by $\bullet$, is a linear binary product on $\mathbb{Q}[S]$. It satisfies, for any permutations σ and ν,

$$\sigma \bullet \nu = \sum_{\pi \in S} \lambda_{\sigma, \nu}^{\pi} \, \pi, \tag{6}$$

where the coefficients $\lambda^\pi_{\sigma,\nu}$ are the ones of (5). We call $\bullet$ the *shuffle product of permutations*. For instance,

$$
\begin{aligned}
12 \bullet 21 = {} & 1243 + 1324 + 2 \cdot 1342 + 2 \cdot 1423 + 3 \cdot \mathbf{1432} + 2134 + 2 \cdot 2314 \\
& + 3 \cdot 2341 + 2413 + 2 \cdot 2431 + 2 \cdot 3124 + 3142 + 3 \cdot 3214 + 2 \cdot 3241 \quad (7) \\
& + 3421 + 3 \cdot 4123 + 2 \cdot 4132 + 2 \cdot 4213 + 4231 + 4312.
\end{aligned}
$$

Observe that the coefficient 3 of the permutation 1432 in (7) comes from the fact that the coefficient of the tensor $12 \otimes 21$ is 3 in (4).

Intuitively, this product shuffles the values and the positions of the letters of the permutations. One can observe that the empty permutation ϵ is a unit for $\bullet$ and that this product is graded by the sizes of the permutations (*i.e.*, the product of a permutation of size n with a permutation of size m produces a sum of permutations of size $n + m$).

We say that a permutation π *appears* in the shuffle $\sigma \bullet \nu$ of two permutations σ and ν if the coefficient $\lambda^\pi_{\sigma,\nu}$ defined above is different from zero. In a more combinatorial way, this is equivalent to say that there are two sets P_1 and P_2 of disjoints indexes of letters of π satisfying $P_1 \sqcup P_2 = [|\pi|]$ such that the subword $\pi_{|P_1}$ is order-isomorphic to σ and the subword $\pi_{|P_2}$ is order-isomorphic to ν.

A permutation π is a *square* if there is a permutation σ such that π appears in $\sigma \bullet \sigma$. In this case, we say that σ is a *square root* of π. Equivalently, π is a square with σ as square root if and only if in the expansion of $\Delta(\pi)$, there is a tensor $\sigma \otimes \sigma$ with a nonzero coefficient. In a more combinatorial way, this is equivalent to saying that there are two sets P_1 and P_2 of disjoints indexes of letters of π satisfying $P_1 \sqcup P_2 = [|\pi|]$ such that the subwords $\pi_{|P_1}$ and $\pi_{|P_2}$ are order-isomorphic. Computer experiments give us the first numbers of square permutations with respects to their size, which are, from size 0 to 10,

$$1, 0, 2, 0, 20, 0, 504, 0, 21032, 0, 1293418. \tag{8}$$

This sequence (and its subsequence obtained by removing the 0's) is for the time being not listed in [13]. The square permutations of sizes 0 to 4 are

Size 0	Size 2	Size 4
ϵ	12, 21	1234, 1243, 1423, 1324, 1342, 4132, 3124, 3142, 3412, 4312,
		2134, 2143, 2413, 4213, 2314, 2431, 4231, 3241, 3421, 4321

4   Binary Square Words and Permutations

In this section, we shall show that the square binary words are in one-to-one correspondence with square permutations avoiding some patterns. This property establishes a link between the shuffle of binary words and our shuffle of permutations and allows to obtain a new description of square binary words.

Let u be a binary word of length n with k occurrences of 0. We denote by btp (Binary word To Permutation) the map sending any such word u to the permutation obtained by replacing from left to right each occurrence of 0 in u by $1, 2, \ldots, k$, and from right to left each occurrence of 1 in u by $k+1, k+2, \ldots, n$. For instance,

$$\text{btp}(100101101000) = \mathbf{C12B3A948567}, \tag{9}$$

where A, B, and C respectively stand for 10, 11, and 12. Observe that for any nonempty permutation π in the image of btp, there is exactly one binary word u such that $\text{btp}(u0) = \text{btp}(u1) = \pi$. In support of this observation, when π has an even size, we denote by $\text{ptb}(\pi)$ (Permutation To Binary word) the word ua such that $|ua|_0$ and $|ua|_1$ are both even, where $a \in \{0, 1\}$.

Proposition 1. *For any $n \geq 0$, the map btp restricted to the set of square binary words of length $2n$ is a bijection between this last set and the set of square permutations of size $2n$ avoiding the patterns 213 and 231.*

Proof (of Proposition 1). The statement of the proposition is a consequence of the following claims implying that ptb is the inverse map of btp over the set of square binary words.

Claim 1. The image of btp is the set of all permutations avoiding 213 and 231.

Proof (of Claim 1). Let us first show that the image of btp contains only permutations avoiding 213 and 231. Let u be a binary word, $\pi = \text{btp}(u)$, and P_0 (resp. P_1) be the set of the positions of the occurrences of 0 (resp. 1) in u. By definition of btp, from left to right, the subword $v = \pi_{|P_0}$ is increasing and the subword $w = \pi_{|P_1}$ is decreasing, and all letters of w are greater than those of v. Now, assume that π admits an occurrence of 213. Then, since v is increasing and w is decreasing, there is an occurrence of 3 (resp. 13, 23) in v and a relative occurrence of 21 (resp. 2, 1). All these three cases contradict the fact that all letters of w are greater than those of v. A similar argument shows that π avoids 231 as well.

Finally, observe that any permutation π avoiding 213 and 231 necessarily starts by the smallest possible letter or the greatest possible letter. This property is then true for the suffix of π obtained by deleting its first letter, and so on for all of its suffixes. Thus, by replacing each letter a of π by 0 (resp. 1) if a has the role of a smallest (resp. greatest) letter, one obtains a binary word u such that $\text{btp}(u) = \pi$. Hence, all permutations avoiding 213 and 231 are in the image of btp. □

Claim 2. If u is a square binary word, $\text{btp}(u)$ is a square permutation.

Proof (of Claim 2). Since u is a square binary word, there is a binary word v such that $u \in v \sqcup\!\sqcup v$. Then, there are two disjoint sets P and Q of positions of letters of u such that $u_{|P} = v = u_{|Q}$. Now, by definition of btp, the words $\text{btp}(u)_{|P}$ and $\text{btp}(u)_{|Q}$ have the same standardized σ. Hence, and by definition of the shuffle product of permutations, $\text{btp}(u)$ appears in $\sigma \bullet \sigma$, showing that $\text{btp}(u)$ is a square permutation. □

Claim 3. If π is a square permutation avoiding 213 and 231, ptb(π) is a square binary word.

Proof (of Claim 3). Let π be a square permutation avoiding 213 and 231. By Claim 1, π is in the image of btp and hence, $u = \mathrm{ptb}(\pi)$ is a well-defined binary word. Since π is a square permutation, there are two disjoint sets P_1 and P_2 of indexes of letters of π such that $\pi_{|P_1}$ and $\pi_{|P_2}$ are order-isomorphic. This implies, by the definitions of btp and ptb, that $u_{|P_1} = u_{|P_2}$, showing that u is a square binary word. □

□

The number of square binary words is Sequence A191755 of [13] beginning by

$$1, 0, 2, 0, 6, 0, 22, 0, 82, 0, 320, 0, 1268, 0, 5102, 0, 020632. \tag{10}$$

According to Proposition 1, this is also the sequence enumerating square permutations avoiding 213 and 231.

5   Algebraic Issues

The aim of this section is to establish some of properties of the shuffle product of permutations •. It is worth to note that, as we will see, algebraic properties of the unshuffling coproduct Δ of permutations defined in Sect. 3 lead to combinatorial properties of •.

Proposition 2. *The shuffle product* • *of permutations is associative and commutative.*

Proof (of Proposition 2). To prove the associativity of •, it is convenient to show that its dual coproduct Δ is coassociative, that is

$$(\Delta \otimes I)\Delta = (I \otimes \Delta)\Delta, \tag{11}$$

where I denotes the identity map. This strategy relies on the fact that a product is associative if and only if its dual coproduct is coassociative. For any permutation π, we have

$$(\Delta \otimes I)\Delta(\pi) = (\Delta \otimes I) \sum_{P_1 \sqcup P_2 = [|\pi|]} \mathrm{s}\left(\pi_{|P_1}\right) \otimes \mathrm{s}\left(\pi_{|P_2}\right)$$

$$= \sum_{P_1 \sqcup P_2 = [|\pi|]} \Delta\left(\mathrm{s}\left(\pi_{|P_1}\right)\right) \otimes I\left(\mathrm{s}\left(\pi_{|P_2}\right)\right)$$

$$= \sum_{P_1 \sqcup P_2 = [|\pi|]} \sum_{Q_1 \sqcup Q_2 = [|P_1|]} \mathrm{s}\left(\mathrm{s}\left(\pi_{|P_1}\right)_{|Q_1}\right) \otimes \mathrm{s}\left(\mathrm{s}\left(\pi_{|P_1}\right)_{|Q_2}\right) \otimes \mathrm{s}\left(\pi_{|P_2}\right)$$

$$= \sum_{P_1 \sqcup P_2 \sqcup P_3 = [|\pi|]} \mathrm{s}\left(\pi_{|P_1}\right) \otimes \mathrm{s}\left(\pi_{|P_2}\right) \otimes \mathrm{s}\left(\pi_{|P_3}\right). \tag{12}$$

An analogous computation shows that $(I \otimes \Delta)\Delta(\pi)$ is equal to the last member of (12), whence the associativity of •.

Finally, to prove the commutativity of •, we shall show that Δ is cocommutative, that is for any permutation π, if in the expansion of $\Delta(\pi)$ there is a tensor $\sigma \otimes \nu$ with a coefficient λ, there is in the same expansion the tensor $\nu \otimes \sigma$ with the same coefficient λ. Clearly, a product is commutative if and only if its dual coproduct is cocommutative. Now, from the definition (1) of Δ, one observes that if the pair (P_1, P_2) of subsets of $[|\pi|]$ contributes to the coefficient of $\mathrm{s}\left(\pi_{|P_1}\right) \otimes \mathrm{s}\left(\pi_{|P_2}\right)$, the pair (P_2, P_1) contributes to the coefficient of $\mathrm{s}\left(\pi_{|P_2}\right) \otimes \mathrm{s}\left(\pi_{|P_1}\right)$. This shows that Δ is cocommutative and hence, that • is commutative. $\square$

Proposition 2 implies that $\mathbb{Q}[S]$ endowed with the unshuffling coproduct Δ is a coassociative cocommutative coalgebra, or in an equivalent way, that $\mathbb{Q}[S]$ endowed with the shuffle product • is an associative commutative algebra.

Lemma 1. *The three linear maps*

$$\phi_1, \phi_2, \phi_3 : \mathbb{Q}[S] \to \mathbb{Q}[S] \tag{13}$$

linearly sending a permutation π to, respectively, $\widetilde{\pi}$, $\bar{\pi}$, and π^{-1} are endomorphims of associative algebras.

We now use the algebraic properties of • exhibited by Lemma 1 to obtain combinatorial properties of square permutations.

Proposition 3. *Let π be a square permutation and σ be a square root of π. Then,*

(i) the permutation $\widetilde{\pi}$ is a square and $\widetilde{\sigma}$ is one of its square roots;
(ii) the permutation $\bar{\pi}$ is a square and $\bar{\sigma}$ is one of its square roots;
(iii) the permutation π^{-1} is a square and σ^{-1} is one of its square roots.

Proof (of Proposition 3). All statements *(i)*, *(ii)*, and *(iii)* are consequences of Lemma 1. Indeed, since π is a square permutation and σ is a square root of π, by definition, π appears in the product $\sigma \bullet \sigma$. Now, by Lemma 1, for any $j = 1, 2, 3$, since ϕ_j is a morphism of associative algebras from $\mathbb{Q}[S]$ to $\mathbb{Q}[S]$, ϕ_j commutes with the shuffle product of permutations •. Hence, in particular, one has

$$\phi_j(\sigma \bullet \sigma) = \phi_j(\sigma) \bullet \phi_j(\sigma). \tag{14}$$

Then, since π appears in $\sigma \bullet \sigma$, $\phi_j(\pi)$ appears in $\phi_j(\sigma \bullet \sigma)$ and appears also in $\phi_j(\sigma) \bullet \phi_j(\sigma)$. This shows that $\phi_j(\sigma)$ is a square root of $\phi_j(\pi)$ and implies *(i)*, *(ii)*, and *(iii)*. $\square$

Let us make an observation about Wilf-equivalence classes of permutations restrained on square permutations. Recall that two permutations σ and ν of the same size are *Wilf equivalent* if $\#S_n(\sigma) = \#S_n(\nu)$ for all $n \geq 0$. The well-known [12] fact that there is a single Wilf-equivalence class of permutations of size 3 together with Proposition 3 imply that 123 and 321 are in the same Wilf-equivalence class of square permutations, and that 132, 213, 231, and 312 are in the same Wilf-equivalence class of square permutations. Computer experiments show us that there are two Wilf-equivalence classes of square permutations of size 3. Indeed, the number of square permutations avoiding 123 begins by

$$1, 0, 2, 0, 12, 0, 118, 0, 1218, 0, 14272, \tag{15}$$

while the number of square permutations avoiding 132 begins by

$$1, 0, 2, 0, 11, 0, 84, 0, 743, 0, 7108. \tag{16}$$

Besides, an other consequence of Proposition 3 is that its makes sense to enumerate the sets of square permutations quotiented by the operations of mirror image, complement, and inverse. The sequence enumerating these sets begins by

$$1, 0, 1, 0, 6, 0, 81, 0, 2774, 0, 162945. \tag{17}$$

All Sequences (15), (16), and (17) (and their subsequences obtained by removing the 0s) are for the time being not listed in [13].

6   Algorithmic Issues

This section is devoted to proving hardness of recognizing square permutations. In the same way as happens with words, we shall use a linear graph framework where deciding whether a permutation is a square reduces to computing some specific matching in the associated linear graph [3, 11]. We have, however, to deal with oriented perfect matchings. The needed properties read as follows (see Fig. 1).

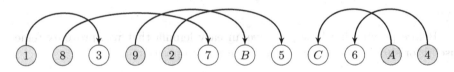

Fig. 1. An oriented perfect matching $\mathcal{M}$ on the permutation $\pi = 183927B5C6A4$ satisfying the properties $\mathbf{P_1}$ and $\mathbf{P_2}$. From $\mathcal{M}$, it follows that π is a square as it appears in the shuffle of $1892A4$ and $37B5C6$, both being order-isomorphic to 145263.

Definition 1 (Property $\mathbf{P_1}$). *Let π be a permutation. An oriented perfect matching $\mathcal{M}$ on π is said to have property $\mathbf{P_1}$ if it avoids all the six patterns*
⌢⌢ , ⌢⌢ , ⌢⌢ , ⌢⌢ , ⌢⌢ , *and* ⌢⌢ .

Definition 2 (Property $\mathbf{P_2}$). *Let π be a permutation. An oriented perfect matching $\mathcal{M}$ on π is said to have property $\mathbf{P_2}$ if, for any two distinct arcs $(\pi(a), \pi(a'))$ and $(\pi(b), \pi(b'))$ in $\mathcal{M}$, we have $\pi(a) < \pi(b)$ if and only if $\pi(a') < \pi(b')$.*

The rationale for introducing properties $\mathbf{P_1}$ and $\mathbf{P_2}$ stems from the following lemma.

Lemma 2. *Let π be a permutation. The following statements are equivalent:*

1. *The permutation π is a square.*
2. *There exists an oriented perfect matching $\mathcal{M}$ on π satisfying $\mathbf{P_1}$ and $\mathbf{P_2}$.*

Let π be a permutation. For the sake of clarity, we will say that a bunch of consecutive positions P of π is *above* (resp. *below*) above another bunch of consecutive positions P' in π if $\pi(i) > \pi(j)$ (resp. $\pi(i) < \pi(j)$) for every $i \in P$ and every $j \in P'$. For example, σ_1 is above σ_2 (in an equivalent manner, σ_2 is below σ_1) in Fig. 2(a), whereas σ_1 is below σ_2 (in an equivalent manner, σ_2 is above σ_1) in Fig. 2(b).

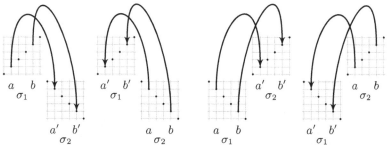

(a) An increasing pattern before and above a decreasing pattern.

(b) A decreasing pattern before and below an increasing pattern.

Fig. 2. Illustration of Lemma 3.

Before proving hardness, we give an easy lemma that will prove extremely useful for simplifying the proof of upcoming Proposition 4.

Lemma 3. *Let $\pi = \pi_1 \sigma_1 \pi_2 \sigma_2 \pi_3$ be a permutation with $|\sigma_1| \geq 2$ and $|\sigma_2| \geq 2$, and $\mathcal{M}$ be an oriented perfect matching on π satisfying $\mathbf{P_1}$ and $\mathbf{P_2}$. The following assertions hold:*

1. *If σ_1 is increasing, σ_2 is decreasing, and σ_1 is above σ_2 (see Fig. 2(a)), then there is at most one arc between σ_1 and σ_2 in $\mathcal{M}$ (this arc can be a (σ_1, σ_2)-arc or a (σ_2, σ_1)-arc).*
2. *If σ_1 is decreasing, σ_2 is increasing, and σ_1 is below σ_2 (see Fig. 2(b)), then there is at most one arc between σ_1 and σ_2 in $\mathcal{M}$ (this arc can be a (σ_1, σ_2)-arc or a (σ_2, σ_1)-arc).*

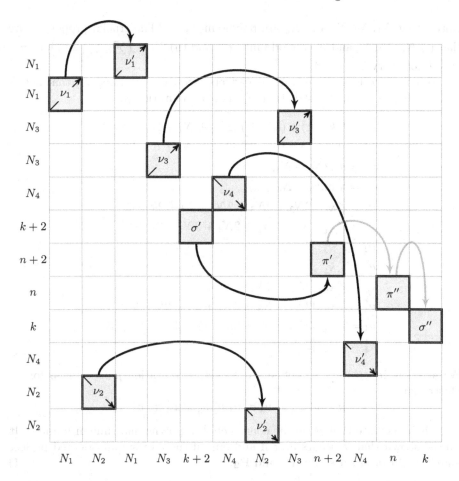

Fig. 3. Schematic representation of the permutation μ used in Proposition 4. Black arcs denote the presence of at least one arc between two bunches of positions in μ. Grey arcs denote edges that are only considered in the forward direction of the proof.

Proposition 4. *Deciding whether a permutation is a square is* **NP***-complete.*

Proof (of Proposition 4). The problem is certainly in **NP**. We propose a reduction from the pattern involvement problem which is known to be **NP**-complete [2]: Given two permutations π and σ, decide whether σ occurs in π (as an order-isomorphic pattern).

Let $\pi \in S_n$ and $\sigma \in S_k$ be two arbitrary permutations. Define

$$N_4 = 2(2n + 2k + 4) + 1 = 4n + 4k + 9$$
$$N_3 = 2(2N_4 + 2n + 2k + 4) + 1 = 20n + 20k + 45$$
$$N_2 = 2(2N_3 + 2N_4 + 2n + 2k + 4) + 1 = 100n + 100k + 225$$
$$N_1 = 2(2N_2 + 2N_3 + 2N_4 + 2n + 2k + 4) + 1 = 1000n + 1000k + 1325.$$

Notice that N_1, N_2, N_3 and N_4 are polynomial in n. The crucial properties are that (i) N_1, N_2, N_3 and N_4 are odd integers and (ii) $N_i > \left(\sum_{i<j\leq 4} 2N_j \right) + 2n + 2k + k$ for every $1 \leq i \leq 4$.

We now turn to defining various gadgets (sequences of integers) that act as building blocks in our construction of a new permutation μ:

$$\sigma' = ((k+1)\ \sigma\ (k+2))\ [2N_2 + N_4 + 2n + k + 2]$$
$$\pi' = ((n+1)\ \pi\ (n+2))\ [2N_2 + N_4 + n + k]$$
$$\sigma'' = \sigma\ [2N_2 + N_4]$$
$$\pi'' = \pi\ [2N_2 + N_4 + k]$$
$$\nu_1 = \nearrow_{N_1}\ [2N_2 + 2N_3 + 2N_4 + 2n + 2k + 4]$$
$$\nu_1' = \nearrow_{N_1}\ [N_1 + 2N_2 + 2N_3 + 2N_4 + 2n + 2k + 4]$$
$$\nu_2 = \nearrow_{N_2}\ [N_2]$$
$$\nu_2' = \searrow_{N_2}$$
$$\nu_3 = \nearrow_{N_3}\ [2N_2 + 2N_4 + 2n + 2k + 4]$$
$$\nu_3' = \nearrow_{N_3}\ [2N_2 + N_3 + 2N_4 + 2n + 2k + 4]$$
$$\nu_4 = \searrow_{N_4}\ [2N_2 + N_4 + 2n + 2k + 4]$$
$$\nu_4' = \searrow_{N_4}\ [2N_2].$$

We are now in position to define our target permutation μ (see Fig. 3 for an illustration):
$$\mu = \nu_1\ \nu_2\ \nu_1'\ \nu_3\ \sigma'\ \nu_4\ \nu_2'\ \nu_3'\ \pi'\ \nu_4'\ \pi''\ \sigma''.$$

It is immediate that μ can be constructed in polynomial-time in n and k. It can be shown that σ occurs in π if and only if there exists an oriented perfect matching $\mathcal{M}$ on μ satisfying $\mathbf{P_1}$ and $\mathbf{P_2}$. $\qquad\square$

7   Conclusion

There are a number of further directions of investigation in this general subject. They cover several areas: algorithmic, combinatorics, and algebra. Let us mention several - not necessarily new - open problems that are, in our opinion, the most interesting. How many permutations of S_{2n} are squares? How many $(213, 231)$-avoiding permutations of S_{2n} are squares? (Equivalently, by Proposition 1, how many binary strings of length $2n$ are squares; see also Problem 4 in [10])? How hard is the problem of deciding whether a $(213, 231)$-avoiding permutation is a square (Problem 4 in [10], see also [3,11])? Given two permutations π and σ, how hard is the problem of deciding whether σ is a square root of π? As for algebra, one can ask for a complete algebraic study of $\mathbb{Q}[S]$ as a graded associative algebra for the shuffle product $\bullet$. Describing a generating family for $\mathbb{Q}[S]$, defining multiplicative bases of $\mathbb{Q}[S]$, and determining whether $\mathbb{Q}[S]$ is free as an associative algebra are worthwhile questions.

References

1. Allauzen, C.: Calcul efficace du shuffle de k mots. Technical report, Institut Gaspard Monge, Université Marne-la-Vallée (2000)
2. Bose, P., Buss, J.F., Lubiw, A.: Pattern matching for permutations. Inf. Process. Lett. **65**(5), 277–283 (1998)
3. Buss, S., Soltys, M.: Unshuffling a square is NP-hard. J. Comput. Syst. Sci. **80**(4), 766–776 (2014)
4. Choffrut, C., Karhumäki, J.: Combinatorics of words. In: Rozenberg, G., Salomaa, A. (eds.) Handbook of Formal Languages. Springer, Heidelberg (1997)
5. Duchamp, G., Hivert, F., Thibon, J.-Y.: Noncommutative symmetric functions. VI. Free quasi-symmetric functions and related algebras. Int. J. Algebr. Comput. **12**(5), 671–717 (2002)
6. Eilenberg, S., Mac Lane, S.: On the groups of $H(\Pi, n)$. I. Ann. of Math. **58**(2), 58:55–58:106 (1953)
7. Grinberg, D., Reiner, V.: Hopf Algebras in Combinatorics (2014). arxiv:1409.8356
8. Joni, S.A., Rota, G.-C.: Coalgebras and bialgebras in combinatorics. Stud. Appl. Math. **61**(2), 93–139 (1979)
9. Mansfield, A.: On the computational complexity of a merge recognition problem. Discrete Appl. Math. **5**, 119–122 (1983)
10. Henshall, D., Rampersad, N., Shallit, J.: Shuffling and unshuffling (2011). http://arxiv.org/abs/1106.5767
11. Rizzi, R., Vialette, S.: On recognizing words that are squares for the shuffle product. In: Bulatov, A.A., Shur, A.M. (eds.) CSR 2013. LNCS, vol. 7913, pp. 235–245. Springer, Heidelberg (2013)
12. Simion, R., Schmidt, F.W.: Restricted permutations. Eur. J. Comb. **6**(4), 383–406 (1985)
13. Sloane, N.J.A.: The On-Line Encyclopedia of Integer Sequences. https://oeis.org/
14. Spehner, J.-C.: Le calcul rapide des melanges de deux mots. Theoret. Comput. Sci. **47**, 181–203 (1986)
15. van Leeuwen, J., Nivat, M.: Efficient recognition of rational relations. Inf. Process. Lett. **14**(1), 34–38 (1982)
16. Y. Vargas. Hopf algebra of permutation pattern functions. In: 26th International Conference on Formal Power Series and Algebraic Combinatorics, pp. 839–850 (2014)
17. Warmuth, M.K., Haussler, D.: On the complexity of iterated shuffle. J. Comput. Syst. Sci. **28**(3), 345–358 (1984)

Generating Random Spanning Trees via Fast Matrix Multiplication

Nicholas J.A. Harvey and Keyulu Xu[✉]

University of British Columbia, Vancouver, BC, Canada
nickhar@cs.ubc.ca, keyulu.x@gmail.com

Abstract. We consider the problem of sampling a uniformly random spanning tree of a graph. This is a classic algorithmic problem for which several exact and approximate algorithms are known. Random spanning trees have several connections to Laplacian matrices; this leads to algorithms based on fast matrix multiplication. The best algorithm for dense graphs can produce a uniformly random spanning tree of an n-vertex graph in time $O(n^{2.38})$. This algorithm is intricate and requires explicitly computing the LU-decomposition of the Laplacian.

We present a new algorithm that also runs in time $O(n^{2.38})$ but has several conceptual advantages. First, whereas previous algorithms need to introduce *directed* graphs, our algorithm works only with undirected graphs. Second, our algorithm uses fast matrix inversion as a black-box, thereby avoiding the intricate details of the LU-decomposition.

Keywords: Uniform spanning trees · Spectral graph theory · Fast matrix multiplication · Laplacian matrices

1 Introduction

Enumerating and sampling spanning trees of a graph is a classic problem in combinatorics dating back to Kirchhoff's celebrated matrix-tree theorem [16] from 1847. From this result, one can fairly easily derive a polynomial-time algorithm to generate a uniformly random spanning tree. Over the past few decades, researchers have developed several startling algorithms for this problem with improved running times.

The existing algorithms fall into three broad classes.

Laplacian-Based Algorithms. Properties of the graph's Laplacian matrix allow one to compute the number of spanning trees in the graph. Similarly, one can compute the probability that a given edge is in a uniformly random spanning tree. A sequence of papers [8,9,12,18] developed improved algorithms following this approach. This culminated in the algorithm of Colbourn, Myrvold and Neufeld which has running time $O(n^\omega)$, where $\omega < 2.373$ is the best-known exponent for matrix multiplication. These algorithms are most efficient on dense graphs.

© Springer-Verlag Berlin Heidelberg 2016
E. Kranakis et al. (Eds.): LATIN 2016, LNCS 9644, pp. 522–535, 2016.
DOI: 10.1007/978-3-662-49529-2_39

Random Walks. Aldous [1], Broder [3] and Wilson [21] showed that remarkably simple algorithms using random walks can be used to generate a uniformly random spanning tree. These algorithms are particularly efficient on graphs whose cover time or mean hitting time is small.

Approximate Algorithms. Recent advances in algorithmic spectral graph theory have led to nearly-linear time algorithms for approximately solving linear systems involving Laplacian matrices [17]. These methods can be used to accelerate the random walk algorithms by identifying regions of the graph where the random walk will be slow [15,20]. These algorithms are most efficient on sparse graphs.

Applications. The interest in enumerating and sampling spanning trees is not only due to its origins as a foundational problem in combinatorics. Random spanning trees have also turned out to be useful in many other contexts in combinatorics and computer science. For example, Colbourn et al. [7] showed how the coefficients of the reliability polynomial can be estimated using random spanning trees. Goyal, Rademacher and Vempala [11] have used random spanning trees to generate expander graphs. Recent breakthroughs on the traveling salesman problem [2,10] involve so-called "λ-random spanning trees", which are essentially uniformly random spanning trees in multigraphs. Other distributions on spanning trees have been used to show results in spectral graph theory [14]. More generally, random distributions on matroid bases have had interesting applications in submodular optimization [6].

1.1 Related Work

Consider the following algorithm for sampling any subgraph [18, Algorithm A]. Consider the edges in order; for each edge, decide if it is in the subgraph or not with probability conditioned on the previous decisions. It is a trivial consequence of the chain rule for conditional probabilities that this generates a random subgraph according to the desired distribution.

This algorithm can be used to generate uniformly random spanning trees if one can determine the probability of an edge being in the tree, conditioned on all previous decisions. It turns out that conditioning on an edge *not* being in the tree is the same as deleting the edge, whereas conditioning on an edge being in the tree is the same as *contracting* the edge. Thus, we may use the matrix-tree theorem to determine the sampling probability for each edge, by considering the graph with all the necessary deletions and contractions. Guenoche [12] and Kulkarni [18] discussed this method and showed that it can be implemented in time $O(n^3 m)$. A more detailed discussion of this method is given in Sect. 3.

Colbourn, Day and Nel [8] showed that the runtime of this method can be improved to $O(n^3)$. Their algorithm is recursive and applies partial Gaussian elimination. Colbourn, Myrvold and Neufeld [9] presented a different algorithm that also has runtime $O(n^3)$. Their first observation is that the desired sampling probabilities can be determined in constant time from the inverse of the (modified) Laplacian matrix (which they call the Kirchhoff matrix). Then, they observe that, after contracting an edge, the new inverse of the Laplacian matrix can be computed in $O(n^2)$ time by the Sherman-Morrison formula. Since the algorithm performs $n-1$ contractions, the total runtime is $O(n^3)$.

The best running time for dense graphs is obtained by another algorithm of Colbourn, Myrvold and Neufeld (CMN) [9]. They show that fast matrix multiplication can be used to give an algorithm with runtime $O(n^\omega)$. This algorithm abandons the Sherman-Morrison formula and instead computes the LU-decomposition of the Laplacian matrix via a "six-way divide-and-conquer algorithm". The rather intricate details of this approach are strongly reminiscent of the Bunch-Hopcroft algorithm [4] for fast matrix inversion.

1.2   Our Techniques

In this paper, we present a new algorithm for sampling a uniformly random spanning tree in $O(n^\omega)$ time. Our approach is different from, and arguably simpler than, the CMN algorithm. We recursively enumerate all edges in the graph, and lazily update the inverse of the Laplacian matrix as edges are chosen to be added to the tree or not. The updates are determined by an extension of the Sherman-Morrison formula and can be performed using fast matrix inversion as a black box. This avoids many of the intricacies of the approach based on LU-decomposition. Our idea for this approach originates from a similar algorithm for non-bipartite matching that also uses fast matrix inversion [13].

Nevertheless, there are numerous challenges that must be addressed in the present work. One challenge is that the Laplacian matrix is not invertible. Previous algorithms dealt with that by deleting the row and column associated with an arbitrary vertex and inverting the resulting matrix instead. We avoid this issue by working with the Moore-Penrose *pseudoinverse* of the Laplacian, which always exists. We must then derive a new extension of the Sherman-Morrison formula for updating the pseudoinverse. Such formulas are known, but quite complicated in general — a standard reference [5, §3.1] describes an algorithm that involves six different cases! Our formulas are much simpler.

Another challenge relates to the contraction of edges. Normally contracting an edge involves decreasing the number of vertices by one. Performing the corresponding operation to the Laplacian and its pseudoinverse is quite cumbersome. The CMN algorithm avoids this issue by working with directed graphs and sampling arboresences. In a directed graph, the analog of this contraction operation is to delete all-but-one incoming arc to a vertex; this does not affect the number of vertices. We adopt a different approach that avoids unnecessarily resorting to directed graphs. We effectively contract an edge by increasing its weight to be a large value k. In the limit $k \to \infty$, this is equivalent to contracting the edge, from the point of view of electrical networks and spanning trees.

2   Preliminaries

The graph G is assumed to be undirected, simple, connected and unweighted.

2.1   Notations

In this section, we explain the notations that we use in the algorithms and theorems.

Definition 1. *Given an unweighted graph $G = (V_G, E_G)$ with $|V_G| = n$, its Laplacian matrix $L_G = (l_{i,j})_{n \times n}$ is defined as $L_G = D - A$, where D is the degree matrix and A is the adjacency matrix, i.e.*

$$l_{i,j} = \begin{cases} deg(v_i) & (\text{if } i = j) \\ -1 & (\text{if } i \neq j \text{ and } v_i v_j \in E_G) . \\ 0 & \text{otherwise} \end{cases}$$

Given any set $E \subseteq E_G$, we may define its Laplacian L_E to be the Laplacian of the subgraph (V_G, E).

We also define the Laplacian of a graph with finite weights. Suppose that $w : E \to \mathbb{R}_{\geq 0}$ assigns weights to the edges of G. Then the weighted Laplacian is $L_w = (l_{i,j})_{n \times n}$ where

$$l_{i,j} = \begin{cases} \sum_{e \text{ incident on } i} w_e & (\text{if } i = j) \\ -w_e & (\text{if } e = \{i, j\} \in E) \\ 0 & (\text{otherwise}) \end{cases}$$

Definition 2. *Let A be a matrix. A submatrix containing rows S and columns T is denoted $A_{S,T}$. A submatrix containing all rows (resp., columns) is denoted $A_{*,T}$ (resp., $A_{S,*}$).*

Remark 1. Throughout this paper we will use the notation of Definition 2 for matrices such as L_G whose notation already involves a subscript. Mathematical correctness would suggest using the notation $(L_G)_{S,T}$ but for typographical clarity we will instead use the notation $L_{G_{S,T}}$.

Definition 3. *Let $A \in \mathbb{M}_{m \times n}$, a pseudoinverse of A is defined as $A^+ \in \mathbb{M}_{n \times m}$ satisfying all of the following criteria: $AA^+A = A$, $A^+AA^+ = A^+$, $(AA^+)^T = AA^+$, $(A^+A)^T = A^+A$.*

Definition 4. *Define $\omega \in \mathbb{R}$ as the infimum over all $c \in \mathbb{R}$ such that multiplying two $n \times n$ matrices takes $O(n^c)$ time. Matrix inverse of an $n \times n$ matrix can also be computed in $O(n^\omega)$ time.*

2.2   Facts

We will use the following basic facts. Proofs of these facts can be found in books on linear algebra and spectral graph theory.

Fact 1 (Sherman-Morrison-Woodbury formula).   *Let $M \in \mathbb{M}_{n \times n}, U \in \mathbb{M}_{n \times k}, V \in \mathbb{M}_{n \times k}$. Suppose M is non-singular. Then $M + UV^T$ is non-singular if and only if $I + V^T M^{-1} U$ is non-singular. If $M + UV^T$ is non-singular, then*

$$(M + UV^T)^{-1} = M^{-1} - M^{-1}U(I + V^T M^{-1}U)^{-1}V^T M^{-1}$$

Fact 2. *For any $L \in \mathbb{M}_{n \times n}$ with kernel* span($\mathbf{1}$), *we have $LL^+ = I - \frac{\mathbf{11}^T}{n}$. We call $I - \frac{\mathbf{11}^T}{n}$ the projection matrix P.*[1]

Fact 3 (Facts about Submatrices)

1. *For any $A, B \in \mathbb{M}_{m \times n}$ and index set S, $(A + B)_{S,S} = A_{S,S} + B_{S,S}$.*
2. *For any matrices C, D, E, F and index set S, if $C = DEF$, then $C_{S,S} = D_{S,*}EF_{*,S}$.*
3. *For any $A \in \mathbb{M}_{m \times n}, B \in \mathbb{M}_{n \times l}$ and index set S, if A or B is only non-zero in S, S, then $(AB)_{S,S} = A_{S,S} \times B_{S,S}$.*
4. *For any matrices $C = DEF$ and index set S. If $D_{*,S^c} = 0$ and $F_{S^c,*} = 0$, then $C = D_{*,S}E_{S,S}F_{S,*}$.*
5. *Suppose $D = \begin{bmatrix} M & 0 \\ 0 & 0 \end{bmatrix}$ and $E = \begin{bmatrix} A & B \\ X & Y \end{bmatrix}$ where M, A are n-by-n and $MA - I$ is non-singular. Then we have*

$$(DE - I)^{-1} = \begin{bmatrix} (MA - I)^{-1} & (MA - I)^{-1}MB \\ 0 & -I \end{bmatrix}$$

Fact 4. *Let $A, B \in \mathbb{M}_{n \times n}$ with B symmetric positive semi-definite. Suppose x is an eigenvector of AB corresponding to eigenvalue λ. Then $B^{1/2}x$ is an eigenvector of $B^{1/2}AB^{1/2}$ corresponding to eigenvalue λ.*

Fact 5. *Let G be a graph with n vertices. Let $\lambda_1 \leq \cdots \leq \lambda_n$ be the eigenvalues of L_G with the corresponding eigenvectors $v_1, \cdots, v_n$. Then L_G is symmetric positive semi-definite. $\lambda_1 = 0$ and $v_1 = \mathbf{1}$. Moreover, $\lambda_2 > 0$ if and only if G is connected, i.e. G is disconnected if and only if $\exists z$ with $z^T\mathbf{1} = 0$ and $z^TL_Gz = 0$. Everything above holds for L_G^+ as well.*

3   The Chain-Rule Algorithm

Given a simple undirected connected graph $G = (V_G, E_G)$, let $\mathcal{T}$ be the set of all spanning trees of G. We want to sample a uniformly random spanning tree $\hat{T} \subseteq E_G$ such that for any $T \in \mathcal{T}$, $\mathbb{P}(\hat{T} = T) = 1/|\mathcal{T}|$.

As described in Sect. 1.1, there is a simple algorithm for generating uniformly random spanning trees based on the chain-rule for conditional probabilities [12] [18, Algorithm A8] [19, §4.2]. The algorithm traverses the graph and samples an edge with the conditional probability of it belonging to the tree. Fact 6 below shows that this conditional probability is determined by effective resistances in the graph where edges are contracted or deleted in accordance with the algorithm's previous decisions. This algorithm is shown in Algorithm 1.

Fact 6. *Given an graph $G = (V_G, E_G)$ with Laplacian L_G, the effective resistance of an edge $e = \{u, v\} \in E_G$ is defined as*

$$R_e^{eff} = (\mathcal{X}_u - \mathcal{X}_v)^T L_G^+ (\mathcal{X}_u - \mathcal{X}_v).$$

---

[1] $P := I - \mathbf{11}^T/n$

Algorithm 1. Sampling a uniformly random spanning tree using the chain-rule.

1: **function** SAMPLESPANNINGTREE($G = (V, E)$)
2:     **for** $e = \{u, v\} \in E$ **do**
3:         $R_e^{\text{eff}} \leftarrow (\mathcal{X}_u - \mathcal{X}_v)^T L_G^+ (\mathcal{X}_u - \mathcal{X}_v)$
4:         Flip a biased coin that turns head with probability R_e^{eff}
5:         **if** head **then**
6:             Add e to the spanning tree
7:             Contract e from G and update L_G^+
8:         **else**
9:             Delete e from G and update L_G^+

---

where $\mathcal{X}_u$ is a unit vector of size $|V_G|$ with $\mathcal{X}_u(u) = 1$ and 0 otherwise. Let $\hat{T}$ be a random variable denoting a uniformly random spanning tree, i.e. $\mathbb{P}(\hat{T} = T) = 1/|\mathcal{T}|$ for any $T \in \mathcal{T}$, where $\mathcal{T}$ is the set of all spanning trees of G. Then for any $e \in E_G$, we have $\mathbb{P}(e \in \hat{T}) = R_e^{\text{eff}}$.

The algorithm involves three key properties that guarantee correctness.

- **P1:** It visits every edge of E_G exactly once.
- **P2:** It examines L_G^+ to compute the correct conditional probability of sampling an edge.
- **P3:** It updates L_G^+ to incorporate the contraction or deletion of that edge.

The naive method to update L_G^+ is to recompute it from scratch, which would require $O(n^3)$ time. There are at most n^2 edges, so overall the algorithm runs in $O(n^5)$ time.

4   A Recursive Algorithm with Lazy Updates

In this section, we present Algorithm 2, which, based on Algorithm 1, provides a faster way to update the Laplacian pseudoinverse and reduces the runtime to $O(n^\omega)$. The only difference between Algorithm 2 and Algorithm 1 is that Algorithm 2 visits the edges in a specific order to exploit lazy updates to L_G^+.

4.1   Update Formulas

In this subsection, we present our update formulas for L_G^+. We first observe that the effective resistance of any edge only depends on four entries of L_G^+. To see that, for any edge $\{u, v\}$, it follows from Fact 3.4 that

$$R_e^{\text{eff}} = (\mathcal{X}_u - \mathcal{X}_v)^T L_G^+ (\mathcal{X}_u - \mathcal{X}_v) = [1, -1] L_{G_{\{u,v\},\{u,v\}}}^+ \begin{bmatrix} 1 \\ -1 \end{bmatrix}$$

Therefore, when we are deciding whether to sample an edge, all we need to ensure is that the value of the corresponding entries in the Laplacian pseudoinverse is correct, which makes lazy updates desirable. Suppose we have made sampling decisions for some edges of a graph G but have not changed L_G^+ to reflect these decisions. Let F be the set of edges sampled and D be the set of edges discarded. We want to (partially) update L_G^+ to the Laplacian pseudoinverse of the graph obtained by contracting edges in F and deleting edges in D from G.

Because the order of updates does not matter, we make the deletion updates all together before making the contraction updates. Theorem 1 and Corollary 1 give update formulas for deletion. Lemma 1 states that these formulas are well-defined.

Lemma 1. *Let $G = (V_G, E_G)$ be a connected graph and $D \subseteq E_G$. $I - L_D L_G^+$ is non-singular iff $G \setminus D$ contains at least one spanning tree.*

Proof. $I - L_D L_G^+$ is singular iff $1 \in eig(L_D L_G^+)$ because I only has eigenvalue 1. $eig(L_D L_G^+) = eig((L_G - L_{G \setminus D})L_G^+)$. By Fact 5, 1 lies in the kernel of L_G^+. Suppose $1 \in eig(L_D L_G^+)$. Let $x \perp \mathbf{1}$ be an eigenvector of $(L_G - L_{G \setminus D})L_G^+$ corresponding to eigenvalue 1. Let $y = (L_G^+)^{1/2} x / \|(L_G^+)^{1/2} x\|$. By Fact 4, y is an eigenvector of $(L_G^+)^{1/2}(L_G - L_{G \setminus D})(L_G^+)^{1/2}$ corresponding to eigenvalue 1. We have

$$y^T (L_G^+)^{1/2}(L_G - L_{G \setminus D})(L_G^+)^{1/2} y = 1$$

Also, it is clear that

$$y^T (L_G^+)^{1/2} L_G (L_G^+)^{1/2} y = y^T L_G^+ L_G y = y^T P y = y^T (I - \mathbf{1}^T \mathbf{1}/n) y = y^T y = 1$$

It follows that $y^T (L_G^+)^{1/2} L_{G \setminus D}(L_G^+)^{1/2} y = 0$. Also, $y^T (L_G^+)^{1/2} \mathbf{1} = x^T L_G^+ \mathbf{1} = 0$. By Fact 5, $G \setminus D$ is disconnected. Hence $L_D L_G^+$ is non-singular if $G \setminus D$ contains at least one spanning tree.

Conversely, suppose $G \setminus D$ is disconnected. Then by Facts 5 and 4, there exists $y \perp \mathbf{1}$ of length 1 such that $y^T (L_G^+)^{1/2} L_{G \setminus D}(L_G^+)^{1/2} y = 0$. Also, $y^T (L_G^+)^{1/2} L_G (L_G^+)^{1/2} y = y^T y = 1$. Hence $y^T (L_G^+)^{1/2}(L_G - L_{G \setminus D})(L_G^+)^{1/2} y = 1$. It follows that $1 \in eig(L_D L_G^+)$ and $I - L_D L_G^+$ is singular.

$(L_G - L_D)^+$ is the Laplacian pseudoinverse of the graph obtained by deleting edges in D from G. The runtime of each update in Theorem 1 is $O(|V_G|^\omega)$.

Theorem 1. *Let $G = (V_G, E_G)$ be a connected graph and $D \subseteq E_G$. If $G \setminus D$ contains at least one spanning tree, then*

$$(L_G - L_D)^+ = L_G^+ - L_G^+ \left(L_D L_G^+ - I\right)^{-1} L_D L_G^+$$

Proof. By Lemma 1, $(L_D L_G^+ - I)^{-1}$ is well-defined. Since G and $G \setminus D$ are connected, by Facts 5 and 2, $(L_G - L_D)(L_G - L_D)^+ = P$. We have

$$(L_G - L_D)(L_G^+ - L_G^+(L_D L_G^+ - I)^{-1} L_D L_G^+)$$
$$= L_G L_G^+ - L_D L_G^+ - ((L_G L_G^+ - L_D L_G^+)(L_D L_G^+ - I)^{-1} L_D L_G^+)$$
$$= P - L_D L_G^+ + ((L_D L_G^+ - I + \mathbf{1} \cdot \mathbf{1}^T/n)(L_D L_G^+ - I)^{-1} L_D L_G^+)$$
$$= P - L_D L_G^+ + L_D L_G^+ + \mathbf{1} \cdot \mathbf{1}^T/n(L_D L_G^+ - I)^{-1} L_D L_G^+$$

We claim $\mathbf{1}^T(L_D L_G^+ - I)^{-1} = -\mathbf{1}^T$. To see that,

$$-\mathbf{1}^T(L_D L_G^+ - I) = \mathbf{1}^T(I - L_D L_G^+)$$
$$= \mathbf{1}^T(I - L_G L_G^+ + L_{G \setminus D} L_G^+)$$
$$= \mathbf{1}^T(\mathbf{1} \cdot \mathbf{1}^T/n + L_{G \setminus D} L_G^+)$$
$$= \mathbf{1}^T + \mathbf{1}^T(L_{G \setminus D} L_G^+) = \mathbf{1}^T$$

It follows from the claim that $1 \cdot \mathbf{1}^T/n(L_D L_G^+ - I)^{-1} L_D L_G^+ = 0$ because $\mathbf{1}^T L_D = 0$. Hence $(L_G - L_D)(L_G^+ - L_G^+(L_D L_G^+ - I)^{-1} L_D L_G^+) = P$.

The formula in Theorem 1 updates the entire L_G^+, which is unnecessary because we will not be using most entries of L_G^+ immediately. Corollary 1 gives a formula that updates a submatrix of L_G^+, using only the values of that submatrix. The updated submatrix has the same value as the submatrix of the Laplacian pseudoinverse of the graph obtained by deleting edges in D from G. The runtime of each update is improved to $O(|S|^\omega)$.

Corollary 1. *Let $G = (V_G, E_G)$ be a connected graph and $D \subseteq G$. Let $S \subseteq V_G$. Define $E[S]$ as the set of edges whose vertices are in S. Suppose $D \subseteq E[S]$ and $G \setminus D$ contains at least one spanning tree, then*

$$(L_G - L_D)_{S,S}^+ = L_{G_{S,S}}^+ - L_{G_{S,S}}^+(L_{D_{S,S}} L_{G_{S,S}}^+ - I)^{-1} L_{D_{S,S}} L_{G_{S,S}}^+.$$

Proof. L_D is only non-zero on the rows and columns indexed by S, since $D \subseteq E[S]$. Fact 3.5 implies that

$$(L_D L_G^+ - I)^{-1} = \begin{bmatrix} (L_{D_{S,S}} L_{G_{S,S}}^+ - I)^{-1} & (L_{D_{S,S}} L_{G_{S,S}}^+ - I)^{-1} L_{D_{S,S}} L_{G_{S,S^c}} \\ 0 & -I \end{bmatrix} \quad (1)$$

and in particular that

$$(L_D L_G^+ - I)_{S,S}^{-1} = (L_{D_{S,S}} L_{G_{S,S}}^+ - I)^{-1}. \quad (2)$$

Combining Theorem 1, Facts 3.1 and 3.3 gives

$$(L_G - L_D)_{S,S}^+ = L_{G_{S,S}}^+ - L_{G_{S,S}}^+(L_D L_G^+ - I)_{S,S}^{-1} L_{D_{S,S}} L_{G_{S,S}}^+.$$

The result now follows from (2).

We present similar update formulas for contraction. As mentioned in Sect. 1.2, algorithms for generating random spanning trees must contract edges but somehow avoid the cumbersome updates to the Laplacian that result from decreasing the number of vertices. Our approach is to increase the edge's weight to a large value k. By Fact 7 below, this is equivalent to contracting the edge in the limit as $k \to \infty$. One must be careful to specify formally what this means, because we have only defined the Laplacian of a weighted graph when the weights are finite. However, this does not matter. The main object of interest to us is L_G^+, and this *does* have a finite limit as $k \to \infty$.

To emphasize the graph under consideration, we use the following notation: $R_e^{\text{eff}}[H]$ denotes the effective resistance of edge e in the graph H.

Fact 7. *Let G be a weighted graph. Let e, f be distinct edges in G. Let G/e be the graph obtained by contracting edge e. Let $G + ke$ be the weighted graph obtained by increasing e's weight by k. Then*

$$R_f^{\textit{eff}}[G/e] \;=\; \lim_{k \to \infty} R_f^{\textit{eff}}[G + ke].$$

Let us make explicit the dependence on k in the graphs and matrices used by the algorithm. For any finite k, define $G(k) := G \setminus D + kF$, the graph obtained by deleting the edges D then increasing the weight of edges in F by k. For any edge $e = \{u, v\}$, we have

$$
\begin{aligned}
R_e^{\textit{eff}}[G \setminus D/F] &= \lim_{k \to \infty} R_e^{\textit{eff}}[G(k)] &&\text{(by Fact 7)} \\
&= \lim_{k \to \infty} (\mathcal{X}_u - \mathcal{X}_v)^T L_{G(k)}^+ (\mathcal{X}_u - \mathcal{X}_v) &&\text{(by Fact 6)} \\
&= (\mathcal{X}_u - \mathcal{X}_v)^T \lim_{k \to \infty} L_{G(k)}^+ (\mathcal{X}_u - \mathcal{X}_v)
\end{aligned}
$$

Thus, if the Laplacian pseudoinverse is updated to $\lim_{k \to \infty} L_{G(k)}^+$, then the algorithm will sample edges with the correct probability. The next few theorems give the update formulas. Let us first give a definition of incidence matrices.

Definition 5. *Let $G = (V_G, E_G)$ be a graph with n vertices. Given an edge $e = \{u, v\} \in E_G$, we define the incidence vector of e as $v_e = (\mathcal{X}_u - \mathcal{X}_v)$. Given a set of edges $E = \{e_1, e_2, \cdots, e_m\} \subseteq E_G$, we define the incidence matrix of E as $V_E = [v_{e_1} | v_{e_2} | \cdots | v_{e_m}]$.*

By the definition of the weighted Laplacian, $L_{G+kF} = L_G + kV_F V_F^T$. The next two lemmas state that our contraction update formulas are well-defined.

Lemma 2. *Let $G = (V_G, E_G)$ be a connected graph. Given $F \subseteq E_G$ with $|F| = r$, let V be the incidence matrix of F. $V^T L_G^+ V$ is non-singular iff F is a forest.*

Proof. Suppose F is a forest. For any $x \in \mathbb{R}^r$, $x \neq 0$, let $y = Vx$. Since F is a forest, V has full column rank. Therefore $y \neq 0$. Clearly $y^T \mathbf{1} = x^T(V^T \mathbf{1}) = 0$. By Fact 5, L_G^+ is PSD and $\ker(L_G^+) = \mathbf{1}$. Thus $y \perp \ker(L_G^+)$. We have

$$x^T V^T L_G^+ V x = y^T L_G^+ y > 0$$

Hence $V^T L_G^+ V$ is positive definite and thus non-singular. The converse is trivial.

Lemma 3. *Let G be a connected graph. Given $F \subseteq E_G$, let V be the incidence matrix of F. If F is a forest, then $I/k + V^T L_G^+ V$ is non-singular for any $k > 0$.*

Proof. By Lemma 2, $V^T L_G^+ V$ is positive definite. Since $k > 0$, I/k is also positive definite. The lemma follows from the sum of two positive definite matrices is positive definite.

Theorem 2 and Corollary 2 give contraction update formulas for a finite k. Corollary 2 improves on Theorem 2 by only updating a submatrix. The runtime of each update in Corollary 2 is $O(|S|^\omega)$.

Theorem 2. *Let $G = (V_G, E_G)$ be a connected graph. Given a forest $F \subseteq E_G$, let V be the incidence matrix of F. For any $k > 0$,*

$$(L_G + k \cdot L_F)^+ = L_G^+ - L_G^+ V(I/k + V^T L_G^+ V)^{-1} V^T L_G^+$$

Proof. Let $M_k = L_G + k \cdot L_F = L_G + k \cdot VV^T$ and $N_k = L_G^+ - L_G^+ V(I/k + V^T L_G^+ V)^{-1} V^T L_G^+$. By Lemma 3, N_k is well-defined. By Fact 5, $\ker(L_G^+) = \operatorname{span}(\mathbf{1})$. By Fact 2, $L_G L_G^+ = P = I - \mathbf{1} \cdot \mathbf{1}^T / |V_E|$. We have

$$
\begin{aligned}
M_k N_k &= (L_G + kVV^T)(L_G^+ - L_G^+ V(I/k + V^T L_G^+ V)^{-1} V^T L_G^+) \\
&= P + kVV^T L_G^+ - (L_G L_G^+ V + kVV^T L_G^+ V)(I/k + V^T L_G^+ V)^{-1} V^T L_G^+ \\
&= P + kVV^T L_G^+ - kV(I/k + V^T L_G^+ V)(I/k + V^T L_G^+ V)^{-1} V^T L_G^+ \quad (3) \\
&= P + kVV^T L_G^+ - kVV^T L_G^+ = P
\end{aligned}
$$

where (3) follows from the sum of any column of an incidence matrix is 0. Since $G + kF$ is connected, we have $M_k^+ = N_k$.

Corollary 2. *Let $G = (V_G, E_G)$ be a connected graph. Given a forest $F \subseteq E_G$, let V be the incidence matrix of F. Suppose $F \subseteq E[S]$, where $S \subseteq V_G$. Then for any $k > 0$,*

$$(L_G + k \cdot L_F)_{S,S}^+ = L_{G_{S,S}}^+ - L_{G_{S,S}}^+ V_{S,*}(I/k + V_{S,*}^T L_{G_{S,S}}^+ V_{S,*})^{-1} V_{S,*}^T L_{G_{S,S}}^+$$

Proof. V is only non-zero in rows in S. By Fact 3.4 $V_{S,*}^T L_{G_{S,S}}^+ V_{S,*} = V^T L_G^+ V$. The corollary then follows from Facts 3.1, 3.2 and 3.3.

Remark 2. Because the set of sampled edges, i.e. contracted edges F is a forest, V has at most $|S|$ columns.

The following theorem extends the result in Theorem 2 to $k = \infty$ and gives a contraction update formula that we use in Algorithm 2.

Theorem 3. *Let G be a graph with finite weights. Let $G(k) = G + kF_1$ for a forest $F_1 \subseteq E_G$. Let $F_2 \subseteq E_G$ be disjoint from F_1 such that $F_1 \cup F_2$ is a forest. Let V be the incidence matrix of F_2. For $k > 0$, define $N = \lim_{k \to \infty} L_{G(k)}^+$. Then*

$$\lim_{k \to \infty} L_{G(k)+kF_2}^+ = N - NV(V^T N V)^{-1} V^T N.$$

Furthermore $\ker(\lim_{k \to \infty} L_{G(k)+kF_2}^+) = \operatorname{span}(V_{F_1 \cup F_2} \cup \mathbf{1})$.

Proof. We first show that $\lim_{k \to \infty} L_{G+kF}^+ = L_G^+ - L_G^+ V(V^T L_G^+ V)^{-1} V^T L_G^+$, where V is the incidence matrix of F. By Lemma 2, $V^T L_G^+ V$ is invertible so the RHS of the formula above is well-defined. Let $N_k = (L_G + k \cdot L_F)^+ = L_G^+ - L_G^+ V(I/k + V^T L_G^+ V)^{-1} V^T L_G^+$ and $N = L_G^+ - L_G^+ V(V^T L_G^+ V)^{-1} V^T L_G^+$. We show as $k \to \infty$, N_k converges to N with respect to any matrix norm. Let $A = V^T L_G^+ V$. We have

$$
\begin{aligned}
\|N_k - N\| &= \|L_G^+ V((I/k + A)^{-1} - A^{-1})V^T L_G^+\| \\
&\leq \|L_G^+\|^2 \cdot \|V\| \cdot \|V^T\| \cdot \|(I/k + A)^{-1} - A^{-1}\| \quad (4)
\end{aligned}
$$

By the Sherman-Morrison-Woodbury formula (Fact 1),

$$
\begin{aligned}
\|(I/k + A)^{-1} - A^{-1}\| &= \|A^{-1} - A^{-1}(I + A^{-1}/k)^{-1}A^{-1}/k - A^{-1}\| \\
&= \|A^{-1}(I + A^{-1}/k)^{-1}A^{-1}/k\| \\
&\leq \|A^{-1}\|^2 \cdot \|(I + A^{-1}/k)^{-1}\|/k \\
&\rightarrow \|A^{-1}\|^2\|I\|/k \quad\quad\quad\quad\quad\quad (5) \\
&\rightarrow 0 \quad\quad\quad\quad\quad\quad\quad\quad\quad\quad\quad (6)
\end{aligned}
$$

where (5) follows from the fact that $I + A^{-1}/k \rightarrow I$ uniformly as $k \rightarrow \infty$, and the facts that matrix norm and matrix inverse are continuous functions for non-singular matrices. Hence, combining (4) and (6), $\|N_k - N\| \rightarrow 0$ as $k \rightarrow \infty$. The theorem then follows from the fact that the order of applying the update formulas does not matter and that applying the formula for F_1 and F_2 is the same as for $F_1 \cup F_2$.

A similar argument as Corollary 1 can show that the submatrix version of Theorem 3 holds as well. The only remaining detail is to establish that $V^T N V$ is non-singular. This follows by the same argument as Lemma 2 because the columns of V_{F_2} are not spanned by the columns of V_{F_1}, since $F_1 \cup F_2$ is a forest.

4.2   The Recursive Algorithm

We say an edge $\{u, v\}$ is in a submatrix if entries (u, v) and (v, u) are inside the submatrix. Corollarys 1 and 2 say that if we have only made sampling decisions for edges in a submatrix, then we can update the submatrix of the Laplacian pseudoinverse with a small cost, using only the values of that submatrix. Algorithm 2 samples the edges in a matrix by diving the matrix into submatrices and recursively samples the edges in each submatrix. Whenever the algorithm returns from a recursive call to a submatrix, it updates the current matrix with the formulas given by Corollary 1 and Theorem 3 to ensure that the next submatrix it enters has been updated, which is enough for the algorithm to correctly sample the edges in that submatrix. Let us formally define the way we recurse on the edges.

Definition 6. *Let* $G = (V_G, E_G)$ *be an graph and* S, R *be disjoint sets of* V_G. *We define the following subsets of edges.*

$$
\begin{aligned}
E[S] &= \{\{u, v\} \in E_G : u, v \in S\} \\
E[R, S] &= \{\{u, v\} \in E_G : u \in R, v \in S\}
\end{aligned}
$$

Remark 3. Suppose that $R = R_1 \cup R_2$ and $S = S_1 \cup S_2$. Then

$$
\begin{aligned}
E[S] &= E[S_1] \cup E[S_2] \cup E[S_1, S_2] \\
E[R, S] &= E[R_1, S_1] \cup E[R_1, S_2] \cup E[R_2, S_1] \cup E[R_2, S_2]
\end{aligned}
$$

Algorithm 2. A Recursive Algorithm

1: **function** SampleSpanningTree($G = (V_G, E_G)$)
2:     $N \leftarrow L_G^+$
3:     SampleEdgesWithin(V_G)
4:     **return** the uniform spanning tree T
5: **function** SampleEdgesWithin(S)
6:     **if** $|S| = 1$ **then return**
7:     Divide S in half: $S = S_1 \cup S_2$
8:     **for** $i \in \{1, 2\}$ **do**
9:         SampleEdgesWithin(S_i)
10:         Restore N_{S_i, S_i} to its value before entering the recursion
11:         $F \leftarrow$ the set of edges contracted in SampleEdgesWithin(S_i)
12:         $D \leftarrow$ the set of edges deleted in SampleEdgesWithin(S_i)
13:         Update(S, F, D)
14:     SampleEdgesCrossing(S_1, S_2)
15: **function** SampleEdgesCrossing(R, S)
16:     **if** $|R| = 1$ **then**
17:         Let $r \in R$ and $s \in S$, $R^{\text{eff}} \leftarrow (\mathcal{X}_r - \mathcal{X}_s)^T N (\mathcal{X}_r - \mathcal{X}_s)$
18:         Flip a biased coin that turns head with probability R^{eff}
19:         **if** head **then**
20:             Add $e_{r,s}$ to the uniform spanning tree T and the set of contracted edges
21:         **else**
22:             Add $e_{r,s}$ to the set of deleted edges
23:     **else**
24:         Divide R and S each in half: $R = R_1 \cup R_2$ and $S = S_1 \cup S_2$
25:         **for** $i \in \{1, 2\}$ and $j \in \{1, 2\}$ **do**
26:             SampleEdgesCrossing(R_i, S_j)
27:             Restore $N_{R_i \cup S_j, R_i \cup S_j}$ to its value before entering the recursion
28:             $F \leftarrow$ the set of edges contracted in SampleEdgesCrossing(R_i, S_j)
29:             $D \leftarrow$ the set of edges deleted in SampleEdgesCrossing(R_i, S_j)
30:             Update($R \cup S, F, D$)
31: **procedure** Update(S, F, D)
32:     Let V be the incidence matrix for F
33:     Let L_D be the Laplacian matrix for D
34:     $N_{S,S} \leftarrow N_{S,S} - N_{S,S} V_{S,*} (V_{S,*}^T N_{S,S} V_{S,*})^{-1} V_{S,*}^T N_{S,S}$
35:     $N_{S,S} \leftarrow N_{S,S} - N_{S,S} (L_{D_{S,S}} N_{S,S} - I)^{-1} L_{D_{S,S}} N_{S,S}$

The formulas in Remark 3 give a recursive way to traverse the graph, visiting each edge exactly once. This is the approach adopted by Algorithm 2. The algorithm samples the edges in $E[S]$ with SampleEdgesWithin(S), where we partition the current vertex set S into $S = S_1 \cup S_2$ and then recurse to visit edges in $E[S_1]$, $E[S_2]$ and $E[S_1, S_2]$, calling SampleEdgesWithin(S_1) and SampleEdgesWithin(S_2) respectively on $E[S_1]$, $E[S_2]$ and calling SampleEdgesCrossing(S_1, S_2) on $E[S_1, S_2]$. In SampleEdgesCrossing(S_1, S_2) We do a similar splitting and recursion. So, Algorithm 2 satisfies the property **P1** mentioned in Sect. 3.

Because Algorithm 2 does lazy updates, in order not to confuse with the true L_G^+, we denote the matrix that Algorithm 2 maintains by N. The way N is updated ensures that the following invariants are satisfied.

Invariant 1: SAMPLEEDGESWITHIN(S) initially has $N_{S,S} = L_{G_{S,S}}^+$. The algorithm restores this property after each recursive call to the functions SAMPLEEDGESWITHIN(S_i) and SAMPLEEDGESCROSSING(S_i, S_j).

Invariant 2: SAMPLEEDGESCROSSING(R, S) initially has $N_{R \cup S, R \cup S} = L_{G_{R \cup S, R \cup S}}^+$. The algorithm restores this property after each recursive call to the function SAMPLEEDGESCROSSING(R_i, S_j).

Since the two invariants guarantee that for any edge $\{r, s\}$, $N_{\{r,s\},\{r,s\}}$ is equal to $L_{G_{\{r,s\},\{r,s\}}}^+$ when we are deciding whether to keep the edge, the values of the effective resistances are correct for all edges. So, Algorithm 2 satisfies the properties **P2** and **P3**.

4.3   Analysis of Runtime

Let $f(n)$ and $g(n)$ respectively denote the runtime of SAMPLEEDGESWITHIN(S) and SAMPLEEDGESCROSSING(R, S), where $n = |R| = |S|$. Updating N requires $O(|S|^\omega)$ time. Therefore, we have

$$f(n) = 2f(n/2) + g(n) + O(n^\omega)$$
$$g(n) = 4g(n/2) + O(n^\omega)$$

By standard theorems on recurrence relations, the solutions of these recurrences are $g(n) = O(n^\omega)$ and $f(n) = O(n^\omega)$. Thus, the runtime of Algorithm 2 is $O(n^\omega)$.

5   Conclusions

In this paper, we have shown a new algorithm for sampling random spanning trees, which is arguably simpler and cleaner than the algorithm of Colbourn, Myrvold and Neufeld (CMN) [9]. Our algorithm uses a similar framework as the algorithm for non-bipartite matching of Harvey [13]. Some open questions are whether the same type of framework can be applied to other graph-theoretic problems, and whether it is possible to bring this line of work and the recent results on the sparse graph case of random spanning trees generation closer together.

References

1. Aldous, D.: The random walk construction of uniform spanning trees and uniform labelled trees. SIAM J. Discrete Math. **3**, 450–465 (1990)
2. Asadpour, A., Goemans, M., Madry, A., Gharan, S.O., Saberi, A.: An $O(\log n/\log \log n)$-approximation algorithm for the asymmetric traveling salesman problem. In: Proceedings of SODA (2010)

3. Broder, A.: Generating random spanning trees. In: Proceedings of FOCS, pp. 442–447 (1989)
4. Bunch, J.R., Hopcroft, J.E.: Triangular factorization and inversion by fast matrix multiplication. Math. Comput. **28**, 231–236 (1974)
5. Campbell, S.L., Meyer, C.D.: Generalized Inverses of Linear Transformations. SIAM (1973)
6. Chekuri, C., Vondrak, J., Zenklusen, R.: Dependent randomized rounding via exchange properties of combinatorial structures. In: Proceedings of FOCS (2010)
7. Colbourn, C.J., Debroni, B.M., Myrvold, W.J.: Estimating the coefficients of the reliability polynomial. Congr. Numer. **62**, 217–223 (1988)
8. Colbourn, C.J., Day, R.P.J., Nel, L.D.: Unranking and ranking spanning trees of a graph. J. Algor. **10**, 271–286 (1989)
9. Colbourn, C.J., Myrvold, W.J., Neufeld, E.: Two algorithms for unranking arborescences. J. Algor. **20**, 268–281 (1996)
10. Gharan, S.O., Saberi, A., Singh, M.: A randomized rounding approach to the traveling salesman problem. In: Proceedings of FOCS (2011)
11. Goyal, N., Rademacher, L., Vempala, S.: Expanders via random spanning trees. In: Proceedings of SODA (2009)
12. Guénoche, A.: Random spanning tree. J. Algor. **4**, 214–220 (1983)
13. Harvey, N.J.A.: Algebraic algorithms for matching and matroid problems. SIAM J. Comput. **39**, 679–702 (2009)
14. Harvey, N.J.A., Olver, N.: Pipage rounding, pessimistic estimators and matrix concentration. In: Proceedings of SODA (2014)
15. Kelner, J.A., Madry, A.: Faster generation of random spanning trees. In: Proceedings of FOCS (2009)
16. Kirchhoff, G.: Über die Auflösung der Gleichungen, auf welche man bei der Untersuchung der linearen Vertheilung galvanischer Ströme geführt wird. Ann. Phys. und Chem. **72**, 497–508 (1847)
17. Koutis, I., Miller, G.L., Peng, R.: A fast solver for a class of linear systems. Commun. ACM **55**(10), 99–107 (2012)
18. Kulkarni, V.G.: Generating random combinatorial objects. J. Algor. **11**(2), 185–207 (1990)
19. Lyons, R., Peres, Y.: Probability on Trees and Networks. Cambridge University Press (in preparation). Current version available at http://pages.iu.edu/~rdlyons/
20. Madry, A., Straszak, D., Tarnawski, J.: Fast generation of random spanning trees and the effective resistance metric. In: Proceedings of SODA (2015)
21. Wilson, D.B.: Generating random spanning trees more quickly than the cover time. In: Proceedings of STOC (1996)

Routing in Unit Disk Graphs

Haim Kaplan[1], Wolfgang Mulzer[2] , Liam Roditty[3], and Paul Seiferth[2]([✉])

[1] School of Computer Science, Tel Aviv University, Tel Aviv, Israel
haimk@post.tau.ac.il
[2] Institut Für Informatik, Freie Universität Berlin, Berlin, Germany
{mulzer,pseiferth}@inf.fu-berlin.de
[3] Department of Computer Science, Bar Ilan University, Ramat Gan, Israel
liamr@macs.biu.ac.il

Abstract. Let $S \subset \mathbb{R}^2$ be a set of n sites. The *unit disk graph* $\mathrm{UD}(S)$ on S has vertex set S and an edge between two distinct sites $s, t \in S$ if and only if s and t have Euclidean distance $|st| \leq 1$.

A routing scheme R for $\mathrm{UD}(S)$ assigns to each site $s \in S$ a *label* $\ell(s)$ and a *routing table* $\rho(s)$. For any two sites $s, t \in S$, the scheme R must be able to route a packet from s to t in the following way: given a *current site* r (initially, $r = s$), a *header* h (initially empty), and the *target label* $\ell(t)$, the scheme R may consult the current routing table $\rho(r)$ to compute a new site r' and a new header h', where r' is a neighbor of r. The packet is then routed to r', and the process is repeated until the packet reaches t. The resulting sequence of sites is called the *routing path*. The *stretch* of R is the maximum ratio of the (Euclidean) length of the routing path of R and the shortest path in $\mathrm{UD}(S)$, over all pairs of sites in S.

For any given $\varepsilon > 0$, we show how to construct a routing scheme for $\mathrm{UD}(S)$ with stretch $1 + \varepsilon$ using labels of $O(\log n)$ bits and routing tables of $O(\varepsilon^{-5} \log^2 n \log^2 D)$ bits, where D is the (Euclidean) diameter of $\mathrm{UD}(S)$. The header size is $O(\log n \log D)$ bits.

1 Introduction

Routing in graphs constitutes a fundamental problem in distributed graph algorithms [7,10]. Given a graph G, we would like to be able to route a packet from any node in G to any other node. The routing algorithm should be *local*, meaning that it uses only information stored with the packet and with the current node, and it should be *efficient*, meaning that the packet does not travel much longer than necessary. There is an obvious solution to this problem: with each node s of G, we store the shortest path tree for s. Then it is easy to route a packet along the shortest path to its destination. However, this solution is very inefficient: we need to store the complete topology of G with each node, leading to quadratic space usage. Thus, the goal of a routing scheme is to store as little information as possible with each node of the graph, while still guaranteeing a routing path that is not too far from optimal.

This work is supported by GIF project 1161 & DFG project MU/3501/1.

© Springer-Verlag Berlin Heidelberg 2016
E. Krankis et al. (Eds.): LATIN 2016, LNCS 9644, pp. 536–548, 2016.
DOI: 10.1007/978-3-662-49529-2_40

For general graphs a plethora of results is available, reflecting the work of almost three decades (see, e.g., [3,11] and the references therein). However, for general graphs, any efficient routing scheme needs to store $\Omega(n^\alpha)$ bits per node, for some $\alpha > 0$ [10]. Thus, it is natural to ask whether improved results are possible for specialized graph classes. For example, for trees it is known how to obtain a routing scheme that follows a shortest path and requires $O(\log n)$ bits of information at each node [5,12,14]. In planar graphs, for any $\varepsilon > 0$ it is possible to store a polylogarithmic number of bits at each node in order to route a packet along a path of length at most $1 + \varepsilon$ times the length of the shortest path [13].

A graph class that is of particular interest for routing problems comes from the study of mobile and wireless networks. Such networks are traditionally modeled as *unit disk graphs* [4]. The nodes are represented as points in the plane, and two nodes are connected if and only if the distance between the corresponding points is at most one. Even though unit disk graphs may be dense, they share many properties with planar graphs, in particular with respect to algorithmic problems. There exists a vast literature on routing in unit disk graphs (cf. [7]), but most known schemes cannot ensure a short routing path in the worst case. Yan, Xiang, and Dragan [15] present a scheme with provable worst case guarantees. They extend a scheme by Gupta et al. [8] for planar graphs to unit disk graphs by using a delicate planarization argument to obtain small-sized balanced separators. Even though the scheme by Yan et al. is conceptually simple, it requires a detailed analysis with an extensive case distinction.

We propose an alternative approach to routing in unit disk graphs. Our scheme is based on the well-separated pair decomposition for unit disk graphs [6]. It stores a polylogarithmic number of bits with each node of the graph, and it constructs a routing path that can be made arbitrarily close to a shortest path (see Sect. 2 for a precise statement of our results). This compares favorably with the scheme by Yan et al. [15] which achieves only a constant factor approximation. Moreover, our scheme is arguably simpler to analyze. However, unlike the algorithm by Yan et al., we require that the packet contain a modifiable *header* with a polylogarithmic number of bits. It is an interesting question whether this header can be removed.

2   The Model and Our Results

Let $S \subset \mathbb{R}^2$ be a set of n *sites* in the plane. We say that S has *density* δ if every unit disk contains at most δ points from S. The density δ of S is *bounded* if $\delta = O(1)$. The *unit disk graph* for S is the graph $UD(S)$ with vertex set S and an edge st between two distinct sites $s, t \in S$ if and only if $|st| \leq 1$, where $|\cdot|$ denotes the Euclidean distance. We define the *weight* of the edge st to be its Euclidean length and use $d(\cdot, \cdot)$ to denote the shortest path distance in $UD(S)$.

We would like to obtain a *routing scheme* for $UD(S)$ with small *stretch* and compact *routing tables*. Formally, this is defined as follows: we can preprocess $UD(S)$ to obtain for each site $s \in S$ (i) a *label* $\ell(s) \in \{0,1\}^*$, and (ii) a *routing* *table* $\rho(s) \in \{0,1\}^*$. Furthermore, we need to define a *routing function* $f : S \times \{0,1\}^* \times \{0,1\}^* \rightarrow S \times \{0,1\}^* \times \{0,1\}^*$. The function f takes as input a *current site* s, the label $\ell(t)$ of a *target site* t, and a *header* $h \in \{0,1\}^*$. The routing function may use its input and the routing table $\rho(s)$ of s to compute a new site s', a modified header h', and the label of an *intermediate target* $\ell(t')$. The new site s' may be either s or a neighbor of s in UD(S). Even though the eventual goal of the packet is the target t, we introduce the intermediate target t' into the notation, since it allows for better presentation of the routing algorithm.

The routing scheme is *correct* if the following holds: let h_0 be the empty header. For any two sites $s, t \in S$, consider the sequence of triples given by $(s_0, \ell_0, h_0) = (s, \ell(t), h_0)$ and $(s_i, \ell_i, h_i) = f(s_{i-1}, \ell_{i-1}, h_{i-1})$ for $i \geq 1$. Then there exists a $k = k(s,t) \geq 0$ such that $s_k = t$ and $s_i \neq t$ for $i < k$, i.e., the routing scheme reaches t after k steps. We call $s_0, s_1, \dots, s_k$ the *routing path* between s and t, and we define the *routing distance* $d_\rho(s,t)$ between s and t as $d_\rho(s,t) = \sum_{i=1}^{k} |s_{i-1}s_i|$. The quality of the routing scheme is measured by several parameters:

- the *label size* $L(n) = \max_{|S|=n} \max_{s \in S} |\ell(s)|$,
- the *table size* $T(n) = \max_{|S|=n} \max_{s \in S} |\rho(s)|$,
- the *header size* $H(n) = \max_{|S|=n} \max_{s \neq t \in S} \max_{i=1,\dots,k(s,t)} |h_i|$,
- and the *stretch* $\varphi(n) = \max_{|S|=n} \max_{s \neq t \in S} d_\rho(s,t)/d(s,t)$.

We show that for any $S \subset \mathbb{R}^2$, $|S| = n$, and any $\varepsilon > 0$ we can construct a routing scheme with $\varphi(n) = 1+\varepsilon$, $L(n) = O(\log n)$, $T(n) = O(\varepsilon^{-5} \log^2 n \log^2 D)$, and $H(n) = O(\log n \log D)$, where D is the weighted diameter of UD(S), i.e., the maximum length of a shortest path between two sites in UD(S).

3   The Well-Separated Pair Decomposition for UD(S)

Our routing scheme uses the well-separated pair decomposition (WSPD) for the unit disk graph metric given by Gao and Zhang [6]. WSPDs provide a compact way to efficiently encode the approximate pairwise distances in a metric space. Originally, WSPDs were introduced by Callahan and Kosaraju [2] in the context of the Euclidean metric, and they have found numerous applications since then (see, e.g., [6,9] and the references therein).

Since our routing scheme relies crucially on the specific structure of the WSPD described by Gao and Zhang, we remind the reader of the main steps of their algorithm and analysis.

First, Gao and Zhang assume that S has bounded density and that UD(S) is connected. They construct the Euclidean minimum spanning tree T for S. It is easy to see that T is a spanning tree for UD(S) with maximum degree 6. Furthermore, T can be constructed in $O(n \log n)$ time [1]. Since T has maximum degree 6, there exists an edge e in T such that $T \backslash e$ consists of two trees with at least $\lceil (n-1)/6 \rceil$ vertices each. By applying this observation recursively, we obtain a *hierarchical decomposition* H of T. The decomposition H is a binary tree. Each node v of H represents a subtree T_v of T with vertex set $S_v \subseteq S$ such that (i) the root of H corresponds to T; (ii) the leaves of H are in one-to-one correspondence with the sites in S; and (iii) let v be an inner node of H with children u and w. Then v has an *associated edge* $e_v \in T_v$ such that removing e_v from T_v yields the two subtrees T_u and T_w represented by u and w (see Fig. 1). Furthermore, we have $|S_u|, |S_w| \geq \lceil (|S_v| - 1)/6 \rceil$.

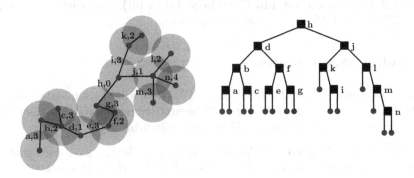

Fig. 1. An EMST of UD(S) (left) where the edges are annotated with their level in the hierarchical decomposition (right).

It follows that H has height $O(\log n)$. The *depth* $\delta(v)$ of a node $v \in H$ is defined as the number of edges on the path from v to the root of H. The *level* of the associated edge e_v of v is the depth of v in H. This uniquely defines a level for each edge in T. Now, for each node $v \in H$, the subtree T_v is a connected component in the forest that is induced in T by the edges of level at least $\delta(v)$.

After computing the hierarchical decomposition, the algorithm of Gao and Zhang essentially uses the greedy algorithm of Callahan and Kosaraju to construct a WSPD, with H in place of the quadtree (or the fair split tree). Let $c \geq 1$ be a separation parameter. The algorithm traverses H and produces a sequence $\Xi = (u_1, v_1), (u_2, v_2), \ldots, (u_m, v_m)$ of pairs of nodes of H, with the following properties:

1. The sets $S_{u_1} \times S_{v_1}, S_{u_2} \times S_{v_2}, \ldots, S_{u_m} \times S_{v_m}$ constitute a partition of $S \times S$. This means that for each ordered pair of sites $(s, t) \in S \times S$, there is exactly one pair $(u, v) \in \Xi$ with $(s, t) \in S_u \times S_v$. We say that (u, v) *represents* (s, t).
2. Each pair $(u, v) \in \Xi$ is *c-well-separated*, i.e., we have

$$(c + 2) \max\{|S_u| - 1, |S_v| - 1\} \leq |\sigma(u)\sigma(v)|, \qquad (1)$$

where $\sigma(u), \sigma(v)$ are arbitrary sites in S_u and S_v chosen by the algorithm.

Since in the unit distance graph metric the diameter diam(S_u) is at most $|S_u| - 1$ and since $|\sigma(u)\sigma(v)| \leq d(\sigma(u), \sigma(v))$, (1) implies that

$$(c + 2) \max\{\text{diam}(S_u), \text{diam}(S_v)\} \leq d(\sigma(u), \sigma(v)), \qquad (2)$$

which is the traditional well-separation condition. However, (1) is easier to check algorithmically and has additional advantages that we will exploit in our routing scheme below.

Gao and Zhang show that their algorithm produces a c-WSPD with $m = O(\delta c^2 n \log n)$ pairs, where δ is the density of S. More precisely, they prove the following lemma:

Lemma 3.1 (Lemma 4.3 and Corollary 4.6 in [6]). *For each node $u \in H$, the WSPD Ξ has $O(\delta c^2 |S_u|)$ pairs that contain u.*   □

4   Preliminary Lemmas

We begin with two technical lemmas on WSPDs that will be useful later on. The proofs can be found in the full version. The first lemma shows that the choice of the sites $\sigma(u)$ for the nodes $u \in H$ is essentially arbitrary.

Lemma 4.1. *Let Ξ be a c-WSPD for S and let s, t be two sites such that the pair $(u, v) \in \Xi$ represents (s, t). Then $c \, diam(S_u) \le c(|S_u| - 1) \le d(s, t)$.*

The next lemma is a direct consequence of Lemma 4.1 and shows that short distances are represented by singletons.

Lemma 4.2. *Let Ξ be a c-WSPD for S and let $s, t \in S$ be two sites with $d(s, t) < c$. If $(u, v) \in \Xi$ represents (s, t), then $S_u = \{s\}$ and $S_v = \{t\}$.*

5   The Routing Scheme

Let δ be the density of S. First we describe a routing scheme whose parameters depend on δ. Then we show how to remove this dependency and extend the scheme to work with arbitrary density. Our routing scheme uses the WSPD described in Sect. 3, and it is based on the following idea: let Ξ be the c-WSPD for $UD(S)$ and let T be the EMST for S used to compute it. We distribute the information about the pairs in Ξ among the sites in S (in a way described later) such that each site stores $O(\delta c^2 \log n)$ pairs in its routing table. To route from s to t, we explore T, starting from s, until we find the site r with the pair (u, v) representing (s, t). Our scheme will guarantee that s and r are sites in S_u, and therefore it suffices to walk along T_u to find r (see Fig. 2). This is called the *local routing*. With (u, v), we store in $\rho(r)$ the *middle site* m on the shortest path from r to $\sigma(v)$, i.e., the vertex "halfway" between r and $\sigma(v)$. We recursively route from r to m and when reaching m from m to t. To keep track of intermediate targets during the recursion, we store a stack in the header. This second step, the recursive routing through the middle site, we call the *global routing*. We now describe our routing scheme in detail. Let $1 + \varepsilon$, $\varepsilon > 0$, be the desired stretch factor.

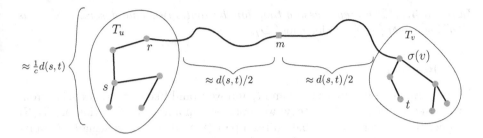

Fig. 2. To route a packet from s to t, we first walk along T_u until we find r. Then we recursively route from r to m and from m to t.

5.1   Preprocessing

The preprocessing phase works as follows. We set $c = (\alpha/\varepsilon) \log D$, where D is the Euclidean diameter of $\mathrm{UD}(S)$ and α is a sufficiently large constant we will fix later. Then we compute a c-WSPD for $\mathrm{UD}(S)$. As explained in Sect. 3, the WSPD consists of a bounded degree spanning tree T of $\mathrm{UD}(S)$, a hierarchical balanced decomposition H of T whose nodes $u \in H$ correspond to subtrees T_u of T, and a sequence $\Xi = (u_1, v_1), (u_2, v_2), \ldots, (u_m, v_m)$ of $m = O(\delta c^2 n \log n) = O(\delta \varepsilon^{-2} n \log n \log^2 D)$ well-separated pairs that represent a partition of $S \times S$.

First, we determine the *labeling* ℓ for the sites in S. For this, we perform a postorder traversal of H. Let l be a counter which is initialized to 1. Whenever we encounter a leaf of H, we set the label $\ell(s)$ of the corresponding site $s \in S$ to l, and we increment l by 1. Whenever we visit an internal node u of H for the last time, we annotate it with the interval I_u of the labels in T_u. Thus, a site $s \in S$ lies in a subtree T_u if and only if $\ell(s) \in I_u$. Each label has $O(\log n)$ bits.

Next, we describe the routing tables. Each routing table consists of two parts, the *local* routing table and the *global* routing table. The local routing table $\rho_L(s)$ of a site s stores the neighbors of s in T, in counterclockwise order, together with the levels in H of the corresponding edges (cf. Sect. 3). Since T has degree at most 6, each local routing table consists of $O(\log n)$ bits. The global routing table $\rho_G(s)$ of a site s is obtained as follows: we go through all $O(\log n)$ nodes u of H that contain s in their subtree T_u. By Lemma 3.1, Ξ contains at most $O(\delta c^2 |S_u|)$ well-separated pairs in which u represents one of the sets. We assign $O(\delta c^2) = O(\delta \varepsilon^{-2} \log^2 D)$ of these pairs to s, such that each pair is assigned to exactly one site in S_u. For each pair (u, v) assigned to s, we store the interval I_v corresponding to S_v. Furthermore, if $\sigma(v)$ is not a neighbor of s, we store the label $\ell(m)$ of the *middle site* m of a shortest path π from s to $\sigma(v)$. Here, m is a site on π that minimizes the maximum distance, $\max\{d(s, m), d(m, \sigma(v))\}$, to the endpoints of π. A site s lies in $O(\log n)$ different sets S_u, at most one for each level of H. For each such set, we store $O(\delta \varepsilon^{-2} \log^2 D)$ pairs in $\rho_G(s)$, each of which requires $O(\log n)$ bits. Thus, ρ_G has $O(\delta \varepsilon^{-2} \log^2 n \log^2 D)$ bits.

Finally, we argue that the routing scheme can be computed efficiently. See the full version for a proof.

Lemma 5.1. *The preprocessing time for the routing scheme described above is*
$O(n^2 \log n + \delta n^2 + \delta \varepsilon^{-2} n \log n \log^2 D)$.

5.2   Routing a Packet

Suppose we are given two sites s and t, and we would like to route a packet from
s to t. Recall our overall strategy: we first perform a local exploration of $UD(S)$
in order to discover a site r that stores a pair $(u, v) \in \Xi$ representing (s, t) in its
global routing table $\rho_G(r)$. To find r, we consider the subtrees of T that contain
s by increasing size, and we perform an Euler tour in each subtree until we find
r. In $\rho_G(r)$ we have stored the middle site m of a shortest path from r to $\sigma(v)$.
We put t into the header, and we recursively route the packet from r to m. Once
we reach m, we retrieve the original target t from the header and recursively
route from m to t, see Algorithm 1 for pseudo-code.

Local Routing: The Euler-Tour. We start at s, and we would like to find the
site r that stores the pair (u, v) representing (s, t). By construction, both s and r
are contained in S_u, and it suffices to perform an Euler tour on T_u to discover r.
Since we do not know u in advance, we begin with the leaf in H that contains s,
and we explore all nodes on the path to the root until we find u (see Fig. 3).

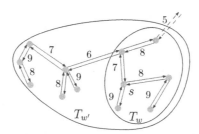

Fig. 3. To find r we do an Euler Tour on T_u, the subtree that contains s whose edges
have level at least 7. Since we do not find r, we search the next larger subtree $T_{u'}$,
where u' is the parent of u in H by decreasing the search level to 6.

We store s as the *start site* in the header h. Let $w \in H$ be the node to be
explored, and let $l = \delta(w)$ be the depth of w in H. We store l in h. Recall that T_w
is a connected component of the forest induced by all edges of level at least l. We
perform an Euler tour on T_w using the local routing tables as follows: starting
at s, we follow the first edge in $\rho_L(s)$ that has level at least l. Every time we
visit a site r, we check for all WSPD-pairs (u, v) in $\rho_G(r)$ whether $\ell(t) \in I_v$,
i.e., whether $t \in S_v$. If so, we clear the local routing information from h, and we
proceed with the global routing. If not, we scan $\rho_L(r)$ for the next edge in $\rho_L(r)$
that has level at least l, going back to the beginning of $\rho_L(r)$ if necessary, and
we follow this edge. For this, we must remember in h the edge through which
we last entered r (note that we must store only the last edge of the tour). Once

we reach s for the last time (i.e., through the last edge in $\rho_L(s)$ with level at least l), we decrease l by one and restart the process. Decreasing l corresponds to proceeding with the parent of w in H.

Global Routing: The WSPD. Suppose we are at a site s such that $\rho_G(s)$ contains the pair (u, v) with the target t being in S_v. If t is not a neighbor of s, then $\rho_G(s)$ also contains the label of a middle site m for (u, v). We push (the label of) t onto the header stack, and we use $\ell(m)$ as the new target. Then we perform a local routing, starting at s, in order to find a pair (u', v') with $m \in S_{v'}$. If t is a neighbor of s, we go directly to t. Since t may be an intermediate target, we pop the next element from the header stack and set it as the new target label. If the header stack is empty, t is our final destination.

Input: currentSite s, targetLabel $\ell(t)$, header h
Output: nextSite, nextTargetLabel, header
1   **if** $\ell(s) = \ell(t)$ **then**   /* intermediate target reached? */
2   **if** h.stack $= \emptyset$ **then**   /* final target? */
3 | **return** $(s, \perp, \perp)$
4   **else**
5 | **return** $(s, h.\text{stack.pop}(), h)$
6   **else if** $\rho(s)$ stores a WSPD-pair (u, v) with $t \in S_v$ **then** /* global routing */
7   h.startSite $\leftarrow \emptyset$
8   **if** s and t are neighbors in $\text{UD}(S)$ **then**
9 | **return** $(t, \ell(t), h)$
10   **else**
11   nextTargetLabel $\leftarrow$ label of middle site for (u, v)
12   h.stack.push$(\ell(t))$
13   **return** $(s, \text{nextTargetLabel}, h)$
14 **else**   /* local routing */
15   **if** h.startSite $= \emptyset$ **then**
16   h.startSite $\leftarrow s$
17   h.level $\leftarrow \delta(s)$
18      $r \leftarrow$ next clockwise neighbor of s with level of edge $sr \geq h$.level
19   **if** $r = \perp$ **then**   /* Euler tour is finished */
20   h.level $\leftarrow h$.level $- 1$
21   **return** $(s, \ell(t), h)$
22   **else**
23 | **return** $(r, \ell(t), h)$

Algorithm 1. The routing algorithm.

5.3   Analysis of the Routing Scheme

We now prove that the described routing scheme is correct and has low stretch, i.e., that for any two sites s and t, it produces a routing path $s = s_0, \ldots, s_k = t$ of length at most $(1 + \varepsilon)d(s, t)$.

Correctness. First, we consider only small distances and show that in this case our routing scheme produces an actual shortest path.

Lemma 5.2. *Let s, t be two sites in S with $d(s,t) < c$. Then, the routing scheme produces a routing path $s_0, s_1, \ldots, s_k$ with the following properties*

(i) $s_0 = s$ and $s_k = t$,
(ii) $d_\rho(s,t) = d(s,t)$, and
(iii) the header stack is in the same state at the beginning and at the end of the routing path.

Proof. We prove that our routing scheme has properties (i)–(iii) by induction on the rank of $d(s,t)$ in the sorted list of the pairwise distances in $\mathrm{UD}(S)$.

For the base case, consider the edges st in $\mathrm{UD}(G)$, i.e., $d(s,t) = |st| \leq 1$. By Lemma 4.2, there exists a pair (u,v) with $S_u = \{s\}$ and $S_v = \{t\}$. Thus, Algorithm 1 correctly routes to t in one step and does not manipulate the header stack. All properties are fulfilled.

Now, consider an arbitrary pair s, t with $1 < d(s,t) < c$. By Lemma 4.2, there is a pair (u,v) with $S_u = \{s\}$ and $S_v = \{t\}$. By construction, (u,v) is stored in $\rho_G(s)$ and the routing algorithm directly proceeds to the global routing phase. Since $d(s,t) > 1$, the routing table contains a middle site m and since S_u and S_v are singletons, m is a middle site on a shortest path from s to t. Algorithm 1 pushes $\ell(t)$ onto the stack and sets m as the new target. By induction, the routing scheme now routes the packet along a shortest path from s to m (i, ii), and when the packet arrives at m, the target label $\ell(t)$ is at the top of the stack (iii). Thus, Algorithm 1 executes line 5, and routes the packet from m to t. Again by induction, the packet now follows a shortest path from m to t (i, ii), and when the packet arrives at t, the stack is in the same state a before pushing $\ell(t)$ (iii). The claim follows. □

Building on Lemma 5.2, we can now prove that our scheme is correct.

Lemma 5.3. *Let s, t be two sites in S. Then, the routing scheme produces a routing path $s_0, s_1, \ldots, s_k$ with the following properties*

(i) $s_0 = s$ and $s_k = t$, and
(ii) the header stack is in the same state at the beginning and at the end of the routing path.

Proof. Again, we use induction on the rank of $d(s,t)$ in the sorted list of pairwise distances in $\mathrm{UD}(S)$. If $d(s,t) < c$, the claim is immediate by Lemma 5.2.

Now, consider an arbitrary pair $s, t \in S$. By construction, our routing scheme will eventually find a site $r \in S$ whose global routing table stores a WSPD-pair (u,v) that represents (s,t), together with a middle site m (m exists for $d(s,t) \geq c$ large enough). So far, the stack remains unchanged. Algorithm 1 pushes $\ell(t)$ onto the stack and sets m as the new target. By induction, the routing scheme routes the packet correctly from s to m (i), and when the packet arrives at m, the target label $\ell(t)$ is at the top of the stack (ii). Thus, Algorithm 1 executes line 5, and routes the packet from m to t. Again by induction, the packet arrives at t, with the stack in the same state as before pushing $\ell(t)$ (i, ii). □

Stretch Factor. The analysis of the stretch factor requires some more technical work. For space reasons, we omit the proofs of Lemmas 5.4 and 5.6 and refer the reader to the full version. We begin with a lemma that justifies the term "middle site".

Lemma 5.4. *Let s,t be two sites in S with $d(s,t) \geq c \geq 14$ and let $(u,v) \in \Xi$ be the WSPD-pair that represents (s,t). If m is a middle site of a shortest path from s to $\sigma(v)$ in $\mathrm{UD}(S)$, then*

(i) $d(s,m) + d(m,t) \leq (1+2/c)d(s,t)$, and
(ii) $d(s,m), d(m,t) \leq (5/8)d(s,t)$.

In the next lemma, we bound the distance traveled during the local routing.

Lemma 5.5. *Let s,t be two sites in S with $d(s,t) \geq c$. Then, the total distance traveled by the packet during the local routing phase before the WSPD-pair representing (s,t) is discovered is at most $(48/c)d(s,t)$.*

Proof. Let (u,v) be the WSPD-pair representing (s,t), and let $u_0, u_1, \ldots, u_k = u$ be the path in H from the leaf u_0 for s to u. Let $T_0, T_1, \ldots, T_k$ and $S_0, S_1, \ldots, S_k$ be the corresponding subtrees of T and sites of S. The local routing algorithm iteratively performs an Euler tour of $T_0, T_1, \ldots, T_k$ (the tour of T_k may stop early). An Euler tour in T_i takes $2|S_i| - 2$ steps, and each edge has length at most 1. As described in Sect. 3, for $i = 0. \ldots, k-1$, the WSPD ensures that

$$|S_i| \leq |S_{i+1}| - \lceil(|S_{i+1}| - 1)/6\rceil \leq (5/6)|S_{i+1}| + 1/6 \leq (11/12)|S_{i+1}|,$$

since $|S_{i+1}| \geq 2$. It follows that the total distance for the local routing is at most

$$\sum_{i=0}^{k}(2|S_i| - 2) \leq 2|S_k| \sum_{i=0}^{k}(11/12)^i \leq 24|S_k|.$$

By Lemma 4.1, we have $d(s,t) \geq c(|S_u|-1)$ and since $S_k = S_u$ the total distance is bounded by $24|S_u| \leq 24(d(s,t)/c+1) \leq (48/c)d(s,t)$, where the last inequality is true for $d(s,t) \geq c$. □

Finally, we can bound the stretch factor:

Lemma 5.6. *For any two sites s and t, we have $d_\rho(s,t) \leq (1+\varepsilon)d(s,t)$.*

Combining Lemmas 5.1 and 5.6 we obtain the following theorem.

Theorem 5.7. *Let S be a set of n sites in the plane with density δ. For any $\varepsilon > 0$, we can preprocess S into a routing scheme for $\mathrm{UD}(S)$ with labels of size $O(\log n)$ bits and routing tables of size $O(\delta\varepsilon^{-2}\log^2 n \log^2 D)$, where D is the diameter of $\mathrm{UD}(S)$. For any two sites s,t, the scheme produces a routing path with $d_\rho(s,t) \leq (1+\varepsilon)d(s,t)$ and during the routing the maximum header size is $O(\log n \log D)$. The preprocessing time is $O(n^2 \log n + \delta n^2 + \delta\varepsilon^{-2}n \log n \log^2 D)$.*

5.4 Extension to Arbitrary Density

Let $1 + \varepsilon$, $\varepsilon > 0$, be the desired stretch factor. To extend the routing scheme to point sets of unbounded density, we follow a strategy similar to Gao and Zhang [6, Sect. 4.2]: we first pick an appropriate $\varepsilon_1 > 0$, and we compute an ε_1-*net* $R \subseteq S$, i.e., a subset of sites such that each site in S has distance at most ε_1 to the closest site in R and such that any two sites in R have distance at least ε_1. It is easy to see that R has density $O(\varepsilon_1^{-2})$, and we would like to represent each site in S by the closest site in R. However, the connectivity in $\mathrm{UD}(R)$ might differ from $\mathrm{UD}(S)$. To rectify this, we add additional sites to R. This is done as follows: two sites $s, t \in R$ are called *neighbors* if $|st| > 1$, but there are $p, q \in S$ such that s, p, q, t is a path in $\mathrm{UD}(S)$ and such that $|sp| \leq \varepsilon_1$ and $|qt| \leq \varepsilon_1$ (possibly, $s = p$ or $q = t$). In this case, p and q are called a *bridge* for s, t. Let R' be a point set that contains an arbitrary bridge for each pair of neighbors in R. Set $Z = R \cup R'$. A simple volume argument shows that Z has density $\delta = O(\varepsilon_1^{-3})$. Furthermore, Gao and Zhang show the following:

Lemma 5.8 (Lemmas 4.8 and 4.9 in [6]). *We can compute Z in $O((n/\varepsilon_1^2) \log n)$ time, and if $d^Z(\cdot, \cdot)$ denotes the shortest path distance in $\mathrm{UD}(Z)$, then, for any $s, t \in R$, we have $d^Z(s, t) \leq (1 + 12\varepsilon_1)d(s, t) + 12\varepsilon_1$.*

Now, our extended routing scheme proceeds as follows: first, we compute R and Z as described above, and we perform the preprocessing algorithm for Z with ε_1 as stretch parameter. We assign arbitrary new labels to the sites in $S \backslash Z$. Then, we extend the label $\ell(s)$ of each site $s \in S$, such that it also contains the label of a site in R closest to s. The label size remains $O(\log n)$.

To route between two sites $s, t \in S$, we first check whether we can go from s to t in one step (we assume that this can be checked locally in the routing function). If so, we route the packet directly. Otherwise, we have $d(s, t) > 1$. Let $s', t' \in R$ be the closest sites in R to s and to t. By construction, we can obtain s' and t' from $\ell(s)$ and $\ell(t)$. Now, we first go from s to s'. Then, we use the low-density algorithm to route from s' to t' in $\mathrm{UD}(Z)$, and finally we go from t' to t in one step. Using the discussion above, the total routing distance is bounded by

$$d_\rho(s, t) \leq |ss'| + d_\rho^Z(s', t') + |t't|,$$

where $d_\rho^Z(\cdot, \cdot)$ is the routing distance in $\mathrm{UD}(Z)$. By Lemmas 5.6 and 5.8, this is

$$\leq \varepsilon_1 + (1 + \varepsilon_1)d^Z(s', t') + \varepsilon_1$$
$$\leq 2\varepsilon_1 + (1 + \varepsilon_1)\big((1 + 12\varepsilon_1)d(s', t') + 12\varepsilon_1\big),$$

and by using the triangle inequality twice this is

$$\leq 2\varepsilon_1 + (1 + \varepsilon_1)\big((1 + 12\varepsilon_1)(d(s, t) + 2\varepsilon_1) + 12\varepsilon_1\big).$$

Rearranging and using $d(s,t) > 1$ yields

$$\leq (1 + 29\varepsilon_1 + 50\varepsilon_1^2 + 24\varepsilon_1^3)d(s,t) \leq (1 + \varepsilon)d(s,t),$$

where the last inequality holds for $\varepsilon_1 \leq \varepsilon/103$. This establishes our main theorem:

Theorem 5.9. *Let S be a set of n sites in the plane. For any $\varepsilon > 0$, we can preprocess S into a routing scheme for $UD(S)$ with labels of $O(\log n)$ bits and routing tables of size $O(\varepsilon^{-5} \log^2 n \log^2 D)$, where D is the diameter of $UD(S)$. For any two sites s,t, the scheme produces a routing path with $d_\rho(s,t) \leq (1 + \varepsilon)d(s,t)$ and during the routing the maximum header size is $O(\log n \log D)$. The preprocessing time is $O(n^2 \log n + \varepsilon^{-3}n^2 + \varepsilon^{-5}n \log n \log^2 D)$.*

Proof. The theorem follows from the above discussion and from the fact that the set Z has density $O(\varepsilon^{-3})$, by our choice of ε_1. □

References

1. de Berg, M., Cheong, O., van Kreveld, M., Overmars, M.: Computational Geometry: Algorithms and Applications, 3rd edn. Springer, Berlin (2008)
2. Callahan, P., Kosaraju, S.: A decomposition of multidimensional point sets with applications to k-nearest-neighbors and n-body potential fields. J. ACM **42**(1), 67–90 (1995)
3. Chechik, S.: Compact routing schemes with improved stretch. In: Proceedings of 32nd ACM Symposium on Principles of Distributed Computing (PODC), pp. 33–41 (2013)
4. Clark, B.N., Colbourn, C.J., Johnson, D.S.: Unit disk graphs. Discrete Math. **86**(1–3), 165–177 (1990)
5. Fraigniaud, Pierre, Gavoille, Cyril: Routing in trees. In: Orejas, F., Spirakis, Paul G., van Leeuwen, Jan (eds.) ICALP 2001. LNCS, vol. 2076, p. 757. Springer, Heidelberg (2001)
6. Gao, J., Zhang, L.: Well-separated pair decomposition for the unit-disk graph metric and its applications. SIAM J. Comput. **35**(1), 151–169 (2005)
7. Giordano, S., Stojmenovic, I.: Position based routing algorithms for ad hoc networks: a taxonomy. In: Cheng, X., Huang, X., Du, D.-Z. (eds.) Ad Hoc Wireless Networking. Network Theory and Applications, vol. 14, pp. 103–136. Springer, New York (2004)
8. Gupta, A., Kumar, A., Rastogi, R.: Traveling with a Pez dispenser (or, routing issues in MPLS). SIAM J. Comput. **34**(2), 453–474 (2004)
9. Narasimhan, G., Smid, M.H.M.: Geometric Spanner Networks. Cambridge University Press, Cambridge (2007)
10. Peleg, D., Upfal, E.: A trade-off between space and efficiency for routing tables. J. ACM **36**(3), 510–530 (1989)
11. Roditty, L., Tov, R.: New routing techniques and their applications. In: Proceedings of 34th ACM Symposium on Principles of Distributed Computing (PODC), pp. 23–32 (2015)
12. Santoro, N., Khatib, R.: Labelling and implicit routing in networks. Comput. J. **28**(1), 5–8 (1985)

13. Thorup, M.: Compact oracles for reachability and approximate distances in planar digraphs. J. ACM **51**(6), 993–1024 (2004)
14. Thorup, M., Zwick, U.: Compact routing schemes. In: Proceedings of 13th ACM Symposium on Parallel Algorithms and Architectures (SPAA), pp. 1–10 (2001)
15. Yan, C., Xiang, Y., Dragan, F.F.: Compact and low delay routing labeling scheme for unit disk graphs. Comput. Geom. **45**(7), 305–325 (2012)

Graph Drawings with One Bend and Few Slopes

Kolja Knauer[1]([✉]) and Bartosz Walczak[2]

[1] Aix-Marseille Université, CNRS, LIF UMR 7279, Marseille, France
kolja.knauer@lif.univ-mrs.fr
[2] Theoretical Computer Science Department, Faculty of Mathematics
and Computer Science, Jagiellonian University, Kraków, Poland
walczak@tcs.uj.edu.pl

Abstract. We consider drawings of graphs in the plane in which edges are represented by polygonal paths with at most one bend and the number of different slopes used by all segments of these paths is small. We prove that $\lceil \frac{\Delta}{2} \rceil$ edge slopes suffice for outerplanar drawings of outerplanar graphs with maximum degree $\Delta \geqslant 3$. This matches the obvious lower bound. We also show that $\lceil \frac{\Delta}{2} \rceil + 1$ edge slopes suffice for drawings of general graphs, improving on the previous bound of $\Delta + 1$. Furthermore, we improve previous upper bounds on the number of slopes needed for planar drawings of planar and bipartite planar graphs.

1 Introduction

A *one-bend drawing* of a graph G is a mapping of the vertices of G into distinct points of the plane and of the edges of G into polygonal paths each consisting of at most two segments joined at the *bend* of the path, such that the polygonal paths connect the points representing their end-vertices and pass through no other points representing vertices nor bends of other paths. If it leads to no confusion, in notation and terminology, we make no distinction between a vertex and the corresponding point, and between an edge and the corresponding path. The *slope* of a segment is the family of all straight lines parallel to this segment. The *one-bend slope number* of a graph G is the smallest number s such that there is a one-bend drawing of G using s slopes. Similarly, one defines the *planar one-bend slope number* and the *outerplanar one-bend slope number* of a planar and respectively outerplanar graphs if the drawing additionally has to be planar and respectively outerplanar. Since at most two segments at each vertex can use the same slope, $\lceil \frac{\Delta}{2} \rceil$ is a lower bound on the one-bend slope number. Here and further on, Δ denotes the maximum degree of the graph considered.

1.1 Results

Our main contribution (Theorem 1) is that the outerplanar one-bend slope number of every outerplanar graph is equal to the above-mentioned obvious lower

K. Knauer—Supported by ANR EGOS grant ANR-12-JS02-002-01 and PEPS grant EROS.
B. Walczak—Supported by MNiSW grant 911/MOB/2012/0.

E. Kranakis et al. (Eds.): LATIN 2016, LNCS 9644, pp. 549–561, 2016.
DOI: 10.1007/978-3-662-49529-2_41

bound of $\lceil \frac{\Delta}{2} \rceil$ except for graphs with $\Delta = 2$ that contain cycles, which need 2 slopes. For general graphs, we show that every graph admits a one-bend drawing using at most $\lceil \frac{\Delta}{2} \rceil + 1$ slopes (Theorem 7), which improves on the upper bound of $\Delta + 1$ by Dujmović et al. [7].

For planar graphs, it was shown by Keszegh et al. [14] that the planar one-bend slope number is always at most 2Δ. In the same paper, it was shown that sometimes $\frac{3}{4}(\Delta - 1)$ slopes are necessary. We improve the upper bound to $\frac{3}{2}\Delta$ (Proposition 4) and bound the planar one-bend slope number of planar bipartite graphs by $\Delta + 1$ (Proposition 5). We also show that there are planar bipartite graphs requiring $\frac{2}{3}(\Delta-1)$ slopes in any planar one-bend drawing (Proposition 6). Furthermore, Keszegh et al. [14] showed that every planar graph admits a planar 2-bend drawing with $\lceil \frac{\Delta}{2} \rceil$ slopes.

Apart from improving upon earlier results, one of our motivations for studying the one-bend slope number is that it arises as a relaxation of the slope number, a measure of "geometric complexity" of a graph studied quite extensively since the 1990s. The one-bend slope number also naturally generalizes problems concerning one-bend orthogonal drawings, which have been of interest in the graph drawing community over the past years. We continue with a short overview of these studies. In addition to that, drawings with few slopes are motivated by the real-world need for maps and diagrams of large networks that are well-understandable for the human eye.

1.2   Related Results: Slope Number

The *slope number* of a graph G, introduced by Wade and Chu [25], is the smallest number s such that there is a *straight-line drawing* of G using s slopes. As for the one-bend slope number, $\lceil \frac{\Delta}{2} \rceil$ is an obvious lower bound on the slope number. Dujmović and Wood [8] asked whether the slope number can be bounded from above by a function of the maximum degree. This was answered independently by Barát et al. [1] and by Pach and Pálvölgyi [23] in the negative: graphs with maximum degree 5 can have arbitrarily large slope number. Dujmović et al. [7] further showed that for all $\Delta \geqslant 5$ and sufficiently large n, there exists an n-vertex graph with maximum degree Δ and slope number at least $n^{\frac{1}{2} - \frac{1}{\Delta-2} - o(1)}$. On the other hand, Mukkamala and Pálvölgyi [20] proved that graphs with maximum degree 3 have slope number at most 4, improving earlier results of Keszegh et al. [15] and of Mukkamala and Szegedy [21]. The question whether graphs with maximum degree 4 have slope number bounded by a constant remains open.

The situation is different for *planar* straight-line drawings. It is well known that every planar graph admits a planar straight-line drawing. The *planar slope number* of a planar graph G is the smallest number s such that there is a planar straight-line drawing of G using s slopes. This parameter was first studied by Dujmović et al. [6] in relation to the number of vertices. They also asked whether the planar slope number of a planar graph is bounded in terms of its maximum degree. Jelínek et al. [12] gave an upper bound of $O(\Delta^5)$ for planar graphs of treewidth at most 3. Lenhart et al. [17] showed that the maximum planar slope number of a graph of treewidth at most 2 lies between Δ and 2Δ. Di Giacomo et al. [5] showed that subcubic planar graphs with at least 5 vertices have planar slope number at most 4. The problem has been solved in full generality by Keszegh et al. [14], who showed (with a non-constructive proof) that the planar slope number is bounded from above by an exponential function of the maximum degree. It is still an open problem whether this can be improved to a polynomial upper bound.

Knauer et al. [16] showed that every outerplanar graph with $\Delta \geqslant 4$ has an outerplanar straight-line drawing using at most $\Delta - 1$ slopes and this bound is best possible. For outerplanar graphs with $\Delta = 2$ or $\Delta = 3$, the optimal upper bound is 3.

1.3   Related Results: Orthogonal Drawings

Drawings of graphs that use only the horizontal and the vertical slopes are called *orthogonal*. Every drawing with two slopes can be made orthogonal by a simple affine transformation of the plane. Felsner et al. [9] proved that a graph G with $\Delta \leqslant 4$ admits a one-bend orthogonal drawing if and only if every induced subgraph H of G satisfies $E(H) \leqslant 2V(H) - 2$. Since outerplanar graphs satisfy the latter condition, it follows that every outerplanar graph with $\Delta \leqslant 4$ admits a one-bend orthogonal drawing (our Theorem 1 gives an outerplanar one-bend orthogonal drawing). Biedl and Kant [2] and Liu et al. [18] showed that every planar graph with $\Delta \leqslant 4$ has a planar 2-bend orthogonal drawing with the only exception of the octahedron, which has a planar 3-bend orthogonal drawing. Kant [13] showed that every planar graph with $\Delta \leqslant 3$ has a planar one-bend orthogonal drawing with the only exception of K_4. Nomura et al. [22] proved that every triangle-free outerplanar graph with $\Delta \leqslant 3$ has an outerplanar straight-line orthogonal drawing.

1.4   Related Results: Upward Drawings

Another setting in which one-bend drawings have been considered are upward one-bend drawings of diagrams of posets. Here an edge from a smaller to a larger element of the poset has to be represented by a y-monotone path with at most one bend. Czyzowicz et al. [4] showed that diagrams with maximum degree Δ have an upward one-bend drawing using at most Δ slopes. On the other hand, Czyzowicz [3] constructed examples showing that in the straight-line setting the number of slopes cannot be bounded by the maximum downward or upward degree (whichever is larger) even in the case of lattices. To our knowledge, it is open whether there exists a function f such that the upward slope number of diagrams of maximum degree Δ is bounded from above by $f(\Delta)$.

2   Outerplanar Graphs

The main contribution of this section is to show the following:

Theorem 1. *Every outerplanar graph with maximum degree Δ admits an outerplanar one-bend drawing using at most $\max\{\lceil\frac{\Delta}{2}\rceil, 2\}$ slopes. Furthermore, the set of slopes can be prescribed arbitrarily.*

The structure of the proof of Theorem 1 will follow the same recursive decomposition of an outerplanar graph into *bubbles* that was used in [16] in the proof that every outerplanar graph has a straight-line outerplanar drawing using at most $\Delta - 1$ slopes. Although this decomposition is very natural, for completeness we present it in detail recalling definitions and lemmas from [16].

Let G be an outerplanar graph provided together with its arbitrary outerplanar drawing in the plane. The drawing determines the cyclic order of edges at each vertex and identifies the *outer face* (which is unbounded and contains all vertices on its boundary) and the *inner faces* of G. The edges on the boundary of the outer face are *outer edges*, and all remaining ones are *inner edges*. A *snip* is a simple closed counterclockwise-oriented curve γ which

- passes through some pair of vertices u and v of G (possibly being the same vertex) and through no other vertex of G,
- on the way from v to u goes entirely through the outer face of G and crosses no edge of G,
- on the way from u to v (considered only when $u \neq v$) goes through inner faces of G possibly crossing some inner edges of G that are not incident to u or v, each at most once,
- crosses no edge of G incident to u or v at a point other than u or v.

Every snip γ defines a *bubble* H in G as the subgraph of G induced by the vertices lying on or inside γ. Since γ crosses no outer edges, H is a connected induced subgraph of G. The *roots* of H are the vertices u and v together with all vertices of H adjacent to $G - H$. The snip γ breaks the cyclic clockwise order of the edges of H around each root of H making it a linear order, which we envision as going from left to right. We call the first edge in this order *leftmost* and the last one *rightmost*. The *root-path* of H is the simple oriented path P in H that starts at u with the rightmost edge, continues counterclockwise along the boundary of the outer face of H, and ends at v with the leftmost edge. If $u = v$, then the root-path consists of that single vertex only. All roots of H lie on the root-path—their sequence in the order along the root-path is the *root-sequence* of H. A *k-bubble* is a bubble with k roots. See Fig. 1 for an illustration.

Except at the very end of the proof where we regard the entire G as a bubble, we deal with bubbles H whose first root u and last root v are adjacent to $G - H$. For such bubbles H, all the roots, the root-path, the root-sequence and the left-to-right order of edges at every root do not depend on the particular snip γ used

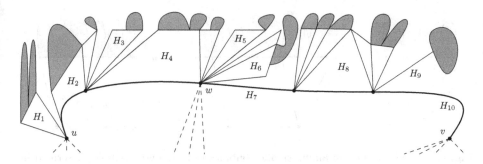

Fig. 1. A 3-bubble H with root-path drawn thick, root-sequence (u, w, v) (connected to the remaining graph by dashed edges), and splitting sequence $(H_1, \ldots, H_{10})$, in which H_1, H_3, H_5, H_6, H_9 are v-bubbles and $H_2, H_4, H_7, H_8, H_{10}$ are e-bubbles.

to define H. Specifically, for such bubbles H, the roots are exactly the vertices adjacent to $G - H$, while the root-path consists of the edges of H incident to inner faces of G that are contained in the outer face of H. From now on, we will refer to the roots, the root-path, the root-sequence and the left-to-right order of edges at every root of a bubble H without specifying the snip γ explicitly.

Lemma 2 ([16, **Lemma 1**]). *Let H be a bubble with root-path $v_1 \ldots v_k$. Every component of $H - \{v_1, \ldots, v_k\}$ is adjacent to either one vertex among $v_1, \ldots, v_k$ or two consecutive vertices from $v_1, \ldots, v_k$. Moreover, there is at most one component adjacent to v_i and v_{i+1} for $1 \leqslant i < k$.*

Lemma 2 allows us to assign each component of $H - \{v_1, \ldots, v_k\}$ to a vertex of P or an edge of P so that every edge is assigned at most one component. For a component C assigned to a vertex v_i, the graph induced by $C \cup \{v_i\}$ is called a *v-bubble*. Such a v-bubble is a 1-bubble with root v_i. For a component C assigned to an edge $v_i v_{i+1}$, the graph induced by $C \cup \{v_i, v_{i+1}\}$ is called an *e-bubble*. Such an e-bubble is a 2-bubble with roots v_i and v_{i+1}. If no component is assigned to an edge of P, then we let that edge alone be a *trivial e-bubble*. All v-bubbles of v_i in H are naturally ordered by their clockwise arrangement around v_i in the drawing. All this leads to a decomposition of the bubble H into a sequence $(H_1, \ldots, H_b)$ of v- and e-bubbles such that the naturally ordered v-bubbles of v_1 precede the e-bubble of $v_1 v_2$, which precedes the naturally ordered v-bubbles of v_2, and so on. We call it the *splitting sequence* of H. The splitting sequence of a single-vertex 1-bubble is empty. Every 1-bubble with more than one vertex is a v-bubble or a bouquet of several v-bubbles. The splitting sequence of a 2-bubble may consist of several v- and e-bubbles. Again, see Fig. 1 for an illustration.

The following lemma provides the base for the recursive structure of the proof of Theorem 1. See Fig. 2 for an illustration.

Lemma 3 ([16, **Lemma 2, statements 2.1 and 2.3**]).

1. *Let H be a v-bubble rooted at u. Let $u^1, \ldots, u^k$ be the neighbors of u in H from left to right. Then $H - \{u\}$ is a bubble with root-sequence $(u^1, \ldots, u^k)$.*

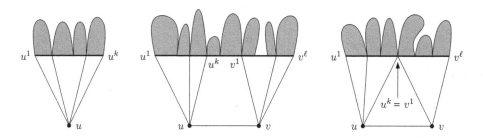

Fig. 2. Various ways of obtaining smaller bubbles from v- and e-bubbles described in Lemma 3. The new bubbles are grayed, and the new root-paths are drawn thick.

2. Let H be an e-bubble with roots u and v. Let $u^1, \ldots, u^k, v$ and $u, v^1, \ldots, v^\ell$ be respectively the neighbors of u and v in H from left to right. Then $H - \{u, v\}$ is a bubble with root-sequence $(u^1, \ldots, u^k, v^1, \ldots, v^\ell)$ in which u^k and v^1 coincide if the inner face of H containing uv is a triangle.

Proof (Theorem 1). We fix $s \geqslant 2$, assume to be given an outerplanar graph G with maximum degree $\Delta \leqslant 2s$, and construct an outerplanar one-bend drawing of G with a prescribed set of s slopes. Actually, for most of the proof, we assume $s \geqslant 3$. The case $s = 2$ is sketched at the very end of the proof.

Let D denote the set of $2s$ *directions*, that is, oriented slopes from the prescribed set of s slopes. For a direction $d \in D$, let d^- and d^+ denote respectively the previous and the next directions in the clockwise cyclic order on D.

We can assume without loss of generality that every vertex of G has degree either 1 or $2s$. Indeed, we can raise the degree of any vertex by connecting it to new vertices of degree 1 placed in the outer face. With this assumption, at each vertex u, the direction in which one edge leaves u determines the directions of the other edges at u. When a vertex u has all edge directions determined, we write $d(uv)$ to denote the direction determined for an edge uv at u.

For an edge uv drawn as a union of two segments ux and xv and for two directions $d_v, d_u \in D$ consecutive in the clockwise order on D, let $Q(uv, d_u, d_v)$ denote the quadrilateral $uxvy$, where y is the intersection point of the rays going out of u and v in directions d_u and d_v, respectively. We express the condition that the point y exists saying that the quadrilateral is *well defined*.

First, consider the setting of Lemma 3 statement 2. Assume that the edge uv is the only predrawn part of H. Assume further that two *leading directions* $d_v, d_u \in D$ that are consecutive in the clockwise order on D and have the following properties are provided:

a. $-d_u \notin \{d(uu^1), \ldots, d(uu^k)\}$ and $-d_v \notin \{d(vv^1), \ldots, d(vv^\ell)\}$,
b. no part of the graph other than the edge uv and some short initial parts of other edges at u and v is predrawn in the ϵ-neighborhood Q_ϵ of the quadrilateral $Q = Q(uv, d_u, d_v)$, for some sufficiently small $\epsilon > 0$.

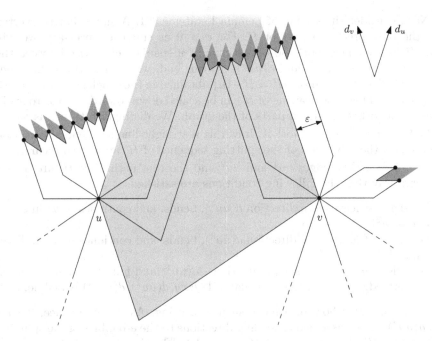

Fig. 3. Drawing bubbles: a v-bubble of Case 3 (left), an e-bubble (middle), and a v-bubble of Case 2 (right). The directions d_u and d_v used to draw the e-bubble are also shown. The target quadrilaterals for e-bubbles are grayed.

We call Q the *target quadrilateral* for H. We will draw H in Q_ϵ in a way that will guarantee that H does not cross the predrawn parts of the graph. To this end, we need to draw the edges $uu^1, \ldots, uu^k, vv^1, \ldots, vv^\ell$ and the bubble $H' = H - \{u, v\}$ obtained in the conclusion of Lemma 3 statement 2.

The edges $uu^1, \ldots, uu^k, vv^1, \ldots, vv^\ell$ and the root-path P of H' are drawn in Q_ϵ in such a way that the following conditions are satisfied:

- each edge uu^i leaves u in direction $d(uu^i)$, bends shortly after (but far enough to avoid crossing other edges at u), and continues to u^i in direction d_u,
- each edge vv^i leaves v in direction $d(vv^i)$, bends shortly after (but far enough to avoid crossing other edges at v), and continues to v^i in direction d_v,
- each edge xy of P leaves x in direction $-d_v^-$ if $x \in \{v^1, \ldots, v^{\ell-1}\}$ or $-d_v$ otherwise, and leaves y in direction $-d_u^+$ if $y \in \{u^2, \ldots, u^k\}$ or $-d_u$ otherwise,
- for every edge xy of P, the quadrilateral $Q(xy, d_u, d_v)$ is well defined.

Figure 3 illustrates how to achieve such a drawing. As a consequence, d_v and d_u can be assigned as leading directions to the e-bubbles of the splitting sequence of H', because (a) at their roots, the directions $-d_v$ and $-d_u$ are occupied by edges of the root-path of H' or by edges going to u and v, and (b) their target quadrilaterals are pairwise disjoint except at their common vertices and are contained in Q_ϵ far enough from u and v. The drawing of H is completed by drawing all bubbles of the splitting sequence of H' recursively.

Now, consider the setting of Lemma 3 statement 1. Assume that the vertex u is the only predrawn part of H. For $\epsilon > 0$ as small as necessary, we will draw H in the ϵ-neighborhood of the cone at u spanned clockwise between the rays in directions $d(uu^1)$ and $d(uu^k)$. To this end, we need to draw the edges $uu^1, \ldots, uu^k$ and the bubble $H' = H - \{u\}$ obtained in the conclusion of Lemma 3 statement 1. Then, the drawing of H can be scaled down towards u so as to avoid crossing the other predrawn parts of the graph. We distinguish three cases:

Case 1: $k = 1$. The edge uu^1 is drawn as a straight-line segment in direction $d(uu^1)$, and the v-bubbles of the splitting sequence of H' are drawn recursively.

Case 2: $k = 2$. The edges uu^1 and uu^2 and the root-path P of H' are drawn in such a way that the following conditions are satisfied:

- the edge uu^1 leaves u in direction $d(uu^1)$, bends, and continues to u^1 in direction $d(uu^2)$,
- the edge uu^2 leaves u in direction $d(uu^2)$, bends, and continues to u^2 in direction $d(uu^1)$,
- each edge xy of P leaves x in direction $-d(uu^1)$ and y in direction $-d(uu^2)$,
- for every edge xy of P, the quadrilateral $Q(xy, d(uu^2), d(uu^1))$ is well defined.

Figure 3 illustrates how to achieve such a drawing. As a consequence, $d(uu^1)$ and $d(uu^2)$ can be assigned as leading directions to the e-bubbles of the splitting sequence of H'. The drawing of H is completed by drawing all bubbles of the splitting sequence of H' recursively.

Case 3: $k \geqslant 3$. Let P denote the root-path of H and $u^{k-1}x_1 \ldots x_m u^k$ denote the part of P between u^{k-1} and u^k. Choose a direction $d \in \{d(uu^1), \ldots, d(uu^k)\}$ so that $-d \notin \{d(uu^1), \ldots, d(uu^k)\}$. The edges $uu^1, \ldots, uu^k$ and the root-path P are drawn in such a way that the following conditions are satisfied:

- each edge uu^i leaves u in direction $d(uu^i)$, bends shortly after, and continues to u^i in direction d,
- each edge xy of P leaves x in direction $-d$ if $x \in \{x_1, \ldots, x_m\}$ or $-d^-$ otherwise, and leaves y in direction $-d^+$ if $y \in \{u^2, \ldots, u^{k-1}, x_1, \ldots, x_m, u^k\}$ or $-d$ otherwise.
- for every edge xy of P, the quadrilateral $Q(xy, d_x^{xy}, d_y^{xy})$ is well defined, where

$$(d_x^{xy}, d_y^{xy}) = \begin{cases} (d, d^-) & \text{if } x, y \in \{u^{k-1}, x_1, \ldots, x_m, u^k\}, \\ (d^+, d) & \text{otherwise.} \end{cases}$$

Again, Fig. 3 illustrates how to achieve such a drawing. As a consequence, d_y^{xy} and d_x^{xy} can be assigned as leading directions to every e-bubble of the splitting sequence of H', where x and y are the roots of the e-bubble (case distinction in the definition of (d_x^{xy}, d_y^{xy}) is needed to ensure property a). The drawing of H is completed by drawing all bubbles of the splitting sequence of H' recursively.

To complete the proof for $s \geqslant 3$, pick any vertex u of G of degree 1, assign an arbitrary direction to the edge at u, and continue the drawing as in Case 1.

The proof for $s = 2$ keeps the same general recursive scheme following from Lemma 3. As before, all e-bubbles are drawn in ϵ-neighborhoods of their target

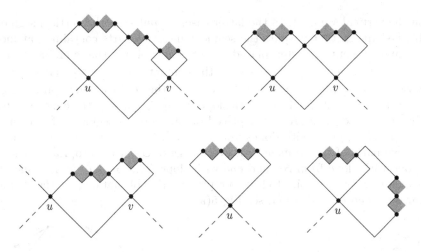

Fig. 4. Various ways of drawing v- and e-bubbles when $s = 2$. The target quadrilaterals for recursive e-bubbles are grayed. The edges of the root-path of H' that form trivial e-bubbles do not have target quadrilaterals.

quadrilaterals, which are always parallelograms when $s = 2$. The details of the drawing algorithm for various possible cases should be clear from Fig. 4.   □

3   Planar Graphs and Planar Bipartite Graphs

Using contact representations as in [14, Theorem 2], where the upper bound of 2Δ on the planar one-bend slope number is shown for planar graphs, we improve the upper bounds on this parameter for planar and bipartite planar graphs.

Proposition 4. *Every planar graph with maximum degree Δ admits a planar one-bend drawing using at most $\Delta + \lceil \frac{\Delta}{2} \rceil - 1$ slopes.*

Proof. Let G be a graph as in the statement. By [11, Theorem 4.1], G can be represented as a contact graph of T-shapes in the plane. Every T-shape consists of a horizontal segment and a vertical segment touching at the upper endpoint of the vertical one. That point, called the *center* of the T-shape, splits the horizontal segment into the *left segment* and the *right segment* of the T-shape. The T-shapes of the contact representation are modified as follows: for each T-shape, considered one by one in the top-down order of horizontal segments, move its vertical segment horizontally so as to make its left segment and its right segment contain at most $\lceil \frac{\Delta}{2} \rceil$ contact points with other T-shapes, and scale accordingly the two bottomless rectangular stripes going down from the left and the right segment. This keeps the contact graph unchanged.

We construct a one-bend drawing of G using a set S_H of $\lceil \frac{\Delta}{2} \rceil$ almost horizontal slopes and a set S_V of $\Delta - 1$ almost vertical slopes. We place each vertex v at the center of the T-shape representing v unless all contact points of the T-shape lie on the vertical segment. In the latter case, we put v on the vertical segment of the T-shape so that it splits the segment into two parts containing at most $\lceil \frac{\Delta}{2} \rceil$ contact points. A vertex placed at the center of a T-shape emits at most $\lceil \frac{\Delta}{2} \rceil$ rays with slopes from S_H towards the contact points on the left segment, at most $\lceil \frac{\Delta}{2} \rceil$ rays with slopes from S_H towards the contact points on the right segment, and at most $\Delta - 1$ rays with slopes from S_V towards the contact points on the vertical segment. A vertex placed on the vertical segment of a T-shape emits at most $\lceil \frac{\Delta}{2} \rceil$ rays with slopes from S_V towards the contact points on either of the two parts of the segment. For every edge of G, two appropriately chosen rays, one with slope from S_H and one with slope from S_V, are joined near the corresponding contact point to form a representation of that edge in the claimed planar one-bend drawing of G, see Fig. 5(a). $\qquad\square$

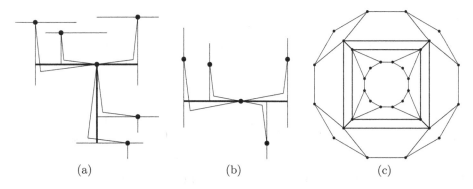

Fig. 5. (a) One-bend drawing of a planar graph (b) One-bend drawing of a bipartite planar graph (c) Graph G_5 constructed in the proof of Proposition 6

Proposition 5. *Every bipartite planar graph with maximum degree Δ admits a planar one-bend drawing using at most $2\lceil \frac{\Delta}{2} \rceil$ slopes.*

Proof. Let G be a graph as in the statement. By [10, Theorem 1.5], G can be represented as a contact graph of horizontal and vertical segments in the plane. We construct a one-bend drawing of G using a set S_H of $\lceil \frac{\Delta}{2} \rceil$ almost horizontal slopes and a set S_V of $\lceil \frac{\Delta}{2} \rceil$ almost vertical slopes. We place every vertex v of G on the segment representing v so that it splits the segment into two parts containing at most $\lceil \frac{\Delta}{2} \rceil$ contact points with other segments. A vertex placed on a horizontal segment emits at most $\lceil \frac{\Delta}{2} \rceil$ rays with slopes from S_H towards the contact points on either of the two parts of the segment. Similarly, a vertex placed on a vertical segment emits at most $\lceil \frac{\Delta}{2} \rceil$ rays with slopes from S_V towards the contact points on either of the two parts of the segment. For every edge of G, two appropriately chosen rays, one with slope from S_H and one with slope from S_V, are joined near the corresponding contact point to form a representation of that edge in the claimed planar one-bend drawing of G, see Fig. 5(b). $\qquad\square$

The following is a straightforward adaptation of [14, Theorem 4], where planar graphs with planar one-bend slope number at least $\frac{3}{4}(\Delta - 1)$ are constructed.

Proposition 6. *For every $\Delta \geqslant 3$, there is a planar bipartite graph with maximum degree Δ and with planar one-bend slope number at least $\frac{2}{3}(\Delta - 1)$.*

Proof. A graph G_Δ of maximum degree Δ is constructed starting from a plane drawing of the 3-dimensional cube. Two opposite faces of the cube are chosen, say, the outer and the central. In either of them, a cycle of $8\Delta - 28$ new vertices is drawn; then, each boundary vertex of the face picks a subpath of $2\Delta - 7$ vertices of the cycle and connects to the $\Delta - 3$ odd vertices of the subpath, see Fig. 5(c).

It is well known that the measures of the interior angles of a simple k-gon sum up to $(k - 2)\pi$. This is a consequence of a more general observation: if P is a simple k-gon (with angles of measure π allowed), then every slope is covered exactly $k - 2$ times by interior angles of P. For the purpose of this statement, at each vertex of P, either of the two directions of a slope is counted separately—once if it points towards the interior of P and $\frac{1}{2}$ times if it points towards the boundary of P. Therefore, if S is a set of slopes and P is a simple k-gon drawn using slopes from S, then every slope from S can be used by at most $k - 2$ segments that are sides of P or go from a vertex of P towards the interior of P.

Suppose we are given a planar one-bend drawing of G_Δ using a set of slopes S. The restriction of the drawing to the starting cube must have one of the two selected faces, call it F, as an inner face. The face F is drawn as a simple octagon (with angles of measure π allowed), and each of the four vertices of the cube that lie on the boundary of F emits $\Delta - 3$ edges towards the interior of F. By the observation above, every slope from S can be used by at most 6 of the $8 + 4(\Delta - 3)$ segments that are sides of the octagon or initial parts of the edges going from the four vertices towards the interior of F. We conclude that $8 + 4(\Delta - 3) \leqslant 6|S|$, which yields $|S| \geqslant \frac{2}{3}(\Delta - 1)$. $\square$

4   General Graphs

The main contribution of this section is to show the following:

Theorem 7. *Every graph with maximum degree Δ admits a one-bend drawing using at most $\lceil \frac{\Delta}{2} \rceil + 1$ slopes. Such a drawing exists with all vertices placed on a common line. Furthermore, the set of slopes can be prescribed arbitrarily.*

Proof. Let G be a graph with maximum degree Δ. By Vizing's theorem [24], G has a proper edge-coloring using at most $\Delta + 1$ colors, and moreover, such a coloring can be obtained in polynomial time [19]. This yields a partition of the edge set of G into $\Delta + 1$ matchings $M_1, \ldots, M_{\Delta+1}$. Let $n = |V(G)| - |M_{\Delta+1}|$, and let $f \colon V(G) \to \{1, \ldots, n\}$ be such that $f(u) = f(v)$ if and only if $uv \in M_{\Delta+1}$.

Let S be a set of $k = \lceil \frac{\Delta}{2} \rceil + 1$ slopes and ℓ be a line with slope not in S. Without loss of generality, we can assume that ℓ is horizontal. Order S as $\{s_1, \ldots, s_k\}$ clockwise starting from the horizontal slope (that is, if $i < j$, then s_i occurs before s_j when rotating a line clockwise starting from the horizontal position). Fix n pairwise disjoint segments $I_1, \ldots, I_n$ in this order on ℓ.

Each vertex v of G is placed on the segment $I_{f(v)}$. Each edge $uv \in M_i$ with $1 \leqslant i \leqslant k-1$ and $f(u) < f(v)$ is drawn above ℓ so that its slope at v is s_i and its slope at u is s_j, where j is the least index in $\{i+1, \ldots, k-1\}$ for which there is no edge $u'u \in M_j$ with $f(u') < f(u)$, or $j = k$ if such an index does not exist. This way, since $M_1, \ldots, M_{k-1}$ are matchings, no two edges of $M_1, \ldots, M_{k-1}$ use the same slope at any vertex. The edges of $M_k, \ldots, M_\Delta$ are drawn in an analogous way below ℓ. At least one slope above ℓ and at least one below ℓ are left free at every vertex.

Now, consider an edge $uv \in M_{\Delta+1}$. In the drawing presented above, u and v have degree at most $\Delta - 1$, so either of them has an additional free slope above or below ℓ. Therefore, either above or below ℓ, there are two distinct slopes, one free at u and the other free at v. They can be used to draw the edge uv if u and v are placed in an appropriate order within $I_{f(u)} = I_{f(v)}$. Occurrence of bend points of some edges on other edges can be fixed by perturbing vertices slightly within their segments on ℓ. ◻

5 Problems

Apart from determining precisely the planar one-bend slope number of planar and planar bipartite graphs, it would be interesting to drop the restriction to the very particular set of slopes in the constructions in Propositions 4 and 5. Are there planar one-bend drawings of planar graphs that work for any given set of few slopes (like all the other constructions in the present paper)?

The second problem concerns one-bend drawings of general graphs. The results of [9] directly yield a characterization of the graphs that require $\lceil \frac{\Delta}{2} \rceil + 1$ slopes for a one-bend drawing when $\Delta \leqslant 4$. Are there graphs with $\Delta \geqslant 5$ that require $\lceil \frac{\Delta}{2} \rceil + 1$ slopes for a one-bend drawing?

Acknowledgment. We are grateful to Piotr Micek for fruitful discussions.

References

1. Barát, J., Matoušek, J., Wood, D.R.: Bounded-degree graphs have arbitrarily large geometric thickness. Electron. J. Combin. **13**(1), #R3, 14 pp. (2006)
2. Biedl, T., Kant, G.: A better heuristic for orthogonal graph drawings. Comput. Geom. **9**(3), 159–180 (1994)
3. Czyzowicz, J.: Lattice diagrams with few slopes. J. Combin. Theory, Ser. A **56**(1), 96–108 (1991)
4. Czyzowicz, J., Pelc, A., Rival, I., Urrutia, J.: Crooked diagrams with few slopes. Order **7**(2), 133–143 (1990)
5. Di Giacomo, E., Liotta, G., Montecchiani, F.: The planar slope number of subcubic graphs. In: Pardo, A., Viola, A. (eds.) LATIN 2014. LNCS, vol. 8392, pp. 132–143. Springer, Heidelberg (2014)
6. Dujmović, V., Eppstein, D., Suderman, M., Wood, D.R.: Drawings of planar graphs with few slopes and segments. Comput. Geom. **38**(3), 194–212 (2007)

7. Dujmović, V., Suderman, M., Wood, D.R.: Graph drawings with few slopes. Comput. Geom. **38**(3), 181–193 (2007)
8. Dujmović, V., Wood, D.R.: On linear layouts of graphs. Discrete Math. Theor. Comput. Sci. **6**(2), 339–358 (2004)
9. Felsner, S., Kaufmann, M., Valtr, P.: Bend-optimal orthogonal graph drawing in the general position model. Comput. Geom. **47**(3), 460–468 (2014)
10. de Fraysseix, H., de Mendez, P.O., Pach, J.: A left-first search algorithm for planar graphs. Discrete Comput. Geom. **13**(3–4), 459–468 (1995)
11. de Fraysseix, H., de Mendez, P.O., Rosenstiehl, P.: On triangle contact graphs. Combin. Prob. Comput. **3**(2), 233–246 (1994)
12. Jelínek, V., Jelínková, E., Kratochvíl, J., Lidický, B., Tesař, M., Vyskočil, T.: The planar slope number of planar partial 3-trees of bounded degree. Graphs Combin. **29**(4), 981–1005 (2013)
13. Kant, G.: Drawing planar graphs using the canonical ordering. Algorithmica **16**(1), 4–32 (1996)
14. Keszegh, B., Pach, J., Pálvölgyi, D.: Drawing planar graphs of bounded degree with few slopes. SIAM J. Discrete Math. **27**(2), 1171–1183 (2013)
15. Keszegh, B., Pach, J., Pálvölgyi, D., Tóth, G.: Drawing cubic graphs with at most five slopes. Comput. Geom. **40**(2), 138–147 (2008)
16. Knauer, K., Micek, P., Walczak, B.: Outerplanar graph drawings with few slopes. Comput. Geom. **47**(5), 614–624 (2014)
17. Lenhart, W., Liotta, G., Mondal, D., Nishat, R.I.: Planar and plane slope number of partial 2-trees. In: Wismath, S., Wolff, A. (eds.) GD 2013. LNCS, vol. 8242, pp. 412–423. Springer, Heidelberg (2013)
18. Liu, Y., Morgana, A., Simeone, B.: A linear algorithm for 2-bend embeddings of planar graphs in the two-dimensional grid. Discrete Appl. Math. **81**(1–3), 69–91 (1998)
19. Misra, J., Gries, D.: A constructive proof of Vizing's theorem. Inform. Process. Lett. **41**(3), 131–133 (1992)
20. Mukkamala, P., Pálvölgyi, D.: Drawing cubic graphs with the four basic slopes. In: van Kreveld, M., Speckmann, B. (eds.) GD 2011. LNCS, vol. 7034, pp. 254–265. Springer, Heidelberg (2012)
21. Mukkamala, P., Szegedy, M.: Geometric representation of cubic graphs with four directions. Comput. Geom. **42**(9), 842–851 (2009)
22. Nomura, K., Tayu, S., Ueno, S.: On the orthogonal drawing of outerplanar graphs. In: Chwa, K.-Y., Munro, J.I. (eds.) COCOON 2004. LNCS, vol. 3106, pp. 300–308. Springer, Heidelberg (2004)
23. Pach, J., Pálvölgyi, D.: Bounded-degree graphs can have arbitrarily large slope numbers. Electron. J. Combin. **13**(1), #N1, 4 pp. (2006)
24. Vizing, V.G.: Ob otsenke khromaticheskogo klassa p-grafa (On an estimate of the chromatic class of a p-graph). Diskret. Analiz **3**, 25–30 (1964)
25. Wade, G.A., Chu, J.H.: Drawability of complete graphs using a minimal slope set. Comput. J. **37**(2), 139–142 (1994)

Edge-Editing to a Dense
and a Sparse Graph Class

Michal Kotrbčík[1], Rastislav Královič[2], and Sebastian Ordyniak[3(✉)]

[1] Faculty of Informatics, Masaryk University, Brno, Czech Republic
kotrbcik@fi.muni.cz
[2] Department of Computer Science, Comenius University, Bratislava, Slovakia
kralovic@dcs.fmph.uniba.sk
[3] Institute of Information Systems, TU Wien, Vienna, Austria
sordyniak@gmail.com

Abstract. We consider a graph edge-editing problem, where the goal is to transform a given graph G into a disjoint union of two graphs from a pair of given graph classes, investigating what properties of the classes make the problem fixed-parameter tractable. We focus on the case when the first class is dense, i.e. every such graph G has minimum degree at least $|V(G)| - \delta$ for a constant δ, and assume that the cost of editing to this class is fixed-parameter tractable parameterized by the cost. Under the assumptions that the second class either has bounded maximum degree, or is edge-monotone, can be defined in MSO_2, and has bounded treewidth, we prove that the problem is fixed-parameter tractable parameterized by the cost. We also show that the problem is fixed-parameter tractable parameterized by degeneracy if the second class consists of independent sets and SUBGRAPH ISOMORPHISM is fixed-parameter tractable for the input graphs. On the other hand, we prove that parameterization by degeneracy is in general W[1]-hard even for editing to cliques and independent sets.

Keywords: Graph modification problems · Clique-editing · Degeneracy · Parameterized complexity · Treewidth

1 Introduction

Graph editing problems ask for modifying an input graph to a graph with a given property using at most k operations, where the allowed operations are usually edge addition, edge deletion, vertex deletion, or their combinations. Variants of graph editing problems received significant attention ever since the seminal Yannakakis's paper [23], and have found use in many application areas including machine learning [2], and social networks [13,18]. Graph editing is similarly

M. Kotrbčík and S. Ordyniak—Research funded by the Employment of Newly Graduated Doctors of Science for Scientific Excellence (CZ.1.07/2.3.00/30.0009) and the Austrian Science Fund (FWF, project P26696).

© Springer-Verlag Berlin Heidelberg 2016
E. Kranakis et al. (Eds.): LATIN 2016, LNCS 9644, pp. 562–575, 2016.
DOI: 10.1007/978-3-662-49529-2_42

interesting from the algorithm design point of view and a considerable effort was invested into determining the (parameterized) complexity of editing into particular graph classes, for example, Eulerian graphs [7,11], regular graphs [16,17], or hereditary classes [6]; for a recent survey we refer to Bodlaender et al. [3].

Recently, variants of graph editing involving two classes were considered. The goal in these problems is to change a given graph into a disjoint union of two graphs, one from each class, using as few edge insertions and deletions as possible. A particular instance of editing to two classes, the CLIQUEEDITING problem, asks for editing into a disjoint union of a clique and an independent set and arises as a clustering problem in noisy data sets [8]. CLIQUEEDITING is fixed-parameter tractable (and solvable in subexponential time) parameterized by the number of edge modifications [8]. Only later it was shown that CLIQUEEDITING is NP-hard, both in the general case and in the class of bipartite graphs [15]. On the other hand, CLIQUEEDITING is solvable in polynomial-time on planar graphs [15], admits a PTAS on bipartite graphs [15], and there is a 3-approximation algorithm on general graphs [14]. A different case of editing to two classes considers editing to a disjoint union of a balanced biclique ($K_{n,n}$) and an independent set [14] — the problem admits a kernelization scheme that for any positive ϵ yields a kernel of size ϵk, where k is the cost of editing.

Our Results. We consider the problem of $(\mathcal{C}_1, \mathcal{C}_2)$-EDITING, where the goal is to transform an input graph G into a disjoint union of two graphs, one from $\mathcal{C}_1$ and the other from $\mathcal{C}_2$, using as few edge insertions and deletions as possible. Editing to two classes is a quite general problem, for example, one can imagine a scenario as follows. A bus network, represented by a graph, is going to be split and sold to two interested parties. The parties are interested only in networks with certain topology, and the network may need to be changed to accommodate the supported topologies. The cost of edge modifications then represents the inherent cost of changing the infrastructure accordingly. The cost of editing to two classes corresponds to the minimum cost of infrastructure changes required to transfer the network to parties (or one of them) while still serving all the nodes in the network.

In this paper we focus on the theoretical aspects of the problem and pursue the question *when is editing to two classes tractable*. In line with the existing research, we focus on the case where $\mathcal{C}_1$ is a dense and $\mathcal{C}_2$ is a sparse class. Technically, we achieve this by requiring all graphs G from $\mathcal{C}_1$ to have minimum degree at least $|V(G)| - \delta$ for some constant δ and requiring $\mathcal{C}_2$ to have bounded either maximum degree, or treewidth. Our main contribution is a general approach which allows us to treat many cases in a unified fashion, using very little class-specific structural information. First, we introduce novel notions of weakly-hereditary and weakly anti-hereditary classes. These properties guarantee that the cost of editing to a graph in $\mathcal{C}$ does not change enormously by omitting only one vertex from the graph, respectively by adding an isolated vertex to a graph, see Sect. 4 for details. Second, we make use of a separation property between dense weakly hereditary and weakly-anti hereditary classes, which implies that the degrees of the induced subgraph that is being edited to the dense class are large in terms of the size of the subgraph. Finally, to obtain a solution, at some point it is necessary to perform computations for individual classes. This is achieved by assuming that the cost of single-class editing is either decidable or fixed-parameter tractable, or by assuming that the class can be defined in MSO_2 logic. In particular, we show that for bounded degree sparse classes $(\mathcal{C}_1, \mathcal{C}_2)$-EDITING is fixed-parameter tractable parameterized by the cost of the editing as long as $\text{cost}_{\mathcal{C}_1}(G)$ and $\text{cost}_{\mathcal{C}_2}(G)$ are fixed-parameter tractable. For bounded treewidth sparse classes, the problem is fixed-parameter tractable parameterized by the cost whenever $\text{cost}_{\mathcal{C}_1}(G)$ is fixed-parameter tractable and $\mathcal{C}_2$ is an edge-monotone and MSO_2-definable class. Our assumptions, in particular on the sparse classes, are weak, since our results hold for example for regular graphs, acyclic graphs, bounded degree trees, and k-colorable or bounded genus graphs with bounded treewidth.

For parameterization by degeneracy we prove that the problem is fixed-parameter tractable assuming only computability of $\text{cost}_{\mathcal{C}_1}(G)$, the tradeoff is that the input graphs have to belong to a class for which SUBGRAPH ISO-MORPHISM is fixed-parameter tractable, and that $\mathcal{C}_2$ contains only independent sets. Since graphs with bounded expansion have bounded degeneracy and admit SUBGRAPH ISOMORPHISM in FPT, we obtain a linear-time algorithm for $(\mathcal{C}, \mathcal{I})$-EDITING when the input graphs have bounded expansion, where $\mathcal{I}$ is the class of all independent sets, which is a great improvement over the existing $O(n^{11})$ algorithm for CLIQUEEDITING on planar graphs. On the other hand, we prove that the parameterization by degeneracy is in general W[1]-hard already for CLIQUEEDITING. Finally, we obtain a kernel if $\text{cost}_{\mathcal{C}_1}(G)$ is computable in polynomial time and the class $\mathcal{C}_2$ contains all graphs with maximum degree Δ. Without the second condition, or a similarly strong assumption about the structure of the graphs in the second class, it would not be possible to guarantee that the instance constructed as a kernel has the desired cost.

2   Preliminaries

We assume that the reader is familiar with the basic concepts and definitions of parameterized complexity; for details we refer to the standard texts on parameterized complexity [9, 10, 20].

2.1   Graphs

We use standard graph-theoretic notation and terminology and consider only finite, undirected, and simple graphs. For disjoint sets of vertices A and B of a graph G, by $E_G(A, B)$ we denote the set of edges of G with one endpoint in A and the other in B; if the graph G is clear from the context we omit the subscript G. For a graph G, by $|G|$ we denote the number of vertices of G and by $\deg_G(v)$ the degree of a vertex v in G. For a set of vertices X and a set of edges Y of a graph G, we denote by $G[X]$ the subgraph of G induced by X, by $G \backslash X$ the subgraph of G induced by $V(G) \backslash X$, and by $G \backslash Y$ the subgraph of G with vertices $V(G)$ and edges $E(G) \backslash Y$. The *degeneracy* of a graph G is defined as the minimum integer r such that every subgraph of G has minimum degree at most r and is denoted by $\mathrm{degen}(G)$. Finally, the isomorphism relation between graphs is denoted by $\cong$.

2.2 Treewidth

A *tree-decomposition* $\mathcal{T}$ of a graph G is a pair (T, χ), where T is a tree and χ is a function that assigns each tree node t a set $\chi(t) \subseteq V(G)$ of vertices such that the following conditions hold: (T1) For every vertex $v \in V(G)$, there is a tree node t such that $v \in \chi(t)$, (T2) For every edge $\{u, v\} \in E(G)$ there is a tree node t such that $u, v \in \chi(t)$, (T3) For every vertex $v \in V(G)$, the set of tree nodes t with $v \in \chi(t)$ forms a subtree of T. The sets $\chi(t)$ are called *bags* of the decomposition $\mathcal{T}$ and $\chi(t)$ is the bag associated with the tree node t. The *width* of a tree-decomposition (T, χ) is the size of a largest bag minus 1. A tree-decomposition of minimum width is called *optimal*. The *treewidth* of a graph G, denoted by $\mathrm{tw}(G)$, is the width of an optimal tree decomposition of G.

Proposition 1. *Let G be a graph and let G' be a graph obtained from G by applying at most k edge-modifications to G. Then $\mathrm{tw}(G) \leq \mathrm{tw}(G') + k$.*

Proposition 2 [4]. *Let G be a graph and ω a natural number. Then the problems of deciding whether G has a tree-decomposition of width at most ω and if yes, constructing such a tree-decomposition, are linear time fixed-parameter tractable parameterized by ω.*

2.3 Monadic Second Order Logic and Monotone Classes

We consider *Monadic Second Order* (MSO_2) logic on graphs in terms of their incidence structure whose universe contains vertices and edges; the incidence between vertices and edges is represented by a binary relation. We assume an infinite supply of *individual variables* $x, x_1, x_2, \ldots$ and of *set variables* $X, X_1, X_2, \ldots$ The *atomic formulas* are Ixy ("vertex x is incident with edge y"), $x = y$ (equality), $x \neq y$ (inequality), and Xx ("vertex or edge x is an element of set X"). *MSO formulas* are built up from atomic formulas using the usual Boolean connectives ($\neg, \wedge, \vee, \rightarrow, \leftrightarrow$), quantification over individual variables ($\forall x, \exists x$), and quantification over set variables ($\forall X, \exists X$).

Let $\Phi(X)$ be an MSO_2 formula with a free set variable X. For a graph G and a set $S \subseteq E(G)$ we write $G \models \Phi(S)$ if the formula Φ holds true on G whenever X is instantiated with S.

The following theorem shows that if G has bounded treewidth, then in linear-time we can verify whether there is an S with $G \models \Phi(S)$ and $|S| \leq \ell$.

Theorem 3 [1]. *Let $\Phi(X)$ be an MSO_2 formula with a free set variable X and let ω be a constant. Then there is a linear-time algorithm that, given a graph G*

of treewidth at most ω, and an integer ℓ, decides whether there is a set $S \subseteq E(G)$ with $|S| \leq \ell$ such that $G \models \Phi(S)$.

While the original version of Theorem 3 [1] requires a tree-decomposition of width at most w to be provided with the input, for a graph of treewidth at most w such a tree decomposition can be found in linear time (Proposition 2) and thus the assumption is not necessary.

To employ Theorem 3, we utilize MSO_2-definable graph classes. A class $\mathcal{C}$ is MSO_2-*definable* if there is an MSO_2 formula $\Phi_{\mathcal{C}}(X)$ such that for any graph G, the formula $\Phi_{\mathcal{C}}(X)$ is satisfiable on G if and only if G belongs to $\mathcal{C}$. The vast majority of well studied graph classes are MSO_2-definable, examples include all graphs with bounded treewidth or bounded genus, bipartite, chordal, and perfect graphs, r-degenerate graphs for any r, trees, cliques, and many others. For our approach we additionally need the graph class to be *monotone* (with respect to edge deletion), that is, if G belongs to the class, then every subgraph of G on $V(G)$ also belongs to the class. Monotone classes are well studied (see for example Rivest and Vuillemin [22]) and examples of monotone MSO_2-definable graph classes include acyclic graphs, bipartite graphs, r-degenerate graphs for any r, or the class of all graphs with genus at most g.

3   Problem Definition

We consider the problem of edge-editing a graph to a disjoint union of two graphs, one from $\mathcal{C}_1$ and the other from $\mathcal{C}_2$. More formally, for a graph class $\mathcal{C}$ and a graph G, by $\text{cost}_{\mathcal{C}}(G)$ we denote the minimum number of edge additions and edge removals required to modify G to a graph in $\mathcal{C}$. In other words, $\text{cost}_{\mathcal{C}}(G)$ is the minimum size of a set of edges F such that the graph with vertex set $V(G)$ and edge set $E(G) \triangle F$ belongs to $\mathcal{C}$, where $\triangle$ denotes the symmetric difference. Note that $\text{cost}_{\mathcal{C}}(G) = 0$ if and only if G belongs to $\mathcal{C}$. If $\mathcal{C}$ does not contain a graph on $|V(G)|$ vertices, then we let $\text{cost}_{\mathcal{C}}(G) = \infty$. Let $\mathcal{C}_1$ and $\mathcal{C}_2$ be two graph classes. Let G be a graph and D and S two induced subgraphs of G whose vertex sets partition $V(G)$. We define $\text{cost}^G_{\mathcal{C}_1, \mathcal{C}_2}(D, S)$ to be the minimum number of edge additions and edge removals required to modify D to a graph in $\mathcal{C}_1$, modify S to a graph in $\mathcal{C}_2$, and to remove all edges between D and S. Formally, $\text{cost}^G_{\mathcal{C}_1, \mathcal{C}_2}(D, S) = \text{cost}_{\mathcal{C}_1}(D) + \text{cost}_{\mathcal{C}_2}(S) + |E_G(D, S)|$. For a graph G, any pair (D, S) of vertex-disjoint induced subgraphs of G such that $V(D) \cup V(S) = V(G)$ is called a *solution* of $(\mathcal{C}_1, \mathcal{C}_2)$-EDITING. *The cost of editing* of G to $(\mathcal{C}_1, \mathcal{C}_2)$ is defined by $\text{cost}_{\mathcal{C}_1, \mathcal{C}_2}(G) = \min_{(D,S)} \text{cost}^G_{\mathcal{C}_1, \mathcal{C}_2}(D, S)$, where the minimum is taken over all solutions. A solution (D, S) of $(\mathcal{C}_1, \mathcal{C}_2)$-EDITING is *optimum* if $\text{cost}^G_{\mathcal{C}_1, \mathcal{C}_2}(D, S) = \text{cost}_{\mathcal{C}_1, \mathcal{C}_2}(G)$. These definitions lead to the following problem.

$(\mathcal{C}_1, \mathcal{C}_2)$-EDITING
Input: Graph G and a natural number k.
Question: Is $\text{cost}_{\mathcal{C}_1, \mathcal{C}_2}(G) \leq k$?

We denote the class of all edgeless graphs by $\mathcal{I}$ and the class of all cliques by $\mathcal{K}$. If $(\mathcal{C}_1, \mathcal{C}_2) = (\mathcal{K}, \mathcal{I})$, then we call $(\mathcal{C}_1, \mathcal{C}_2)$-EDITING problem CLIQUEEDITING and omit $(\mathcal{C}_1, \mathcal{C}_2)$ from the cost functions.

Consider the special case of $(\mathcal{C}, \mathcal{I})$-EDITING. Clearly, for any solution (D, S) we have $\text{cost}^G_{\mathcal{C}, \mathcal{I}}(D, S) = \text{cost}_{\mathcal{C}}(D) + |E(D, S)| + |E(S)|$. Since $|E(D)| + |E(D, S)| + |E(S)| = |E(G)|$, we have

$$\text{cost}^G_{\mathcal{C}, \mathcal{I}}(D, S) = \text{cost}_{\mathcal{C}}(D) + |E(G)| - |E(D)|. \tag{1}$$

For a class of graphs $\mathcal{C}$, the $\mathcal{C}$-COST problem asks whether, given a graph G and a natural number k, it holds that $\text{cost}_{\mathcal{C}}(G) \le k$. For the remainder of the paper, we implicitly assume that the problems $(\mathcal{C}_1, \mathcal{C}_2)$-EDITING and $\mathcal{C}$-COST are parameterized by the cost of editing unless stated otherwise.

4   Graph Classes

In general, we are concerned with editing a given graph to a disjoint union of a dense and a sparse graph; in this section we make the technical requirements on the classes precise.

A class of graphs $\mathcal{D}$ is called a $\mathcal{D}(d, \delta)$-class, or a *dense class*, denoted by $\mathcal{D} \in \mathcal{D}(d, \delta)$, if each graph G from $\mathcal{D}$ satisfies the following two conditions:

($\mathcal{D}1$) For each vertex v of G, the degree of v is at least $|V(G)| - \delta$;
($\mathcal{D}2$) For each vertex v of G we have $\text{cost}_{\mathcal{D}}(G - v) \le d$.

In the absence of significant structural information about the class, the condition ($\mathcal{D}2$) or a similar one seems to be necessary for our approach. In particular, without condition ($\mathcal{D}2$), it would be possible that $\mathcal{C}$ contains a graph H with n vertices, but does not contain any graph on $n - 1$ vertices (or all graphs from $\mathcal{C}$ on $n - 1$ vertices are very far from H in terms of editing cost). In such cases, an optimum solution might be forced to include a costly vertex just to raise the number of vertices in the dense part to n. Consequently, it would not be possible to prove a separation property analogous to the one we will prove in Lemma 7, which is crucial for our results.

On the other hand, the condition ($\mathcal{D}2$) is not particularly restrictive. Indeed, all hereditary graph classes (and in particular all minor-closed graph classes) satisfy ($\mathcal{D}2$) with $d = 0$, which leads us to call classes satisfying ($\mathcal{D}2$) *weakly hereditary*. The common property shared by the sparse classes considered in this paper is the following property, which is in a sense complementary to ($\mathcal{D}2$). A class of graphs $\mathcal{S}$ is called *weakly anti-hereditary* if there is an integer s such that for each graph G from $\mathcal{S}$ and for a vertex v not in G we have $\text{cost}_{\mathcal{S}}(G + v) \le s$, where $G + v$ is the disjoint union of G and the single-vertex graph $\{v\}$. We call a class $\mathcal{S}$ s-weakly anti-hereditary to indicate the smallest integer s for which $\mathcal{S}$ satisfies the definition of weakly anti-hereditary class. Again, being weakly anti-hereditary is not particularly restrictive, since many well studied graph classes are weakly anti-hereditary. Examples include connected, bipartite, r-regular for any r, k-colorable for any k, chordal, perfect, and bounded-genus graphs.

Informally, to guarantee that a weakly anti-hereditary class is indeed sparse, we additionally require either bounded maximum degree, or bounded treewidth (graphs with treewidth ω have at most ωn edges). Specific basic examples of weakly anti-hereditary classes of bounded degree Δ include independent sets ($\Delta = 0$, $s = 0$), matchings covering all but at most one vertex ($\Delta = 1$, $s = 1$), paths ($\Delta = 2$, $s = 1$), cycles ($\Delta = 2$, $s = 3$), disjoint unions of paths ($\Delta = 2$, $s = 0$), disjoint unions of cycles ($\Delta = 2$, $s = 3$), forests with bounded maximum degree Δ ($s = 0$), trees with bounded maximum degree Δ ($s = 1$), and r-regular graphs for any even r ($\Delta = r$, $s = 2r$).[1] Further examples of sparse classes may be obtained by considering any weakly anti-hereditary class and restricting it to graphs with maximum degree Δ for suitably chosen Δ. Similarly, examples of weakly anti-hereditary classes of bounded treewidth include acyclic graphs and any weakly anti-hereditary class restricted to graphs with treewidth at most ω. Examples of dense classes may be obtained by taking complements of bounded degree sparse classes (see the next section for the precise definition). While we postpone verification of the fact that these classes have all the required properties to Sect. 7, the preceding classes can be used by the reader as specific illustrative examples of classes covered by our main theorems in Sect. 6.

5 Editing to a Single Class

To design parameterized algorithms for editing to two graph classes, it is necessary to assume that the cost of single-class editing can be determined efficiently. In this section we introduce the notation for the complexity of single-class editing and present several tools assuring efficient computation of the cost of single-class editing. These results are then used in Sect. 7 to establish the required complexity of editing to the basic examples of graph classes from the previous section. Except for the introduction of our notation in the following paragraph, the reader interested only in the framework for editing into two classes may skip the remainder of this section.

To discriminate among the complexity classes of the cost of single-class editing, we introduce additional notation as follows. By $\mathcal{D}_{\mathbf{P}}(d, \delta)$ we denote the set of all $\mathcal{D}(d, \delta)$ classes $\mathcal{C}$ such that $\mathcal{C}$-COST can be computed in polynomial time. Similarly, by $\mathcal{D}_{\mathbf{FPT}}(d, \delta)$ we denote the set of all $\mathcal{D}(d, \delta)$ classes $\mathcal{C}$ such that $\mathcal{C}$-COST is fixed-parameter tractable. Finally, by $\mathcal{D}_{\mathbf{C}}(d, \delta)$ we denote the set of all $\mathcal{D}(d, \delta)$ classes $\mathcal{C}$ such that $\mathcal{C}$-COST is computable. Observe that $\mathcal{D}_{\mathbf{P}}(d, \delta) \subseteq \mathcal{D}_{\mathbf{FPT}}(d, \delta) \subseteq \mathcal{D}_{\mathbf{C}}(d, \delta)$ and that if $\mathcal{C} \in \mathcal{D}_{\mathbf{FPT}}(d, \delta)$, then it is possible to decide the membership in $\mathcal{C}$ in polynomial time.

The following lemma allows us to considerably weaken the requirements on the complexity of computation of the single-class editing cost function, and thus it may be interesting on its own.

---

[1] Note that in the case of r-regular graphs the smallest graph in the class has $r + 1$ vertices.

Lemma 4. *Let C be a class of graphs with maximum degree at most Δ such that C-COST is fixed-parameter tractable when the input is restricted to graphs with maximum degree at most Δ. Then C-COST is fixed-parameter tractable.*

In particular, we use the preceding lemma to prove Theorem 16 that establishes the complexity of single-class editing of several of the examples given in Sect. 4. While the proof of Theorem 16 treats each class separately and it is rather difficult to envision a general proof, we are able to show in a uniform way that S-COST is fixed-parameter tractable for MSO_2-definable sparse classes of bounded treewidth

Theorem 5. *Let S be a monotone, MSO_2-definable class of bounded treewidth. Then S-COST is fixed-parameter tractable.*

We say that a class C_1 is *complement* of a class C_2 if a graph G lies in C_1 if and only if the complement of G lies in C_2.

Proposition 6. *Let S be a weakly hereditary class of bounded maximum degree. Then the complement $\mathcal{D}$ of S is a $\mathcal{D}(d, \delta)$ class for some d and δ. Furthermore, if S-COST is computable in polynomial time, fixed-parameter tractable, respectively computable, then so is $\mathcal{D}$-COST.*

6   Editing to Two Classes

In this section we give our fixed-parameter (in-)tractability results for variants of the (C_1, C_2)-EDITING problem parameterized by treewidth, the cost of editing, respectively the degeneracy of the input graph.

We make use of the following separation lemma, which provides a lower bound on the minimum degree in the dense part of any optimal solution. While the lemma may seem intuitively clear, it is not necessarily the case. For instance, it is not immediately obvious why the bound should not be a function of degrees in S. However, even for weakly anti-hereditary classes of bounded degree Δ the lemma would be false with the bound $\deg_D(v) \geq (|D| + \Delta - \delta - d - s)/2$.

Lemma 7. *Assume that $\mathcal{D} \in \mathcal{D}(d, \delta)$, S is an s-weakly anti-hereditary class, and let (D, S) be an optimum solution of $(\mathcal{D}, S)$-EDITING on a graph G. Then $\deg_D(v) \geq (|D| - \delta - d - s)/2$ for each vertex v of D.*

6.1   Parameterization by Treewidth

Theorem 8. *Let $\mathcal{D} \in \mathcal{D}_C(d, \delta)$ and let S be a weakly anti-hereditary class. If S is monotone and MSO_2-definable, then $(\mathcal{D}, S)$-EDITING is fixed-parameter tractable parameterized by treewidth.*

Let (G, k) be an instance of $(\mathcal{D}, S)$-EDITING satisfying the assumptions of Theorem 8 and let (D, S) be an optimum solution of (G, k). To prove the theorem we first use Lemma 7 and the fact that a graph with treewidth ω has degeneracy at most ω to show that $|D| \leq 2\omega + d + \delta + s$. Since $\mathcal{S}$ is monotone, no edge is added to S. Therefore, to prove the theorem it is sufficient to show that the problem of finding the dense part D with at most $2\omega + d + \delta + s$ vertices and the set of edges to be deleted from S is fixed-parameter tractable parameterized by treewidth. The proof is concluded by constructing an MSO_2-formula deciding the last problem and using Theorem 3.

6.2   Parameterization by Editing Cost

In this section we show that $(\mathcal{D}, \mathcal{S})$-EDITING is fixed-parameter tractable parameterized by the cost of editing for every $\mathcal{D} \in \mathcal{D}_{\mathbf{FPT}}(d, \delta)$ provided that $\mathcal{S}$ is weakly anti-hereditary and either has bounded maximum degree, or is monotone, MSO_2-definable, and has bounded treewidth. We also obtain polynomial kernels for several cases if $\mathcal{D} \in \mathcal{D}_{\mathbf{P}}(d, \delta)$ and $\mathcal{S}$ contains all graphs with maximum degree at most Δ. Our starting point is the following result for sparse classes with bounded treewidth.

Theorem 9. *Let* $\mathcal{D} \in \mathcal{D}_{\mathbf{FPT}}(d, \delta)$ *and let* $\mathcal{S}$ *be a weakly anti-hereditary class. If* $\mathcal{S}$ *is monotone,* MSO_2*-definable, and has bounded treewidth, then* $(\mathcal{D}, \mathcal{S})$*-* EDITING *is fixed-parameter tractable.*

The proof of Theorem 9 proceeds as follows. Let $\mathcal{D}$ and $\mathcal{S}$ be classes satisfying the assumptions of the theorem, let (G, k) be an instance of $(\mathcal{D}, \mathcal{S})$-EDITING and let (D, S) be an optimum solution of (G, k). We separately treat two cases, namely $|D| \geq 2k + \delta + d + s + 1$ and $|D| \leq 2k + \delta + d + s$. In the first case we construct the largest set of vertices X such that the minimum degree of the subgraph of G induced by X is more than k. We further show that $V(D) \subseteq X$ and that $|X \backslash V(D)|$ is constant (with respect to $\mathcal{D}$ and $\mathcal{S}$). It follows that there are only polynomially many choices for the dense part $\mathcal{D}$ and because $\mathcal{D} \in \mathcal{D}_{\mathbf{FPT}}(d, \delta)$ and Theorem 5, we can compute the cost for each of them in fpt-time.

In the second case, we show that the treewidth of any YES-instance with $|D| \leq 2k + \delta + d + s$ is at most $3k + \delta + d + s + \omega_{\mathcal{S}}$, where $\omega_{\mathcal{S}}$ is the bound on the treewidth of graphs in $\mathcal{S}$. To see this, let (D', S') be a pair of graphs on $V(D)$, respectively $V(S)$, such that there is a set of $cost_{\mathcal{D},\mathcal{S}}(G)$ edge modifications that turn G into disjoint union of D' and S'. Observe that treewidth of D' is at most $2k + \delta + d + s$, the treewidth of S' is at most $\omega_{\mathcal{S}}$, and G differs from the disjoint union of D' and S' by at most k edges. Using Proposition 1 we obtain that the treewidth of G is at most $3k + \delta + d + s + \omega_{\mathcal{S}}$ and we can solve $(\mathcal{D}, \mathcal{S})$-EDITING with the help of Theorem 8.

Theorem 9 implies fixed-parameter tractability of $(\mathcal{D}, \mathcal{S})$-EDITING for sparse classes such as acyclic graphs and series-parallel graphs, as well as for k-colorable graphs and bounded genus graphs of bounded treewidth.

The following theorem, which is our main result for sparse classes of bounded maximum degree, is proved directly by first decomposing the graph according to degrees, using Lemma 7 to restrict the dense part, and then employing the fact that the cost of editing to the sparse class is fixed-parameter tractable.

Theorem 10. *Assume that $\mathcal{D} \in \mathcal{D}_{\mathbf{FPT}}(d, \delta)$ and $\mathcal{S}$ is a weakly anti-hereditary class. If $\mathcal{S}$ has bounded maximum degree and $\mathcal{S}$-COST is fixed-parameter tractable, then $(\mathcal{D}, \mathcal{S})$-EDITING is fixed-parameter tractable.*

In a more restricted setting, we are able to obtain polynomial kernels, as shown by the following theorem.

Theorem 11. *Let $\mathcal{D} \in \mathcal{D}_{\mathbf{P}}(d, \delta)$ and let $\mathcal{S}_\Delta$ be the class containing all graphs with maximum degree at most Δ. If $\Delta = 0$, then $(\mathcal{D}, \mathcal{S}_\Delta)$-EDITING admits a kernel with $O(k)$ vertices and $O(k^2)$ edges. If $\Delta \leq 1$ or $\Delta \geq \max\{\delta, 2d+1, 6\}$, then $(\mathcal{D}, \mathcal{S}_\Delta)$-EDITING admits a kernel with $O(k^2)$ vertices and $O(k^3)$ edges.*

6.3 Parameterization by Degeneracy

In this section we consider $(\mathcal{C}, \mathcal{I})$-EDITING parameterized by the degeneracy of the input graph. Before we can state our results, we need to introduce the SUBGRAPH ISOMORPHISM problem, where given two graphs G and H, one asks whether G contains a subgraph isomorphic to H. In the following we will assume that SUBGRAPH ISOMORPHISM is parameterized by the size of H. We will denote by SI-FPT the set of all classes of graphs such that SUBGRAPH ISOMORPHISM is fixed-parameter tractable whenever the graph G is restricted to come from some class in SI-FPT. We will show that $(\mathcal{C}, \mathcal{I})$-EDITING is fixed-parameter tractable parameterized by degeneracy if the input graph comes from some class in SI-FPT and also if the input graph is bipartite. We also show that already CLIQUEEDITING is W[1]-hard parameterized by degeneracy on general graphs.

Theorem 12. *For any $\mathcal{D}_{\mathbf{C}}(d, \delta)$-class $\mathcal{D}$, $(\mathcal{D}, \mathcal{I})$-EDITING is fixed-parameter tractable parameterized by the degeneracy for input graphs restricted to a class in SI-FPT.*

We want to note here that the above theorem is obtained via a linear time fpt-reduction from $(\mathcal{D}, \mathcal{I})$-EDITING parameterized by degeneracy to SUBGRAPH ISOMORPHISM. This, in particular, implies that if the SUBGRAPH ISOMORPHISM is linear time fixed-parameter tractable, then so is the $(\mathcal{D}, \mathcal{I})$-EDITING problem.

Note that restricting the input graphs to a class in SI-FPT is not particularly restrictive. Indeed, it is known that the SUBGRAPH ISOMORPHISM problem is fpt-equivalent to the model checking problem of existential first-order logic parameterized by the length of the formula [5, Proposition 1]. Therefore, SI-FPT contains also all classes for which the first-order model checking problem parameterized by the length of the formula is fixed-parameter tractable and, in particular, the very general class of nowhere-dense graphs [12]. A particular example of a nowhere-dense class is formed by graphs with bounded expansion, which contain for example all planar graphs, all classes with bounded treewidth, and all classes defined by a finite set of forbidden minors. Since we do not need the precise definition of bounded expansion, which is rather lengthy and technical, we only collect the following two results about classes with bounded expansion: (i) they belong to SI-FPT, and (ii) they have bounded degeneracy.

We refer to a book by Nešetřil and Ossona de Mendez [19] for more details on classes with bounded expansion and related classes in the context of nowhere-dense/somewhere-dense dichotomy. In the light of the preceding discussion, it follows that if input graphs are restricted to any class with bounded expansion, both conditions in Theorem 12 are satisfied, and since SUBGRAPH ISOMORPHISM is linear-time fixed-parameter tractable on graphs of bounded expansion [12], we obtain that $(\mathcal{D}, \mathcal{I})$-EDITING is solvable in linear time.

Corollary 13. *For any class $\mathcal{C}$ of graphs of bounded expansion and any $\mathcal{D} \in \mathcal{D}_{\mathbf{C}}(d, \delta)$, $(\mathcal{D}, \mathcal{I})$-EDITING can be solved in linear time for input graphs restricted to $\mathcal{C}$.*

While on bipartite graphs CLIQUEEDITING is still NP-hard, it appears that requiring the input graphs to be bipartite leads to a somewhat simpler problem, as there is a polynomial-time approximation scheme [15]. We observe a similar behavior also with respect to parameterization — we prove that while CLIQUEEDITING parameterized by degeneracy is in general W[1]-hard, it is fixed-parameter tractable when the input graphs are bipartite.

Theorem 14. CLIQUEEDITING *is linear-time fixed-parameter tractable parameterized by the degeneracy for input graphs restricted to being bipartite.*

Our final result, proved by a reduction from CLIQUE, which is W[1]-hard [21], is that editing to two classes is W[1]-hard parameterized by the degeneracy even for the case of cliques and independent sets.

Theorem 15. CLIQUEEDITING *is* W[1]-*hard parameterized by the degeneracy.*

Proof. To show W[1]-hardness for CLIQUEEDITING parameterized the degeneracy of the input graph, we use a parameterized reduction from the CLIQUE problem, which is well-known to be W[1]-complete [21].

CLIQUE   **Parameter:** k
Input: A graph G and a natural number k.
Question: Does G have a clique of size at least k?

Given G and k we construct a graph H with $\operatorname{degen}(H) \leq 1 + 3\binom{k}{2} + k$ such that G has a clique of size at least k if and only if $\operatorname{cost}(H) \leq b$, where $b = |E(H)| - (3\binom{k}{2} + k - 1) - 8\binom{k}{2}$.

Let G' be the graph obtained from G after subdividing every edge of G three times, i.e., replacing every edge of G by a path consisting of three novel internal vertices. Then the graph H consists of a copy of G' together with a set I of $3\binom{k}{2} + k - 1$ vertices, which are completely connected to all vertices in the copy of G'. Then $\operatorname{degen}(H) \leq 1 + 3\binom{k}{2} + k$ and it remains to show that G has a clique of size at least k if and only if $\operatorname{cost}(H) \leq b$, the proof of which can be found in the full version of the paper.

7   Applications

In this section we outline applications of our results by showing that our examples of sparse and dense classes have the properties required to employ Theorems 8 and 10. Let $\mathcal{G}$ be the set of the following graph classes: independent sets, matchings covering all but at most one vertex, paths, cycles, disjoint unions of paths, disjoint unions of cycles, forests with bounded maximum degree, trees with bounded maximum degree, and r-regular graphs for any even r.

The cost function $\mathrm{cost}_C(G)$ is computable in polynomial time for the class of independent sets, the class of matchings covering all but at most one vertex, and the class of graphs with bounded maximum degree [24]. It is also known that computing $\mathrm{cost}_C(G)$ is fixed-parameter tractable if C is the class of all r-regular graphs [16]. To show that the same holds also for the remaining classes in $\mathcal{G}$, we employ Lemma 4.

Theorem 16. *Let C be a class in $\mathcal{G}$. Then C-COST is fixed-parameter tractable.*

Proof. It is easy to see that all the classes given in $\mathcal{G}$ have bounded maximum degree and are weakly anti-hereditary. It hence remains to show that for all these classes computing the cost of editing is fixed-parameter tractable w.r.t. the editing cost. Because of Lemma 4 it is sufficient to show that for all these classes (having maximum degree at most Δ) computing the cost of editing is fixed-parameter tractable w.r.t. the editing cost if the input graph has maximum degree at most Δ.

Here we only show that this is indeed the case if C is the class of all connected graphs of maximum degree at most $d = \Delta$. The proofs for other cases are similar and can be found in the full version of the paper. Let c_1 be the number of components of G with minimum degree less than Δ and c_2 be the number of components of G with minimum degree at least Δ. Since the maximum degree of G is Δ, each component that is contributing to c_2 is a Δ-regular graph. To edit a graph with $c_1 + c_2$ components into a connected graph $c_1 + c_2 - 1$ edge additions are necessary and sufficient, and to maintain maximum degree Δ it is necessary and sufficient to delete one edge from each component that is contributing to c_2. Therefore, the cost of editing into C is $\mathrm{cost}_C(G) = c_1 + c_2 - 1 + c_2 = c_1 + 2c_2 - 1$. □

To the best of our knowledge, the exact (parameterized) complexity status of computing the cost for the classes in $\mathcal{G}$ is not known. In particular, none of the classes in $\mathcal{G}$ fall into the general setting of classes characterized by finitely many forbidden induced subgraphs considered previously [6]. However, at least for the class of all paths and the class of all cycles there is a straightforward reduction from the HAMILTONIAN PATH or HAMILTONIAN CYCLE problem, respectively, which shows NP-completeness of computing the cost of editing to these classes.

Recall that a class C_1 is the complement of a class C_2 if a graph G lies in C_1 if and only if the complement of G lies in C_2. Let $\mathcal{G}'$ be the set $\mathcal{G}$ without the classes of regular graphs, cycles, and disjoint unions of cycles. Since all the examples of weakly anti-hereditary classes in $\mathcal{G}'$ are also weakly hereditary,

Proposition 6 yields that their complements are examples of dense classes. For the complementary classes of $2r$-regular graphs, cycles, or disjoint unions of cycles to be weakly hereditary, they would need to contain at least one graph on n vertices for each integer n. It is easy to see that if we add to the class of $2r$-regular graphs any graph on n vertices for each $n \leq 2r$, the class becomes weakly hereditary and thus its complement is a dense class. Similarly, if we add to the classes of cycles and disjoint union of cycles a graph with one vertex and a graph with two vertices, then the resulting classes are weakly hereditary and its complement is a $\mathcal{D}(d, \delta)$ class.

Finally, since $\mathcal{S}$-COST is fixed-parameter tractable for each class $\mathcal{S}$ in $\mathcal{G}$ by Theorem 16, Proposition 6 yields that $\mathcal{C}_{\mathcal{S}}$-COST is fixed-parameter tractable for the complement $\mathcal{C}_{\mathcal{S}}$ of each such class $\mathcal{S}$. Therefore, Theorem 8 holds with any class from $\mathcal{G}$ in the role of sparse class and the complement of any class from $\mathcal{G}$' (resp. complements of the extended classes of $2r$-regular graphs, cycles, or disjoint unions of cycles) in the role of dense class. Similarly, Theorem 10 also holds for the same set of complements in the role of dense class.

8   Conclusion

This paper introduces the $(\mathcal{C}_1, \mathcal{C}_2)$-EDITING problem, where the goal is to edit an input graph into a disjoint union of two graphs, one from each class. We investigate for which classes the problem is fixed-parameterized tractable using only limited class-specific structural information. This is achieved by focusing on novel relevant properties, weakly hereditary and weakly anti-hereditary, and a separation property between such classes. While as far as we know weakly hereditary classes were not considered before, they may prove to be an useful concept also for other editing problems, extending the results from hereditary classes. Our results allow us to prove fixed-parameter tractability for a number of interesting sparse classes such as acyclic graphs, bounded degree trees or forests, series-parallel graphs, or k-colorable graphs of bounded treewidth, for instance, bipartite graphs of bounded treewidth. Since this is the first attempt to solve the problem in general, there are many open problems, including other conceivable approaches to choosing the technical requirements on the dense class. Another particular problem left is extending our results for parameterization by the degeneracy beyond independent sets. Our approach uses the equality in Eq. (1) in an essential way, and it seems to be very difficult, and necessary, to calculate the cost of editing to the class exactly when the class is different from independent sets.

References

1. Arnborg, S., Lagergren, J., Seese, D.: Easy problems for tree-decomposable graphs. J. Algorithms **12**(2), 308–340 (1991)
2. Bansal, N., Blum, A., Chawla, S.: Correlation clustering. Mach. Learn. **56**(1–3), 89–113 (2004)

3. Bodlaender, H., Heggernes, P., Lokshtanov, D.: Graph modification problems (Dagstuhl seminar 14071). Dagstuhl Rep. **4**(2), 38–59 (2014)
4. Bodlaender, H.L.: A linear-time algorithm for finding tree-decompositions of small treewidth. SIAM J. Comput. **25**(6), 1305–1317 (1996)
5. Bova, S., Ganian, R., Szeider, S.: Model checking existential logic on partially ordered sets. In: CSL-LICS. ACM (2014)
6. Cai, L.: Fixed-parameter tractability of graph modification problems for hereditary properties. Inform. Process. Lett. **58**(4), 171–176 (1996)
7. Cygan, M., Marx, D., Pilipczuk, M., Pilipczuk, M., Schlotter, I.: Parameterized complexity of eulerian deletion problems. Algorithmica **68**(1), 41–61 (2014)
8. Damaschke, P., Mogren, O.: Editing simple graphs. J. Graph Algorithms Appl. **18**(4), 557–576 (2014)
9. Downey, R.G., Fellows, M.R.: Parameterized Complexity. Monographs in Computer Science. Springer, New York (1999)
10. Flum, J., Grohe, M.: Parameterized Complexity Theory. Texts in Theoretical Computer Science. An EATCS Series, vol. 41. Springer, Heidelberg (2006)
11. Fomin, F., Golovach, P.: Long circuits and large euler subgraphs. SIAM J. Discrete Math. **28**(2), 878–892 (2014)
12. Grohe, M., Kreutzer, S., Siebertz, S.: Deciding first-order properties of nowhere dense graphs. In: STOC. ACM (2014)
13. Hartung, S., Nichterlein, A., Niedermeier, R., Suchý, O.: A refined complexity analysis of degree anonymization in graphs. In: Fomin, F.V., Freivalds, R., Kwiatkowska, M., Peleg, D. (eds.) ICALP 2013, Part II. LNCS, vol. 7966, pp. 594–606. Springer, Heidelberg (2013)
14. Hüffner, F., Komusiewicz, C., Nichterlein, A.: Editing graphs into few cliques: complexity, approximation, and kernelization schemes. In: Dehne, F., Sack, J.-R., Stege, U. (eds.) WADS 2015. LNCS, vol. 9214, pp. 410–421. Springer, Heidelberg (2015)
15. Kováč, I., Selečéniová, I., Steinová, M.: On the clique editing problem. In: Csuhaj-Varjú, E., Dietzfelbinger, M., Ésik, Z. (eds.) MFCS 2014, Part II. LNCS, vol. 8635, pp. 469–480. Springer, Heidelberg (2014)
16. Mathieson, L., Szeider, S.: Editing graphs to satisfy degree constraints: a parameterized approach. J. Comput. Syst. Sci. **78**(1), 179–191 (2012)
17. Moser, H., Thilikos, D.M.: Parameterized complexity of finding regular induced subgraphs. J. Discrete Algorithms **7**(2), 181–190 (2009)
18. Nastos, J., Gao, Y.: Familial groups in social networks. Soc. Netw. **35**(3), 439–450 (2013)
19. Nešetřil, J., de Mendez, P.O.: Sparsity: Graphs, Structures, and Algorithms. Algorithms and Combinatorics, vol. 28. Springer, Heidelberg (2012)
20. Niedermeier, R.: Invitation to Fixed-Parameter Algorithms. Oxford University Press, Oxford (2006)
21. Pietrzak, K.: On the parameterized complexity of the fixed alphabet shortest common supersequence and longest common subsequence problems. J. Comput. Syst. Sci. **67**(4), 757–771 (2003)
22. Rivest, R.L., Vuillemin, J.: On recognizing graph properties from adjacency matrices. Theor. Comput. Sci. **3**(3), 371–384 (1976)
23. Yannakakis, M.: Node-and edge-deletion NP-complete problems. In: STOC, pp. 253–264. ACM (1978)
24. Yannakakis, M.: Edge-deletion problems. SIAM J. Comput. **10**(2), 297–309 (1981)

Containment and Evasion
in Stochastic Point Data

Nirman Kumar$^{(\boxtimes)}$ and Subhash Suri

Department of Computer Science, University of California, Santa-Barbara, USA
{nirman,suri}@cs.ucsb.edu

Abstract. Given two disjoint and finite point sets A and B in $\mathbb{R}^d$, we say that B is *contained* in A if all the points of B lie within the convex hull of A, and that B *evades* A if no point of B lies inside the convex hull of A. We investigate the containment and evasion problems of this type when the set A is stochastic, meaning each of its points a_i is present with an independent probability $\pi(a_i)$. Our model is motivated by situations in which there is uncertainty about the set A, for instance, due to randomized strategy of an adversarial agent or scheduling of monitoring sensors. Our main results include the following: (1) we can compute the exact probability of containment or evasion in two dimensions in worst-case $O(n^4 + m^2)$ time and $O(n^2 + m^2)$ space, where $n = |A|$ and $m = |B|$, and (2) we prove that these problems are #P-hard in 3 or higher dimensions.

1 Introduction

Geometric containment and evasion problems are useful abstractions for a variety of applications, including robotics, pursuit evasion, computer vision and graphics, and security and anomaly detection among others [9,11,13,23]. In the containment problem, we are interested in ensuring that a target set B is *contained* within the convex hull of another point set A, while the evasion problem models the situation from the target's perspective—how to exclude all members of B from the convex hull of A. In pursuit evasion or security-related applications, for instance, B might represent a set of *valuable assets* that require monitoring, and A represents the positions of guarding agents tasked to ensure that all points of B remain surrounded. Conversely, in evasion problems, the goal is to ensure that no member of B is captured or surrounded.

In this paper, we explore the containment and evasion problems of this type when the containing set A is stochastic, meaning each of its points exists with an arbitrary but known probability. Formally, we are given two sets of points $A = \{a_1, \ldots, a_n\}$ and $B = \{b_1, \ldots, b_m\}$ in a d-dimensional space. The set B is deterministic, but A is stochastic, meaning that each of its points a_i is associated with an existence probability $\pi(a_i)$. (The probabilities of different points are

This research was partially supported by NSF grants CCF-1161495 and CCF-1525817.

E. Kranakis et al. (Eds.): LATIN 2016, LNCS 9644, pp. 576–589, 2016.
DOI: 10.1007/978-3-662-49529-2_43

independent, but otherwise arbitrary valued.) We can interpret the stochasticity of A as a randomized strategy of the monitoring agents or, from B's perspective, a probabilistic belief about A's planned deployment. With this input, we want to evaluate the *probability* of B's containment or evasion. In other words, what is the probability that B evades the convex hull of A, or the probability that the convex hull of A contains B?

If B were a singleton point $\{b\}$, then the stochastic containment measures the probability that b lies in the convex hull of the stochastic set A. We recently showed that this *membership* probability can be computed *exactly* in time $O(n \log n)$ when $d = 2$ and in time $O(n^d)$ when $d \geq 3$ [17]. Unfortunately, the containment probabilities for different points of B are not independent, and so we cannot solve the *set containment* problem by solving multiple instances of point containment. In fact, as our results below show, the complexity of the set containment (or evasion) differs sharply from the point containment.

Our Results. We show that the stochastic set containment and set evasion problems are both #P-hard in dimensions $d \geq 3$. In two dimensions, however, we show that both problems admit a polynomial-time algorithm. In particular, we present a dynamic programming algorithm that runs in $O(n^4 + m^2)$ worst-case time and uses $O(n^2 + m^2)$ space.

Related Work. The set containment and disjointness problems are well-studied in computational geometry [9,10,22,23]. When A and B are deterministic sets, these problems can be solved trivially by performing $|B|$ membership queries in the convex hull of A, where each query takes $O(|A|)$ time using fixed-dimensional linear programming algorithm of Megiddo or Clarkson [12,21], although more specialized algorithms are known in 2 or 3 dimensions.

Our work is a contribution to the growing body of research dealing with *uncertain* data, which has received a great deal of attention in recent years in the research communities of databases, machine learning, AI, algorithms and computational geometry. While much of the research in the database community primarily focuses on models, schema, and query evaluation [7,14,24], there is significant algorithmic overlap as well, especially on problems such as indexing, clustering, range searching and skyline computation over uncertain data [2,3,7,8]. Within computational geometry, a number of problems have been addressed, including convex hulls, range searching, skylines, Voronoi diagrams, nearest neighbor searching, and minimum spanning trees [1–5,18–20,25,26].

The work most closely related to the present paper is the problem of computing the probability that a point lies in the convex hull of uncertain points [6,17]. The main result of these papers is that the point-membership probability can be computed in polynomial time $O(n^d)$, in any fixed dimension d. As mentioned earlier, however, the membership probabilities of different points of B being contained in the convex hull of A are *not independent* and therefore these result cannot be applied to our *set containment* or set disjointness problem. Indeed, as our results show these problems are #P-hard, for $d \geq 3$.

Paper Organization. In Sect. 2 we present our dynamic programming algorithm for 2 dimensions. In Sect. 3 we present our hardness results. In Sect. 4 we present a simple Monte-Carlo sampling scheme for estimating the containment and evasion probabilities. We conclude in Sect. 5.

2   The 2-Dimensional Case

In this section we present a polynomial time algorithm to evaluate the evasion and containment probabilities in two dimensions. Since the algorithms for the two problems are quite similar, we focus mainly on the evasion problem, with only a brief discussion of the modifications needed for the containment problem. For ease of presentation, we assume the input points $A \cup B$ are in general position, meaning that no 3 are collinear, but our results hold for the degenerate case as well, with some additional technical details. Without loss of generality, we also assume that no two points have the same x or y coordinates—otherwise, we can rotate the coordinate axes to achieve this.

2.1   Algorithm for the Evasion Problem

Consider a random outcome S of the stochastic point set A. The probability of this outcome is $\mathbf{Pr}[S] = \prod_{a \in S} \pi(a) \prod_{a' \notin S}(1 - \pi(a'))$. For this outcome, it is easy to decide whether any point of B lies inside the convex hull of A. Unfortunately though, there are an exponential number of outcomes, so we cannot afford to enumerate them all, to compute the total probability of B's evasion. As a first simplification, we observe that all the outcomes with the same convex hull P are essentially the same for our purpose—if P does not intersect B then all these outcomes are favorable outcomes, while if P does intersect B, none of these outcomes are favorable. Although we still cannot afford to enumerate all different convex hulls, since their number may be exponential, the idea helps us explain the basic structure of our polynomial time algorithm.

The convex hull of any non-empty outcome is a convex polygon P with vertices among A. Throughout, we let P denote a convex polygon with vertices in A. In order to group over all samples that have a given P as their convex hull, we define the *realization probability* of a polygon P, denoted $h(P)$, as the sum of the probability $\mathbf{Pr}[S]$ of all samples S with convex hull P:

$$h(P) = \sum_{S \text{ s.t. } \mathsf{CH}(S)=P} \mathbf{Pr}[S].$$

In general, there can be an exponential number of samples S that satisfy $\mathsf{CH}(S) = P$ but fortunately there is a simple way to compute $h(P)$. Suppose that the vertices of P are $a_1, \ldots, a_r$. The necessary and sufficient condition for S to satisfy $\mathsf{CH}(S) = P$ is the following:

> The set S includes *all* the vertices of P, namely, $a_1, \ldots, a_r$, and *none* of the points in A lying outside P.

We can use this observation to write our probability as

$$h(P) = \phi\left(\overline{P}\right) \prod_{i=1}^{r} \pi(a_i).$$

where $\overline{P}$ is the shorthand for the region of the plane outside P, and $\phi\left(\overline{P}\right)$ is probability that no point of A lies in it. Throughout the paper, we will use the notation $\phi(R)$ for the probability that no point of A contained in region R is present, with the convention that $\phi(\emptyset) = 1$.

The quantity that we wish to compute is the sum of $h(P)$ over all polygons P that do not intersect B, plus the probability of the empty sample—because in the latter case the hull trivially evades B. This motivates us to define the *likelihood of evasion* of a polygon P, denote $\mathcal{L}(P)$, as $h(P)$ if P does not intersect B and 0 otherwise. Then, the desired probability is given by the following equation, where the first term accounts for the sample S being empty:

$$\prod_{i=1}^{n}(1 - \pi(a_i)) + \sum_{P} \mathcal{L}(P).$$

Evasion Probabilities. We show that $\mathcal{L}(P)$ can be expressed as the product of certain probabilities associated with the edges of P such that these probabilities are independent of the polygon composed from these edges, *subject to the condition that the lowest vertex of P is fixed.* We call these the *evasion probabilities*. In order to keep the notation simple, we assume without loss of generality that a_1 is the lowest point and $a_2, a_3, \ldots, a_n$, are the points above a_1, sorted in counter-clockwise order around a_1. We precede this sequence by a_1 and add a new point a_{n+1} which is a copy of a_1 but is considered "ahead" of the other a_i in counter-clockwise order. Given a polygon P, we first show how to decompose the space outside it, namely, $\overline{P}$, into regions that only depend on the lowest vertex a_1, and the edges. See Fig. 1 for illustration. Consider a polygon P with the lowest vertex a_1. Any edge of P has endpoints a_i and a_j, for some $i < j$. We denote such an edge by the simpler notation $i \to j$. For such an edge, consider the wedge shaped region Z_i^j as shown in Fig. 1. The wedge Z_i^j is a half open wedge defined as follows. It is the polygonal region to the left of the half ray $\overrightarrow{a_1 a_i}$ not including its boundary, to the right of the half ray $\overrightarrow{a_1 a_j}$ including its boundary, and beyond the polygon boundary, i.e., the segment $a_i a_j$ but excluding it. The bounding ray $\overrightarrow{a_1 a_i}$ precedes $\overrightarrow{a_1 a_j}$ in the counter-clockwise order around a_1, and so we can define such a half open wedge by requiring its earlier bounding ray to be excluded, its second bounding ray to be included, and the part on P to be excluded. The same definition extends in fact to the regions Z_1^1, Z_1^j or Z_i^{n+1}. For example the region Z_1^1 associated with the bottom vertex a_1 is the region below a_1 which excludes the horizontal half ray, left of a_1, includes the half ray right of a_1 and excludes a_1. All these regions are disjoint, and Z_i^j only depends on the edge $i \to j$ and the bottom vertex a_1. We, therefore, have the decomposition:

$$\overline{P} = Z_1^1 \cup \bigcup_{i \to j \text{ edge of } P} Z_i^j.$$

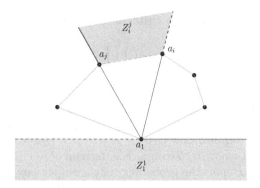

Fig. 1. The regions Z_i^j for a polygon P. For the shaded regions, the excluded boundary portions are shown dashed.

If R_1, R_2 are disjoint regions, then their emptiness probability are independent, meaning that $\phi(R_1 \cup R_2) = \phi(R_1)\phi(R_2)$. We, therefore, have the emptiness probability for $\overline{P}$:

$$\phi(\overline{P}) = \phi(Z_1^1) \times \prod_{i \to j \text{ edge of } P} \phi(Z_i^j). \tag{1}$$

Let us denote the triangle formed by the vertices a_1, a_i, a_j as $\mathsf{T}(i, j)$. A necessary and sufficient condition that $i \to j$ occurs as an edge of a polygon P evading B is that (1) $\mathsf{T}(i, j)$ does not contain any points from B, and, (2) the region Z_i^j does not contain any points from A. Indeed, the first condition is necessary if the polygon P is disjoint from B, and the second condition certifies that no points outside of the triangle can exist if $i \to j$ bounds the convex hull P. The sufficiency is trivial as the triangle $\mathsf{T}(i, j)$ is itself a convex polytope. This observation motivates us to define the evasion probability $L(i, j)$, of the edge $i \to j$ as,

$$L(i, j) = \begin{cases} \pi(a_i) \times \phi(Z_i^j) & \text{if } \mathsf{T}(i, j) \cap B = \emptyset \\ 0 & \text{otherwise.} \end{cases}$$

This definition is valid for $2 \le i < j \le n$, but requires a minor modification when the endpoints include either the index 1 or $n + 1$. In particular, $L(1, j) = \pi(a_1)\phi(Z_1^j)$ and $L(i, n + 1) = \pi(a_i)\phi(Z_i^{n+1})$, because there are no triangles involved here. (Notice that $L(1, j)$ and $L(i, n + 1)$ are never 0.) The evasion probabilities can be used to compute the likelihood of P as the following lemma shows. We assume that P is non-degenerate, meaning that it has at least three vertices. The degenerate case when P consists of two or fewer vertices is easier to handle, and is discussed in the algorithm.

Lemma 1. *Let P be a convex polygon with vertices $a_{\alpha(1)}, a_{\alpha(2)}, \ldots, a_{\alpha(r)}$, for $r \ge 3$, in counter-clockwise order around $a_1 = a_{\alpha(1)}$, and write $a_{\alpha(r+1)} = a_{n+1}$*

which is a copy of a_1. Then, the likelihood $\mathcal{L}(P)$ equals the product of the evasion probabilities of its edges and $\phi\left(Z_1^1\right)$:

$$\mathcal{L}(P) = \phi\left(Z_1^1\right) \times \prod_{i=1}^{r} L(\alpha(i), \alpha(i+1)).$$

Proof. First suppose that P intersects B. Since P can be decomposed via the bottom-vertex triangulation into the triangles, $\mathsf{T}(\alpha(2), \alpha(3)), \ldots, \mathsf{T}(\alpha(r-1), \alpha(r))$, at least one of them must contain a point of B, say $\mathsf{T}(\alpha(j), \alpha(j+1))$. By definition, the evasion probability of this edge $\alpha(j) \to \alpha(j+1)$ is 0, which renders the entire product $\phi\left(Z_1^1\right) \times \prod_{i=1}^{r} L(\alpha(i), \alpha(i+1))$ to zero. On the other hand, if P does not intersect B, then the evasion probabilities of the edges are $\pi(a_{\alpha(j)}) \times \phi\left(Z_{\alpha(j)}^{\alpha(j+1)}\right)$ for $j = 1 \ldots r$. Thus, the product $\phi\left(Z_1^1\right) \times \prod_{i=1}^{r} L(\alpha(i), \alpha(i+1))$ evaluates to

$$\phi\left(Z_1^1\right) \times \prod_{i=1}^{r} \pi(a_{\alpha(i)}) \times \prod_{i=1}^{r} \phi\left(Z_{\alpha(i)}^{\alpha(i+1)}\right).$$

By Eq. 1 this product equals $\phi\left(\overline{P}\right) \prod_{i=1}^{r} \pi(a_{\alpha(i)})$, which is equal to $h(P) = \mathcal{L}(P)$. This completes the proof. □

The Algorithm. Let $\ell(P)$ denote the lowest vertex of a polygon. The probability of evasion that we want to compute is equal to the value of the expression,

$$\prod_{i=1}^{n}(1 - \pi(a_i)) + \sum_{P} \mathcal{L}(P) = \prod_{i=1}^{n}(1 - \pi(a_i)) + \sum_{a_i \in A} \left(\sum_{\ell(P)=a_i} \mathcal{L}(P) \right),$$

where the first term is to account for the empty sample which trivially evades B. The algorithm evaluates the inner sum,

$$\sum_{\ell(P)=a_i} \mathcal{L}(P),$$

for each point a_i as the lowest point, which we now show. For notational simplicity, we use a_1 as the lowest point, with all the remaining points $a_2, \ldots, a_n$ lying above it. We first observe that

$$\sum_{\substack{P \\ \ell(P)=a_1}} \mathcal{L}(P) = \sum_{\substack{\text{degenerate } P \\ \ell(P)=a_1}} \mathcal{L}(P) + \sum_{\substack{\text{non-degenerate } P \\ \ell(P)=a_1}} \mathcal{L}(P).$$

To evaluate the first term we notice that it is equal to the probability of all 1 point and 2 point subsets that include a_1—by our general position assumptions the convex hull of any such subset always evade B. Thus we have

$$\sum_{\substack{\text{degenerate } P \\ \ell(P)=a_1}} \mathcal{L}(P) = \phi\left(Z_1^1\right)\left(\pi(a_1)\prod_{i=2}^{n}(1-\pi(a_i)) + \sum_{i=2}^{n}\pi(a_1)\pi(a_i)\prod_{\substack{j=2 \\ j\neq i}}^{n}(1-\pi(a_j))\right),$$

where the outer $\phi\left(Z_1^1\right)$ is to account for the fact that no points below a_1 can exist in the sample. We now turn to evaluating the sum over the non-degenerate polygons P. In this case, by Lemma 1, the likelihood $\mathcal{L}(P)$ is the product of the evasion probabilities of all the edges and $\phi\left(Z_1^1\right)$. The term $\phi\left(Z_1^1\right)$ is computed easily in $O(n\log n)$ time using plane sweep. The main part of the algorithm is to compute the sum over all non-degenerate polygons, with lowest vertex a_1, the product of edge evasion probabilities, for which we use a dynamic programming algorithm. Assume that the evasion probabilities of all the edges $i \rightarrow j$ for $1 \leq i < j \leq n+1$ have been computed. For a non-degenerate polygon P we can order the triangles of its bottom-vertex triangulation counter-clockwise in a natural fashion. We define the function $F(i,j)$, for $2 \leq i < j \leq n$, as follows,

$F(i,j) =$ the sum of $\mathcal{L}(P)$ over all polygons P with last triangle $\mathsf{T}(i,j)$.

We now set up a recursive definition for this function. Suppose first that $i = 2$, i.e., the last triangle has as a vertex the point a_2, which is first in the counter-clockwise order around a_1. In this case, it is easy to see that the only such polygon is itself the triangle $\mathsf{T}(i,j)$, and so we have

$$F(2,j) = \mathcal{L}(\mathsf{T}(2,j)) = \phi\left(Z_1^1\right) \times L(1,2) \times L(2,j) \times L(j,n+1).$$

For other i, consider any polygon P with last triangle $\mathsf{T}(i,j)$. Either the entire polygon is just this triangle, i.e., $P = \mathsf{T}(i,j)$, or else, such a polygon is made up of a polygon P' suffixed with the last triangle $\mathsf{T}(i,j)$. See Figure on the right. Indeed, given such a P, we can let P' be the (convex) polygon without the last triangle $\mathsf{T}(i,j)$. Observe that in this case, the triangle $\mathsf{T}(k,i)$ that precedes $\mathsf{T}(i,j)$, must be such that the edge $k \rightarrow i$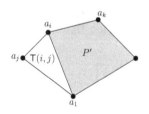
makes a left turn into the edge $i \rightarrow j$. Conversely, given any polygon P' with last triangle $\mathsf{T}(k,i)$, such that $k \rightarrow i$ turns left into $i \rightarrow j$, we can suffix the triangle $\mathsf{T}(i,j)$ to make up a polygon P with last triangle $\mathsf{T}(i,j)$. Thus, there is a one-to-one correspondence between polygons P with last two triangles as $\mathsf{T}(i,j)$ and $\mathsf{T}(k,i)$, and polygons P' with last triangle $\mathsf{T}(k,i)$, whenever $k \rightarrow i$ turns left into $i \rightarrow j$. Observe that the edge $i \rightarrow (n+1)$ is the last edge of P' but this is not present in P, which has two additional edges $i \rightarrow j$ and $j \rightarrow (n+1)$. By using Lemma 1 on P and P' we observe that,

$$\mathcal{L}(P) = \frac{\mathcal{L}(P')}{L(k,n+1)} \times L(i,j) \times L(j,n+1).$$

It follows that when $i > 2$ we have the following recursive definition for $F(i,j)$.

$$F(i,j) = \mathcal{L}(\mathsf{T}(i,j)) + L(j,n+1) \times L(i,j) \times \sum_{k \rightarrow i \rightarrow j \text{ is a left turn}} \frac{F(k,i)}{L(k,n+1)}.$$

The function $F(i, j)$ can be evaluated using dynamic programming if we consider edges successively in the order where edges with smaller i occur earlier—for a fixed i they can be ordered arbitrarily. We omit the routine technical details. The evaluation of $\mathcal{L}(\mathsf{T}(i, j))$ can be done using Lemma 1. Summing over all possible last triangles of non-degenerate polygons P, we have the desired sum:

$$\sum_{\substack{\text{non-degenerate } P \\ \ell(P)=a_1}} \mathcal{L}(P) = \sum_{2 \le i < j \le n} F(i, j).$$

Finally, we now show how to compute the edge evasion probabilities. In order to compute $\phi\left(Z_i^j\right)$, essentially we need to be able to answer queries of the following type: given a triangle (with vertices among the points of A), find the product of $(1 - \pi(a))$ for all points a inside the triangle. This follows because because the regions Z_i^j can be decomposed into a constant number of such triangles and we can compute $\phi\left(Z_i^j\right)$ by taking the product of the numbers obtained for the corresponding triangles. The half openness of the edges does not really present us with problems as there are no points of A on these edges (except those belonging to the edge $i \to j$, and these points are known—a_i and a_j), so we might as well consider them as closed edges for our purposes. We can compute all these quantities in $O(1)$ time per triangle after a one-time $O(n^2)$ preprocessing using ideas from [16, 26]. To determine which edges have evasion probability 0 we use triangle emptiness queries, for triangles $\mathsf{T}(i, j)$ and for points of B. Each query takes $O(1)$ time, after a one-time $O((n + m)^2) = O(n^2 + m^2)$ preprocessing, using the data structure of [16].

Complexity Analysis. The computation of the $\phi\left(Z_i^i\right)$ can be done by first sorting the set A by the x_2 coordinate, and then performing a linear scan. Consider a fixed lowest point a_i, and the computation of the evasion probabilities of the edges among points above a_i. For each such edge, we require $O(1)$ queries on the data structure [16, 26] and 1 query to the triangle emptiness data structure from [16]; in total $O(1)$ per edge . Thus, in $O(n^2)$ total time we can determine all the edge evasion probabilities. The dynamic programming algorithm then takes $O(n^3)$ time. Thus for a fixed lowest point a_i, our algorithm takes $O(n^3)$ time, and so the overall time over all choices of the lowest point is $O(n^4 + m^2)$. The total space requirement is $O(n^2 + m^2)$, dominated by the data structures from [16], which need to be prepared only once. In each iteration with a fixed lowest point a_i we need $O(n^2)$ space to store the evasion probabilities of the edges, which is reused for different a_i.

We have established the following result.

Theorem 1. *Given a set A of n stochastic points, and a set B of m points in the plane, all in general position, we can compute in $O(n^4 + m^2)$ time and $O(n^2 + m^2)$ space the probability that A evades B.*

2.2   The Containment Problem

The algorithm for the containment problem is similar but some key concepts need to be redefined. Specifically,

(I) the region Z_i^i needs to be empty of points in B. Otherwise, we may ignore the point a_i and any higher points in the overall summation.

(II) the "containment probability" of an edge $i \rightarrow j$ is set to 0 if the region Z_i^j contains a point of B. This ensures that $\overline{P}$ is disjoint from B, which is equivalent to $B \subseteq P$.

Our result on the stochastic containment can be stated as follows.

Theorem 2. *Given a set A of n stochastic points and a set B of m points in the plane, all in general position, we can compute in $O(n^4 + m^2)$ time and $O(n^2 + m^2)$ space the probability that A contains B.*

3   Hardness in Higher Dimensions

In this section, we show that the evasion problem is #P-hard in three or higher dimensions. The stochastic set containment problem is also hard, as shown in Appendix A.

Our reduction is from the #P-hard problem of counting independent sets in planar graphs [27]. Let $G = (V, E)$ be an instance of the independent set problem, with $V = [n]$ and $m = |E|$. Corresponding to each vertex $i \in [n]$, we create a point $a_i \in A$, with associated probability $\pi(a_i) = 1/2$, and for each edge $(i, j) \in E$, we create a a corresponding point $b_{ij} \in B$. For a subset $V' \subseteq V$, we let $A' \subseteq A$ denote the corresponding set of points. The crucial property of the reduction is that V' is an independent set in G if and only if $\mathsf{CH}(A') \cap B = \emptyset$.

Suppose for now that the point set A can be constructed in such a way that all the points a_i are vertices of their convex hull P (as such all of them are also in general position), and that for each $(i, j) \in E$, the segment $a_i a_j$ is an edge of P. (In other words, the polytope P is an embedding of G.) If we choose the points of B as $b_{ij} = (a_i + a_j)/2$, then we can prove the following result.

Lemma 2. *A subset $V' \subseteq V$ is an independent set of G iff the corresponding set of points A' satisfies $\mathsf{CH}(A') \cap B = \emptyset$.*

Proof. If $\mathsf{CH}(A') \cap B = \emptyset$, then clearly both a_i, a_j cannot be in A' for any edge (i, j) of G—otherwise the midpoint $b_{ij} = (a_i + a_j)/2$ will lie inside $\mathsf{CH}(A')$. Conversely, let V' be an independent set in G, with A' the corresponding subset of points in A. We show that any point $b_{ij} \in B$ must lie outside $\mathsf{CH}(A')$. Since $a_i a_j$ is an edge of the polytope P there is a hyperplane H, that is a supporting hyperplane of $\mathsf{CH}(A')$ and intersects $\mathsf{CH}(A')$ in the edge $a_i a_j$, i.e., $P \cap H = a_i a_j$, and all other vertices lie in one of the open halfspaces of H. Suppose both i and j are not in V'. Then, H shifted slightly towards $\mathsf{CH}(A')$ will act as a separating hyperplane, separating b_{ij} from $\mathsf{CH}(A')$. If V' includes one of them, say i, then H

rotated slightly (while still containing b_{ij}), so that i, j remain on the appropriate sides (i.e., i on the same side as P, and j on the other side), and then shifted towards $\mathsf{CH}(A')$ will act as a separating hyperplane.   □

Constructing the Polytope. We now discuss how to construct the desired set of points A. In 4 dimensions, this is possible for any graph G—simply choose the points a_i on the moment curve $(\gamma, \gamma^2, \gamma^3, \gamma^4)$, for $\gamma = 1, 2, \ldots, n$, and verify that every pair is an edge of the convex hull [28]. In 3 dimensions, such a construction is not feasible for an arbitrary graph G, but can be achieved when G is *planar*. In particular, we can assume that G is maximally planar—otherwise add edges until it becomes maximal—and then use the well-known theorem of Steinitz to realize a 3-dimensional embedding of this 3-connected graph [28]. Furthermore, using a result of [15], we can also construct such a 3-dimensional polytope in polynomial time because our graph is a *triangulation* (maximally planar). Moreover, in this construction, there is a one-to-one correspondence between vertices of G and the constructed polytope.

Since each point of A occurs with probability $1/2$, the probability of any sample S is precisely $1/2^n$. Therefore, we now have a one-to-one correspondence between independent sets in G and samples A' with $\mathsf{CH}(A') \cap B = \emptyset$. We can therefore count the number of independent sets in G, as follows:

$$\mathbf{Pr}[\mathsf{CH}(A') \cap B = \emptyset] = \frac{1}{2^n} \times (\text{Number of independent sets in } G).$$

We have established the following result.

Theorem 3. *Given a stochastic point set A in $\mathbb{R}^d$, for $d \geq 3$, the problem of computing the evasion probability of another (deterministic) point set $B \subset \mathbb{R}^d$ is #P-hard.*

Remark 1. The point sets constructed in our hardness proof are not in general position (for example the point b_{ij} is collinear with a_i and a_j), but they can be modified to achieve non-degeneracy. The details are mostly technical, and omitted from this abstract.

4   Approximation

Given the #P-hardness of both the containment and the evasion problems, it is natural to explore efficient approximation schemes. In the following we briefly discuss a Monte-Carlo algorithm for approximating the probabilities of the complement of the evasion problem and that of the containment problem, under the assumption that the the probabilities of the stochastic points are not too small, namely, each point occurs with probability at least α, for some $\alpha > 0$.

Lemma 3. *Let $A = \{a_1, \ldots, a_n\}$ be a set of n stochastic points, and B be a set of m points. Let p denote the probability of the complement of the evasion problem (resp. the containment problem) , i.e., $p = \mathbf{Pr}[B \cap \mathsf{CH}(S) \neq \emptyset]$ (resp.,*

$p = \mathbf{Pr}[B \subseteq \mathsf{CH}(S)]$, for a random sample S of A. Suppose, $\pi(a_i) \geq \alpha$ for some constant $\alpha > 0$, for every $1 \leq i \leq n$. Then, for $0 < \varepsilon \leq 1, 0 < \delta \leq 1$ we can compute in time $O(1/\varepsilon^2 \log(1/\delta)mnM)$, a number q, such that $(1 - \varepsilon)p \leq q \leq (1 + \varepsilon)p$ with probability at least $1 - \delta$, where $M = O(1/\alpha^{d+1})$ for the evasion problem and $M = O(1/\alpha^{m(d+1)})$ for the containment problem.

Proof. The algorithm generates $O(1/\varepsilon^2 \log(1/\delta)M)$ samples independently, where in a sample S we choose the $a_i \in A$ independently with probability $\pi(a_i)$, and for each of these samples S, tests in time $O(mn)$ whether $B \cap \mathsf{CH}(S) \neq \emptyset$, for the complement of the evasion (resp. $B \subseteq \mathsf{CH}(S)$, for the containment) problem. These tests are easy to do as one can test each point b of B for containment in the convex hull $\mathsf{CH}(S)$ by using a $O(n)$ time algorithm for fixed dimension linear programming [12,21]. Consider the indicator random variable X which assumes 1 if the relevant condition for the evasion (resp. the containment) problem holds, and 0 otherwise, and we output the mean over all our samples as our estimate. If $p = 0$, clearly we output 0, as all tests will have $X = 0$. On the other hand if $p \neq 0$, then observe that for the complement of the evasion problem, the probability that $\mathsf{CH}(S) \cap B \neq \emptyset$ is at least the (nonzero) probability that for some i, $a_i \in \mathsf{CH}(S)$. This probability is at least $\mu \geq \alpha^{d+1}$ as a successful draw consists of choosing all the points in some simplex containing a_i (such a simplex must exist). Similarly, for the containment problem there must be m such simplices, one for each of the m points, and the probability of a successful draw is at least $\mu \geq \alpha^{m(d+1)}$. In each draw, we have $E[X] = \mu$. Finally, by an application of Chernoff bound we have the result.   □

5   Conclusions

An immediate open problem is to improve the running time and space requirements of our algorithms in 2 dimensions. Another open problem is to achieve an efficient *multiplicative factor* approximation for the containment or evasion probabilities in dimensions $d \geq 3$.

A   The Containment Problem Is #P-hard for $d \geq 3$

Theorem 4. *Given a stochastic point set A, and another point set B, in $\mathbb{R}^d$ for $d \geq 3$, the problem of computing the probability that A contains B is #P-hard.*

Proof. We reduce from the #P-hard problem of counting vertex covers in planar graphs. In any graph, a subset of vertices is a vertex cover iff its complement is an independent set. As such the problem of counting vertex covers is equivalent to counting independent sets, and moreover, the problem is hard on any class of graphs on which the problem of counting independent sets is hard. In particular, it is hard on planar graphs. We use the transformation used in the proof of Theorem 3. More specifically, given a planar graph $G = (V, E)$ on $V = [n]$, it constructs a graph $G = (V, E')$ where $E \subseteq E'$ and a set A of n points

$\{a_1, \ldots, a_n\}$, which are the vertices of a convex body, such that every edge (i,j) in E' corresponds to an edge (a_i, a_j) of the polytope. We assign the probability $1/2$ to each of these stochastic points. Let $P = \mathsf{CH}(A)$ denote the polytope. We now construct a new polytope $P' \subseteq P$, where each edge e of P will have a corresponding (2 dimensional) facet f_e in P' "close" to e, i.e., lying inside P and in the close vicinity of e. To this end, we consider a supporting plane H_e such that $H_e \cap P = e$. We now move all such planes H_e, for each edge e slightly, shifting it parallel to itself, inside the polytope by a distance that is smaller than $1/2$ of the minimum nonzero distance of any vertex of P to any of the planes H_e. Let H'_e denote the shifted plane. As a result, we get a new polytope, $\bigcap H'_e$ which lies inside P and where now for each edge e of P, the portion of H'_e lying inside P' is a facet. This is the facet f_e corresponding to e. Clearly, P' as defined, can be computed in polynomial time given P.

Consider an edge $(i,j) \in E$, and the corresponding edge (a_i, a_j) of P. We now add 3 points to A and one point to B, for each such edge, as follows. Choose 3 points $b'_{ij}, b''_{ij}, b'''_{ij}$ in general position on the facet f_e of P'.

We add them to A and assign them the probability 1 each. The simplices $b'_{ij} b''_{ij} b'''_{ij} a_i$, and $b'_{ij} b''_{ij} b'''_{ij} a_j$ share a base, and are on the same side of it, and therefore they share a point that is interior to both of them. We choose such a point b_{ij} and include it in B. See Figure on the right for an example of this construction in 2 dimensions.

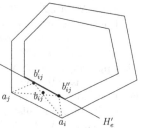

It is clear that if $A' \subseteq A$ excludes both a_i, a_j, then H'_e is a hyperplane that separates b_{ij} from $\mathsf{CH}(A')$. Conversely, if at least one of a_i, a_j are included, then b_{ij} must be in $\mathsf{CH}(A')$. Therefore, if $\mathsf{CH}(A')$ contains all such points b_{ij}, the corresponding vertices forms a vertex cover in G. It may be observed that since all the new points added in A have probability 1 they do not affect our counting argument.   □

References

1. Abdullah, A., Daruki, S., Phillips, J.M.: Range counting coresets for uncertain data. In: Proceedings of the 29th Annual Symposium Computational Geometry, pp. 223–232. ACM (2013)
2. Afshani, P., Agarwal, P.K., Arge, L., Larsen, K.G., Phillips, J.M.: (Approximate) uncertain skylines. Theory Comput. Syst. **52**(3), 342–366 (2013)
3. Agarwal, P.K., Aronov, B., Har-Peled, S., Phillips, J.M., Yi, K., Zhang, W.: Nearest neighbor searching under uncertainty II. In: Proceedings of the 32nd ACM Symposium Principles Database Systems, pp. 115–126 (2013)
4. Agarwal, P.K., Cheng, S.W., Yi, K.: Range searching on uncertain data. ACM Trans. Algorithms **8**(4), 43:1–43:17 (2012)
5. Agarwal, P.K., Efrat, A., Sankararaman, S., Zhang, W.: Nearest-neighbor searching under uncertainty. In: Proceedings of the 31st ACM Symposium Principles Database Systems, pp. 225–236. ACM (2012)

6. Agarwal, P.K., Har-Peled, S., Suri, S., Yıldız, H., Zhang, W.: Convex hulls under uncertainty. In: Proceedings of the 22nd Annual European Symposium on Algorithms, pp. 37–48 (2014)
7. Aggarwal, C.C.: Managing and Mining Uncertain Data. Springer, US (2009)
8. Aggarwal, C.C., Yu, P.S.: A survey of uncertain data algorithms and applications. IEEE Trans. Knowl. Data Eng. **21**(5), 609–623 (2009)
9. de Berg, M., Cheong, O., van Kreveld, M., Overmars, M.: Computational Geometry: Algorithms and Applications, 3rd edn. Springer, Heidelberg (2008)
10. Chazelle, B.: The polygon containment problem. In: Preparata, F.P. (ed.) Advances in Computing Research, vol. 1, pp. 1–33. JAI Press (1983)
11. Chung, T.H., Hollinger, G.A., Isler, V.: Search and pursuit-evasion in mobile robotics-A survey. Auton. Robots **31**(4), 299–316 (2011)
12. Clarkson, K.L.: Linear programming in $o(n3^{d^2})$ time. Inform. Process. Lett. **22**, 21–24 (1986)
13. Costa, G.B.P., Ponti, M., Frery, A.C.: Partially supervised anomaly detection using convex hulls on a 2D parameter space. In: Zhou, Z.-H., Schwenker, F. (eds.) PSL 2013. LNCS, vol. 8183, pp. 1–8. Springer, Heidelberg (2013)
14. Dalvi, N., Ré, C., Suciu, D.: Probabilistic databases: diamonds in the dirt. Commun. ACM **52**(7), 86–94 (2009)
15. Das, G., Goodrich, M.T.: On the complexity of optimization problems for 3-dimensional convex polyhedra and decision trees. Comput. Geom. Theory Appl. **8**(3), 123–137 (1997)
16. Eppstein, D., Overmars, M., Rote, G., Woeginger, G.: Finding minimum area k-gons. Discrete Comput. Geom. **7**(1), 45–58 (1992)
17. Fink, M., Hershberger, J., Kumar, N., Suri, S.: Hyperplane separability and convexity of probabilistic points (2015) (unpublished manuscript)
18. Jørgensen, A., Löffler, M., Phillips, J.M.: Geometric computations on indecisive and uncertain points. CoRR abs/1205.0273 (2012)
19. Kamousi, P., Chan, T.M., Suri, S.: Stochastic minimum spanning trees in Euclidean spaces. In: Proceedings of the 27th Annual Symposium Computational Geometry, pp. 65–74 (2011)
20. Kamousi, P., Chan, T.M., Suri, S.: Closest pair and the post office problem for stochastic points. Comput. Geom. Theory Appl. **47**(2), 214–223 (2014)
21. Megiddo, N.: Linear programming in linear time when the dimension is fixed. J. ACM **31**(1), 114–127 (1984)
22. Milenkovic, V.J.: Translational polygon containment and minimal enclosure using linear programming based restriction. In: Proceedings of the 28th Annual ACM Symposium on Theory of Computing, pp. 109–118. ACM (1996)
23. Preparata, F.P., Shamos, M.I.: Computational Geometry: An Introduction. Springer, New York (1985)
24. Sarma, A.D., Benjelloun, O., Halevy, A.Y., Nabar, S.U., Widom, J.: Representing uncertain data: models, properties, and algorithms. VLDB J. **18**(5), 989–1019 (2009)
25. Suri, S., Verbeek, K.: On the most likely voronoi diagram and nearest neighbor searching. In: Ahn, H.-K., Shin, C.-S. (eds.) ISAAC 2014. LNCS, vol. 8889, pp. 338–350. Springer, Heidelberg (2014)
26. Suri, S., Verbeek, K., Yıldız, H.: On the most likely convex hull of uncertain points. In: Bodlaender, H.L., Italiano, G.F. (eds.) ESA 2013. LNCS, vol. 8125, pp. 791–802. Springer, Heidelberg (2013)

27. Vadhan, S.P.: The complexity of counting in sparse, regular, and planar graphs. SIAM J. Comput. **31**(2), 398–427 (2001)
28. Ziegler, G.M.: Lectures on Polytopes. Graduate Texts in Mathematics. Springer, New York (1995)

Tree Compression Using String Grammars

Moses Ganardi[(✉)], Danny Hucke, Markus Lohrey, and Eric Noeth

University of Siegen, Siegen, Germany
{ganardi,hucke,lohrey,eric.noeth}@eti.uni-siegen.de

Abstract. We study the compressed representation of a ranked tree by a straight-line program (SLP) for its preorder traversal string, and compare it with the previously studied representation by straight-line context-free tree grammars (also known as tree straight-line programs or TSLPs). Although SLPs may be exponentially more succinct than TSLPs, we show that many simple tree queries can still be performed efficiently on SLPs, such as computing the height of a tree, tree navigation, or evaluation of Boolean expressions. Other problems like pattern matching and evaluation of tree automata become intractable.

1 Introduction

The idea of *grammar-based compression* is to represent a given string s by a small context-free grammar that generates only s; such a grammar is also called a *straight-line program* (SLP) for s. By repeated doubling, it is easy to produce a string of length 2^n by an SLP of size n (measured as the total length of all right-hand sides of the productions), i.e., exponential compression can be achieved in the best case. The goal of grammar-based compression is to construct from a given string s a small SLP for s. Whereas computing a smallest SLP for a given string is not possible in polynomial time unless $\mathsf{P} = \mathsf{NP}$ [9,28], there exist several linear time algorithms that produce grammars that are at worst $\mathcal{O}(\log(N/g))$ larger than the size of a smallest SLP, where N is the length of the input string s and g is the size of a smallest SLP for s [9,18,26].

Motivated by applications like XML processing, where large tree-structured data occur, grammar-based compression has been extended to trees, see [24] for a survey. Unless otherwise specified, a tree in this paper is always a rooted ordered tree over a ranked alphabet, i.e., every node is labelled with a symbol and the rank of this symbol is equal to the number of children of the node. This class of trees occurs in many different contexts like term rewriting, expression evaluation and tree automata. A tree over a ranked alphabet is uniquely represented by its preorder traversal. For instance, the preorder traversal of the tree $f(g(a), f(a, b))$ is the string $fgafab$. It is now a natural idea to apply a string compressor to this preorder traversal. In this paper we study the compression of ranked trees by SLPs for their preorder traversals. This idea is very similar to [6], where unranked unlabelled trees are compressed by SLPs for their balanced parenthesis representations.

The third and fourth author are supported by the DFG-project LO 748/10-1.

© Springer-Verlag Berlin Heidelberg 2016
E. Krankis et al. (Eds.): LATIN 2016, LNCS 9644, pp. 590–604, 2016.
DOI: 10.1007/978-3-662-49529-2_44

In Sect. 3 we compare the size of SLPs for preorder traversals with other grammar-based compressed tree representations. SLPs for strings can also be generalized directly to trees, using context-free tree grammars that produce a single tree (so called tree straight-line programs, briefly TSLPs). TSLPs generalize dags (directed acyclic graphs), which are widely used as a compact tree representation. Whereas dags only allow to share repeated subtrees, TSLPs can also share repeated internal tree patterns. The algorithm from [13] produces for every tree over a fixed ranked alphabet a TSLP of size $\mathcal{O}(N/\log N)$, which is worst-case optimal. A grammar-based tree compressor using TSLPs with an approximation ratio of $\mathcal{O}(\log N)$ can be found in [19]. It was shown in [7] that from a given TSLP $\mathbb{A}$ of size m for a tree t one can efficiently construct an SLP of size $\mathcal{O}(m \cdot r)$ for the preorder traversal of t, where r is the maximal rank occurring in t (i.e. the maximal number of children of a node). Hence a smallest SLP for the traversal of t cannot be much larger than a smallest TSLP for t. Our first main result shows that SLPs can be exponentially more succinct than TSLPs: We construct a family of binary trees t_n ($n \geq 0$) such that the size of a smallest SLP for the traversal of t_n is polynomial in n but the size of a smallest TSLP for t_n is $\Omega(2^{n/2})$. Moreover, we also construct a family of binary trees t_n ($n \geq 0$) such that the size of a smallest SLP for the preorder traversal of t_n is polynomial in n but the size of a smallest SLP for the balanced parenthesis representation is $\Omega(2^{n/2})$. It remains open whether a family of trees with the opposite behavior exists.

We also study algorithmic problems for SLP-compressed trees. We extend some of the results from [6] on querying SLP-compressed balanced parenthesis representations to our context. Specifically, we show that after a linear time preprocessing we can navigate (i.e., move to the parent node and to the kth child), compute lowest common ancestors and subtree sizes in time $\mathcal{O}(\log N)$, where N is the size of the tree represented by the SLP. For a couple of other problems (computation of the tree's height, the depth of a node and evaluation of Boolean expressions) we provide polynomial time algorithms for the case that the input tree is given by an SLP for the preorder traversal. On the other hand, there exist problems that are polynomial time solvable for TSLP-compressed trees but intractable for SLP-compressed trees: examples for such problems are pattern matching, evaluation of max-plus expressions, and membership for tree automata. Looking at tree automata is also interesting when compared with the situation for explicitly given (i.e., uncompressed) preorder traversals. For these, evaluating Boolean expressions (which is the membership problem for a particular tree automaton) is NC^1-complete by a famous result of Buss [8], and the NC^1 upper bound was generalized to every fixed tree automaton [21]. If we compress the preorder traversal by an SLP, the problem is still solvable in polynomial time for Boolean expressions (Theorem 13), but there is a fixed tree automaton with a PSPACE-complete evaluation problem (Theorem 16).

Missing proofs can be found in the long version [14].

Related Work on Tree Compression. There are also tree compressors based on other grammar formalisms. In [1] so called elementary ordered tree grammars are used, and a polynomial time compressor with an approximation ratio of $\mathcal{O}(N^{5/6})$ is presented. Also the *top dags* from [5] can be seen as a variation of TSLPs for unranked trees. Recently, in [15] it was shown that for every tree of size N with σ many node labels, the top dag has size $\mathcal{O}(N \cdot \log\log_\sigma N/\log_\sigma N)$, which improved the bound from [5]. An extension of TSLPs to higher order tree grammars was proposed in [20].

Another class of tree compressors use succinct data structures for trees. Here, the goal is to represent a tree in a number of bits that asymptotically matches the information theoretic lower bound, and at the same have efficient querying. For unlabelled (resp., node-labelled) unranked trees of size N there exist representations with $2N + o(N)$ bits (resp., $(2 + \log\sigma) \cdot N + o(N)$ bits, where σ is the number of node labels) that support navigation and some other tree queries in time $\mathcal{O}(1)$ [3, 12, 16, 17, 25].

2   Preliminaries

Let Σ be a finite alphabet. For a string $w = a_1 \cdots a_N \in \Sigma^*$ we define $|w| = N$, $w[i] = a_i$ and $w[i : j] = a_i \cdots a_j$ where $w[i : j] = \varepsilon$, if $i > j$. Let $w[: i] = w[1 : i]$ and $w[i :] = w[i : |w|]$. With $\mathrm{rev}(w) = a_N \cdots a_1$ we denote w reversed. For $u, v \in \Sigma^*$, the *convolution* $u \otimes v \in (\Sigma \times \Sigma)^*$ is the string of length $\min\{|u|, |v|\}$ defined by $(u \otimes v)[i] = (u[i], v[i])$ for $1 \le i \le \min\{|u|, |v|\}$.

We assume familiarity with basic complexity classes like P, NP and PSPACE. The counting class #P contains all functions $f : \Sigma^* \to \mathbb{N}$ for which there is a nondeterministic polynomial time machine M such that for all $x \in \Sigma^*$, $f(x)$ is the number of accepting computation paths of M on input x. The class PP contains all problems A for which there is a nondeterministic polynomial time machine M such that for all inputs x: $x \in A$ iff more than half of all computation paths of M on input x are accepting. When referring to linear time algorithms, we assume the standard RAM model of computation, where registers can hold numbers with $\mathcal{O}(\log n)$ bits for n the input size, and arithmetic operations on register values can be done in constant time.

A *ranked alphabet* $\mathcal{F}$ is a finite set of symbols, where every $f \in \mathcal{F}$ has a rank $\mathrm{rank}(f) \in \mathbb{N}$. By $\mathcal{F}_n$ we denote the symbols of $\mathcal{F}$ of rank n. We assume that $\mathcal{F}_0 \ne \emptyset$. Later we will also allow ranked alphabets where $\mathcal{F}_0$ is infinite. For the purpose of this paper, it is convenient to define trees as particular strings over the alphabet $\mathcal{F}$ (namely as preorder traversals). The set $\mathcal{T}(\mathcal{F})$ of all *trees* over $\mathcal{F}$ is the subset of $\mathcal{F}^*$ defined inductively as follows: If $f \in \mathcal{F}_n$ with $n \ge 0$ and $t_1, \ldots, t_n \in \mathcal{T}(\mathcal{F})$, then also $f t_1 \cdots t_n \in \mathcal{T}(\mathcal{F})$ (we denote this tree also with $f(t_1, \ldots, t_n)$, which corresponds to the standard term notation). A string $s \in \mathcal{F}^*$ is a *fragment* if there exist a tree $t \in \mathcal{T}(\mathcal{F})$ and a non-empty string $x \in \mathcal{F}^+$ such that $sx = t$. Note that the empty string ε is a fragment. Intuitively, a fragment is a tree with gaps. For every non-empty fragment $s \in \mathcal{F}^+$ there is a unique $n \ge 1$ such that $\{x \in \mathcal{F}^* \mid sx \in \mathcal{T}(\mathcal{F})\} = (\mathcal{T}(\mathcal{F}))^n$; this n is denoted with $\mathrm{gaps}(s)$. We set $\mathrm{gaps}(\varepsilon) = 0$. Since $\mathcal{T}(\mathcal{F})$ is prefix-free we have:

Fig. 1. The tree t from Example 2 and the tree fragment corresponding to $ffaafff$.

Lemma 1. *For every $w \in \mathcal{F}^*$ there exist unique $n \geq 0$, $t_1, \ldots, t_n \in \mathcal{T}(\mathcal{F})$ and a unique fragment $s \in \mathcal{F}^*$ such that $w = t_1 \cdots t_n s$.*

Let $w \in \mathcal{F}^*$ and let $w = t_1 \cdots t_n s$ as in Lemma 1. We define $c(w) = (n, \mathsf{gaps}(s))$. The number n counts the number of full trees in w and $\mathsf{gaps}(s)$ is the number of trees that are missing in order to make the fragment s a tree.

We also consider trees in their graph-theoretic interpretation where the set of nodes of a tree t is the set of positions $\{1, \ldots, |t|\}$ of the string t. The root node is 1. If t factorizes as $uft_1 \cdots t_n v$ for $u, v \in \mathcal{F}^*$, $f \in \mathcal{F}_n$, and $t_1, \ldots, t_n \in \mathcal{T}(\mathcal{F})$, then the n children of node $|u| + 1$ are $|u| + 2 + \sum_{i=1}^{k} |t_i|$ for $0 \leq k \leq n - 1$. We define the depth of a node in t (number of edges from the root to the node) and the height of t (maximal depth of a node) as usual. Note that the tree t as a string is simply the preorder traversal of the tree t seen in its standard graph-theoretic interpretation. Since for a ranked tree the number of children of a node is uniquely determined by the node label, a tree (in the above graph-theoretic interpretation) is uniquely determined by its preorder traversal and vice versa.

Example 2. Let $t = ffaafffaaaa = f(f(a,a), f(f(f(a,a),a),a))$ be the tree depicted in Fig. 1 with $f \in \mathcal{F}_2$ and $a \in \mathcal{F}_0$. Its height is 4. All prefixes (including the empty word, excluding the full word) of t are fragments. The fragment $s = ffaafff$ is also depicted in Fig. 1 in a graphical way. The dashed edges visualize the gaps. We have $\mathsf{gaps}(s) = 4$. For the factor $u = aafffa$ of t we have $c(u) = (2, 3)$. The children of node 5 (the third f-labelled node) are 6 and 11.

A *straight-line program*, briefly SLP, is a context-free grammar that produces a single string. Formally, it is a tuple $\mathbb{A} = (N, \Sigma, P, S)$, where N is a finite set of nonterminals, Σ is a finite set of terminals such that $\Sigma \cap N = \emptyset$, $S \in N$ is the start nonterminal, and P is a finite set of productions (or rules) of the form $A \to w$ for $A \in N$, $w \in (N \cup \Sigma)^*$ such that: (i) For every $A \in N$, there exists exactly one production of the form $A \to w$, and (ii) the binary relation $\{(A, B) \in N \times N \mid (A \to w) \in P,\ B \text{ occurs in } w\}$ is acyclic. Every nonterminal $A \in N$ produces a unique string $\mathsf{val}_{\mathbb{A}}(A) \in \Sigma^*$. The string defined by $\mathbb{A}$ is $\mathsf{val}(\mathbb{A}) = \mathsf{val}_{\mathbb{A}}(S)$. We omit the subscript $\mathbb{A}$ when it is clear from the context. The *size* of the SLP $\mathbb{A}$ is $|\mathbb{A}| = \sum_{(A \to w) \in P} |w|$. An SLP for a nonempty word can be transformed in linear time into *Chomsky normal form*, i.e., for each production $A \to w$, either $w \in \Sigma$ or $w = BC$ where $B, C \in N$. The following lemma summarizes known results about SLPs which we will use throughout the paper, see e.g. [23].

Lemma 3. *Let* $\mathbb{A}$ *be an SLP. There are algorithms running in time* $\mathcal{O}(|\mathbb{A}|)$ *for the following problems (the numbers i and j are given in binary encoding):*

1. *Compute the set of symbols occurring in* $\mathsf{val}(\mathbb{A})$.
2. *Let Σ be the terminal set of $\mathbb{A}$ and let $\Gamma \subseteq \Sigma$. Compute the number of occurrences of symbols from Γ in* $\mathsf{val}(\mathbb{A})$.
3. *Let Σ be the terminal set of $\mathbb{A}$ and let $\Gamma \subseteq \Sigma$. Given a number i, compute the position of the ith occurrence of a symbol from Γ in* $\mathsf{val}(\mathbb{A})$ *(if it exists).*
4. *Given $1 \leq i, j \leq |\mathsf{val}(\mathbb{A})|$, compute an SLP of size $\mathcal{O}(|\mathbb{A}|)$ for* $\mathsf{val}(\mathbb{A})[i:j]$.

We want to compress trees (viewed as particular strings) by SLPs. This leads to the question whether a given SLP produces a tree, which is also known as the compressed membership problem for the language $\mathcal{T}(\mathcal{F}) \subseteq \mathcal{F}^*$. By computing bottom-up for each nonterminal A the pair $c(\mathsf{val}(A))$, we can show:

Theorem 4. *Given an SLP $\mathbb{A}$, one can check in time $\mathcal{O}(|\mathbb{A}|)$ whether* $\mathsf{val}(\mathbb{A}) \in \mathcal{T}(\mathcal{F})$.

Note that $\mathcal{T}(\mathcal{F})$ is context-free. In general the compressed membership problem for context-free languages belongs to PSPACE and there is a deterministic context-free language with a PSPACE-complete compressed membership problem [22].

Tree straight-line programs (briefly TSLPs) generalize SLPs to trees [13,19]. In addition to terminals and nonterminals, the productions of a TSLP also contain so called parameters $x_1, x_2, x_3, \ldots$, which are treated as symbols of rank zero (i.e., they only label leaves). Formally, a TSLP is a tuple $\mathbb{A} = (\mathcal{V}, \mathcal{F}, P, S)$, where $\mathcal{V}$ (resp., $\mathcal{F}$) is a ranked alphabet of nonterminals (resp., terminals), $S \in \mathcal{V}_0$ is the start nonterminal and P is a finite set of productions of the form $A(x_1, \ldots, x_n) \to t$ (which is also briefly written as $A \to t$), where $n \geq 0$, $A \in \mathcal{V}_n$ and $t \in \mathcal{T}(\mathcal{F} \cup \mathcal{V} \cup \{x_1, \ldots, x_n\})$ is a tree in which every parameter x_i ($1 \leq i \leq n$) occurs at most once, such that: (i) For every $A \in \mathcal{V}_n$ there exists exactly one production of the form $A(x_1, \ldots, x_n) \to t$, and (ii) the binary relation $\{(A, B) \in \mathcal{V} \times \mathcal{V} \mid (A \to t) \in P, B \text{ is a label in } t\}$ is acyclic. These conditions ensure that exactly one tree $\mathsf{val}_{\mathbb{A}}(A) \in \mathcal{T}(\mathcal{F} \cup \{x_1, \ldots, x_n\})$ is derived from every nonterminal $A \in \mathcal{V}_n$ by using the rules as rewriting rules in the usual sense. As for SLPs, we omit the subscript $\mathbb{A}$ when the context is clear. The tree defined by $\mathbb{A}$ is $\mathsf{val}(\mathbb{A}) = \mathsf{val}_{\mathbb{A}}(S)$. The *size* $|\mathbb{A}|$ of a TSLP is the total number of non-parameter nodes in all right-hand sides of productions; see [13] for a justification of this. TSLPs in which every nonterminal has rank 0 correspond to dags (the nodes of the dag are the nonterminals of the TSLP).

3   Relative Succinctness of SLP-Compressed Trees

In [7] it is shown that a TSLP $\mathbb{A}$ for a tree t can be transformed into an SLP of size $\mathcal{O}(|\mathbb{A}| \cdot r)$ for (the traversal of) t, where r is the maximal rank of a label in t. In this section we discuss the other direction, i.e., transforming an SLP into a

TSLP. For tree families of unbounded maximal rank, SLPs can trivially achieve exponentially better compression: The size of the smallest TSLP for $t_n = f_n a^n$ (with $f_n \in \mathcal{F}_n$) is $n+1$, whereas the size of the smallest SLP for t_n is in $\mathcal{O}(\log n)$. Note that this does not contradict the $\mathcal{O}(\frac{n}{\log n})$ bound from [13] since the trees t_n have unbounded rank. It is less obvious that such an exponential gap can also occur with trees of bounded rank. To show this, we use the following result:

Theorem 5. ([4, Theorem 2]). *For every $n > 0$, there exist words $u_n, v_n \in \{0,1\}^*$ with $|u_n| = |v_n|$ such that u_n and v_n have SLPs of size $n^{\mathcal{O}(1)}$, but the smallest SLP for the convolution $u_n \otimes v_n$ has size $\Omega(2^{n/2})$.*

For two words $u = i_1 \cdots i_n \in \{0,1\}^*$ and $v = j_1 \cdots j_n \in \{0,1\}^*$ we define the *comb tree* $t(u,v) = f_{i_1}(f_{i_2}(\ldots f_{i_n}(\$, j_n)\ldots j_2), j_1)$ over the ranked alphabet $\{f_0, f_1, 0, 1, \$\}$ where f_0, f_1 have rank 2 and $0, 1, \$$ have rank 0.

Theorem 6. *For every $n > 0$ there exists a tree t_n such that the size of a smallest SLP for t_n is polynomial in n, but the size of a smallest TSLP for t_n is in $\Omega(2^{n/2})$.*

Proof sketch. Let $t_n = t(u_n, v_n)$ be the comb tree, where u_n, v_n are from Theorem 5. These words have SLPs of size $n^{\mathcal{O}(1)}$, which yield an SLP of size $n^{\mathcal{O}(1)}$ for t_n. On the other hand, one can transform a TSLP for t_n of size m into an SLP of size $\mathcal{O}(m)$ for $u_n \otimes v_n$, which implies the result.   □

Note that the height of the tree t_n in Theorem 6 is linear in the size of t_n. By the following result, large height and rank are always responsible for the exponential succinctness gap between SLPs and TSLPs.

Theorem 7. *Let $t \in \mathcal{T}(\mathcal{F})$ be a tree of height h and maximal rank r, and let $\mathbb{A}$ be an SLP for t. Then there exists a TSLP $\mathbb{B}$ with $\mathsf{val}(\mathbb{B}) = t$ such that $|\mathbb{B}| \in \mathcal{O}(|\mathbb{A}| \cdot h \cdot r)$, which can be constructed in time $\mathcal{O}(|\mathbb{A}| \cdot h \cdot r)$.*

Proof sketch. Without loss of generality we assume that $\mathbb{A}$ is in Chomsky normal form. Consider a nonterminal A of $\mathbb{A}$ with $c(A) = (a_1, a_2)$. This means that $\mathsf{val}(A) = t_1 \cdots t_{a_1} s$, where the t_i is a full tree and s is a fragment with a_2 many gaps. For the TSLP $\mathbb{B}$, we introduce (i) a_1 nonterminals $A_1, \ldots, A_{a_1}$ of rank 0, which produce the trees $t_1, \ldots, t_{a_1}$, and (ii), if $a_2 > 0$, one nonterminal A' of rank a_2 for the fragment s. For every rule of the form $A \to f$ with $f \in \mathcal{F}_n$ we add to $\mathbb{B}$ the TSLP-rule $A_1 \to f$ if $n = 0$ or $A'(x_1, \ldots, x_n) \to f(x_1, \ldots, x_n)$ if $n \geq 1$. For a rule of the form $A \to BC$ with $c(B) = (b_1, b_2)$ and $c(C) = (c_1, c_2)$, one has to distinguish the cases $b_2 = 0$, $0 < b_2 \leq c_1$, and $b_2 > c_1$. In each of these cases it is straightforward to define the rules in such a way that A_i derives t_i and, in case $a_2 > 0$), A' produces the fragment s. Finally, it is easy to achieve the size bound $\mathcal{O}(|\mathbb{A}| \cdot h \cdot r)$ in the construction for $\mathbb{B}$.   □

Balanced parenthesis sequences are widely used as a succinct representation of ordered unranked unlabelled trees [25]. One defines the balanced parenthesis sequence $\mathsf{bp}(t)$ of such a tree t inductively as follows. If t consists of a single node, then $\mathsf{bp}(t) = ()$. If the root of t has n children in which the subtrees $t_1, \ldots, t_n$ are rooted (from left to right), then $\mathsf{bp}(t) = (\mathsf{bp}(t_1) \cdots \mathsf{bp}(t_n))$. Using a construction similar to the proof of Theorem 6 we can show:

Theorem 8. *For every $n > 0$ there exists a binary tree $t_n \in \mathcal{T}(\{a, f\})$ (where f has rank 2 and a has rank 0) such that the size of a smallest SLP for t_n is polynomial in n, but the size of a smallest SLP for $\mathsf{bp}(t_n)$ is in $\Omega(2^{n/2})$.*

It remains open whether there is also a family of trees where the opposite situation arises, i.e., where a smallest SLP for the balanced parenthesis sequence is exponentially smaller than a smallest SLP for the preorder traversal.

4 Algorithmic Problems on SLP-Compressed Trees

For trees given by TSLPs or other compressed representations, various algorithmic questions have been studied in the literature [5, 15, 24, 27]. Here, we study the complexity of several basic algorithmic problems on trees that are represented by SLPs. In this context the main difficulty for SLPs in contrast to TSLPs is that the tree structure is only given implicitly by the ranked alphabet.

4.1 Tree Navigation and Pattern Matching

In [6] it is shown that from an SLP of size n that produces the balanced parenthesis representation of an unranked tree t of size N, one can compute in time $\mathcal{O}(n)$ a data structure of size $\mathcal{O}(n)$ that supports navigation as well as other important computations (e.g. lowest common ancestors) in time $\mathcal{O}(\log N)$. Here, the word RAM model is used, where memory cells can store numbers with $\log N$ bits and arithmetic operations on $\log N$-bit numbers can be carried out in constant time. An analogous result was shown in [5] for top dags. Here, we show the same result for SLPs that produce (preorder traversals of) ranked trees. Recall that we identify the nodes of a tree t with the positions $1, \ldots, |t|$ in the string t. The proof of the following result combines results from [3, 6, 17] and uses a correspondence between preorder traversals of ranked trees and the DFUDS (depth-first unary-degree sequence) representation of unranked trees from [3].

Theorem 9. *Given an SLP of size n for a tree t of size N, one can produce in time $\mathcal{O}(n)$ a data structure of size $\mathcal{O}(n)$ that allows to do the following computations in time $\mathcal{O}(\log N) \leq \mathcal{O}(n)$, where $i, j, k \in \mathbb{N}$ with $1 \leq i, j \leq N$ are given in binary notation:*

(a) Compute the parent node of node $i > 1$ in t.
(b) Compute the kth child of node i in t, if it exists.
(c) Compute the number k such that $i > 1$ is the kth child of its parent node.
(d) Compute the size of the subtree rooted at node i.
(e) Compute the lowest common ancestor of node i and j in t.

The data structure of [6] allows to compute the height and the depth of a given tree node in time $\mathcal{O}(\log N)$ as well. It is not clear to us whether this result also can be extended to our setting. On the other hand, in Sect. 4.2, we show that the height and the depth of a given node of an SLP-compressed tree can be computed in polynomial time.

In contrast to navigation, simple pattern matching problems are intractable for SLP-compressed trees. The *pattern matching problem for SLP-compressed trees* is defined as follows: Given a tree $s \in T(\mathcal{F} \cup \mathcal{X})$ (the *pattern*), where every variable $x \in \mathcal{X}$ (a symbol of rank zero) occurs at most once, and an SLP $\mathbb{A}$ producing a tree $t \in T(\mathcal{F})$, is there a substitution $\sigma : \mathcal{X} \to T(\mathcal{F})$ such that $\sigma(s)$ is a subtree of t? Here, $\sigma(s) \in T(\mathcal{F})$ denotes the tree obtained from s by substituting each variable $x \in \mathcal{X}$ by $\sigma(x)$. Note that the pattern is given uncompressed. If the tree t is given by a TSLP, the corresponding problem can be solved in polynomial time [27].[1] For SLP-compressed trees we have:

Theorem 10. *The pattern matching problem for SLP-compressed trees is* NP-*complete. Moreover,* NP-*hardness holds for a fixed pattern of the form* $f(x, a)$.

NP-hardness is shown by a reduction from the question whether $(1, 1)$ appears in the convolution of two SLP-compressed strings over $\{0, 1\}$ [23, Theorem 3.13].

4.2   Tree Evaluation Problems

The algorithmic difficulty of SLP-compressed trees already becomes clear when computing the height. For TSLPs it is easy to see that the height of the produced tree can be computed in linear time: Compute bottom-up for each nonterminal the height of the produced tree and the depths of the parameter nodes. However, this direct approach fails for SLPs since each nonterminal encodes a possibly exponential number of trees. The crucial observation to solve this problem is that one can store and compute the required information for each nonterminal in a compressed form.

In the following we present a general framework to define and solve evaluation problems on SLP-compressed trees. We assign to each alphabet symbol of rank n an n-ary operator which defines the value of a tree by evaluating it bottom-up. This approach includes natural tree problems like computing the height of a tree, evaluating a Boolean expression or determining whether a fixed tree automaton accepts a given tree. We only consider operators on $\mathbb{Z}$ but other domains with an appropriate encoding of the elements are also possible. To be able to consider arbitrary arithmetic expressions properly, it is necessary to allow the set $\mathcal{F}_0 \subseteq \mathcal{F}$ of constants to be an infinite subset of $\mathbb{Z}$. If such a constant $a \in \mathcal{F}_0$ appears in an SLP for a tree, then its contribution to the SLP size is the number of bits of the binary representation of a.

---

[1] In fact, there is a polynomial time algorithm that checks whether a TSLP-compressed pattern tree s occurs in a TSLP-compressed tree t [27]. But for this, it is important that every variable x occurs at most once in the pattern s. For the case that variables are allowed to occur repeatedly in the pattern, the precise complexity is open.

Let $\mathcal{D} \subseteq \mathbb{Z}$ be a possibly infinite set of integers and let $\mathcal{F}$ be a ranked alphabet with $\mathcal{F}_0 = \mathcal{D}$. An *interpretation* $\mathcal{I}$ *of* $\mathcal{F}$ *over* $\mathcal{D}$ assigns to each symbol $f \in \mathcal{F}_n$ an n-ary function $f^{\mathcal{I}} : \mathcal{D}^n \to \mathcal{D}$ with the restriction that $a^{\mathcal{I}} = a$ for all $a \in \mathcal{D}$. We lift the definition of $\mathcal{I}$ to $\mathcal{T}(\mathcal{F})$ inductively by $(f\, t_1 \cdots t_n)^{\mathcal{I}} = f^{\mathcal{I}}(t_1^{\mathcal{I}}, \ldots, t_n^{\mathcal{I}})$, where $f \in \mathcal{F}_n$ and $t_1, \ldots, t_n \in \mathcal{T}(\mathcal{F})$. The problem $\mathcal{I}$*-evaluation for SLP-compressed trees* is: Given an SLP $\mathbb{A}$ over $\mathcal{F}$ with $\mathsf{val}(\mathbb{A}) \in \mathcal{T}(\mathcal{F})$, compute $\mathsf{val}(\mathbb{A})^{\mathcal{I}}$.

In a first step, we reduce $\mathcal{I}$-evaluation for SLP-compressed trees to the corresponding problem for SLP-compressed caterpillar trees. A tree $t \in \mathcal{T}(\mathcal{F})$ is called a *caterpillar tree* if every node has at most one child which is not a leaf. Let $s \in \mathcal{F}^*$ be an arbitrary string. Then $s^{\mathcal{I}} \in \mathcal{F}^*$ denotes the unique string obtained from s by replacing every maximal substring $t \in \mathcal{T}(\mathcal{F})$ of s by its value $t^{\mathcal{I}}$. By Lemma 1 we can factorize s uniquely as $s = t_1 \cdots t_n u$ where $t_1, \ldots, t_n \in \mathcal{T}(\mathcal{F})$ and u is a fragment. Hence $s^{\mathcal{I}} = m_1 \cdots m_n u^{\mathcal{I}}$ with $m_1, \ldots, m_n \in \mathcal{D}$. Since u is a fragment, the string $u^{\mathcal{I}}$ is the fragment of a caterpillar tree (briefly, *caterpillar fragment*). For instance, with the standard interpretation of $+$ and $\times$ on integers, we have $(0, 2, +, 2, +, +, \times, 2, +, 2, 1, +, \times)^{\mathcal{I}} = 0, 2, +, 2, +, +, 6, +, \times$ (commas are added for better readability).

Our reduction to caterpillar trees only works for interpretations $\mathcal{I}$ that are *polynomially bounded* in the following sense: There exist constants $\alpha, \beta \geq 0$ such that for every tree $t \in \mathcal{T}(\mathcal{F})$, $\mathsf{abs}(t^{\mathcal{I}}) \leq \left(\beta \cdot |t| + \sum_{i \in L} \mathsf{abs}(t[i])\right)^{\alpha}$, where $\mathsf{abs}(z)$ is the absolute value of $z \in \mathbb{Z}$ (we write $\mathsf{abs}(z)$ instead of $|z|$ in order to not get confused with the size $|t|$ of a tree) and $L \subseteq \{1, \ldots, |t|\}$ is the set of leaves of t. The purpose of this definition is to ensure that for every SLP $\mathbb{A}$ for a tree t, the length of the binary encoding of $t^{\mathcal{I}}$ is polynomially bounded in $|\mathbb{A}|$ and the binary lengths of the integer constants that appear in $\mathbb{A}$.

Theorem 11. *Let* $\mathcal{I}$ *be a polynomially bounded interpretation. Then the* $\mathcal{I}$*-evaluation for SLP-compressed trees is polynomial time Turing-reducible to the* $\mathcal{I}$*-evaluation for SLP-compressed caterpillar trees.*

Proof. In the proof we use an extension of SLPs by the cut-operator, called *composition systems*. A *composition system* $\mathbb{A} = (N, \Sigma, P, S)$ is an SLP where P may also contain rules of the form $A \to B[i : j]$ where $A, B \in N$ and $i, j \geq 0$. Here we let $\mathsf{val}(A) = \mathsf{val}(B)[i : j]$. It is known (see e.g. [23]) that a given composition system can be transformed in polynomial time into an SLP with the same value. We may also use more complex rules like for instance $A \to B[i : j]C[k : l]$. Such rules can be easily reduced to the above format.

Let $\mathbb{A} = (N, \mathcal{F}, P, S)$ be the input SLP in Chomsky normal form. We compute a composition system, which contains for each nonterminal $A \in N$ two nonterminals A_1 and A_2 such that the following holds: Assume that $\mathsf{val}(A) = t_1 \cdots t_n s$, where $t_1, \ldots, t_n \in \mathcal{T}(\mathcal{F})$ and s is a fragment (hence $c(\mathsf{val}(A)) = (n, \mathsf{gaps}(s))$). Then we will have $\mathsf{val}(A_1) = t_1^{\mathcal{I}} \cdots t_n^{\mathcal{I}} \in \mathcal{D}^*$ and $\mathsf{val}(A_2) = s^{\mathcal{I}}$. In particular, $\mathsf{val}(A_1)\mathsf{val}(A_2) = \mathsf{val}(A)^{\mathcal{I}}$ and $\mathsf{val}(\mathbb{A})^{\mathcal{I}}$ is given by a single number in $\mathsf{val}(S_1)$. The fact that $\mathcal{I}$ is polynomially bounded ensures that all numbers $t_i^{\mathcal{I}}$ as well as all numbers that appear in the caterpillar tree $s^{\mathcal{I}}$ have polynomially many bits in the input length $|\mathbb{A}|$.

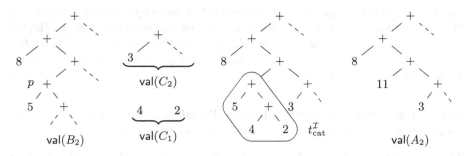

Fig. 2. An example for the case $b_2 > c_1$ in the proof of Theorem 11 ($+$ is interpreted as addition). Inserting the values from $\mathsf{val}(C_1) = 4\,2$ into the caterpillar fragment $\mathsf{val}(B_2) = + + 8 + + 5 +$ produces a caterpillar subtree t_{cat}, which evaluates to 11. Then, the fragment $\mathsf{val}(C_2) = +3$ is appended, which yields $\mathsf{val}(A_2) = + + 8 + 11 + 3$.

The computation is straightforward for rules $A \to f$ with $A \in N$ and $f \in \mathcal{F}$: If $\mathrm{rank}(f) = 0$, then $\mathsf{val}(A_1) = f$ and $\mathsf{val}(A_2) = \varepsilon$. If $\mathrm{rank}(f) > 0$, then $\mathsf{val}(A_1) = \varepsilon$ and $\mathsf{val}(A_2) = f$. For a nonterminal $A \in N$ with the rule $A \to BC$ we make a case distinction depending on $c(\mathsf{val}(B)) = (b_1, b_2)$ and $c(\mathsf{val}(C)) = (c_1, c_2)$.

Case $b_2 \le c_1$: Then concatenating $\mathsf{val}(B)$ and $\mathsf{val}(C)$ yields a new tree t_{new} (or ε if $b_2 = 0$) in $\mathsf{val}(A)$. Notice that $t_{\mathrm{new}}^{\mathcal{I}}$ is the value of the tree $\mathsf{val}(B_2)\,\mathsf{val}(C_1)[: b_2]$. Hence we can compute $t_{\mathrm{new}}^{\mathcal{I}}$ in polynomial time by computing an SLP that produces $\mathsf{val}(B_2)\,\mathsf{val}(C_1)[: b_2]$ and querying the oracle for caterpillar trees. We add the rules $A_1 \to B_1 t_{\mathrm{new}}^{\mathcal{I}} C_1[b_2 + 1 : c_1]$, $A_2 \to C_2$ to the composition system.

Case $b_2 > c_1$: Then all trees and the fragment produced by C are inserted into the gaps of the fragment encoded by B. If $c_1 = 0$ (i.e., $\mathsf{val}(C_1) = \varepsilon$), then we add the productions $A_1 \to B_1$ and $A_2 \to B_2C_2$. Now assume that $c_1 > 0$. Consider the fragment $s = \mathsf{val}(B_2)\,\mathsf{val}(C_1)\,\mathsf{val}(C_2)$. Intuitively, this fragment s is obtained by taking the caterpillar fragment $\mathsf{val}(B_2)$, where the first c_1 many gaps are replaced by the constants from the sequence $\mathsf{val}(C_1)$ and the $(c_1 + 1)^{\mathrm{st}}$ gap is replaced by the caterpillar fragment $\mathsf{val}(C_2)$, see Fig. 2 for an example. If s is not already a caterpillar fragment, then we have to replace the (unique) largest factor of s which belongs to $T(\mathcal{F})$ by its value under $\mathcal{I}$ to get $s^{\mathcal{I}}$. To do so we proceed as follows: Consider the tree $t' = \mathsf{val}(B_2)\,\mathsf{val}(C_1)\diamond^{b_2 - c_1}$, where $\diamond$ is an arbitrary symbol of rank 0, and let $r = |\mathsf{val}(B_2)| + c_1 + 1$ (the position of the first $\diamond$ in t'). Let q be the parent node of r, which can be computed in polynomial time by Theorem 9. Using Lemma 3 we compute the position p (which is marked in the left tree in Fig. 2) of the first occurrence of a symbol in $t'[q + 1 :]$ with rank > 0. If no such symbol exists, then s is already a caterpillar fragment and we add the rules $A_1 \to B_1$ and $A_2 \to B_2C_1C_2$ to the composition system. Otherwise p is the first symbol of the largest factor from $T(\mathcal{F})$ described above. Using Theorem 9(d), we can compute in polynomial time the last position p' of the subtree of t' that is rooted in p. Note that the position p must belong to $\mathsf{val}(B_2)$ and that p' must belong to $\mathsf{val}(C_1)$ (since $c_1 > 0$). The string

$t_{\text{cat}} = (\mathsf{val}(B_2)\,\mathsf{val}(C_1))[p : p']$ is a caterpillar tree for which we can compute an SLP in polynomial time by the above remark on composition systems. Hence, using the oracle we can compute the value $t_{\text{cat}}^{\mathcal{I}}$. We then add the rules $A_1 \to B_1$, $A' \to B_2 C_1$, and $A_2 \to A'[: p - 1]\, t_{\text{cat}}^{\mathcal{I}}\, A'[p' + 1 :]\, C_2$ to the composition system. This completes the proof.   □

Polynomial Time Solvable Evaluation Problems. Next, we present several applications of Theorem 11. We start with the height of a tree.

Theorem 12. *The height of a tree $t \in \mathcal{T}(\mathcal{F})$ given by an SLP and the depth of a given node in t can be computed in polynomial time.*

Proof. We can assume that t is not a single constant. We replace every symbol in $\mathcal{F}_0$ by the integer 0. Then the height of t is given by its value under the interpretation $\mathcal{I}$ with $f^{\mathcal{I}}(a_1, \ldots, a_n) = 1 + \max\{a_1, \ldots, a_n\}$ for symbols $f \in \mathcal{F}_n$ with $n > 0$. Clearly $\mathcal{I}$ is polynomially bounded. By Theorem 11 it is enough to show how to evaluate a caterpillar tree t given by an SLP $\mathbb{A}$ in polynomial time under the interpretation $\mathcal{I}$. But note that arbitrary natural numbers may occur at leaf positions in this caterpillar tree.

Let $\mathcal{D}_t = \{d \in \mathbb{N} \mid d \text{ labels a leaf of } t\}$. The size of this set is bounded by $|\mathbb{A}|$. For $d \in \mathcal{D}_t$ let v_d be the deepest node such that d is the label of a child of node v_d (in particular, v_d is not a leaf). Let us first argue that v_d can be computed in polynomial time: Let k be the maximal position in t where a symbol of rank larger than zero occurs. The number k is computable in polynomial time by Lemma 3 (point 2 and 3). Again using Lemma 3 we compute the position of d's last (resp., first) occurrence in $t[: k]$ (resp., $t[k + 1 :]$). Then using Theorem 9 we compute the parent nodes of those two nodes. The larger (i.e., deeper one) is v_d.

Assume that $\mathcal{D}_t = \{d_1, \ldots, d_m\}$, where w.l.o.g. $v_{d_1} < v_{d_2} < \cdots < v_{d_m}$ (if $v_{d_i} = v_{d_j}$ for $d_i < d_j$, then we simply ignore d_i in the following consideration). Note that v_{d_m} is the maximal position in t where a symbol of rank at least one occurs (called k above) and that all children of v_{d_m} are labelled with d_m. Let t_i be the subtree rooted at v_{d_i}. Then $t_m^{\mathcal{I}} = d_m + 1$. We claim that from the value $t_{i+1}^{\mathcal{I}}$ we can compute in polynomial time the value $t_i^{\mathcal{I}}$. The crucial point is that all constants that appear in the interval $[v_{d_i} + 1, v_{d_{i+1}} - 1]$ except for d_i have a deeper occurrence in the tree and therefore can be ignored for the evaluation under $\mathcal{I}$. More precisely, if a is the number of occurrences of symbols of rank at least one in the interval $[v_{d_i} + 1, v_{d_{i+1}} - 1]$ (which can be computed in polynomial time by Lemma 3), then $t_i^{\mathcal{I}} = 1 + \max\{t_{i+1}^{\mathcal{I}} + a, d_i\}$. Finally, using the same argument, we can compute $t^{\mathcal{I}}$ from $t_1^{\mathcal{I}}$.

For the second part of the theorem, the computation of the depth of a given node can be easily reduced to a height computation.   □

In the full version [14], we show with similar arguments that also the *Horton-Strahler number* [11] of an SLP-compressed tree can be computed in polynomial time. It can be defined as the value $t^{\mathcal{I}}$ under the interpretation $\mathcal{I}$ over $\mathbb{N}$ which interprets constant symbols $a \in \mathcal{F}_0$ by $a^{\mathcal{I}} = 0$ and each symbol $f \in \mathcal{F}_n$

with $n > 0$ as follows: Let $a_1, \ldots, a_n \in \mathbb{N}$ and $a = \max\{a_1, \ldots, a_n\}$. We set $f^{\mathcal{I}}(a_1, \ldots, a_n) = a$ if exactly one of $a_1, \ldots, a_n$ is equal to a, and otherwise $f^{\mathcal{I}}(a_1, \ldots, a_n) = a + 1$.

If the interpretation $\mathcal{I}$ is clear from the context, we also speak of the problem of *evaluating SLP-compressed $\mathcal{F}$-trees*. In the following theorem the interpretation is given by the Boolean operations $\wedge$ and $\vee$ over $\{0, 1\}$.

Theorem 13. *SLP-compressed $\{\wedge, \vee, 0, 1\}$-trees can be evaluated in polynomial time.*

Difficult Arithmetical Evaluation Problems. Assume that $\mathcal{I}$ is the interpretation that assigns to the binary symbols max, $+$, and $\times$ their standard meanings over $\mathbb{Z}$. We consider the problem of evaluating SLP-compressed expressions over $\{\max, +\}$ or $\{+, \times\}$. For circuits, these problems are well-studied. Circuits over max and $+$ can be evaluated bottom-up in polynomial time, since all values that arise in the circuit only need polynomially many bits. Circuits are dags, and the latter correspond to TSLPs where all nonterminals have rank 0. Moreover, it was shown in [13] that a TSLP that evaluates to an expression over a semiring can be transformed in polynomial time into an equivalent circuit over the same semiring. Hence, TSLPs over max and $+$ can be evaluated in polynomial time. In contrast, for SLP-compressed expressions we can show the following result. The counting hierarchy CH is a hierarchy of complexity classes within PSPACE, and it is conjectured that CH $\subsetneq$ PSPACE, see [2] for more details.

Theorem 14. *The evaluation of SLP-compressed $(\{\max, +\} \cup \mathbb{Z})$-trees belongs to* CH *and is $\#$P-hard (even for SLP-compressed $(\{\max, +\} \cup \mathbb{N})$-trees).*

For expressions over $+$ and $\times$ the situation is more difficult. Clearly a circuit of size $\mathcal{O}(n)$ can produce the number 2^{2^n} which has 2^n bits. Hence, we cannot evaluate a circuit over $+$ and $\times$ in polynomial time. In [2] it was shown that the problem BitSLP of computing the kth bit (k is given in binary) of the number to which a given arithmetic circuit evaluates to belongs to CH and is $\#$P-hard. By [13] these results also hold for TSLPs. For the related problem PosSLP of deciding, whether a given arithmetic circuit computes a positive number, no non-trivial lower bound is known, see also [2]. For SLP-compressed expressions over $+$ and $\times$ we can show the following:

Theorem 15. *The problem of computing for a given binary encoded number k and an SLP $\mathbb{A}$ over $\{+, \times\} \cup \mathbb{Z}$ the kth bit of $\mathsf{val}(\mathbb{A})^{\mathcal{I}}$ belongs to* CH. *Moreover, the problem of checking $\mathsf{val}(\mathbb{A})^{\mathcal{I}} \geq 0$ is* PP-*hard.*

Tree Automata. A *(deterministic) tree automaton* $\mathcal{A} = (Q, \mathcal{F}, \Delta, F)$ consists of a finite set of *states* Q, a ranked alphabet $\mathcal{F}$, a set of *final states* $F \subseteq Q$ and a set Δ of *transition rules*, which contains for all $f \in \mathcal{F}_n$, $q_1, \ldots, q_n \in Q$ exactly one rule $f(q_1, \ldots, q_n) \to q$. A tree $t \in \mathcal{T}(\mathcal{F})$ is *accepted* by $\mathcal{A}$ if $t \xrightarrow{*}_{\Delta} q$ for some $q \in F$ where $\to_{\Delta}$ is the rewriting relation defined by Δ as usual. See [10] for more details on tree automata. One can also define nondeterministic tree automata, but the above deterministic model fits better into our framework: A tree automaton as defined above can be seen as a finite algebra (i.e., an interpretation $\mathcal{I}$, where the domain $\mathcal{D}$ is finite): The domain of the algebra is the set of states, and the operations of the algebra correspond to the transitions of the automaton. Then, the membership problem for the tree automaton corresponds to the evaluation problem in the finite algebra. The uniform membership problem for tree automata asks whether a given tree automaton accepts a given tree. In [21] it was shown that this problem belongs to LogDCFL $\subseteq$ P (for nondeterministic tree automata it becomes LogCFL-complete). For every fixed tree automaton, the membership problem belongs to NC^1 [21] if the tree is represented by its traversal string. If the input tree is given by a TSLP, the uniform membership problem becomes P-complete [24]. For SLP-compressed trees we have:

Theorem 16. *Uniform membership for tree automata is* PSPACE*-complete if the input tree is given by an SLP. Moreover,* PSPACE*-hardness holds for a fixed tree automaton.*

Proof Sketch. For the upper bound one uses the fact that the uniform membership problem for explicitly given trees is in LogDCFL $\subseteq$ DSPACE$(\log^2(n))$. Given an SLP $\mathbb{A}$ for the tree $t = \text{val}(\mathbb{A})$, one can run the DSPACE$(\log^2(n))$-algorithm on the tree t without producing the whole tree t (which does not fit into polynomial space) before. This leads to a polynomial space algorithm.

For the lower bound we use a fixed regular language $L \subseteq (\{0,1\}^2)^*$ from [22] such that the following problem is PSPACE-complete: Given SLPs $\mathbb{A}$ and $\mathbb{B}$ over $\{0,1\}$ with $|\text{val}(\mathbb{A})| = |\text{val}(\mathbb{B})|$, is $\text{val}(\mathbb{A}) \otimes \text{val}(\mathbb{B}) \in L$? It is straightforward to transform a finite automaton for the fixed language L into a tree automaton $\mathcal{A}$ such that $\mathcal{A}$ accepts the comb tree $t(\text{rev}(u), \text{rev}(v))$ with $u = \text{val}(\mathbb{A})$ and $v = \text{val}(\mathbb{B})$ iff $u \otimes u \in L$. □

Theorem 16 implies that there exists a fixed finite algebra for which the evaluation problem for SLP-compressed trees is PSPACE-complete. This is somewhat surprising if we compare the situation with dags or TSLP-compressed trees. For these, membership for tree automata is still doable in polynomial time [24], whereas the evaluation problem of arithmetic expressions (in the sense of computing a certain bit of the output number) belongs to the counting hierarchy and is #P-hard. In contrast, for SLP-compressed trees, the evaluation problem for finite algebras (i.e., tree automata) is harder than the evaluation problem for arithmetic expressions (PSPACE versus the counting hierarchy).

References

1. Akutsu, T.: A bisection algorithm for grammar-based compression of ordered trees. Inf. Process. Lett. **110**(18–19), 815–820 (2010)
2. Allender, E., Bürgisser, P., Kjeldgaard-Pedersen, J., Miltersen, P.B.: On the complexity of numerical analysis. SIAM J. Comput. **38**(5), 1987–2006 (2009)

3. Benoit, D., Demaine, E.D., Munro, J.I., Raman, R., Raman, V., Rao, S.S.: Representing trees of higher degree. Algorithmica **43**(4), 275–292 (2005)
4. Bertoni, A., Choffrut, C., Radicioni, R.: Literal shuffle of compressed words. In: Ausiello, G., Karhumäki, J., Mauri, G., Ong, L. (eds.) Fifth IFIP International Conference on Theoretical Computer Scienc – TCS 2008. IFIP, vol. 273, pp. 87–100. Springer, Boston (2008)
5. Bille, P., Gørtz, I.L., Landau, G.M., Weimann, O.: Tree compression with top trees. Inform. Comput. **243**, 166–177 (2015)
6. Bille, P., Landau, G.M., Raman, R., Sadakane, K., Satti, S.R., Weimann, O.: Random access to grammar-compressed strings and trees. SIAM J. Comput. **44**(3), 513–539 (2015)
7. Busatto, G., Lohrey, M., Maneth, S.: Efficient memory representation of XML document trees. Inform. Syst. **33**(4–5), 456–474 (2008)
8. Buss, S.R.: The boolean formula value problem is in ALOGTIME. In: Proceedings of STOC 1987, pp. 123–131. ACM Press (1987)
9. Charikar, M., Lehman, E., Lehman, A., Liu, D., Panigrahy, R., Prabhakaran, M., Sahai, A., Shelat, A.: The smallest grammar problem. IEEE Trans. Inf. Theory **51**(7), 2554–2576 (2005)
10. Comon, H., Dauchet, M., Gilleron, R., Jacquemard, F., Lugiez, D., Löding, C., Tison, S., Tommasi, M.: Tree automata techniques and applications. tata.gforge.inria.fr/
11. Esparza, J., Luttenberger, M., Schlund, M.: A brief history of strahler numbers. In: Dediu, A.-H., Martín-Vide, C., Sierra-Rodríguez, J.-L., Truthe, B. (eds.) LATA 2014. LNCS, vol. 8370, pp. 1–13. Springer, Heidelberg (2014)
12. Ferragina, P., Luccio, F., Manzini, G., Muthukrishnan, S.: Compressing and indexing labeled trees, with applications. J. ACM **57**(1), 4 (2009)
13. Ganardi, M., Hucke, D., Jeż, A., Lohrey, M., Noeth, E.: Constructing small tree grammars and small circuits for formulas. arXiv.org (2014). arxiv.org/abs/1407.4286
14. Ganardi, M., Hucke, D., Lohrey, M., Noeth, E.: Tree compression using string grammars. arXiv.org (2014). arxiv.org/abs/1504.05535
15. Hübschle-Schneider, L., Raman, R.: Tree compression with top trees revisited. In: Bampis, E. (ed.) SEA 2015. LNCS, vol. 9125, pp. 15–27. Springer, Heidelberg (2015)
16. Jacobson, G.: Space-efficient static trees and graphs. In: Proceedings of FOCS 1989, pp. 549–554. IEEE Computer Society (1989)
17. Jansson, J., Sadakane, K., Sung, W.-K.: Ultra-succinct representation of ordered trees with applications. J. Comput. Syst. Sci. **78**(2), 619–631 (2012)
18. Jeż, A.: Approximation of grammar-based compression via recompression. In: Fischer, J., Sanders, P. (eds.) CPM 2013. LNCS, vol. 7922, pp. 165–176. Springer, Heidelberg (2013)
19. Jeż, A., Lohrey, M.: Approximation of smallest linear tree grammars. In: Proceedings of STACS 2014. LIPIcs, vol. 25, pp. 445–457. Schloss Dagstuhl - Leibniz-Zentrum für Informatik (2014)
20. Kobayashi, N., Matsuda, K., Shinohara, A.: Functional programs as compressed data. In: Proceedings of PEPM 2012, pp. 121–130. ACM Press (2012)
21. Lohrey, M.: On the parallel complexity of tree automata. In: Middeldorp, A. (ed.) RTA 2001. LNCS, vol. 2051, pp. 201–215. Springer, Heidelberg (2001)
22. Lohrey, M.: Leaf languages and string compression. Inform. Comput. **209**(6), 951–965 (2011)

23. Lohrey, M.: The Compressed Word Problem for Groups. Springer, New York (2014)
24. Lohrey, M.: Grammar-based tree compression. In: Potapov, I. (ed.) DLT 2015. LNCS, vol. 9168, pp. 46–57. Springer, Heidelberg (2015)
25. Munro, J.I., Raman, V.: Succinct representation of balanced parentheses and static trees. SIAM J. Comput. **31**(3), 762–776 (2001)
26. Rytter, W.: Application of Lempel-Ziv factorization to the approximation of grammar-based compression. Theor. Comput. Sci. **302**(1–3), 211–222 (2003)
27. Schmidt-Schauß, M.: Linear compressed pattern matching for polynomial rewriting. In: Proceedings of TERMGRAPH 2013. EPTCS, vol. 110, pp. 29–40 (2013)
28. Storer, J.A., Szymanski, T.G.: The macro model for data compression. In: Proceedings of STOC 1978, pp. 30–39. ACM (1978)

Trees and Languages with Periodic Signature

Victor Marsault[1]([✉]) and Jacques Sakarovitch[2]

[1] LIAFA, Université Denis Diderot, 8 place Aurélie Nemours, 75013 Paris, France
`Victor.Marsault@liafa.univ-paris-diderot.fr`
[2] Telecom-ParisTech and CNRS, 46 rue Barrault, 75013 Paris, France

Abstract. The *signature* of a labelled tree (and hence of its prefix-closed branch language) is the sequence of the degrees of the nodes of the tree in the *breadth-first traversal*. In a previous work, we have characterised the signatures of the regular languages. Here, the trees and languages that have the simplest possible signatures, namely the periodic ones, are characterised as the sets of representations of the integers in rational base numeration systems.

1 Introduction

Rational base numeration systems were defined in a joint work of the second author with S. Akiyama and Ch. Frougny [1] and allowed to make some progress in a number theoretic problem, by means of automata theory and combinatorics of words. At the same time, it raised the problem of understanding the structure of the sets of the representations of the integers in these systems from the point of view of formal language theory.

At first sight, these sets look rather chaotic and do not fit in the classical Chomsky hierarchy of languages. They all enjoy a property that makes them defeat, so to speak, any kind of iteration lemma. On the other hand, the most common example given by the set of representations in the base $\frac{3}{2}$ exhibits a remarkable regularity. The set $L_{\frac{3}{2}}$ of representations, which are words written with the three digits $\{0, 1, 2\}$, is prefix-closed and thus naturally represented as a subtree of the full ternary tree which is shown in Fig. 1. It is then easily observed that the *breadth-first* traversal of that tree yields an infinite *periodic* sequence of degrees: $2, 1, 2, 1, 2, 1, \ldots = (2\,1)^\omega$. Moreover, the sequence of labels of the arcs in the same breadth-first search is also a purely periodic sequence $0, 2, 1, 0, 2, 1, \ldots = (0\,2\,1)^\omega$.[1]

Let us call *signature* of a tree (or of the corresponding prefix-closed language) the sequence of degrees in a breadth-first traversal of the tree. With this example, we are confronted with a situation where a regular process, a periodic signature, give birth to the highly non regular language, $L_{\frac{3}{2}}$. This paradox was the incentive to look at the breadth-first traversal description of languages in general. We have

---

[1] The sequence of degrees observed on the tree in the figure begins indeed with a 1 instead of a 2, the sequence of labels begins at the second term. These discrepancies will be explained later.

© Springer-Verlag Berlin Heidelberg 2016
E. Kranakis et al. (Eds.): LATIN 2016, LNCS 9644, pp. 605–618, 2016.
DOI: 10.1007/978-3-662-49529-2_45

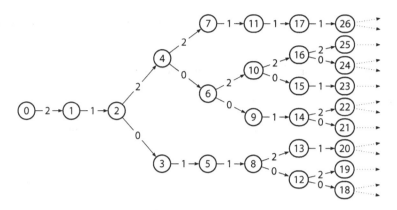

Fig. 1. The tree $\mathcal{T}_{\frac{3}{2}}$, representation of the language $L_{\frac{3}{2}}$

shown in [11] that regular languages are characterised by signatures belonging to a special class of morphic words. The purpose of this paper is to establish that a periodic signature is *characteristic* of the languages of representations of the integers in rational base numeration systems (roughly speaking and up to very simple and rational transformations).

Let us be more specific in order to state more precisely the characterisation results. An ordered tree of finite degree $\mathcal{T}$ is characterised by the infinite sequence of the degrees of its nodes visited in the order given by the breadth-first search, which we call the *signature* s of $\mathcal{T}$. Such a signature s, together with an infinite sequence λ of letters taken in an ordered alphabet form a *labelled signature* (s, λ) and characterises then a *labelled* tree $\mathcal{T}$. The breadth-first search of $\mathcal{T}$ corresponds to the enumeration in the *radix order* of the prefix-closed language $L_{\mathcal{T}}$ of branches of $\mathcal{T}$.

We call *rhythm* of *directing parameter* (q, p) a q-tuple $\mathbf{r}$ of integers whose sum is p: $\mathbf{r} = (r_0, r_1, \ldots, r_{q-1})$ and $p = r_0 + r_1 + \cdots + r_{q-1}$. With $\mathbf{r}$, we associate sequences γ of p letters that meet some consistency conditions. And we consider the languages that are determined by the labelled signature $(\mathbf{r}^{\omega}, \gamma^{\omega})$. The characterisation announced above splits in two parts.

We first determine (Theorem 1) the remarkable labelled signature $(\mathbf{r}_{\frac{p}{q}}^{\omega}, \gamma_{\frac{p}{q}}^{\omega})$ of the languages $L_{\frac{p}{q}}$. The rhythm $\mathbf{r}_{\frac{p}{q}}$ of $L_{\frac{p}{q}}$ corresponds roughly to *the most equitable way of partitioning p objects into q parts*. We call it the *Christoffel rhythm* associated with $\frac{p}{q}$, as it can be derived from the more classical notion of Christoffel word of slope $\frac{p}{q}$ (*cf.* [2]), that is, the canonical way to approximate the line of slope $\frac{p}{q}$ on a $\mathbb{Z} \times \mathbb{Z}$ lattice. The labelling $\gamma_{\frac{p}{q}}$ is induced by the generation of $\mathbb{Z}/p\mathbb{Z}$ by q.

The converse is more convoluted but its complexity is confined in the definition of a *special labelling* $\gamma_{\mathbf{r}}$ associated with every rhythm $\mathbf{r}$ (Definition 5). It is then established (Theorem 2) that the language $L_{\mathbf{r}}$ generated by the labelled signature $(\mathbf{r}^{\omega}, \gamma_{\mathbf{r}}^{\omega})$ is a non-canonical representation of the integers in the base which is the growth ratio of the rhythm **r**. The properties of alphabet conversion in rational base numeration systems (*cf.* [1] or [5]) allow to conclude that for every rhythm **r**, the language $L_{\mathbf{r}}$ is as complicated (or as simple, in the degenerate case where the growth ratio happens to be an integer) as these languages $L_{\frac{p}{q}}$.

The same techniques allow to treat the generalisation to *ultimately periodic* which raises no special difficulties and the results readily extend.

The languages with periodic labelled signature keep most of their mystery. But we have at least established that they are all alike, essentially similar to the representation languages of rational base numeration systems, and that variations in the rhythm and labelling do not really matter.

Due to space constraints, some proofs are only sketched and some figures have been removed. A complete version may be found on arXiv [9].

2 Rythmic Trees and Languages

Trees and I-trees. Classically, a tree is an undirected graph in which any two vertices are connected by exactly one path (*cf.* [3], for instance). Our point of view differs in two respects (as already discussed in [11]).

First, a tree is a *directed* graph $\mathcal{T} = (V, \Gamma)$ such that there exists a *unique* vertex, called *root*, which has no incoming arc, and there is a *unique (oriented) path* from the root to every other vertex. In the figures, we draw trees with the root on the left, and arcs rightwards.

Second, our trees are *ordered*, that is, the set of children of every node is totally ordered. The order will be implicit in the figures, with the convention that lower children are smaller (according to this order).

It will prove to be convenient to have a slightly different look at trees and to consider that the root of a tree is also a *child of itself*, that is, bears a loop onto itself. We call such a structure an *i-tree*. It is so close to a tree that we pass from one to the other with no further ado. Nevertheless, some definitions or results are easier or more straightforward when stated for i-trees, and others when stated for trees: it is then handy to have both available. A tree will usually be denoted by $\mathcal{T}_x$ for some index x and the associated i-tree by $\mathcal{I}_x$. Figure 1 shows a tree and Fig. 2a shows an i-tree.

The degree of a node is the number of its children. In the sequel, we consider infinite (i-)trees of finite degree, that is, all nodes of which have finite degree. The breadth-first traversal of such a tree defines a total ordering of its nodes. We then consider that the set of nodes of an (i-)tree is always the set of integers $\mathbb{N}$. The root is 0 and n is the $(n+1)$-th node visited by the search. We write $n \xrightarrow{\mathcal{T}} m$ if and only if m is a child of n in $\mathcal{T}$.

Let $\mathcal{I}$ be an (infinite) i-tree (of finite degree). The sequence $\boldsymbol{s}$ of the degrees of the nodes of $\mathcal{I}$ visited in the breadth-first search of $\mathcal{I}$ is called the *signature* of $\mathcal{I}$ and is *characteristic* of $\mathcal{I}$, that is, one can compute $\mathcal{I}$ from $\boldsymbol{s}$ (*cf.* Proposition 1). By convention, *the signature of a tree $\mathcal{T}$ is always that of the corresponding i-tree $\mathcal{I}$.*

In this paper, we are interested in signatures that are purely periodic. We call the period of a periodic signature *a rhythm*.

Rhythms. Given two integers n and m such that $m > 0$, we denote by $\frac{n}{m}$ their division in $\mathbb{Q}$; by $n \div m$ and $n \% m$ respectively the quotient and the remainder of the Euclidean division of n by m, that is verifying $n = (n \div m) m + (n \% m)$ and $0 \leqslant (n \% m) < m$. We also denote the integer interval $\{n, (n+1), \ldots, m\}$ by $[\![n, m]\!]$.

Definition 1. *Let p and q be two integers with $p > q \geqslant 1$.*

(i) *We call* rhythm *of directing parameter (q, p), a q-tuple $\mathbf{r}$ of non-negative integers whose sum is p:*

$$\mathbf{r} = (r_0, r_1, \ldots, r_{q-1}) \quad and \quad \sum_{i=0}^{q-1} r_i = p.$$

(ii) *We say that a rhythm $\mathbf{r}$ is* valid *if it satisfies the following equation:*

$$\forall j \in [\![0, q-1]\!] \quad \sum_{i=0}^{j} r_i > j + 1. \tag{1}$$

(iii) *We call* growth ratio *of $\mathbf{r}$ the rational number $z = \frac{p}{q}$, also written $z = \frac{p'}{q'}$ where p' and q' are coprime; it is always greater than 1.*

Examples of rhythms of growth ratio $\frac{5}{3}$ are $(2, 2, 1)$, $(3, 1, 1)$, $(1, 2, 2)$, $(3, 0, 2)$, $(2, 1, 3, 0, 0, 4)$; all but the third one are valid; the directing parameter is $(3,5)$ for the first four, and $(6,10)$ for the last one.

In the following, whenever the reference to a rhythm $\mathbf{r} = (r_0, r_1, \ldots, r_{q-1})$ is clear, we denote simply by R_j the partial sum of the first j components of $\mathbf{r}^\omega$:

$$\forall j \in \mathbb{N} \quad R_j = \sum_{i=0}^{j-1} r_{i \% q} \quad \left(= R_{j-1} + r_{(j-1) \% q} \quad \text{if } j > 0 \right).$$

Generating Trees by Rhythm. An (i-)tree can be 'reconstructed' from its signature s (*cf.* [11]), hence in the present case, from its rhythm.

Proposition 1. *Let $\mathbf{r} = (r_0, r_1, \ldots, r_{q-1})$ be a (valid) rhythm. Then, there exists a unique i-tree $\mathcal{I}_\mathbf{r}$ whose signature is $\mathbf{r}^\omega$.*

Proof (Sketch). The i-tree $\mathcal{I}_\mathbf{r}$ is built from $\mathbf{r}$ by a kind of procedure which maintains two integers, n and m, both initialised to 0: n is *the node to be processed* and m is *the next node to be created*. At every step of the procedure, $\mathbf{r}_{(n \% q)}$ nodes are created: the nodes m, $(m+1)$, $\ldots$, $(m + \mathbf{r}_{(n \% q)} - 1)$, and the corresponding arcs from n to every new node are created. Then n is incremented by 1, and m by $\mathbf{r}_{(n \% q)}$. It is verified by induction that at every step, m is equal to R_n. In particular, since R_0 is an empty sum hence equal to 0, the root 0 of $\mathcal{I}_\mathbf{r}$ is a child of itself. The next equation then gives an explicit definition of $\mathcal{I}_\mathbf{r}$:

$$\forall n, m \in \mathbb{N} \quad n \xrightarrow[\mathcal{I}_\mathbf{r}]{} m \iff R_n \leqslant m < R_{n+1}. \tag{2}$$

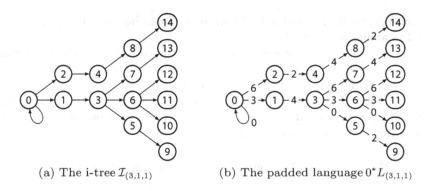

(a) The i-tree $\mathcal{I}_{(3,1,1)}$ (b) The padded language $0^* L_{(3,1,1)}$

Fig. 2. Tree and language generated by the rhythm $(3,1,1)$

We denote by $\mathcal{T}_\mathbf{r}$ the tree resulting from the removal from $\mathcal{I}_\mathbf{r}$ of the loop on its root and call respectively $\mathcal{T}_\mathbf{r}$ and $\mathcal{I}_\mathbf{r}$ the tree and i-tree *generated by* $\mathbf{r}$. Figure 2a shows $\mathcal{I}_{(3,1,1)}$ and Fig. 1 shows $\mathcal{T}_{(2,1)}$ (if one forgets the labels on the arcs).

The *validity* of the rhythm is the necessary and sufficient condition for m to always be greater than n in the course of the execution of the procedure, that is, a node is always 'created' before being 'processed', or, equivalently, for the i-tree described in Proposition 1 be infinite.

A direct consequence of the proof is that q consecutive nodes of $\mathcal{I}_\mathbf{r}$ (in the breadth-first traversal) have p (consecutive) children, hence the name *growth ratio* given to the *number* $\frac{p}{q}$. More precisely, the following holds.

Lemma 1. *Let $\mathcal{I}_\mathbf{r}$ be the i-tree generated by the rhythm $\mathbf{r}$ of directing parameter (q,p). Then, for all n, m in $\mathbb{N}$:* $n \xrightarrow[\mathcal{I}_\mathbf{r}]{} m \quad \Longleftrightarrow \quad (n+q) \xrightarrow[\mathcal{I}_\mathbf{r}]{} (m+p)$.

Generating Languages by Rhythm and Labelling. If the arcs of an i-tree $\mathcal{I}$ are labelled then $\mathcal{I}$ also defines the sequence $\boldsymbol{\lambda}$ of the labels of the arcs as they are visited in the breadth-first search; conversely, $\mathcal{I}$ as well as its branch language, will be determined by the pair $(\boldsymbol{s}, \boldsymbol{\lambda})$. In this paper, labels are digits, that is, integers, hence naturally ordered. The labelling of $\mathcal{I}$ has to be consistent with the order of $\mathcal{I}$, that is, the children of every node are in the same order as the labels of their incoming arcs.

We consider here periodic signatures $\boldsymbol{s} = \mathbf{r}^\omega$ where $\mathbf{r}$ is a rhythm of directing parameter (q,p). We then will consider pairs $(\boldsymbol{s}, \boldsymbol{\lambda})$ with $\boldsymbol{\lambda} = \boldsymbol{\gamma}^\omega$ where $\boldsymbol{\gamma}$ is a sequence of letters (digits) *of length p*.

It follows from Lemma 1 that the labelling is consistent on the whole tree if and only if it is consistent on the first q nodes, hence on the first p arcs. Let $\boldsymbol{\gamma} = u_0 u_1 \cdots u_{q-1}$ be the factorisation of $\boldsymbol{\gamma}$ *induced by* $\mathbf{r}$, that is, satisfying $|u_i| = r_i$ for every i, $0 \leqslant i < q$. Note that $u_i = \varepsilon$ if $r_i = 0$. The labelling $\boldsymbol{\gamma}$ is then *consistent with* $\mathbf{r}$ if and only if each u_i is increasing[2] and the pair $(\mathbf{r}, \boldsymbol{\gamma})$ is *valid* if in addition $\mathbf{r}$ is valid.

For instance, the labelling $\boldsymbol{\gamma} = (0, 3, 6, 4, 2)$ is consistent with the rhythm $\mathbf{r} = (3, 1, 1)$ since $u_0 = (0, 3, 6)$, $u_1 = (4)$ and $u_2 = (2)$ are all increasing and $u_0 \, u_1 \, u_2$ is the factorisation of $\boldsymbol{\gamma}$ induced by $\mathbf{r}$.

We denote by $\mathcal{I}_{(\mathbf{r}, \boldsymbol{\gamma})}$ the labelled i-tree *generated by* a rhythm $\mathbf{r}$ of directing parameter (q, p) and a labelling $\boldsymbol{\gamma} = (\gamma_0, \gamma_1, \ldots, \gamma_{p-1})$ consistent with $\mathbf{r}$. The labels of the arcs of $\mathcal{I}_{(\mathbf{r}, \boldsymbol{\gamma})}$ are determined by

$$\forall n, m \in \mathbb{N} \qquad n \xrightarrow[\mathcal{I}_{\mathbf{r}}]{a} m \quad \text{implies} \quad a = \gamma_{(m \, \% \, p)} \text{ which belongs to } u_{(n \, \% \, q)}. \quad (3)$$

By convention, we denote by $L_{(\mathbf{r}, \boldsymbol{\gamma})}$ the branch language of the **tree** $\mathcal{T}_{(\mathbf{r}, \boldsymbol{\gamma})}$ rather than the one of **i-tree** $\mathcal{I}_{(\mathbf{r}, \boldsymbol{\gamma})}$, and we call it *the language generated by* $(\mathbf{r}, \boldsymbol{\gamma})$. The branch language of $\mathcal{I}_{(\mathbf{r}, \boldsymbol{\gamma})}$ is thus $z^* L_{(\mathbf{r}, \boldsymbol{\gamma})}$ where $z = \gamma_0$ is the label of the loop $0 \longrightarrow 0$ in $\mathcal{I}_{(\mathbf{r}, \boldsymbol{\gamma})}$ and we call it the *padded* language generated by $(\mathbf{r}, \boldsymbol{\gamma})$.

For instance, the language generated by $\mathbf{r} = (2, 1)$ and $\boldsymbol{\gamma} = (0, 2, 1)$ is shown in Fig. 1 and the padded language generated by $\mathbf{r} = (3, 1, 1)$ and $\boldsymbol{\gamma} = (0, 3, 6, 4, 2)$ in Fig. 2b.

Let L be a prefix-closed language over an ordered alphabet A and $\mathcal{T}_L$ its associated labelled tree (whose set of nodes is then $\mathbb{N}$). The enumeration of L in the radix order is then equivalent to the breadth-first traversal of $\mathcal{T}_L$. This ordering of L is precisely the idea underlying the notion of Abstract Numeration System (ANS) as defined by Lecomte and Rigo (*cf.* [7,8]). An ANS is a language L over an ordered alphabet and in this system every integer n is represented by the $(n+1)$-th word of L in the radix order; this word is denoted by $\langle n \rangle_L$. The integer representations in the ANS L and the nodes of the tree $\mathcal{T}_L$ are thus linked by: $\langle 0 \rangle_L = \varepsilon$ and

$$\forall n \in \mathbb{N}, \, \forall m \in \mathbb{N}^+, \, \forall a \in A \qquad \langle n \rangle_L \, a = \langle m \rangle_L \quad \Longleftrightarrow \quad n \xrightarrow[\mathcal{T}_L]{a} m. \quad (4)$$

3   From Rational Base Numeration Systems to Rhythms

Integer and Rational Base Numeration Systems. Let p be an integer, $p \geqslant 2$, and $A_p = [\![0, p-1]\!]$ the alphabet of the first p digits. Every word $w = a_n \, a_{n-1} \cdots a_0$ of A_p^* is given a value n in $\mathbb{N}$ by the *evaluation function* π_p: $\pi_p(a_n \, a_{n-1} \cdots a_0) = \sum_{i=0}^n a_i \, p^i$, and w is a p-development of n. Every n in $\mathbb{N}$ has a unique p-*development* without leading 0's in A_p^*: it is called the p-*representation* of n and is denoted by $\langle n \rangle_p$. The p-*representation* of n can be computed from left-to-right by a greedy algorithm, and also from *right-to-left* by iterating the Euclidean division of n by p, the digits a_i being the successive remainders. The language of the p-representations of the integers is the regular language $L_p = \{\langle n \rangle_p \mid n \in \mathbb{N}\} = (A_p \setminus 0) \, A_p^*$.

---

[2] A word $a_0 \, a_1 \, a_2 \cdots a_n$ is *increasing* if $a_0 < a_1 < a_2 < \cdots < a_n$.

Let p and q be two co-prime integers, $p > q > 1$. In [1], these classical statements have been generalised to the case of *numeration system with rational base* $\frac{p}{q}$. The $\frac{p}{q}$-evaluation function $\pi_{\frac{p}{q}}$ is defined by:

$$\forall a_n a_{n-1} \cdots a_0 \in A_p^* \qquad \pi_{\frac{p}{q}}(a_n a_{n-1} \cdots a_0) = \sum_{i=0}^{n} \frac{a_i}{q} \left(\frac{p}{q}\right)^i,$$

and it is shown that every integer n has a unique $\frac{p}{q}$-*representation* $\langle n \rangle_{\frac{p}{q}}$, that is, a word of A_p^* such that $\pi_{\frac{p}{q}}\left(\langle n \rangle_{\frac{p}{q}}\right) = n$. This representation is computed (from right to left) by the *modified Euclidean division algorithm* as follows: let $N_0 = n$ and, for all $i > 0$,

$$q N_i = p N_{(i+1)} + a_i, \tag{5}$$

where a_i is the remainder of the Euclidean division of $q N_i$ by p, hence belongs to $A_p = [\![0, p-1]\!]$. Since $p > q$, the sequence $(N_i)_{i \in \mathbb{N}}$ is strictly decreasing and eventually stops at $N_{k+1} = 0$. The $\frac{p}{q}$-representation of n is then the word $\langle n \rangle_{\frac{p}{q}} = a_k a_{k-1} \cdots a_0$ of A_p^*.

The set $L_{\frac{p}{q}} = \{\langle n \rangle_{\frac{p}{q}} \mid n \in \mathbb{N}\}$ of $\frac{p}{q}$-representations of integers is 'far' from being a regular language. It has a property that we have later called *FLIP*[3] (for *Finite Left Iteration Property*, cf. [10]) and which is equivalent (for *prefix-closed* languages) to the fact that it contains no infinite regular subsets (IRS condition of [6]). This implies that $L_{\frac{p}{q}}$ does not meet any kind of iteration lemma and in particular that it is not context-free. It is also shown in [1] that the numeration system with rational base $\frac{p}{q}$ coincide with the ANS $L_{\frac{p}{q}}$.

In many respects, the case of integer base can be seen as a special case of rational base numeration system. The definitions of $\pi_{\frac{p}{q}}$, $\langle n \rangle_{\frac{p}{q}}$ and $L_{\frac{p}{q}}$ coincide with those of π_p, $\langle n \rangle_p$ and L_p respectively, when $q = 1$. In the sequel, we consider the base $\frac{p}{q}$ where p and q are two coprime integers verifying $p > q \geqslant 1$, that is, indifferently one numeration system or the other. In particular, the following holds in both integer or rational cases:

$$\forall n \in \mathbb{N}, \ \forall m \in \mathbb{N}^+, \ \forall a \in A_p \qquad \langle m \rangle_{\frac{p}{q}} = \langle n \rangle_{\frac{p}{q}} a \quad \Longleftrightarrow \quad a = q m - p n. \tag{6}$$

Geometric Representations of Rhythms. Rhythms are given a very useful geometric representation as *paths* in the $(\mathbb{Z} \times \mathbb{Z})$-lattice and such paths are coded by *words* of $\{x, y\}^*$ where x denotes an horizontal unit segment and y a vertical unit segment. Hence the name *path* given to a *word* associated with a rhythm.

Definition 2. *Let* $\mathbf{r} = (r_0, r_1, \ldots, r_{q-1})$ *be a rhythm of directing parameter* (q, p). *With* $\mathbf{r}$, *we associate the word* $\mathtt{path}(\mathbf{r})$ *of* $\{x, y\}^*$:

$$\mathtt{path}(\mathbf{r}) = y^{r_0} x \, y^{r_1} x \, y^{r_2} \cdots x \, y^{r_{q-1}} x$$

which corresponds to a path from $(0, 0)$ *to* (q, p) *in the* $(\mathbb{Z} \times \mathbb{Z})$-*lattice.*

---

[3] This property was introduced in [10] under the unproper name of *Bounded Left Iteration Property*, or *BLIP* for short.

Figure 3 shows the paths associated with three rhythms of directing parameter $(3,5)$. It then appears clearly that Definition 1 (ii) can be restated as 'a rhythm is valid if and only if the associated path is strictly above the line of slope 1 passing through the origin'.

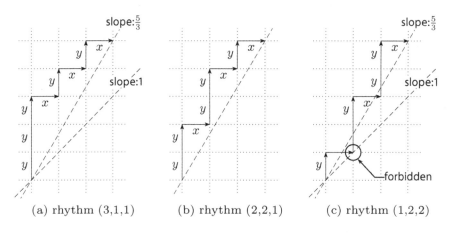

(a) rhythm (3,1,1)   (b) rhythm (2,2,1)   (c) rhythm (1,2,2)

Fig. 3. Words and paths associated with rhythms of directing parameter $(3,5)$

Rhythm and Labelling of Rational Base. We introduce $r_{\frac{p}{q}}$, a particular *rhythm* of directing parameter (q,p) associated with a canonical *labelling* $\gamma_{\frac{p}{q}}$. The former relates to the classical notion of *Christoffel words* while the later results from the generation of $\mathbb{Z}/p\mathbb{Z}$ by q. The remarkable fact is then that the representation language in the $\frac{p}{q}$-numeration system is generated by $(r_{\frac{p}{q}}, \gamma_{\frac{p}{q}})$.

Christoffel words code the 'best (upper) approximation' of segments the $\mathbb{Z} \times \mathbb{Z}$-lattice and have been studied in the field of combinatorics of words (*cf.* [2]).

Definition 3 ([2]). *The (upper) Christoffel word of slope $\frac{p}{q}$, denoted by $\mathbf{w}_{\frac{p}{q}}$, is the label of the path from $(0,0)$ to (q,p) on the $(\mathbb{Z} \times \mathbb{Z})$-lattice, such that*

- *the path is above the line of slope $\frac{p}{q}$ passing through the origin;*
- *the region enclosed by the path and the line contains no point of $\mathbb{Z} \times \mathbb{Z}$.*

We translate then Christoffel words into rhythms.

Definition 4. *The Christoffel rhythm associated with $\frac{p}{q}$, and denoted by $\mathbf{r}_{\frac{p}{q}}$, is the rhythm whose path is $\mathbf{w}_{\frac{p}{q}}$: $\mathtt{path}(\mathbf{r}_{\frac{p}{q}}) = \mathbf{w}_{\frac{p}{q}}$, hence its directing parameter is (q,p).*

Figure 3b shows the path of $\mathbf{w}_{\frac{5}{3}} = \overline{yy}\, x\, \overline{yy}\, x\, \overline{y}\, x$, the Christoffel word associated with $\frac{5}{3}$; then, $\mathbf{r}_{\frac{5}{3}} = (2,2,1)$. Other instances of Christoffel rhythms are $\mathbf{r}_{\frac{3}{2}} = (2,1)$, $\mathbf{r}_{\frac{4}{3}} = (2,1,1)$ and $\mathbf{r}_{\frac{12}{5}} = (3,2,3,2,2)$. The definition of Christoffel words yields the following proposition on rhythms.

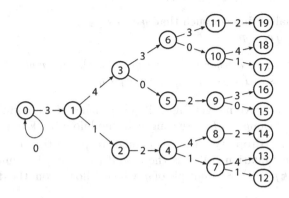

Fig. 4. The padded language $0^* L_{\frac{5}{3}}$ of the representation of integers in base $\frac{5}{3}$

Proposition 2. *Given a base $\frac{p}{q}$ of rhythm $\mathbf{r}_{\frac{p}{q}} = (r_0, r_1, \cdots, r_{q-1})$, for every integer $k \leqslant q$, the partial sum R_k of the first k components of $\mathbf{r}$ is equal to the smallest integer greater than $k\frac{p}{q}$.*

Since p and q are coprime integers, q is a generator of the group $\mathbb{Z}/p\mathbb{Z}$ (additive). We denote by $\boldsymbol{\gamma}_{\frac{p}{q}}$ the sequence induced by this generation process:

$$\boldsymbol{\gamma}_{\frac{p}{q}} = (0, (q\%p), (2q\%p), \ldots, ((p-1)q\%p)). \tag{7}$$

Theorem 1. *Let p and q be two coprime integers, $p > q \geqslant 1$. The language $L_{\frac{p}{q}}$ of the $\frac{p}{q}$-representations of the integers is generated by the rhythm $\mathbf{r}_{\frac{p}{q}}$ and the labelling $\boldsymbol{\gamma}_{\frac{p}{q}}$.*

For instance, $L_{\frac{3}{2}}$, shown in Fig. 1, is built with the rhythm $\mathbf{r}_{\frac{3}{2}} = (2, 1)$ and the labelling $\boldsymbol{\gamma}_{\frac{3}{2}} = (0, 2, 1)$ while the padded language $0^* L_{\frac{5}{3}}$, shown in Fig. 4, is built with the rhythm $\mathbf{r}_{\frac{5}{3}} = (2, 2, 1)$ and the labelling $\boldsymbol{\gamma}_{\frac{5}{3}} = (0, 3, 1, 4, 2)$.

The proof of Theorem 1 requires additional definitions and statements. We define the sequence of integers $e_0, e_1, \ldots, e_{q-1}$ such that e_j is the difference between the approximation $R_k = (r_0 + r_1 + \cdots + r_{k-1})$ and the point of the associated line of the respective abscissa, that is $(k\,\frac{p}{q})$. This difference is a rational number smaller than 1 and whose denominator is q, in order to obtain an integer we multiply it by q:

$$\forall k \in [\![0, q-1]\!] \qquad e_k = q\,R_k - k\,p. \tag{8}$$

Below are compiled basic properties of the r_j's and e_j's that follow directly from Proposition 2 and Equation (8).

Property 1. Let $\mathbf{r}_{\frac{p}{q}} = (r_0, r_1, \ldots, r_{q-1})$ be the Christoffel rhythm of slope $\frac{p}{q}$. For every integer k in $[\![0, q-1]\!]$, it holds:

(a) e_k belongs to $[\![0, q-1]\!]$;

(b) r_k is the smallest integer such that $q\,r_k + e_k \geqslant p$;
(c) $e_{k+1} = e_k + q\,r_k - p$.

Lemma 2. *For every integer $n > 0$ (resp. $n = 0$), the smallest letter a of A_p such that $\langle n \rangle_{\frac{p}{q}}\, a$ is in $L_{\frac{p}{q}}$ is $e_{(n\,\%\,q)}$ (resp. $e_0 + q$).*

Proof. Let n be positive integer and k its congruence class modulo q. Letters a such that $\langle n \rangle_{\frac{p}{q}}\, a$ belongs to $L_{\frac{p}{q}}$ are congruent modulo q (*cf.* Equation (6)). Since e_k is in $[\![0, q-1]\!]$ (Property 1a), it is enough to prove that e_k is an outgoing label of n. From Equation (6), it is the case if $(n\,p + e_k)$ is a multiple of q or, equivalently if $(k\,p + e_k)$ is a multiple of q, which follows from the definition of e_k (Equation (8)).

For $n = 0$, $e_0 = 0$ and although the equation $e_0 = q\,m - p\,n$ is verified for some integer m, that integer is $m = 0$. It then follows from Equation (6) that $\langle 0 \rangle_{\frac{p}{q}}\, e_0$ does **not** belong to $L_{\frac{p}{q}}$ (since m is not positive). The reasoning of the previous paragraph then works for $(e_0 + q)$. $\qquad\square$

Proposition 3. *For every integer $n > 0$ (resp. $n = 0$), there are exactly $r_{(n\,\%\,q)}$ (resp. $(r_0 - 1)$) letters a of A_p such that $\langle n \rangle_{\frac{p}{q}}\, a$ belongs to $L_{\frac{p}{q}}$.*

Proof. Let n be positive integer and k its congruence class modulo q. From Property 1b, r_k is the smallest integer such that $q\,r_k + e_k > p$. It follows that for all k in $[\![0, r_k - 1]\!]$ $(e_k + q\,k) < p$ and that $e_k + q\,r_k > p$.

The set $S = \{e_k, (e_k + q), \ldots, (e_k + q\,(r_k - 1))\}$ contains all the letters of A_p that are congruent to e_k modulo q. Since $\langle n \rangle_{\frac{p}{q}}\, e_k$ belongs to $L_{\frac{p}{q}}$ (Lemma 2), it follows from Equation (6) that

$$S \;=\; \{\, a \in A_p \mid \langle n \rangle_{\frac{p}{q}}\, a \in L_{\frac{p}{q}} \,\}.$$

The set S is of cardinal r_k, concluding the case $n > 0$.

The proof is similar in the case where $n = 0$, except that the smallest letter is $(e_0 + q)$ instead of e_0 (Lemma 2). $\qquad\square$

The next proposition follows directly from Equation (6).

Proposition 4. *For every positive integer m, the rightmost letter of $\langle m \rangle_{\frac{p}{q}}$ is equal to $(q\,m)\,\%\,p$.*

Proposition 3 yields that the rhythm of $L_{\frac{p}{q}}$ is indeed $\mathbf{r}_{\frac{p}{q}}$ and Proposition 4 that its labelling is $\boldsymbol{\gamma}_{\frac{p}{q}}$, hence concluding the proof of Theorem 1.

The next statement gives a different way to compute $\boldsymbol{\gamma}_{\frac{p}{q}}$; its generalisation in the next section (Definition 5) to arbitrary rhythms will be instrumental in the proof of Theorem 2.

Proposition 5. *Let $\mathbf{r}_{\frac{p}{q}}$ be a Christoffel rhythm and $\boldsymbol{\gamma}_{\frac{p}{q}} = \gamma_0 \gamma_1 \cdots \gamma_{(p-1)}$ the associated labelling. We denote by $\boldsymbol{\gamma}_{\frac{p}{q}} = u_0 u_1 \cdots u_{q-1}$ the factorisation of $\boldsymbol{\gamma}_{\frac{p}{q}}$ induced by $\mathbf{r}_{\frac{p}{q}}$. Then, $\gamma_0 = 0$ and, for all integer i, $0 \leqslant i < (p - 1)$,*

- *if the letters γ_i and $\gamma_{(i+1)}$ belong to the same factor u_j then $\gamma_{(i+1)} = \gamma_i + q$;*
- *otherwise, $\gamma_{(i+1)} = \gamma_i + q - p$.*

Proof. We denote by $c_0\, c_1 \,\cdots\, c_{p-1}$ the integers computed by the recursive algorithm of the proposition, that is:

$$\forall i,\, 0 \leqslant i < q \qquad c_i = q\,i - p\,j \qquad \text{if } \gamma_i \text{ is a letter of the factor } u_j.$$

It should be noted that $c_i \equiv i\,q\ [p]$, hence that $c_i \equiv \gamma_i\ [p]$ from Equation (7); it is then enough to show that $0 \leqslant c_i < p$ for every integer $i < p$.

Let us take $i, j > 0$ such that $\gamma_0 \gamma_1 \cdots \gamma_{i-1} = u_0 u_1 \cdots u_{j-1}$, a word of length $i = R_j$. It follows from Proposition 2 that $i = \lceil j\frac{p}{q} \rceil$, or, in other word, that $j\,p - q \leqslant q\,(i-1) < j\,p$. Since γ_{i-1} is the last letter of u_{j-1}

$$c_{i-1} = q\,(i-1) - p\,(j-1) \qquad \text{hence} \qquad (p-q) \leqslant c_{(i-1)} < p,$$

and since γ_i is the first letter of u_j

$$c_i = (c_{(i-1)} + q - p) \qquad \text{hence} \qquad 0 \leqslant c_i < (q-1).$$

We have just shown that the first letter of every factor u_k is non-negative and that its last letter is strictly smaller than p. Since every factor is increasing (each letter being equal to the previous letter plus q), every letter a of every factor satisfies $0 \leqslant a < p$.

4 From Rhythms Back to Rational Bases

We now establish a kind of converse of Theorem 1. With an arbitrary rhythm is associated a rational base (its growth ratio) and a *special* labelling. We consider the language generated by this rhythm and labelling as an abstract numeration system and show that it features a rule much like Equation (6). We finally show that this abstract numeration system is simply a rational base on a non-canonical alphabet (Theorem 2).

In this section, p and q are two integers, $p > q \geqslant 1$, *not necessarily coprime*, and $\mathbf{r}$ is a rhythm of directing parameter (q, p). As in Definition 1, we denote by p' and q' their respective quotient by their gcd.

Special Labelling. The next definition is a generalisation of the labelling of rational base for arbitrary rhythms; it is based on the characterisation given by Proposition 5 but is more complicated in order to take into accounts the possible components equal to 0 appearing in the rhythm.

Definition 5. *We call* special labelling *(associated with $\mathbf{r}$), and denote by $\boldsymbol{\gamma_r} = (\gamma_0, \gamma_1, \ldots, \gamma_{p-1})$, the sequence of digits of length p defined as follows. First $\gamma_0 = 0$. Second, we denote by $\boldsymbol{\gamma_r} = u_0\, u_1 \,\cdots\, u_{q-1}$ the factorisation of $\boldsymbol{\gamma_r}$ induced by $\mathbf{r}$ (for all i, $0 \leqslant i < p$, $|u_i| = r_i$). Then, for every i, $0 \leqslant i < p - 1$, if k and j are the indices such that γ_i belongs to u_k and γ_{i+1} belongs to u_{k+j}, then $\gamma_{i+1} = \gamma_i + q' - j\,p'$.*

Example 1. Let $\mathbf{r} = (3, 1, 1)$; its directing parameter is $(3, 5)$, hence $p = p' = 5$, $q = q' = 3$ and the computation of $\boldsymbol{\gamma_r}$ is given below, on the left. Within a factor u_i, the difference between two consecutive digits is $3(= q')$, otherwise it is $-2\, (= q' - p')$.

$$\mathbf{r} = \quad (\underbrace{3,}_{u_0} \quad \underbrace{1,}_{u_1} \quad \underbrace{1)}_{u_2} \qquad\qquad (\underbrace{4,}_{u_0} \quad \underbrace{0,}_{u_1} \quad \underbrace{0,}_{u_2} \quad \underbrace{2)}_{u_3}$$

$$\boldsymbol{\gamma_r} = \quad (\overbrace{0, 3, 6,}\ \overbrace{4}\ ,\ \overbrace{2}\ ) \qquad\qquad (\overbrace{0, 2, 4, 6,}\ \overbrace{3}\ \ \overbrace{3}\ \overbrace{-1, 1}\ )$$

Let now $\mathbf{r} = (4, 0, 0, 2)$; its directing parameter is $(4, 6)$, $p' = 3$, $q' = 2$ and the computation of $\boldsymbol{\gamma_r}$ is given above, on the right. Within a factor u_i, the difference between two consecutive digits is $+2(= +q')$; the fourth digit belongs to u_0 and the fifth to u_3: the difference between the two is $-7(= +q' - 3\, p')$.

It directly follows from Definition 5 that $\boldsymbol{\gamma_r}$ *is always consistent with* $\mathbf{r}$. *Notation.* We denote by $L_\mathbf{r}$ the language generated by a rhythm $\mathbf{r}$ and the associated special labelling $\boldsymbol{\gamma_r}$, that is, $L_\mathbf{r} = L_{(\mathbf{r}, \boldsymbol{\gamma_r})}$.

Non-canonical Representation of Integers. If $\mathbf{r}$ happens to be a Christoffel rhythm, then, by Theorem 1, $L_\mathbf{r}$ is equal to $L_{\frac{p'}{q'}}$ (which, in this case, is also $L_{\frac{p}{q}}$). The key result of this work is that $L_\mathbf{r}$ and $L_{\frac{p'}{q'}}$ are indeed of the same kind.

Theorem 2. *Let* $\mathbf{r}$ *be a rhythm of directing parameter* (q, p) *and* $\frac{p'}{q'}$ *the reduced fraction of* $\frac{p}{q}$. *Then, the language* $L_\mathbf{r}$ *is a set of representations of the integers in the rational base* $\frac{p'}{q'}$ *using a non-canonical set of digits.*

The proof of Theorem 2 is sketched below. Let us call $\mathbf{r}$-representation of an integer n, and denote it by $\langle n \rangle_\mathbf{r}$, the representation of n in the abstract numeration system $L_\mathbf{r}$. We know from Equation (4) that $\langle n \rangle_\mathbf{r}$ labels the path from the root 0 to the node n in the labelled tree defined by $L_\mathbf{r}$. First we show that the existence of arcs in $L_\mathbf{r}$ has a necessary condition similar to those of $L_{\frac{p'}{q'}}$ (*cf.* Equation (6)).

Lemma 3. *Let* $\mathbf{r}$ *be a rhythm of directing parameter* (q, p) *and* $\frac{p'}{q'}$ *the reduced fraction of* $\frac{p}{q}$. *Then, for every integers* n *and* $m > 0$, *it holds:*
$$\langle n \rangle_\mathbf{r}\, a \;=\; \langle m \rangle_\mathbf{r} \quad \Longrightarrow \quad a \;=\; q'\, m - p'\, n.$$

The converse of Lemma 3 does not hold in general; it holds only for rhythms (of directing parameter (q, p)) such that p and q are coprime, and for powers of such rhythms. Otherwise, the alphabet of the letters appearing in $\boldsymbol{\gamma_r}$ contains at least two different digits congruent modulo p'; the incoming arc of a given node then depends on its congruence class modulo p (and not only modulo p').

Theorem 2 is then equivalent to the following statement.

Proposition 6. *Let* $\mathbf{r}$ *be a rhythm of directing parameter* (q, p), $\frac{p'}{q'}$ *the reduced fraction of* $\frac{p}{q}$ *and* $\pi_{\frac{p'}{q'}}$ *the evaluation function in the* $\frac{p'}{q'}$-*numeration system. Then, for every integer* n, $\pi_{\frac{p'}{q'}}(\langle n \rangle_\mathbf{r}) = n$ *holds.*

Proof. By induction on the length of $\langle n \rangle_{\mathbf{r}}$. The equality is obviously verified for $\langle 0 \rangle_{\mathbf{r}} = \varepsilon$. Let m be a positive integer and $\langle m \rangle_{\mathbf{r}} = a_{k+1} \, a_k \, a_{k-1} \cdots a_1 \, a_0$ its $\mathbf{r}$-representation, that is, a word of $L_{\mathbf{r}}$. The word $a_{k+1} \, a_k \, a_{k-1} \cdots a_1$ is also in $L_{\mathbf{r}}$; it is the $\mathbf{r}$-representation of an integer n strictly smaller than m, verifying $\langle n \rangle_{\mathbf{r}} \, a_0 = \langle m \rangle_{\mathbf{r}}$, hence $n \xrightarrow[L_{\mathbf{r}}]{a_0} m$. On the right hand, by induction hypothesis, $n = \pi_{\frac{p'}{q'}}(\langle n \rangle_{\mathbf{r}})$ and on the other hand, it follows from the previous Lemma 3 that $a_0 = q'm - p'n$, or, equivalently, that $m = \frac{np' + a_0}{q'}$, hence

$$m \;=\; \frac{p'}{q'} \pi_{\frac{p'}{q'}}(\langle n \rangle_{\mathbf{r}}) + \frac{a_0}{q} \;=\; \pi_{\frac{p}{q}}(\langle n \rangle_{\mathbf{r}} \, a_0) \;=\; \pi_{\frac{p'}{q'}}(\langle m \rangle_{\mathbf{r}}).$$

It is shown in [1] that in spite of this 'complexity' of $L_{\frac{p}{q}}$, the *conversion* from any digit-alphabet B into the canonical alphabet A_p is realised by a *finite transducer* exactly as in the case of an integer numeration system (*cf.* also [5]). More precisely:

Theorem 3 ([1]). *For all digit alphabets B, the function $\chi \colon B^* \to A_{p'}^*$ which maps every word w of B^* onto the word of $A_{p'}^*$ which has the same value in the $\frac{p'}{q'}$-numeration system — hence $\pi_{\frac{p'}{q'}}(w) = \pi_{\frac{p'}{q'}}(\chi(w))$ — is a (right sequential) rational function.*

If we write B for the set of digits appearing in $\boldsymbol{\gamma}_{\mathbf{r}}$, Theorem 3 implies in particular that $\chi(L_{\mathbf{r}}) = L_{\frac{p'}{q'}}$. Hence, that the complexity of $L_{\frac{p'}{q'}}$ extends to $L_{\mathbf{r}}$.

Corollary 1. *Let $\mathbf{r}$ be a rhythm of directing parameter (q,p) and $L_{\mathbf{r}}$ the language generated by the pair $(\mathbf{r}, \boldsymbol{\gamma}_{\mathbf{r}})$. If $\frac{p}{q}$ is an integer, then $L_{\mathbf{r}}$ is a regular language, otherwise, $L_{\mathbf{r}}$ is a FLIP language.*

Example 2. Given a directing parameter (q,p), let $\mathbf{r}$ be the *extreme* rhythm where all components are 0 but one which is p. The validity condition implies that the positive digit is necessarily the first one: $\mathbf{r} = (p, 0, \ldots, 0)$ and the associated special labelling is then $\boldsymbol{\gamma}_{\mathbf{r}} = (0, q, (2\,q), \ldots, (p-1)\,q)$. Since every letter of $\boldsymbol{\gamma}_{\mathbf{r}}$ is a multiple of q, we perform a component-wise division of $\boldsymbol{\gamma}_{\mathbf{r}}$ by q and obtain $\boldsymbol{\gamma} = (0, 1, 2, \ldots, (p-1))$.

The language $L_{(\mathbf{r}, \boldsymbol{\gamma})}$ generated by $(\mathbf{r}, \boldsymbol{\gamma})$ is then the language of the representations of the integers in a variant (that we call FK after its authors) of $\frac{p}{q}$-numeration systems considered in [4]. In the variant FK, the value of a word u, denoted by $\pi_{FK}(u)$, is q times its standard evaluation: $\pi_{FK}(u) = q \times \pi_{\frac{p}{q}}(u)$. This is exactly the behaviour described by Proposition 6, since all digits have been divided by q. This example highlights the soundness of the relationship between rational base numeration system and periodic signature.

5   Extension, Future Work and Conclusion

For sake of simplicity, we have considered here purely periodic signatures and the periodic labellings that go with them. The same techniques as the ones developed in Sect. 4 allow to treat the generalisation to *ultimately periodic* which raises no special difficulties and the results established here readily extend. One may even generalise these results to every aperiodic signature whose path (as defined in Sect. 2) is confined to a strip between two parallel lines of slope $\frac{p}{q}$.

Using rhythm often sheds light on problems related to rational base. It is the case for the question of representation of the negative integers, tackled in [4], that may be given a new approach in terms of Christoffel words and their properties.

There is certainly still much to be understood on the relationship between the 'high regularity' of periodic signatures and the apparent disorder or complexity of trees that are generated by these periodic signatures. Some questions, such as statistics of labels along infinite branches, are indeed related to identified problems in number theory that are recognised as very difficult.

We have established in this paper that the infinite trees or languages generated by periodic signatures are completely determined (up to very simple transformations — that is, rational sequential functions) by the growth ratio of the period only and independent of the actual components of the period. This first step was somehow unexpected. It makes the scenery simpler but the call for further investigations on the subject even stronger.

References

1. Akiyama, S., Frougny, C., Sakarovitch, J.: Powers of rationals modulo 1 and rational base number systems. Isr. J. Math. **168**, 53–91 (2008)
2. Berstel, J., Lauve, A., Reutenauer, C., Saliola, F.: Combinatorics on Words: Christoffel Words and Repetition in Words, vol. 27 of CRM Monograph Series. American Math. Soc., Providence, Rhode Island, USA (2008)
3. Diestel, R.: Graph Theory. Springer, New York (1997)
4. Frougny, C., Klouda, K.: Rational base number systems for p-adic numbers. RAIRO Theor. Inf. Appl. **46**(1), 87–106 (2012)
5. Frougny, C., Sakarovitch, J.: Number representation and finite automata. In: Berthé, V., Rigo, M. (eds.) Combinatorics, Automata and Number Theory. Cambridge University Press, Cambridge (2010)
6. Greibach, S.A.: One counter languages and the IRS condition. J. Comput. Syst. Sci. **10**(2), 237–247 (1975)
7. Lecomte, P., Rigo, M.: Numeration systems on a regular language. Theor. Comput. Syst. **34**, 27–44 (2001)
8. Lecomte, P., Rigo, M.: Abstract numeration systems. In: Berthé, V., Rigo, M. (eds.) Combinatorics, Automata and Number Theory. Cambridge University Press, Cambridge (2010)
9. Marsault, V., Sakarovitch, J.: Rhythmic generation of infinite trees, languages (full version). In preparation. Preprint available at arXiv:1403.5190
10. Marsault, V., Sakarovitch, J.: On sets of numbers rationally represented in a rational base number system. In: Muntean, T., Poulakis, D., Rolland, R. (eds.) CAI 2013. LNCS, vol. 8080, pp. 89–100. Springer, Heidelberg (2013)
11. Marsault, V., Sakarovitch, J.: Breadth-first serialisation of trees and rational languages. In: Shur, A.M., Volkov, M.V. (eds.) DLT 2014. LNCS, vol. 8633, pp. 252–259. Springer, Heidelberg (2014)

Rank Reduction of Directed Graphs by Vertex and Edge Deletions

Syed Mohammad Meesum[1(✉)] and Saket Saurabh[1,2]

[1] Institute of Mathematical Sciences, Chennai, India
{meesum,saket}@imsc.res.in
[2] University of Bergen, Bergen, Norway

Abstract. In this paper we continue our study of graph modification problems defined by reducing the rank of the adjacency matrix of the given graph, and extend our results from undirected graphs to directed graphs. An instance of a graph modification problem takes as input a graph G, a positive integer k and the objective is to delete k vertices (edges) so that the resulting graph belongs to a particular family, $\mathcal{F}$, of graphs. Given a fixed positive integer r, we define $\mathcal{F}_r$ as the family of directed graphs where for each $G \in \mathcal{F}_r$, the rank of the adjacency matrix of G is at most r. Using the family $\mathcal{F}_r$ we do algorithmic study, both in classical and parameterized complexity, of the following graph modification problems: r-RANK VERTEX DELETION, r-RANK EDGE DELETION. We first show that both the problems are NP-Complete. Then we show that these problems are fixed parameter tractable (FPT) by designing an algorithm with running time $2^{\mathcal{O}(k \log r)} n^{\mathcal{O}(1)}$ for r-RANK VERTEX DELETION, and an algorithm for r-RANK EDGE DELETION running in time $2^{\mathcal{O}(f(r)\sqrt{k} \log k)} n^{\mathcal{O}(1)}$. We complement our FPT result by designing polynomial kernels for these problems. Our main structural result, which is the fulcrum of all our algorithmic results, is that for a fixed integer r the size of any "reduced graph" in $\mathcal{F}_r$ is upper bounded by 3^r. This result is of independent interest and generalizes a similar result of Kotlov and Lovász regarding reduced undirected graphs of rank r.

1 Introduction

Editing a graph by either deleting vertices or deleting edges or adding edges such that the resulting graph satisfies certain properties or becomes a member of some well-understood graph class is one of the basic problems in graph theory and graph algorithms. These problems are called graph modification problems. However, most of the graph modification problems are NP-Complete [21,27] and thus they are subjected to intensive study in algorithmic paradigms that are meant for coping with NP-completeness [16,18,22,24]. These paradigms among

S. Saurabh—The research leading to these results has received funding from the European Research Council under the European Union's Seventh Framework Programme (FP7/2007–2013) / ERC grant agreement no. 306992.

E. Kranakis et al. (Eds.): LATIN 2016, LNCS 9644, pp. 619–633, 2016.
DOI: 10.1007/978-3-662-49529-2_46

others include applying restrictions on inputs, approximation algorithms and parameterized complexity.

Graph modification problems have been at forefront of research in parameterized complexity and several interesting and important results have been obtained recently. In fact, just over the course of the last couple of years there have been results on parameterized algorithms for CHORDAL EDITING [8], UNIT INTERVAL EDITING [6], INTERVAL VERTEX (EDGE) DELETION [7,9], PROPER INTERVAL COMPLETION [4], INTERVAL COMPLETION [5], CHORDAL COMPLETION [17], CLUSTER EDITING [15], THRESHOLD EDITING [12], CHAIN EDITING [12], TRIVIALLY PERFECT EDITING [13,14] and SPLIT EDITING [19]. Even more recently, a theory for lower bounds for these problems has also been proposed [3]. We would like to mention that the above list is not comprehensive but rather illustrative.

All the articles on graph modification problems in parameterized complexity (except one) is about modifying the input graph to a graph in a family, $\mathcal{F}$, defined by either forbidden induced subgraphs or minors. In our earlier paper [25] we studied graph modification problems on undirected graph defined by reducing the rank of the adjacency matrix of the input graph. Our main goal of this article is to continue our study of graph modification problems defined by some algebraic properties like rank of a given adjacency matrix. In particular we extend our results to directed graphs from undirected graphs in the realm of classical and parameterized complexity. In parameterized complexity, each instance of the problem is *parameterized*, i.e. assigned a number k, which is called the *parameter*, representing some property of the instance. For example, a typical parameter is the size of the optimum solution to the instance. The problem is called *fixed parameter tractable* (FPT), if a parameterized instance (x, k) of the problem is solvable in time $f(k)n^{\mathcal{O}(1)}$, where n is the size of the instance. A parameterized problem is said to admit a *polynomial kernel* if there is a polynomial time algorithm that given an instance of the problem, outputs an equivalent instance whose size is bounded by a polynomial in the parameter.

Given a fixed positive integer r, we define $\mathcal{F}_r$ as the family of directed graphs where for each $G \in \mathcal{F}_r$, the rank of the adjacency matrix of G is at most r. In particular we study the following problems in this paper.

---

r-RANK VERTEX DELETION   **Parameter:** k
Input: A directed graph D and a positive integer k
Question: Does there exist a set $S \subseteq V(D)$ of size at most k such that $\mathsf{rank}(A_{D \setminus S}) \le r$?

---

r-RANK EDGE DELETION   **Parameter:** k
Input: A directed graph D and a positive integer k
Question: Does there exist a set $F \subseteq E(D)$ of size at most k such that $\mathsf{rank}(A_{D'}) \le r$, where $D' = (V(D), E(D) \setminus F)$?

These problems are also related to some well known problems in graph algorithms. If $\mathsf{rank}(A_D) = 0$, then D is a disjoint union of isolated vertices. So for $r = 0$,

r-RANK VERTEX DELETION is the well known VERTEX COVER problem. Similarly for $r = 2$, after converting an undirected graph to a directed graph, a solution to r-RANK EDGE DELETION is a complement of a solution to the MAXIMUM EDGE BICLIQUE problem. In the MAXIMUM EDGE BICLIQUE problem the goal is to find a complete bipartite subgraph of the given graph with maximum number of edges [26] possible. These problems are also related to the concept of "rigidity of matrices" [23].

Our Results and Methods. In this paper, we obtain the following results about these problems.

1. We first show that both the problems are NP-Complete.
2. Then we show that these problems are FPT by designing an algorithm with running time $2^{\mathcal{O}(k \log r)} n^{\mathcal{O}(1)}$ for r-RANK VERTEX DELETION, and an algorithm for r-RANK EDGE DELETION running in time $2^{\mathcal{O}(f(r)\sqrt{k} \log k)} n^{\mathcal{O}(1)}$. Note that the edge deletion problem admits a sub-exponential FPT algorithm.
3. Finally, we design polynomial kernels for these problems.

The main difficulty in extending our result from undirected to directed is that the entries of an adjacency matrix of an undirected graph are from $\{0, 1\}$, while the entries of an adjacency matrix of a directed graph are from $\{0, 1, -1\}$. Also, the adjacency matrix of an undirected graph is symmetric while for a directed graph it is only skew-symmetric. Our algorithmic results are based on a structural result regarding directed graphs whose adjacency matrices have low rank. In particular we define an equivalence relation on the neighborhood of directed graphs and show that the size of directed graphs, in which each equivalence class has size one, is upper bounded by a function of the rank of its adjacency matrix. In case of undirected graphs, such results were proved by Kotlov and Lovász [20]. In particular they showed that if a graph is reduced (if no two vertices have the same set of neighbours) and its adjacency matrix has rank r then its size is upper bounded by $2^{r/2}$. See Akbari et al. [2] for more details. We show that the "reduced" directed graphs of rank r have size at most 3^r. This result could be of independent interest. We use this result to define a subfamily of $\mathcal{F}_r$ (which is sufficient for our purposes) whose size is upper bounded by a function of r alone. This together with standard methods in parameterized complexity easily imply FPT and kernel result for r-RANK VERTEX DELETION. To obtain sub-exponential algorithm for r-RANK EDGE DELETION we do a modification to an algorithm of Damaschke et. al. [11] and use it as a subroutine.

2   Preliminaries

We use $\mathbb{R}$ to denote the set of real number. For a positive integer n, $[n]$ denotes the set of integers $\{1, \ldots, n\}$. For two sets X and Y, we define $X \bigtriangleup Y = (X \setminus Y) \cup (Y \setminus X)$, i.e. the set of elements which belong to X or Y but not both. A vector v of length n is an ordered sequence of n values from $\mathbb{R}$. A collection of vectors $\{v_1, v_2, \ldots, v_k\}$ is said to be linearly dependent if there exist numbers $a_1, a_2, \ldots, a_k \in \mathbb{R}$, not all zeros, such that $\sum_{i=1}^{k} a_i v_i = 0$; otherwise these vectors are called linearly independent. The span of a set of vectors $\{v_1, \ldots, v_m\}$, denoted as $\mathrm{span}(v_1, \ldots, v_m)$, is defined as the set $\{a_1 v_1 + \cdots + a_m v_m : a_1, \ldots, a_m \in \mathbb{R}\}$. A matrix A of dimension $n \times m$, is a sequence of values (a_{ij}). The i-th row of A is defined as the vector $(a_{i1}, a_{i2}, \ldots, a_{im})$ and is denoted using A_i. The j-th column of A is defined as the vector $(a_{1j}, a_{2j}, \ldots, a_{nj})$ and is represented using A^j. The row set and the column set of A are denoted by $\mathbf{R}(A)$ and $\mathbf{C}(A)$ respectively. For $I \subseteq \mathbf{R}(A)$ and $J \subseteq \mathbf{C}(A)$, we define $A[I, J] = (a_{ij} : i \in I, \ j \in J)$, i.e. it is the submatrix of A with the row set I and the column set J. When $I = J$, the submatrix $A[I, I]$ is called as a principal submatrix of A. The *rank* of a matrix is the cardinality of the maximum sized collection of columns which are linearly independent. Equivalently, the rank of a matrix is the cardinality of the maximum sized collection of rows which are linearly independent. It is denoted by $\mathrm{rank}(A)$.

For a graph G, we use $V(G)$ and $E(G)$ to denote the vertex set and the edge set of G. Let the vertex set be ordered as $V(G) = \{v_1, v_2, \ldots, v_n\}$. We say a graph G is undirected if whenever $(u, v) \in E(G)$ then $(v, u) \in E(G)$, otherwise it will be referred to as a directed graph or a digraph. Whenever we refer to an edge $(u, v) \in E(G)$ of an undirected graph G we mean it to be the pair $(u, v), (v, u) \in E(G)$. The *adjacency matrix* of an undirected graph G, denoted by A_G, is defined as the $n \times n$ symmetric matrix with $\{0, 1\}$ as entries whose rows and columns are indexed by $V(G)$ such that $A_G(i, j) = 1$ if and only if $(v_i, v_j) \in E(G)$.

We use D to denote directed graphs and its edges will be referred as arcs. In this paper we will be exclusively concerned with loopless directed (or undirected) graphs (*simple graphs*) which have at most one arc between any two vertices i.e. for two vertices $u, v \in V(D)$ only one the following conditions holds; (i) $(u, v) \in E(D)$, (ii) $(v, u) \in E(D)$ or (iii) there is no arc between u and v. For a directed graph D, given a vertex $v \in V(D)$ we denote the set of vertices with an outgoing arc to v by $N^-(v) = \{u : (u, v) \in E(D)\}$. Similarly, we define the set of vertices with an incoming arc from v as $N^+(v) = \{u : (v, u) \in E(D)\}$. The *adjacency matrix* of a directed graph D, denoted by A_D, is a $|V(D)| \times |V(D)|$ skew-symmetric matrix $(A = -A^{\mathbf{T}})$ with entries from $\{-1, 0, +1\}$ whose columns and rows are indexed using vertices and $A_D(u, v) = -A_D(v, u) = 1$ if and only if (u, v) is an arc in D. A *directed bi-clique* is a directed graph D, where $V(D) = V_1 \uplus V_2$ and for every pair of vertices $v_1 \in V_1$ and $v_2 \in V_2$, there is an arc $(v_1, v_2) \in E(G)$. Note that in a directed bi-clique all the arcs are directed from one part to the other.

For a vertex $v \in V(G)$, we use $R(v)$ to denote the row vector of A_G corresponding to v. Similarly we use $C(v)$ to denote the column vector of A_G corresponding to v. For $X, Y \subseteq V(G)$, we use $A_G[X, Y]$ to denote the submatrix of A_G corresponding to rows in $R(X)$ and columns in $C(Y)$. An *independent set* in a graph G is a set of vertices X such that for every pair of vertices $u, v \in X$,

$(u, v) \notin E(G)$. For a set of vertices U and a graph G, $G[U]$ denotes the induced subgraph of G, and $G \setminus U$ denotes the graph $G[V(G) \setminus U]$.

3   A Structural Result on Directed Graphs of Rank r

In this section we give our main structural result about directed graphs of low rank. This result will be exploited heavily in all the algorithms presented here. We start by defining an equivalence relation on the neighborhoods of directed graphs. Using this we show that the number of vertices in a digraph of rank r, where each equivalence class of this relation has size exactly one, is upper bounded by 3^r.

Given a directed graph D we define a relation $\sim$ on $V(D)$. Two vertices $u, v \in V(D)$ are $u \sim v$ if they have the *same neighborhood*. That is, $u \sim v$ if and only if $N^+(u) = N^+(v)$ and $N^-(u) = N^-(v)$. From the definition of the equivalence relation $\sim$ the following properties regarding it easily follow.

Lemma 1. *Let D be a digraph and $\sim$ be the relation as defined above. Then,*

(i) $u \sim v$ if and only if $R(u) = R(v)$.
(ii) $\sim$ partitions $V(D)$ as $V_1,,\ldots,V_\ell$ and each V_i is an independent set in D.
(iii) For each pair of distinct V_i and V_j, either there are no edges between V_i and V_j or $D[V_i \uplus V_j]$ is a directed bi-clique.

When the graph is clear from the context, an equivalence class of $\sim$ will be referred to as a *module* or an *equivalence class*. Given a directed graph D, we define $D^\sim$ to be the *reduced graph* of D obtained as follows. Let $V_1, \ldots, V_\ell$ denote the equivalence classes of $\sim$ in D. The vertex set of $D^\sim$ is $V(D^\sim) = \{V_i : i \in [\ell]\}$. The edge set of $D^\sim$ is $E(D^\sim) = \{(V_i, V_j) : D[V_i \uplus V_j]$ is a directed bi-clique$\}$. For a digraph D the operation of obtaining the reduced graph can also be thought of as selecting one vertex from each equivalence class arbitrarily and then constructing the induced subgraph of D over them. This gives us the following.

Observation 1. *$D^\sim$ is an induced subgraph of D and the columns of adjacency matrix of $D^\sim$ are distinct.*

The operation of obtaining a reduced graph preserves the rank.

Lemma 2. $(\star)$.[1] $\mathsf{rank}(A_D) = \mathsf{rank}(A_{D^\sim})$.

In the rest of the section we prove that the number of directed graphs whose adjacency matrix has rank r is bounded by a function of r. Similar result exists for the case of undirected graphs [20].

Lemma 3. $(\star)$. *Any skew-symmetric matrix A has a principal submatrix of rank equal to $\mathsf{rank}(A)$.*

---

[1] Due to space constraints proofs of results marked $\star$ have been omitted. They appear in the full version of the paper.

Using the proof above and noting that the determinant of an odd sized skew-symmetric matrix is zero we can derive the following lemma.

Lemma 4. *The rank of a skew-symmetric matrix is an even number.*

The following corollary is a consequence of Lemma 3.

Corollary 1. *Any rank r skew-symmetric matrix A, with full rank principal submatrix $B = A\,[[r],[r]]$, can be written in the following form for an appropriate sized matrix X,*

$$\begin{bmatrix} B & BX \\ -X^T B & -X^T BX \end{bmatrix}. \tag{1}$$

Theorem 2. *Any rank r skew-symmetric matrix A with entries from $\{-1, 0, +1\}$ has at most 3^r distinct columns.*

Proof. For an $n \times n$ matrix A, let $R = [r]$ and assume without loss of generality that the principal submatrix $B = A[R, R]$ is of full rank. By Corollary 1, matrix A can be written in the form as shown in Eq. 1. We prove that no two columns in the matrix $A[R, [n]]$ are repeated. As B is of full rank it has no repeated columns. Assume that the columns i and j in BX are repeated. As A is skew-symmetric, the rows i and j are repeated in $-X^T B$. Therefore, the product $(-X^T B)X$ has rows i and j repeated which again, as A is skew-symmetric, implies that the columns i and j are repeated in $-X^T BX$. Hence, the columns i and j are repeated in A. Therefore it suffices to enforce that no column in $A[R, [n]]$ is repeated. For an upper bound notice that the maximum possible number of distinct vectors of length r with entries from $\{-1, 0, +1\}$ is 3^r.   □

As a consequence of the theorem above we have the following.

Theorem 3. *Let D be a digraph having adjacency matrix A of rank r. Then the number of vertices in $D^\sim$ is at most 3^r.*

The following lemma provides an operation which does not change the rank of adjacency matrix of a given graph.

Lemma 5. *Given a digraph D and a vertex $u \in V(D)$, let D' be the graph obtained by adding a new vertex u' in D which has exactly the same neighborhood as u. Then $\mathsf{rank}(A_D) = \mathsf{rank}(A_{D'})$.*

Proof. The adjacency matrix of D can be obtained from that of D' by deleting the column and row corresponding to u' from the matrix $A_{D'}$ which does not change the rank.   □

4   Reducing Rank by Deleting Vertices

We will be considering r-RANK VERTEX DELETION in this section. For a graph D, we denote the graph obtained after deleting a set of vertices $S \subseteq V[D]$ by $D \setminus S$. We begin with some structural lemma about modules of induced graphs.

Lemma 6 ($\star$). *Let D be a digraph, $S \subseteq V(D)$ and $u, v \in V(D) \setminus S$. If u and v have the same neighborhood in D then they have the same neighborhood in $D \setminus S$ and* $\mathsf{rank}(A_{D \setminus S}) \leq \mathsf{rank}(A_D)$.

We will be working with solutions of a special form, we define S to be a *minimal solution* to an instance (D, k) of r-RANK VERTEX DELETION if it either contains all the vertices of an equivalence class defined by $\sim$ on $V(D)$ or none of it.

Lemma 7 ($\star$). *Let (D, k) be an input instance of r-RANK VERTEX DELETION and $\sim$ be the equivalence relation on $V(D)$. If $S \subseteq V(D)$ is a solution for the instance (D, k) then there exists a* minimal *solution S' such that $S' \subseteq S$.*

As an immediate consequence of the lemma above, we have the following.

Corollary 2. *Any* minimal *solution S for an instance (D, k) of r-RANK VERTEX DELETION is disjoint from a $\sim$-module of size strictly more than k.*

4.1 Complexity of r-Rank Vertex Deletion

The r-RANK VERTEX DELETION problem generalizes VERTEX COVER. In this section we give a reduction from VERTEX COVER to r-RANK VERTEX DELETION which also proves that no parameterized algorithm is possible for r-RANK VERTEX DELETION which uses rank alone as the parameter.

Theorem 4 ($\star$). r-RANK VERTEX DELETION *is* NP-Complete.

4.2 A Parameterized Algorithm for r-Rank Vertex Deletion

We start by recalling that the rank of any skew-symmetric matrix is always even. In the rest of the paper we always assume that rank is even. We will be using the following result that characterizes skew-symmetric matrices that have rank less than or equal to r.

Theorem 5 ($\star$). *For an even positive integer r, a skew-symmetric matrix A has rank less than or equal to r if and only if all it's submatrices of size $(r+2) \times (r+2)$ have rank less than or equal to r.*

We obtain an algorithm and a kernel by reducing r-RANK VERTEX DELETION to $(2r + 4)$-HITTING SET, which has a parameterized algorithm and a polynomial kernel [1]. Given a set family $\mathcal{F}$ over a universe U, a set S is said to be a *hitting set* of $\mathcal{F}$ if $S \cap F \neq \varnothing, \forall F \in \mathcal{F}$. An input to d-HITTING SET consists of $(U, \mathcal{F}, k)$ where $\mathcal{F}$ is a family of subsets of U of size at most d and the objective is to check if there exists a set $S \subseteq U$ of size at most k which is a hitting set of $\mathcal{F}$. Given a graph D with adjacency matrix A_D, define the set family $\mathcal{H}_r(D)$ as follows,

$$\mathcal{H}_r(D) \triangleq \{X \cup Y : X, Y \subseteq V(D), |Y| = |X| = \mathsf{rank}(A_D[X, Y]) = r + 2\} \quad (2)$$

Lemma 8 ($\star$). *An input instance* (D, k) *is a yes instance of* r-RANK VERTEX DELETION *if and only if* $(\mathcal{H}_r(D), k)$ *is a yes instance of* $(2r + 4)$-HITTING SET.

Lemma 8 allows us to reformulate our problem as $(2r + 4)$-HITTING SET and thus using the known algorithm and kernel for the problem we get the following result [1,10]. Also see Theorems 6 and 7 in [25].

Theorem 6. r-RANK VERTEX DELETION *admits a kernel of size* $\mathcal{O}(k^{2r+4})$ *and an algorithm with running time* $2^{\mathcal{O}(k \log r)} n^{\mathcal{O}(1)}$.

5   Deleting Arcs to Reduce Rank

This section considers the problem of rank reduction of the adjacency matrix of a directed graph D by deletion of arcs. We prove several properties for the general case of editing arcs. In the case of editing a graph, a solution F is a subset of $V(D) \times V(D)$. Note that deletion corresponds to the case when $F \subseteq E(D)$. Let the modules of $\sim$ in D be $V_1, \ldots, V_t$. We call F a *minimal* edit if it either contains all the arcs in $V_i \times V_j$ or none of them, for all $i, j \in [t]$. We call a vertex $v \in V(D)$ an *affected* vertex if it receives an edit i.e. there exists an $e \in F$ such that v is contained in the arc e. For simplicity we use $D \triangle F$ to denote the graph obtained after performing edits in F on D. Let F be a set of edits on D, we use $J_{in}(u) = \{w : w \in V(D), (w, u) \in F\}$ and $J_{out}(u) = \{w : w \in V(D), (u, w) \in F\}$ to denote the set of *affected* vertices due to edits on arcs incident on $u \in V(D)$.

Lemma 9. *Let the equivalence classes of* $\sim$ *in* D *be* $V_1, \ldots, V_t$, *and the equivalence classes of* $\sim$ *in* $D' = D \triangle F$ *be* $U_1, \ldots, U_s$, *for some set* $F \subseteq V(D) \times V(D)$. F *is a* minimal *edit of* D *if and only if for all* $i \in [t]$ *there exists a unique* $j \in [s]$ *such that* $V_i \subseteq U_j$.

Proof. For the forward implication, if $|V_i| = 1$ then it is true, otherwise let $u, v \in V_i$ be any two vertices. If F does not *affect* u or v then they both go in the same module in D' as they have the *same neighborhood*. If F *affects* u but it does not *affect* v then there is an element $e \in F$ such that u is adjacent to e, without loss of generality let $e = (u, w)$. Since v is not *affected*, (v, w) is not present in F, a contradiction. In the case that F *affects* both u and v, let $J_{in}[u]$ and $J_{out}[u]$ be the set of *affected* vertices having edited arcs to and from u as defined earlier. If $J_x[u] = J_x[v]$ for $x \in \{in, out\}$ then both u and v recieve same edits and have the *same neighborhood* in D'. Otherwise, suppose $J_x[u] \neq J_x[v]$ for some $x \in \{in, out\}$. Both $J_x[u] \setminus J_x[v]$ and $J_x[v] \setminus J_x[u]$ cannot be empty, assume that $J_x[u] \setminus J_x[v]$ is non-empty and let $x = in$, other cases can be handled similarly. Let $w \in J_x[u] \setminus J_x[v]$, since $(w, v) \notin F$, F is not *minimal*, a contradiction. Since u and v are arbitrary, the statements above hold for all vertices in V_i therefore all the vertices of V_i have the *same neighborhood* in D' and hence belong to the same module in D'. Moreover the proof holds for all $i \in [t]$.

For the reverse direction, we construct a set of *minimal* edits which results in the graph D'. Pick any two modules V_i and V_j which have arcs between them, note that all directed arcs are present between them, for clarity assume the set of arcs is $V_i \times V_j$. If V_i and V_j are in the same module of D' then all the arcs in $V_i \times V_j$ get deleted, hence $V_i \times V_j \subseteq F$. If V_i and V_j are in different modules of D' which have an arc then no arc gets deleted; if they do not have arcs then all the arcs in $V_i \times V_j \subseteq F$ get deleted. Similar argument can be made for the case when V_i and V_j do not have any arcs between them. This proves that F is a set of *minimal* edits. □

Lemma 10. *Let D be a digraph with $V_1, \ldots, V_t$ as modules of $\sim$ in D. For any vertex $v \in V_i$, let $D_{i,j}^v$ be the graph obtained by moving vertex v from module V_i to any other module V_j and performing necessary edits to make its neighborhood same as any other vertex in V_j. Then $\mathsf{rank}(A_{D_{i,j}^v}) \leq \mathsf{rank}(A_D)$.*

Proof. The operation of moving a vertex consists of three operations, deleting vertex v from D, copying a vertex $u \in V_j$ and relabeling the copy as v. Relabeling a vertex does not change the rank, using Lemma 6 and 5, the result follows. □

Lemma 11. *Any set of edits F on a graph D can be transformed into a minimal set of edits F' of D such that $|F'| \leq |F|$ and $\mathsf{rank}(A_{D \triangle F'}) \leq \mathsf{rank}(A_{D \triangle F})$.*

Proof. Let the modules of $\sim$ in D be $V_1, \ldots, V_t$. Denote the modules of $\sim$ in $D' = D \triangle F$ by $U_1, \ldots, U_s$. Suppose F is not a *minimal* edit. By Lemma 9, there exists a V_i such that $V_i \cap U_h \neq \emptyset$ and $V_i \cap U_j \neq \emptyset$ for some $h \neq j$, $h, j \in [s]$. Let $v \in V_i$ be a vertex receiving minimum number of edits i.e. $|J_{in}(x) \cup J_{out}(x)|$ is minimum for $x = v$ among all the vertices $x \in V_i$. Suppose $v \in U_h$, pick a vertex $w \in U_j \cap V_i$. Obtain a new graph D'' by moving the vertex w from module U_j to U_h. Let F'' be the new set of edits required to get D'' from D, clearly $|F''| \leq |F|$ by the minimality of edits performed on v and observing that if u ends up in the same module as v then same number of edits as v have been performed on u in D. Also, by Lemma 10, $\mathsf{rank}(A_{D''}) \leq \mathsf{rank}(A_{D'})$.

We reuse the names $U_1, \ldots, U_s$ for the modules of D'', only two modules get affected, one of them possibly becoming empty due to moving. Apply the above procedure for every vertex in V_i but not in U_h. Inductively, the rank and the number of edits performed on D never increase. The process stops when there are no such vertices, when this happens $V_i \subseteq U_h$.

Apply the procedure above for all $i \in [t]$ violating the condition of F being minimal. Finally all the modules satisfy the condition that there exists a unique $j \in [s]$ such that $V_i \subseteq U_j$, by the reverse direction of Lemma 9 we get that the set of edits performed on D is *minimal*. □

Using Lemma 11 we get the following.

Corollary 3. *Any* minimal *solution F of an instance (D, k) of r-RANK EDGE DELETION is disjoint from the arcs incident on any module having more than k vertices.*

5.1   NP Completeness

In this section we prove the hardness result for the arc deletion problem r-RANK EDGE DELETION. The problem of r-RANK EDGE DELETION is NP-Complete for $r \geq 2$. We will first prove a lemma which characterises "bipartite" digraphs of rank 2.

Definition 1. A *bi-digraph* is a digraph D whose vertex set consists of two disjoint sets V_1 and V_2 and its arc set $E(D)$ is a subset of $(V_1 \times V_2) \uplus (V_2 \times V_1)$. A *simple-digraph* is a *bi-digraph* such that for $i \in [2]$ and any two non-isolated vertices $a, b \in V_i$, exactly one of the following conditions holds,

1. $N^-(a) = N^-(b)$ and $N^+(a) = N^+(b)$.
2. $N^-(a) = N^+(b)$ and $N^+(a) = N^-(b)$.

Note that a directed bi-clique is a *bi-digraph* as well as a *simple-digraph*.

Lemma 12. *A* bi-digraph *is a* simple-digraph *if and only if the rank of it's adjacency matrix is 2. In addition, a* simple-digraph *is the disjoint union of isolated vertices and a directed graph D such that the underlying undirected graph of D is a complete bi-partite graph.*

Proof. A *bi-digraph* D, by construction, has an adjacency matrix of the following form

$$A_D = \begin{bmatrix} \mathbf{0} & A \\ -A^T & \mathbf{0} \end{bmatrix}, \tag{3}$$

where A is a matrix whose upper half of rows are indexed by V_1 and the lower half by V_2 and $\mathbf{0}$ is the all zeroes matrix of appropriate size.

For the forward implication, observe that $\mathsf{rank}(A_D) = 2 \times \mathsf{rank}(A)$. Therefore it suffices to prove that $\mathsf{rank}(A) = 1$. By Definition 1, for any two non-isolated vertices $a, b \in V_2$, the columns A^a and A^b are related as $A^a = \pm A^b$. Therefore, a column of A is either all-zero or one of $\pm A^a$, which implies that $\mathsf{rank}(A) = 1$.

For the backward implication, suppose D is a *bi-digraph* with $\mathsf{rank}(A_D) = 2$, which implies that $\mathsf{rank}(A) = 1$. Therefore, there exists a non-zero column A^j such that every column of A is contained in its span. As the entries of A are restricted to take values from $\{-1, 0, 1\}$, each column of A is equal to one of A^j, $-A^j$ or the all-zero column. Thus D is a *simple-digraph*.

To prove the last property, observe that deleting the all-zero columns and rows (which correspond to isolated vertices) from A results in a matrix which consists exclusively of $+1$ or -1. The adjacency matrix of the underlying undirected graph is obtained by mapping each -1 to $+1$ in the adjacency matrix of a directed graph, in this case it results in the adjacency matrix of an undirected complete bi-partite graph.   □

The problem r-RANK EDGE DELETION is polynomial time solvable for $r = 0$ as the solution consists of deleting all the arcs. We prove that r-RANK EDGE DELETION is NP-Complete for any $r \geq 2$.

Theorem 7. r-RANK EDGE DELETION *is* NP-Complete *for $r \geq 2$.*

Proof. We give a reduction from the MAXIMUM BI-CLIQUE problem which was shown to be NP-Complete by Peeters [26]. The MAXIMUM BI-CLIQUE problem takes an undirected graph G and a positive integer K as input and outputs "yes" if and only if there exists a bi-clique (a complete bi-partite graph) on vertex sets $A, B \subseteq V(G)$ such that $|A| \times |B| \geq K$.

Given an undirected graph G construct the *bi-digraph* D as follows; the vertex set $V(D) = \{u_i : u \in V(G), i \in [2]\}$ and arc set $E(D) = \{(u_1, v_2) : (u, v) \in E(G)\}$. For any $a_i \in V_i$ for $i \in [2]$, we use the un-subscripted symbol a to denote the corresponding vertex in the undirected graph G. Denote the number of arcs in D by $m = 2 \times |E(G)|$. Let (G, K) be an instance of the MAXIMUM BI-CLIQUE problem then $(D, m - K)$ is the input instance of 2-RANK ARC DELETION.

For the forward implication, assume (G, K) has a bi-clique on the sets $A, B \subset V(G)$. Let $S = \{(a_1, b_2) : a \in A, b \in B\}$ be a set of arcs; since (A, B) is a solution, $|S| \geq K$. Now $S' = E(D) \setminus S$ is a solution of $(D, m - K)$ as $D \setminus S'$ is a directed bi-clique from $A_1 = \{a_1 : a \in A\}$ to $B_2 = \{b_2 : b \in B\}$ and is also a *simple-digraph*, by Lemma 12 rank of its adjacency matrix is 2.

For the backward implication, let S be a solution for $(D, m - K)$. By construction, the remaining number of arcs in $D \setminus S$ is more than K. Since rank$(A_{D \setminus S}) = 2$ and $D \setminus S$ is a *bi-digraph*, by Lemma 12 it is a *simple-digraph* and consists of disjoint union of isolated vertices and a directed bi-clique over sets $A_1 \subseteq V_1$ and $B_2 \subseteq V_2$. Let $A = \{a : a_1 \in A_1\}$ and $B = \{b : b_2 \in B_2\}$ be subsets of $V(G)$. We claim that (A, B) is a solution of (G, K). Observe that $A \cap B = \emptyset$, otherwise let $x \in A \cap B$. As A_1 and B_2 induce a directed bi-clique in D, (x_1, x_2) is an arc in D which is not possible as that implies (x, x) is an edge in G which contradicts the assumption that G is a simple undirected graph. Therefore, (A, B) is a solution of (G, K).

The statements above prove that r-RANK EDGE DELETION is NP-Complete for $r = 2$. To prove it NP-Complete for arbitrary r, a trick similar to Theorem 4 can be employed.   □

5.2   A Parameterized Algorithm for r-Rank Edge Deletion

In this section we design a polynomial kernel and a sub-exponential parameterized algorithm for r-RANK EDGE DELETION.

Lemma 13. *For any instance (D, k) of the r-RANK EDGE DELETION problem. If $D^{\sim}$ has more than $3^r + 2k$ vertices then (D, k) is a no instance.*

Proof. Let (D, k) be a yes instance and D' be the graph obtained from D after deleting a *minimal* solution having at most k arcs. The graph D' has rank r, by Theorem 3 it has at most 3^r modules. Any *minimal* solution having size at most k can *affect* at most $2k$ modules of D. If two distinct modules of D end up in the same module of D' then at least one of them must be *affected*. Every module of D ends up in some module of D', therefore there can be at most $3^r + 2k$ modules in D.   □

An application of Lemma 13 and Corollary 3 results in the following kernel for r-RANK EDGE DELETION.

Theorem 8. r-RANK EDITING *admits a kernel with* $\mathcal{O}((3^r + 2k) \cdot (k+1))$ *vertices.*

Proof. Let (D, k) be an input instance of r-RANK EDGE DELETION. If D has more than $3^r + 2k$ modules then by Lemma 13 it is a no instance. Otherwise, construct the graph D' by removing all but $k + 1$ vertices from each module of D. We will show that the instances (D, k) and (D', k) are equivalent.

Let F be a minimal solution of (D', k). Let D'_{k+1} and D_{k+1} be the graphs obtained after deletion of any module having more than k vertices. Observe that D_{k+1} and D'_{k+1} are the same graph (isomorphic) and hence have the same rank implying $\mathsf{rank}(A_{D_{k+1} \setminus F}) = \mathsf{rank}(A_{D'_{k+1} \setminus F})$. It is also easy to see that both $D_{k+1} \setminus F$ and $D'_{k+1} \setminus F$ have the same reduced graph. Let V_i be a module which was deleted. Now, add $|V_i|$ vertices in $D_{k+1} \setminus F$ and $k+1$ vertices in $D'_{k+1} \setminus F$ such that they have the same induced neighborhood as any other vertex in V_i; both the new graphs have the same reduced graph and hence have the same rank. To simplify presentation we denote the graph obtained by the addition of vertices as $(D_{k+1} \setminus F) \cup V_i$. The graphs $(D_{k+1} \setminus F) \cup V_i$ and $(D_{k+1} \cup V_i) \setminus F$ are the same graph because $F \subseteq E(D_{k+1})$ by Corollary 3. Therefore adding back the deleted modules keeps the rank of both the graphs equal i.e. $\mathsf{rank}(A_{D \setminus F}) = \mathsf{rank}(A_{D' \setminus F})$. We can repeat this argument with the assumption that F is a minimal solution of (D, k) to prove the equivalence.

We output (D', k) as the desired kernel. The instance (D', k) has at most $3^r + 2k$ equivalence classes and each module has at most $k + 1$ vertices, so the claimed bound on the size $V(D')$ follows.   □

The family of undirected $\sim$-reduced graphs of rank equal to r can be constructed in time which is a function of r alone [2]. With very minor changes in the algorithm (Sect. 3, [2]), we have the following theorem.

Theorem 9. *The family* $\mathcal{D}_r$ *of directed reduced graphs of rank at most* r *can be constructed in time which is a function of* r *alone.*

We will be making use of an algorithm for the DIRECTED H-BAG DELETION problem which is defined as follows.

DIRECTED H-BAG DELETION   **Parameter:** k
Input: Directed graphs D, H and an integer k, where H is a $\sim$-reduced graph.
Question: Does there exist a set F having at most k arcs such that the $\sim$-reduced graph of $D' = (V(D), E(D) \setminus F)$ is an induced subgraph of H ?

The rest of the section gives an algorithm for DIRECTED H-BAG DELETION and shows how to use it for solving r-RANK EDGE DELETION. Our algorithm for

DIRECTED H-BAG DELETION is similar to the bag-editing algorithm of Damaschke et. al. [11] but is novel in the sense that it handles a different equivalence relation and works with directed graphs. We now present the algorithm for DIRECTED H-BAG DELETION.

Theorem 10 ($\star$). *The* DIRECTED H-BAG DELETION *problem with a fixed graph H and an input instance (D, k) can be solved in $2^{\mathcal{O}(\sqrt{k}\log k)}n^{\mathcal{O}(1)}$ time.*

Making use of the algorithm above we get the algorithm for r-RANK EDGE DELETION.

Theorem 11. *An instance (D, k) of r-RANK EDGE DELETION can be solved in $2^{\mathcal{O}(\sqrt{k}\log k)}n^{\mathcal{O}(1)}$ time.*

Proof. We generate family $\mathcal{D}_r$ of $\sim$-reduced graphs using Theorem 9. For each graph $H \in \mathcal{D}_r$, we check if arc deletions can convert D into a graph having H as reduced graph using Theorem 10. If the answer is "no" for all H, we output "no", otherwise we output "yes". For the correctness of algorithm observe that after k deletions the reduced graph of D is a member of $\mathcal{D}_r$ if and only if it's rank is at most r. □

6 Conclusion

In this paper we presented algorithms for reducing rank of the adjacency matrix of a directed graph by vertex and edge deletions. We note that the algorithm for edge deletion can be modified easily to work for rank reduction by edge editing as well. This paper leaves several questions open. Can the bound for maximum number of vertices in a reduced directed graph of rank r be improved? Is the problem NP-Complete if we allow deletion as well as addition of edges? We believe that the editing version is NP-Complete for $r \geq 4$. Apart from that, can the algorithms and kernels be improved for these problems? In particular, is it possible to get rid of the $\log k$ factor in the exponent of running time for r-RANK EDGE DELETION?

References

1. Abu-Khzam, F.N.: A kernelization algorithm for d-hitting set. J. Comput. Syst. Sci. **76**(7), 524–531 (2010)
2. Akbari, S., Cameron, P. J., Khosrovshahi, G. B.: Ranks and signatures of adjacency matrices (2004)
3. Bliznets, I., Cygan, M., Komosa, P., Mach, L., Pilipczuk, M.: Lower bounds for the parameterized complexity of minimum fill-in and other completion problems. In: SODA 16 (2016, to appear)
4. Bliznets, I., Fomin, F.V., Pilipczuk, M., Pilipczuk, M.: A subexponential parameterized algorithm for proper interval completion. In: Schulz, A.S., Wagner, D. (eds.) ESA 2014. LNCS, vol. 8737, pp. 173–184. Springer, Heidelberg (2014)

5. Bliznets, I., Fomin, F.V., Pilipczuk, M., Pilipczuk, M.: A subexponential parameterized algorithm for interval completion. In: SODA 16 (2016, to appear)
6. Cao, Y.: Unit interval editing is fixed-parameter tractable. In: Halldórsson, M.M., Iwama, K., Kobayashi, N., Speckmann, B. (eds.) ICALP 2015. LNCS, vol. 9134, pp. 306–317. Springer, Heidelberg (2015)
7. Cao, Y.: Linear recognition of almost interval graphs. In: SODA 16 (2016, to appear)
8. Cao, Y., Marx, D.: Chordal editing is fixed-parameter tractable. In: STACS, vol. 25, pp. 214–225. Schloss Dagstuhl - Leibniz-Zentrum fuer Informatik (2014)
9. Cao, Y., Marx, D.: Interval deletion is fixed-parameter tractable. ACM Trans. Algorithms 11(3), 21:1–21:35 (2015)
10. Cygan, M., Fomin, F.V., Kowalik, L., Lokshtanov, D., Marx, D., Pilipczuk, M., Pilipczuk, M., Saurabh, S.: Parameterized Algorithms. Springer, Switzerland (2015)
11. Damaschke, P., Mogren, O.: Editing the simplest graphs. In: Pal, S.P., Sadakane, K. (eds.) WALCOM 2014. LNCS, vol. 8344, pp. 249–260. Springer, Heidelberg (2014)
12. Drange, P.G., Dregi, M.S., Lokshtanov, D., Sullivan, B.D.: On the threshold of intractability. In: Bansal, N., Finocchi, I. (eds.) ESA 2015. LNCS, vol. 9242, pp. 411–423. Springer, Heidelberg (2015)
13. Drange, P.G., Fomin, F.V., Pilipczuk, M., Villanger, Y.: Exploring subexponential parameterized complexity of completion problems. In: STACS, LIPIcs, vol. 25, pp. 288–299. Schloss Dagstuhl - Leibniz-Zentrum fuer Informatik (2014)
14. Drange, P.G., Pilipczuk, M.: A polynomial kernel for trivially perfect editing. In: Bansal, N., Finocchi, I. (eds.) ESA 2015. LNCS, vol. 9294, pp. 424–436. Springer, Heidelberg (2015)
15. Fomin, F.V., Kratsch, S., Pilipczuk, M., Pilipczuk, M., Villanger, Y.: Tight bounds for parameterized complexity of cluster editing with a small number of clusters. J. Comput. Syst. Sci. 80(7), 1430–1447 (2014)
16. Fomin, F.V., Lokshtanov, D., Misra, N., Saurabh, S., F-deletion, P.: Approximation, kernelization and optimal FPT algorithms. In: FOCS (2012)
17. Fomin, F.V., Villanger, Y.: Subexponential parameterized algorithm for minimum fill-in. SIAM J. Comput. 42(6), 2197–2216 (2013)
18. Fujito, T.: A unified approximation algorithm for node-deletion problems. Discrete Appl. Math. 86, 213–231 (1998)
19. Ghosh, E., Kolay, S., Kumar, M., Misra, P., Panolan, F., Rai, A., Ramanujan, M.S.: Faster parameterized algorithms for deletion to split graphs. Algorithmica 71(4), 989–1006 (2015)
20. Kotlov, A., Lovász, L.: The rank and size of graphs. J. Graph Theor. 23(2), 185–189 (1996)
21. Lewis, J.M., Yannakakis, M.: The node-deletion problem for hereditary properties is NP-complete. J. Comput. Syst. Sci. 20(2), 219–230 (1980)
22. Lund, C., Yannakakis, M.: On the hardness of approximating minimization problems. J. ACM 41, 960–981 (1994)
23. Mahajan, M., Sarma, J.: On the complexity of matrix rank and rigidity. Theory Comput. Syst. 46(1), 9–26 (2010)
24. Marx, D., O'Sullivan, B., Razgon, I.: Finding small separators in linear time via treewidth reduction. ACM Trans. Algorithms 9(4), 30 (2013)
25. Meesum, S.M., Misra, P., Saurabh, S.: Reducing rank of the adjacency matrix by graph modification. In: Xu, D., Du, D., Du, D. (eds.) COCOON 2015. LNCS, vol. 9198, pp. 361–373. Springer, Heidelberg (2015)

26. Peeters, R.: The maximum edge biclique problem is np-complete. Discrete Appl. Math. **131**(3), 651–654 (2003)
27. Yannakakis, M.: Node-and edge-deletion NP-complete problems. In: STOC 1978, pp. 253–264. ACM, New York (1978)

New Deterministic Algorithms for Solving Parity Games

Matthias Mnich, Heiko Röglin, and Clemens Rösner[✉]

Department of Computer Science, University of Bonn, Bonn, Germany
mmnich@uni-bonn.de, {roeglin,roesner}@cs.uni-bonn.de

Abstract. We study parity games in which one of the two players controls only a small number k of nodes and the other player controls the $n - k$ other nodes of the game. Our main result is a fixed-parameter algorithm that solves bipartite parity games in time $k^{O(\sqrt{k})} \cdot O(n^3)$ and general parity games in time $(p+k)^{O(\sqrt{k})} \cdot O(pnm)$, where p denotes the number of distinct priorities and m denotes the number of edges. For all games with $k = o(n)$ this improves the previously fastest algorithm by Jurdziński, Paterson, and Zwick (SICOMP 2008).

We also obtain novel kernelization results and an improved deterministic algorithm for graphs with small average degree.

1 Introduction

A parity game [4] is a two-player game of perfect information played on a directed graph G by two players, *even* and *odd*, who move a token from node to node along the edges of G so that an infinite path is formed. The nodes of G are partitioned into two sets V_0 and V_1; the even player moves if the token is at a node in V_0 and the odd player moves if the token is at a node in V_1. The nodes of G are labeled by a *priority function* $p : V \to \mathbb{N}_0$, and the players compete for the parity of the highest priority occurring infinitely often on the infinite path $v_0, v_1, v_2 \dots$ describing a play: the even player wins if $\limsup_{i \to \infty} p(v_i)$ is even, and the odd player wins if it is odd.

The winner determination problem for parity games is the algorithmic problem to determine for a given parity game $G = (V_0 \uplus V_1, E, p)$ and an initial node $v_0 \in V_0 \cup V_1$, whether the even player has a winning strategy in the game if the token is initially placed on node v_0. We say that an algorithm for this problem *solves* parity games. Parity games have various applications in computer science and the theory of formal languages and automata in particular. They are closely related to other games of infinite duration, such as mean payoff games, discounted payoff games, and stochastic games [9]. Solving parity games is linear-time equivalent to the model checking problem for the modal μ-calculus [18]. Hence, any parity game solver is also a model checker for the μ-calculus (and vice versa).

This research was supported by ERC Starting Grant 306465 (BeyondWorstCase).

E. Kranakis et al. (Eds.): LATIN 2016, LNCS 9644, pp. 634–645, 2016.
DOI: 10.1007/978-3-662-49529-2_47

Many algorithms have been suggested for solving parity games [3,11,19,20], yet none of them is known to run in polynomial time. McNaughton [13] showed that the winner determination problem belongs to the class NP ∩ coNP, and Jurdziński [9] strengthened this to UP ∩ coUP. It is a long-standing open question whether parity games can be solved in polynomial time. The fastest known deterministic algorithm is due to Jurdziński, Paterson, and Zwick [11] and it has a run time of $n^{O(\sqrt{n})}$ for general parity games and of $n^{O(\sqrt{n/\log n})}$ for parity games in which every node has out-degree at most two. The fastest known randomized algorithm for general parity games is due to Björklund et al. [3] and it has a run time of $n^{O(\sqrt{n/\log n})}$.

As a polynomial-time algorithm for solving parity games has remained elusive, researchers have started to consider which restrictions on the game allow for polynomial-time algorithms. One such well-studied restriction is the treewidth t of the underlying undirected graph G of the game. Obdržálek [15] found an algorithm solving parity games on n nodes in time $n^{O(t^2)}$. Later, Fearnley and Lachish [5] gave an algorithm solving parity games in time $n^{O(t \log n)}$. Another well-studied parameter for parity games is the number p of distinct priorities by which the nodes of the game are labeled. The progress-measure lifting algorithm by Jurdziński [10] solves parity games in time $O(pm(2n/p)^{p/2})$, where m denotes the number of edges of G. This run time has been improved by Schewe [17] to $O(m((2e)^{3/2}n/p)^{p/3})$. Fearnley and Schewe [6] presented an algorithm for solving parity games with run time $O(n(t+1)^{t+5}(p+1)^{3t+5})$, assuming that a tree decomposition of G with width t is given.

For a given parameter κ, one usually aims for *fixed-parameter algorithm* algorithms, i.e., algorithms that run in time $f(\kappa) \cdot n^c$ for some computable function f and some constant c that is independent of κ. Such an algorithm can be practical for large instances if f grows moderately and c is small. From the previously mentioned algorithms only the algorithm by Fearnley and Schewe [6] is a fixed-parameter algorithm for the combined parameter (t,p). It is not known if fixed-parameter algorithms exist for the parameter t or the parameter p alone.

Further parameters for which polynomial-time algorithms for parity games have been suggested include DAG-width [1], clique-width [16], and entanglement [2]; none of these are fixed-parameter algorithms.

1.1   Our Contributions

We study as parameter the number k of nodes that belong to the player who controls the smaller number of nodes in the parity game. Our first result is a *subexponential* fixed-parameter algorithm for solving general parity games for parameters p and k and for parameter only k for bipartite parity games (where players alternate between their moves).

Theorem 1. *There is a deterministic algorithm that solves any parity game G on n nodes and m edges in time $(p+k)^{O(\sqrt{k})} \cdot O(pnm)$, where k denotes the minimum number of nodes owned by one of the players and p the number of distinct priorities. If G is bipartite, the algorithm runs in time $k^{O(\sqrt{k})} \cdot O(n^3)$.*

Thus, our algorithm is particularly efficient if the game is unbalanced, in the sense that one player owns only k nodes and the other player owns the remaining $n - k \gg k$ nodes.

Let us remark that it is not very hard to show fixed-parameter tractability for parameter $p + k$; indeed McNaughton's algorithm [13] can be shown to run in time $p^k \cdot n^{O(1)}$, and this was improved to $p^{\log k} \cdot 4^k \cdot n^{O(1)}$ by Gajarský et al. [7]. Our key contribution here is to reduce the dependence of k to a *subexponential* function. Indeed, this improvement allows us to derive the following immediate corollary of Theorem 1 to expedite the run time for solving *general* parity games.

Corollary 2. *There is a deterministic algorithm that solves parity games in time $n^{O(\sqrt{k})}$.*

Our algorithm is asymptotically always at least as fast as the fastest known deterministic parity game solver by Jurdziński, Paterson, and Zwick [11], which runs in time $n^{O(\sqrt{n})}$. For the case $k = o(n)$, our algorithm is asymptotically faster than theirs and constitutes the fastest known deterministic solver for such games.

We also prove the existence of a small kernel, as our second result. For a parameterized problem, a *kernelization algorithm* takes as input an instance x with parameter κ and computes in time $(|x| + \kappa)^{O(1)}$ an equivalent instance x' with parameter κ' (a *kernel*) with size $|x'| \leq g(\kappa)$, for some computable function g; here, equivalent means that an optimal solution for x can be derived in polynomial time from an optimal solution of x'.

Theorem 3. *Parity games can be kernelized in time $O(pmn)$ to at most $(p + 1)^k + (p + 1)k$ nodes, and bipartite parity games can be kernelized in time $O(n^3)$ to at most $k + 2^k \cdot \min\{k, p\}$ nodes and at most $k2^k \cdot \min\{k, p\}$ edges.*

This kernelization result is not only interesting for its own sake, but it is also an important ingredient in the proof of Theorem 1.

As our third result, we generalize the algorithm by Jurdziński, Paterson, and Zwick [11] for parity games with maximum out-degree 2 to arbitrary out-degree Δ.

Theorem 4. *There is a deterministic algorithm that solves parity games on n nodes out of which s_j nodes have out-degree at most j in time*

$$n^{O\left(\min_{1 \leq j \leq n}\left\{\sqrt{n - s_j} + \sqrt{\frac{s_j}{\log_j s_j}}\right\}\right)} .$$

Corollary 5. *There is a deterministic algorithm that solves parity games on n nodes with maximum out-degree Δ in time $n^{O(\sqrt{\log(\Delta) \cdot n / \log(n)})}$ and parity games on n nodes with average out-degree Δ in time $n^{O(\sqrt{\log(\log(n)\Delta) \cdot n / \log(n)})}$.*

Due to space constraints, all proofs are deferred to the full version [14] of this paper.

1.2   Detailed Comparison with Previous Work

Let us discuss in detail how our results compare to previous work. It is well-known (cf. [12, Lemma 3.2]) and easy to prove that the treewidth of a complete bipartite graph equals the size of the smaller side. Since the treewidth of a graph can only decrease when deleting edges, the graph underlying a bipartite parity game in which one player owns k nodes has a treewidth of at most k. However, as it is not known if there exists a fixed-parameter algorithm for parameter treewidth, the result in Theorem 1 for the bipartite case does not follow from previous work about parity games with bounded treewidth. As a parity game in which one player owns k nodes can have up to n different priorities, also the fixed-parameter algorithm for the combined parameter (t, p) by Fearnley and Schewe [6] does not imply our result.

The algorithm of Jurdziński, Paterson, and Zwick [11] for parity games with maximum out-degree two with run time $n^{O(\sqrt{n/\log n})}$ can easily be generalized to arbitrary parity games at the expense of its run time. For this, one only needs to observe that every parity game can be transformed into a game with maximum out-degree two by replacing each node with a higher out-degree by an appropriate binary tree. This transformation increases the number of nodes from n to $\Theta(m)$ where m denotes the number of edges in the original parity game. Hence, the run time becomes $m^{O(\sqrt{m/\log m})} = n^{O(\sqrt{m/\log n})}$. For graphs with average out-degree $\Delta = \omega(\log \log n)$ the resulting run time of $n^{O(\sqrt{\Delta n/\log n})}$ is asymptotically worse than the run time we obtain in Corollary 5.

For graphs in which the variance of the out-degrees is large, our algorithm can even be better than stated in Corollary 5. If, for example, there are $n^{1-\varepsilon}$ nodes with an arbitrary out-degree for some $\varepsilon > 0$ and all remaining nodes have constant out-degree at most c then our algorithm has a run time of $n^{O(\sqrt{\frac{n}{\log n}})}$ (the minimum in Theorem 4 is assumed for $j = c$). This matches the best known bound for randomized algorithms.

Gajarský et al. [7] present an algorithm that solves parity games in time $w^{O(\sqrt{w})} \cdot n^{O(1)}$, where w denotes the modular width of G. Since the modular width of a bipartite graph can be exponential in the size of the smaller side, Theorem 1 does not follow from this result.

2   Fundamental Properties of Parity Games

A parity game $G = (V_0 \uplus V_1, E, p)$ consists of a directed graph $(V_0 \uplus V_1, E)$, where V_0 is the set of *even* nodes and V_1 is the set of *odd* nodes, and a priority function $p : V_0 \cup V_1 \to \mathbb{N}_0$. We often abuse notation and also refer to $(V_0 \uplus V_1, E)$ as the graph G. For each node $v \in V(G)$, we denote by $N_G^+(v)$ and $N_G^-(v)$ the set of out-neighbors and in-neighbors of v in G, respectively.

Two standard assumptions about parity games are (1) that G is bipartite with $E \subseteq (V_0 \times V_1) \cup (V_1 \times V_0)$, and (2) that each node $u \in V$ has at least one outgoing edge $(u, v) \in E$. The first assumption is often made because it is easy to transform a non-bipartite instance into a bipartite instance. However, the usual transformation increases the number of nodes in V_i by an amount of $|\{v \in V_{1-i} \mid N_G^-(v) \cap V_{1-i} \neq \emptyset\}|$, and can therefore increase the parameter $k = \min\{|V_0|, |V_1|\}$ significantly. We therefore consider bipartite and non-bipartite instances separately in Theorem 1.

We write $n = |V(G)|$, $m = |E|$ and $p = |\{p(v) \mid v \in V(G)\}|$. The game is played by two players, the *even* player (or player 0) and the *odd* player (or player 1). The game starts at some node $v_0 \in V(G)$. The players construct an infinite path (a *play*) as follows. Let u be the last node added so far to the path. If $u \in V_0$, then player 0 chooses an edge $(u, v) \in E$. Otherwise, if $u \in V_1$, then player 1 chooses an edge $(u, v) \in E$. In either case, node v is added to the path and a new edge is then chosen by either player 0 or player 1. As each node has at least one outgoing edge, the path constructed can always be continued. Let $v_0, v_1, v_2, \ldots$ be the infinite path constructed by the two players and let $p(v_0), p(v_1), p(v_2), \ldots$ be the sequence of the priorities of the nodes on the path. Player 0 *wins* the game if the largest priority seen infinitely often is even, and player 1 wins if the largest priority seen infinitely often is odd.

We will define $p_1(v)$ as $p(v)$ if $p(v)$ is odd and as $-p(v)$ if $p(v)$ is even. This allows us to say that, in case $p_1(v) > p_1(u)$ for some $v, u \in V$, player 1 *prefers* $p(v)$ over $p(u)$. Observe that removing an arbitrary finite prefix of a play in a parity game does not change the winner; we refer to this property of parity games as *prefix independence*. A *strategy* for player $i \in \{0, 1\}$ in a game G specifies, for every finite path $v_0, v_1, \ldots, v_k$ in G that ends in a node $v_k \in V_i$, an edge $(v_k, v_{k+1}) \in E$. A strategy is *positional* if the edge $(v_k, v_{k+1}) \in E$ chosen depends only on the last node v_k visited and is independent of the prefix path $v_0, v_1, \ldots, v_{k-1}$. A strategy for player $i \in \{0, 1\}$ is *winning* (for player i) from a start node v_0 if following this strategy ensures that player i wins the game, regardless of which strategy is used by the other player.

The fundamental determinacy theorem for parity games [4,8] says that for every parity game G and every start node v_0, either player 0 has a winning strategy or player 1 has a winning strategy. Furthermore, if a player has a winning strategy from some node in a parity game, then she also has a winning positional strategy from this node. From now on we will therefore, unless stated differently, assume every strategy to be positional. Given positional strategies s_0 on V_0 and s_1 on V_1 and a start node $v_0 \in V$ the infinite path starting in v_0 corresponding to these strategies consists of a finite prefix and an infinite recurrence of a cycle $C = C(s_0, s_1, v_0)$. We call C the cycle *corresponding* to s_0, s_1, v_0 and say that s_0 and s_1 *create* C. The parity of the highest priority $p(u)$ of all nodes $u \in V(C)$ in cycle C then determines the winner of the game. The *winning set* of player $i \in \{0, 1\}$ is the set $\mathrm{win}_i(G) \subseteq V$ of nodes of the game G from which player i has a winning strategy.

For $i \in \{0, 1\}$, an *i-dominion* is a set of nodes $D \subseteq V$ so that player i can win from every node of D, without leaving D and without allowing the other player to leave D. An example of an i-dominion is the set $\mathrm{win}_i(G)$, but there may be smaller subsets of $\mathrm{win}_i(G)$ that are i-dominions as well. Although finding i-dominions can be just as hard as finding $\mathrm{win}_i(G)$, searching only for dominions with certain properties (e.g. small dominions) can be easier. In our algorithm we will use the fact that once an i-dominion is found, it can easily be removed from the graph, leaving a smaller game to be solved.

Next, we recall some well-known results about parity games that form the basis of the algorithms for solving parity games by McNaughton [13] and Zielonka [20]. We include them here as our algorithm relies on them as well; for a detailed exposition we refer to Grädel et al. [8]. Fix a parity game $G = (V_0 \uplus V_1, E, p)$.

For $i \in \{0, 1\}$, a set $B \subseteq V(G)$ is i-closed if for every $u \in B$ the following holds (we use the notation $\neg i$ for the element $1 - i \in \{0, 1\}$):

- If $u \in V_i$, then there exists some $(u, v) \in E$ such that $v \in B$; and
- if $u \in V_{\neg i}$, then for every $(u, v) \in E$, we have $v \in B$.

In other words, a set B is i-closed if player i can always choose to stay in B while simultaneously player $\neg i$ cannot escape from it, i.e., B is a "trap" for player $\neg i$.

Lemma 6. *For each $i \in \{0, 1\}$, the set $\text{win}_i(G)$ is i-closed.*

Let $A \subseteq V(G)$ be a set of nodes and let $i \in \{0, 1\}$. The i-*reachability* set of A is the set $\text{reach}_i(A)$ of nodes in A together with all nodes in $V(G)\backslash A$ from which player i has a strategy σ to enter A at least once (regardless of the strategy of the other player); we call such a strategy σ an i-*reachability strategy* to A.

Lemma 7. *For $A \subseteq V(G)$ and $i \in \{0, 1\}$, the set $V(G)\backslash\text{reach}_i(A)$ is $(\neg i)$-closed.*

We will from now on assume that the graph of the parity game we operate on is encoded as an adjacency list.

Lemma 8. *For every set $A \subseteq V(G)$ and $i \in \{0, 1\}$, the set $\text{reach}_i(A)$ can be computed in $O(m)$ time, where $m = |E|$ is the number of edges in the game.*

If $B \subseteq V(G)$ is such that for each node $u \in V(G)\backslash B$ there is an edge (u, v) with $v \in V(G)\backslash B$, then the sub-game $G - B$ is the game obtained from G by removing the nodes of B. We will only be using B's for which $V(G)\backslash B$ is an i-closed set for some i. In this case every node in $v \in V(G)\backslash B$ has at least one out-going edge (v, w) with $w \in V(G)\backslash B$ and $G - B$ will therefore be well-defined. The next lemmas show some useful properties of sub-games.

Lemma 9. *Let G' be a sub-game of G and let $i \in \{0, 1\}$. If the node set of G' is i-closed in G, then $\text{win}_i(G') \subseteq \text{win}_i(G)$.*

The next lemma shows that if we know some non-empty subset U of the winning set of some player $\neg i$ in a game G, then computing the winning sets of both players in G can be reduced to computing their winning sets in the smaller game $G - \text{reach}_{\neg i}(U)$.

Lemma 10. *For any parity game G and $i \in \{0, 1\}$, if $U \subseteq \text{win}_{\neg i}(G)$ and $U^* = \text{reach}_{\neg i}(U)$, then $\text{win}_{\neg i}(G) = U^* \cup \text{win}_{\neg i}(G - U^*)$ and $\text{win}_i(G) = \text{win}_i(G - U^*)$.*

The next lemma complements Lemma 10 by providing a way to find a non-empty subset of the winning set of player $\neg i$ in a parity game G or to conclude that player i can win from every node in G.

Lemma 11. *Let G be a parity game with largest priority $p_{\max}$ and let $V_{p_{\max}} \subseteq V(G)$ be the set of nodes with priority $p_{\max}$. Let $i = p_{\max} \pmod 2$ and let $G' = G - \mathsf{reach}_i(V_{p_{\max}})$. Then $\mathsf{win}_{\neg i}(G') \subseteq \mathsf{win}_{\neg i}(G)$. Also, if $\mathsf{win}_{\neg i}(G') = \emptyset$, then $\mathsf{win}_i(G) = V$, i.e., player i wins from every node of G.*

3   Kernelization of Parity Games

In this section, we describe some reduction rules for parity games. Theses rules are such that we can efficiently compute the winning sets of the original parity game once we know the winning sets of the reduced game.

3.1   Non-bipartite Games

Lemma 12. *Any parity game $G = (V_0 \uplus V_1, E, p)$ can be transformed in time $O(pmn)$ to a parity game $G' = (V_0' \uplus V_1', E', p')$ with $V_1' \subseteq V_1$ such that*

- *there are no edges inside V_1', and*
- *for each node $v \in V_0'$ either $N_G^+(v) \subseteq V_1'$ or $N_G^-(v) \subseteq V_1'$, and*
- *$|V_0'| \leq \min\{n + pk, (p+1)^k + pk\}$, where $k = |V_1|$.*

Moreover, G and G' have the same winning sets on V_1' and the winner of the remaining nodes of G can be computed either during the transformation or from the winning sets of G' in linear time.

3.2   Bipartite Games

In this section we give some reduction rules that efficiently reduce any bipartite game $G = (V_0 \uplus V_1, E, p)$ to a structurally simpler bipartite game $G' = (V_0' \uplus V_1', E', p')$, such that the winning sets of G can be efficiently recovered from the winning sets of G'. After exhaustive application of the reduction rules, the reduced game G' will have size bounded by some function of k and p only, independent of the size of G.

The digraphs of our underlying parity game may have self-loops and bidirected edges, but (without loss of generality) no parallel edges between the same two nodes. Thus, whenever parallel edges arise during the application of one of the reduction rules, we remove one of them without explicit mention.

Lemma 13. *Let $G = (V_0 \uplus V_1, E, p)$ be a bipartite parity game, and let $u, v \in V_0$ be such that $N_G^+(v) \subseteq N_G^+(u)$ and $p_1(v) \geq p_1(u)$. Let G' be the parity game obtained from G by deleting the edges $\{(w, u) \in E \mid (w, v) \in E\}$. Then the winning sets of G and G' are equal.*

Lemma 14. *Let $G = (V_0 \uplus V_1, E, p)$ be a bipartite parity game, and let $u, v \in V_0$ be nodes with $N_G^+(u) = N_G^+(v)$ and $p(v) = p(u)$. Let G' be the parity game obtained from G by contracting u and v into a new node v' with priority $p(v)$. Then u and v belong to the same winning set $\mathsf{win}_i(G)$ in G and v' belongs to the winning set $\mathsf{win}_i(G')$ of the same player in G'. For all other nodes the winning sets of G and G' coincide.*

Lemma 15. *Let $G = (V_0 \uplus V_1, E, p)$ be a bipartite parity game, and let $v \in V(G)$ be such that $N_G^-(v) = \emptyset$. Then for the parity game $G' = G - v$ and for $i \in \{0, 1\}$, any node $v' \neq v$ is winning for player i in G if and only if it is winning for player i in G'.*

Lemma 16. *Let $G = (V_0 \uplus V_1, E, p)$ be a parity game with largest priority $p_{\max} = \max\{p(v) \mid v \in V(G)\}$. If $p^{-1}(z) = \emptyset$ for some $z \in \{1, \ldots, p_{\max}\}$ then let $G' = (V_0 \uplus V_1, E, p')$ be the parity game obtained from G by setting $p'(v) = p(v) - 2$ for all $v \in V$ with $p(v) > z$ and $p'(v) = p(v)$ for all $v \in V$ with $p(v) < z$. Then the winning sets of the games G and G' coincide.*

Corollary 17. *In any parity game with maximum priority $p_{\max}$ to which the reduction rule described in Lemma 16 cannot be applied anymore, the set of priorities is either $\{0, 1, \ldots, p_{\max}\}$ or $\{1, \ldots, p_{\max}\}$.*

Lemma 18. *Let $G = (V_0 \uplus V_1, E, p)$ be a bipartite parity game with $|V_1| = k$ that is reduced according to Lemmas 13–15. Then $|V_0| \leq 2^k \cdot \min\{k, p\}$.*

Lemma 19. *There exists a sequence of applications of the reduction rules described in Lemmas 13–16 with a total run time of $O(n^3)$ that leads to a game in which none of these rules applies anymore.*

4   A Simple Exponential-Time Algorithm

A simple algorithm with run time $O(2^n)$ for the solution of parity games originates from the work of McNaughton [13] and was first presented for parity games by Zielonka [20]; see also Grädel et al. [8]. Algorithm $\mathbf{win}(G)$ receives a parity game G and returns the pair of winning sets $(\mathsf{win}_0(G) = W_0, \mathsf{win}_1(G) = W_1)$.

Algorithm $\mathbf{win}(G)$ is based on Lemmas 10 and 11. Let $p_{\max}$ be the largest priority in G and let $V_{p_{\max}}$ be the set of nodes with priority $p_{\max}$. Let $i = p_{\max}$ (mod 2) be the player who owns the highest priority. The algorithm first finds the winning sets (W_0', W_1') of the smaller game $G' = G - \mathsf{reach}_i(V_{p_{\max}})$ in a first recursive call. If $W_{\neg i}' = \emptyset$, then by Lemma 11 player i wins from all nodes of G and we are done. Otherwise, again by Lemma 11 we know that $W_{\neg i}' \subseteq \mathsf{win}_{\neg i}(G)$. The algorithm then finds the winning sets (W_0'', W_1'') of the smaller game $G'' = G - \mathsf{reach}_{\neg i}(W_{\neg i}')$ by a second recursive call. By Lemma 10, $\mathsf{win}_i(G) = W_i''$ and $\mathsf{win}_{\neg i}(G) = \mathsf{reach}_{\neg i}(W_{\neg i}') \cup W_{\neg i}'' = V(G) \backslash W_i''$. The pseudocode of $\mathbf{win}(G)$ can be found in the full version [14] of this paper.

5   Overview of the New Algorithms

Before we describe our new algorithms that lead to Theorems 1 and 4 in detail in Sect. 7 and the full version [14] of this paper, we present an overview of the main ideas. The algorithm **new-win**(G) by Jurdziński, Paterson, and Zwick [11] with run time $n^{O(\sqrt{n})}$ is a slight modification of the just described algorithm **win**(G). At the beginning of each recursive call it tests in time $O(n^\ell)$ if the parity game contains a dominion D of size at most $\ell = \lceil\sqrt{2n}\rceil$. If this is the case then D is removed and the remaining game is solved recursively. Else, the parity game is solved by the algorithm **win**(G), except that the recursive calls in lines 4 and 8 are made to **new-win**(G). Since this happens only when G does not contain a dominion of size at most ℓ, the dominion $\mathsf{reach}_j(W'_j)$ that is removed in line 8 has size greater than ℓ and hence, the second recursive call is to a substantially smaller game. Overall, this leads to the improved run time of $n^{O(\sqrt{n})}$.

Our new algorithms are based on a similar idea. Instead of simply searching for a dominion of size at most ℓ, our algorithm **new-win**$_1(G)$ that leads to Theorem 1 searches for a dominion that contains at most $\ell = \lfloor\sqrt{2k}\rfloor$ nodes of the odd player, assuming without loss of generality that the odd player controls fewer nodes, i.e., $k = |V_1|$. If such a dominion is found then we remove it from the game and solve the remaining game recursively. Otherwise, we use the algorithm **win**(G) to solve the parity game, except that the recursive calls in lines 4 and 8 are made to **new-win**$_1(G)$. It can happen that in the game to which the first recursive call in line 4 is made, the odd player controls again k nodes. We will show that in bipartite instances this cannot happen in two consecutive calls. For general instances we use that the observation that at least the number of different priorities decreases by one in the recursive call. Searching efficiently for a dominion that contains at most $\ell = \lfloor\sqrt{2k}\rfloor$ nodes of the odd player is more involved than simply searching for dominions whose total size is at most ℓ. We use multiple recursive calls of **new-win**$_1$ to test if such a dominion exists, which makes the recursion of our algorithm and its analysis more complicated.

Our second algorithm leading to Theorem 4 is based on the same approach and inspired by the algorithm of Jurdziński, Paterson, and Zwick [11]. In this case we let s_j, for some $j \in \mathbb{N}$, equal the number of vertices with out-degree at most j. We separate the nodes into s_j nodes with out-degree at most j and $n - s_j$ nodes with out-degree larger than j and, at the beginning of each iteration, search for and remove dominions that contain at most $\ell = \lceil\sqrt{2(n - s_j)}\rceil$ nodes with out-degree larger than j and at most $s = \lceil\sqrt{s_j \cdot \log_j s_j}\rceil$ nodes with out-degree at most j. This algorithm has a run time of $n^{O\left(\sqrt{n - s_j} + \sqrt{\frac{s_j}{\log_j s_j}}\right)}$, which implies Theorem 4. Since our second algorithm is very similar to our first algorithm, we moved the detailed description and analysis to the full version [14] of this paper.

6   Finding Small Dominions

We now describe how dominions with the previously discussed properties can be found. Let $G = (V_0 \uplus V_1, E, p)$ be a parity game. Recall that for $i \in \{0, 1\}$, a set $D \subseteq V$ is an i-dominion if player i can win from every node of D without ever leaving D, regardless of the strategy of player $\neg i$. Note that any i-dominion must be i-closed. A set $D \subseteq V$ is a *dominion* if it is either a 0-dominion or a 1-dominion. By prefix independence of parity games, the winning set $\mathrm{win}_i(G)$ of player i is an i-dominion.

For $k, p \in \mathbb{N}$, let $T(k)$ denote the maximum number of steps needed to solve a bipartite parity game $G = (V_0 \uplus V_1, E, p)$ and let $T(k, p)$ denote the maximum number of steps needed to solve a general parity game $G = (V_0 \uplus V_1, E, p)$ with $|V_1| = k$ and $p = |\{p(v) \mid v \in V\}|$ using some fixed algorithm. For $k, p, \ell \in \mathbb{N}$, let $\mathrm{dom}_k(\ell)$ denote the maximum number of steps required to find a dominion D with $|V_1 \cap D| \le \ell$ in a bipartite parity game $G = (V_0 \uplus V_1, E, p)$ with $|V_1| = k$ and let $\mathrm{dom}_{k,p}(\ell)$ denote the maximum number of steps required to find a dominion D with $|V_1 \cap D| \le \ell$ in a general parity game $G = (V_0 \uplus V_1, E, p)$ with $|V_1| = k$ and $p = |\{p(v) \mid v \in V\}|$, or to determine that no such dominion exists.

We will in the analysis of run times make the assumption that computation and removal of reachability sets as well as kernelization are elementary operation and can therefore be performed in time $O(1)$. To obtain the actual run times of our algorithms we will in the end multiply the computed run times by a factor corresponding to the time needed for these operations.

Lemma 20. *For* $k \ge 4$, $\mathrm{dom}_k(\ell) = O(k^\ell \cdot T(\ell))$ *and* $\mathrm{dom}_{k,p}(\ell) = O(k^\ell \cdot T(\ell, p))$.

With the algorithm described in Lemma 20 we can find a dominion D such that $|D \cap V_1| \le \ell$ if such a dominion exists. We denote this algorithm by **dominion$_1$**(G, ℓ) and assume that it returns either the pair (D, i) if an i-dominion D is found, or $(\emptyset, -1)$ if not.

7   New Algorithms for Solving Parity Games

We present the algorithm **new-win$_1$**(G) discussed in Sect. 5 in detail. Let $G = (V_0 \uplus V_1, E, p)$ with $|V_1| = k$ be a parity game with p distinct priorities.

The algorithm **new-win$_1$** starts by trying to find a "small" dominion D, where small means $|D \cap V_1| \le \ell$, where $\ell = \lfloor \sqrt{2k} \rfloor$ is a parameter chosen to minimize the run time of the algorithm. If such an i-dominion is found, then we remove it together with its i-reachability set from the game and solve the remaining game recursively. If no small dominion is found, then **new-win$_1$** simply calls algorithm **old-win$_1$**, which is almost identical to algorithm **win**. The only difference between **old-win$_1$** and **win** is that its recursive calls are made to **new-win$_1$** and not to itself.

The recursion stops once the number of odd nodes is at most 4, in which case we will test each of the at most $((p+1)^4)^4$ (due to the size of our kernel) different strategies for player 1 in constant time. We will call this brute force method **solve$_1$**(G). We will also kernelize using the reduction rules described in Sect. 3. We will call the kernelization subroutine **kernel**(G). The pseudocode of **new-win$_1$**(G) can be found in Sect. 8.

The correctness of the algorithm follows analogously to the correctness of $\mathbf{win}(G)$. We analyze the run time of $\mathbf{new\text{-}win}_1(G)$ and prove Theorem 1 in the full version [14] of this paper.

8   Pseudocode for Algorithm new-win

We will now give the pseudocode for algorithm $\mathbf{new\text{-}win}_1(G)$.

Algorithm 1. $\mathbf{new\text{-}win}_1(G)$

Input: A parity game $G = (V_0 \uplus V_1, E, p)$.
Output: A partition (W_0, W_1) of V, where W_i is the winning set of player $i \in \{0, 1\}$.
1: $k \leftarrow |V_1|; \ell \leftarrow \left\lfloor \sqrt{2k} \right\rfloor$; $G = \mathbf{kernel}(G)$
2: **if** $k \leq 4$ **then return** $\mathbf{solve}_1(G)$
3: $(D, i) \leftarrow \mathbf{dominion}_1(G, \ell)$
4: **if** $D = \emptyset$ **then**
5: $\quad (W_0, W_1) \leftarrow \mathbf{old\text{-}win}_1(G)$
6: **else**
7: $\quad (W_0', W_1') \leftarrow \mathbf{new\text{-}win}_1(G - \mathrm{reach}_i(D))$
8: $\quad (W_{\neg i}, W_i) \leftarrow (W_{\neg i}', V \setminus W_{\neg i}')$
9: **return** (W_0, W_1)

Algorithm 2. $\mathbf{old\text{-}win}_1(G)$

Input: A parity game $G = (V_0 \uplus V_1, E, p)$.
Output: A partition (W_0, W_1) of V, where W_i is the winning set of player $i \in \{0, 1\}$.
1: $G = \mathbf{kernel}(G)$
2: $i \leftarrow p_{\max} \pmod 2$
3: $(W_0', W_1') \leftarrow \mathbf{new\text{-}win}_1(G - \mathrm{reach}_i(V_{p_{\max}}))$
4: **if** $W_{\neg i}' = \emptyset$ **then**
5: $\quad (W_i, W_{\neg i}) \leftarrow (V, \emptyset)$
6: **else**
7: $\quad (W_0'', W_1'') \leftarrow \mathbf{new\text{-}win}_1(G - \mathrm{reach}_{\neg i}(W_{\neg i}'))$
8: $\quad (W_i, W_{\neg i}) \leftarrow (W_i'', V \setminus W_i'')$
9: **return** (W_0, W_1)

Acknowledgements. M.M. thanks Lászlo Végh for introducing him to parity games, and the authors of [7] for sending us a preprint.

References

1. Berwanger, D., Dawar, A., Hunter, P., Kreutzer, S.: DAG-width and parity games. In: Durand, B., Thomas, W. (eds.) STACS 2006. LNCS, vol. 3884, pp. 524–536. Springer, Heidelberg (2006)

2. Berwanger, D., Grädel, E., Kaiser, L., Rabinovich, R.: Entanglement and the complexity of directed graphs. Theoret. Comput. Sci. **463**, 2–25 (2012)
3. Björklund, H., Sandberg, S., Vorobyov, S.: A discrete subexponential algorithm for parity games. In: Alt, H., Habib, M. (eds.) STACS 2003. LNCS, vol. 2607, pp. 663–674. Springer, Heidelberg (2003)
4. Emerson, E.A., Jutla, C.S.: Tree automata, mu-calculus and determinacy. In: Proceedings of FOCS 1991, pp. 368–377 (1991)
5. Fearnley, J., Lachish, O.: Parity games on graphs with medium tree-width. In: Murlak, F., Sankowski, P. (eds.) MFCS 2011. LNCS, vol. 6907, pp. 303–314. Springer, Heidelberg (2011)
6. Fearnley, J., Schewe, S.: Time and parallelizability results for parity games with bounded treewidth. In: Czumaj, A., Mehlhorn, K., Pitts, A., Wattenhofer, R. (eds.) ICALP 2012, Part II. LNCS, vol. 7392, pp. 189–200. Springer, Heidelberg (2012)
7. Gajarský, J., Lampis, M., Makino, K., Mitsou, V., Ordyniak, S.: Parameterized algorithms for parity games. In: Italiano, G.F., Pighizzini, G., Sannella, D.T. (eds.) MFCS 2015. LNCS, vol. 9235, pp. 336–347. Springer, Heidelberg (2015)
8. Grädel, E., Thomas, W., Wilke, T. (eds.): Automata, Logics, and Infinite Games: A Guide to Current Research. LNCS, vol. 2500. Springer, Heidelberg (2002)
9. Jurdziński, M.: Deciding the winner in parity games is in UP ∩ co-UP. Inform. Process. Lett. **68**(3), 119–124 (1998)
10. Jurdziński, M.: Small progress measures for solving parity games. In: Reichel, H., Tison, S. (eds.) STACS 2000. LNCS, vol. 1770, pp. 290–301. Springer, Heidelberg (2000)
11. Jurdziński, M., Paterson, M., Zwick, U.: A deterministic subexponential algorithm for solving parity games. SIAM J. Comput. **38**(4), 1519–1532 (2008)
12. Kloks, T., Bodlaender, H.L.: On the treewidth and pathwidth of permutation graphs (1992). http://www.cs.uu.nl/research/techreps/repo/CS-1992/1992-13.pdf
13. McNaughton, R.: Infinite games played on finite graphs. Ann. Pure Appl. Logic **65**(2), 149–184 (1993)
14. Mnich, M., Röglin, H., Rösner, C.: New deterministic algorithms for solving parity games. arXiv.org, cs.CC, December 2015. http://arxiv.org/abs/1512.03246
15. Obdržálek, J.: Fast μ-calculus model checking when tree-width is bounded. In: Hunt Jr., W.A., Somenzi, F. (eds.) CAV 2003. LNCS, vol. 2725, pp. 80–92. Springer, Heidelberg (2003)
16. Obdržálek, J.: Clique-width and parity games. In: Duparc, J., Henzinger, T.A. (eds.) CSL 2007. LNCS, vol. 4646, pp. 54–68. Springer, Heidelberg (2007)
17. Schewe, S.: Solving parity games in big steps. In: Arvind, V., Prasad, S. (eds.) FSTTCS 2007. LNCS, vol. 4855, pp. 449–460. Springer, Heidelberg (2007)
18. Stirling, C.: Local model checking games. In: Lee, I., Smolka, S.A. (eds.) CONCUR 1995. LNCS, vol. 962, pp. 1–11. Springer, Heidelberg (1995)
19. Vöge, J., Jurdziński, M.: A discrete strategy improvement algorithm for solving parity games. In: Emerson, E.A., Sistla, A.P. (eds.) CAV 2000. LNCS, vol. 1855, pp. 202–215. Springer, Heidelberg (2000)
20. Zielonka, W.: Infinite games on finitely coloured graphs with applications to automata on infinite trees. Theoret. Comput. Sci. **200**(1–2), 135–183 (1998)

Computing a Geodesic Two-Center of Points in a Simple Polygon

Eunjin Oh[1]([✉]), Sang Won Bae[2], and Hee-Kap Ahn[1]

[1] Department of Computer Science and Engineering,
POSTECH, Pohang, South Korea
{jin9082,heekap}@postech.ac.kr
[2] Department of Computer Science, Kyonggi University, Suwon, South Korea
swbae@kgu.ac.kr

Abstract. Given a simple polygon P and a set Q of points contained in P, we consider the geodesic k-center problem in which we seek to find k points, called *centers*, in P to minimize the maximum geodesic distance of any point of Q to its closest center. In this paper, we focus on the case for $k = 2$ and present the first exact algorithm that efficiently computes an optimal 2-center of Q with respect to the geodesic distance in P.

1 Introduction

Computing the centers of a point set in a metric space is a fundamental algorithmic problem in computational geometry, which has been extensively studied with numerous applications in science and engineering. This family of problems is also known as the *facility location problem* in operations research that asks an optimal placement of facilities to minimize transportation costs. A historical example is the *Weber problem* in which one wants to place one facility to minimize the (weighted) sum of distances from the facility to input points. In cluster analysis, the objective is to group input points in such a way that the points in the same group are relatively closer to each other than to those in other groups. A natural solution finds a few number of centers and assign the points to the nearest center, which relates to the well known *k-center problem*.

The k-center problem is formally defined as follows: for a set Q of m points, find a set C of k points that minimizes $\max_{q \in Q} \{\min_{c \in C} d(q, c)\}$, where $d(x, y)$ denotes the distance between x and y. The k-center problem has been investigated for point sets in two-, three-, or higher dimensional Euclidean spaces. For the special case where $k = 1$, the problem is equivalent to finding the smallest enclosing ball containing all points. It can be solved in $O(m)$ time for any fixed dimension [5,7,12]. The case of $k = 2$ can be solved in deterministic

Work by Oh and Ahn was supported by the NRF grant 2011-0030044 (SRC-GAIA) funded by the government of Korea. Work by S.W. Bae was supported by Basic Science Research Program through the National Research Foundation of Korea (NRF) funded by the Ministry of Science, ICT & Future Planning (2013R1A1A1A05006927) and by the Ministry of Education (2015R1D1A1A01057220).

© Springer-Verlag Berlin Heidelberg 2016
E. Kranakis et al. (Eds.): LATIN 2016, LNCS 9644, pp. 646–658, 2016.
DOI: 10.1007/978-3-662-49529-2_48

$O(m \log^2 m \log^2 \log m)$ time [4] in $\mathbb{R}^2$. If $k > 2$ is part of input, it is NP-hard to approximate the Euclidean k-center within approximation factor 1.822 [8], while an $m^{O(\sqrt{k})}$-time exact algorithm is known for points in $\mathbb{R}^2$ [11].

There has been studied several variants of the k-center problem. One variant is the problem for finding k smallest congruent disks in the presence of obstacles. More specifically, the problem takes a set of pairwise disjoint simple polygons (obstacles) with a total of n edges in addition to a set S of m points as inputs. It aims to find k smallest congruent disks whose union contains S and whose centers do not lie on the interior of the obstacles. Here, the obstacles do not affect the distance between two points. For $k = 2$, Halperin et al. [10] gave an expected $O(n \log^2(mn) + mn \log^2 m \log(mn))$-time algorithm for this problem.

In this paper, we consider another variant of the k-center problem in which the set Q of m points are given in a simple n-gon P and the centers are constrained to lie in P. Here the boundary of the polygon P is assumed to act as an obstacle and the distance between any two points in P is thus measured by the length of the geodesic (shortest) path connecting them in P in contrast to [10]. We call this constrained version *the geodesic k-center problem* and its solution *a geodesic k-center* or simply *an optimal k-center* of Q with respect to P.

This problem has been investigated for the simplest case $k = 1$. The geodesic one-center of Q with respect to P is proven to coincide with the geodesic one-center of the geodesic convex hull of Q with respect to P [2], which is the smallest subset $C \subseteq P$ containing Q such that for any two points $p, q \in C$, the geodesic path between p and q is also contained in C. Thus, the geodesic one-center can be computed by first computing the geodesic convex hull of Q in $O((m + n) \log(m + n))$ time [15] and second finding its geodesic one-center. The geodesic convex hull of Q forms a (weakly) simple polygon with $O(m + n)$ vertices. The problem of finding the geodesic one-center of a (weakly) simple polygon was introduced by Asano and Toussaint [3]. The first algorithm by Asano and Toussaint takes $O(n^4 \log n)$ time, where n denotes the number of vertices of the input polygon. It was improved to $O(n \log n)$ in [14] and finally improved again to $O(n)$ in [1]. Consequently, the geodesic one-center of Q with respect to P can be computed in $O((m + n) \log(m + n))$ time.

However, even for $k = 2$, finding a geodesic k-center of Q with respect to P is not equivalent to finding a geodesic k-center of a (weakly) simple polygon, which was addressed in [13]. One can easily construct an example of P and Q in which the two geodesic disks corresponding to a geodesic 2-center of Q do not contain the geodesic convex hull of Q. See Fig. 1.

In this paper, we consider the geodesic k-center problem for $k = 2$ and present the first exact algorithm to compute a geodesic two-center, that is, a pair (c_1, c_2) of points in P such that $\max_{q \in Q}\{\min\{d(q, c_1), d(q, c_2)\}\}$ is minimized, where $d(x, y)$ denote the length of the geodesic path connecting x and y in P. Our algorithm takes $O(m(m + n) \log^3(m + n) \log m)$ time using $O(mn)$ space. A simple modification of the algorithm allows us to improve the space complexity by sacrificing the running time. The algorithm takes $O(m^2(m + n) \log^3(m + n))$ time when only $O(m + n)$ space is allowed. If n and m are asymptotically equal, then our algorithms take $O(n^2 \log^4 n)$ and $O(n^3 \log^3 n)$ time, respectively, using $O(n^2)$ and (n) space, respectively.

All missing proofs can be found in the full version of this paper.

2 Preliminaries

Let P be a simple polygon with n vertices. The *geodesic path* between x and y contained in P, denoted by $\pi(x, y)$, is the unique shortest path between x and y inside P. We often consider $\pi(x, y)$ directed from x to y. The length of $\pi(x, y)$ is called the *geodesic distance* between x and y, and we denote it by $d(x, y)$.

A subset A of P is *geodesically convex* if it holds that $\pi(x, y) \subseteq A$ for any $x, y \in A$. For a set Q of m points contained in P, the common intersection of all the geodesically convex subsets of P that contain Q is also geodesically convex and it is called the *geodesic convex hull* of Q. It is already known that the geodesic convex hull of any set of m points in P is a weakly simple polygon and can be computed in $O((m + n) \log(m + n))$ time [15].

Note that once the geodesic convex hull of Q is computed, our algorithm regards the geodesic convex hull as a new polygon and never consider the points lying outside of the geodesic convex hull. Thus, we simply use $\mathcal{C}_Q$ to denote the geodesic convex hull of Q. Each point $q \in Q$ lying on the boundary of $\mathcal{C}_Q$ is called *extreme*.

For a set A, we use ∂A to denote the boundary of A. Since the boundary of $\mathcal{C}_Q$ is not necessarily simple, the clockwise order of $\partial \mathcal{C}_Q$ is not defined naturally in contrast to a simple curve. Aronov et al. [2] presented a way to label the extreme points of Q with $v_1, \ldots, v_k \in Q$ such that the circuit $\pi(v_1, v_2)\pi(v_2, v_3) \cdots \pi(v_k, v_1)$ is a closed walk of the boundary of $\mathcal{C}_Q$ allowing repetitions only for extreme points. We use this labeling of extreme points for our problem. The circuit $\pi(v_1, v_2)\pi(v_2, v_3) \cdots \pi(v_k, v_1)$ is called the *clockwise traversal* of $\partial \mathcal{C}_Q$ from v_1. The *clockwise order* follows from the clockwise traversal along $\partial \mathcal{C}_Q$.

Let $\mathcal{C}_Q(v, w)$ denote the portion of $\partial \mathcal{C}_Q$ from v to w in clockwise order (including v and w) for $v, w \in \partial \mathcal{C}_Q$. For any two extreme points v_i and v_j, we use $\mathcal{C}_Q(i, j)$ to denote the chain $\mathcal{C}_Q(v_i, v_j)$ for simplicity. The subpolygon bounded by $\pi(w, v)$ and $\mathcal{C}_Q(v, w)$ is denoted by $\mathcal{P}_Q(v, w)$. Clearly, $\mathcal{P}_Q(v, w)$ is a weakly simple polygon.

The *geodesic disk* centered at $c \in P$ with radius $r \in \mathbb{R}$, denoted by $D_r(c)$, is the set of points whose geodesic distance from c is at most r. We call a connected set of points of $D_r(c)$ that are at distance exactly r from c an *extreme arc* of $D_r(c)$. A set of geodesic disks with the same radius satisfies the *pseudo-disk property*. An extended form of the pseudo-disk property of geodesic disks can be stated as follows.

Lemma 1 ([13, Lemma 8]). *Let $\mathcal{D} = \{D_1, \ldots, D_k\}$ be a set of geodesic disks with the same radius and let I be the common intersection of all disks in $\mathcal{D}$. Let $S = \langle s_1, \ldots, s_t \rangle$ be the cyclic sequence of the extreme arcs of geodesic disks appearing on ∂I along its boundary in clockwise order. For any $i \in \{1, \ldots, k\}$, the extreme arcs in $\partial I \cap \partial D_i$ are consecutive in S.*

3 Bipartition by Two Centers

We first compute the geodesic convex hull $\mathcal{C}_Q$ of the point set Q using the algorithm by Toussaint [15]. Let Q_B be the set of extreme points in Q and let $Q_I := Q \setminus Q_B$. Note that each $q \in Q_B$ lies on $\partial\mathcal{C}_Q$ while each $q' \in Q_I$ lies in the interior of $\mathcal{C}_Q$. The points of Q_B are readily sorted along the boundary of $\mathcal{C}_Q$, being labeled by $v_1, \ldots, v_k$ following the notion of Aronov et al. [2].

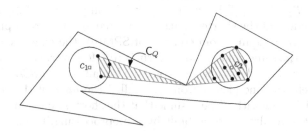

Fig. 1. The gray region is the geodesic convex hull $\mathcal{C}_Q$. The center c_1 lies outside of $\mathcal{C}_Q$, while the center c_2 lies inside $\mathcal{C}_Q$.

Note that it is possible that an optimal two-center has one of its two centers lying outside of $\mathcal{C}_Q$ (See Fig. 1). However, there always exists an optimal two-center of Q with respect to P such that both the centers are contained in $\mathcal{C}_Q$ as stated in the following lemma. Thus we may search only $\mathcal{C}_Q$ to find an optimal two-center of Q with respect to P.

Lemma 2. *There is an optimal two-center (c_1, c_2) of Q with respect to P such that both c_1 and c_2 are contained in the geodesic convex hull $\mathcal{C}_Q$ of Q.*

Let $c_1, c_2 \in \mathcal{C}_Q$ such that $Q \subset D_r(c_1) \cup D_r(c_2)$. By Lemma 1, the boundaries of the two geodesic disks cross each other at most twice. If every extreme point on $\mathcal{C}_Q(j+1, i)$ is contained in $D_r(c_1)$, so is the whole chain $\mathcal{C}_Q(j+1, i)$ since $D_r(c_1)$ is geodesic convex and therefore $\pi(v, v') \subset D_r(c_1)$ for any two extreme points v, v' on $\mathcal{C}_Q(j+1, i)$. Thus there exists a pair (i, j) of indices such that $\mathcal{C}_Q(j+1, i) \subset D_r(c_1)$ and $\mathcal{C}_Q(i+1, j) \subset D_r(c_2)$. We call such a pair (i, j) of indices a *partition pair* of $D_r(c_1)$ and $D_r(c_2)$. If (c_1, c_2) is an optimal two-center and $r = \max_{q \in Q}\{\min\{d(q, c_1), d(q, c_2)\}\}$, then the partition pair of $D_r(c_1)$ and $D_r(c_2)$ is called an *optimal partition pair*.

For a pair (i, j) of indices, an *optimal (i, j)-restricted two-center* is defined as a pair of points (c_1, c_2) that minimizes $r > 0$ satisfying $\mathcal{C}_Q(j+1, i) \subset D_r(c_1)$, $\mathcal{C}_Q(i+1, j) \subset D_r(c_2)$, and $Q \subset D_r(c_1) \cup D_r(c_2)$. Let r_{ij}^* denote the radius of an optimal (i, j)-restricted two-center.

In this paper, we give an algorithm for computing an optimal two-center of Q with respect to P. The overall algorithm is described in Sect. 6. As subprocedures, we use the decision and the optimization algorithms described in Sects. 4 and 5, respectively. The decision algorithm determines whether $r \geq r_{ij}^*$ for a given triple (i, j, r) and the optimization algorithm computes r_{ij}^* for a given pair (i, j). While executing the whole algorithm, we call the decision and the optimization algorithms repeatedly with different inputs.

4   Decision Algorithm for a Partition Pair

In this section, we present an algorithm that decides whether or not $r \geq r_{ij}^*$ given a partition pair (i, j) and a radius $r \geq 0$.

The *shortest path map* $\mathsf{SPM}(q)$ of q is the decomposition of $\mathcal{C}_Q$ obtained by extending the edges of the shortest path tree rooted at q towards opposite to the root. For every point x in a common cell of $\mathsf{SPM}(q)$, the geodesic path $\pi(q, x)$ from q to x has the same combinatorial structure. Moreover, each cell of $\mathsf{SPM}(q)$ forms a triangle with at least one side contained in an edge of $\partial \mathcal{C}_Q$. Among three vertices of a cell, the one closest from q is called the *apex* of the cell. For each call of our decision algorithm, we assume that the shortest path maps $\mathsf{SPM}(q)$ of q for all $q \in Q$ have already been built by a preprocessing. Computing $\mathsf{SPM}(q)$ for all $q \in Q$ can be done in $O(mn)$ time [9].

Note that $r \geq r_{ij}^*$ if and only if there is a pair (c_1, c_2) of points in $\mathcal{C}_Q$ such that $D_r(c_1)$ contains $\mathcal{C}_Q(j + 1, i)$, $D_r(c_2)$ contains $\mathcal{C}_Q(i + 1, j)$ and $D_r(c_1) \cup D_r(c_2)$ contains Q. We call such a pair (c_1, c_2) an (i, j, r)-*restricted two-center*. As discussed above, the set Q_B is partitioned by the partition pair (i, j) into two subsets, $Q_1 = Q_B \cap \mathcal{C}_Q(j + 1, i)$ and $Q_2 = Q_B \cap \mathcal{C}_Q(i + 1, j)$.

If r is sufficiently large, the decision can be made relatively easy. If r is at least the radius of the smallest geodesic disk containing $\mathcal{C}_Q$, which can be computed in time linear to the complexity of $\mathcal{C}_Q$, then our decision algorithm surely returns "yes." Another easy case is when r is large enough so that at least one of the four vertices v_i, v_{i+1}, v_j, v_{j+1} is contained in both $D_r(c_1)$ and $D_r(c_2)$ for some (i, j, r)-restricted two-center (c_1, c_2). In this case, an (i, j, r)-restricted two-center can be computed in $O((m + n) \log^2(m + n))$ time as follows. Assume without loss of generality that v_i is contained in both $D_r(c_1)$ and $D_r(c_2)$. Let $Q' := Q \setminus \{v_i\}$. Then, we observe that Q' can be bipartitioned into Q_1' and Q_2' by a geodesic path $\pi(v_i, w)$ from v_i to some $w \in \partial \mathcal{C}_Q$ such that $Q_1' \cup \{v_i\} \subset D_r(c_1)$ and $Q_2' \cup \{v_i\} \subset D_r(c_2)$. We thus search the boundary $\partial \mathcal{C}_Q$ for a $w \in \partial \mathcal{C}_Q$ implying such an optimal bipartition (Q_1', Q_2') of Q'. For the purpose, we decompose each cell $\triangle$ of $\mathsf{SPM}(v_i)$ into smaller triangular cells by the line ℓ_q through q and the apex of $\triangle$ for each $q \in Q \cap \triangle$. Now, we have $O(m + n)$ triangular cells in the resulting refined map M. Note that all the vertices of the refined map M lie on $\partial \mathcal{C}_Q$ and every $q \in Q$ lies on the geodesic $\pi(v_i, w)$ for some vertex w of M. We sort the vertices of the cells along $\partial \mathcal{C}_Q$ in clockwise order from v_i, and then apply a binary search on them to find a vertex w that minimizes the radius of the larger of smallest geodesic disks containing $(Q' \cap \mathcal{P}_Q(v_i, w)) \cup \{v_i\}$ and $(Q' \setminus \mathcal{P}_Q(v_i, w)) \cup \{v_i\}$, respectively. And an (i, j, r)-restricted two-center corresponds to the optimal bipartition obtained in the above binary search. This takes $O((m + n) \log^2(m + n))$ time, and by the same procedure one can also decide if this is the case where there is an (i, j, r)-restricted two-center (c_1, c_2) such that $v_i \in D_r(c_1) \cap D_r(c_2)$.

In the following, we thus assume that there is no (i, j, r)-restricted two-center (c_1, c_2) such that any of the four vertices $v_i, v_{i+1}, v_j, v_{j+1}$ is contained in both $D_r(c_1)$ and $D_r(c_2)$. This also means that $D_r(c_1) \cap \{v_i, v_{i+1}, v_j, v_{j+1}\} = \{v_i, v_{j+1}\}$ and $D_r(c_2) \cap \{v_i, v_{i+1}, v_j, v_{j+1}\} = \{v_{i+1}, v_j\}$ for any (i, j, r)-restricted two-center (c_1, c_2).

4.1   Intersection of Geodesic Disks and Events

We first compute the common intersection of the geodesic disks of radius r centered at extreme points on each subchain. Let $I_1 = \bigcap_{q \in Q_1} D_r(q)$ and $I_2 = \bigcap_{q \in Q_2} D_r(q)$. Let t be 1 or 2 in the following. Given the farthest-point geodesic Voronoi diagram of Q_t, I_t can be computed in $O(m + n)$ time. The boundary ∂I_t of I_t consists of points that are at distance at most r from all points Q_t. Recall that each extreme arc in ∂I_t is from an extreme arc on the boundary of a geodesic disk centered at a point in Q_t. On the other hand, any straight portion of ∂I_t is a subset of the polygon boundary ∂P. We denote the union of the extreme arcs of ∂I_t by A_t. Note that $r < r_{ij}^*$ if $I_1 = \emptyset$ or $I_2 = \emptyset$. Our decision algorithm returns "no" immediately if this is the case. Otherwise, both A_1 and A_2 are nonempty because r is smaller than the radius of a smallest disk containing C_Q. Also, if $Q_I = \emptyset$, then our algorithm returns "yes" immediately; so, in the following, we also assume that $Q_I \neq \emptyset$.

Lemma 3. *There is an (i, j, r)-restricted two-center (c_1, c_2) such that $c_1 \in A_1$ and $c_2 \in A_2$, provided that $r \geq r_{ij}^*$.*

Since the boundary of I_t is a simple closed curve, we can define the clockwise and the counterclockwise directions of ∂I_t. We choose a *reference point* o_t from ∂I_t such that for all $q \in Q_I$, either the interior of $D_r(q)$ avoids o_t or $I_t \subseteq D_r(q)$. Such reference points o_1 and o_2 can be found easily.

Lemma 4. *For each $t \in \{1, 2\}$, there exists a reference point o_t on A_t, and a reference point can be found in $O(m)$ time.*

Using the reference point o_t, we define the ordering $\prec_t$ on I_t. We write $x \prec_t y$ for two points $x, y \in \partial I_t$ if x comes before y as we traverse ∂I_1 in clockwise order from the reference point o_t. We also write $x \preceq_t y$ if either $x = y$ or $x \prec_t y$. Since $A_t \subseteq I_t$, the order $\prec_t$ on A_t is naturally inherited.

We then consider the intersection of $\partial D_r(q)$ with A_t for each $q \in Q_I$. By Lemma 1, the intersection $\partial D_r(q) \cap A_t$ consists of at most two points. Moreover, for each $x \in \partial D_r(q) \cap A_t$, there are no two $y, y' \in D_r(q) \cap A_t$ such that $y \prec_t x \prec_t y'$ by our choice of the reference points o_t on A_t. We call each intersection point $x \in \partial D_r(q) \cap A_t$ an *event* of q on A_t. Each event x of q is associated with its defining point $def(x) = q$ and a boolean value $io(x) = $ IN or OUT. More specifically, if there are two events x, x' of q on A_t with $x \prec_t x'$, then $io(x) = $ IN and $io(x') = $ OUT; if there is a unique event x of q on A_t, then $D_r(q)$ is tangent to I_t at x, so we regard x as two distinct events x and x' at a common position with $io(x) = $ IN and $io(x') = $ OUT. Note that if x and x' are two events of $q \in Q_I$

on A_t with $x \preceq_t x'$, then we have $y \in D_r(q) \cap A_t$ for any y with $x \preceq_t y \preceq_t x'$. Let $M_t := \bigcup_{q \in Q_I} (\partial D_r(q) \cap A_t)$ be the set of events on A_t. Clearly, the number of events is $|M_t| = O(m)$.

Lemma 5. *Suppose that both M_1 and M_2 are nonempty. Them, there is an (i, j, r)-restricted two-center (c_1, c_2) such that $c_1 \in M_1$ and $c_2 \in M_2$, if $r \geq r_{ij}^*$.*

If either M_1 or M_2 is empty, then it becomes easier. If $M_1 = M_2 = \emptyset$, then no $q \in Q_I$ can be contained in any disk of radius r centered at A_1 or A_2, so our decision algorithm returns "no." If one of them is empty, say $M_1 = \emptyset$, and (c_1, c_2) is an (i, j, r)-restricted two-center, then we must have $Q_I \subset D_r(c_2)$. Thus, this case can be handled by computing the smallest geodesic disk containing $Q_2 \cup Q_I$ and testing if its radius is at most r.

Hence, we further assume that both M_1 and M_2 are nonempty. Then, by Lemma 5, we can decide if $r \geq r_{ij}^*$ by finding a pair (c_1, c_2) of points such that $c_1 \in M_1, c_2 \in M_2$ and $Q_I \subset D_r(c_1) \cup D_r(c_2)$. For the purpose, we traverse A_1 and A_2 simultaneously by handling the events in M_1 and M_2 in order, after sorting events in M_1 and M_2 with respect to $\prec_1$ and $\prec_2$, respectively.

Lemma 6. *The sets A_1, A_2, M_1, and M_2 can be computed in $O((m + n) \log^2(m + n))$ time.*

4.2   Traversing A_1 and A_2 by Scanning Events

We scan M_1 once by moving a pointer c_1 from the reference point in clockwise order. We also scan M_2 from the reference point o_2 of ∂I_2 by moving one pointer c_2^c in clockwise order and another pointer c_2^{cc} in counterclockwise order at the same time. We continue to scan and handle the events until c_1 points to the last event of M_1 or c_2^c and c_2^{cc} point at the same event of M_2. We often regard the three pointers as events which they point to. For example, we write $D_r(c_2^c)$ to indicate the set of points whose geodesic distance from the event in M_1 which c_2^c points is at most r.

Whenever we handle an event, we apply two operations, which we call DECISION and UPDATE. We maintain the sets $D_r(c_1) \cap Q_I$, $D_r(c_2^c) \cap Q_I$, and $D_r(c_2^{cc}) \cap Q_I$. Operation UPDATE updates the sets, and operation DECISION checks whether $Q_I \subset D_r(c_1) \cup D_r(c_2^c)$ or $Q_I \subset D_r(c_1) \cup D_r(c_2^{cc})$.

In the following paragraphs, we describe (1) how to handle the events in $M_1 \cup M_2$, and (2) how the two operations work.

(1) How to Handle the Events in $M_1 \cup M_2$. We move the three pointers c_1, c_2^c and c_2^{cc} as follows. First, we scan M_1 until it reaches an event x with $io(x) = $ OUT. If $D_r(c_2^c)$ does not contain $def(x)$, then we scan M_2 from c_2^c in clockwise order until we reach the event y with $def(x) = def(y)$. If $D_r(c_2^{cc})$ does not contain $def(x)$, then we also scan M_2 from c_2^{cc} in counterclockwise order until we reach the event y with $def(x) = def(y)$. Afterwards, we scan again M_1 from c_1 in clockwise order until we reach an event with the attribute OUT. Whenever we

reach an event in $M_1 \cup M_2$, we check whether $Q_I \subset D_r(c_1) \cup D_r(c_2^s)$ for $s = c, cc$. If this test passes at some event, we stop traversing and return a solution (c_1, c_2^c) or (c_1, c_2^{cc}) accordingly. If the pointer c_1 goes back to the reference point or c_2^c, c_2^{cc} meet each other, our decision algorithm returns "no." Clearly, this algorithm terminates and we consider $O(m)$ event points in total. If UPDATE and DECISION take constant time, the total running time for this step is $O(m)$.

(2) How the Two Operations Work. To apply DECISION and UPDATE in constant time, we use five arrays of points in Q_I. Each element of the arrays is a Boolean value corresponding to each point in Q_I. For the first array, each element indicates whether $D_r(c_1)$ contains its corresponding point in Q_I. Similarly, the second and the third arrays have Boolean values for $D_r(c_2^c)$ and $D_r(c_2^{cc})$, respectively. Each element of the remaining two arrays indicates whether its corresponding point in Q_I is contained in $D_r(c_1) \cup D_r(c_2^c)$ and $D_r(c_1) \cup D_r(c_2^{cc})$, respectively. In addition to the five arrays, we also maintain the number of points of Q_I contained in each set; $D_r(c_1)$, $D_r(c_2^c)$, $D_r(c_2^{cc})$, $D_r(c_1) \cup D_r(c_2^c)$, and $D_r(c_1) \cup D_r(c_2^{cc})$.

At reference points, we initialize the five arrays and the five numbers in $O(m)$ time. For DECISION, we just check whether the number of points contained in either $D_r(c_1) \cup D_r(c_2^c)$ or $D_r(c_1) \cup D_r(c_2^{cc})$ is equal to the number of points in Q_I, which takes constant time. To apply UPDATE when c_1 reaches an event $x \in M_1$ with $def(x) = q$, we first change Boolean values of the elements in the arrays assigned for $D_r(c_1)$, $D_r(c_1) \cup D_r(c_2^c)$ and $D_r(c_1) \cup D_r(c_2^{cc})$ according to $io(x)$. When we change Boolean values, we also update the number of points contained in the sets accordingly. These procedures can be done in constant time.

We are now ready to conclude the following.

Theorem 1. *Given a partition pair (i, j) and a nonnegative real r, our decision algorithm correctly decides in $O((m+n) \log^2(m+n))$ time whether or not $r \geq r_{ij}^*$. If so, it also returns an (i, j, r)-restricted two-center.*

5   Optimization Algorithm for a Partition Pair

In this section, we present an optimization algorithm for a given partition pair (i, j) that computes r_{ij}^* and an optimal (i, j)-restricted two-center.

Our optimization algorithm will work with an *assistant interval* $[r_L, r_U]$ which will be given also as part of input and satisfy the following condition: $r^* \in [r_L, r_U]$ and the combinatorial structure of $\partial D_r(q)$ for each $q \in Q$ remains the same for all $r \in [r_L, r_U]$, where $r^* = \min_{i,j} r_{ij}^*$ denotes the radius of an optimal two-center of Q. Also, we assume that the combinatorial structure of $\partial D_r(q)$ for all $q \in Q$ has already been computed before the first call of the optimization algorithm. The algorithm will return exactly r_{ij}^* if $r_{ij}^* \leq r_U$; otherwise, it just reports that $r_{ij}^* > r_U$. The latter case means that (i, j) is never an optimal partition pair, as we have assured that $r_{ij}^* > r_U \geq r^*$. Testing whether $r_{ij}^* > r_U$ or $r_{ij}^* \leq r_U$ can be done by running the decision algorithm for input (i, j, r_U).

In the following, we thus assume that $r^*_{ij} \in [r_L, r_U]$, and search for r^*_{ij} in the assistant interval $[r_L, r_U]$.

As in the decision algorithm, we consider the intersection of geodesic disks and events on extreme arcs. For each $r \in [r_L, r_U]$ and each $t \in \{1, 2\}$, let $I_t(r) := \bigcap_{q \in Q_t} D_r(q)$, and $A_t(r)$ be the union of extreme arcs of I_t. Also, let $M_t(r)$ be the set of events of each $q \in Q_I$, if any, on $A_t(r)$, as defined in Sect. 4. Here, we identify each event $x \in M_t(r)$ by as a pair $x = (def(x), io(x))$, not by its exact position on A_t. Note that the set $M_t(r)$ and the combinatorial structure of $\partial I_t(r)$ may not be constant over $r \in [r_L, r_U]$. In order to fix them also, we narrow the assistant interval $[r_L, r_U]$ into $[\rho_L, \rho_U]$ as follows.

Lemma 7. *In $O((m + n) \log^3(m + n))$ time, one can find an interval $[\rho_L, \rho_U] \subseteq [r_L, r_U]$ containing r^*_{ij} such that the combinatorial structure of each of the following remains the same over $r \in [\rho_L, \rho_U]$: $\partial I_t(r)$ and $M_t(r)$ for $t = 1, 2$.*

We proceed with the interval $[\rho_L, \rho_U]$ described as in Lemma 7. For any $r \in [\rho_L, \rho_U]$, $M_1(r)$ and $M_2(r)$ are fixed, so we write $M_1 = M_1(r)$ and $M_2 = M_2(r)$. The sets M_1 and M_2 can be computed by Lemma 6. Note that $M_1 = M_1(r^*_{ij})$ and $M_2 = M_2(r^*_{ij})$ since $r^*_{ij} \in [\rho_L, \rho_U]$. We then pick a reference point $o_t(r)$ on $\partial I_t(r)$ as done in Sect. 4 such that the trace of $o_t(r)$ over $r \in [\rho_L, \rho_U]$ is a simple curve. This is always possible because the combinatorial structure of $\partial I_t(r)$ is constant. (See also the proof of Lemma 4.) Such a choice of references $o_t(r)$ ensures that the order on the events in M_t stays the same as r continuously increases unless the positions of two distinct events in $M_t(r)$ coincides.

We are now interested in the order $\prec^*_t$ on the events in M_t at $r = r^*_{ij}$. In the following, we obtain a sorted list of events in M_t with respect to $\prec^*_t$ without knowing the exact value of r^*_{ij}.

Deciding Whether or Not $x \preceq^*_t x'$ for $x, x' \in M_t$. This is a primitive operation to sort M_t with respect to $\prec^*_t$. Let $q = def(x)$ and $q' = def(x')$. The order of x and x' over $r \in [\rho_L, \rho_U]$ may change only when we have a nonempty intersection of $A_t(r) \cap \partial D_r(q) \cap \partial D_r(q')$. Let $\rho_t(q, q')$ be such a radius $r > 0$ that $A_t(r) \cap \partial D_r(q) \cap \partial D_r(q')$ is nonempty for any two distinct $q, q' \in Q_I$. Note that the intersection $A_t(r) \cap \partial D_r(q) \cap \partial D_r(q')$ at $r = \rho_t(q, q')$ forms a single point c and $D_{\rho_t(q,q')}(c)$ is the smallest-radius geodesic disk containing $Q_t \cup \{q, q'\}$. Thus, the value $\rho_t(q, q')$ is uniquely determined.

Lemma 8. *Let $q_1, q_2 \in Q_I$ be two distinct points. For $t \in \{1, 2\}$, we can decide whether or not $\rho_t(q_1, q_2) \in [\rho_L, \rho_U]$ in $O(\log(m+n))$ time. If $\rho_t(q_1, q_2) \in [\rho_L, \rho_U]$, the value of $\rho_t(q_1, q_2)$ can be computed in $O(\log(m + n) \log n)$ time.*

If $\rho_t(q, q') \notin [\rho_L, \rho_U]$, then the order of x and x' can be determined by computing their positions at $r = \rho_L$ or ρ_U. Otherwise, we can decide whether or not $x \preceq^*_t x'$ by running the decision algorithm for input $(i, j, \rho_t(q, q'))$, once we know the value $\rho_t(q, q')$.

Sorting the Events in M_t with Respect to $\prec_t^*$. This can be done in $O(T_c \cdot m \log m)$ time, where T_c denotes the time needed to compare two events as above. A more efficient method applies a parallel sorting algorithm due to Cole [6]. They gave a parallel algorithm for sorting N elements in $O(\log N)$ time using $O(N)$ processors. In Cole's algorithm, we need to apply $O(m)$ comparisons at each iteration, while comparisons in each iteration are independent of one another. For each iteration, we compute the values of $\rho_t(def(x), def(x'))$ that are necessary for the $O(m)$ comparisons of $x, x' \in M_t$, and sort them in increasing order. On the sorted list of the values, we apply binary search using the decision algorithm. Then we complete the comparisons in each iteration in time $O(m \log(m + n) \log n + T_d \log m)$ by Lemma 8, where T_d denotes the time taken by the decision algorithm. Since Cole's algorithm requires $O(\log m)$ iterations in total, the total running time for sorting the events in M_t is $O(m \log (m + n) \log m \log n + T_d \log^2 m)$.

Computing r_{ij}^* and a Corresponding Two-Center. For any two neighboring events x and x' in M_t with respect to $\prec_t^*$, we call the value of $\rho_t(def(x), def(x'))$ a *critical radius* if it belongs to $[\rho_L, \rho_U]$. Let R be the set of all critical radii, including ρ_L and ρ_U.

Lemma 9. *The set R contains r_{ij}^*.*

Hence, r_{ij}^* is exactly the smallest value $\rho \in R$ such that there exists an (i, j, ρ)-restricted two-center. The last step of our optimization algorithm thus performs a binary search on R using the decision algorithm.

This completes the description of the optimization algorithm and we conclude the following.

Theorem 2. *Given a partition pair (i, j) and an assistant interval $[r_L, r_U]$, an optimal (i, j)-restricted two-center of Q can be computed in $O((m + n) \log^3 (m + n) \log m)$ time, provided that $r_L \leq r_{ij}^* \leq r_U$.*

6   Computing an Optimal Two-Center of Points

Finally, we present an algorithm that computes an optimal two-center of Q with respect to P. As the optimization algorithm described in Sect. 5 works with a fixed partition pair, trying all partition pairs (i, j) already implies such an algorithm, once an assistant interval $[r_L, r_U]$ is computed. In the following, we show how to choose $O(m)$ partition pairs that are guaranteed to include an optimal pair.

6.1   Finding Candidate Pairs

Let $\gamma(v, w)$ denote the radius of the smallest geodesic disk containing $\mathcal{P}_Q(v, w)$ for $v, w \in \partial \mathcal{C}_Q$. We define the function f which maps each index i of an extreme vertex v_i of $\mathcal{C}_Q$ to the index j of the first clockwise extreme vertex v_j of $\mathcal{C}_Q$ from v_i that minimizes $\max\{\gamma(v_i, v_j), \gamma(v_j, v_i)\}$.

Lemma 10. *Let v_i be any extreme vertex of C_Q and v_j be an extreme vertex of C_Q lying on $C_Q(v_i, v_{f(i)})$. Then $v_{f(j)} \in C_Q(f_{cc}, v_i)$, where f_{cc} is the neighboring extreme vertex of $v_{f(i)}$ in counterclockwise order.*

An index pair (i, j) is called *a candidate pair* if (1) $j - 1 \le f(i + 1) \le j + 1$, (2) $j - 1 \le f(i) \le j + 1$, or (3) both v_j and v_{j+1} lie on $C_Q(v_{f(i)-2}, v_{f(i+1)+2})$ when $v_{f(i)}$ appears ahead of $v_{f(i+1)}$ along the boundary in clockwise from v_i.

By the definition of candidate pairs, we can bound the number of candidate pairs as follows.

Lemma 11. *The number of candidate pairs is $O(m)$.*

The following is the key observation on candidate pairs.

Lemma 12. *There exists an optimal partition pair that is a candidate pair.*

Now we describe a procedure that finds the set of all candidate pairs. First, we compute the index $f(1)$ by traversing all extreme vertices of C_Q in clockwise order. Afterwards, we find $f(i)$ for all indices larger than 1. Suppose that we have already computed $f(i - 1)$ and we want to find $f(i)$. By Lemma 10, we do not need to consider the vertices lying in the interior of $C_Q(v_{i-1}, f_{cc})$, where f_{cc} is the neighboring extreme vertex of $v_{f(i-1)}$ in counterclockwise direction. Thus we traverse the vertices from f_{cc} in clockwise order and check whether the current vertex is $v_{f(i)}$. To do this, we consider three vertices: the current vertex v_c and two neighbor vertices v_{k_1}, v_{k_2} of v_c. If $\max\{\gamma(v_c, v_i), \gamma(v_i, v_c)\} \le \min\{\max\{\gamma(v_{k_1}, v_i), \gamma(v_i, v_{k_1})\}, \max\{\gamma(v_{k_2}, v_i), \gamma(v_i, v_{k_2})\}\}$, the vertex $v_{f(i)}$ is the current vertex v_c by the monotonicity of the functions $\gamma(v, \cdot)$ and $\gamma(\cdot, v)$, where v is a fixed extreme vertex of C_Q. Otherwise, v_c is not $f(i)$, so we move to the extreme vertex next to v_c. Hence, we can find $f(i)$ for all indices i by traversing ∂C_Q twice. For each vertex we visit during the traversal, we compute $\max\{\gamma(v, w), \gamma(w, v)\}$ for three different pairs (v, w) of extreme vertices of C_Q, each of which takes $O(m + n)$ time by the algorithm in [1].

Afterwards, we compute the set of all candidate pairs based on the information we have just computed. For each index i, we traverse the vertices lying between $v_{f(i)-2}$ and $v_{f(i+1)+2}$. It takes time proportional to the number of candidate pairs, which is $O(m)$ by Lemma 11.

Thus, we conclude the following lemma.

Lemma 13. *The set of all candidate index pairs can be computed in $O(m(m + n))$ time.*

6.2 Applying the Optimization Algorithm for Candidate Pairs

To compute the optimal radius r^*, we apply the optimization algorithm in Sect. 5 on the set of all candidate pairs. Let C be the set of candidate pairs. By Lemma 12, we have $r^* = \min_{(i,j) \in C} r_{ij}^*$. To apply the optimization algorithm, we have to compute an assistant interval $[r_L, r_U]$ satisfying that $r^* \in [r_L, r_U]$ and the combinatorial structure of $\partial D_r(q)$ for each $q \in Q$ remains the same for all $r \in [r_L, r_U]$.

Lemma 14. *An assistant interval* $[r_L, r_U]$, *together with the combinatorial struc-ture of* $\partial D_r(q)$ *for all* $q \in Q$ *and any* $r \in [r_L, r_U]$, *can be computed in* $O(m$ $(m + n) \log^3(m + n))$ *time.*

Now, we are ready to execute our optimization algorithm. We run the opti-mization algorithm for each $(i, j) \in C$ and find the minimum of r_{ij}^* over $(i, j) \in C$.

Theorem 3. *An optimal two-center of* m *points with respect to a simple* n-*gon can be computed in* $O(m(m + n) \log^3(m + n) \log m)$ *time using* $O(mn)$ *space.*

We can improve the space complexity by sacrificing the running time. Instead of building and storing $\mathsf{SPM}(q)$ and $\partial D_{r_L}(q)$ for $q \in Q_I$ in the preprocessing step, we can compute them from scratch whenever necessary. Then the running time of the algorithm gains into $O(m^2(m + n) \log^3(m + n))$.

Corollary 1. *An optimal two-center of* m *points with respect to a simple* n-*gon can be computed in* $O(m^2(m + n) \log^3(m + n))$ *time using* $O(m + n)$ *space.*

References

1. Ahn, H.K., Barba, L., Bose, P., De Carufel, J.L., Korman, M., Oh, E.: A linear-time algorithm for the geodesic center of a simple polygon. In: Proceedings of the 31st Symposium Computational Geometry (SoCG), vol. 34, pp. 209–223 (2015)
2. Aronov, B., Fortune, S., Wilfong, G.: The furthest-site geodesic voronoi diagram. Discrete Comput. Geom. **9**(1), 217–255 (1993)
3. Asano, T., Toussaint, G.T.: Computing geodesic center of a simple polygon. Tech-nical report SOCS-85.32, McGill University (1985)
4. Chan, T.M.: More planar two-center algorithms. Comput. Geom. **13**(3), 189–198 (1999)
5. Chazelle, B., Matoušek, J.: On linear-time deterministic algorithms for optimiza-tion problems in fixed dimension. J. Algorithms **21**(3), 579–597 (1996)
6. Cole, R.: Parallel merge sort. SIAM J. Comput. **17**(4), 770–785 (1988)
7. Dyer, M.E.: On a multidimensional search technique and its application to the euclidean one-centre problem. SIAM J. Comput. **15**(3), 725–738 (1986)
8. Feder, T., Greene, D.H.: Optimal algorithms for approximate clustering. In: Pro-ceedings of the 20th ACM Symposium Theory Computing (STOC), pp. 434–444 (1988)
9. Guibas, L., Hershberger, J., Leven, D., Sharir, M., Tarjan, R.: Linear-time algo-rithms for visibility and shortest path problems inside triangulated simple poly-gons. Algorithmica **2**(1–4), 209–233 (1987)
10. Halperin, D., Sharir, M., Goldberg, K.Y.: The 2-center problem with obstacles. J. Algorithms **42**(1), 109–134 (2002)
11. Hwang, R., Lee, R., Chang, R.: The slab dividing approach to solve the Euclidean p-center problem. Algorithmica **9**(1), 1–22 (1993)
12. Megiddo, N.: Linear-time algorithms for linear programming in R^3 and related problems. SIAM J. Comput. **12**(4), 759–776 (1983)
13. Oh, E., De Carufel, J.-L., Ahn, H.-K.: The 2-center problem in a simple polygon. In: Elbassioni, K., Makino, K. (eds.) ISAAC 2015. LNCS, vol. 9472, pp. 307–317. Springer, Heidelberg (2015). doi:10.1007/978-3-662-48971-0_27

14. Pollack, R., Sharir, M., Rote, G.: Computing the geodesic center of a simple polygon. Discrete Comput. Geom. **4**(1), 611–626 (1989)
15. Toussaint, G.T.: Computing geodesic properties inside a simple polygon. Revue D'Intelligence Artificielle **3**, 9–42 (1989)

Simple Approximation Algorithms for Balanced MAX 2SAT

Alice Paul, Matthias Poloczek[(⊠)], and David P. Williamson

School of Operations Research and Information Engineering,
Cornell University, Ithaca, NY 14850, USA
{ajp336,poloczek}@cornell.edu, dpw@cs.cornell.edu

Abstract. We study simple algorithms for the balanced MAX 2SAT problem, where we are given weighted clauses of length one and two with the property that for each variable x the total weight of clauses that x appears in equals the total weight of clauses for $\bar{x}$. We show that such instances have a simple structural property in that any optimal solution can satisfy at most the total weight of the clauses minus half the total weight of the unit clauses. Using this property, we are able to show that a large class of greedy algorithms, including Johnson's algorithm, gives a $\frac{3}{4}$-approximation algorithm for balanced MAX 2SAT; a similar statement is false for general MAX 2SAT instances. We further give a spectral 0.81-approximation algorithm for balanced MAX E2SAT instances (in which each clause has exactly 2 literals) by a reduction to a spectral algorithm of Trevisan for the maximum colored cut problem. We provide experimental results showing that this spectral algorithm performs well and is slightly better than Johnson's algorithm and the Goemans-Williamson semidefinite programming algorithm on balanced MAX E2SAT instances.

1 Introduction and Overview

In the MAX SAT problem we are given a set of Boolean variables and a set of clauses. Each clause consists of a disjunction of literals and is associated with a nonnegative weight. The goal is to find a Boolean assignment to the variables that maximizes the weight of satisfied clauses. Well-studied special cases include MAX kSAT, in which each clause has at most k literals, and MAX EkSAT, in which each clause has exactly k literals. In this paper, we will consider both MAX 2SAT and MAX E2SAT.

MAX 2SAT is known to be NP-hard, and hence approximation algorithms have been studied for this problem. We say an efficient algorithm A is an α-approximation algorithm, if A obtains an assignment that satisfies clauses with a total weight of at least α times the optimum for every instance; the value α is the

A. Paul—Supported by an NDSEG fellowship.

M. Poloczek—Supported by the Alexander von Humboldt Foundation within the Feodor Lynen program and by NSF grant CCF-1115256.

D.P. Williamson—Supported in part by NSF grant CCF-1115256.

© Springer-Verlag Berlin Heidelberg 2016
E. Kranakis et al. (Eds.): LATIN 2016, LNCS 9644, pp. 659–671, 2016.
DOI: 10.1007/978-3-662-49529-2_49

performance guarantee of the algorithm. In the case of a randomized algorithm, we require the guarantee to hold in expectation. In a seminal paper Goemans and Williamson [8] used semidefinite programming to give a 0.878-approximation algorithm for MAX 2SAT. Concluding a series of papers [6,12], the (currently) best algorithm for MAX 2SAT due to Lewin, Livnat, and Zwick [11] achieves a 0.94-approximation. Austrin [2] showed that no polynomial-time algorithm can achieve a better approximation ratio assuming the Unique Games Conjecture. Without using semidefinite programming, the best known approximation algorithm is a $\frac{3}{4}$-approximation algorithm. Such an algorithm was first given by Yannakakis [21]; he uses a network flow computation to reduce MAX 2SAT instances to an equivalent MAX E2SAT instance, then applies a greedy algorithm of Johnson [9], which gives a $\frac{3}{4}$-approximation algorithm for MAX E2SAT instances. Chan, Lee, Raghavendra, and Steurer [5] have shown that having an integrality gap larger than $\frac{3}{4}$ for MAX 2SAT requires superpolynomially-sized linear programs, suggesting that something stronger than linear programming is required to get better than a $\frac{3}{4}$-approximation algorithm for the problem. Note that their bound holds for balanced MAX E2SAT instances.

One recent theme of work in the area of approximation algorithms has been that of finding *simple* approximation algorithms that achieve the same or nearly the same performance guarantee as complicated or computationally intensive algorithms currently in the literature. One such stream of work has been for the MAX SAT problem; in 2011, Poloczek and Schnitger [16] gave a randomized, greedy $\frac{3}{4}$-approximation algorithm for MAX SAT. Previously, the solution to a linear program had been needed to achieve such a performance guarantee [7,21]. There were several subsequent variants of this algorithm considered [4,13,18,22]; most recently Poloczek et al. gave a deterministic, two-pass $\frac{3}{4}$-approximation algorithm [17].

In this paper, we continue the theme of finding simple approximation algorithms for MAX SAT by focusing on balanced MAX 2SAT. A set of clauses is called *balanced* if for each Boolean variable x the total weight of clauses containing literal x equals the total weight of all clauses containing literal $\bar{x}$. Better approximation algorithms are known for balanced MAX 2SAT: Khot, Kindler, Mossel, and O'Donnell [10] showed that the semidefinite programming algorithm of Goemans and Williamson [8] achieves at least a 0.943-approximation on balanced MAX E2SAT; note that this guarantee is slightly better than the 0.940-approximation algorithm of Levin, Livnat, and Zwick [11] for MAX 2SAT and the hardness bound of Austrin [2].

Here we show that a broad class of deterministic and randomized greedy algorithms, called *majority-preserving* algorithms by Poloczek [14], achieve a $\frac{3}{4}$-approximation for balanced MAX 2SAT. This class includes Johnson's original greedy algorithm. To achieve this result, we prove a simple but interesting structural result for balanced MAX 2SAT instances: if W is the total weight of all clauses, and W_1 the total weight of all unit clauses (clauses with just one literal), then any assignment can satisfy weight at most $W - \frac{1}{2}W_1$. The existence of deterministic and greedy $\frac{3}{4}$-approximation algorithms for balanced MAX 2SAT stands in contrast to recent work by Poloczek [13], who showed that randomness seems to be essential for greedy algorithms for general MAX 2SAT by giving a 0.729-inapproximability bound for adaptive priority algorithms. Adaptive priority algorithms capture the characteristics of deterministic, greedy-like algorithms. The MAX 2SAT instances used in the inapproximability bound are intriniscally unbalanced.

We further show that it is possible to do better than a $\frac{3}{4}$-approximation algorithm for balanced MAX E2SAT without solving a semidefinite program, by performing an eigenvalue computation; in particular, we are able to achieve a 0.81-approximation algorithm for balanced MAX E2SAT by using an algorithm of Trevisan. Trevisan [20] gave a spectral 0.531-approximation algorithm for the maximum cut problem; Soto [19] generalized and improved Trevisan's analysis to a 0.614-approximation algorithm for the *maximum colored cut problem*. In the maximum colored cut problem, we are given an undirected graph $G = (V, E)$, with weights $w_{ij} \geq 0$ on $(i, j) \in E$, and a partition of the edge set E into R and B. The goal is to find a subset S of vertices that maximizes the total weight of edges in R with an odd number of endpoints in S and the weight of edges in B with an even number of endpoints in S. Trevisan's algorithm works by computing an eigenvector for the Laplacian of the graph, then computing a tripartition of the vertices into those in S, those not in S, and an unassigned set of vertices; the algorithm then recurses on the unassigned set. We achieve our 0.81-approximation algorithm by a reduction of the balanced MAX E2SAT problem to the maximum colored cut problem, and a generalization of the analysis of Soto.

Again, this spectral algorithm stands in contrast to recent work of Poloczek [14], which shows that one cannot hope to achieve a performance guarantee better than $\frac{3}{4}$ for MAX SAT via any online priority algorithm; this class generalizes majority-preserving algorithms and captures a class of deterministic and randomized greedy algorithms. The instances used to prove this result are balanced MAX E2SAT instances, and the result implies that no greedy-like algorithm will be able to improve on a performance guarantee of $\frac{3}{4}$ even for balanced MAX E2SAT.

Finally, we perform an empirical evaluation of our algorithms: in particular, we compare Johnson's greedy algorithm, the Goemans-Williamson semidefinite programming algorithm, and our spectral algorithm on 162 balanced MAX E2SAT instances drawn from the MAX SAT 2014 Competition. All algorithms satisfy about the same fraction of clauses on average, with the spectral algorithm slightly outperforming the others. In particular, the spectral algorithm really shines on a certain family of instances describing the maximum cut in random graphs. The average running times of the spectral algorithm and the SDP-based algorithm are comparable to each other and are orders of magnitude higher than the greedy algorithm. We propose a variation that speeds up the spectral algorithm by more than a factor of three, while affecting the quality of the solution negligibly.

The structure of our paper is as follows. In Sect. 2, we prove our structural result about balanced MAX 2SAT instances. In Sect. 3, we use this structural result to show that the class of majority-preserving algorithms are $\frac{3}{4}$-approximation algorithms, including Johnson's algorithm. In Sect. 4, we show how to achieve our spectral 0.81-approximation algorithm for balanced MAX E2SAT via a reduction to the maximum colored cut problem. We discuss our experimental work in Sect. 5, and conclude in Sect. 6.

2   The Structure of Balanced MAX 2SAT

We begin by proving a lemma about the structure of balanced MAX 2SAT instances.

Lemma 1. *Let C be a balanced clause set with at most two literals per clause. Furthermore, we denote the total weight of all clauses by W and the total weight of all unit clauses by W_1.*

Then any Boolean assignment leaves clauses with a total weight of at least $\frac{W_1}{2}$ unsatisfied on C.

Note that the lemma implies that if there is a Boolean assignment that satisfies clauses with a total weight of at least $(1 - \varepsilon) \cdot W$, then the total weight of unit clauses of C is at most $2 \cdot \varepsilon \cdot W$. In particular, if C is satisfiable, then the set contains no unit clauses.

Proof of Lemma 1. Let C be a balanced set of unit clauses and clauses of length two over the set of variables V. A nonnegative weight w_c is associated with each clause $c \in C$.

Suppose that there is a Boolean assignment b that satisfies clauses with a total weight larger than $W - \frac{1}{2}W_1$. For the sake of simplicity, we flip signs of literals such that b is the all-ones assignment. Note that this does not affect the balance of C. Then exactly those clauses without a positive literal are those not satisfied by b, and their total weight is bounded by

$$\sum_{x \in V} w_{(\overline{x})} + \frac{1}{2} \sum_{x \in V} \sum_{y \in V,\, x \neq y} w_{(\overline{x} \vee \overline{y})} < \frac{1}{2} \sum_{x \in V} \left(w_{(x)} + w_{(\overline{x})} \right), \tag{1}$$

where $\frac{1}{2} \sum_{x \in V} \left(w_{(x)} + w_{(\overline{x})} \right) = \frac{1}{2} W_1$ holds by definition. Observe that the coefficient $\frac{1}{2}$ on the LHS compensates for the fact that each 2-clause contributes twice to the sum. We rewrite Eq. (1) as

$$\sum_{x \in V} \sum_{y \in V,\, x \neq y} w_{(\overline{x} \vee \overline{y})} < \sum_{x \in V} \left(w_{(x)} - w_{(\overline{x})} \right). \tag{2}$$

The assumption that C is balanced implies

$$\sum_{x \in V} \left[\left(w_{(x)} - w_{(\overline{x})} \right) + \sum_{y \in V,\, x \neq y} \left(w_{(x \vee y)} + w_{(x \vee \overline{y})} - w_{(\overline{x} \vee y)} - w_{(\overline{x} \vee \overline{y})} \right) \right] = 0.$$

Now the heart of the proof is that $\sum_{x\in V}\sum_{y\in V,\,x\neq y}\left[w_{(x\vee\overline{y})}-w_{(\overline{x}\vee y)}\right]=0$ holds, since each of these clauses occurs exactly once with a positive sign and exactly once with a negative one. Hence, we have

$$\sum_{x\in V}\left[\left(w_{(x)}-w_{(\overline{x})}\right)+\sum_{y\in V,\,x\neq y}\left(w_{(x\vee y)}-w_{(\overline{x}\vee\overline{y})}\right)\right]=0.$$

But this contradicts Eq. (2), because $\sum_{x\in V}\sum_{y\in V,\,x\neq y}w_{(x\vee y)}$ is nonnegative. Thus, every Boolean assignment to V leaves clauses with a total weight of at least $\frac{W_1}{2}$ unsatisfied. □

This bound is tight; consider for instance a pair of contradictory unit clauses of same weight.

3 Majority-Preserving Algorithms for Balanced MAX 2SAT

In this section, we define a large class of randomized and deterministic greedy algorithms, called majority-preserving algorithms, and use the structure lemma of the previous section to prove that they are $\frac{3}{4}$-approximation algorithms for balanced MAX 2SAT.

In his fundamental work, Johnson [9] presented a greedy algorithm that processes variables in an arbitrary order. To favor shorter clauses which are in greater danger of being falsified, the modified weight μ is introduced.

Definition 1 (Modified weight μ). *For a clause c we denote its weight by w_c and the number of unfixed literals by $|c|$. The modified weight, also referred to as Johnson measure, of a clause c that is not yet satisfied is*

$$\mu(c)=w_c\cdot 2^{-|c|}.$$

As a convention we set $\mu(c)=0$ if c has already been satisfied. Moreover, we define the modified weight of literal x as

$$\mu_x=\sum_{c,\,x\in c}\mu(c).$$

Observe that the modified weight of a clause increases in the course of the computation, as some of its literals evaluate to zero. If x is the currently processed variable, then Johnson's algorithm sets x to one iff $\mu_x\geq\mu_{\overline{x}}$ holds.

A *majority-preserving algorithm* is a randomized algorithm which processes the variables according to a fixed order, but utilizes randomization to decide the assignment. In particular, the currently processed variable, say x, is assigned the value *true* with probability $p\left(\frac{\mu_x}{\mu_x+\mu_{\overline{x}}}\right)$; we demand the function $p:[0,1]\to[0,1]$ to be *majority-preserving*, i.e. $p(z)\geq\frac{1}{2}$ holds whenever $z\geq\frac{1}{2}$. In particular, any monotone increasing function p with $p\left(\frac{1}{2}\right)\geq\frac{1}{2}$ is majority-preserving. Thus, we cover a broad class of approaches, including the deterministic algorithm of Johnson, as well as the natural randomizations that assign one with probability $p\left(\frac{\mu_x}{\mu_x+\mu_{\overline{x}}}\right) = \frac{\mu_x}{\mu_x+\mu_{\overline{x}}}$ or with probability $p\left(\frac{\mu_x}{\mu_x+\mu_{\overline{x}}}\right) = \frac{1}{2}$. Note that we may assume $\mu_x + \mu_{\overline{x}} > 0$, since the respective variable can be skipped otherwise.

Assume that variable x is to be decided. Then we denote the discrepancy of the decision for x by its slack

$$\mathrm{slack}_x = p\left(\frac{\mu_x}{\mu_x + \mu_{\overline{x}}}\right) \cdot (\mu_x - \mu_{\overline{x}}) + p\left(\frac{\mu_{\overline{x}}}{\mu_x + \mu_{\overline{x}}}\right) \cdot (\mu_{\overline{x}} - \mu_x).$$

In [14] it is shown that the slack is always nonnegative if p is a majority-preserving function. Let $\mathbb{E}[\mathrm{slack}]$ be the expected slack accumulated during the computation. The following result relates the satisfied weight, denoted by Sat, to the total weight of a CNF formula C and the expected slack. We point out that the theorem does not require C to be balanced.

Theorem 1 (Chapter 2.3 in [14]**).** *Let W_j denote the total weight of all clauses of initial length j. Any majority-preserving algorithm satisfies clauses with an expected weight of*

$$\mathbb{E}[\mathrm{Sat}] = \sum_j \left(1 - 2^{-j}\right) W_j + \mathbb{E}[\mathrm{slack}].$$

We can now prove the following.

Theorem 2. *Any majority-preserving algorithm achieves a $\frac{3}{4}$-approximation (in expectation) when invoked on a balanced MAX 2SAT instance. In particular, Johnson's deterministic algorithm guarantees a $\frac{3}{4}$-approximation in this case.*

Proof. Given an instance of balanced MAX 2SAT, let W_1 be the weight of the unit clauses and W_2 the weight of the length two clauses. Then Lemma 1 shows that an optimal solution can satisfy at most total weight $\frac{1}{2}W_1 + W_2$. From Theorem 1, we know that for any majority-preserving algorithm satisfies clauses with expected weight of at least $\frac{1}{2}W_1 + \frac{3}{4}W_2$, since $\mathbb{E}[\mathrm{slack}]$ is nonnegative for a majority-preserving algorithm. The theorem statement follows.   □

4   Beating $\frac{3}{4}$ for Balanced MAX E2SAT

In this section we give an approximation-preserving reduction from balanced MAX E2SAT to the maximum colored cut problem (MAX CC). Recall that in the maximum colored cut problem, we are given an undirected graph $G = (V, E)$, with weights $w_{ij} \geq 0$ on $(i, j) \in E$, and a partition of the edge set E into R and B. The goal is to find a subset S of vertices that maximizes the total weight of edges in R with an odd number of endpoints in S and the weight of edges in B with an even number of endpoints in S. We combine this reduction with Trevisan's algorithm to obtain a 0.81-approximation algorithm.

We state our reduction in a slightly more general form: consider now any problem that can be written in the form

$$\max_{y\in\{-1,1\}^n} \sum_{(i,j)\in E^+} w_{ij}(\alpha + \beta \cdot y_i y_j) + \sum_{(i,j)\in E^-} w_{ij}(\alpha - \beta \cdot y_i y_j),$$

where E^+ and E^- are sets of pairs of n variables with values in $\{-1,+1\}$, and $\alpha, \beta \geq 0$ (we assume $\alpha + \beta > 0$). Let $W = \sum_{(i,j)\in E^+\cup E^-} w_{ij}$ be the total weight of pairs. Then we can rewrite the problem as

$$\max_{y\in\{-1,1\}^n} (\alpha - \beta)\cdot W + \sum_{(i,j)\in E^+} w_{ij}(\beta + \beta \cdot y_i y_j) + \sum_{(i,j)\in E^-} w_{ij}(\beta - \beta \cdot y_i y_j)$$

or equivalently,

$$\max_{y\in\{-1,1\}^n} (\alpha - \beta)\cdot W$$
$$+ 2\cdot\beta\cdot\left(\frac{1}{4}\sum_{(i,j)\in B=E^+} w_{ij}(y_i + y_j)^2 + \frac{1}{4}\sum_{(i,j)\in R=E^-} w_{ij}(y_i - y_j)^2\right).$$

To see that we have reduced the problem to a MAX CC instance, observe that a MAX CC instance $G = (V, R\cup B)$ with $|V| = n$ can be expressed as quadratic form by introducing a variable $x_i \in \{-1,1\}$ for all $i \in V$ that represents which side of the cut i lies on. Then the problem of finding a maximum colored cut can be represented as:

$$\max_{x\in\{-1,1\}^n} \frac{1}{4}\sum_{(i,j)\in R} w_{ij}(x_i - x_j)^2 + \frac{1}{4}\sum_{(i,j)\in B} w_{ij}(x_i + x_j)^2. \tag{3}$$

In a moment, we will show that balanced MAX E2SAT can be expressed in the quadratic form in 3. Let G be from now on the MAX CC instance that is created by our reduction. Our proposed algorithm is to run Trevisan's spectral algorithm [20] on the resulting instance and return the result. Consider an optimal assignment to the MAX CC instance on G. Suppose this assignment achieves weight $\mathrm{OPT}_G(\varepsilon) = (1 - \varepsilon)W$ for some $0 \leq \varepsilon \leq \frac{1}{2}$. Then, an optimal assignment for the original problem has weight

$$\mathrm{OPT}(\varepsilon) = (\alpha - \beta)\cdot W + 2\beta\cdot\mathrm{OPT}_G(\varepsilon) = (\alpha - \beta)\cdot W + 2\beta(1 - \varepsilon)\cdot W.$$

Let $\mathrm{LB}_G(\varepsilon)$ be a lower bound on the fraction of weight achieved by the spectral algorithm on G. Then we obtain a lower bound, denoted by $\mathrm{ALG}(\varepsilon)$, on the total weight of clauses satisfied by our proposed algorithm by

$$\mathrm{ALG}(\varepsilon) = (\alpha - \beta)\cdot W + 2\beta\cdot\mathrm{LB}_G(\varepsilon)\cdot W.$$

Hence its approximation ratio is at least

$$\frac{(\alpha - \beta) + 2\beta\cdot\mathrm{LB}_G(\varepsilon)}{(\alpha - \beta) + 2\beta\cdot(1 - \varepsilon)}$$

on the instance. To find the overall approximation ratio, we must find the minimum of the above expression over $0 \leq \varepsilon \leq \frac{1}{2}$. Without loss of generality, we may assume that $\alpha + \beta = 1$ (otherwise divide by $\alpha + \beta > 0$ in the objective function which does not change the approximation ratio). Using the closed form of $\mathrm{LB}_G(\varepsilon)$ (cp. Sect. A in the appendix) and Matlab's *fminbnd* we can plot a lower bound on the approximation ratio for all (α, β) pairs (see Fig. 1 in Sect. B in the appendix).

Now suppose we have an instance of balanced MAX E2SAT with clauses $C = \{c_1, c_2, \ldots, c_m\}$ over Boolean variables $V = \{x_1, x_2, \ldots, x_n\}$. Each clause c_k consists of a disjunction of exactly two literals and is associated with a nonnegative weight w_k (with $k = 1, 2, \ldots, m$). For each clause c_k, let i_k and j_k be the indices of the variables in c_k and let $\mathrm{sgn}(\ell, k)$ be the sign of variable x_ℓ in clause c_k.

Let $y_i \in \{-1, +1\}$ for $1 \leq i \leq n$ be a variable indicating the assignment of x_i where $y_i = +1$ corresponds to setting x_i to one and -1 to zero, then an instance of balanced MAX E2SAT can be written in the following quadratic form (see [10]):

$$
\max_{y \in \{-1, +1\}^n} \left(\sum_{k: \, \mathrm{sgn}(i_k, k) = \mathrm{sgn}(j_k, k)} w_k \left(\frac{3}{4} - \frac{1}{4} y_{i_k} y_{j_k} \right) \right.
$$

$$
\left. + \sum_{k: \, \mathrm{sgn}(i_k, k) \neq \mathrm{sgn}(j_k, k)} w_k \left(\frac{3}{4} + \frac{1}{4} y_{i_k} y_{j_k} \right) \right)
$$

Thus, by setting $\alpha = \frac{3}{4}$ and $\beta = \frac{1}{4}$ we can see that this yields approximation ratio at least

$$
H(\varepsilon) := \frac{\frac{1}{2} + \frac{1}{2} \mathrm{LB}_G(\varepsilon)}{\frac{1}{2} + \frac{1}{2}(1 - \varepsilon)} = \frac{1 + \mathrm{LB}_G(\varepsilon)}{2 - \varepsilon}
$$

for all $0 \leq \varepsilon \leq \frac{1}{2}$. This function has a unique minimum at $\varepsilon^* \approx 0.0912$ with value $H(\varepsilon^*) \approx 0.8173$. This yields the following theorem.

Theorem 3. *Trevisan's algorithm is a 0.8173-approximation algorithm for balanced MAX E2SAT.*

5   Experimental Results

In this section, we perform computational experiments in order to study the performance of our algorithms in practice. In particular, we compare the performance of Johnson's deterministic greedy algorithm [9], the spectral algorithm we proposed in Sect. 4 that is based on Trevisan's algorithm [20], and the algorithm of Goemans and Williamson [8] based on semidefinite programming (SDP). The testbed is 162 balanced MAX E2SAT instances taken from the MAX SAT 2014 Competition [1]. They all belong to the MS-CRAFTED category: while most sets encode maximum cut instances, the benchmark also contains instances representing vertex cover or applications in fault diagnosis, for example. The instances are rather small, having less than 150 variables in at most 1500 clauses. All clauses are unweighted. We remark that the SAT 2014 competition [3] does not contain any balanced clause sets.

The Setup. The spectral algorithm and the SDP-based algorithm were implemented using julia 0.38. In order to solve the semidefinite programs, we relied on CSDP (A C Library for Semidefinite Programming) version 6.1.1. The experiments were conducted on a Dell Precision 490 workstation (Intel Xeon 5140 2.33 GHz with 8 GB RAM) under Debian wheezy. Every algorithm was run ten times on each input instance.

For comparison we present experimental data for Johnson's algorithm that is taken from [15]: this implementation was done in C++ using GCC 4.7.2 and run on the same machine. The performance indicator that we are interested in is the average fraction of satisfied clauses, since in most cases the value of the optimum is unknown. Moreover, we also measured the average running time.

Our Findings. The general picture is that all three algorithms satisfy typically a very similar number of clauses. Looking closer, we see that the average fraction of satisfied clauses for our spectral algorithm is typically larger by 1.54 % than the average of Johnson's algorithm, and still better by 0.54 % when compared to the SDP-based algorithm. However, no algorithm strictly dominates another algorithm on every instance.

Since the SDP-based algorithm performs randomized rounding, it might benefit from multiple iterations on a single instance. Indeed, this is what we observe in our experiments: the average fraction of satisfied clauses is increased by 0.73 % if we consider the best solution found in ten iterations.

The majority of the testbed clause sets are MAX CUT instances in random graphs with a high girth; their filenames start with "maxcut". One hundred of the 162 instances are of this type. A closer examination reveals that our spectral algorithm and the SDP-based algorithm perform particularly well on these instances: here Johnson's algorithm trails behind the spectral algorithm by 2.07 % and behind the Goemans-Williamson algorithm by 1.44 %.

Even more interestingly, when we focus on the remaining instances, then the differences in the performances of the algorithms become almost negligible: our spectral method exceeds Johnson's algorithm by 0.68 % and the SDP algorithm by 0.40 % in terms of average satisfied clauses.

Considering the running times, Johnson's algorithm outperforms the others by orders of magnitude. On average it is faster by a factor of 10^5. Comparing the other algorithms based on more sophisticated techniques, we see that our implementation of the spectral algorithm in julia is about twice as fast as our implementation of the Goemans-Williamson algorithm using CSDP. Our findings are summarized in Table 1.

Improving the Runtime of the Spectral Algorithm. Trevisan's algorithm repeatedly computes a tripartition of variables set to false, to true, and those it recurses

Table 1. A summary of the experimental results for different algorithms.

	Johnson's algorithm		Spectral algorithm		GW algorithm	
	% sat	∅ time	% sat	∅ time	% sat	∅ time
All Instances	81.97	$4.9 \cdot 10^{-5}$ s	83.51	0.1999 s	82.97	0.5562 s
only MAX CUT	83.63	0.0001 s	85.70	0.2779 s	85.07	0.8401 s
w/o MAX CUT	79.28	0.0000 s	79.96	0.0741 s	79.56	0.0984 s

on in the subsequent iteration. In order to obtain such a tripartition, the algorithm determines a *threshold* t and considers for each variable x_i the corresponding component e_i of the principal eigenvector: if $e_i \geq \sqrt{t}$ then x_i is set to true, and if $e_i \leq -\sqrt{t}$ then is fixed to false. Any variable not fixed will be considered again in subsequent iterations.

Two computational tasks seem to dominate the running time of Trevisan's algorithm. The first is the search for the best threshold. The second factor is the recursion depth: the more variables are set early on, the smaller the overall work seems to be. Trevisan's implementation of his algorithm simply tries each value e_i^2 as threshold t, thereby aiming to minimize the second factor.

We propose a hybrid approach: Instead of evaluating all possible cutoff points, we sample k values $t_1, \ldots, t_k \in (0, 1]$ uniformly at random and try each t_j as threshold. Observe that worst case analyses of [19,20] rely on a randomly chosen threshold, hence the approximation guarantee is still valid for our variants. We evaluate experimentally different choices for k. Note that the differences in the average fraction of satisfied clauses are negligible, but we see a big impact on the running times. When the number of random thresholds is reduced, the fluctuation of the running times increases significantly: for example, if $k = 3$ then the standard deviation in the running times is more than half the average running time. Therefore, the choice $k = |V|/4$ seems preferable, since it combines a low average running time with a small fluctuation. This parameter setting performs 3.4 times faster than the vanilla version of Trevisan's algorithm, and 9.5 times faster than the Goemans-Williamson algorithm (Table 2).

Table 2. The effects of the number of random thresholds. The first column summarizes the vanilla version of Trevisan's algorithm.

	vanilla	$k = |V|/4$	$k = 20$	$k = 5$	$k = 3$	$k = 1$
% sat	0.8351	0.8344	0.8344	0.8331	0.8324	0.8312
∅ time	0.1999 s	0.0588 s	0.0508 s	0.0524 s	0.0716 s	0.1338 s
SD time	0.0026 s	0.0043 s	0.0063 s	0.0372 s	0.0546 s	0.0784 s

6   Conclusions and Future Work

We have shown that balanced instances of MAX 2SAT allow better approximation guarantees: on the one hand, a large class of greedy algorithms gives a $\frac{3}{4}$-
approximation, whereas it is known that they do not achieve this performance for
general MAX 2SAT. On the other hand, we can go even beyond the $\frac{3}{4}$-barrier and
obtain a 0.81-approximation using our reduction from balanced MAX E2SAT
to the maximum colored cut problem. We wonder if a similar approach can be
used for the general case of MAX 2SAT, especially since the result of Chan et
al. [5] implies that linear programming alone is not sufficient to get better than
a $\frac{3}{4}$-approximation algorithm. Note that the straightforward reduction applied
to unbalanced sets of 2-clauses yields an approximation guarantee of only $\frac{2}{3}$.
Nevertheless, it would be interesting how this reduction performs in practice.

We are also interested in the approximability of balanced instances that contain clauses of length greater than two. While the statement of Lemma 1 is
not valid for longer clauses, we do not know of any balanced clause set that is
approximated worse than $\frac{3}{4}$ by Johnson's algorithm.

An interesting aspect of our experimental evaluation is that the spectral
algorithm performs better than the Goemans-Williamson algorithm, although
its approximation guarantee is worse. In addition, by sampling a set of random thresholds instead of testing all possibilities, we have found a better trade-
off between the two computationally intense tasks for the spectral algorithms.
This allows the hybrid algorithm to run considerably faster than the SDP-based
algorithm. We leave it as future work to explore this connection both from an
experimental and a theoretical point of view.

A   Soto's Bound for MAX CC

Recall from Sect. 4 that $\mathrm{LB}_G(\varepsilon)$ is a lower bound on the fraction of weight
achieved by Trevisan's spectral algorithm on G, where G is the MAX CC instance
that was created by our reduction on the balanced set of 2-clauses C.

Lemma 2 (Sect. 3.1 in [19]). *Let ε_0 be the unique solution of the equa-
tion* $\frac{1}{1+2\sqrt{\varepsilon(1-\varepsilon)}} = \frac{-1+\sqrt{4\varepsilon^2-8\varepsilon+5}}{2(1-\varepsilon)}$. *Then,*
If $\varepsilon \geq \frac{1}{3}$,

$$\mathrm{LB}_G(\varepsilon) := \frac{1}{2}.$$

If $\varepsilon_0 \leq \varepsilon \leq \frac{1}{3}$,

$$\mathrm{LB}_G(\varepsilon) := \frac{1}{2} \cdot \left(\varepsilon - 1 + \sqrt{4\varepsilon^2 - 8\varepsilon + 5} - \varepsilon \ln \left(\frac{1 + \sqrt{4\varepsilon^2 - 8\varepsilon + 5}}{8\varepsilon} \right) \right.$$
$$\left. + \frac{\sqrt{5}}{5} \varepsilon \ln \left(\frac{5 - 4\varepsilon + \sqrt{5(4\varepsilon^2 - 8\varepsilon + 5)}}{(11 + 5\sqrt{5})\varepsilon} \right) \right).$$

If $\varepsilon \leq \varepsilon_0$,

$$
\begin{aligned}
\mathrm{LB}_G(\varepsilon) := \frac{1}{2} \cdot \Bigg(& \varepsilon \left(1 - \frac{3}{\varepsilon_0} \right) + 2 + \frac{\varepsilon}{\varepsilon_0} \sqrt{4\varepsilon_0^2 - 8\varepsilon_0 + 5} \\
& - \varepsilon \ln \left(\frac{1 + \sqrt{4\varepsilon_0^2 - 8\varepsilon_0 + 5}}{8\varepsilon_0} \right) \\
& + \frac{\sqrt{5}}{5} \varepsilon \ln \left(\frac{5 - 4\varepsilon_0 + \sqrt{5(4\varepsilon_0^2 - 8\varepsilon_0 + 5)}}{(11 + 5\sqrt{5})\varepsilon_0} \right) \\
& + 16\varepsilon \ln \left(\frac{\sqrt{\varepsilon} + \sqrt{1 - \varepsilon}}{\sqrt{\varepsilon} + \sqrt{\frac{\varepsilon}{\varepsilon_0} - \varepsilon}} \right) + 8\varepsilon \frac{\sqrt{\varepsilon_0(1 - \varepsilon_0)} + 1 - 2\varepsilon_0}{\varepsilon_0 + \sqrt{\varepsilon_0(1 - \varepsilon_0)}} \\
& - 8\sqrt{\varepsilon} \frac{\sqrt{\varepsilon(1 - \varepsilon)} + 1 - 2\varepsilon}{\sqrt{\varepsilon} + \sqrt{\varepsilon(1 - \varepsilon)}} \Bigg).
\end{aligned}
$$

B   Dependency of the Approximation Ratio on α and β

Fig. 1. Approximation ratio for (α, β) pairs, where $\beta = 1 - \alpha$. α is given on the horizontal axis and the approximation ratio on the vertical axis.

References

1. Argelich, J., Li, C.M., Manyà, F., Planes, J.: MAX-SAT 2014: Ninth Max-SAT evaluation. www.maxsat.udl.cat/14/. Accessed 9 January 2015
2. Austrin, P.: Balanced MAX 2-SAT might not be the hardest. In: STOC, pp. 189–197 (2007)
3. Belov, A., Diepold, D., Heule, M.J., Järvisalo, M.: Proceedings of the SAT COMPETITION 2014: solver and benchmark descriptions (2014)
4. Buchbinder, N., Feldman, M., Naor, J., Schwartz, R.: A tight linear time (1/2)-approximation for unconstrained submodular maximization. In: FOCS, pp. 649–658 (2012)
5. Chan, S.O., Lee, J., Raghavendra, P., Steurer, D.: Approximate constraint satisfaction requires large LP relaxations. In: FOCS, pp. 350–359 (2013)
6. Feige, U., Goemans, M.X.: Approximating the value of two prover proof systems, with applications to MAX 2SAT and MAX DICUT. In: ISTCS, pp. 182–189 (1995)
7. Goemans, M.X., Williamson, D.P.: New 3/4-approximation algorithms for the maximum satisfiability problem. SIAM J. Discrete Math. **7**(4), 656–666 (1994)
8. Goemans, M.X., Williamson, D.P.: Improved approximation algorithms for maximum cut and satisfiability problems using semidefinite programming. J. ACM **42**(6), 1115–1145 (1995)

9. Johnson, D.S.: Approximation algorithms for combinatorial problems. J. Comput. Syst. Sci. **9**(3), 256–278 (1974)
10. Khot, S., Kindler, G., Mossel, E., O'Donnell, R.: Optimal inapproximability results for MAX-CUT and other 2-variable CSPs? SIAM J. Comput. **37**(1), 319–357 (2007)
11. Lewin, M., Livnat, D., Zwick, U.: Improved rounding techniques for the MAX 2-SAT and MAX DI-CUT problems. In: Cook, W.J., Schulz, A.S. (eds.) IPCO 2002. LNCS, vol. 2337, pp. 67–82. Springer, Heidelberg (2002)
12. Matuura, S., Matsui, T.: 0.935-approximation randomized algorithm for MAX-2SAT and its derandomization. Technical report METR 2001-03, Department of Mathematical Engineering and Physics, the University of Tokyo, Japan (2001)
13. Poloczek, M.: Bounds on greedy algorithms for MAX SAT. In: Demetrescu, C., Halldórsson, M.M. (eds.) ESA 2011. LNCS, vol. 6942, pp. 37–48. Springer, Heidelberg (2011)
14. Poloczek, M.: Greedy Algorithms for MAX SAT and Maximum Matching: Their Power and Limitations. Ph.D. thesis, Johann Wolfgang Goethe-Universitaet, Frankfurt am Main (2012)
15. Poloczek, M.: An experimental evaluation of fast approximation algorithms for the maximum satisfiability problem (2015) (in preparation)
16. Poloczek, M., Schnitger, G.: Randomized variants of Johnson's algorithm for MAX SAT. In: SODA, pp. 656–663 (2011)
17. Poloczek, M., Schnitger, G., Williamson, D.P., van Zuylen, A.: Greedy algorithms for the maximum satisfiability problem: simple algorithms and inapproximability bounds (2015) (in preparation)
18. Poloczek, M., Williamson, D.P., van Zuylen, A.: On some recent approximation algorithms for MAX SAT. In: Pardo, A., Viola, A. (eds.) LATIN 2014. LNCS, vol. 8392, pp. 598–609. Springer, Heidelberg (2014)
19. Soto, J.A.: Improved analysis of a Max-Cut algorithm based on spectral partitioning. SIAM J. Discrete Math. **29**(1), 259–268 (2015)
20. Trevisan, L.: Max Cut and the smallest eigenvalue. SIAM J. Comput. **41**(6), 1769–1786 (2012)
21. Yannakakis, M.: On the approximation of maximum satisfiability. J. Algorithms **17**(3), 475–502 (1994)
22. van Zuylen, A.: Simpler 3/4-approximation algorithms for MAX SAT. In: WAOA, pp. 188–197 (2011)

A Parameterized Algorithm for MIXED-CUT

Ashutosh Rai[1]([✉]), M.S. Ramanujan[2], and Saket Saurabh[1,3]

[1] The Institute of Mathematical Sciences, Chennai, India
{ashutosh,saket}@imsc.res.in
[2] Vienna Institute of Technology, Vienna, Austria
ramanujan@ac.tuwien.ac.at
[3] University of Bergen, Bergen, Norway

Abstract. The classical Menger's theorem states that in any undirected (or directed) graph G, given a pair of vertices s and t, the maximum number of vertex (edge) disjoint paths is equal to the minimum number of vertices (edges) needed to disconnect s from t. This min-max result can be turned into a polynomial time algorithm to find the maximum number of vertex (edge) disjoint paths as well as the minimum number of vertices (edges) needed to disconnect s from t. In this paper we study a mixed version of this problem, called MIXED-CUT, where we are given an undirected graph G, vertices s and t, positive integers k and l and the objective is to test whether there exist a k sized vertex set $S \subseteq V(G)$ and an l sized edge set $F \subseteq E(G)$ such that deletion of S and F from G disconnects from s and t. Apart from studying a generalization of classical problem, one of our main motivations for studying this problem comes from the fact that this problem naturally arises as a subproblem in the study of several graph editing (modification) problems. We start with a small observation that this problem is NP-complete and then study this problem, in fact a much stronger generalization of this, in the realm of parameterized complexity. In particular we study the MIXED MULTIWAY CUT-UNCUT problem where along with a set of terminals T, we are also given an equivalence relation $\mathcal{R}$ on T, and the question is whether we can delete at most k vertices and at most l edges such that connectivity of the terminals in the resulting graph respects $\mathcal{R}$. Our main result is a fixed parameter algorithm for MIXED MULTIWAY CUT-UNCUT using the method of recursive understanding introduced by Chitnis et al. (FOCS 2012).

1 Introduction

Given a graph, a typical *cut problem* asks for finding a set of vertices or edges such that their removal from the graph makes the graph satisfy some separation property. The most fundamental version of the cut problems is MINIMUM CUT, where given a graph and two vertices, called *terminals*, we are asked to find

The research leading to these results has received funding from the European Research Council under the European Union's Seventh Framework Programme (FP7/2007-2013)/ERC grant agreement no. 306992.

© Springer-Verlag Berlin Heidelberg 2016
E. Kranakis et al. (Eds.): LATIN 2016, LNCS 9644, pp. 672–685, 2016.
DOI: 10.1007/978-3-662-49529-2_50

the minimum sized subset of vertices (or edges) of the graph such that deleting them separates the terminals. The MINIMUM CUT problem is known to be polynomial time solvable for both edge and vertex versions and both in undirected and directed graphs. The core of the polynomial time solvability of the MINIMUM CUT problem is one of the classical min-max results in graph theory – the Menger's theorem. Menger's theorem states that in any undirected (or directed) graph G, given a pair of vertices s and t, the maximum number of vertex (edge) disjoint paths is equal to the minimum number of vertices (edges) needed to be deleted to disconnect s from t.

While MINIMUM CUT is polynomial time solvable; even a slight generalization becomes NP-hard. Two of the most studied generalizations of MINIMUM CUT problem which are NP-hard are MULTIWAY CUT and MULTICUT. In the MULTIWAY CUT problem, we are given a set of terminals, and we are asked to delete minimum number of vertices (or edges) to separate the terminals from each other. This problem is known to be NP-hard even when the number of terminals is at least three. In the MULTICUT problem, given pairs of terminals, we are asked to delete minimum number of vertices (or edges) so that it separates all the given terminal pairs. The MULTICUT problem is known to be NP-hard when the number of pairs of terminals is at least three. The mixed version of the problem, which is the central topic of this paper, namely MIXED CUT is also NP-hard. In this problem we are given an undirected graph G, vertices s and t, positive integers k and l and the objective is to test whether there exist a k sized vertex set $S \subseteq V(G)$ and an l sized edge set $F \subseteq E(G)$ such that deletion of S and F from G disconnects from s and t. In this paper we study MIXED CUT, in fact a stronger generalization of it, in the realm of parameterized complexity.

The field of parameterized complexity tries to provide efficient algorithms for NP-complete problems by going from the classical view of single-variate measure of the running time to a multi-variate one. It aims at getting algorithms of running time $f(k)n^{\mathcal{O}(1)}$, where k is an integer measuring some aspect of the problem. These algorithms are called fixed parameter tractable (FPT) algorithms and the integer k is called the *parameter*. In most of the cases, the solution size is taken to be the parameter, which means that this approach gives faster algorithms when the solution is of small size. For more background on parameterized complexity, the reader is referred to the monographs [9,10,18]. In this paper we study the problem called MIXED MULTIWAY CUT-UNCUT (MMCU) where given a graph G, $T \subseteq V(G)$, and equivalence relation $\mathcal{R}$ on T and integers k and l, we are asked whether there exist $X \subseteq (V(G) \setminus T)$ and $F \subseteq E(G)$ such that $|X| \leq k$, $|F| \leq l$ and for all $u, v \in T$, u and v belong to the same connected component of $G - (X, F)$ if and only if $(u, v) \in \mathcal{R}$. We start by giving a brief overview of related work and then give our results and methods.

Related Works. MULTIWAY CUT was one of the first cut problems to be explored under the realm of parameterized complexity. It was known to be FPT using graph minors, but Marx [15] was the first one to give a constructive algorithm to show that MULTIWAY CUT is FPT when parameterized by the solution size. He also showed that MULTICUT is FPT when parameterized by the solution size plus the number of terminals. Subsequently, a lot of work has been done on cut problems in the field of parameterized complexity [3,5,8,13,14,16,17]. Recently, Chitnis et al. [7] introduced the technique of *randomized contractions* and used that to solve the UNIQUE LABEL COVER problem. They also show that the same techniques can be applied to solve a generalization of MULTIWAY CUT problem, namely MULTIWAY CUT-UNCUT, where an equivalence relation $\mathcal{R}$ is also supplied along with the set of terminals and we are to delete minimum number of vertices (or edges) such that the terminals lie in the same connected of the resulting graph if and only if they lie in the same equivalence class of $\mathcal{R}$. The MULTIWAY CUT-UNCUT problem was first shown to FPT by Marx et al. [16]. It is easy to see that MMCU not only generalized MIXED CUT and MIXED MULTIWAY CUT, but also both edge and vertex versions of MULTIWAY CUT and MULTIWAY CUT-UNCUT problems. MIXED CUT is studied and mentioned in the books [2,11] and is also a useful subroutine in parameterized graph editing problems [4]. Cao and Marx [4] studied this problem during their study on CHORDAL EDITING problem and gave an algorithm with running time $2^{\mathcal{O}(k+l)}n^{\mathcal{O}(1)}$ on chordal graphs. Algorithms for cut-problems can be applied to several problems, which at first do not look like cut problems. Examples include well studied problems such as FEEDBACK VERTEX SET [6] and ODD CYCLE TRANSVERSAL [19]. Thus, MMCU is not only an interesting combinatorial problem in itself but it is also useful in designing other parameterized algorithms (for example in editing problems, *cf.* [4]). Hence, it is natural and timely to obtain a parameterized algorithms for MMCU.

Our Results and Methods. Even though the vertex and edge versions of MINIMUM CUT problem are polynomial time solvable, we show that allowing deletion of both, the vertices and the edges, makes the MIXED CUT problem NP-hard. To show that, we use a simple reduction from the BIPARTITE PARTIAL VERTEX COVER problem which was recently shown to be NP-hard [1,12]. Then we show that MMCU is FPT when parameterized by $k + l$. In particular we prove the following theorem.

Theorem 1. MIXED MULTIWAY CUT-UNCUT *is FPT with an algorithm running in time* $2^{(k+l)^{\mathcal{O}(1)}} \cdot n^{\mathcal{O}(1)}$.

There are two ways to approach our problem – one is via treewidth reduction technique of Marx et al. [16] and the second is via the method of recursive understanding introduced by Chitnis et al. [7]. However, the method of treewidth reduction technique would lead to an algorithm for MMCU that has double exponential dependence on $k + l$ and thus we did not pursue this method. We use recursive understanding introduced by Chitnis et al. [7] to solve the problem. The main observation is that if there is a small vertex separation which divides the graph into big parts, then we can recursively reduce the size of one of the big parts. Otherwise, the graph is highly connected, and the structure of the graph can be exploited to obtain a solution. We follow the framework given in [7] and design our algorithm. In particular we utilise the recursive understanding technique to first find a small separator in the graph which separates the graph into two parts, each of sufficiently large size and then recursively solve a 'border' version of the same problem on one of the two sides. The border version of the problem is a generalization which also incorporates a special bounded set of vertices, called *border terminals*. During the course of our algorithm, we will attempt to solve the border problem on various subgraphs of the input graph. The objective in the border problem is to find a bounded set of vertices contained within a particular subgraph such that any vertex in this subgraph *not* in the computed set is not required in *any* solution for the given instance irrespective of the vertices chosen outside this subgraph. The algorithm in [7] returns the minimum solutions in the recursive steps. Since we allow both edge and vertex deletion, there is no clear ordering on the solutions, and hence we need to look for solutions of all possible sizes while making the recursive call.

This leaves us with the base case of the recursion, that is when we are unable to find a separator of the required kind. This is called high connectivity phase and this is the place where one needs a problem specific algorithm in the framework given in [7]. Since the solution we are looking for contains both edges and vertices, we need some additional work, as the good node separation framework gives bound only for vertices that can be part of the solution. Once we have done that, the frameworks lends itself for our use, and we can use a separating set family to get to the solution. This results in an extra factor of $k + l$ in the exponent as compared to the algorithm in [7], as the separating family needs to take care of both vertices and edges.

2   Preliminaries

In this section, we first give the notations and definitions which are used in the paper. Then we state some basic properties of mixed-cuts and some known results which will be used later in the paper.

Notations and Definitions: A tuple $G = (V, E)$ is a multigraph if V is a set (called *vertices*) and E is a multiset of 2-element subsets of V (called *edges*). For a multigraph G, we denote the set of vertices of the multigraph by $V(G)$ and the set of edges of the multigraph by $E(G)$. In slight abuse of terminology, we will be calling multigraphs also as graphs in the rest of the paper. We denote $|V(G)|$ and $|E(G)|$ by n and m respectively, where the graph is clear from context. For a set $S \subseteq V(G)$, the *subgraph of G induced by* S is denoted by $G[S]$ and it is defined as the subgraph of G with vertex set S and edge set $\{(u, v) \in E(G) : u, v \in S\}$ and the subgraph obtained after deleting S is denoted as $G - S$. For $F \subseteq E(G)$, by $V(F)$ we denote the set $\{v \mid \exists u \text{ such that } uv \in F\}$. For a set $Z = V' \cup E'$ where $V' \subseteq V(G)$ and $E' \subseteq E(G)$, by $G(Z)$ we denote the subgraph $G' = (V' \cup V(E'), E')$. For a tuple $\mathcal{X} = (X, F)$ such that $X \subseteq V(G)$ and $F \subseteq E(G)$, by $G - \mathcal{X}$ we denote the graph $G' = (V(G) \backslash X, E(G) \backslash F)$ and by $V(\mathcal{X})$ we denote the vertex set $X \cup V(F)$. All vertices adjacent to a vertex v are called *neighbours* of v and the set of all such vertices is called *open neighbourhood* of v, denoted by $N_G(v)$. For a set of vertices $S \subseteq V(G)$, we define $N_G(S) = (\cup_{v \in S} N(v)) \backslash S$. We drop the subscript G when the graph is clear from the context.

We define the MIXED MULTIWAY CUT-UNCUT problem as follows.

MIXED MULTIWAY CUT-UNCUT (MMCU)

Input: A multigraph G, a set of terminals $T \subseteq V(G)$, an equivalence relation $\mathcal{R}$ on the set T and integers k and l.

Parameters: k, l

Question: Does there exist $X \subseteq (V(G) \setminus T)$ and $F \subseteq E(G)$ such that $|X| \leq k$, $|F| \leq l$ and for all $u, v \in T$, u and v belong to the same connected component of $G - (X, F)$ if and only if $(u, v) \in \mathcal{R}$?

We say that a tuple $\mathcal{X} = (X, F)$, where $X \subseteq V(G) \setminus T$ and $F \subseteq E(G)$, is a solution to an MMCU instance $\mathcal{I} = (G, T, \mathcal{R}, k, l)$ if $|X| \leq k$, $|F| \leq l$ and for all $u, v \in T$, u and v belong to the same connected component of $G - (X, F)$ if and only if $(u, v) \in \mathcal{R}$. We define a partial order on the solutions of the instance $\mathcal{I}$. For two solutions $\mathcal{X} = (X, F)$ and $\mathcal{X}' = (X', F')$ of an MMCU instance $\mathcal{I}$, we say that $\mathcal{X}' \leq \mathcal{X}$ if $X' \subseteq X$ and $F' \subseteq F$. We say that a solution $\mathcal{X}$ to an MMCU instance $\mathcal{I}$ is *minimal* if there does not exist another solution $\mathcal{X}'$ to $\mathcal{I}$ such that $\mathcal{X}' \neq \mathcal{X}$ and $\mathcal{X}' \leq \mathcal{X}$. For a solution $\mathcal{X} = (X, F)$ of an MMCU instance $\mathcal{I} = (G, T, \mathcal{R}, k, l)$ and $v \subseteq V(G)$, we say that $\mathcal{X}$ *affects* v if either $v \in X$ or there exists $u \in V(G)$ such that $uv \in F$.

Observation 1. *If $\mathcal{X} = (X, F)$ is a minimal solution to an MMCU* instance $\mathcal{I} = (G, T, \mathcal{R}, k, l)$, then none of the edges in F are incident to X.*

Now we state the definitions of good node separations and flower separations from [7]. Then we state the lemmas that state the running time to find such separations and the properties of the graph if such separations do not exist.

Lemma 2 (1.1 in [7]). *Given a set U of size n together with integers $0 \leq a, b \leq n$, one can in $\mathcal{O}(2^{\mathcal{O}(\min(a,b) \log(a+b))} n \log n)$ time construct a family $\mathcal{F}$ of at most $\mathcal{O}(2^{\mathcal{O}(\min(a,b) \log(a+b))} \log n)$ subsets of U, such that the following holds: for any sets $A, B \subseteq U$, $A \cap B = \emptyset$, $|A| \leq a$, $|B| \leq b$, there exists a set $S \in \mathcal{F}$ with $A \subseteq S$ and $B \cap S = \emptyset$.*

Definition 3 (C.1 in [7]). *Let G be a connected graph and $V^\infty \subseteq V(G)$ a set of undeletable vertices. A triple (Z, V_1, V_2) of subsets of $V(G)$ is called a (q, k)-good node separation, if $|Z| \leq k$, $Z \cap V^\infty = \emptyset$, V_1 and V_2 are vertex sets of two different connected components of $G - Z$ and $|V_1 \setminus V^\infty|, |V_2 \setminus V^\infty| > q$.*

Definition 4 (C.2 in [7]). *Let G be a connected graph, $V^\infty \subseteq V(G)$ a set of undeletable vertices, and $T_b \subseteq V(G)$ a set of border terminals in G. A pair $(Z, (V_i)_{i=1}^r)$ is called a (q, k)-flower separation in G (with regard to border terminals T_b), if the following holds:*

- *$1 \leq |Z| \leq k$ and $Z \cap V^\infty = \emptyset$; the set Z is the core of the flower separation $(Z, (V_i)_{i=1}^r)$;*
- *V_i are vertex sets of pairwise different connected components of $G - Z$, each set V_i is a petal of the flower separation $(Z, (V_i)_{i=1}^r)$;*

- $V(G) \setminus (Z \cup \bigcup_{i=1}^{r} V_i)$, called a stalk, contains more than q vertices of $V \setminus V^\infty$;
- for each petal V_i we have $V_i \cap T_b = \emptyset$, $|V_i \setminus V^\infty| \leq q$ and $N_G(V_i) = Z$;
- $|(\bigcup_{i=1}^{r} V_i) \setminus V^\infty| > q$.

Lemma 5 (C.3 in [7]). *Given a connected graph G with undeletable vertices $V^\infty \subseteq V(G)$ and integers q and k, one may find in $\mathcal{O}(2^{\mathcal{O}(\min(q,k) \log(q+k))} n^3 \log n)$ time a (q,k)-good node separation of G, or correctly conclude that no such separation exists.*

Lemma 6 (C.4 in [7]). *Given a connected graph G with undeletable vertices $V^\infty \subseteq V(G)$ and border terminals $T_b \subseteq V(G)$ and integers q and k, one may find in $\mathcal{O}(2^{\mathcal{O}(\min(q,k) \log(q+k))} n^3 \log n)$ time a (q,k)-flower separation in G w.r.t. T_b, or correctly conclude that no such flower separation exists.*

Lemma 7 (C.5 in [7]). *If a connected graph G with undeletable vertices $V^\infty \subseteq V(G)$ and border terminals $T_b \subseteq V(G)$ does not contain a (q,k)-good node separation or a (q,k)-flower separation w.r.t. T_b then, for any $Z \subseteq V(G) \setminus V^\infty$ of size at most k, the graph $G - Z$ contains at most $(2q + 2)(2^k - 1) + |T_b| + 1$ connected components containing a vertex of $V(G) \setminus V^\infty$, out of which at most one has more than q vertices not in V^∞.*

3   NP-Completeness of Mixed Cut

We prove that MIXED CUT in NP-complete by giving a reduction from the BIPARTITE PARTIAL VERTEX COVER problem which is defines as follows.

BIPARTITE PARTIAL VERTEX COVER (BPVC)	
Input:	A bipartite graph $G = (X \uplus Y, E)$, integers p and q.
Output:	Does there exist $S \subseteq V(G)$ such that $|S| \leq p$ and at least q edges in E are incident on X?

Theorem 8 ([1,12]). BPVC *is NP-complete.*

For an instance of BPVC, we assume that the given bipartite graph does not have any isolated vertices, as a reduction rule can be applied in polynomial time which takes care of isolated vertices and produces an equivalent instance. Given an instance (G, p, q) of BPVC where $G = (X \uplus Y, E)$ is a bipartite graph, we get an instance (G', s, t, k, l) of MIXED CUT as follows. To get the graph G', we introduce two new vertices s and t and add all edges from s to X and t to Y. More formally, $G' = (V', E')$ where $V' = V(G) \cup \{s, t\}$ and $E' = E \cup \{sx \mid x \in X\} \cup \{ty \mid y \in Y\}$. Then we put $k = p$ and $l = m - q$, where $m = |E|$. It is easy to see that (G, p, q) is a YES instance of BPVC if and only if (G', s, t, k, l) is a YES instance of MIXED CUT, and hence we get the following theorem.

Theorem 9. MIXED CUT *is NP-complete even on bipartite graphs.*

4   An Algorithm for MMCU

In this section, we describe the FPT algorithm for MMCU. In fact, we will give an algorithm, which when provided with an instance $(G, T, \mathcal{R}, k, l)$ of MMCU, not only decides whether there exists a solution (X, F) such that $|X| \leq k$ and $|F| \leq l$, but also outputs such a solution that is also minimal. To that end, we assume that the graph is connected and that the number of equivalence classes is bounded by $(k + l)(k + l + 1)$.

4.1   Operations on the Graph

Definition 10. *Let* $\mathcal{I} = (G, T, \mathcal{R}, k, l)$ *be an* MMCU *instance and let* $v \in V(G) \setminus T$. *By bypassing a vertex* v *we mean the following operation: we delete the vertex* v *from the graph and, for any* $u_1, u_2 \in N_G(v)$, *we add an edge* (u_1, u_2) *if it is not already present in* G.

Definition 11. *Let* $\mathcal{I} = (G, T, \mathcal{R}, k, l)$ *be an* MMCU *instance and let* $u, v \in T$ *such that* $(u, v) \in \mathcal{R}$. *By identifying vertices* u *and* v *in* T, *we mean the following operation: we make a new set* $T' = (T \setminus \{u, v\}) \cup \{x_{uv}\}$, *for each edge of the form* $uw \in E(G)$ *or* $vw \in E(G)$, *we add an edge* $x_{uv}w$ *to* $E(G)$ *and we put the new vertex* x_{uv} *in the same equivalence class as vertices* u *and* v. *Observe that the operation might add parallel edges.*

Lemma 12. *Let* $\mathcal{I} = (G, T, \mathcal{R}, k, l)$ *be an* MMCU *instance, let* $v \in V(G) \setminus T$ *and let* $\mathcal{I}' = (G', T, \mathcal{R}, k, l)$ *be the instance* $\mathcal{I}$ *with* v *bypassed. Then:*

- *if* $\mathcal{X} = (X, F)$ *is a solution to* $\mathcal{I}'$, *then* $\mathcal{X}$ *is a solution to* $\mathcal{I}$ *as well;*
- *if* $\mathcal{X} = (X, F)$ *is a solution to* $\mathcal{I}$ *and* $v \notin X$ *and for all* $u \in N(v)$ *$vu \notin F$ then* $\mathcal{X}$ *is a solution to* $\mathcal{I}'$ *as well.*

Lemma 13. *Let* $\mathcal{I} = (G, T, \mathcal{R}, k, l)$ *be an* MMCU *instance and let* $u, v \in T$ *be two different terminals with* $(u, v) \notin \mathcal{R}$, *such that* $uv \in E(G)$, *then for any solution* $\mathcal{X} = (X, F)$ *of* $\mathcal{I}$, *we have* $uv \in F$.

The proof of the Lemma 13 follows from the fact that any solution must delete the edge uv to disconnect u from v. The proof of the next lemma follows by simple observation that u and v have at least $k+l+1$ internally vertex disjoint paths or from the fact that $(u, v) \in \mathcal{R}$ and thus after deleting the solution they must belong to the same connected component and thus every minimal solution does not use the edge $uv \in E(G)$.

Lemma 14. *Let* $\mathcal{I} = (G, T, \mathcal{R}, k, l)$ *be an* MMCU *instance and let* $u, v \in T$ *be two different terminals with* $(u, v) \in \mathcal{R}$, *such that* $uv \in E(G)$ *or* $|N_G(u) \cap N_G(v)| > k + l$. *Let* $\mathcal{I}'$ *be instance* $\mathcal{I}$ *with terminals* u *and* v *identified. Then the set of minimal solution of* $\mathcal{I}$ *and* $\mathcal{I}'$ *is the same.*

Lemma 15. *Let* $\mathcal{I} = (G, T, \mathcal{R}, k, l)$ *be an* MMCU *instance and let* $U = \{v_1, v_2, \ldots, v_t\} \subseteq T$ *be different terminals of the same equivalence class of* $\mathcal{R}$, *pairwise nonadjacent and such that* $N_G(u_1) = N_G(u_2) = \cdots = N_G(u_t) \subseteq V(G) \setminus T$ *and* $t > l + 2$. *Let* $\mathcal{I}'$ *be obtained from* $\mathcal{I}$ *by deleting all but* $l + 2$ *terminals in* U *(and all pairs that contain the deleted terminals in* $\mathcal{R}$). *Then the sets of minimal solutions to* $\mathcal{I}$ *and* $\mathcal{I}'$ *are equal.*

Lemma 16. *Let* $\mathcal{I} = (G, T, \mathcal{R}, k, l)$ *be an* MMCU *instance and let* $uv \in E(G)$ *be an edge with multiplicity more than* $l + 1$. *Then for any minimal solution* $\mathcal{X} = (X, F)$ *of* $\mathcal{I}$, F *does not contain any copies of* uv.

Proof. If $\{u, v\} \cap X \neq \emptyset$, then by Observation 1, we have that none of the copies of uv are in F. Otherwise, F contains at most l copies of edge uv. Let $\mathcal{X}' = (X, F \setminus \{uv\})$. Then we have that for any two $x, y \in V(G)$, x and y are adjacent in $G - \mathcal{X}$ if and only if they are adjacent in $G - \mathcal{X}'$, contradicting the minimality of $\mathcal{X}$.

4.2   Borders and Recursive Understanding

In this section, we define the bordered problem and describe the recursive phase of the algorithm. Let $\mathcal{I} = (G, T, \mathcal{R}, k, l)$ be an MMCU instance and let $T_b \subseteq V(G) \setminus T$ be a set of border terminals, where $|T_b| \leq 2(k + l)$. Define $\mathcal{I}_b = (G, T, \mathcal{R}, k, l, T_b)$ to be an instance of the bordered problem. By $\mathbb{P}(\mathcal{I}_b)$ we define the set of all tuples $\mathcal{P} = (X_b, E_b, \mathcal{R}_b, k', l')$, such that $X_b \subseteq T_b$, E_b is an equivalence relation on $T_b \setminus X_b$, $\mathcal{R}_b$ is an equivalence relation on $T \cup (T_b \setminus X_b)$ such that $E_b \subseteq \mathcal{R}_b$ and $\mathcal{R}_b|_T = \mathcal{R}$, $k' \leq k$ and $l' \leq l$. For a tuple $\mathcal{P} = (X_b, E_b, \mathcal{R}_b, k', l')$, by $G_\mathcal{P}$ we denote the graph $G \cup E_b$, that is the graph G with additional edges E_b.

The intuition behind defining the tuple $\mathcal{P}$ is as following. The set X_b denotes the intersection of the solution with the border terminals. The equivalence relation E_b tells which of the border terminals can be connected from outside the graph considered. This can be looked at as analogous to *torso* operation on the graph. The equivalence relation $\mathcal{R}_b$ tells how the terminals and border terminals are going to get partitioned in different connected components after deleting the solution. Since deletion of any solution respects the relation $\mathcal{R}$, we have that $\mathcal{R}_b|_T = \mathcal{R}$. The numbers k' and l' are guesses for how much the smaller graph is going to contribute to the solution.

We say that a tuple $\mathcal{X} = (X, F)$ is a solution to $(\mathcal{I}_b, \mathcal{P})$ where $\mathcal{P} = (X_b, E_b, \mathcal{R}_b, k', l')$ if $|X| \leq k'$, $|F| \leq l'$ and for all $u, v \in T \cup (T_b \setminus X_b)$, u and v belong to the same connected component of $G_\mathcal{P} - (X, F)$ if and only if $(u, v) \in \mathcal{R}_b$. We also say that $\mathcal{X}$ is a solution to $\mathcal{I}_b = (G, T, \mathcal{R}, k, l, T_b)$ whenever $\mathcal{X}$ is a solution to $\mathcal{I} = (G, T, \mathcal{R}, k, l)$. Now we define the bordered problem as follows.

BORDER-MIXED MULTIWAY CUT-UNCUT(B-MMCU)

Input:	An MMCU instance $\mathcal{I} = (G, T, \mathcal{R}, k, l)$ with G being connected and a set $T_b \subseteq V(G) \setminus T$ such that $|T_b| \le 2(k+l)$; denote $\mathcal{I}_b = (G, T, \mathcal{R}, k, l, T_b)$.
Output:	For each $\mathcal{P} = (X_b, E_b, \mathcal{R}_b, k', l') \in \mathbb{P}(\mathcal{I}_b)$, output a $\text{sol}_\mathcal{P} = \mathcal{X}_\mathcal{P}$ being a minimal solution to $(\mathcal{I}_b, \mathcal{P})$, or $\text{sol}_\mathcal{P} = \bot$ if no solution exists.

It is easy to see that MMCU reduces to B-MMCU, by putting $T_b = \emptyset$. Also, in this case, any answer to B-MMCU for $\mathcal{P} = (\emptyset, \emptyset, \mathcal{R}, k, l)$ returns a solution for MMCU instance. To bound the size of the solutions returned for an instance of B-MMCU we observe the following.

$$|\mathbb{P}(\mathcal{I}_b)| \le (k+1)(l+1)(1+|T_b|(|T_b|+(k+l)(k+l+1)))^{|T_b|}$$
$$\le (k+1)(l+1)(1+2(k+l)^2(k+l+3))^{2(k+l)}$$
$$= 2^{\mathcal{O}((k+l)\log(k+l))}$$

This is true because $\mathcal{R}_b$ has at most $(k+l)(k+l+1)+|T_b|$ equivalence classes, E_b has at most T_b equivalence classes, each $v \in T_b$ can either go to X_b or choose an equivalence class in $\mathcal{R}_b$ and E_b, and k' and l' have $k+1$ and $l+1$ possible values respectively. Let $q = (k+2l)(k+1)(l+1)(1+2(k+l)^2(k+l+3))^{2(k+l)}+k+l$, then all output solutions to a B-MMCU instance $\mathcal{I}_b$ affect at most $q - (k+l)$ vertices in total. Now we are ready to state the lemma which is central for the recursive understanding step.

Lemma 17. *Assume we are given a B-MMCU instance $\mathcal{I}_b = (G, T, \mathcal{R}, k, l, T_b)$ and two disjoint sets of vertices $Z, V^* \subseteq V(G)$, such that $|Z| \le k+l$, $Z \cap T = \emptyset$, $Z_W := N_G(V^*) \subseteq Z$, $|V^* \cap T_b| \le k+l$ and the subgraph of G induced by $W := V^* \cup Z_W$ is connected. Denote $G^* = G[W]$, $T_b^* = (T_b \cup Z_W) \cap W$, $T^* = T \cap W$, $\mathcal{R}^* = \mathcal{R}|_{T \cap W}$ and $\mathcal{I}^* = (G^*, T^*, R^*, k, l, T_b^*)$. Then $\mathcal{I}^*$ is a proper B-MMCU instance. Moreover, if we denote by $(\text{sol}_{\mathcal{P}^*}^*)_{\mathcal{P}^* \in \mathbb{P}(\mathcal{I}_b^*)}$ an arbitrary output to the B-MMCU instance $\mathcal{I}_b^*$ and*

$$U(\mathcal{I}_b^*) = T_b^* \cup \{v \in V(G) \mid \mathcal{P}^* \in \mathbb{P}(\mathcal{I}_b^*), \text{sol}_{\mathcal{P}^*}^* = \mathcal{X}_{\mathcal{P}^*}^* \ne \bot \text{ and } \mathcal{X}_{\mathcal{P}^*}^* \text{ affects } v\},$$

then there exists a correct output $(\text{sol}_\mathcal{P})_{\mathcal{P} \in \mathbb{P}(\mathcal{I}_b)}$ to the B-MMCU instance $\mathcal{I}_b$ such that whenever $\text{sol}_\mathcal{P} = \mathcal{X}_\mathcal{P} \ne \bot$ and $\mathcal{X}_\mathcal{P}$ is a minimal solution to $(\mathcal{I}_b, \mathcal{P})$ then $V(\mathcal{X}_\mathcal{P}) \cap V^* \subseteq U(\mathcal{I}_b^*)$.

The proof of lemma basically says that for any solution (X, F) of $\mathcal{I}_b, \mathcal{P}$ we can replace its intersection with the graph G^* with one of the solutions of the recursive calls and it still remains a solution. Now we describe the recursive step of the algorithm.

Step 1. *Assume we are given a* B-MMCU *instance* $\mathcal{I}_b = (G, T, \mathcal{R}, k, l, T_b)$. *Invoke first the algorithm of Lemma 5 in a search for* $(q, k + l)$-*good node separation (with* $V^\infty = T$). *If it returns a good node separation* (Z, V_1, V_2), *let* $j \in \{1, 2\}$ *be such that* $|V_j \cap T_b| \le k + l$ *and denote* $Z^* = Z$, $V^* = V_j$. *Otherwise, if it returns that no such good node separation exists in* G, *invoke the algorithm of Lemma 6 in a search for* $(q, k + l)$-*flower separation w.r.t.* T_b *(with* $V^\infty = T$ *again). If it returns that no such flower separation exists in* G, *pass the instance* $\mathcal{I}_b$ *to the next step. Otherwise, if it returns a flower separation* $(Z, (V_i)_{i=1}^r)$, *denote* $Z^* = Z$ *and* $V^* = \bigcup_{i=1}^r V_i$.

In the case we have obtained Z^* *and* V^* *(either from Lemma 5 or Lemma 6), invoke the algorithm recursively for the* B-MMCU *instance* $\mathcal{I}_b^*$ *defined as in the statement of Lemma 17 for separator* Z^* *and set* V^*, *obtaining an output* $(\mathsf{sol}_{\mathcal{P}^*}^*)\mathcal{P}^* \in \mathbb{P}(\mathcal{I}_b^*)$. *Compute the set* $U(\mathcal{I}_b^*)$. *Bypass (in an arbitrary order) all vertices of* $V^* \setminus (T \cup U(\mathcal{I}_b^*))$. *Recall that* $T_b^* \subseteq U(\mathcal{I}_b^*)$, *so no border terminal gets bypassed. After all vertices of* $V^* \setminus U(\mathcal{I}_b^*)$ *are bypassed, perform the following operations on terminals of* $V^* \cap T$:

1. *As long as there exist two different* $u, v \in V^* \cap T$ *such that* $(u, v) \notin \mathcal{R}$, *and* $uv \in E(G)$, *then delete the edge* uv *and decrease* l *by 1; if* l *becomes negative by this operation, return* $\bot$ *for all* $\mathcal{P} \in \mathbb{P}(\mathcal{I}_b)$.
2. *As long as there exist two different* $u, v \in V^* \cap T$ *such that* $(u, v) \in \mathcal{R}$ *and either* $uv \in E(G)$ *or* $|N_G(u) \cap N_G(v)| > k + l$, *identify* u *and* v.
3. *If the above two rules are not applicable, then, as long as there exist pairwise distinct terminals* $u_1, u_2, \ldots, u_t \in T$ *of the same equivalence class of* $\mathcal{R}$ *that have the same neighbourhood and* $t > l + 2$, *delete* u_i *for* $i > l + 2$ *from the graph (and delete all pairs containing* u_i *from* $\mathcal{R}$).

Let $\mathcal{I}_b'$ *be the outcome instance.*

Finally, restart this step on the new instance $\mathcal{I}_b'$ *and obtain a family of solutions* $(\mathsf{sol}_{\mathcal{P}})_{\mathcal{P} \in \mathbb{P}(\mathcal{I}_b')}$ *and return this family as an output to the instance* $\mathcal{I}_b$.

After the bypassing operations, we have that V^* contains at most q vertices that are not terminals (at most $k + l$ border terminals and at most $q - (k + l)$ vertices which are neither terminals nor border terminals). Let us now bound the number of terminal vertices once Step 1 is applied. Note that, after Step 1 is applied, for any $v \in T \cap V^*$, we have $N_G(v) \subseteq (V^* \setminus T) \cup Z$ and $|(V^* \setminus T) \cup Z| \le (q + k + l)$. Due to the first and second rule in Step 1, for any set $A \subseteq (V^* \setminus T) \cup Z$ of size $k + l + 1$, at most one terminal of $T \cap V^*$ is adjacent to all vertices of A. Due to the third rule in Step 1, for any set $B \subseteq (V^* \setminus T) \cup Z$ of size at most $k + l$ and for each equivalence class of $\mathcal{R}$, there are at most $l + 2$ terminals of this equivalence class with neighbourhood exactly B. Let $q' := |T \cup V^*|$, then we have the following.

$$q' \le (q + k + l)^{k+l+1} + (l + 2)(k + l)(k + l + 1) \sum_{i=1}^{k+l} (q + k + l)^i = 2^{\mathcal{O}((k+l)^2 \log(k+l))}$$

Lemma 18. *Assume that we are given a* B-MMCU *instance* $\mathcal{I}_b$ = $(G, T, \mathcal{R}, k, l, T_b)$ *on which Step 1 is applied, and let* $\mathcal{I}_b'$ *be an instance after Step 1 is applied. Then any correct output to the instance* $\mathcal{I}_b'$ *is a correct output to the instance* $\mathcal{I}_b$ *as well. Moreover, if Step 1 outputs* $\perp$ *for all* $\mathcal{P} \in \mathbb{P}(\mathcal{I}_b')$, *then this is a correct output to* $\mathcal{I}_b$.

Proof. We first note that by Lemma 17, for all $\mathcal{P} \in \mathbb{P}(\mathcal{I}_b)$, for all the vertices $v \notin U(\mathcal{I}_b^*)$, there exists a minimal solution to $(\mathcal{I}_b, \mathcal{P})$ that does not affect v, hence by Lemma 12, the bypassing operation is justified. The second and third rules are justified by Lemmas 14 and 15 respectively. The first rule is justified by Lemma 13, and if application of this rule makes l negative then for any $\mathcal{P} \in \mathbb{P}(\mathcal{I}_b)$, there is no solution to $(\mathcal{I}_b, \mathcal{P})$.

A careful running time analysis of Step 1 gives us the following recurrence.

$$T(n) \leq \max_{q+1 \leq n' \leq n-q-1} \left(\mathcal{O}(2^{\mathcal{O}((k+l)^2 \log(k+l))} n^3 \log n) + T(n' + k + l) \right.$$
$$\left. + T(\min(n - 1, n - n' + q + q'))) \right)$$

The base case for the recursive calls is the high connectivity phase, which takes time $\mathcal{O}(2^{\mathcal{O}((k+l)^3 \log(k+l))} n^3 \log n)$ as we will argue later. Solving the recurrence for the worst case gives $T(n) = \mathcal{O}(2^{\mathcal{O}((k+l)^3 \log(k+l))} n^4 \log n)$, which is the desired upper bound for the running time of the algorithm.

4.3   High Connectivity Phase

In this section we describe the high connectivity phase for the algorithm. Assume we have a B-MMCU instance $\mathcal{I}_b = (G, T, \mathcal{R}, k', l', T_b)$ where Step 1 is not applicable. Let us fix $\mathcal{P} = (X_b, E_b, \mathcal{R}_b, k, l) \in \mathbb{P}(\mathcal{I}_b)$. We iterate through all possible values of $\mathcal{P}$ and try to find a minimal solution to $(\mathcal{I}_b, \mathcal{P})$. Since $|\mathbb{P}(\mathcal{I}_b)| = 2^{\mathcal{O}((k+l) \log(k+l))}$ it results in a factor of $2^{\mathcal{O}((k+l) \log(k+l))}$ in the running time. For a graph G, by $\mathcal{L}(G)$ we denote the set $V(G) \cup E(G)$. Similarly, for a tuple $\mathcal{X} = (X, F)$, by $\mathcal{L}(\mathcal{X})$ we denote the set $X \cup F$. We once again need to use Lemmas 13–15 to bound number of terminals. We also need to apply Lemma 16 to bound the number of edges.

Step 2. *Apply Lemmas 13, 14 and 15 exhaustively on the set T of terminals in the graph (as done in rules 1–3 of Step 1, but doing it for all of T instead of just $T \cap V^*$). Apply Lemma 16 to reduce multiplicity of all edges in the graph to at most $l + 1$.*

The running time analysis of applying Lemmas 13–15 in this step is exactly the same as the one done in Step 1. Also, Lemma 16 can be applied in $\mathcal{O}(n^2 l)$ time. Hence, the step takes $\mathcal{O}(n^3(k + l + \log n))$ time. After applying Step 2 exhaustively, we know that no two terminals are adjacent, and hence for any solution $\mathcal{X} = (X, F)$, we have that $F \cap E(G[T]) = \emptyset$.

Now we look at what can happen after deleting a set $\mathcal{X} = (X, F)$ from the graph G such that $X \subseteq V(G) \setminus T$, $F \subseteq E(G)$, $|X| \leq k$ and $|F| \leq l$. Since we have assumed that Step 1 is not applicable, for any $\mathcal{X} = (X, F)$ where $X \subseteq V(G) \setminus T$, $F \subseteq E(G)$, $|X| \leq k$ and $|F| \leq l$, Lemma 7 implies that the graph $G - \mathcal{X}$ contains at most $t := (2q + 2)(2(k + l)1) + 2(k + l) + 1$ connected components containing a non-terminal out of which at most one can contain more than q vertices outside T. Let us denote its vertex set by $\mathsf{big}(\mathcal{X})$ (observe that this can possibly be the empty set, in case such a component does not exist). Now we define the notion of interrogating a solution, which will help us in highlighting the solution.

Definition 19. *Let $\mathcal{Z} = (Z, F')$ where $Z \subseteq V(G) \setminus T$, $F' \subseteq E(G) \setminus E(G[T])$, $|Z| \leq k$ and $|F'| \leq l$ and let $S \subseteq \mathcal{L}(G) \setminus T$. We say that S interrogates $\mathcal{Z}$ if the following holds:*

– $S \cap \mathcal{L}(\mathcal{Z}) = \emptyset$;
– for any connected component C of $G - \mathcal{Z}$ with at most q vertices outside T, all vertices and edges of C belong to $S \cup T$.

Lemma 20. *Let $q'' = (qt + k + l)^{k+l+1} + (l+2)(k+l)(k+l+1) \sum_{i=1}^{k+l}(qt+k+l)^i$. Let $\mathcal{F}$ be a family obtained by the algorithm of Lemma 7 for universe $U = \mathcal{L}(G) \setminus T$ and constants $a = qt + (l+1)\binom{q''+qt}{2}$ and $b = k + l$, Then, for any $\mathcal{Z} = (Z, F')$ where $Z \subseteq V(G) \setminus T$, $F' \subseteq E(G) \setminus E(G[T])$, $|Z| \leq k$ and $|F'| \leq l$, there exists a set $S \in \mathcal{F}$ that interrogates $\mathcal{Z}$.*

The proof of this lemma follows from the observation that since we can bound the number of vertices and edges in the small components, there exists a set family of desired size.

Step 3. *Compute the family $\mathcal{F}$ from Lemma 20 and branch into $|\mathcal{F}|$ subcases, indexed by sets $S \in \mathcal{F}$. In a branch S we seek for a minimal solution $\mathcal{X}_\mathcal{P}$ to $(\mathcal{I}_b, \mathcal{P})$, which is interrogated by S.*

Note that since we have $q'' = 2^{\mathcal{O}((k+l)^2 \log(k+l))}$ and $q, t = 2^{\mathcal{O}((k+l) \log(k+l))}$, the family $\mathcal{F}$ of Lemma 2 is of size $\mathcal{O}(2^{\mathcal{O}((k+l)^3 \log(k+l))} \log n)$ and can be computed in $\mathcal{O}(2^{\mathcal{O}((k+l)^3 \log(k+l))} n \log n)$ time. The correctness of Step 3 is obvious from Lemma 20. As discussed, it can be applied in $\mathcal{O}(2^{\mathcal{O}((k+l)^3 \log(k+l))} n \log n)$ time and gives rise to $\mathcal{O}(2^{\mathcal{O}((k+l)^3 \log(k+l))} \log n)$ subcases. The size of the separating family here is the reason for the extra factor of $k + l$ in the exponent, as compared to the algorithm in [7]. There, it was needed to only consider the vertices of the small connected components while looking for a solution, while for our problem, we needed to find a separating family for both the vertices and edges of small components, which has a larger size, and hence a larger separating family is needed.

Lemma 21. *Let $\mathcal{X}_\mathcal{P} = (X, F)$ be a solution to $(\mathcal{I}_b, \mathcal{P})$ interrogated by S. Then there exists a set $T^{\mathsf{big}} \subseteq T \cup (T_b \setminus X_b)$ that is empty or contains all vertices of exactly one equivalence class of $\mathcal{R}_b$, such that $X \subseteq (X_b \cup N_G(S(T^{\mathsf{big}}))$ and*

$F = A_{G,X}(S(T^{\mathsf{big}}))$, where $S(T^{\mathsf{big}})$ is the union of vertex sets of all connected components of $G(S \cup T \cup (T_b \setminus X_b))$ that contain a vertex of $(T \cup (T_b \setminus X_b)) \setminus T^{\mathsf{big}}$ and $A_{G,X}(S(T^{\mathsf{big}}))$ is set of edges in G which have at least one end point in $S(T^{\mathsf{big}})$ but do not belong to any of the connected components of $G[S(T^{\mathsf{big}})]$ and are not incident on X.

Step 4. *For each branch, where S is the corresponding guess, we do the following. For each set T^{big} that is empty or contains all vertices of one equivalence class of $\mathcal{R}_b$, if $|N_G(S(T^{\mathsf{big}}))| \leq k + l$, then for each $X \subseteq X_b \cup N_G(S(T^{\mathsf{big}}))$ such that $|X| \leq k$, and $F = A_{G,X}(S(T^{\mathsf{big}}))$, check whether (X, F) is a solution to $(\mathcal{I}_b, \mathcal{P})$ interrogated by S. For each $\mathcal{P}$, output a minimal solution to $(\mathcal{I}_b, \mathcal{P})$ that is interrogated by S. Output $\perp$ if no solution is found for any choice of S, T^{big} and X.*

The correctness of the step follows from Lemma 21 and the fact that if S interrogates a solution $\mathcal{X}$ to $(\mathcal{I}_b, \mathcal{P})$, then $|N_G(S(T^{\mathsf{big}}))| \leq k + l$. Note that $\mathcal{R}$ has at most $(k+l)(k+l+1)$ equivalence classes. As $|T_b| \leq 2(k+l)$, we have $\mathcal{R}_b$ has at most $(k+l)(k+l+3)$ equivalence classes, and hence there are at most $(k+l)(k+l+3)+1$ choices of the set T^{big}. For each T^{big}, computing $N_G(S(T^{\mathsf{big}}))$ and checking whether $|N_G(S(T^{\mathsf{big}}))| \leq k + l$ takes $\mathcal{O}(n^2)$ time. Since $X_b \leq k$, there are at most $(k+1)(2k+l)^k$ choices for X, and then computing $F = A_{G,X}(S(T^{\mathsf{big}}))$ and checking whether (X, F) is a solution to $(\mathcal{I}_b, \mathcal{P})$ interrogated by S take $\mathcal{O}(n^2)$ time each. Finally, checking whether the solution is minimal or not and computing a minimal solution takes additional $\mathcal{O}((k + l)n^2)$ time. Therefore Step 4 takes $\mathcal{O}(2^{\mathcal{O}((k+l)^3 \log(k+l))} n^2 \log n)$ time for all subcases.

This finishes the description of fixed-parameter algorithm for MMCU and we get the following theorem.

Theorem 22. MMCU *can be solved in* $\mathcal{O}(2^{\mathcal{O}((k+l)^3 \log(k+l))} n^4 \log n)$ *time.*

References

1. Apollonio, N., Simeone, B.: The maximum vertex coverage problem on bipartite graphs. Discrete Appl. Math. **165**, 37–48 (2014)
2. Beineke, L.W., Wilson, R.J. (eds.): Topics in Structural Graph Theory. Cambridge University Press, Cambridge (2013)
3. Bousquet, N., Daligault, J., Thomassé, S.: Multicut is FPT. In: Proceedings of the 43rd ACM Symposium on Theory of Computing, STOC 2011, San Jose, CA, USA, 6–8 June 2011, pp. 459–468 (2011)
4. Cao, Y., Marx, D.: Chordal editing is fixed-parameter tractable. In: 31st International Symposium on Theoretical Aspects of Computer Science, STACS 2014, Lyon, France, 5–8 March 2014, pp. 214–225 (2014)
5. Chen, J., Liu, Y., Lu, S.: An improved parameterized algorithm for the minimum node multiway cut problem. Algorithmica **55**(1), 1–13 (2009)
6. Chen, J., Liu, Y., Lu, S., O'Sullivan, B., Razgon, I.: A fixed-parameter algorithm for the directed feedback vertex set problem. J. ACM **55**(5), 1–19 (2008)

7. Chitnis, R.H., Cygan, M., Hajiaghayi, M., Pilipczuk, M., Pilipczuk, M.: Designing FPT algorithms for cut problems using randomized contractions. In: 53rd Annual IEEE Symposium on Foundations of Computer Science, FOCS 2012, New Brunswick, NJ, USA, 20–23 October 2012, pp. 460–469 (2012)
8. Chitnis, R.H., Hajiaghayi, M., Marx, D.: Fixed-parameter tractability of directed multiway cut parameterized by the size of the cutset. SIAM J. Comput. **42**(4), 1674–1696 (2013)
9. Downey, R.G., Fellows, M.R.: Fundamentals of Parameterized Complexity. Texts in Computer Science. Springer, London (2013)
10. Flum, J., Grohe, M.: Parameterized Complexity Theory. Texts in Theoretical Computer Science. An EATCS Series. Springer, Berlin (2006)
11. Frank, A.: Connections in combinatorial optimization. Discrete Appl. Math. **160**(12), 1875 (2012)
12. Joret, G., Vetta, A.: Reducing the rank of a matroid. CoRR, abs/1211.4853 (2012)
13. Kawarabayashi, K., Thorup, M.: The minimum k-way cut of bounded size is fixed-parameter tractable. In: IEEE 52nd Annual Symposium on Foundations of Computer Science, FOCS 2011, Palm Springs, CA, USA, 22–25 October 2011, pp. 160–169 (2011)
14. Kratsch, S., Pilipczuk, M., Pilipczuk, M., Wahlström, M.: Fixed-parameter tractability of multicut in directed acyclic graphs. In: Czumaj, A., Mehlhorn, K., Pitts, A., Wattenhofer, R. (eds.) ICALP 2012, Part I. LNCS, vol. 7391, pp. 581–593. Springer, Heidelberg (2012)
15. Marx, D.: Parameterized graph separation problems. Theor. Comput. Sci. **351**(3), 394–406 (2006)
16. Marx, D., O'Sullivan, B., Razgon, I.: Finding small separators in linear time via treewidth reduction. ACM Trans. Algorithms **9**(4), 30 (2013)
17. Marx, D., Razgon, I.: Fixed-parameter tractability of multicut parameterized by the size of the cutset. SIAM J. Comput. **43**(2), 355–388 (2014)
18. Niedermeier, R.: Invitation to Fixed Parameter Algorithms. Oxford Lecture Series in Mathematics and Its Applications. Oxford University Press, Oxford (2006)
19. Reed, B.A., Smith, K., Vetta, A.: Finding odd cycle transversals. Oper. Res. Lett. **32**(4), 299–301 (2004)

$(k, n - k)$-MAX-CUT: An $\mathcal{O}^*(2^p)$-Time Algorithm and a Polynomial Kernel

Saket Saurabh[1,2] and Meirav Zehavi[3(✉)]

[1] University of Bergen, Bergen, Norway
saket.saurabh@ii.uib.no
[2] The Institute of Mathematical Sciences, Chennai, India
saket@imsc.res.in
[3] Tel Aviv University, Tel Aviv, Israel
meizeh@post.tau.ac.il

Abstract. MAX-CUT is a well-known classical NP-hard problem. This problem asks whether the vertex-set of a given graph $G = (V, E)$ can be partitioned into two disjoint subsets, A and B, such that there exist at least p edges with one endpoint in A and the other endpoint in B. It is well known that if $p \leq |E|/2$, the answer is necessarily positive. A widely-studied variant of particular interest to parameterized complexity, called $(k, n - k)$-MAX-CUT, restricts the size of the subset A to be exactly k. For the $(k, n - k)$-MAX-CUT problem, we obtain an $\mathcal{O}^*(2^p)$-time algorithm, improving upon the previous best $\mathcal{O}^*(4^{p+o(p)})$-time algorithm, as well as the first polynomial kernel. Our algorithm relies on a delicate combination of methods and notions, including independent sets, depth-search trees, bounded search trees, dynamic programming and treewidth, while our kernel relies on examination of the closed neighborhood of the neighborhood of a *certain* independent set of the graph G.

1 Introduction

MAX-CUT is a widely-studied classical NP-hard problem. Here, the input consists of a graph G and a positive integer p, and the objective is to check whether there is a cut of size at least p. A cut of a graph is a partition of the vertices of the graph into two disjoint subsets. The size of the cut is the number of edges whose endpoints belong to different subsets of the partition. MAX-CUT is NP-hard and has been the focus of extensive study, from the algorithmic perspective in computer science as well as the extremal perspective in combinatorics.

A problem is *fixed-parameter tractable (FPT)* with respect to a parameter t if it can be solved in time $\mathcal{O}^*(f(t))$ for some function f, where $\mathcal{O}^*$ hides factors polynomial in the input size. In the context of MAX-CUT, it is well known that if $p \leq |E(G)|/2$, the answer is necessarily positive. Indeed, given an arbitrary partition of the vertex-set of G, by repeatedly moving vertices from one subset of the partition to the other as long as the size of the cut increases, it is easy to see that one obtains a cut of size at least $|E(G)|/2$. A variant of particular interest to parameterized complexity restricts the size of one of the subsets of

© Springer-Verlag Berlin Heidelberg 2016
E. Kranakis et al. (Eds.): LATIN 2016, LNCS 9644, pp. 686–699, 2016.
DOI: 10.1007/978-3-662-49529-2_51

the partition to be exactly k. More precisely, the problem we consider in this paper is as follows.

$(k, n - k)$-MAX-CUT   **Parameters:** k, p
Input: An undirected graph G such that $|V(G)| = n$, and positive integers k and p.
Question: Does there exist a subset $A \subseteq V(G)$ of size exactly k such that $E(G)$ contains at least p edges with exactly one endpoint in A?

Related Work. It is well known that MAX-CUT is APX-hard [14], and that one can always obtain a cut of size at least $|E(G)|/2$ in polynomial time. A breakthrough result by Goemans and Williamson [11] gave a 0.878-approximation algorithm, which is optimal under the Unique Games Conjecture [12]. There has been extensive study of MAX-CUT from the viewpoint of parameterized complexity [6,7,13,15]. A notable one is the parameterized algorithm for an above-guarantee version of MAX-CUT [7].

The $(k, n - k)$-MAX-CUT problem is a well-known adaptation of MAX-CUT to the realm of parameterized complexity. For this specific variant, there exists a 0.5-approximation algorithm [1], which has been slightly improved in [10], as well as a parameterized approximation scheme with respect to the parameter k [3]. Note that, with respect to the parameter k, Cai [4] proved that $(k, n - k)$-MAX-CUT is W[1]-hard. With respect to the parameter p, Bonnet et al. [3] showed that $(k, n - k)$-MAX-CUT is solvable in time $\mathcal{O}^*(p^p)$, to which end they first showed that it is solvable in time $\mathcal{O}^*(\Delta^k)$, where Δ is the maximum degree of a vertex in G. Bonnet et al. [3] also gave an algorithm that solves $(k, n - k)$-MAX-CUT in time $\mathcal{O}^*(2^{tw})$, where tw is the treewidth of G. Recently, by relying on a derandomization of the method of random separation [5], Shachnai et al. [16] showed that $(k, n - k)$-MAX-CUT is solvable in time $\mathcal{O}^*(\binom{p+k}{p} 2^{o(p)})$, which, in particular, implies that it is solvable in time $\mathcal{O}^*(4^{p+o(p)})$.

Our Contribution. Our contribution is twofold. First, we obtain a fast algorithm for $(k, n - k)$-MAX-CUT, which runs in time $\mathcal{O}^*(2^p)$, thus significantly improving upon the previous best $\mathcal{O}^*(4^{p+o(p)})$-time algorithm for $(k, n - k)$-MAX-CUT. Second, we show that $(k, n - k)$-MAX-CUT admits an $\mathcal{O}(\frac{p^3}{k^2})$-vertex kernel, thus obtaining the *first* polynomial kernel for this problem (with respect to the parameter p). That is, we present a polynomial-time algorithm that given an instance (G, k, p) of $(k, n - k)$-MAX-CUT, constructs an equivalent instance (G', k, p) of $(k, n' - k)$-MAX-CUT where $n' = \mathcal{O}(\frac{p^3}{k^2})$. We note that the analysis of the kernel is quite intuitive, relying on examination of the closed neighborhood of the neighborhood of a *certain* independent set of the graph G (see Sect. 4).

Our $\mathcal{O}^*(2^p)$-time algorithm relies on a delicate combination of methods and notions, including independent sets, depth-search trees, bounded search trees, dynamic programming and treewidth (see Sect. 3). In case $k \leq p/4$, we solve

$(k, n - k)$-MAX-CUT by calling the above mentioned $\mathcal{O}^*((\binom{p+k}{p})2^{o(p)})$-time algorithm, while in case $k > p/2$, we show that the problem can be solved in an elegant manner by using a depth-search tree as well as the above mentioned $\mathcal{O}^*(2^{tw})$-time algorithm. Then, in case $p/4 < k \le p/2$, we turn to compute a *certain* independent set I. In case the size of the neighborhood of I is small, we show that we can again rely on the $\mathcal{O}^*(2^{tw})$-time algorithm. Otherwise, we further distinguish between two cases, which require more careful analysis. First, if $|I|$ is small, we turn to compute yet another independent set. Otherwise, we take a divide-and-conquer approach. We partition $V(G)$ into four subsets, A^*, B^*, C^* and D^*, where the most "difficult" set among these subsets is A^*, and the subset that "connects" A^* to the rest of the graph is B^*. We handle the subset B^* by using an exhaustive search, thus isolating the subset A^*. Afterwards, we can handle A^*, C^* and D^* by using a somewhat technical procedure that is based on the method of bounded search trees as well as dynamic programming.

2   Preliminaries

Standard Notation. Given a graph G, let $V(G)$ and $E(G)$ denote its vertex set and edge set, respectively. Given a vertex $v \in V(G)$, let $N_G(v)$ and $N_G[v]$ denote the open and closed neighborhoods of v, respectively. Let Δ_G denote the maximum degree of a vertex in G. Given a set $U \subseteq V(G)$, let $N_G(U)$ and $N_G[U]$ denote the open and closed neighborhoods of U, respectively. That is, $N_G(U) = (\bigcup_{v \in U} N_G(v)) \setminus U$ and $N_G[U] = N_G(U) \cup U$. Moreover, let $G[U]$ denote the subgraph of G induced by U. Given a partition (A, B) of $V(G)$, let $E_G(A, B) = \{\{v, u\} \in E(G) : v \in A, u \in B\}$. We omit the subscript G when it is clear from the context. The definition of treewidth is given in the appendix (we only use known results that rely on this concept).

Bounded Search Trees: The *bounded search trees* method is fundamental in the design of recursive FPT algorithms (see, e.g., [8,9]). In applying this method, one defines a list of rules. Each rule is of the form Rule X. [condition] action, where X is the number of the rule in the list. At each recursive call (i.e., a node in the search tree), the algorithm performs the action of the first rule whose condition is satisfied. If, by performing an action, the algorithm recursively calls itself at least twice, the rule is a *branching rule*; otherwise, it is a *reduction rule*. We only consider polynomial-time actions that increase neither the parameter nor the size of the instance, and decrease at least one of them.

The running time of an algorithm that uses bounded search trees can be analyzed as follows (see, e.g., [2]). Suppose that the algorithm executes a branching rule which has ℓ branching options (each leading to a recursive call with the corresponding parameter value), such that, in the ith branch option, the current value of the parameter decreases by b_i. Then, $(b_1, b_2, \ldots, b_\ell)$ is called the *branching vector* of this rule. We say that α is the *root* of $(b_1, b_2, \ldots, b_\ell)$ if it is the (unique) positive real root of $x^{b^*} = x^{b^* - b_1} + x^{b^* - b_2} + \cdots + x^{b^* - b_\ell}$, where $b^* = \max\{b_1, b_2, \ldots, b_\ell\}$. If $r > 0$ is the initial value of the parameter, and the algorithm (a) returns a result when (or before) the parameter is negative, and (b) only executes branching rules whose roots are bounded by a constant $c > 0$, then its running time is bounded by $\mathcal{O}^*(c^r)$.

Known Results and Simple Observations: Next, we restate known results and present simple observations relevant to the following sections.

Lemma 1 [3]. $(k, n - k)$-MAX-CUT *is solvable in time* $\mathcal{O}^*(2^{tw})$.

Given a vertex cover of size vc, it is well known that one can compute (in polynomial time) a tree decomposition of width at most vc (see, e.g., [9]). Thus, we obtain the following result.

Corollary 1. *Given a vertex cover U of G, $(k, n - k)$-MAX-CUT is solvable in time* $\mathcal{O}^*(2^{|U|})$.

Lemma 2 [16]. $(k, n - k)$-MAX-CUT *is solvable in time* $\mathcal{O}^*(\binom{p+k}{p} 2^{o(p)})$.

In case $k \leq p/4$, it holds that $\mathcal{O}^*(\binom{p+k}{p} 2^{o(p)}) = \mathcal{O}^*(2^p)$. Thus, we have the following result.

Corollary 2. *In case $k \leq p/4$, $(k, n - k)$-MAX-CUT is solvable in time* $\mathcal{O}^*(2^p)$.

Now, observe that to solve $(k, n - k)$-MAX-CUT, one can actually solve $(n - k, k)$-MAX-CUT. Thus, we can assume that $2k \leq n$. Moreover, if $p \leq \Delta$ and $p + k \leq n$, the input instance is a yes-instance (simply let v denote a vertex of degree at least p in G, and define the solution A as an arbitrary set of k vertices that contains v and excludes at least p vertices in $N(v)$). Thus, we can also assume that either $\Delta < p$ or $n < p + k$. Lemma 3 in [3] shows that if G does not contain isolated vertices and $p \leq \min\{k, n - k\}$, the input instance is a yes-instance. Now, let U be the set of isolated vertices in G, and note that (G, k, p) is a yes-instance if and only if there exists $k' \in \{k - |U|, \ldots, k\}$ such that $(G' = (V(G) \setminus U, E(G)), k', p)$ is a yes-instance. Clearly, we can assume that $|U| < k$. Thus, in Sect. 3, at a cost of a factor of $\mathcal{O}(k)$ in the running time, we can assume that $k < p$. We also get that if there exists $k' \in \{k - |U|, \ldots, k\}$ such that $p \leq \min\{k', n - |U| - k'\}$, the input instance is a yes-instance. In particular, if either $p \leq \min\{\frac{k}{2}, n - \frac{3k}{2}\} \leq \min\{k - \frac{|U|}{2}, n - |U| - (k - \frac{|U|}{2})\}$ or $p \leq \min\{k, n - 2k\}$, the input instance is a yes-instance. The first case allows us to assume that $p > \frac{k}{2}$. Now, if $n < 3k$, then $n < 6p$, and no computation is necessary to obtain the kernel we desire. Thus, the second case allows us to assume, also in Sect. 4, that $k < p$. Thus, we have the following assumption.

Assumption 3. *From now on, we can assume that $2k \leq n$, either $\Delta < p$ or $n < p + k$, and $k < p$.*

Finally, observe that if $n < p + k$, then $n < 2p$, and no computation is necessary to obtain a kernel of the size we desire. Thus, we further have the following assumption.

Assumption 4. *In Sect. 4, we can assume that* $\max\{k, \Delta\} < p$.

3   An Algorithm for $(k, n - k)$-Max-Cut

In this section, we prove the following result.

Theorem 5. $(k, n - k)$-MAX-CUT *is solvable in time* $\mathcal{O}(2^p \cdot \mathrm{poly}(p) + pn)$.

To obtain this result, we first apply the kernelization algorithm in Sect. 4, which runs in time $\mathcal{O}(pn)$, after which we have that $\mathcal{O}^*(2^p) = \mathcal{O}(2^p \cdot \mathrm{poly}(p))$. Thus, it is next sufficient to show that $(k, n - k)$-MAX-CUT is solvable in time $\mathcal{O}^*(2^p)$, to which end we follow the overview given in the introduction. We start by handling the case where $k > p/2$ via examination of a depth-first spanning tree (Lemma 6).

Lemma 6. *The* $(k, n - k)$-MAX-CUT *problem, restricted to instances where* $k > p/2$, *is solvable in time* $\mathcal{O}^*(2^p)$.

Proof. Compute a depth-first spanning tree for each connected component of G. It is well known that if there is a back edge that "stretches" a distance greater than p, we can compute (in polynomial time) a path P in G on $p+1$ vertices, and otherwise we can compute (in polynomial time) a tree decomposition of width at most p (see, e.g., [9]). By Lemma 1, $(k, n - k)$-MAX-CUT is solvable in time $\mathcal{O}^*(2^{tw})$. Thus, we can next focus on the case where we have a path P on $p + 1$ vertices.

Denote $V(P) = \{v_1, v_2 \ldots, v_{p+1}\}$, where $\{v_{i-1}, v_i\} \in E(P)$ for all $2 \leq i \leq p$. If p is an even number, let $U_A = \{v_2, v_4, \ldots, v_p\}$ and $U_B = \{v_1, v_3, \ldots, v_{p+1}\}$, and otherwise let $U_A = \{v_2, v_4, \ldots, v_{p+1}\}$ and $U_B = \{v_1, v_3, \ldots, v_p\}$. Note that $k \geq \lceil \frac{p+1}{2} \rceil$ (since $k > p/2$), as well as $n \geq 2k$ (Assumption 3). Therefore, $|V(G) \setminus V(P)| = n - |U_A| - |U_B| \geq 2k - |U_A| - \lceil \frac{p+1}{2} \rceil \geq k - |U_A|$. Moreover, $|U_A| = \lceil \frac{p}{2} \rceil \leq k$. Thus, we can choose an arbitrary set, S, of $k - |U_A|$ elements from $V(G) \setminus V(P)$. Denote $A = U_A \cup S$. Observe that A is computed in polynomial time. Moreover, $|A| = k$, and every edge in $E(P)$ connects a vertex in $U_A \subseteq A$ to a vertex in $U_B \subseteq V(G) \setminus A$. Since $|E(P)| = p$, set A is a solution.   □

Thus, by Corollary 2 and Lemma 6, we can next focus on the case where $p/4 < k \leq p/2$, as stated in the following assumption.

Assumption 7. *From now on, we can assume that* $p/4 < k \leq p/2$.

To handle this case, we rely on the following construction of an independent set I. Initially, let I be an empty set. Then, as long as there is a vertex in $V(G)$ without any neighbor in I, insert into I such a vertex of maximum degree. Clearly, at the end of this process, I is a *maximal* independent set (in particular, $V(G) = N[I]$). We consider three subcases that cover the case where $p/4 < k \leq p/2$, and prove that each of them is solvable in time $\mathcal{O}^*(2^p)$. Thus, we overall conclude that Theorem 5 is correct. For the sake of clarity of presentation, each subcase is presented in a separate subsection.

3.1   The Subcase where $|N(I)| \leq p$

Since I is a maximal independent set, $N(I)$ is a vertex cover. Thus, by Corollary 1, $(k, n - k)$-MAX-CUT is solvable in time $\mathcal{O}^*(2^{|N(I)|})$. In this subcase, $|N(I)| \leq p$, and thus we have that $(k, n - k)$-MAX-CUT is solvable in time $\mathcal{O}^*(2^p)$.

3.2   The Subcase where $|N(I)| > p$ and $|I| \leq k$

Since $|N(I)| > p$, there are more than p edges in $E(I, N(I))$. As long as $|I| < k$ and there is a vertex that can be added to I such that $|E(I, N(I))|$ remains greater than or equal to p, we add such a vertex. Let U be the set obtained at the end of this process. If $|U| = k$, we are done (U is a solution). Thus, we next assume that $|U| < k$. Observe that $V(G) = N[U]$.

If every vertex in $N(U)$ has at least four neighbors in U, then we can choose an arbitrary set, S, of k vertices from $N(U)$, and S will be a solution (since $|E(S, N(S))| \geq 4|S| = 4k$ and, by Assumption 7, $4k > p$). Therefore, we next assume that there is a vertex in $N(U)$ with at most three neighbors in U. Since the above process did not insert this vertex into U, we have that $p \leq |E(U, N(U))| \leq p + 2$.

Compute (in polynomial time) a maximal independent set, I^*, of $G[N(U)]$. Note that $U \cup (N(U) \setminus I^*)$ is a vertex cover. Thus, if $|U \cup (N(U) \setminus I^*)| \leq p$, we can solve the problem in time $\mathcal{O}^*(2^p)$ by using the idea described in Sect. 3.1. Now, assume that $|U \cup (N(U) \setminus I^*)| \geq p + 1$. Since I^* is a maximal independent set of $G[N(U)]$, every vertex in $N(U) \setminus I^*$ has at least one neighbor in I^*. Therefore, since every vertex in $N(U) \setminus I^*$ could not have been added to U, we have that every vertex in $N(U) \setminus I^*$ has at least *two* neighbors in U. This implies that $|N(U) \setminus I^*| \leq \dfrac{|E(U, N(U))|}{2} \leq \dfrac{p + 2}{2}$, which, in turn, implies that $|U \cup (N(U) \setminus I^*)| \leq (k - 1) + \frac{p+2}{2}$. Since $k \leq p/2$ (Assumption 7), we get that $|U \cup (N(U) \setminus I^*)| \leq p$, which is a contradiction.

3.3   The Subcase where $|N(I)| > p$ and $|I| > k$

To handle this subcase, we take a divide-and-conquer approach. Let A^* be the set that contains the first k vertices that were inserted into I, $B^* = N(A^*)$, $C^* = I \setminus A^*$, and $D^* = N(C^*) \setminus B^*$. Clearly, (A^*, B^*, C^*, D^*) is a partition of $V(G)$. If the set A^* is a solution, we are done. Thus, we can next assume that $|E(A^*, B^*)| < p$, which implies that $|B^*| < p$ and that there is a vertex in A^* of degree *smaller* than p/k. By the construction of I, we have that the degree of each vertex in $C^* \cup D^*$ is smaller than p/k. A rough sketch of this situation is illustrated in Fig. 1.

Now, in a sense, we would like to "isolate" A^* from $C^* \cup D^*$ by deciding the "roles" of all of the vertices in B^* (i.e., for each vertex in B^*, we would like to know whether we should insert it into the solution we shall attempt to construct). To this end, we iterate over every subset of B^* of size at most k.

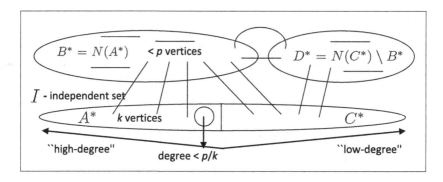

Fig. 1. The partition (A^*, B^*, C^*, D^*) of $V(G)$.

Next, consider some iteration, corresponding to a subset $B' \subseteq B^*$. Informally, B' aims to capture exactly the vertices in B^* that should be inserted into the desired solution.

We concisely present the bounded search tree-based procedure we use to solve the subcase (recall that the method of bounded search trees was explained in Sect. 2). We first give a subroutine based on dynamic programming that is used by this procedure. The subroutine is summarized in the following result.

Lemma 8. *Let U_A and U_B be disjoint subsets of $A^* \cup C^* \cup D^*$ such that $G[V(G) \setminus (B^* \cup U_A \cup U_B)]$ is a graph of maximum degree at most 2. Denote $X = B' \cup U_A$. Then, there is a polynomial-time subroutine that accepts the input if and only if there exists a solution that contains X and is disjoint from $(B^* \setminus B') \cup U_B$.*

Proof. Observe that the set of connected components of $G[V(G) \setminus (B^* \cup U_A \cup U_B)]$, $\mathcal{C} = \{C_1, C_2, \ldots, C_{|\mathcal{C}|}\}$, consists only of simple cycles and paths (an isolated vertex is treated as a simple path). For each component $C_i \in \mathcal{C}$, denote $V(C_i) = \{v_1^i, v_2^i, \ldots, v_{|V(C_i)|}^i\}$ such that $\{v_{j-1}^i, v_j^i\} \in E(C_i)$ for all $j \in \{2, \ldots, |V(C_i)|\}$. Furthermore, let $V(\mathcal{C}^{>i}) = V(C_{i+1}) \cup \cdots \cup V(C_{|\mathcal{C}|})$, and given $j \in \{1, \ldots, |V(C_i)|\}$, denote $V(C_i^{>j}) = \{v_{j+1}^i, v_{j+2}^i, \ldots, v_{|V(C_i)|}^i\}$

We use a matrix M with an entry $[k', i, j, b_{first}, b_{next}]$ for all $k' \in \{0, \ldots, k - |X|\}, i \in \{1, \ldots, |\mathcal{C}|\}, j \in \{1, \ldots, |V(C_i)|\}$, and $b_{first}, b_{next} \in \{0, 1\}$ such that if $j = |V(C_i)|$, then $b_{next} = 0$. Such an entry should store the maximum value $|E(A, B)|$ of a cut (A, B) that satisfies the following conditions, where an undefined entry has value $-\infty$.

- $|A| = |X| + k'$.
- $X \subseteq A$, and $(B^* \setminus B') \cup U_B \cup V(\mathcal{C}^{>i}) \cup V(C_i^{>j+1}) \subseteq B$.
- $v_1^i \in A \Leftrightarrow b_{first} = 1$, and $v_{j+1}^i \in A \Leftrightarrow b_{next} = 1$.

The computation of the matrix M is straightforward. For the sake of completeness, we present the formulas below; correctness can be verified by standard induction on the order of computation. Having computed M, the subroutine accepts if and only if $p \leq \max\limits_{b_{first}, b_{next} \in \{0,1\}} \{M[k - |X|, |\mathcal{C}|, |V(C_{|\mathcal{C}|})|, b_f, b_n]\}$.

Base Cases: $i = j = 1$.

- $M[0, 1, 1, 0, 0] \leftarrow |E(X, V(G) \setminus X)|$.
- $M[1, 1, 1, 1, 0] \leftarrow |E(X \cup \{v_1^1\}, V(G) \setminus (X \cup \{v_1^1\}))|$.
- $|V(C_1)| \geq 2$: $M[1, 1, 1, 0, 1] \leftarrow |E(X \cup \{v_2^1\}, V(G) \setminus (X \cup \{v_2^1\}))|$.
- $|V(C_1)|, k - |X| \geq 2$: $M[2, 1, 1, 1, 1] \leftarrow |E(X \cup \{v_1^1, v_2^1\}, V(G) \setminus (X \cup \{v_1^1, v_2^1\}))|$.
- Otherwise: $M[k', 1, 1, b_f, b_n] \leftarrow -\infty$.

Next, we denote $w(v, Y) = |N(v) \setminus (X \cup Y)| - |N(v) \cap (X \cup Y)|$ and $w(v) = w(v, \emptyset)$.

Steps: $i > 1$ or $j > 1$.

- $j = 1$: $M[k', i, 1, b_f, b_n] \leftarrow$
  $\max\limits_{b \in \{0,1\}} \{M[k' - b_f - b_n, i - 1, |V(C_{i-1})|, b, 0] + t\}$, where
  - $b_f = b_n = 0$: $t = 0$.
  - $b_f = 1$ and $b_n = 0$: $t = w(v_1^i)$.
  - $|V(C_i)| \geq 2$, $b_f = 0$ and $b_n = 1$: $t = w(v_2^i)$.
  - $|V(C_i)| \geq 2$ and $b_f = b_n = 1$: $t = w(v_1^i) + w(v_2^i) - 2$.
  - Otherwise: $t = -\infty$.
- $j \geq 2$:
  - $M[k', i, j, b_f, 0] \leftarrow \max\{M[k', i, j - 1, b_f, 0], M[k', i, j - 1, b_f, 1]\}$.
  - $|V(C_i)| \geq j + 1$: $M[k', i, j, 0, 1] \leftarrow$
    $\max\{M[k' - 1, i, j - 1, 0, 0] + w(v_{j+1}^i), M[k' - 1, i, j - 1, 0, 1] + w(v_{j+1}^i, \{v_j^i\})\}$.

  - $|V(C_i)| \geq j + 1$: $M[k', i, j, 1, 1] \leftarrow$
    $\max\{M[k' - 1, i, j - 1, 1, 0] + w(v_{j+1}^i, \{v_1^i\}), M[k' - 1, i, j - 1, 1, 1] + w(v_{j+1}^i, \{v_1^i, v_j^i\})\}$.

Observe that the decision of the subroutine is computed in polynomial time by calculating the entries of M in lexicographic order of the pair (i, j). □

We proceed by developing MaxCut$(G, k, p, A^*, B^*, C^*, D^*, B', U_A, U_B)$, our bounded search tree-based procedure, where U_A and U_B are disjoint subsets of $A^* \cup C^* \cup D^*$. Recall that $X = B' \cup U_A$. Since U_A and U_B are the only arguments changed during the execution, we use the abbreviation MaxCut(U_A, U_B). At the first call, $U_A = U_B = \emptyset$. The measure is $k - |X|$. We will ensure that once the measure drops to a non-positive value, MaxCut returns a decision in polynomial time (see Rule 1), and that the following claim is correct.

Lemma 3. MaxCut *accepts if and only if there is a solution that contains X and is disjoint from $(B^* \setminus B') \cup U_B$. Moreover, if $k \geq p/3$, MaxCut performs only reduction rules; otherwise, MaxCut performs either reduction rules or branching rules associated with roots smaller than 2.31.*

In case $k \geq p/3$, for each subset of B^*, we perform a polynomial-time computation (see Lemma 3). Recall that $|B^*| < p$. Thus, in case $k \geq p/3$, the overall running time is bounded by $\mathcal{O}^*(2^{|B^*|}) = O^*(2^p)$. Otherwise, for each subset $B' \subseteq B^*$ such that $|B'| \leq k$, Lemma 3 and the discussion in Sect. 2 imply the the running time is bounded by $2.31^{k-|B'|}$ (since, initially, the measure is $k - |B'|$). Thus, in case $k < p/3$, we get that the overall running time is bounded by the following expression:

$$\mathcal{O}^*(\sum_{B' \subseteq B^* \text{ s.t. } |B'| \leq k} 2.31^{k-|B'|})$$

$$= \mathcal{O}^*(\sum_{t=0}^{p/3} \binom{p}{t} \cdot 2.31^{p/3-t})$$

$$= \mathcal{O}^*\left(\max_{0 \leq \alpha \leq 1/3}\{\binom{p}{\alpha p} \cdot 2.31^{(1/3-\alpha)p}\}\right)$$

$$= \mathcal{O}^*\left((\max_{0 \leq \alpha \leq 1/3}\{\frac{2.31^{1/3-\alpha}}{\alpha^\alpha \cdot (1-\alpha)^{1-\alpha}}\})^p\right)$$

$$= \mathcal{O}^*(2^p).$$

Observe that the first equality follows from the fact that $k < p/3$, and the third equality follows from Stirling's formula. Now, at the fourth formula, the maximum is obtained at $\alpha \cong 0.3021$, where the value of the expression is bounded by $\mathcal{O}^*(1.8942^p)$; for our purpose, it is sufficient to state that the expression is bounded by $\mathcal{O}^*(2^p)$.

Thus, it remains to give the list of rules of $\mathsf{MaxCut}(U_A, U_B)$, and prove that they indeed satisfy the properties mentioned in Lemma 3.

Reduction Rule 1 $[|X| \geq k]$
If $|X| = k$ and $|E(X, V(G) \setminus X)| \geq p$, accept; otherwise, reject.

If $|X| \geq k$, then if there exists a solution that contains X, it is necessarily X (since the size of a solution is exactly k), which means that $|X| = k$ and $|E(X, V(G) \setminus X)| \geq p$. Moreover, the rule is performed in polynomial time. Thus, we preserve the correctness of Lemma 3. Note that this rule ensures that once the measure drops to (or below) 0, MaxCut returns an answer.

Reduction Rule 2 $[V(G) = B^* \cup U_A \cup U_B]$
Reject.

Since Rule 1 was not applied, $|X| < k$. Thus, by the condition of this rule, there is no solution that contains X and is disjoint from $(B^* \setminus B') \cup U_B$ (there are simply no vertices outside $X \cup B^* \cup U_B$ that can be added to X to obtain a solution). The rule is performed in polynomial time, and thuse it preserves the correctness of Lemma 3.

Reduction Rule 3 $[G[V(G) \setminus (B^* \cup U_A \cup U_B)]$ is a graph of maximum degree at most 2]
Use the subroutine in Lemma 8 to decide whether to accept or reject.

In the context of this rule, the preservation of the correctness of Lemma 3 follows directly from Lemma 8.

Recall that we have shown that the degree of each vertex in $C^* \cup D^*$ is *smaller* than p/k, which implies that, if $k \geq p/3$, the degree of each vertex in $C^* \cup D^*$ is at most 2. **Therefore, if $k \geq p/3$, the condition of Rule 3 is necessarily true.** That is, for the following rules, it is sufficient to preserve the correctness of the following claim.

Lemma 4. MaxCut *accepts if and only if there is a solution that contains X and is disjoint from* $(B^* \setminus B') \cup U_B$. *Moreover, besides Rules 1–3,* MaxCut *consists only of reduction rules and branching rules associated with roots smaller than 2.31.*

Define $w(v) = |N(v) \setminus X| - |N(v) \cap X|$ for each vertex $v \in V(G) \setminus (B^* \cup U_A \cup U_B)$. Let v^* be a vertex that maximizes $w(v^*)$. Recall that $B^* = N(A^*)$ and $k \geq p/4$ (Assumption 7), and that the degree of each vertex in $C^* \cup D^*$ is *smaller* than p/k. Therefore, $|N(v^*) \setminus (B^* \cup U_A \cup U_B)| \leq 3$.

Reduction Rule 4 $[N(v^*) \setminus (B^* \cup U_A \cup U_B) = \emptyset]$
Return MaxCut$(U_A \cup \{v^*\}, U_B)$.

To prove the correctness of Lemma 4, we need to show that if there is a solution A that contains X and is disjoint from $(B^* \setminus B') \cup U_B \cup \{v^*\}$, then there is a solution A' that contains $X \cup \{v^*\}$ and is disjoint from $(B^* \setminus B') \cup U_B$. Now, suppose that such a solution A exists. Let x be a vertex in $A \setminus X$ (since Rule 1 was not applied, $A \setminus X \neq \emptyset$). Denote $A' = (A \setminus \{x\}) \cup \{v^*\}$. Due to the condition of this rule, $|E(A', V(G) \setminus A')| \geq |E(A, V(G) \setminus A)| + w(v^*) - w(x)$. Thus, by the choice of v^*, $|E(A', V(G) \setminus A')| \geq p$.

Branching Rule 5. $[N(v^*) \setminus (B^* \cup U_A \cup U_B) = \{v\}]$
Accept if and only if at least one of the following branches accepts.

1. MaxCut$(U_A \cup \{v^*\}, U_B)$.
2. MaxCut$(U_A \cup \{v\}, U_B \cup \{v^*\})$.

The branching vector is $(1, 1)$ (in each branch, $|U_A|$ increases by 1), whose root is 2. By the definition of the branches, if there is a solution A that contains $X \cup \{v^*\}$ (resp. $X \cup \{v\}$) and is disjoint from $(B^* \setminus B') \cup U_B$ (resp. $(B^* \setminus B') \cup U_B \cup \{v^*\}$), it is examined in the first (resp. second) branch. Thus, to prove the correctness of Lemma 4, it is sufficient to show that if there is a solution A that contains X and is disjoint from $(B^* \setminus B') \cup U_B \cup \{v^*, v\}$, then there is a solution A' that contains $X \cup \{v^*\}$ and is disjoint from $(B^* \setminus B') \cup U_B$. By letting x be a vertex in $A \setminus X$, the correctness of this claim follows from the arguments given in Rule 4.

Branching Rule 6. $[N(v^*) \setminus (B^* \cup U_A \cup U_B) = \{v, u\}]$
Accept if and only if at least one of the following branches accepts.

1. MaxCut($U_A \cup \{v^*\}, U_B$).
2. MaxCut($U_A \cup \{v, u\}, U_B \cup \{v^*\}$).

The branching vector is $(1, 2)$ (in the first branch, $|U_A|$ increases by 1, and in the second branch, it increases by 2), whose root is smaller than 2. By the definition of the branches, to prove the correctness of Lemma 4, it is sufficient to show that if there is a solution A that contains X and *at most* one vertex from $\{v, u\}$, and is disjoint from $(B^* \setminus B') \cup U_B \cup \{v^*\}$, then there is a solution A' that contains $X \cup \{v^*\}$ and is disjoint from $(B^* \setminus B') \cup U_B$. Now, suppose that such a solution A exists. If $A \cap \{v, u\} \neq \emptyset$, let x be the vertex in $A \cap \{v, u\}$, and otherwise let x be a vertex in $A \setminus X$. Observe that $A \setminus \{x\}$ does not contain any vertex from $N(v^*) \setminus (B^* \cup U_A \cup U_B))$. Thus, the correctness of the claim follows from the arguments given in Rule 4.

Branching Rule 7. $[N(v^*) \setminus (B^* \cup U_A \cup U_B) = \{v, u, r\}]$
Accept if and only if at least one of the following branches accepts.

1. MaxCut($U_A \cup \{v^*\}, U_B$).
2. MaxCut($U_A \cup \{v, u\}, U_B \cup \{v^*\}$).
3. MaxCut($U_A \cup \{v, r\}, U_B \cup \{v^*\}$). ·
4. MaxCut($U_A \cup \{u, r\}, U_B \cup \{v^*\}$).

Recall that $|N(v^*) \setminus (B^* \cup U_A \cup U_B)| \leq 3$, and therefore this rule handles all of the cases that were not handled by previous rules. The branching vector is $(1, 2, 2, 2)$ (in the first branch, $|U_A|$ increases by 1, and in each of the other three branches, it increases by 2), whose root is smaller than 2.31. By the definition of the branches, to prove the correctness of Lemma 4, it is sufficient to show that if there is a solution A that contains X and *at most* one vertex from $\{v, u, r\}$, and is disjoint from $(B^* \setminus B') \cup U_B \cup \{v^*\}$, then there is a solution A' that contains $X \cup \{v^*\}$ and is disjoint from $(B^* \setminus B') \cup U_B$. Now, suppose that such a solution A exists. If $A \cap \{v, u, r\} \neq \emptyset$, let x be the vertex in $A \cap \{v, u, r\}$, and otherwise let x be a vertex in $A \setminus X$. Again, $A \setminus \{x\}$ does not contain any vertex from $N(v^*) \setminus (B^* \cup U_A \cup U_B))$. Thus, the correctness of the claim follows from the arguments given in Rule 4.

4   Kernel

In this section, we prove that $(k, n - k)$-MAX-CUT admits a polynomial kernel. To this end, let I be the empty set. As long as both $|I| < p$ and there is a vertex in $V(G)$ without any neighbor in I, insert into I such a vertex of maximum degree. Clearly, at the end of this process, which can be performed in time $\mathcal{O}(pn)$, I is an independent set of size at most p. We denote $I = \{v_1, v_2, \ldots, v_{|I|}\}$, such that for all $2 \leq i \leq |I|$, v_{i-1} was inserted before v_i into I. Moreover, we denote $M = p + \lceil \frac{p}{k} \rceil \cdot p + \lceil \frac{\lceil \frac{p}{k} \rceil \cdot p}{k} \rceil \cdot p$, and proceed by proving two claims. Recall that in this section, we rely on Assumption 4 (i.e., $\max\{k, \Delta\} < p$).

Observation 9. *If $n > M$, then either $|I| = p$ or the input is solvable in time $\mathcal{O}(pn)$.*

Proof. Recall that $|I| \leq p$, and suppose that $n > M$. If $|I| < p$, then the construction of I implies that $V(G) = N[I]$. First, suppose that $|I| < k$. Then, let $\widetilde{I}$ denote the set I to which we add $k - |I|$ vertices arbitrarily chosen from $N(I)$. If $|N(\widetilde{I})| \geq p$, then the set $\widetilde{I}$ is a solution, and we are done. Thus, next suppose that this is not the case. Then, $n = |N[I]| < |\widetilde{I}| + |N(\widetilde{I})| < k + p < M$, which is a contradiction. Now, suppose that $k \leq |I| < p$. If $\sum_{i=1}^{k} |N(v_i)| \geq p$, then the set $\{v_1, v_2, \ldots, v_k\}$ is a solution, and we are done. Thus, next suppose that $\sum_{i=1}^{k} |N(v_i)| \leq p-1$. The construction of I implies that $|N(v_{i-1})| \geq |N(v_i)|$ for every $i \in \{2, 3, \ldots, |I|\}$. Therefore, $|N(v_i)| < p/k$ for every $i \in \{k, \ldots, p\}$. We get that $n = |N[I]| \leq |I| + \sum_{i=1}^{k} |N(v_i)| + \sum_{i=k+1}^{p} |N(v_i)| < (p - 1) + (p - 1) + (p/k)(p - k) < M$, which is a contradiction. $\square$

Lemma 10. *If $|N[N[I]]| > M$, then the input is solvable in time $\mathcal{O}(pn)$.*

Proof. By Observation 9, $|I| = p$. Let $\widetilde{I}$ denote an arbitrary subset of $\lfloor \frac{p}{k} \rfloor k$ vertices from I, and let $I_1, I_2, \ldots, I_{\lfloor \frac{p}{k} \rfloor}$ be a partition of $\widetilde{I}$ into subsets of k vertices. Moreover, if $\lceil \frac{p}{k} \rceil \neq \lfloor \frac{p}{k} \rfloor$, let $I_{\lceil \frac{p}{k} \rceil}$ be the set of vertices in $I \setminus \widetilde{I}$ to which we add $k - |I \setminus \widetilde{I}|$ other vertices arbitrarily chosen from I (since $|I| = p > k$, $I_{\lceil \frac{p}{k} \rceil}$ is well-defined). If there exists $i \in \{1, 2, \ldots, \lceil \frac{p}{k} \rceil\}$ such that $|N(I_i)| \geq p$, we are done (the set I_i is a solution). Therefore, we next assume that this is not the case, which implies that $|N(I)| \leq \sum_{i=1}^{\lceil \frac{p}{k} \rceil} |N(I_i)| < \lceil \frac{p}{k} \rceil p$.

Next, let $\widetilde{N}$ denote an arbitrary subset of $\lfloor \frac{|N(I)|}{k} \rfloor k$ vertices from $N(I)$, and let $N_1, N_2, \ldots, N_{\lfloor \frac{p}{k} \rfloor}$ be a partition of $\widetilde{N}$ into subsets of k vertices. Moreover, if $\lceil \frac{|N(I)|}{k} \rceil \neq \lfloor \frac{|N(I)|}{k} \rfloor$, let $N_{\lceil \frac{|N(I)|}{k} \rceil}$ be the set of vertices in $N(I) \setminus \widetilde{N}$ to which we add $k - |N(I) \setminus \widetilde{N}|$ other vertices arbitrarily chosen from I. If there exists $i \in \{1, 2, \ldots, \lceil \frac{|N(I)|}{k} \rceil\}$ such that $|N(N_i)| \geq p$, we are done (the set N_i is a solution). Therefore, we next assume that this is not the case, which implies that $|N(N(I))| \leq \sum_{i=1}^{\lceil \frac{|N(I)|}{k} \rceil} |N(N_i)| < \lceil \frac{|N(I)|}{k} \rceil p = \lceil \frac{\lceil \frac{p}{k} \rceil p}{k} \rceil p$.

Overall, we have that $|N[N[I]]| \leq |I| + |N(I)| + |N(N(I))| < p + \lceil \frac{p}{k} \rceil p + \lceil \frac{\lceil \frac{p}{k} \rceil p}{k} \rceil p = M$, which concludes the correctness of the lemma. $\square$

We are now ready to compute the kernel.

Theorem 11. *$(k, n - k)$-Max-Cut admits an $\mathcal{O}(\frac{p^3}{k^2})$-vertex kernel, which can be computed in time $\mathcal{O}(pn)$.*

Proof. Assume that $n > M$, since otherwise we can just return the input instance. By Observation 9 and Lemma 10, we can further assume that $|I| = p$ and $|N[N[I]]| \leq M$. If $|N[N[I]]| \geq k + p$, let $S = N[N[I]]$, and otherwise let

$S = N[N[I]] \cup X$, where X is an arbitrary set of $k + p - |N[N[I]]|$ vertices from $V(G) \setminus N[N[I]]$ (since $n > M > 2p \geq k + p$, such a choice is possible). Since $|S| \leq M = \mathcal{O}(\frac{p^3}{k^2})$, it is sufficient to show that $(G[S], k, p)$ is a yes-instance if and only if (G, k, p) is a yes-instance. Clearly, since $G[S]$ is a subgraph of G, the forward direction is correct.

Now, we prove the reverse direction. We can assume that there is no isolated vertex in I, else it is immediate that the claim holds (by the choice of I). Suppose that (G, k, p) has a solution, and let A be a solution that maximizes $A \cap N[I]$. If $|A \cap N[I]| = k$ (i.e., $A \subseteq N[I]$), then A is a solution for $(G[S], k, p)$ (since $N[A] \subseteq S$), and we are done. Thus, we next suppose that $|A \cap N[I]| < k$, and show that this supposition leads to a contradiction. First, this supposition implies that there is a vertex v in $A \setminus N[I]$. By our choice of A, we cannot replace v by a vertex in $N[I] \setminus A$ and obtain a solution for (G, k, p). However, by the construction of I, the degree of every vertex in I is at least as large as the degree of every vertex in $V(G) \setminus N[I]$. Therefore, every vertex in I either belongs to A or has at least one neighbor that belongs to A. That is, $I \subseteq N[A]$.

Initialize A^* to be $A \cap N(I)$ (note that $I \setminus A \subseteq N(A^*)$). Then, $|A^*| < k - |A \cap I|$. As long as $|A^*| < k$ and there is a vertex $u \in I$ without a neighbor in A^*, add to A^* a vertex from $N(u)$. At the end of this process, every vertex in I has a neighbor that belongs to A^* (since initially it was true that $|A^*| < k - |A \cap I|$). Moreover, by the construction of A^*, $A^* \cap I = \emptyset$. If at the end of this process $|A^*| < k$, insert $k - |A^*|$ additional vertices arbitrarily chosen from $S \setminus I$ into A^* (since $|S| \geq k + p$, this is possible). Thus, $|A^*| = k$ and $|N(A^*)| \geq |I| = p$. Observe that $A^* \subseteq N[I]$, while $A \setminus N[I] \neq \emptyset$. We get that A^* is a solution for $(G[S], k, p)$ (and therefore it is a solution for (G, k, p)) such that $|A \cap N[I]| < |A^* \cap N[I]|$. This is a contradiction to the choice of A. $\square$

Appendix

Treewidth: A *tree decomposition* of a graph G is a pair (D, β), where D is a rooted tree and $\beta : V(D) \rightarrow 2^{V(G)}$ is a mapping that satisfies the following conditions.

- For each vertex $v \in V(G)$, the set $\{d \in V(D) : v \in \beta(d)\}$ induces a nonempty and connected subtree of D.
- For each edge $\{v, u\} \in E(G)$, there exists $d \in V(D)$ such that $\{v, u\} \subseteq \beta(d)$.

The set $\beta(d)$ is called the bag at d, and the width of (D, β) is the size of the largest bag minus one (i.e., $\max_{d \in V(D)} |\beta(d)| - 1$). The *treewidth* of G, tw_G, is the minimum width among all possible tree decompositions of G.

References

1. Ageev, A.A., Sviridenko, M.: Approximation algorithms for maximum coverage and max cutwith given sizes of parts. In: IPCO, pp. 17–30 (1999)
2. Binkele-Raible, D.: Amortized analysis of exponential time and parameterized algorithms: Measure & conquer and reference search trees. Ph.D. thesis, Universität Trier (2010)
3. Bonnet, E., Escoffier, B., Paschos, V.T., Tourniaire, E.: Multi-parameter analysis for local graph partitioning problems: using greediness for parameterization. Algorithmica **71**(3), 566–580 (2015)
4. Cai, L.: Parameter complexity of cardinality constrained optimization problems. Comput. J. **51**(1), 102–121 (2008)
5. Cai, L., Chan, S.M., Chan, S.O.: Random separation: a new method for solving fixed-cardinality optimization problems. In: Bodlaender, H.L., Langston, M.A. (eds.) IWPEC 2006. LNCS, vol. 4169, pp. 239–250. Springer, Heidelberg (2006)
6. Crowston, R., Gutin, G., Jones, M., Muciaccia, G.: Maximum balanced subgraph problem parameterized above lower bound. Theor. Comput. Sci. **513**, 53–64 (2013)
7. Crowston, R., Jones, M., Mnich, M.: Max-cut parameterized above the Edwards-Erdős bound. Algorithmica **72**(3), 734–757 (2015)
8. Cygan, M., Fomin, F.V., Kowalik, L., Lokshtanov, D., Marx, D., Pilipczuk, M., Saurabh, S.: Parameterized Algorithms. Springer, Switzerland (2015)
9. Downey, R.G., Fellows, M.: Fundamentals of Parameterized Complexity. Springer, London (2013)
10. Feige, U., Langberg, M.: Approximation algorithms for maximization problems arising in graph partitioning. J. Algorithms **41**(2), 174–211 (2001)
11. Goemans, M.X., Williamson, D.P.: Improved approximation algorithms for maximum cut and satisfiability problems using semidefinite programming. J. ACM **42**(6), 1115–1145 (1995)
12. Khot, S., Kindler, G., Mossel, E., O'Donnell, R.: Optimal inapproximability results for MAX-CUT and other 2-variable CSPs? SICOMP **37**(1), 319–357 (2007)
13. Mahajan, M., Raman, V.: Parameterizing above guaranteed values: MaxSat and MaxCut. J. Algorithms **31**(2), 335–354 (1999)
14. Papadimitriou, C.H., Yannakakis, M.: Optimization, approximation, and complexity classes. J. Comput. Syst. Sci. **43**(3), 425–440 (1991)
15. Raman, V., Saurabh, S.: Improved fixed parameter tractable algorithms for two "edge" problems: MAXCUT and MAXDAG. Inf. Process. Lett. **104**(2), 65–72 (2007)
16. Shachnai, H., Zehavi, M.: Parameterized algorithms for graph partitioning problems. In: WG, pp. 384–395 (2014)

Independent Set of Convex Polygons: From n^ϵ to $1 + \epsilon$ via Shrinking

Andreas Wiese[(✉)]

Max Planck Institute for Computer Science, Saarbrücken, Germany
awiese@mpi-inf.mpg.de

Abstract. In the Independent Set of Convex Polygons problem we are given a set of weighted convex polygons in the plane and we want to compute a maximum weight subset of non-overlapping polygons. This is a very natural and well-studied problem with applications in many different areas. Unfortunately, there is a very large gap between the known upper and lower bounds for this problem. The best polynomial time algorithm we know has an approximation ratio of n^ϵ and the best known lower bound shows only strong NP-hardness.

In this paper we close this gap completely, assuming that we are allowed to shrink the polygons a little bit, by a factor $1 - \delta$ for an arbitrarily small constant $\delta > 0$, while the compared optimal solution cannot do this (resource augmentation). In this setting, we improve the approximation ratio from n^ϵ to $1 + \epsilon$ which matches the above lower bound that still holds if we can shrink the polygons.

1 Introduction

Maximum Weight Independent Set of Convex Polygons (MWISCP) is a natural but algorithmically very challenging problem. We are given a set of weighted convex polygons $\mathcal{P}$ in the plane and our goal is to select a subset $\mathcal{P}' \subseteq \mathcal{P}$ such that the polygons in $\mathcal{P}'$ are pairwise non-overlapping. The objective is to maximize the total weight of the selected polygons. The problem and its special cases arise in many settings such as map labeling [5,12,23], cellular networks [11], unsplittable flow [6,8], chip manufacturing [18], or data mining [16,20,21].

On the one hand, the best known polynomial time approximation algorithm for the problem has an approximation ratio of n^ϵ [15]. On the other hand, the best complexity result shows only strong NP-hardness [14,19] which leaves an enormous gap. Even more, there is a QPTAS [3,17] which suggests that much better polynomial time approximation results are possible.

When dealing with a very difficult problem it is useful to first study simplified settings or relaxations of the original question in order to gain understanding. In this paper, we consider a relaxation of MWISCP in which we are allowed to shrink the input polygons slightly while the compared optimal solution cannot do this: we assume that there is a small constant $\delta > 0$ such that we can shrink each polygon by a factor $1 - \delta$ and the new polygon lies in the center of the original one (see Fig. 1). The reader may think of editing the polygons in a

© Springer-Verlag Berlin Heidelberg 2016
E. Kranakis et al. (Eds.): LATIN 2016, LNCS 9644, pp. 700–711, 2016.
DOI: 10.1007/978-3-662-49529-2_52

vector graphics program like Adobe InDesign or Inkscape and shrinking them by dragging two opposite corners of their respective bounding boxes slightly towards the center point. This yields the Maximum Weight Independent Set of δ-Shrinkable Convex Polygons problem (δ-MWISCP).

We believe that allowing to shrink the input polygons does not change the nature of the problem very much and, thus, insights for δ-MWISCP can be useful for the general case as well. Also, in many applications it is justified to shrink the input objects slightly without losing much benefit, e.g., in map labeling.

1.1 Our Contribution

We present a $(1 + \epsilon)$-approximation algorithm for δ-MWISCP. This generalizes a previous result for the special case of axis-parallel rectangles [1] to the much larger class of arbitrary convex polygons. Thus, we show that if we are allowed to shrink the input polygons by a little bit then we can improve the best known approximation ratio from n^ϵ to $1 + \epsilon$. This is the best possible result since δ-MWISCP is NP-hard, even for unit squares [1].

Core of our reasoning is that there exists a $(1 + \epsilon)$-approximative shrunk solution for which there is a special cut sequence. This sequence recursively cuts the input plane into smaller and smaller pieces until each piece either coincides with a polygon from the solution (i.e., the polygon is "cut out") or it has empty intersection with all polygons from this solution. Importantly, each piece arising in this sequence and each recursive cut has only constant complexity, i.e., a constant number of vertices and edges. This allows us to design a dynamic program that recursively guesses the above cut sequence and then outputs the corresponding $(1 + \epsilon)$-approximative shrunk solution.

A key difficulty when approximating independent set in the geometric setting is that the input objects can have very different angles. Note that for independent set of axis-parallel rectangles there is a polynomial time $O(\log n / \log \log n)$-approximation algorithm [10] but for straight line segments (with possibly very different angles) we know only an n^ϵ-approximation [15]. Also in our argumentation we need to control the angles of the polygons, or more precisely the angles of the polygon's edges with an underlying grid that guides the construction of our cut sequence. We need that these angles are bounded away from $\pi/2$. To achieve this we give our grid a random rotation. We are not aware of any prior work in which a randomly rotated grid was used and in our setting it turns out to be exactly the right tool to address one of our key difficulties.

1.2 Other Related Work

Many cases of geometric independent set have been studied, being distinguished by the types of the arising input objects. For axis-parallel squares of arbitrary sizes there is a PTAS due to Erlebach et al. [13]. For axis-parallel rectangles Chan and Har-Peled presented a $O(\log n / \log \log n)$-approximation algorithm [10], improving on several previously known $O(\log n)$-approximation

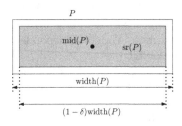
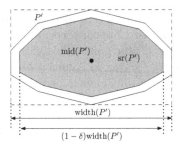

Fig. 1. The outer black lines denote the boundaries of the input polygons P and P'. The gray areas denote their shrunk counterparts $\mathrm{sr}(P)$ and $\mathrm{sr}(P')$. The dashed lines denote the rectangular bounding box of P' (for P the polygon and its rectangular bounding box coincide).

algorithms [5,7,20,22]. In the unweighted case there is even a $O(\log \log n)$-approximation by Chalermsook and Chuzhoy [9]. For arbitrary curves in the plane Fox and Pach give an n^ϵ-approximation, assuming that any two curves intersect only $O(1)$ times [15]. This improves and generalizes an earlier $n^{1/2+o(1)}$-approximation due to Agarwal and Mustafa for straight line segments [4].

Going beyond polynomial time results, for independent set of arbitrary polygons there is a QPTAS [3,17], i.e., a $(1+\epsilon)$-approximation with a running time of $n^{(\log n)^{O_\epsilon(1)}}$, building on an earlier QPTAS for axis-parallel rectangles [2]. This implies that all the above problems are not APX-hard, unless it is true that $\mathsf{NP} \subseteq \mathsf{DTIME}(n^{\mathrm{poly}(\log n)})$.

2   Shrinking Model and Preliminaries

We assume that there is a value $N \in \mathbb{N}$ such that each of the n given input polygons $P_i \in \mathcal{P}$ is specified by vertices $v_{i,1}, v_{i,2}, ... \in \{0, ..., N\}^2$ and a weight $w_i \in \mathbb{N}$. For each polygon $P \in \mathcal{P}$ we define its *midpoint* $\mathrm{mid}(P)$ to be the centroid of its rectangular bounding box, see Fig. 1. For any two points p, p' we define by $\ell(p, p')$ the line segment connecting p and p' and we define $\mathrm{dist}(p, p') := \|\ell(p, p')\|_2$. In our shrinking model for each polygon $P_i \in \mathcal{P}$ we define a new polygon $\mathrm{sr}(P_i)$ defined by vertices $v'_{i,1}, v'_{i,2}, ...$ such that $v'_{i,k} \in \ell(v_{i,k}, \mathrm{mid}(P))$ for each k and such that $\mathrm{dist}(\mathrm{mid}(P), v'_{i,k}) = (1 - \delta)\mathrm{dist}(\mathrm{mid}(P), v_{i,k})$. Observe that if P is convex then $\mathrm{sr}(P) \subseteq P$ and also $\mathrm{sr}(P)$ is convex.

In δ-MWISCP our task is to compute a set of polygons $\mathcal{P}' \subseteq \mathcal{P}$ such that for any two polygons $P, P' \in \mathcal{P}'$ we have that $\mathrm{sr}(P)$ and $\mathrm{sr}(P')$ are disjoint. We compare the value of our (almost feasible) solution to the value of an optimal feasible solution $\mathrm{OPT}(\mathcal{P}) \subseteq \mathcal{P}$ which can *not* shrink the polygons, i.e., with the property that $P \cap P' = \emptyset$ for any two polygons $P, P' \in \mathrm{OPT}(\mathcal{P})$. Thus, an α-approximation algorithm for δ-MWISCP computes a solution $\mathcal{P}' \subseteq \mathcal{P}$ such that $w(\mathcal{P}') \geq \alpha^{-1} \cdot w(\mathrm{OPT}(\mathcal{P}))$ and $\mathrm{sr}(P) \cap \mathrm{sr}(P') = \emptyset$ for all $P, P' \in \mathcal{P}'$, where for any set of polygons $\mathcal{P}''$ we define $w(\mathcal{P}'') := \sum_{P_i \in \mathcal{P}''} w_i$.

Note that for a non-convex polygon P we cannot guarantee that $\mathrm{sr}(P) \subseteq P$. Thus, for arbitrary polygons we no longer obtain a relaxation to the original problem. In particular, the optimal solution for the shrunk polygons might be worse than the optimal solution for the original polygons. Therefore, in this paper we allow only convex polygons.

For technical reasons we assume w.l.o.g. that the width of the rectangular bounding box of each input polygon is larger than its height. This can be ensured by stretching the input plane horizontally. Note that also in our shrinking model this yields an equivalent instance.

3   Preprocessing and Shrinking

In this section we describe preprocessing steps in which we remove some of the input polygons and shrink the remaining ones. While doing this, we lose at most a factor $1 + \epsilon$ in our approximation ratio. Also, we ensure that the shrunk polygons are "well-behaved" so that our main algorithm (described in the next section) has an easier task.

Let $\epsilon > 0$ and $\delta > 0$. First, we ensure that each polygon has only few, i.e., constantly many vertices (and thus also constantly many edges).

Lemma 1. *There exists a constant $K = O_\delta(1)$ such that for each polygon P we can compute a polygon P' with at most K vertices such that $\mathrm{sr}(P) \subseteq P' \subseteq P$.*

We group the polygons according to their diameters. For each polygon P denote by $\mathrm{diam}(P)$ its diameter, i.e., the largest distances between two vertices of P. We do our grouping to achieve two goals: we want that within each group the diameters of the polygons differ by at most a factor $O_{\delta,\epsilon}(1)$ and for two different groups they differ by at least a factor $\sin(\epsilon/K^2)\frac{1}{\epsilon\delta}$.

Lemma 2. *By losing a factor of $1 + \epsilon$ in the value of the optimal solution, we can assume that there is a partition of the polygons $\mathcal{P}$ into $O_{\delta,\epsilon}(\log N)$ groups $\mathcal{P}_i$ and values $\mu_i', \mu_i \in \mathbb{N}$ for each group $\mathcal{P}_i$ such that*

- $\mu_i' \leq \mathrm{diam}(P) < \mu_i$ *for each $P \in \mathcal{P}_i$ and*
- $\delta\epsilon \cdot \sin(\epsilon/K^2) \cdot \mu_i' = \mu_{i+1}$ *and $\mu_i/\mu_i' = \left(\frac{1}{\delta\epsilon\cdot\sin(\epsilon/K^2)}\right)^{1/\epsilon}$ for each i.*

3.1   Hierarchical Grids

We define a family of vertical grids $G_0, G_1, ..., G_m$ with $m = O(\log N)$. They are used in a similar way as in [1]. For each $i \in \{0, ..., m\}$ we define $G_i := \{\{x\} \times \mathbb{R} | \exists k \in \mathbb{N} \,\mathrm{s.t.}\, x = k \cdot g_i\}$ with $g_i := \frac{\delta}{4} \cdot \sin(\epsilon/K^2) \cdot \mu_i'$. Observe that the grids are hierarchical, i.e., each grid line of G_i is also grid line of $G_{i'}$ for each $i' > i$. We give these grids a random rotation such that the lines of all grids have exactly the same angle. This angle is drawn uniformly at random from the range $[\pi/4, \pi/2]$, measured with respect to the x-axis. Let ℓ be a line of the grids. We are interested in the angle between ℓ and the edges of the polygons. We say that

ℓ and a line segment ℓ' have a *good angle* if the angle between ℓ and the line containing ℓ' have an angle of at least ϵ/K^2 and at most $\frac{\pi}{2} - \epsilon/K^2$, otherwise we say that they have a *bad* angle.

Lemma 3. *Let $P \in \mathcal{P}$. With probability at least $1 - O(\epsilon)$ all line segments connecting two vertices of P and all line segments connecting $\mathrm{mid}(P)$ with a vertex of P have a good angle with all grid lines.*

We delete each polygon P that has two vertices v, v' such that $\ell(v, v')$ or $\ell(v, \mathrm{mid}(P))$ has a bad angle with the grid lines. By Lemma 3 this costs only a factor $1 + O(\epsilon)$ in the objective. Next, we give the grids a random shift upwards, drawn uniformly at random from the range $[0, g_0)$, without changing their angles.

Lemma 4. *Let $P \in \mathcal{P}_{i+1}$. Then P intersects a grid line of G_i with probability at most 2ϵ.*

For each $i \in \mathbb{N}$ we delete all polygons $P \in \mathcal{P}_{i+1}$ that intersect a grid line of G_i. Due to Lemma 4 this costs at most a factor of $1 + O(\epsilon)$ in the objective. Since the grids are hierarchical, if a polygon $P \in \mathcal{P}_{i+1}$ does not intersect a grid line of G_i then it does not intersect a grid line of $G_{i'}$ for any $i' \geq i$.

3.2   Shrinking

Next, we want to shrink the polygons. For each polygon $P \in \mathcal{P}_i$ let $v^\uparrow(P)$ and $v^\downarrow(P)$ denote top-most and bottom-most vertices "relative to the grid lines". Formally, we define $v^\uparrow(P)$ and $v^\downarrow(P)$ to be two vertices of P for which there exists a line ℓ with the same angle as the grid lines that intersects $v^\uparrow(P)$ (intersects $v^\downarrow(P)$) and no point in the interior of P.

We shrink P to a polygon P' such that $v^\uparrow(P')$ and $v^\downarrow(P')$ lie on grid lines of G_i. The next lemma shows that this is indeed possible by shrinking P by at most a factor $1 - \delta$. Heart of this reasoning is that there are at least $1/\delta$ grid lines of G_i between $v^\uparrow(P)$ and $v^\downarrow(P)$. We do this operation with all input polygons.

Lemma 5. *Let $P \in \mathcal{P}_i$. In polynomial time we can compute a polygon P' with at most $K + 2$ edges such that $\mathrm{sr}(P) \subseteq P' \subseteq P$ and $v^\uparrow(P')$ and $v^\downarrow(P')$ lie on grid lines of G_i. Furthermore, all edges of P' crossing a grid line of G_i in a non-zero angle have a good angle with this grid line.*

3.3   Horizontal Grids

From now on we do not shrink the polygons any further. Let us assume w.l.o.g. that the grid lines are exactly vertical and that there is an integer N' such that the input polygons are contained in the area $[0, N'] \times [0, N']$ for some integer $N' = O(N)$. We add a hierarchical family of horizontal grids $\bar{G}_0, \bar{G}_1, ..., \bar{G}_m$ to the vertical grids $G_0, G_1, ..., G_m$. For each $i \in \{0, ..., m\}$ we define $\bar{G}_i := \{\mathbb{R} \times \{y\} | \exists k \in \mathbb{N} \text{ s.t. } y = k \cdot g_i\}$ with as before $g_i = \frac{\delta}{2} \cdot \sin(\epsilon/K^2) \cdot \mu_i'$. Thus, for each i the grid $\bar{G}_i$ has exactly the same spacing as G_i. We give the horizontal grids $\bar{G}_i$ a random shift upwards, drawn uniformly at random from the range $[0, g_0)$. Then, for each $i \in \{0, ..., m\}$ we delete all remaining polygons from $\cup_{j>i+1} \mathcal{P}_i$ that intersect a grid line in $\bar{G}_i$. The following lemma can be proven similarly as Lemma 4.

Lemma 6. *Let $P \in \mathcal{P}_{i+1}$. Then P intersects a grid line of $\bar{G}_i$ with probability at most 2ϵ.*

Denote by $\mathcal{P}'$ the resulting set of shrunk polygons. For each integer i we define $\mathcal{P}'_i$ to be the sets of polygons that we obtain when we shrink each polygon in $\mathcal{P}_i$. Note that we lost only a factor of $(1 + O(\epsilon))$ in our approximation ratio (see Lemmas 2, 4 and 6).

4   Dynamic Program

Our algorithm is a geometric divide-and-conquer algorithm similar to the algorithm used in [1,3]. It recursively divides the area containing the input polygons into smaller and smaller pieces. When it makes a recursive call for a piece $A \subseteq [0, N']^2$ then it computes a (near-optimal) solution to the subproblem given by all input polygons that are contained in A. To do this, it tries all possibilities to partition A into at most $k = O_{\delta, \epsilon}(1)$ subpieces such that the boundary of each of them consists of at most k line segments out of a suitable set $\mathcal{L}$ defined below. Then, it makes a recursive call on each of these subpieces and obtains a (near-optimal) solution for each of those. By putting them together, it obtains a candidate solution for the original piece A. Additionally, it checks what profit it can obtain by selecting only one polygon that is contained in A. Eventually, it returns the best solution out of all candidate solutions stemming from all partitions of A and all single polygons contained in A. We will show that if the parameter k is sufficiently large then our algorithm will output a set that is at least as profitable as the optimal solution for $\mathcal{P}'$.

We embed the whole procedure into a dynamic program (DP). Let $k = O_{\delta, \epsilon}(1)$ be a parameter to be defined later. Our DP table has one cell for each (not necessarily convex) piece $A \subseteq [0, N']^2$ whose boundary consists of at most k lines segments such that

– each line segment is a subset of an edge of a polygon in $\mathcal{P}'$ or a subset of a grid line in $\mathcal{G} := \cup_{i=0}^m G_i \cup \bar{G}_i$ and
– the endpoint of each line segment is
  - the vertex of a polygon in $\mathcal{P}'$, or
  - the intersection of an edge of a polygon in $\mathcal{P}'$ with a grid line in $\mathcal{G}$, or
  - the intersection of two grid lines in $\mathcal{G}$.

Denote by $\mathcal{L}$ the set of all line segments that arise on the boundaries of the pieces defined above, see Fig. 2 for an example. Denote by GEO-DP our overall algorithm. As the following lemma shows, it has pseudo-polynomial running time. We will explain later how to improve this to polynomial time.

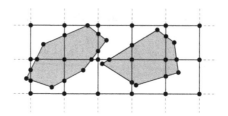

Fig. 2. An instance of δ-MWISCP. The black circles denote the endpoints of the line segments according to our definition. The bold lines denote the lines in $\mathcal{L}$ resulting from them.

Lemma 7. *The number of DP-cells is bounded by* $(n+N)^{O_{\delta,\epsilon}(k)}$. *If* $k = O_{\epsilon,\delta}(1)$ *then the overall running time of GEO-DP is bounded by* $(n+N)^{O_{\delta,\epsilon}(1)}$.

There is a piece containing all input polygons whose subproblem corresponds to the original problem we want to solve. We want to show that GEO-DP outputs a solution that is at least as profitable as the optimal solution for $\mathcal{P}'$. In order to show this we can assume w.l.o.g. that the input to GEO-DP consists only of this optimal solution. We denote it by $\mathcal{P}''$. We define $\mathcal{P}_i'' := \mathcal{P}_i' \cap \mathcal{P}''$ for each i.

4.1 Cutting Sequence

We describe a sequence of cuts that recursively subdivides the whole area $[0, N'] \times [0, N']$ and "cuts out" all polygons in $\mathcal{P}''$. Also for the algorithm in [1] for easier case of shrinkable rectangles such a sequence is used in the analysis. Our construction ensures that for each piece arising in the sequence there exists a DP-cell according to the above definition and that each cut for such a piece partitions it into at most k smaller pieces. Formally, we describe the above sequence of cuts by a tree T where each node v is associated with a piece A_v in the plane. We say that a tree T is a $(k, \mathcal{P}'')$-*region decomposition* if the following holds:

- For each node v in T and each polygon $P \in \mathcal{P}''$ we have that if P does not coincide with A_v, i.e., $P \neq A_v$, then either P is contained in A_v or P is disjoint from A_v.
- For tree nodes u and v such that v is a parent of u we have that $A_u \subseteq A_v$. Each node $v \in T$ has at most $k' \leq k$ children $u_1, ..., u_{k'}$ in T and $\bigcup_{i=1}^{k'} A_{u_i} = A_v$.
- For each leaf node v of T the piece A_v contains at most one polygon in $\mathcal{P}''$ and it has empty intersection with all other polygons in $\mathcal{P}''$.
- For each node v in T the area A_v is connected and its boundary can be described by at most k line segments from $\mathcal{L}$.

Lemma 8. *If there exists a* $(k, \mathcal{P}'')$-*region decomposition then GEO-DP will output a solution of weight at least* $w(\mathcal{P}'')$ *when it is parametrized by* k.

4.2   Existence of Region Decomposition

We prove now that a $(k, \mathcal{P}'')$-region decomposition exists for some $k = O_{\delta,\epsilon}(1)$. Our argumentation proceeds in *levels* with one level for each grid $G_i \cup \bar{G}_i$. We describe it inductively. Our induction hypothesis is that we are given a vertex $v \in T$ whose piece A_v is described as follows: there is a cell C of the grid $G_{i-1} \cup \bar{G}_{i-1}$ and up to two polygons $P_1, P_2 \in \cup_{j=0}^{i-1} \mathcal{P}_j''$ such that each of them intersects both the left and the right grid line of C. Then, A_v is the connected component of $C \setminus \{P_1, P_2\}$ that is adjacent to P_1 and P_2 (see Fig. 3 for a sketch). Note that there exists a DP-cell for A_v. Assume w.l.o.g. that P_1 crosses the left and right grid lines of C below P_2. One or both polygons P_1 and P_2 might be undefined and in this case the bottom and/or the top boundary of C takes the role of P_1 and/or P_2. For the base case, we can assume that there is one cell of an (artificial) grid $G_{-1} \cup \bar{G}_{-1}$ that contains all input polygons. We assume by induction that A_v does not intersect any polygon in $\cup_{j=0}^{i-1} \mathcal{P}_j''$.

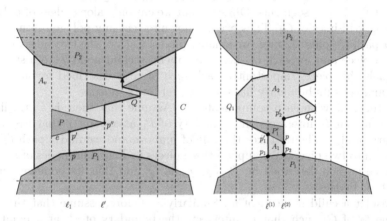

Fig. 3. Left: The first cut Q that separates the piece A_v (given by the area in the cell C between the polygons P_1 and P_2) into two smaller pieces. The point p is defined as the bottom-most point of $\ell_1 \cap A_v$ and ℓ_1 intersects the boundary of P_1 on p. Right: a piece (consisting of $A_1 \cup P \cup A_2$) for which there is no grid line that intersects P_1 on the boundary of the piece (first case in the proof of Lemma 10).

Suppose now that we are given a piece A_v as defined above. We define a cut through A_v. Let ℓ_1 be a grid line of G_i such that ℓ_1 intersects the interior of A_v. Let p be the bottom-most point of $\ell_1 \cap A_v$. Note that at this point ℓ_1 intersects the boundary of P_1. Our cut starts in p and moves up along ℓ_1. If we do not hit any polygon of $\mathcal{P}''$ contained in A_v on the way up then we are done with our cut. Otherwise, suppose that we hit an edge e of a polygon P. Let p' denote the point on e that is hit by ℓ_1. Due to Lemma 3 we know that e has a good angle with ℓ_1. Also, $P \in \mathcal{P}_i''$ since otherwise it would have been deleted before as it intersects a grid line of G_i. Our cut moves along e in the direction that goes up. We continue along the boundary of P in the same direction until we arrive at a leftmost or rightmost vertex of P. Let p'' denote this point. Due to our shrinking, p'' lies on a grid line ℓ' of G_i. See Fig. 3 for a sketch. Since all edges of P have a good angle with the grid, we can prove the following lemma.

Lemma 9. *When moving from p' to p'' along the edge of P, we move up by at least* $\sin(\epsilon^2/K) \cdot g_i = \Omega_{\delta,\epsilon}(1) \cdot g_i$ *units.*

Proof. Since all edges of P have a good angle with the grid lines, each of these angles is at least ϵ^2/K. Therefore, when we move from p' to p'' we move up by at least $\mathrm{dist}(p',p'') \cdot \sin(\epsilon^2/K) \geq \sin(\epsilon^2/K) \cdot g_i$ units.   □

Note that the constructed path from p via p' to p'' consists of at most $K + 1 \leq O_{\delta,\epsilon}(1)$ line segments. We continue iteratively where now p'' takes the role of p. We stop when we hit the upper boundary of A_v (defined by P_2 or the top boundary of C). Denote by Q the constructed path. The height of C is bounded by $g_{i-1} = O_{\delta,\epsilon}(g_i)$. In every iteration we move up by at least $\Omega_{\delta,\epsilon}(1) \cdot g_i$ units. Thus, there are at most $O_{\delta,\epsilon}(1)$ iterations and Q can be described with $O_{\delta,\epsilon}(1) \cdot (K+1)$ line segments. Observe that we cut only along edges of polygons and along grid lines of G_i. Thus, we did not intersect any polygon from $\mathcal{P}''$.

Our path Q splits A_v into two smaller pieces. Each of the two sides of $A_v \setminus Q$ defines a piece and for each of them we append a child node u_i to v such that A_{u_i} equals this piece. Importantly, each such piece is described by only P_1, P_2 and Q and thus its boundary has only $O_{\delta,\epsilon}(1)$ line segments.

We continue with each component A_{u_i}. Assume that there is a grid line ℓ_2 of G_i such that ℓ_2 intersects the boundary of P_1 at a point $\bar{p}$ that lies on the boundary of A_{u_i}. Then $\bar{p}$ takes the role of p above and we find a path Q' that split A_{u_i} into two pieces. We append these pieces in T as child nodes of u_i. Each such piece is then described by P_1, P_2, Q and Q' and thus its boundary has at most $O_{\delta,\epsilon}(1)$ edges.

Consider a child node $\bar{u}_j$ of u_i. Similarly as before, assume that there is a grid line ℓ_3 of G_i such that ℓ_3 intersects the boundary of P_1 at a point that lies on the boundary of $A_{\bar{u}_j}$. We compute a path Q'' through $A_{\bar{u}_j}$ as above. Now, each connected component of $A_{\bar{u}_j} \setminus Q''$ can be described by P_1, P_2 and at most *two* of the paths Q, Q', Q''. Similarly, when we continue further like above in the recursion each resulting piece can be described by P_1, P_2 and two paths Q_1, Q_2 through A_v where each of the latter can be described with only $O_{\delta,\epsilon}(1)$ line segments.

We can apply the above reasoning as long as there is a grid line ℓ_k of G_i such that the boundary of P_1 intersects ℓ_k at a point that lies on the boundary of the considered piece. Suppose now that this is not possible, i.e., we have a piece $A_{\tilde{v}} \subseteq A_v$ such that at the boundary of $A_{\tilde{v}}$ there is no grid line of G_i that intersects P_1. As the next lemma shows, this piece $A_{\tilde{v}}$ can then be partitioned into two smaller pieces A_1, A_2 and one polygon P_1', see Fig. 3. When we continue, for the piece A_1 the polygon P_1' takes the role of P_2, and for the piece A_2 the polygon P_1' takes the role of P_1.

Lemma 10. *Assume that in the above construction we obtain a piece $A_{\tilde{v}}$ such that on the boundary of $A_{\tilde{v}}$ there is no point where a grid line of G_i intersects* the boundary of P_1. Then either $A_{\tilde{v}}$ is contained in a grid column of G_i or $A_{\tilde{v}}$ can be partitioned into two pieces A_1, A_2 and one polygon P'_1 such that

- the boundary of A_1 consists of two upward monotone paths $Q_1^{(1)}, Q_1^{(2)}$ with at most $O_{\delta,\epsilon}(1)$ edges each that both connect P_1 and P'_1, and
- the boundary of A_2 consists of two upward monotone paths $Q_2^{(1)}, Q_2^{(2)}$ with at most $O_{\delta,\epsilon}(1)$ edges each that both connect P'_1 and P_2.

Proof. Assume that $A_{\tilde{v}}$ is not contained in a grid column of G_i since otherwise there is nothing to show. By construction $A_{\tilde{v}}$ is described by two paths Q_1, Q_2 that both connect P_1 and P_2. There is no point on the boundary of $A_{\tilde{v}}$ in which a grid line of G_i intersects P_1. Thus, for the points p_1 and p_2 on which the paths Q_1 and Q_2 start, there must be two consecutive grid lines $\ell^{(1)}, \ell^{(2)}$ of G_i such that $\ell^{(1)}$ intersects P_1 on p_1 and $\ell^{(2)}$ intersects P_1 on p_2, see Fig. 3. Assume that $\ell^{(1)}$ is on the left of $\ell^{(2)}$. Since $A_{\tilde{v}}$ is not contained in a grid column of G_i one of the paths Q_1, Q_2 is not completely vertical.

Assume that both paths are not completely vertical (the other case can be proven with similar arguments). Let p'_1, p'_2 be the points on which Q_1 and Q_2 deviate from being only vertical. Assume w.l.o.g. that the y-coordinate of p'_1 is not larger than the y-coordinate of p'_2.

Assume that on p'_1 the path Q_1 turns left. Then on p'_1 the path Q_1 hits a polygon $P \in \mathcal{P}''_i$. Thus, the boundary of P must intersect $\ell^{(2)}$ at a point p. Then the y-coordinate of this point p must be lower than the y-coordinate of p'_2 (since Q_1 goes monotonously upwards). Then we set $P'_1 := P$ and A_1 consists of the quadrilateral described by p_1, p'_1, p, p_2, and $A_2 = A_{\tilde{v}} \setminus \{A_1, P\}$. The paths $Q_1^{(1)}, Q_1^{(2)}, Q_2^{(1)}, Q_2^{(2)}$ consist of the parts of Q_1 and Q_2 surrounding A_1 and A_2, respectively.

Assume now that on p'_1 the path Q_1 turns right after hitting a polygon P'. Then Q_1 must cross $\ell^{(2)}$ at a point p'. If the y-coordinate of p' is smaller than the y-coordinate of p'_2 then we define $P'_1 := P'$ and we define the pieces A_1, A_2 and the paths $Q_1^{(1)}, Q_1^{(2)}, Q_2^{(1)}, Q_2^{(2)}$ similarly as in the previous case. Finally, suppose that the y-coordinate of p' is not smaller than the y-coordinate of p'_2. Then on p'_2 the path Q_2 hits a polygon P'' and it must turn right (since otherwise $P' \cap P'' \neq \emptyset$). Then the polygon P'' must cross $\ell^{(1)}$ underneath p'_1 and we define $P'_1 := P''$ and the pieces A_1, A_2 and the paths $Q_1^{(1)}, Q_1^{(2)}, Q_2^{(1)}, Q_2^{(2)}$ accordingly. $\qquad\square$

We continue until we obtain pieces that are contained in a grid column of G_i. Let $A_{v'}$ be such a piece. It might not fulfill the induction hypothesis yet since it might still span many grid cells of $G_i \cup \bar{G}_i$. We know that each polygon $P \in \mathcal{P}''_i$ intersects at least two grid columns of G_i. Thus, there can be no polygon $P \in \mathcal{P}''_i$ with $P \subseteq A_{v'}$. Thus, the boundary of $A_{v'}$ consists of two consecutive grid lines in G_i and (parts of) the boundary edges of at most two polygons of $\cup_{j=0}^i \mathcal{P}''_j$, defining the upper and lower boundary of $A_{v'}$.

As long as $A_{v'}$ is not contained in a grid cell of $G_i \cup \bar{G}_i$ there must be a grid row r of $\bar{G}_i$ such that r has non-empty intersection with the interior of $A_{v'}$.

Note that r does not intersect any of the remaining polygons in $\cup_{j \geq i+1} \mathcal{P}_j''$ since we have removed such polygons before already. We split $A_{v'}$ along r into two pieces $A_{v''}, A_{v'''}$ and add the corresponding vertices v'' and v''' to T as children of v'. We continue this process until each piece is contained in a grid cell of G_i. Thus, our resulting pieces fulfill the induction hypothesis. The above reasoning proves the following lemma.

Lemma 11. *There exists a universal constant $k = O_{\delta,\epsilon}(1)$ such that a $(k, \mathcal{P}'')$-region decomposition exists.*

For our main result we parametrize GEO-DP by the constant $k = O_{\delta,\epsilon}(1)$ due to Lemma 11. Then together with Lemma 8 this implies that GEO-DP outputs a solution of weight at least $w(\mathcal{P}'')$. Due to our reasoning in Sect. 3 we have that $w(\mathcal{P}'') \geq (1 - O(\epsilon)) \cdot \text{OPT}(\mathcal{P})$.

It remains to address the fact that in the above form our algorithm has a running time that might be exponential in the input size (since N might be exponential). We argue similarly as in [1]. First observe that there are only $O(\log N)$ recursion levels, which is polynomial in the length of the input encoding. In each level of the grids, it suffices to introduce only grid cells C for which there exists a polygon $P \in \mathcal{P}'$ with $P \subseteq C$. There can be only n such grid cells for each grid $G_i \cup \bar{G}_i$ and thus in total there are only $O(n \cdot \log N)$ such cells. Hence, the total number of grid lines is also bounded by $O(n \cdot \log N)$. This reduces our running time to $(n + \log N)^{O_{\delta,\epsilon}(1)}$.

Theorem 1. *For any constants $\epsilon, \delta > 0$ there is a polynomial time $(1 + \epsilon)$-approximation algorithm for the maximum independent set of δ-shrinkable convex polygons problem.*

Acknowledgments. The author would like to thank Parinya Chalermsook for helpful discussions on the topic of this paper.

References

1. Adamaszek, A., Chalermsook, P., Wiese, A.: How to tame rectangles: solving independent set and coloring of rectangles via shrinking. In: Approximation, Randomization, and Combinatorial Optimization. Algorithms and Techniques (APPROX/RANDOM). Leibniz International Proceedings in Informatics (LIPIcs), vol. 40, pp. 43–60. Schloss Dagstuhl-Leibniz-Zentrum für Informatik, Dagstuhl (2015)
2. Adamaszek, A., Wiese, A.: Approximation schemes for maximum weight independent set of rectangles. In: Proceedings of the 54th Annual IEEE Symposium on Foundations of Computer Science (FOCS), pp. 400–409. IEEE (2013)
3. Adamaszek, A., Wiese, A.: A QPTAS for maximum weight independent set of polygons with polylogarithmically many vertices. In: Proceedings of the Twenty-Fifth Annual ACM-SIAM Symposium on Discrete Algorithms (SODA), pp. 645–656. SIAM (2014)

4. Agarwal, P.K., Mustafa, N.H.: Independent set of intersection graphs of convex objects in 2D. Comput. Geom. **34**(2), 83–95 (2006)
5. Agarwal, P.K., van Kreveld, M., Suri, S.: Label placement by maximum independent set in rectangles. Comput. Geom. **11**, 209–218 (1998)
6. Anagnostopoulos, A., Grandoni, F., Leonardi, S., Wiese, A.: Constant integrality gap LP formulations of unsplittable flow on a path. In: Goemans, M., Correa, J. (eds.) IPCO 2013. LNCS, vol. 7801, pp. 25–36. Springer, Heidelberg (2013)
7. Berman, P., DasGupta, B., Muthukrishnan, S., Ramaswami, S.: Efficient approximation algorithms for tiling and packing problems with rectangles. J. Algor. **41**(2), 443–470 (2001)
8. Bonsma, P., Schulz, J., Wiese, A.: A constant factor approximation algorithm for unsplittable flow on paths. In: Proceedings of the 52th Annual IEEE Symposium on Foundations of Computer Science (FOCS), pp. 47–56 (2011)
9. Chalermsook, P., Chuzhoy, J.: Maximum independent set of rectangles. In: Proceedings of the 20th Annual ACM-SIAM Symposium on Discrete Algorithms (SODA 2009), pp. 892–901. SIAM (2009)
10. Chan, T.M., Har-Peled, S.: Approximation algorithms for maximum independent set of pseudo-disks. Discrete & Comput. Geom. **48**(2), 373–392 (2012)
11. Clark, B.N., Colbourn, C.J., Johnson, D.S.: Unit disk graphs. Discrete Math. **86**(1), 165–177 (1990)
12. de Floriani, L., Magillo, P., Puppo, E.: Applications of computational geometry to geographic information systems. In: Handbook of Computational Geometry, pp. 333–388. North Holland (2000)
13. Erlebach, T., Jansen, K., Seidel, E.: Polynomial-time approximation schemes for geometric intersection graphs. SIAM J. Comput. **34**(6), 1302–1323 (2005)
14. Fowler, R.J., Paterson, M.S., Tanimoto, S.L.: Optimal packing and covering in the plane are NP-complete. Inf. Process. Lett. **12**(3), 133–137 (1981)
15. Fox, J., Pach, J.: Computing the independence number of intersection graphs. In: Proceedings of the Twenty-Second Annual ACM-SIAM Symposium on Discrete Algorithms (SODA), pp. 1161–1165. SIAM (2011)
16. Fukuda, T., Morimoto, Y., Morishita, S., Tokuyama, T.: Data mining with optimized two-dimensional association rules. ACM Trans. Database Syst. (TODS) **26**(2), 179–213 (2001)
17. Har-Peled, S.: Quasi-polynomial time approximation scheme for sparse subsets of polygons. In: Proceedings of the Thirtieth Annual Symposium on Computational Geometry (SoCG), pp. 120–129. ACM (2014)
18. Hochbaum, D.S., Maass, W.: Approximation schemes for covering and packing problems in image processing and VLSI. J. ACM **32**, 130–136 (1985)
19. Imai, H., Asano, T.: Finding the connected components and a maximum clique of an intersection graph of rectangles in the plane. J. Algor. **4**(4), 310–323 (1983)
20. Khanna, S., Muthukrishnan, S., Paterson, M.: On approximating rectangle tiling and packing. In: Proceedings of the 9th Annual ACM-SIAM Symposium on Discrete Algorithms (SODA), pp. 384–393. SIAM (1998)
21. Lent, B., Swami, A., Widom, J.: Clustering association rules. In: Proceedings of the 13th International Conference on Data Engineering, pp. 220–231. IEEE (1997)
22. Nielsen, F.: Fast stabbing of boxes in high dimensions. Theor. Comp. Sc. **246**, 53–72 (2000)
23. Verweij, B., Aardal, K.: An optimisation algorithm for maximum independent set with applications in map labelling. In: Nešetřil, J. (ed.) ESA 1999. LNCS, vol. 1643, pp. 426–437. Springer, Heidelberg (1999)

Author Index

Printed in the United States
By Bookmasters